SP Pvt. Ltd.

Vegetable Processing and Bioactive Compounds

Dattatreya M Kadam
Monika Sharma
Devinder Kaur

2017

Studium Press (India) Pvt. Ltd.

Vegetable Processing and Bioactive Compounds

ISBN: 978-93-85046-08-7

Published by:

Studium Press (India) Pvt. Ltd.
4735/22, 2nd Floor, Prakash Deep Building
(Near Delhi Medical Association),
Ansari Road, Darya Ganj, New Delhi-110 002
Tel.: + 91-11-43240200-15 (15 lines); Fax: 91-11-43240215
E-mail: studiumpress@gmail.com

Printed in India

Dr. S.M. Ilyas

Foreword

The Book titled "*Vegetable Processing and Bioactive Compounds*" edited by Dattatreya M Kadam, Monika Sharma and Devinder Kaur is an important addition to technical literature in the area of vegetable processing and bioactive compounds. The importance of vegetables in human diet cannot be overemphasized since these are valuable sources of nutrition, especially in Indian vegetarian diet. The production and consumption of vegetable in the world in general and India in particular is constantly on rise and India is already Second largest producer of vegetables in the World. Many farmers in India have diversified to growing vegetables in place of much less remunerative cereals.The high water containing vegetables, being highly perishable, require immediate interventions after production and since the huge quantities are being produced, these cannot be consumed right away needing suitable storage and/or processing. Failure of adequate marketing support and dearth of cool/cold storage forces gluts and resultant distress sale and farmers are terribly short changed. Consumers at other places continue to pay higher price for the same produce.

Efforts have been on to save the perishable vegetables through innovative low cost evaporative coolers, building new cold rooms and strengthening cold chain. But, obviously much more needs to be done. Immediate post-production processing has also been attempted with mixed success. If taken up on solid scientific basis, with concept of value chain brought in, it will lead to generation of rural income and employment.

The presences of bio-active compounds in the vegetables are of great economic significance since their extraction adds huge value. In order to maximize gain from the extraction of these valuable compounds, there is a need to develop protocols and infrastructure for size reduction including cryogenic grinding, ultra low cooling, freeze drying and storage etc which appear costly but will give high dividends. The compounds, found in vegetables can be used as ingredients in cosmetics, pharmaceuticals, nutraceuticals and food. For application in food, these can be used as anti-oxidents to prevent oxidation of food containing high amounts of liquid. Obviously, there is a great need to extract these valuable compounds proving that the bye-products will become more valuable than the main product.

The Authors have done a commendable job in bringing out this very useful book which will go a long way in enriching the scarce literature and thus deserve high praise. I am sanguine that the researchers, students, processors and other stakeholders will find it very useful and be immensely benefited.

S.M. Ilyas

About the Editor(s)

Dattatreya M. Kadam

Dr. Dattatreya M. Kadam is Principal Scientist (APE) and Head HCP (Ad. Ch.) at ICAR-CIPHET (GoI), Ludhiana who is actively involved in research and development activities of Agricultural/Food Process Engineering. Presently working on a cutting edge Nano technology projects. He obtained his B. Tech (Agril. Engg), M. Tech (PAS) and Ph. D (Engg) from CAE Raichur (UAS, Dharwad), PAU, Ludhiana and IARI, New Delhi, respectively. He has published more than 35 research papers in National and Internationally reputed journals, 2 booklets, 15 book chapters, 7 bulletins, 4 extension leaflets, 13 popular articles and filled 2 patents. He is working on design and development of agro-processing based machines/ equipments, tools and value added food products. Prepared many workable project profiles to encourage upcoming entrepreneurs and youths to come forward and establish food processing industries. He has received many awards such as The Young Engineer Award; Young Scientist Award; Prestigious BOYSCAST FELLOW; Mother Teresa Sadhavan Award; IFCON 2013 BEST POSTER award; Host Scientist for 'CV Raman International Fellowship for African Researcher; ICAR-CIPHET Silver Jubilee Best Scientist Award. He has undergone an advanced research training on Developed biodegradable biopolymers embedded with nanoparticles at Iowa State University, Ames, U.S.A. as DST's BOYSCAST Fellow award. He has vast experience to handle ITMU (IPR), Business Planning and Development (BPD), Agri-Business Incubation (ABI). He is Officer In charge of Post-Harvest Machine and Equipment (PHME) Testing Centre at ICAR-CIPHET, Ludhiana approved by Mechanization & Technology Division, Department of Agriculture and Farmers Welfare, Ministry of Agriculture and Farmers Welfare, Govt. of India.

Monika Sharma

Dr. Monika Sharma started her career as Scientist in ICAR in 2010 and was posted at ICAR-CIPHET, Ludhiana, Punjab, India till Nov. 2015. She is currently working as Scientist (Food Science & Technology) at Southern Regional Station, ICAR-NDRI, Bengaluru and actively involved in teaching and research activities. She holds a Ph.D (Dairy Technology) from National Dairy Research Institute (NDRI), Karnal, Haryana and has completed M.Sc. (Food Technology) from GBPUA&T, Pantnagar, India. She has worked in the area of convenience and ready to eat foods, functional foods, quality evaluation, composite dairy foods, starch modification etc. She has so far 20 research papers, 2 technical bulletins, 2 technology inventory books, 4 book chapters, about 20 popular articles, more than 15 conference papers to her credit. She has successfully guided two post graduate students. She has worked as PI and Co-PI in about 07 research projects and has developed about 20 technologies. She has also conducted entrepreneurship development program for some of developed technologies and also transferred one technology to two entrepreneurs.

Devinder Kaur

Dr. Devinder Kaur is Assistant Professor at Centre of Food Technology, University of Allahabad, Allahabad. She received her degrees (B.Sc, M.Sc and PhD) in Food Technology from Guru Nanak Dev University, Amritsar. She has published more than 25 research papers in National and Internationally reputed journals. Presently she is working on the area of dietary fiber, plant pigments and food waste management.

Preface

The public perception and expectations from agriculture are also changing fast with the increase in awareness for health, environment and employment generation. Therefore agriculture is also turning to become knowledge-based industry. Consequently, the dimension of agricultural research is bound to change to generate new competitive technologies. Technologies changes are making rapid strides in all fields of science and they directly or indirectly influence our lives. Many new developments in processing, value addition and high value bioactive compounds of vegetables have also taken place.

Vegetable processing and bio-compound extraction technology, an integral part of agricultural system, needs to be developed at par with developments in agriculture. There is need to adopt holistic approach in selection of research and development projects and their execution to reduce losses, and to promote value addition for export and domestic consumption. Development and adoption of eco-friendly techniques for processes of vegetables such as harvesting, cleaning, grading, transportation, packaging, preservation, processing and product development and techniques for extraction of high value and bioactive compounds need to be analyzed and strategies formulated for catching up with contemporary developments in socio-economic and technical sectors. The need of the hour is to concentrate on improved technologies on vegetable processing and bio-compound extractions.

Determination of the nature of the compounds in different commodities by different methods and the influence of pre-harvest, postharvest and processing treatments is well know. This book summarizes the vegetable processing of different crops and wide variety of extraction and characterization techniques which are used to quantify individual or total antioxidants, as well as other bioactive compounds.

Keeping the above in view, we have decided to work on compilation and editing of Book entitled "Vegetable Processing and Bioactive Compound Extraction" by Dr. Dattatreya M. Kadam, Dr. Monika Sharma and Dr. Devinder Kaur. The book contains 22 chapters covering different vegetables i.e. tomatoes, cauliflower, cabbage, broccoli, brinjal/egg plant, mushroom, Bitter gourd, bamboo, chilli, potatoes, leafy vegetables, lady finger, root crops, onion, curry leaves. The chapters are contributed by different authors/

multiple authors from different organizations, institutes and universities. This edited book will be a good source of text/ reference book for teachers, students and researchers of different universities, colleges and institutes. The book targets the areas of Food Processing, Food Process Engineering, Agricultural Processing, Food Science and Technology, Horticulture, Post-harvest Technology, Food and Nutrition, Food Biochemistry, Food Microbiology etc.

We are privileged to express our sincere gratitude to all authors of this book for valuable contribution in the form of chapters. We would also like to place on record our appreciation to Studium Publishers for believing in us and taking this activity to bring out this book in timely manner. Authors wish to acknowledge the sources of information taken from public domain and with consent to bring this edited book in present form.

Dattatreya M. Kadam
Monika Sharma
Devinder Kaur

Table of Contents

1

Overview of Extraction Technology

RAVINDRA NAIK[1], K.P. SUDHEER[2] AND J.S. RUTRAPRIYA[3]

ABSTRACT

A wide range of technologies are available for the extraction of active components from plant and vegetable. This include traditional methods like maceration, infusion, digestion, decoction, percolation, hot continuous extraction, aqueous alcoholic extraction, counter-current extraction, Some of the latest extraction methods include headspace trapping, solid phase micro-extraction, simultaneous distillation extraction, microwave distillation, membrane extraction, phytosol extraction, protoplast extraction, microdistillation, thermomicrodistillation, controlled instantaneous decompression etc. The choice depends on the economic feasibility and suitability of the process in a particular situation. The various processes of extraction of plant and vegetable extracts are briefed in this chapter.

***Key words*:** Vegetables plants, Extraction technology, Economic suitability

I. INTRODUCTION

Extraction, as the term is used, involves the separation of active portions of plant/vegetable tissues from the inactive or inert components by using selective solvents in standard extraction procedures. Extraction forms the first basic step because the preparation of crude extracts is the starting point for the isolation and purification of chemical constituents present.

[1] Post-Harvest Processing, ICAR-Central Institute of Agricultural Engineering, Regional centre Coimbatore-641 007, Tamil Nadu, India

[2] Post Harvest Technology Dept. of Post Harvest Technology & Agrl. Processing, Kelappaji College of Agricultural Engineering and Technology (KCAET) Tavanur, Kerala Agricultural University, India

[3] ICAR-Central Institute of Agricultural Engineering, Regional centre Coimbatore-641 007, Tamil Nadu, India

**Corresponding author* E-mail: naikravindra@gmail.com

Yet the extraction step remains often a neglected area, which over the years has received much less attention and research. Such extraction techniques separate the soluble plant metabolites and leave behind the insoluble cellular mass. The products so obtained from plants are relatively complex mixtures of metabolites, in liquid or semisolid state or in dry powder form, and are intended for oral or external use. The bioactive component extraction starts with the extraction of the active components using various technologies. The general techniques of plant/vegetable extraction include maceration, infusion, percolation, digestion, decoction, hot continuous extraction (Soxhlet), aqueous-alcoholic extraction by fermentation, counter-current extraction, microwave-assisted extraction, ultrasound extraction, supercritical fluid extraction, and phytonic extraction. For leafy vegetables, hydrodistillation techniques (water distillation, steam distillation, water and steam distillation), hydrolytic maceration followed by distillation, expression and enfleurage (cold fat extraction) may be employed. Some of the latest extraction methods for bioactive component extraction include headspace trapping techniques, solid phase microextraction, supercritical fluid extraction, microwave distillation/extraction, membrane extraction, phytosol extraction, protoplast extraction technique, microdistillation, thermomicrodistillation, controlled instantaneous decompression etc[1,2,3].

II. GENERAL TRADITIONAL METHODS OF EXTRACTION INCLUDES

(1) ***Maceration:*** In this process, the whole or coarsely powdered material is placed in a stoppered container with the solvent and allowed to stand at room temperature for a period of at least 3 days with frequent agitation until the soluble matter has dissolved. The mixture then is strained, the marc (the damp solid material) is pressed, and the combined liquids are clarified filtration or decantation after standing.

(2) ***Infusion:*** Fresh infusions are prepared by macerating the material for a short period of time with cold or boiling water. These are dilute solutions of the readily soluble constituents.

(3) ***Digestion:*** This is a form of maceration in which gentle heat is used during the process of extraction. It is used when moderately elevated temperature is not objectionable.

(4) ***Decoction:*** The material is boiled in a specified volume of water for a defined time, it is then cooled and strained or filtered. This procedure is suitable for extracting water-soluble and heat-stable constituents.

(5) ***Percolation:*** The procedure used most frequently to extract active ingredients. The solid ingredients are moistened with an appropriate amount of the solvent and allowed to stand for approximately 24–48 h in a well closed container. The mixed liquid is clarified by filtration or by standing followed by decanting.

(6) ***Hot Continuous Extraction (Soxhlet):*** In this method, the finely ground crude sample is placed in a porous bag or "thimble" made of strong filter paper, which is placed in chamber of the Soxhlet apparatus. This is the most common and widely followed method and can be taken as the control while comparing the modern methods of extraction.

(7) ***Aqueous Alcoholic Extraction by Fermentation:*** Some preparations adopt the technique of fermentation for extracting the active principles. The extraction procedure involves soaking the crude substrate, in the form of either a powder or a decoction for a specified period of time, during which it undergoes fermentation and generates alcohol *in situ*; this facilitates the extraction of the active constituents contained in the plant/vegetable material.

(8) ***Counter-current Extraction:*** In counter-current extraction (CCE), wet raw material is pulverized to produce fine slurry. In this process, the material to be extracted is moved in one direction within a extractor where it comes in contact with extraction solvent. Complete extraction is thus possible when the quantities of solvent and material and their flow rates are optimized. The process is highly efficient, requiring little time and posing no risk from high temperature

The traditional techniques of solvent extraction of plant materials are mostly based on the correct choice of solvents and the use of heat or/and agitation to increase the solubility of the desired compounds and improve the mass transfer. Usually the traditional technique requires longer extraction time, thus running a severe risk of thermal degradation for most of the phyto-constituents. The fact that leafy plant material can contain up to several thousand secondary metabolites, makes the need for the development of high performance and rapid extraction methods an absolute necessity. Keeping in pace with such requirements recent times has witnessed the use and growth of new extraction techniques with shortened extraction time, reduced solvent consumption, increased pollution prevention concern and with special care for thermo labile constituents[4,5,6].

III. MODERN METHOD OF EXTRACTION (NON-TRADITIONAL) HEADSPACE TRAPPING TECHNIQUES

Headspace trapping is essentially a gas extraction technique permitting the direct analysis of volatile compounds present in a non-volatile matrix. Schematic view of head space process is given in Fig. 1. This technique is needed because the aromas of living plant materials are different from those of the extracted oil. Headspace trapping permits getting closer to the natural aroma of the living plant, and gives a clearer view of the differences in volatile constituents between the living plant and the extracted phase. There are two types of headspace trapping: static and dynamic, which is also called the purge-and-trap method[7].

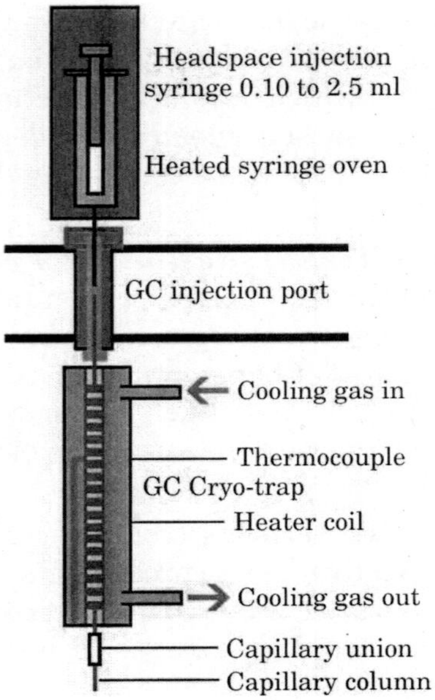

Fig. 1: Schematic view of static head space process.

Types of Headspace Trapping

Headspace trapping can be static or dynamic, which is generally called the purge-and-trap method. In static headspace trapping, gas extraction is carried out in a single step or in a limited number of steps.

(i) *Static headspace*

Principle

If a complex material, such as a piece of vegetable, is placed into a sealed vessel and allowed to stand, some of the more volatile compounds in the sample matrix will leave the sample and pass into the headspace around it. If the concentration of such a compound reaches about 1 ppm in the headspace, then it may be assayed by a simple injection of an aliquot of the atmosphere in the vessel.

First, the sample – a liquid or a solid – is placed in a vessel having a gas volume above it, and the vessel – usually a vial – is closed. This vial is then thermostatted at a constant temperature until equilibrium is reached between the two phases. Then an aliquot of the vials gas phase (the headspace) is introduced into the carrier gas stream which carries it into the column, where it is analyzed in the usual way.

Equipment

This is a single-step gas extraction. By thermosetting the sample for a certain time at a pre-selected temperature, equilibrium is reached between the sample phase and the gas phase of the sample vial. Subsequently, a single aliquot of the headspace is introduced into the carrier gas flow, which then carries it to the column where the volatile compounds are separated in the usual way. The equilibrium of the two phases in the sample vial is characterized by a partition coefficient representing the ratio of the analytes concentration in the sample phase and in the gas phase[8].

Advantages

(a) The sample analytes are not extracted from the sample material using a solvent, there is no need to deal with solvent reduction by evaporation either into the air, with its concerns about pollution, or by recondensing.

(b) It has ability to analyze a sample for low molecular weight volatiles without the presence of a solvent peak.

(c) This is especially important because many samples analyzed by static head space are actually being assayed for residual solvent content.

(d) Packaging, pharmaceuticals, and many other processed materials incorporate the use of solvents in some step of their production, and the amount of those solvents retained in the finished product must be determined.

(e) It is relatively low cost per analysis. Also the sample preparation is simple and the reagents can be eliminated easily.

Disadvantages

(a) Head space analysis can inject only a fraction of the compound of interest to the chromatograph, since the concentration in the headspace is in equilibrium with in the sample matrix, and only a portion of the head space is withdrawn and transferred.

(b) Static headspace techniques may lack the sensitivity required for the determination when the analyte concentration is low.

(c) Analytes at fairly low temperatures also limit the usefulness of static head space for analytes with higher boiling points.

(d) Reproducibility depends on analyzing a sample after it has reached equilibration, and the time required to achieve this point may, especially for less volatile compounds, be a drawback for some analyses.

(ii) Dynamic headspace

Principle

The dynamic headspace technique is principally a continuous method of gas extraction and separates volatile sample constituents from the matrix by a constituents from the matrix by a continuous flow of an inert gas either above a solid or liquid sample, preferably an aqueous one; this technique is known as purge and trap method.

It involves moving the analytes away from the sample matrix in the headspace phase. Instead of allowing sample volatiles to come to equilibrium between the sample matrix and the surrounding headspace, the atmosphere around the sample material is constantly swept away by a flow of carrier gas, taking the volatile analytes with it. This performs two functions relative to the concentrations of the volatiles. First, it prevents the establishment of an equilibrium state, causing more of the volatile dispersed in the sample matrix to leave the sample and pass into the headspace. Second, it increases the size of the headspace sample used beyond the limit of the actual sample vessel.

Equipment

Dynamic headspace sampling involves sweeping the analyte with a stream of air or gas and adsorption of the volatiles from the gas stream on an adsorbent trap. Dynamic headspace sampling techniques can be applied in one of the following ways:

1. Closed-loop Stripping Method
2. Direct Sampling Method

In this technique, the sample is continuously purged with an inert gas (the purge gas), until all volatile compounds are removed. During this step, the gas effluent leaving the sample vessel is conducted through a trap, either cooled to low temperature or containing an adsorbent. This trap retards the volatile analytes purged from the sample. When gas extraction is complete, the condensed or adsorbed analytes by rapid heating of the trap now get purged with the carrier gas. Then the desorbed analytes are conducted directly into the gas chromatograph. Thermal desorption from an adsorbent is not instantaneous: thus, the sample "slug" might be too long, creating broad peaks, with tailing. This is particularly the case when a capillary column is used in the gas chromatograph. For this reason, usually a second, small trap, cooled to low temperature, is placed in the carrier gas line between the primary trap and the column. When desorption is finished, this small trap is then heated very rapidly: in this way, a sharp band of the analytes enters the column[9].

Advantages

(a) It offers same advantages of static headspace, including elimination of the solvent peak, analysis of just the volatiles, automation, and easy sample preparation.
(b) In addition, the trapping stage of the analysis offers increased sensitivity, permitting the analysis of volatiles present at the parts per billion (ppb) level routinely. With careful attention to contaminants and instrument background, they are capable of routine application in the parts per trillion (ppt) range.

Disadvantages

(a) This is more complex and more expensive to purchase than other types of sample introduction systems, because of several steps of monitoring, valving, heating zones and so on.
(b) Because of the functioning of the instrument, there are many opportunities for malfunctions, including heater damage, valve leaking, contamination, and cold spots.

SOLID PHASE MICRO-EXTRACTION

Solid phase microextraction (SPME) is a solvent-free extraction technique that presents some of the characteristics outline before as primordial in new sample preparation strategies. In SPME, a fused silica fiber is coated with a stationary phase. The fiber is exposed to a sample matrix until equilibrium is established between the analyte in the sample and on the fiber (Analyte extraction and pre-concentration are combined in single step). The analyte is then desorbed from the fiber at a high temperature in a chromatograph injector, usually GC, and subsequently analyzed by GC. The schematic diagram of SPME device is given in Fig. 2.

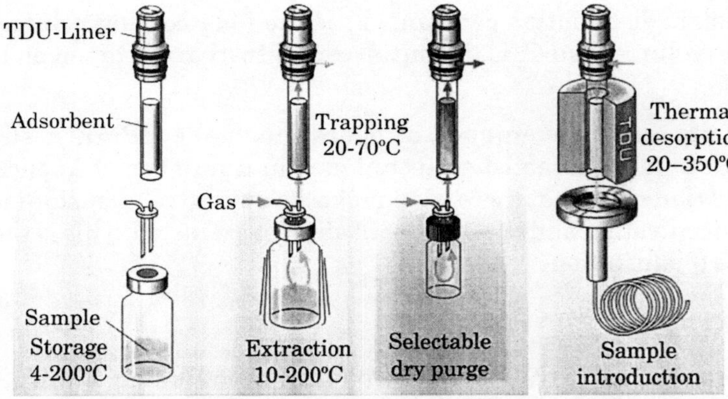

Fig. 2: Schematic view of dynamic head space process.

SPME was developed to provide a quick and solventless technique for the isolation of analytes from a sample matrix. The traditional methods by which the analytes of interest were isolated are typically time and labor-intensive and involve multistep procedures, which could reduce sensitivity. Also, the use of solvents can be hazardous to the operator's health and can damage the environment.

SPME was developed from the technique of solid phase extraction, but the sorbing material is permanently attached to the fiber, allowing reuse of the extracting phase. SPME uses a small volume of sorbent, typically dispersed on the surface of small fibers, to isolate and concentrate analytes from the sample matrix. After contact with the sample, analytes are absorbed or adsorbed by the fiber phase (depending on the nature of the coating). After the extraction step, the fibers are transferred, with a syringe like handling device, to the analytical instrument, for separation and quantification of the analytes. This technique integrates sampling, extraction and sample introduction, and is a simple way of performing on-site monitoring[10].

Principle

SPME is based on the partition equilibrium of target analytes between a polymeric stationary phase, which is a coated fused silica fiber, and the sample matrix. In order to extract analytes SPME does not require organic solvents, which are expensive and may be harmful to health and the environment. The mass extracted and the linear range depend on the partition coefficient (K_{fS}.) and the volume of the stationary phase. The transport of analytes from the matrix into the coating begins as soon as the coated fiber has been placed in contact with the sample. The equilibrium conditions can be described as:

$$n = \frac{K_{fs}V_f V_s C_o}{K_{fs}V_f \quad V_s} \qquad ...(1)$$

where n is the amount extracted by the coating, K_{fS}. is a fiber coating/ sample matrix distribution constant, V_f is the fiber coating volume, V_S is the sample volume, and C_o is the initial concentration of a given analyte in the sample.

Simplicity and convenience of operation make SPME a superior alternative to more established techniques in a number of applications. Equation (1) indicates that there is a direct proportional relationship between sample concentration and the amount of analyte extracted. This is the basis for analyte quantitation.

Extraction modes

Two basic types of extractions can be performed using SPME, direct extraction and headspace extraction. In the direct extraction mode, the coated fiber is inserted directly into the sample and the analytes are transported directly from the sample matrix to the extracting phase. To facilitate rapid extraction, some level of agitation is required to transport analytes from the bulk of the solution to the vicinity of the fiber.

For gaseous samples, natural convection of air is sufficient to facilitate rapid equilibration. For aqueous matrices, more efficient agitation techniques, such as fast sample flow, rapid fiber or vial movement, stirring or sonication is required.

In the headspace mode, the analytes need to be transported through the barrier of air before they can reach the coating;this modification serves primarily to protect fiber coating from damage by high molecular weight and other non-volatile interferences present in the sample matrix, such as humic materials. This headspace mode also allows modification of the matrix, such as a change of the pH, without damaging the fiber.

Amounts of analyte extracted into the coating from the same vial at equilibrium using direct and headspace sampling are identical as long as sample and gaseous headspace volumes are the same. This is caused by the fact that the equilibrium concentration is independent of fiber location in the sample/headspace system. If the above condition is not satisfied, a significant sensitivity difference between the direct and headspace approaches exists only for very volatile analytes[11].

SPME equipment

The configuration of SPME is a small, fused silica fiber, usually coated with a polymeric phase. The fiber is mounted for protection in syringe like device. The analytes are absorbed or adsorbed by the fiber phase until equilibrium

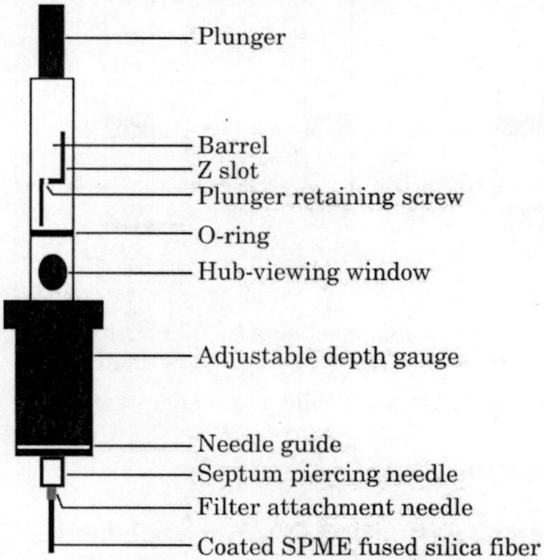

Fig. 3: Solid phase microextraction (SPME) device.

is reached in the system. The amount of an analyte extracted by the coating at equilibrium is determined by the magnitude of the partition coefficient of the analyte between the sample matrix and the coating material.The schematic diagram of soild phase microextraction device is given as Fig. 3.

Extraction procedure

Extraction steps of SPME is given in Fig. 4. The coated fibre is immersed directly in the sample or the Head Space (HS) of the sample, where the analytes are concentrated. After equilibrium has been reached (from a few minutes to several hours depending on the properties of the analytes measured) or after a defined time the fibre is withdrawn and transferred either to a GC injection port or a modified Rheodyne or HPLC valve. The fibre is exposed and the analyte is desorbed either thermally in the hot GC injector porter (in the case of HPLC) eluted by the mobile phase and subsequently chromatographed in a conventional manner. With very complex matrices such as sludges, biological fluids and food products or using solid samples, the SPME technique is mainly applied for the extraction of analytes from the HS of the sample. In both sampling (immersion or HS) modes the agitation of the sample matrix (*e.g.* stirring, sonication) improves transport of analytes from the bulk sample phase to the vicinity of the fibre[12].

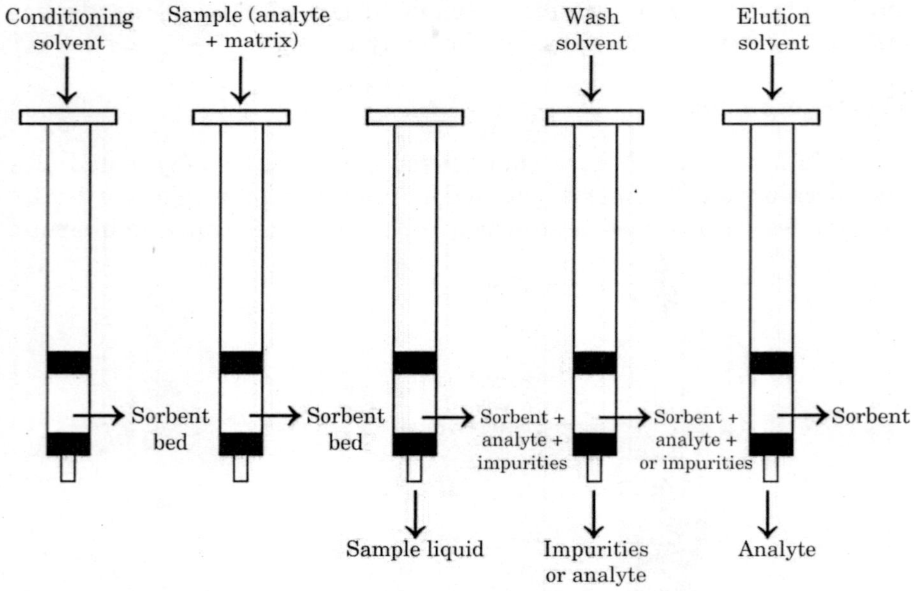

Fig. 4: Extraction steps of solid phase micro-extraction (SPME).

Solid phase extraction steps

SPME sampling can be performed in three basic modes: direct extraction, headspace trapping, and extraction with membrane protection.

Direct extraction: In direct extraction, the coated fiber is inserted into the sample and the analytes are transported directly from the sample matrix to the extracting phase. To facilitate rapid extraction, some agitation is required to transport the analytes from the bulk of the sample to the vicinity of the fiber. For gaseous samples, natural flow of air (*e.g.* convection) is usually sufficient to facilitate rapid equilibrium for volatile analytes.

***Headspace mode*:** In headspace mode, the analytes are extracted from the gas phase equilibrated with the sample. The primary reason for this modification is to protect the fiber from the adverse effects caused by non-volatile, high molecular weight substances present in the sample matrix (*e.g.* vegetable acids or proteins). Here, the amount of an analyte extracted by the fiber coating does not depend on the location of the fiber, in the liquid or gas phase, therefore, the sensitivity of headspace trapping is the same as that of direct sampling as long as the volumes of the two phases are the same in both sampling modes[13,14].

Membrane protection mode: In membrane protection mode, the fiber is separated from the sample with a selective membrane, which lets the analytes through while blocking the interferences. The main purpose for the use of the membrane barrier is to protect the fiber against adverse effects caused by high molecular weight compounds when dirty samples are analyzed. While headspace trapping serves the same purpose, membrane protection enables the analysis of less volatile compounds. Use of thin membranes and an increase in extraction temperature result in shorter extraction times[15].

Advantages

(a) During desorption of the analyte, the polymeric phase is cleaned and ready for reuse.

(b) Absence of solvent makes SPME environmentally friendly, faster separation, increases throughput and allows for use of simpler instruments.

(c) Smaller size which is compactable for field work. Amount of extracting phase is small and equilibrium of system is not disturbed. Further, very small objects can be studied.

(d) High sensitivity, direct sample analysis and limit of determination.

(e) Ranges of analytes that can be analyzed include volatile, semivolatile, nonvolatile, and inorganic species.

(f) Coupled with other instruments besides GC like CE, LC, and MS.

(g) When compared to similar extraction methods, SPME has a better detection limit, precision, cost, time, solvent use, and simple

Disadvantages

(a) Can get relatively expensive if one is not careful with fibers due to the high cost.

(b) Polymer coating is fragile, easily broken, and have limited lifetime.

(c) Its main limitation is its reduced concentration capability due to the small volume of polymer coating on the fiber.

SUPERCRITICAL FLUID EXTRACTION

Supercritical fluid extraction (SFE) is the process of separating one component (the extractant) from another (the matrix) using supercritical fluids as the extracting solvent. Extraction is usually from a solid matrix, but it can also be from liquids. SFE can be used as a sample preparation step for analytical purposes, or on a larger scale to either strip unwanted material from a product or collect a desired product (*e.g.* essential vegetable oils). Carbon dioxide (CO_2) is the most used supercritical fluid, sometimes modified by co-solvents such as ethanol or methanol. Extraction conditions for supercritical CO_2 are above the critical temperature of 31°C and critical pressure of 7400 kPa. Addition of modifiers may slightly alter this. Supercritical extraction mostly uses carbon dioxide at high pressure to extract the high value products from natural materials. Unlike other processes, the extraction process leaves no solvent residue behind. Moreover the CO_2 is non-toxic, nonflammable, odorless, tasteless, inert, and inexpensive. Due to its low critical temperature 31°C, carbon dioxide is known to be perfectly adapted in aromas, essential oils and nutraceutical industries.

Supercritical Fluid

A supercritical fluid is any substance at a temperature and pressure above its critical point. It can diffuse through solids like a gas, and dissolve materials like a liquid. Additionally, close to the critical point, small changes in pressure or temperature result in large changes in density, allowing many properties of a supercritical fluid to be "fine-tuned". Supercritical fluids are suitable as a substitute for organic solvents in a range of industrial and laboratory processes. Carbon dioxide and water are the most commonly used supercritical fluids, being used for decaffeination and power generation, respectively. CO_2 is the kind of extraction solvents for botanicals. It leaves no toxic residue behind. Its extraction properties can be widely and precisely manipulated with subtle changes in pressure and temperature[16].

There are 2 types of CO_2 extraction

1. ***Low pressure cold extraction:*** It involves chilling CO_2 at 1.7–12.7°C and pumping it through the plant material at between 5500–10300 kPa.

2. ***Supercritical extraction:*** It involves heating the CO_2 to above 30.6°C and pumping it above 7580 kPa. Usually, this is between 41400-68950 kPa. Supercritical fluid CO_2 can best be described as a dense fog when CO_2 is used in a dense liquid state. Low-pressure CO_2 is often the best method for producing high quality botanical extracts. CO_2 loading rate in this state means that you have to pump many volumes of CO_2 through botanical.

Properties of Supercritical Fluid

1. Supercritical fluids have highly compressed gases, which combine properties of gases and liquids in an intriguing manner.
2. Supercritical fluids can lead to reactions, which are difficult or even impossible to achieve in conventional solvents.
3. Supercritical fluids have solvent power similar to light hydrocarbons for most of the solutes. However, fluorinated compounds are often more soluble in supercritical CO_2 than in hydrocarbons; this increased solubility is important for polymerization.
4. Solubility increases with increasing density (*i.e.* with increasing pressure). Rapid expansion of supercritical solutions leads to precipitation of a finely divided solid. This is a key feature of flow reactors.
5. The fluids are commonly miscible with permanent gases (*e.g.* N_2 or H_2) and this leads too much higher concentrations of dissolved gases than can be achieved in conventional solvents[17].

Supercritical fluid extraction process: The system must contain a pump for the CO_2, a pressure cell to contain the sample, a means of maintaining pressure in the system and a collecting vessel. The liquid is pumped to a heating zone, where it is heated to supercritical conditions. It then passes into the extraction vessel, where it rapidly diffuses into the solid matrix and dissolves the material to be extracted. The dissolved material is swept from the extraction cell into a separator at lower pressure, and the extracted material settles out. The CO_2 can then be cooled, recompressed and recycled, or discharged to atmosphere. The schematic diagram of supercritical fluid extraction apparatus is given as Fig. 5.

Pumps: Carbon dioxide is usually pumped as a liquid, usually below 5°C and a pressure of about 5000 kPa. The solvent is pumped as a liquid as it is then almost incompressible. As a supercritical fluid, much of the pump stroke will be "used up" in compressing the fluid, rather than pumping it. For small-scale extractions (up to a few grams/minute), reciprocating CO_2 pumps or syringe pumps are often used. For larger scale extractions, diaphragm pumps are most common. The pump heads will usually require cooling, and the CO_2 will also be cooled before entering the pump.

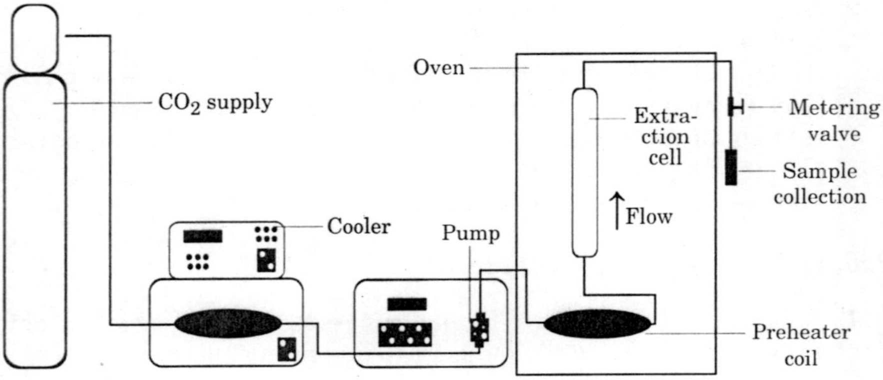

Fig. 5: Schematic diagram of supercritical fluid extraction apparatus.

Fluid reservoir: A gas cylinder provides a source of Supercritical Fluid (*e.g.*, CO_2). Both syringe and reciprocating pumps can be used as solvent delivery systems.

Pressure vessels: Pressure vessels can range from simple tubing to more sophisticated purpose built vessels with quick release fittings. The pressure requirement is at least 7400 kPa, and most extractions are conducted at less than 35000 kPa. However, sometimes, higher pressures will be needed, such as extraction of vegetable oils, where pressures of 80000 kPa are sometimes required for complete miscibility of the two phases. The vessel must be equipped with a means of heating. It can be placed inside an oven for small vessels, or oil or electrically heated jacket for larger vessels. Care must be taken if rubber seals are used on the vessel, as the CO_2 may dissolve in the rubber, causing swelling, and the rubber will rupture on depressurization.

Pressure maintenance: The pressure in the system must be maintained from the pump right through the pressure vessel. In smaller systems (up to about 10 mL/min), a simple restrictor can be used. This can be either a capillary tube cut to length, or a needle valve, which can be adjusted to maintain pressure at different flow rates. In larger systems a backpressure regulator will be used, which maintains pressure upstream of the regulator by means of a spring, compressed air, or electronically driven valve. Whichever is used, heating must be supplied, as the adiabatic expansion of the CO_2 results in significant cooling. This is problematic, if water or other extracted material is present in the sample, as this may freeze in the restrictor or valve and cause blockages.

Collection: The supercritical solvent is passed into a vessel at lower pressure than the extraction vessel. The density, and thus, dissolving power, of supercritical fluids varies sharply with pressure, and hence, the solubility in the lower density CO_2 is much lower, and the material precipitates for collection. It is possible to fractionate the dissolved material using a series of vessels at reducing pressure. The CO_2 can be recycled or depressurized to atmospheric pressure and vented. For analytical SFE, the pressure is usually dropped to atmospheric, and now the gaseous carbon dioxide is bubbled through the solvent to trap the precipitated components.

Heating and cooling: This is an important aspect. The fluid is cooled before pumping to maintain liquid conditions, and then heated after pressurization. As the fluid is expanded into the separator, heat must be provided to prevent excessive cooling. For small-scale extractions, such as for analytical purposes, it is usually sufficient to pre-heat the fluid in a length of tubing inside the oven containing the extraction cell. The restrictor can be electrically heated, or even heated with a hair dryer. For larger systems, the energy required during each stage of the process can be calculated using the thermodynamic properties of the supercritical fluid[18].

Optimization: The optimum will depend on the purpose of the extraction. For an analytical extraction is to determine, say, antioxidant content of a polymer, then the essential factors are complete extraction in the shortest time. However, for production of an essential oil extract from a plant, then quantity of CO_2 used will be a significant cost, and "complete" extraction is not required; a yield of 70–80% perhaps being sufficient to provide economic returns. In another case, the selectivity may be more important, and a reduced rate of extraction will be preferable if it provides greater discrimination.

Maximizing diffusion: This can be achieved by increasing the temperature, swelling the matrix, or reducing the particle size. Matrix swelling can sometimes be increased by increasing the pressure of the solvent, and by adding modifiers to the solvent. Some polymers and elastomers in particular are swelled dramatically by CO_2, with diffusion being increased by several orders of magnitude in some cases.

Maximizing solubility: Generally, higher pressure will increase solubility. The effect of temperature is less certain, as close to the critical point, increase in the temperature causes a decrease in density, and hence, dissolving power. At pressures well above the critical pressure, solubility is likely to increase with temperature. Addition of low levels of modifiers (sometimes called entrainers), such as methanol and ethanol, can also significantly increase solubility, particularly of more polar compounds.

Optimizing flow rate: The flow rate of CO_2 should be measured in terms of mass flow rather than by volume because the density of the CO_2 changes according to the temperature both; before entering the pump heads and during compression. Coriolis flow meters are best used to achieve such flow confirmation. To maximize the rate of extraction, the flow rate should be high enough for the extraction to be completely diffusion limited (but this will be very wasteful of solvent). However, to minimize the amount of solvent used, the extraction should be completely solubility limited (which will take a very long time). Flow rate must therefore be determined depending on the competing factors of time and solvent costs, and also capital costs of pumps, heaters and heat exchangers. The optimum flow rate will probably be somewhere in the region where both solubility and diffusion are significant factors

Equipment

Instrument components include a fluid source, commonly a tank of carbon dioxide followed by a syringe pump having a pressure rating of at least 350000 kPa, a valve to control the flow of the critical fluid into a heated extraction cell having a capacity of a few ml, and lastly an exit valve leading to a flow restrictor that depressurizes the fluid and transfers it into a collection device. The supercritical fluid extraction model is given in Fig. 6.

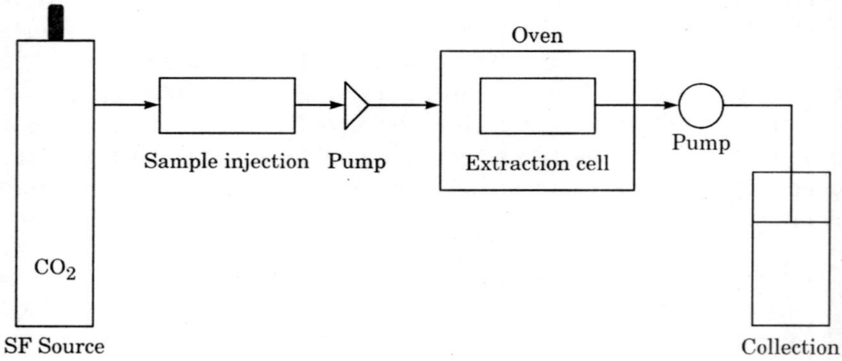

Fig. 6: Supercritical fluid extraction process model.

Procedure

The feed, containing the substance of interest comes in contact with supercritical CO_2, at suitable temperature and pressure, in an extraction device. In this simple scheme, the component is selectively extracted and must be recovered from the supercritical solution, which is usually a dilute one. Product recovery occurs in the separation section, whose temperature and pressure can be adjusted in order to optimize the amount of the component produced. Due to the low solubility in supercritical CO_2, after recovery of the product of interest the solvent must be recycled and pumped back to the extractor, in order to minimize operating costs. It is also noteworthy that the separator can be operated either at the same temperature or at the same pressure of the extractor, the best condition resulting from an economic analysis of the overall production costs. If the temperature is kept constant, product separation is achieved by depressurization and mechanical energy has to be provided to the system to raise the CO_2 pressure from the separator to the extractor conditions. On the other hand, extracted products can be separated from CO_2 by increasing the temperature, and thermal energy must be supplied, where the circulation of the solvent can be done at nearly isobaric conditions. The separation of products from CO_2 is achieved can be more complex: for instance, both temperature and pressure can be varied when passing from the extractor to the separator sectors, or a solid can be used to promote separation by adsorption[19].

Mobile phase: The most widely used mobile phase for SFE is carbon dioxide. It is an excellent solvent for a variety of organic molecules. In addition, it transmits in the ultraviolet and is odorless, nontoxic, readily available, and remarkably inexpensive when compared with other chromatographic mobile phases. Sample Matrix Parameters of Supercritical Fluid Extraction is influenced by particle size and shape, surface area and porosity, moisture content, changes in morphology, sample size, extractable level etc.

The Parameters Effect on Solubility

- The vapor pressure of the component
- Interaction with the supercritical fluid
- Temperature, pressure, density and additives

Detectors: A major advantage of SFC/SFE over HPLC is that the flame ionization detector of gas chromatography can be employed. Mass spectrometers are also more easily adapted as detectors for SFE than HPLC. The following detectors can be adopted

(i) UV detector
(ii) Fluorescence detector
(iii) Flame ionization detector
(iv) Electron capture detector
(v) Mass spectrometric detector

Advantages

(a) It can be easily recycled
(b) Chemically inert and Non flammable
(c) Has a good solvent characteristic for non-polar and slightly polar solutes.
(d) It is a “natural” substance, present in mineral waters and part of the life cycle

(e) It is easily removed from the product; the dissolving power and selectively can be controlled by selection of suitable pressure/ temperature combination.

(f) It has a convenient critical temperature (31.04ºC). This enables extractions to be carried out at comparatively low temperature (often as low as 40 or 50ºC), decreasing the risk of damage of thermalabile compounds.

(g) Extraction of natural raw material with supercritical CO_2, allows the obtaining of extracts which flavour and taste are perfectly respected and reproducible

(h) The supercritical fluid ability to vaporise non-volatile components (at moderate temperatures) reduces the energy spent, when comparing to distillation.

(i) The number of solvents possible to be used on supercritical extraction is superior of classic organic solvents

(j) Supercritical fluids have a superior selectivity although they have an inferior solvent power than classic organic solvents

(k) Supercritical extraction can be connected directly to a chromatograph, allowing analysis immediately after extraction.

Disadvantages

(a) Modelling is inaccurate, expensive and prolonged diffusion rate of solute from the solid to SCF.

(b) Scale is not possible (due to absence of fundamental, molecular-based model of solutes in SCF)

(c) Consistency and reproducibility may vary in continuous production

SIMULTANEOUS DISTILLATION EXTRACTION

Simultaneous distillation extraction (SDE), introduced in 1964 by Likens and Nickerson, is one of most widely employed. This method has been successfully applied in the extraction of essential vegetable oils, aroma compounds and other volatile products from numerous matrices. In the flavour area, this technique is usually considered to be superior to classical ones, such as distillation or solvent extraction. This technique combines steam distillation together with continuous extraction with a solvent or a mixture of solvents. This one-step extraction technique is less time consuming and allows, due to the continuous recycling, a greater reduction of solvent volumes. Under certain conditions a higher concentration factor can be achieved and the direct analysis of the extract can be performed without further concentration. Moreover, given its particular characteristics, this technique allows to carry out the analysis without a sample clean-up step. The extracts obtained by SDE are free from non-volatile materials such as chlorophylls. Because of the high extraction efficiencies associated to the high reproducibility, SDE has also been used to quantify volatile compounds in several matrices. The recoveries are however dependent on the operation conditions and therefore the best experimental conditions should be investigated[20,21].

Optimization of SDE Extraction

SDE exploits the differences in volatility and polarity among the analytes and other non-volatile components present in the matrix. The extracting solvent plays a key role in the process; it has to be chosen to facilitate the extraction of the components that are important as well as to exclude or limit the components that could interfere with the analysis. Another parameter of strong influence is the extraction time. For these reasons the influence of the quoted parameters (solvent nature and extraction time) on the SDE selectivity and extraction yield was investigated in order to find the best experimental conditions to recover the selected compounds[20].

An amount of fragrance generating sample along with distilled water for dry samples, is contained in the reactor, and aqueous phase receives a convenient volume of an extracting solvent denser than water (dichloromethane, chloroform, etc.). The schematic diagram of simultaneous distillation extraction is given in Fig. 7. They are heated, the water and solvent vapors are conducted to the extractor body, where they are allowed to condense over the surface of the cold finger. In this operation, aroma analytes are removed from the matrix by the water vapour and transferred to the organic phase when the liquids condense together on the cooler. Both water and solvent are collected in the extractor body after their condensation and return to the corresponding flasks, allowing continuous reflux. A modified apparatus allows use of solvents lighter than water, such as pentane or ethyl acetate, as extractors[22,23].

Advantages

(a) Efficient stripping of volatiles from food.

(b) This method is more efficient for extracting phenols, free fatty acids, lactones and heavier aldehydes, ketones, alcohols and esters.

(c) Aroma extract will be obtained in a solvent; so many injections can be performed in a single extraction. Hence, one sample can provide material for GC, GC-MS and GC-O[24].

Disadvantages

(a) Possible occurrences of undesirable ester hydrolysis, Maillard reaction and sugar degradation might happen due to its high temperature process.

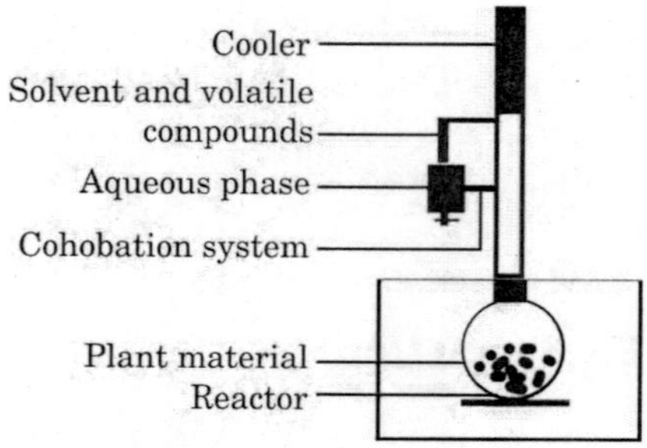

Fig. 7: Simultaneous distillation extraction (SDE) setup.

(b) Loss of low boiling volatiles will occur.

MICROWAVE DISTILLATION

***Microwave Energy*:** Microwaves are a form on non-ionizing electromagnetic energy. This electromagnetic energy can be absorbed, reflected or transmitted to a material in the form of waves at a frequency ranging from approximately 300 MHz to 300 GHz. Domestic microwave units typically operate at a frequency of 2450 MHz while the Federal Communications Commission (FCC) has approved a frequency of 915 MHz for industrial applications. Microwave power that is transmitted through the bulk of a material can be absorbed and converted to heat. There are two proposed mechanisms of microwave heating: dipole rotation and ionic polarization[25].

The mechanism of dipole rotation is based on the fact that many molecules exist as electric dipoles, meaning that the molecule has a negatively charged end and a positively charged end. Water, for example, consists of two hydrogen atoms carrying a slightly positive charge and one oxygen atom having a slightly negative charge. When placed in an electromagnetic field, dipoles attempt to orient themselves according to the polarity of the field, which changes about 4.9 × 109 times per second. The constant rotation of the molecules causes frictional heat. Ionic polarization occurs when ionic compounds such as salt move at an accelerated pace in response to an alternating electric field. The kinetic energy generated by the movement of the ions is converted to thermal energy. The mechanisms mentioned above clearly indicate that dielectric materials and solvents having a permanent dipole moment heat most efficiently in microwaves[26].

Principle

The fundamentals of the microwave extraction (MAE) process are different from those of conventional methods (solid–liquid or simply extraction) because the extraction occurs as the result of changes in the cell structure caused by electromagnetic waves. The solvent penetrates into the solid matrix by diffusion (effective), and the solute is dissolved until reaching a concentration limited by the characteristics of the solid. The solution containing the solute diffuses to the surface by effective diffusion. Finally, by natural or forced convection, the solution is transferred from the surface to the bulk solution. The extraction process takes place in three different steps: an equilibrium phase where the phenomena of solubilisation and partition intervene, in which the substrate is removed from the outer surface of the particle at an approximately constant velocity. Then, this stage is followed by an intermediary transition phase to diffusion. The resistance to mass transfer begins to appear in the solid–liquid interface; in this period the mass transfer by convection and diffusion prevails. In the last phase, the solute must overcome the interactions that bind it to the matrix and diffuse into the extracting solvent. The extraction rate in this period is low, characterized by the removal of the extract through the diffusion mechanism. This point is an irreversible step of the extraction process; it is often regarded as the limiting step of the process[26].

Many forces, such as the physicochemical interactions and relationships, can be exposed during the extraction (dispersion forces, interstitial diffusion, driving forces, and chemical interactions), and the persistence and strength of these phenomena may be closely tied to the properties of the solvent (solubilisation power, solubility in water, purity, polarity, etc.).

Mechanism of microwave heating: In the microwave heating process, energy transfer occurs by two mechanisms: dipole rotation and ionic conduction through reversals of dipoles and displacement of charged ions present in the solute and the solvent. The details of Microwave heating is given in Fig. 8. In many applications these two mechanisms occur simultaneously. Ionic conduction is the electrophoretic migration of ions when an electromagnetic field is applied, and the resistance of the solution to this flow of ions results in friction that heats the solution. Dipole rotation means rearrangement of dipoles with the applied field. Energy transfer is the main characteristic of microwave heating. Traditionally, in heat transfer of the conventional process, the energy is transferred to the material by convection, conduction, and radiation phenomena through the external material surface in the presence of thermal gradients. In contrast, in MAE, the microwave energy is delivered directly to materials through molecular interactions with the electromagnetic field *via* conversions of electromagnetic energy into thermal energy [27].

Factors Affecting Microwave-Assisted Extraction

Solvent nature and volume: Solvent choice for MAE is dictated by the solubility of the extracts of interest, by the interaction between solvent and plant matrix, and finally by the microwave absorbing properties of the solvent

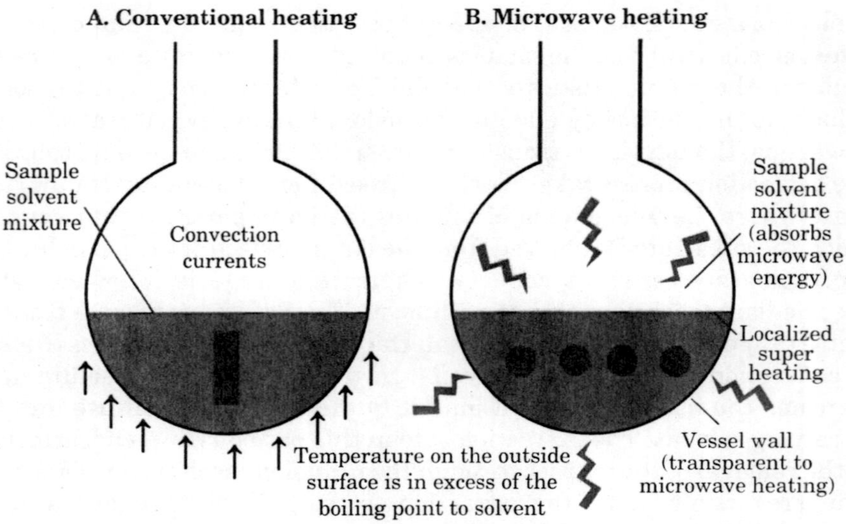

Fig. 8: Principle of microwave heating

determined by its dielectric constant. Both efficacy and selectivity of MAE depend significantly on the dielectric constant of the extraction solvent mixture. Usually, the chosen solvent should possess a high dielectric constant and strongly absorb microwave energy. Solvents such as ethanol, methanol and water are sufficiently polar to be heated by microwave energy. Non-polar solvents with low dielectric constants such as hexane and toluene are not potential solvents for MAE. Using mixtures of solvents can modulate the extracting selectivity and the ability of the solvent to interact with microwaves.

Extraction time: As in other extraction technique, time is another parameter whose influence needs to be taken into account. Generally, by increasing the extraction time, the quantity of analytes extracted is increased, although there is the risk that degradation may occur. Often 15 – 20 min is sufficient, but even 40 sec have been demonstrated to give excellent recovery. A proper study on optimisation of extraction time is vital because extraction time may vary with different plant part used. Irradiation time is also influenced by the dielectric properties of the solvent. Solvents like water, ethanol, and methanol may heat up tremendously on longer exposure thus risking the future of thermolabile constituents.

Microwave power: Microwave power and irradiation time are two such factors, which influences each other to a great extent. A combination of low or moderate power with longer exposure may be a wise approach. Rapid rupture of cell wall takes place at higher temperature when kept at higher power, as a result together with the desired analytes impurities are also leached out into the solvent. Whereas at low power levels the cell wall rupture might take place gradually this enables selective MAE. In closed vessel systems, the chosen power settings depends on the number of samples to be extracted during a single extraction run, as up to 12 vessels can be treated in a single run. The power must be chosen correctly to avoid excessive temperature, which could lead to solute degradation and overpressure inside the vessel.

***Matrix characteristics*:** The plant particle size and the status in which it is presented for MAE can have a profound effect on the recoveries of the compounds. The particle sizes of the extracted materials are generally in the range of 100 µm – 2 mm. Fine powders can enhance the extraction by providing larger surface area, which provides better contact between the plant matrix, and the solvent, also finer particles will allow improved or much deeper penetration of the microwave. One of the disadvantages associated with the use of finer particles is difficulty of separation of the matrix from the solvent after microwave irradiation. Generally, centrifugation or filtration is applied to satisfy the above purpose and use of very fine particles may pose some technical problems. The status of the plant matrix presented for MAE also needs to be evaluated during the extraction process. Sample pre-treatment prior to MAE can bring about effective and selective heating of the plant matrix. The plant matrix may be selectively heated by microwave with the extracting solvent surrounding the sample transparent to microwave. Samples pre-treated with solvents with higher microwave absorbing capacity when coupled with extracting solvents like ethanol or methanol bring about heating by at least two competing mechanisms namely, direct heating from the interaction of microwaves with ethanol and heating from the diffusion of excess heat resulting from the interaction of the microwaves with the pre-treated matrix.

***Temperature*:** Microwave power and temperature are very interrelated to each other and needs to be given special attention particularly when working with closed vessel system. In closed vessel systems, temperature may reach well above the boiling point of the solvent. This elevated temperature does indeed result in improved extraction efficiencies since desorption of analytes from active sites in the matrix will increase. Additionally solvents have higher capacity to solubilizing analytes at higher temperature while surface tension and solvent viscosity decreases with temperature, which will improve sample wetting and matrix penetration respectively. Increase in temperature is also associated with increase in pressure in closed systems, which can raise safety concerns. Temperature can be effectively controlled in open vessel system by proper combinations of extracting solvents which heat up differently[28,29,30].

Principle of Extraction Under Microwave

Microwave-assisted extraction (MAE) offers a rapid delivery of energy to a total volume of solvent and solid plant matrix with subsequent heating of the solvent and solid matrix, efficiently and homogeneously. Because water within the plant/food matrix absorbs microwave energy, cell disruption is promoted by internal superheating, which facilitates desorption of chemicals from the matrix, improving the recovery of neutraceuticals. The change in the plant tissue due to microwave heating gives a considerable increase in the yield of extraction. Furthermore, the migration of dissolved ions increased solvent penetration into the matrix and thus facilitated the release of the chemicals. The effect of microwave energy is thus strongly dependent on the dielectric susceptibility of both the solvent and the solid plant matrix[31].

The moisture when heated up inside the cell due to microwave effect evaporates and generates tremendous pressure on the cell wall due to swelling of the plant cell. The pressure pushes the cell wall from inside, stretching and ultimately rupturing it, which facilitates leaching out of the active constituents from the ruptures cells to the surrounding solvent thus improving the yield of phyto- constituents. This phenomenon can even be more intensified if the plant matrix is impregnated with solvents with higher heating efficiency under microwave[31].

Equipment

Application of Microwave Assisted Extraction (MAE) has increased in recent years due to its inherent advantages (reduction in extraction time and solvent volume) over more traditional extraction techniques(Fig. 9). There are two types of commercially available MAE systems: closed extraction vessels and focused microwave ovens (Fig. 10 and 11). The closed extraction performs extraction under controlled pressure and temperature. The merits and demerits of the closed system is given in Table 1. The focused microwave ovens are also named as focused microwave assisted Soxhlet or solvent extraction (FMASE), in which only a part of the extraction vessel containing the sample is irradiated with microwave. However, both the above-mentioned systems are available as multi- mode and single- mode or focused systems (Fig. 12). A multimode system allows random dispersion of microwave radiation within the microwave cavity, so every zone in the cavity and sample it contains is evenly irradiated. Single mode or focused systems allows focused microwave radiation on a restricted zone where the sample is subjected to a much stronger electric field than in the previous case. Even a modified multimode domestic microwave oven operates as an open vessel extraction system[32].

The use of atmospheric pressure provides certain advantages over pressurized vessels, as follows.

(a) Increased safety results from operating at atmospheric pressure with open vessels.

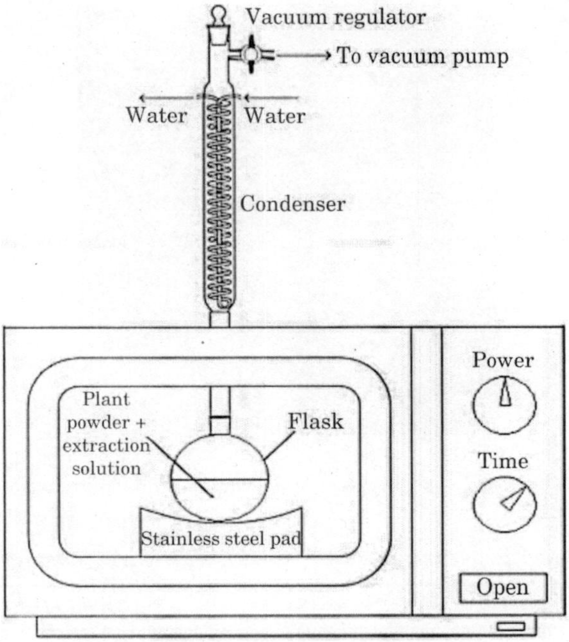

Fig. 9: Microwave assisted extraction setup.

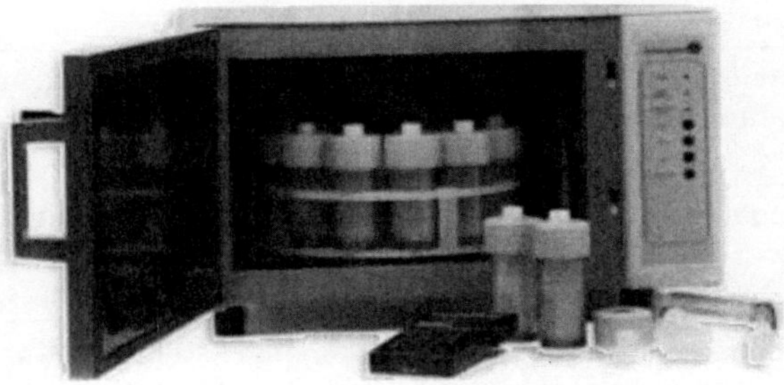

Fig. 10: Microwave assisted extraction unit (*Source*: CEM model, USA).

(b) The ability to add reagents at any time during the treatment.

(c) The ability to use vessels made of various materials, including PTFE, glass and quartz.

(d) The ease with which excess solvent can be removed.

(e) The ability to process large samples.

(f) The absence of any requirement for cooling down or depressurisation.

(g) The low cost of the equipment required.

(h) The ability to go through leaching cycles until quantitative removal of the target species is achieved.

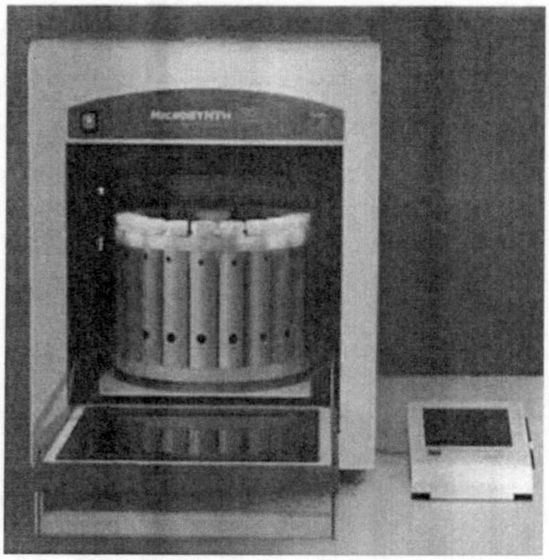

Fig. 11: Closed vessel type of microwave assisted extraction unit.

Table 1: Merits and demerits of closed-vessel systems can be summarized as follows.

Merits	*Demerits*
Can reach higher temperatures than open vessel systems because the increased pressure inside the vessel raises the boiling point of the solvents used. Thereby reducing the time	The high pressures used pose safety (explosion) risks
Loss of volatile substances during microwave irradiation is completely avoided.	The amount of sample that can be processed is limited
Less solvent is required.	The usual constituent material of the vessel, PTFE (polytetrafluoroethylene), does not allow high solution temperatures
Risk of contamination is avoided, as a result there is little or no risk of airborne contamination.	The single-step procedure excludes the addition of reagents or solvents during operation.
Because no evaporation occurs, there is no need to continually add solvent to maintain the volume	The vessel must be cooled down before it can be opened after the treatment to avoid loss of volatile constituents
The fumes produced during an acid microwave extraction are contained within the vessel	

(i) Open-vessel operation is more suitable with thermo labile species as it uses low temperatures relative to closed- vessel systems.

Despite their many advantages, open-vessel systems are also subject to several shortcomings, as follows.

(a) The ensuing methods are usually less precise than those developed using closed-vessel systems.

(b) The sample throughput is lower, as most open systems cannot process many samples simultaneously, while closed-vessel systems can handle up to 20 samples at a time.

(c) The operation times required to obtain results similar to those of closed-vessel systems are usually longer.

Microwave digestion: Microwave digestion is one of the standard sample preparation procedures for elemental determination in analytical chemistry. In closed microwave-transparent pressure vessels the samples are heated together with a suitable acid solution to temperatures that are typically between 200°C and 260°C so that the sample is completely degraded and dissolved. The great advantage of using a microwave field is the speed of heating the sample and the acid solution. However, the precise details of the heating process depend on the nature of the sample, its weight, etc [33].

MAE Instrumentation: There are two types of commercially available MAE systems: Closed extraction vessels and focused microwave ovens (Fig. 12). Both systems are available as multi-mode and single-mode. A multi-mode system allows evenly radiation of sample by random dispersion of microwaves. Single-mode allows focused radiation on a restricted zone by a much stronger electric field. Both multi-mode and focused system comprises of the following four components:

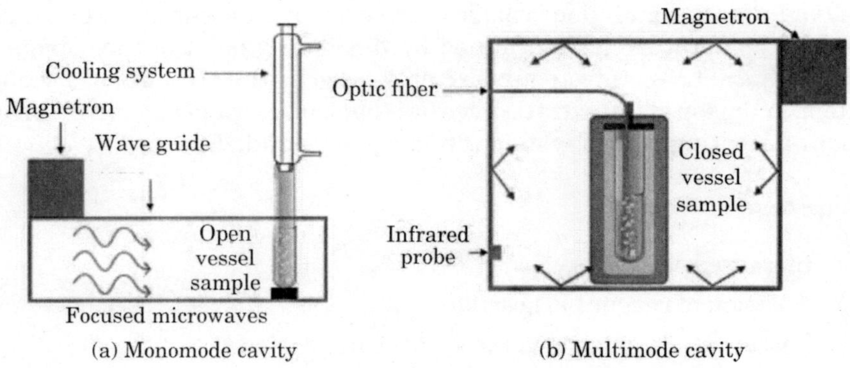

Fig. 12: Multimode and single mode focused MAE system.

(a) *Microwave generator*: Magnetron which generates microwave energy

(b) *Wave guide*: Propagate microwave to microwave cavity

(c) *The applicator*: Where the sample is placed

(d) *Circulator*: Allow microwave to move in forward direction

Closed vessel: With closed vessel extraction, pressurized microwave assisted extraction is performed under pressure (with or without regulation). The pressure allows temperatures above boiling points of the solvents to be reached enhancing speed and efficiency. To avoid overpressure, power, temperature and pressure can be controlled. When dealing with thermolabile compounds, high temperature causes degradation of analytes. Therefore, the open system will be choosing often[33].

Advantages

(a) Decrease in extraction time
(b) Loss of volatile substances is avoided
(c) Less solvent is required because no evaporation occurs
(d) No hazardous fumes during acid microwave since it is a closed vessel

Disavantages

(a) High pressure used pose safety risks
(b) The usual constituent material of the vessel does not allow high solution temperatures
(c) Addition of reagents is impossible since it is a single step procedure
(d) Vessel must be cooled down before it can be opened to prevent loss of volatile constituents.

Open vessel: Extraction is made at atmospheric pressure. The maximum possible temperature is determined by the boiling point of the solvent at that pressure. Losses of vapours are prevented by the presence of a cooling system on the top of the extraction vessel that causes condensation of solvent vapours. Heating in this system is homogenous and efficient.

Advantages

(a) Increased safety
(b) Addition of reagent is possible
(c) Vessels made of various materials can be used
(d) Excess solvent can be removed easily
(e) Ability to process large samples
(f) No requirement for cooling down or depressurization
(g) Low cost of equipment
(h) Suitable for thermolabile products

Disadvantages

(a) The ensuing methods are less precise than in close-vessel system

(b) The sample throughput is lower as open system cannot process many samples simultaneously

(c) Require longer time to achieve same results as for closed system

MEMBRANE EXTRACTION

Membrane extraction is a family of techniques that can be applied to many extraction problems. These techniques require very little solvent and provide remarkable clean-up efficiency. Membrane extraction can also provide very high concentration enrichment factors of hundreds or more. Systems based upon membrane extraction can be automated readily and connected on-line to chromatography systems and other instruments. Both inorganic and organic analytes can be extracted in a wide polarity range.

Membrane extraction has emerged as a promising alternative to conventional sample preparation techniques. It has undergone significant developments in the last two decades and is still evolving. It has been used for the extraction of a wide variety of analytes from different matrices. The sample is in contact with one side of the membrane, which is referred to as the feed (or donor) side. The membrane serves as a selective barrier. The analytes pass through to the other side, referred to as the permeate side. The principle of membrane extraction is given in Fig. 13[34].

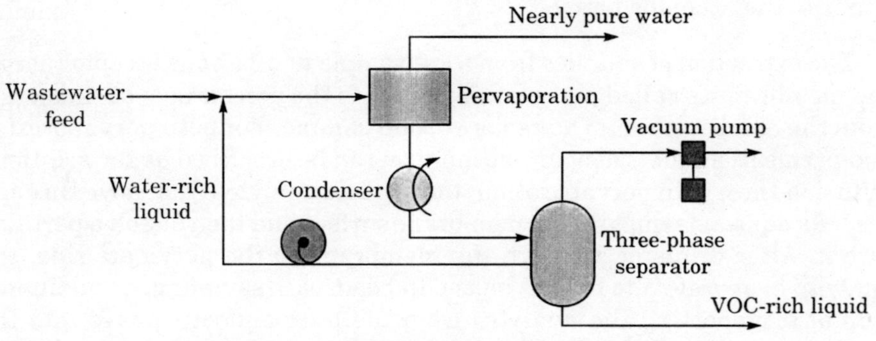

Fig. 13: Flow chart for membrane extraction.

Membrane Systems

Supported liquid membrane extraction is the most versatile membrane extraction technique for analytical sample preparation. In this three-phase extraction technique, analytes are extracted from an aqueous sample into another aqueous phase through an organic liquid. The organic phase is held between the aqueous phases by a porous, hydrophobic supporting membrane. Capillary forces in the pores hold the organic liquid. Thus, two different equilibria are involved, which makes the system chemically analogous to extraction and back-extraction in classic liquid–liquid extraction. The back-extraction step considerably increases the selectivity of the extraction. Supported liquid membrane extraction is applicable to analytes of high or moderate polarity with log Kow (octanol–water partition coefficient) in the range 21 to 13.9 Another type of membrane extraction is based upon a two-phase system, with one aqueous phase and one organic phase; thus, it is equivalent to a single-step liquid–liquid extraction. A microporous hydrophobic membrane separates the two phases, and the organic phase also fills the pores of the membrane to provide a direct contact through a liquid–liquid interface without mixing the phases. This principle is called microporous membrane liquid–liquid extraction.

Liquid Membrane

The pH of the sample can be adjusted to a sufficiently high value, at which the amines are uncharged and therefore capable of being extracted into the organic membrane phase. The sample is pumped through the donor channel, and the amines will partition into the membrane. The acceptor channel on the other side of the membrane is filled with a stagnant acidic buffer, in which an amine molecule that has diffused through the organic membrane is immediately protonated. Because it is protonated, it is prevented from re-entering the membrane and is trapped in the acceptor phase. The resultant driving force continually transports amine molecules from the donor to the acceptor phase.

The extraction of volatiles from a gas sample to a gaseous acceptor across the membrane is called permeation, which is the mechanism of extraction from the headspace of an aqueous or solid sample. For both pervaporation and permeation, the transport mechanism can be described by the solution–diffusion theory. In pervaporation, the organic analytes first move through the bulk aqueous sample to the membrane surface and then dissolve/partition into it. After diffusing through the membrane to the permeate side, the analytes evaporate into the gas phase. In headspace sampling, an additional step of transporting the analytes from the bulk aqueous phase into the headspace is involved. In both cases, the extraction is driven by the concentration gradient across the membrane.

Membranes can be classified as porous and nonporous based on the structure or as flat sheet and hollow fiber based on the geometry. Membranes used in pervaporation and gas permeation are typically hydrophobic, nonporous silicone (polydimethylsiloxane or PDMS) membranes. Organic compounds in water dissolve into the membrane and get extracted, while the aqueous matrix passes unextracted. The use of microporous membrane (made of polypropylene, cellulose, or Teflon) in pervaporation has also been reported, but this membrane allows the passage of large quantities of water. Usually, water has to be removed before it enters the analytical instrument, except when it is used as a chemical ionization reagent gas in MS. It has been reported that permeation is faster across a composite membrane, which has a thin (*e.g.*, 1 mm) siloxane film deposited on a layer of microporous polypropylene[35]. A major advantage of membrane extraction is that it can be coupled to an instrument for continuous online analysis.

PHYTOSOL EXTRACTION

These methods called "Advanced Phytonics", which uses 1, 1, 1, 2-tetrafluoroethane with or without modifiers are known as Phytosol technique. The solvents are named "Phytosols". Phytosol A consists only of the gas, Phytosol B is a mixture of the gas with butane/isobutene, and Phytosol C consists of the gas with dimethyl ether as modifier. Boiling point of 1, 1, 1, 2-tetrafluoroethane or HFC 134 as known in trade is –26.2°C[3]. The flowchart for phytosol extraction is given in given in Fig. 14. The process is aptly applicable to those vegetable compounds that are sensitive to thermal or UV energy, oxygen and acidic conditions.

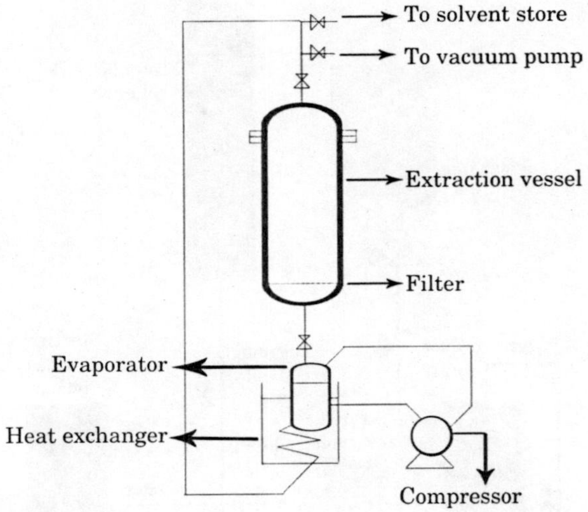

Fig. 14: Phytosol extractor.

PROTOPLAST EXTRACTION TECHNIQUE

Liquid-liquid extraction can be defined as a partition technique where the constituents of a solution are separated due to their different partition coefficients between the two immiscible liquids. Porous support which holds the stationary phase is teflon powder made into a sponge through a special heat treatment. Aqueous dispersion is pumped into the column. Aromatic constituents are transferred from the aqueous phase to the organic phase[3]. The flow chart of Protoplast Extraction Unit is given in Fig. 15.

MICRODISTILLATION

Microdistillation is a micro scale distillation technique which allows rapid and programmable distillation (Fig. 16). To separate a liquid feed distillation process is used, that involves heating a liquid mixture to the point of boiling, whereby the components separate through evaporation and condensation. During this process, different products boil off and are recovered at different temperatures, separating the vegetable samples into broad categories of its component hydrocarbons. The distillation characteristics (*i.e.*, volatility) of hydrocarbons impact the safety and performance of end products, particularly with fuels and solvents. The boiling range gives information on the composition, properties and behavior of the fuel during storage and use[36].

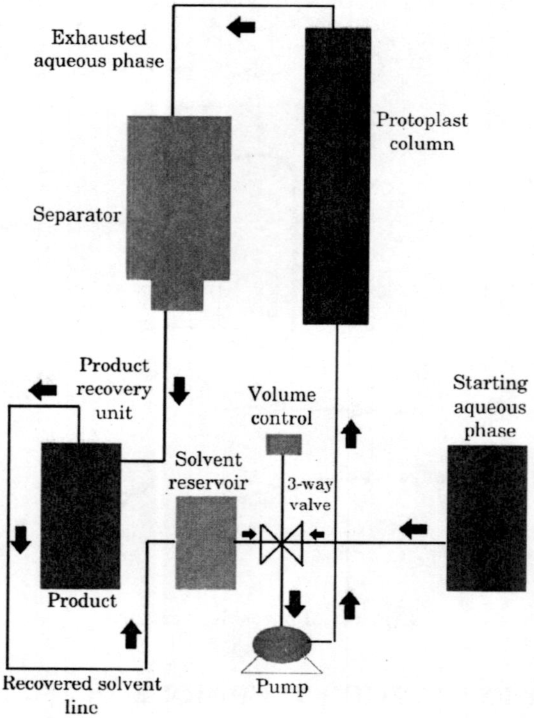

Fig. 15: Protoplast extraction unit.

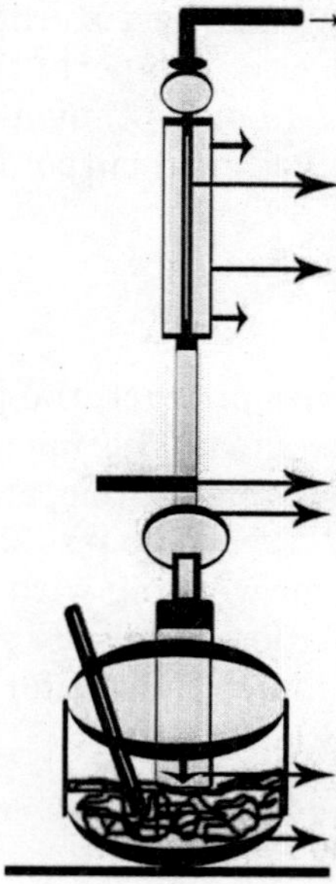

Fig. 16: Micro distillation device.

Equipment

The system has heating and cooling units for parallel operations. Small amount of material to be distilled (100-500 mg) is placed in a vial with water and sealed with a cap having a small slit on top. The key to successful microscale distillations is in avoiding long distillation paths, since this is the main factor leading to loss of material during distillation. Short-path microscale distillations are carried out using the **Hickman distillation head** as the receiving device for the distilled liquid. There are two types of Hickman head, 'ported' and 'unported'. The complete apparatus consists of a flask or vial containing the liquid and a magnetic spin vane or boiling stone, attached to the bottom joint of the Hickman head. If desired, a condenser is attached to the top joint. A thermometer can be suspended down the middle in order to record the distilling temperature, with the bottom of the thermometer in the lower part of the Hickman head just below the circular well. The vapours of the heated liquid rise upward and are cooled and condensed on either the inside walls of the Hickman head or on the walls of the condenser. As liquid drains downward, it will be collected at the circular well, located at the bottom of the still[37].

Collection of fractions is easiest with the ported Hickman head; the port is opened and the liquid in the well removed with a Pasteur pipet. If a condenser or internal thermometer is used, the distilling apparatus must be partially disassembled in order to do this.

Cutpoint optimization: Cutpoints between overhead products and side-cuts can be con-trolled on the basis of temperature, but doing so can downgrade the end product. Since re-finery optimization objectives change daily, re-finers need the ability to optimize cutpoint in consideration of these changing variables[2].

Advantages

By monitoring the boiling point of the product, the processor can get the maximum amount of high-value product. The microdistillation process, provides faster results using small sample volume, and eliminates much of the operator time and subjectivity. It requires less than 1 gram or 1 ml of material, which makes it excellent for working with high value vegetable samples. More over preparation, isolation, readiness and set-up for analysis will be fast. Also, it is cheaper and more efficient for long term usage and easy to operate.

THERMOMICRODISTILLATION

A simple thermo micro procedure is described for the separation of many substances from solid materials and the direct transfer of these substances to the starting point of a thin-layer chromatographic plate (Fig. 17). The sample is introduced into a glass cartridge with a conical tip, and heated rapidly for a short time at a pre-set temperature. The emerging vapours are deposited as a spot on the thin-layer chromatographic plate, which is then chromatographed in the usual way. There are numerous variations of the standard procedure, thus extending the range of application, and this is demonstrated with examples from the fields of phytochemistry, analysis of vegetable additives and residues, and from organic and inorganic chemistry.

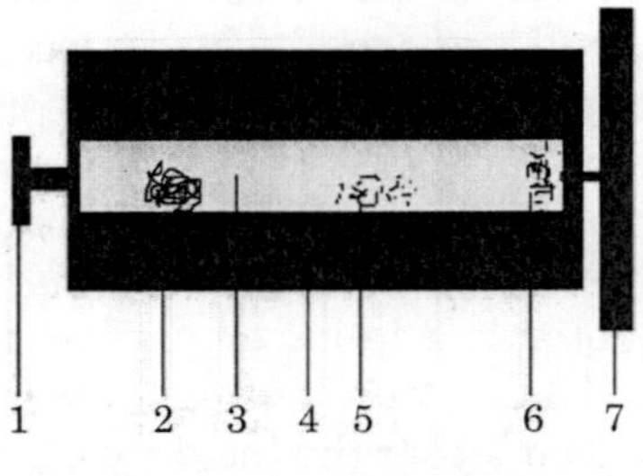

1. Seal; 2. Indicator silica; 3. Glass tube (Pasteur pipette); 4. Oven (220°C); 5. Sample; 6. Glass-wool; 7. TLC Plate.

Fig. 17: Thermomicrodistillation unit.

CONTROLLED INSTANTANEOUS DECOMPRESSION (CID)

The process involves subjecting the partially humidified plant material for a short period of time to a steam pressure varying from 50 to 300 kPa followed by a rapid decompression to a vacuum for 200 msec each time. The vapour in the plant material created by auto vaporization produces a mechanical strength which ruptures the oil cells. Due to pressure differences in the extractor and vacuum chamber, in each opening of the pressure valve oil-rich vapour is instantaneously sucked into the vacuum chamber where it instantly condenses.

The DIC machine has three main parts (Fig. 18).

(a) A processing chamber in which the products are treated at high temperature (up to 170°C) and high pressure (up to 8×10^2 kPa) with a suitable gas (air, inert gas and more commonly steam).

(b) Large vacuum reservoir (50 times the volume of the processing chamber)

(c) Valve with a large diameter (more than 200 mm) that can be opened rapidly (in less than 1/10th of a second)[38].

The DIC technology was applies instant pressure-drop to modify the texture of the material and intensify functional behaviour. DIC treatment usually starts by creating a vacuum condition, followed by injecting steam to the material keeping such a contact for several seconds, proceeds then the sudden pressure drop toward vacuum (about 5 kPa with a rate higher than 0.5 MPa/s). This treatment is also categorized as a HTST (High Temperature Short Time) process. By suddenly dropping pressure, rapid auto vaporisation of the moisture from the material will occur, the material will swell, and lead to texture change which results in higher porosity. It increases the material porosity as well as the specific surface area and reduces the diffusion resistance of moisture during the final dehydration step[38].

Instantaneous controlled pressure-drop consists in heating wet material in an autoclave for a short period of time with saturated steam or microwaves, followed by a rapid expansion to a final pressure lower than 10 kPa. The vapour that is produced from the superheated liquid mechanically strains the material. This stress creates a porous structure, breaks the cell walls and also destroys microorganisms. The exposure of material to a high temperature is limited to a short heating period, after which the temperature decreases rapidly during the expansion and adiabatic auto-vaporization. Hence DIC is suitable for the treatment of thermo-sensitive materials[2].

Instantaneous Controlled Pressure Drop

DIC is a hydro-thermo-mechanical treatment which is mainly used with biological materials like vegetables. DIC treatment consists in placing the moistened product in a processing chamber and exposing it to steam pressure

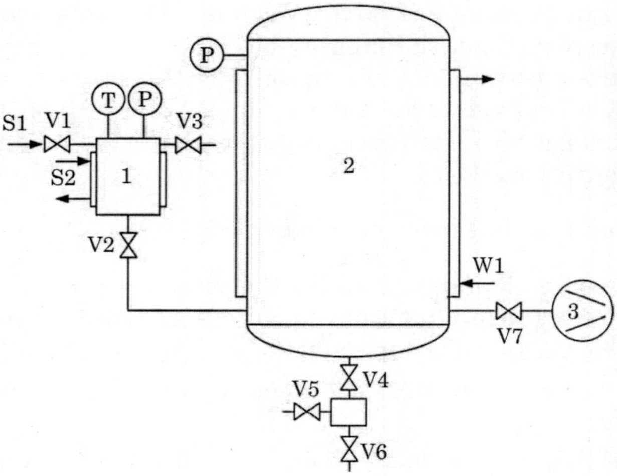

1. Autoclave, 2. Vacuum tank, 3. Vacuum pump, 4. Trap, V1–V7- valves, S1, S2-saturated vapour, W1-cooling water, P-pressure gauge, T- thermocouples

Fig. 18: Steam DIC set-up.

(up to 8-105 Pa) at high temperature (up to 170°C) over a relatively short time (a few seconds to 1 min). The end of the thermal treatment involves a pressure drop which is achieved by the instantaneous opening of a valve between the treatment chamber and a vacuum tank (50-102 Pa) with a volume which is at least 50 times greater. There is a sharp fall in chamber pressure which leads to a partial self vaporisation of water in the product. This produces a porous structure, which varies according to the properties of the treated product and the operating conditions (*e.g.* temperature, amplitude of the pressure drop, processing time). The short duration of the treatment (a few seconds) and the immediate drop in temperature after the pressure drop prevent further thermal deterioration and provide a good quality end. A measuring system consisting of numerous pressure and temperature sensors, linked to a personal computer, ensured data acquisition together with the manual and/or automatic control of the system[39].

The Instantaneous Controlled Pressure Drop (DIC) process seems to be a good means to modify the structure of the product to obtain a better quality. This process has proven its ability to expand the structure of different products improved the drying kinetics of expanded vegetables like potato.

REFERENCES

[1] Wells M.J.M., "Principles of extraction and the extraction of semivolatile organics from liquids", Center for the Management, Utilization and Protection of Water

Resources and Department of Chemistry, Tennessee Technological University, Cookeville, Tennesse (2003).

[2] Handa S.S., Khanuja S.P.S., Longo G., Rakesh R.D., "Extraction technologies for medicinal and aromatic plants", International Centre for Science and High Technology Trieste (2008).

[3] Temelozek., "Extraction, Separation and Analysis Techniques for MAPs: from Field to Market", Anadolu University, Faculty of Pharmacy, 26470-Eskisehir/ Turkey (2013).

[4] Tameshia S.B., Mallikarjunan P., Kequan Z., Keefe S.F.O., "Optimizing the extraction of phenolic antioxidants from peanut skins using response surface methodology", *J Agric Food Chem.,* **57**, 3064–3072 (2009).

[5] Buttery R.G., Ling L.C., Takeoka G.R., Teranishi R., Williams P.J., Kobayashi, "Methods for isolating food and plant volatiles- Bio improved foods and flavours", *ACS Symposium Series.*, **637**, 240–248 (1996).

[6] Ghosh D., Laddha, K.S., Raghavan R., "Herbal Drug Extraction", Chemical Weekly, Sevak Publications. 602-B, K. J. Somaiya Hospital Road, Sion, Mumbai India, p. 185 (2005).

[7] Rouseff R.L., Cadwallader K.R., "Headspace Analysis of Foods and Flavors: Theory and Practice", Kleur Publishing: New York, NY, p. 212 (2001)

[8] Kolb B., Leslie S.E., "Static Headspace-Gas chromatography: Theory and practice", Second Edition, John wiley and sons, (2006).

[9] Stuff R.J., Whitecavage J.A., "Automated Dynamic Headspace Sampling of Aqueous Samples using Replaceable Adsorbent Traps", (2008).

[10] Lord H., Pawliszyn J., "Evolution of solid-phase microextraction technology", *J Chrom.*, **885**, 153–193 (2000).

[11] Chinglin P., "Comparison of simultaneous distillation and extraction (SDE) and headspace solid phase microextraction (SPME) for determination of volatiles of muscadine grapes (vitis rotundifolia)", *Clemson University,* (2014) http://tigerprints.clemson.edu/all_theses/1947.

[12] Kataoka H., "Recent Advances in Solid-Phase Microextraction and Related Techniques for Pharmaceutical and Biomedical Analysis", *Curr Pharm Analy.*, **1,** 65–84 (2005).

[13] Pragst F., "Application of solid-phase microextraction in analytical toxicology", *Anal Bioanal Chem.*, **388,** 1393–1414 (2007).

[14] Heydari S., Haghayegh G.H., "Extraction and Microextraction Techniques for the Determination of Compounds from Saffron", *Canad Chem* Trans., **2**, 221–247 (2014).

[15] Vuckovic D., Cudjoe E., Hein D., Pawliszyn J., "Automation of Solid-Phase Microextraction in High-Throughput Format and Application to Drug Analysis", *Anal Chem.,* **80**, 6870–6880 (2008).

[16] Abraham M.A., Sunol A.K., "Supercritical Fluids: Extraction and Pollution Prevention", ACS Symposium Series., p. 670 (1997).

[17] Carlson D.L.H.C., Lopes T.J., Machado R.A.F., "Comparison of extraction of patchouli (*Pogostemon cablin*) essential oil with supercritical CO_2 and by steam distillation", *J Supercrit Fluid.*, **48**, 15–20 (2009).

[18] Filly, Fernandez X., Minuti M., "Solvent-free microwave extraction of essential oil from aromatic herbs: From laboratory to pilot and industrial scale", *J Food Chem.*, **150**, 193–198 (2014).

[19] Fincan M., Vito D.F., Dejmek P., "Pulsed electric field treatment for solid-liquid extraction of red beetroot pigment", *J Food Eng.,* **64**, 381–388 (2004).

[20] Caia J.J., Liu B., Qingde S., "Comparison of simultaneous distillation extraction and solid-phase microextraction for the determination of volatile flavor components", *J Chrom.*, pp. 1–7 (2001).

[21] Denny E.F.K., "Field Distillation for Herbaceous Oils (2nd Edition.)", Denny-McKenzie, Assoc., Lilydale, Tasmania, p. 265 (1991).

[22] Teixeira S., Mendes A., Alves A., Santos L., "Simultaneous distillation–extraction of high-value volatile compounds from *Cistus ladanifer* L.," *Analytica chimica acta.*, **584**(2), 439–446 (2007).

[23] Dung N.X., Dinh T., "Extraction and Distillation of Essential Oils, Processing, Analysis and Application of Essential Oils", 1st Edition, Har Krishan Bhalla & Sons Book Company, p. 59 (2005).

[24] Ormeno E., Goldstein A., Niinemets U., "Extracting and trapping biogenic volatile organic compounds stored in plant species", *Trends in Anal Chem.*, **30**, 7 (2011).

[25] Alfaro M.J., Belanger J.M.R., Padilla F.C., Pare J.R.R., "Influence of solvent, matrix dielectric properties and applied power on the liquid-phase microwave assistedextraction of ginger *(Zingiberofficinale)", Food Res Int.*, **36,** 499–504 (2003).

[26] Wang H.W., Liu Y.Q., Wei S.L., Yan Z.J., Lu K., "Comparison of Microwave-Assisted and Conventional Hydrodistillation in the Extraction of Essential Oils from Mango (*Mangifera indica* L.) Flowers", *J Mole.,* **15**, 7715–7723 (2010).

[27] Kiss G.A.C., Forgacs E., Serati T.C., Mota T., Morais H., Ramos A., "Optimisation of the microwave assisted extraction of pigments from paprika (*Capsicum annum* L.) powders", *J Chromr A.*, **889**, 41–49 (2000).

[28] Chemat F., Lucchesi M., "Microwave-assisted extraction of essential oils" *In: Microwaves in organic synthesis*", pp. 959–985 (2006).

[29] Barbero G.F., Palma M., Barroso C.G., "Determination of capsaicinoids in peppers by microwave assisted extraction high performance liquid chromatography with fluorescence detection", *Anal Chimca Acta.,* **578**, 227–33 (2006).

[30] Williams O.J., Raghavan.G.S.V., Orsat.V., Dai.J., "Microwave-assisted extraction of capsaicinoids from capsicum fruit", *J Food Bchem.*, **28,** 113–122 (2004).

[31] Cravotto G., Boffa L., Mantegna S., Perego P., Avogadro M., Cintas P., "Improved extraction of vegetable oils under high-intensity ultrasound and/or microwaves", *Ultra Sonochem.,* **15**, 898–902 (2008).

[32] Naik R., Kumar M., Bargale P.C., "Microwave Assisted extraction- Techniques and equipment-Non thermal Non chemical Processing in Food System (ISBN 978-81-909305-2-9)", pp. 99–116 (2012).

[33] Wei L., Zhang Y., Jiang B., "Comparison of Microwave-assisted Hydrodistillation with the Traditional Hydrodistillation Method in the Extraction of Essential Oils from *Dwarfed Cinnamomum Camphora var. Linaolifera Fujita* Leaves and Twigs", *Adv J Fd Sci Tech.*, **5**(11), 1436–1442 (2013).

[34] Jonsson J.A., Mathiasson L., "Liquid membrane extraction in environmental chemical analysis", Gdansk centre of excellence in environmental analysis and monitoring, pp. 84–109 (2003).

[35] Liu J.F., Jonsson J.A., Mayer P., "Equilibrium Sampling through membranes of freely dissolved chlorophenols in water samples with hollow fiber supported liquid membrane", *Anal Chem.,* **77**, 4800–4809 (2005).

[36] Lam K.F., Sorensen E., Gavriilidis A., "On-chip microscale distillation for acetone-water separation", (2010) http://discovery.ucl.ac.uk/id/eprint/1319926.

[37] Fleisher A., "The poroplast extraction technique in the flavour and fragrance industry", *Perfumer Flavor.*, **15**, 27–36 (1999).

[38] Rezzoug S.A., Rezzoug Z.M., Mazoyer J., Jeannin M., Allaf K., "Effect of instantaneous controlled pressure drop process on the hydration capacity of scleroglucan: Optimisation of operating conditions by response surface methodology", *Carbohy Pol.*, **41**(1), 73–84 (2000).

[39] Haddad H.J., Allaf K.K., "A study of the impact of instantaneous controlled pressure drop on the trypsin inhibitors of soybean", *J Food Engg.*, **79,** 353–357 (2007).

2

Potato – Nutritional and Technological Potential

MANJU BALA[1*], SURYA TUSHIR[1] AND ALKA SHARMA[1]

ABSTRACT

Potato is the fourth most important non-cereal food crop of the world, after rice, wheat and maize, with a global production of 365 million metric tons. Globally India stands third in terms of area and second in terms of production (41.5 million metric tons) after China. Potato is an economically important staple crop prevailing all across the world with successful large scale production, consumption and affordability with easy availability in the open market. Potatoes provide basic nutrients such as carbohydrates, dietary fiber through peel, several vitamins and minerals viz. potassium, magnesium. Besides, potatoes are rich source of bioactive compounds. However, retention of bioactive compounds in potatoes is affected by various cooking and processing methods. In India, potato is predominantly consumed as vegetable although potato processing sector has shown immense growth during the recent past. Among the processed products French fries, chips, starch, potato flakes, and flour represent the majority of consumption. This chapter describes regarding potato its production, nutritional composition, bioactive compounds, effect of processing on bioactive compounds, potato processing, product development and byproduct utilization.

***Key words*:** Antioxidant, Nutrient, Phytochemicals, Post-harvest, Potato products, Potato, Processing, Phenolic compounds

[1] ICAR-Central Institute of Post-Harvest Engineering and Technology (ICAR-CIPHET) Ludhiana -141 004, Punjab, India.

**Corresponding author* E-mail: avimithi@gmail.com

I. INTRODUCTION

Potato (*Solanum tuberosum* L., Solanaceae) is the fourth most important non-cereal food crop of the world, after rice, wheat and maize, with a global production of 365 *million metric* tons. Potato biodiversity is vast, approximately 5000 varieties of potatoes are known. Most cultivated varieties belong to the species *Solanum tuberosum*. However, about 10 other *Solanum* species are cultivated, and 200 wild species have been recorded. Potato is a very efficient food crop, produces more dry matter, protein and minerals per unit area in comparison to cereals and plays an important role in human nutrition. Because of its potential to produce highest food, energy and protein per unit area and time, FAO, 2008[1] declared potato as a 'Future Food Crop' and the year 2008 as the International Year of the Potato (IYP) to increase awareness of the relationship that exists between poverty, food security, malnutrition and the potential contribution of the potato to defeating hunger.

Potato is an economically important staple crop prevailing all across the world with successful large scale production, consumption and affordability with easy availability in the open market. Potatoes being the staple food in the developed countries account for 130 kcal of energy per person per day. Countries in the Commonwealth of Independent States (CIS) rely most heavily on potatoes for their energy intakes, with Belarus topping the list at 300 kcal per person per day. Seven other CIS (Commonwealth of Independent States) countries and Portugal follow closely behind, each averaging more than 200 kcal per person per day from potatoes[2]. In the developing countries potato accounts for 41 kcal of energy per person per day, where it is still considered as vegetable[3].

Potato is native to Andean hills of Peru and Bolivia, from where it was first introduced into Europe towards the end of 16th century through Spanish conquerors. There potato developed as a temperate crop and was later distributed throughout the world largely as a consequence of the colonial expansion of European countries. Irish brought it to England in 1718 and was introduced to India by early 17th century probably through British missionaries or Portuguese traders, who were the first to open trade routes to the East.

China is the world's largest potato-producing country, followed by India, Russia, USA and Ukraine (Table 1). China and India together account for about one-third of global potato production. Globally India stands third in terms of area and second in terms of production (41.5 MT) after China[4]. Potato in India is presently grown over 1.97 million ha (2013-14) under diversified agro-climatic conditions. India produced 41.55 million tons with an average yield of 22.72 t/ha[5]. Potato is grown almost in all states of India. Major potato growing states are Uttar Pradesh, Himachal Pradesh, Punjab, Madhya Pradesh, Gujarat, Maharashtra, Karnataka, West Bengal,

Bihar and Assam. The Indian potato belt stretches from Punjab (N.W.) to West Bengal in East India. Uttar Pradesh, West Bengal, Bihar, Madhya Pradesh, Gujarat and Punjab together account for about 87% of India's production. Uttar Pradesh is the largest producer of potatoes for India with about 13.81 million tons production in 2013-14 from an area of about 0.56 million hectares.

Table 1: Major potato producing countries in the world during 2012-2013[5]

Country	*Area in ha*	*Production in metric tones*	*Productivity metric, tones/ha*
China	5429000	85860000	15.8
India	1992211	45343590	22.8
Russian Federation	2197200	29532530	13.4
Ukraine	1444100	23250200	16.1
United States of America	458388	19165865	41.8
Germany	238300	10665600	44.8
Poland	373000	9091900	24.4
Bangladesh	430446	8205470	19.1
Belarus	332255	6910945	20.8
Netherlands	149770	6765618	45.2
Other	6249623	120360640	19.3
World (Total)	**19294293**	**365152358**	**18.9**

Though potato yield has been erratic across the developing world during the past decade (16.3 to 18.9 t/ha during 2000-2012), the Indian-potato productivity has increased steadily during this period (18.8 to 23.6 t/ha)[6]. Diversity in Indian potato varieties for attributes like adaptability, maturity, quality of produce and tolerance to biotic and abiotic factors makes it ideal for inclusion in wide range of cropping systems in different agro-climatic conditions[7]. In India, it is used only as a vegetable. Sometimes it is also mixed with other vegetables. Potato is popularly known as 'The king of vegetables'. At present, roughly two-thirds of the domestic supply of potatoes is consumed as fresh and around 7.5% is consumed in processed form. A considerable amount of potato is wasted due to post-harvest losses (16%) and the remaining production (about 8–10%) is used as seed. In processed form, it is used for making chips, French fries, *chat,* and *tikki* etc. In processing, 10–25% of the raw product is discharged as waste. Peels are the major waste of potato processing that is perishable and cause many management problems in terms of disposal and sanitation.

Classification

- Domain: Eukaryota
- Kingdom: Planta
- Phylum: Magnoliophyta
- Class: Magnoliopsida

- Order: Solanales
- Family: Solanaceae
- Genus: *Solanum*
- Species: *tuberosum*

II. NUTRITIONAL COMPOSITION

Potato is recognized as a source of carbohydrates, proteins, vitamin C, vitamin B6, vitamin B3 and minerals such as potassium, phosphorous and magnesium[8]. More than 95% of the energy in uncooked/unprocessed potatoes is provided by carbohydrates. Potato tuber consists of about 76% moisture, 18% starch, 2% high-quality proteins and other health beneficial ingredient (Table 2). Researchers have reported wide range of starch contents in different varieties of potato from 9.1% for an Indian cultivar Kufri Bahar[9] and as high as 22.6% for another Argentine variety, Imilla Negra *S. tuberosum* ssp. Andigena[10]. Protein content has been reported to vary from less than 1% for Argentine cultivar Revolution[10], to a high of 4.2% for the Spanish Roja Rinon[11] with an average value of 2%. Other important constituents in potatoes include crude fibre (0.7%) and ash content of 1.1%.

Besides these basic nutrients, potatoes contain significant amounts of phytonutrients like phenolic compounds while some colored potato varieties are known to contain anthocyanins also. Another study cites that potatoes serve as one of the low-fat foods with a spectrum of nutrients and phytochemicals present and are particularly rich in vitamin C, vitamin B6, potassium and manganese[12]. Within the macro minerals, the highest contribution to the intakes was observed for potassium, while iron was the trace element with the largest contribution to the proposed intake[13].

Potatoes range in size, shape, color, starch content and flavor. Although nutrient content depends on a number of factors, variety is thought to be among the most significant factors[15]. So, it is apparent that potato is a useful source of nutrition and can deliver virtually all of the nutrients that the body needs ranging from the macro-components that give energy, good levels of the essential amino acids[16] to micro-nutrients such as vitamins and minerals.

III. POTATO HARVESTING & POST-HARVEST HANDLING

(a) Harvesting and Yield

Harvesting of potatoes is done before the maximum soil temperature rises above 26–30°C. It is completed by end of January in central and eastern plains and by mid February in the western plains to avoid rotting of tubers due to high temperatures in March/April. The crop is harvested 10-15 days

Table 2: Composition of potato tuber[2, 14]

Name of parameters	*Content (average values)*
Moisture (%)	76.0
Protein (%)	2.0
Fat (%)	0.2
Starch (%)	18
Total Sugar (%)	0.5
Crude fibre (%)	0.7
Ash (%)	1.1
Thiamine (B1), mg/100 g	0.09
Riboflavin (B2), mg/100 g	0.07
Niacin (B3), mg/100 g	1.26
Vitamin (B6), mg/100 g	0.24
Vitamin C, mg/100 g	17
Folic acid, mcg/100 g	12.42*
Calcium, mg/100 g	10.59
Phosphorus, mg/100 g	71.43
Sodium, mg/100 g	2.87
Potassium, mg/100 g	443.3
Magnesium, mg/100 g	20.21
Manganese, mg/100 g	0.13
Zinc, mg/100 g	0.41
Copper, mg/100 g	0.11
Energy value, kJ/100 g	126

after the last irrigation. This allows tuber skin to become firm and tubers do not get bruised on harvesting. Harvesting is done manually with the help of spade or khurpi or by bullock drawn single row digger/plough. It is also done mechanically with the help of 1-4 row potato digger and sophisticated precision potato harvester.

(b) Post-harvest Handling

Nearly one fifth of the total potato production in the country is used as planting material in the following season. Therefore, post-harvest handling particularly of seed stocks becomes very important. After harvesting, potatoes are kept in heaps in cool places for another l0-15 days for drying and further curing of skin. Heaps of 3-4 metre long, wide at the base and about 1 metre wide at the top are the best. In hills the harvested potatoes are spread in well-ventilated rooms for curing before ready for marketing.

(c) Sorting and Grading

During sorting, ground, stones, vegetal wastes, cut or rotten tubers are separated. Sorting is achieved manually or with sorting machines. Grading of tubers is done both by hand as well as by graders. The tubers are then graded and packed in gunny bags according to sizes preferably in 4 sizes, *e.g.* small (less than 25 g), medium (25-50 g), large (50-75 g) and extra large (above 75 g). Tubers 75 g in weight are kept as table purpose category, small sized tubers are kept as seed. Seed Tubers less than 25 mm and more than 65 mm in size are considered as undersized and oversized, respectively. For processing suitability of tubers is decided by size, shape, depth of eyes, chemical constituents such as dry matter and reducing sugar content.

(d) Storage

Storage requirements of potato vary with the purpose for which potatoes are stored. The methods generally used for storage of potatoes are - room storage, pit storage and cold storage. The room should have a high plinth and good ventilation. Potatoes can be stored in single layers on sand. Frequent examination is necessary to discard tubers showing rotting symptoms. In pit storage system, the tubers are stored in pits 60-75 cm deep and 2.5 metre long and 1 metre wide. These pits are made in a cool shady place. Potatoes in the cold storage are spread over the shelf to a thickness of not more than 30 cm. Individual shelves are about 45 cm high and 15 cm space is allowed between two layers.

Potatoes meant for use as seed during the next year are treated with 3% boric acid solution for 30 minutes for protecting against soil-borne pathogens, *e.g.*, black scurf, common scab, etc. before storing in bags.

IV. BIOACTIVE COMPONENTS

Apart from being a rich source of starch, potatoes contain good quantity of small molecules and secondary metabolites also known as phytochemicals, which play an important role in a number of processes[17]. Phytochemicals are a group of metabolites found in many plants and their presence in fruits and vegetables are positively correlated with many health benefits[18]. Potato tubers contain plenty of bioactive compounds including proteins, peptides, carotenoids, polyamines, polyphenols, suberins, glycoalkaloids and dietary fibre. Vitamin C has been also reported in potato[19]. Most of them enhance human health and they all are useful components for food industry[11]. Indeed, potatoes provide an ideal delivery system for beneficial compounds.

(a) Phenolic Compounds

Polyphenols comprise more than 8000 identified substances, which can be divided into groups according to their chemical structure, such as phenolic acids, stilbenes, coumarins, lignins and flavonoids. Polyphenols are recognized as the most abundant antioxidants in our diet[20]. Further, phenolics have a wide-array of health providing characteristics, therefore, have potential for use as functional food for improving human health.

Phenolic compounds are an extremely heterogeneous class of secondary metabolites originating from phenylalanine and, in some plants from tyrosine *via* the action of phenylalanine ammonia lyase (PAL) and tyrosine ammonia lyase (TAL), respectively[21]. Broadly they can be classified in phenolic acids (C_6-C_1 and C_6-C_3 structures) and flavonoids (C_6-C_3-C_6 backbone). The cinnamic acid and *p*-couramic acid are the main phenylpropanoid (C_6-C_3) compounds which lead to the formation of benzoic acid (C_6-C_1) derivatives by loss of a 2-carbon moiety as well as to chalcones (C_6-C_3-C_6), stilbenes (C_6-C_2-C_6), and flavonoids (C_6-C_3-C_6) by combination with 3 molecules of malonyl coenzyme A. Among the flavonoids, the most widespread compounds are flavonols, flavan-3-ols, flavones, flavanones, and anthocyanins. Apart from modifications to the C_6-C_3-C_6 core, the unmarked structural variety of the flavonoids is also due to their conjugation to sugars at different sites of the molecule, usually to one or more hydroxyl groups or, less frequently, C-glycosidically to an aromatic carbon atom. Additional variations may occur through acylation of the sugar moiety with organic acids. The basic structure of polyphenols is composed of one or more phenolic rings that are substituted with several hydroxyl groups and these are highly correlated with their strong antioxidant activity[22, 23].

Phenolic compounds protect plants against biotic stress caused by herbivores, insects, and pathogens such as bacteria, fungi, and viruses, and also against tissue damages caused by excessive UV light and free radicals[17]. Phenolic compounds are among phytochemicals that may render their effects *via* antioxidation and relief from oxidative stress and its consequences. In general, free radicals are part of life as we consume some 3.5 kg of oxygen each day and some of this is not completely reduced, thus leading to the formation of free radicals and other reactive oxygen species (ROS) such as superoxide, hydroxyl radical, peroxyl radical, and alkoxy radical, as well as hydrogen and other peroxides. Therefore, several kilograms of peroxides and other ROS may be produced in our body each year.

Polyphenols (cinnamic acid derivatives, flavonols, anthocyanins) and vitamins are present in vegetables, fruits, berries, and herbs, which are the main source of natural antioxidants in our daily diet. Potato tubers are one of the richest sources of polyphenols. These compounds are distributed mostly between the cortex and skin (peel) tissues of the potato. About 50% of the compounds are located in the potato peel and adjoining tissues, which often go waste, and their concentration decreases from outside to centre of potato[17]. Potatoes were considered the third most important source of phenols after apples and oranges[24].

Out of phenolic compounds (Table 3), chlorogenic, caffeic acid and catechin, have been reported to be the main compounds present in potato[19].

Total phenolic content of 0.3 g/L (180 mg/100 g), and flavonoid content of 0.06 g/L (36 mg/100 g) has been reported[25]. Total phenolic content varied in the peeled potato samples from 191 to 1864 mg per 100 g DM meanwhile these parameters varied from 345 to 2852 mg per 100 g DM in unpeeled samples in the studied varieties of Chile[26].

Table 3: Phenolic compounds in potatoes[27]

Phenolics/Phenolic acids				***Flavonoids***		***Polyamines***
Hydroxy cinnamic acids	***Hydroxy benzoic acids***	***Flavonols***	***Flavanols***	***Flavanones***	***Anthocyanidins***	***Dihydrocaffeoyl***
5-*O*-Caffeoylquinic acid (chlorogenic acid)	Gallic acid	Kaempferol glycosides	Catechin	Eriodyctiol glycosides	Pelargonidin	*N*[1],*N*[12]-Bis (dihydro caffeoyl) spermine (kukoa mine A)
4-*O*-Caffeo ylquinic acid (*crypto*-chlorogenic acid)	Proto catechuic acid	Quercetin glycosides	Epicatechin	Naringenin	Peonidin glycosides	*N*[1],*N*[8]-Bis (dihydro caffeoyl) sper midine
3-*O*-Caffeo ylquinic acid (*neo*-chlorogenic acid)	Vanillic acid	Rutin			Malvidin glycosides	*N*[1],*N*[4], *N*[12]-Tris (dihydro caffeoyl) spermine
Caffeic acid	Salicylic acid				Petunidin glycosides	*N*[1],*N*4, *N*8-Tris (dihydro caffeoyl) sper midine
p-Coumaric acid						
Ferulic acid						

In potato, main phenolic acid is chlorogenic acid or hydroxycinnamate derivative 5-*O*-caffeoylquinic acid, which is formed from caffeic and quinic acid. It can be hydrolysed to caffeic acid under alkaline conditions[17]. Chlorogenic acid (Fig. 1) constitutes 80-90% of the phenolic compounds showing concentration variation from 0.1-0.2 mg/g on varietal basis[28,29]. A wider range for chlorogenic acid from 13.2 to 68.3 mg/100 gfw potato, has also been reported when measured using HPLC[30, 31]. Stushnoff *et al.*[32] showed that pigmented cultivars such as Mountain Rose and Purple Majesty, and non-pigmented cultivars such as Yukon Gold had an approximately 10-fold difference in their chlorogenic acid content. Navarre *et al.*[33] reported a 20-fold difference in the chlorogenic acid content between the lowest reported (*S. bulbocastanum*) and highest reported (C097226-2R/R) cultivars. Isomers of this compound, *i.e. neo*-chlorogenic acid (3-caf- feroylquinic acid) and *crypto*-chlorogenic acid (4-caffeoyl-quinic acid) have also been reported in small quantities.

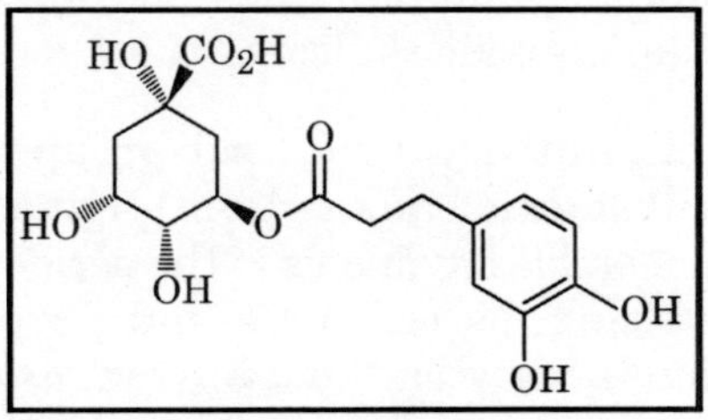

Fig. 1: Structure of chlorogenic acid.

Apart from chlorogenic acid and its isomers, caffeic, *p*-coumaric and ferulic acids as well as various benzoic acid derivatives such as gallic, protocatechuic, vanillic, and salicylic acids (Table 1) have also been reported however, usually in lower amounts[34, 35]. Caffeic acid is the second most abundant phenolic acid in the potato. The caffeic acid content ranged from 310 to 420 µg per 100 gfw potatoes[30, 31]. Phenolic acids in potato peels are not only present in their free form but occur also in bound form, as recently shown for ferulic acid[36]. Hye and Kyung[37] reported that color-fleshed potatoes have higher antioxidant and stronger antiproliferative activities than white-fleshed potatoes. The TPC of the analyzed cultivars varied between 113.3 mg/100g powder ('Superior') and 153.3 mg/100g powder ('Blue').

(b) Flavonoids

Flavonoid content in potatoes was reported in the range from 200 to 300 µg/g fresh weight, with variations in their antioxidant capacity[38]. In skin extracts of coloured potato cultivars, catechin, epicatechin, eriodictyol, kaempferol 3-rutinoside and naringenin at levels between 10 and 150 µg per g fresh weight, and lower concentrations of quercetin glycosides had been reported[34]. However, in wild *Solanum* species the major flavonoids were found to be catechin, epicatechin, eriodictyol and naringin in the skin and flesh[39]. Authors also reported that purple or red-fleshed cultivars had twice the flavonoid concentration of white-fleshed cultivars and their concentrations were considerably higher in skin, approaching 900 mg in purple-fleshed and 500 mg in red-fleshed types per 100 g fresh weight basis.

Flavonol such as rutin have also been reported in potato. The flavonol content of potatoes is not significantly high, but these could be considered as a valuable source of these compounds because of their high consumption[40]. Potato peel contained quercetin, a flavonol with antioxidant activity and such activity in flavonols had been attributed to their action as free radical acceptors. Quercetin was reported to be three times more effective as an antioxidant than kaempferol and eridictyol, and was twice as effective as catechin. The flavonoids and flavones extracts possessed high scavenging activities toward oxygen radicals[41]. Potatoes showed 94% scavenging activity towards hydroxyl radicals, and, along with onions, almost complete inhibition of super-oxide radicals.

Apart from flavonoids, Anthocyanins, a sub-group within the flavonoids, are also present in substantial amounts in pigmented flesh potatoes. Anthocyanins are water soluble flavanoids in the periderm of potatoes, which confer different colors such as red, blue and purple to the plants[42]. Chemically, anthocyanins are glycosylated polyhydroxy and/or polymethoxy derivatives of the 2-(flavylium) salt[43]. Glycosylated anthocyanins can appear acylated with aromatic acids such as *para*-coumaric, caffeic, ferulic, sinapic, gallic or *para*-hydroxybenzoic acids, and/or aliphatic acids such as malonic, acetic, malic, succinic or oxalic acids[44].

Many glycosylated anthocyanins acylated with *para*-coumaric and ferulic acid have been reported in colored potatoes. 3-rutinoside-5-glucoside and 3-rutinoside derivatives of pelargonidin, petunidin, malvidin, cyanidin, peonidin and delphinidin have been reported in colored potatoes[45]. Among colored potatoes, purple-fleshed potatoes contain anthocyanins such as petunidin- and malvidin-3-rutinoside-5-glycosides acylated with *para*-coumaric and ferulic acid while red-fleshed potatoes contain pelargonidin- and peonidin-3-rutinoside-5-glycosides acylated with *para*-coumaric and ferulic acid[46]. The acylation of the saccharide moiety with hydroxycinnamates is considered the main reason for the increased stability of these anthocyanins compared to non-acylated pigments[47].

In colored potatoes, anthocyanins ranged from 17-20 mg *per* 100 g in red-fleshed potatoes and 20-38 mg *per* 100 g in purple-fleshed potatoes[48]. Anthocyanin levels between 5.5 and 35 mg/100 g fw in potatoes have been reported[49]. The pigmented potatoes may serve as a potential source of natural anthocyanin pigments, since these are low-cost crops[50] and also a powerful source of antioxidant micronutrients[8].

Purple potatoes contained higher overall levels of total phenolics and anthocyanins and had a greater antioxidant capacity compared with white potatoes. Domestic processing significantly reduced the total phenolic content of purple potatoes but did not influence anthocyanins or the overall antioxidant capacity. Purple potatoes could provide a rich source of antioxidants in the diet and they compare favourably to other ‘purple vegetables', for example, purple potatoes contained 105.5 mg GAE/100 g fresh weight compared to red cabbage, 439 mg GAE/100 g fresh weight[51].

(c) Folate

Potatoes do not have high folate content, but are a major source of this due to their higher consumption. Potatoes supply about 10% of the total folate intake of the people in European countries such as Netherlands, Norway and Finland[52]. Folate concentrations in potato varied between 12 and 37 μg/100 g fw[53] and 10 to 15 μg/100 g fw[2]. Folate content in more than 70 potato cultivars, advanced hybrids and wild species has been reported to range from 11 to 35 μg/100 g fw. Folate concentrations were higher in younger tubers and substantially higher in some cultivars[54].

(d) Kukoamines

Tuber polyamines have been suggested to play a role in the regulation of starch biosynthesis[55] and making the tubers resistant to diseases[56]. The dihydrocaffeoyl polyamines, kukoamine and related compounds, have also been detected in potato tubers during metabolite profiling. Kukoamines are polyamine conjugates[3] and considered to have health promoting effects. Further studies are needed to understand their stability, beneficial effects and their role in other functions.

(e) Carotenoids

Potatoes are a good source of carotenoids also, which are lipophilic compounds synthesized in plastids from isoprenoids[57]. Because of their high carotenoids content potatoes have been reported to be beneficial for eye health[58]. Lutein, zeaxanthin, violaxanthin and neoxanthin (all of which are xanthophylls) are the major carotenoids present in potatoes and β-carotene has been reported in traces. Potato cultivars with white flesh contained less carotenoids as compared to cultivars with yellow or orange flesh. The orange and yellow colour of the tuber flesh is due to zeaxanthin and lutein, respectively. Total carotenoid content has been reported between 2.83 and 36.21 potato on dry weight basis[8]. However, values for total carotenoid content in white and yellow-fleshed potato cultivars were 50–350 μg/100 g and 800–2000 μg/100 g on fresh weight basis, respectively[49].

Using biotechnological interventions[59] over expression of a bacterial phytoene synthase gene resulted in the increase in the levels of individual carotenoids, β-carotene (more than 11 fold) and lutein (19 fold). As per reports[60], approximately 50% of the RDA of vitamin 'A' can be met by consuming 250 g of carotenoids enriched genetically engineered potatoes. Carotenoid content was significantly affected by cultivation environments and varieties[61].

(f) Glycoalkaloids

Processing of potatoes involves the generation of large amounts of waste especially in the form of peel. However, the potato peel is a rich source of a group of chemical compounds known as steroidal glycoalkaloids (SGAs). Extracted glycoalkaloids from potato skin may be commercially important based on their pharmaceutical activities because of the structural resemblance to the human sex hormones. Particular SGAs can be transformed into steroidal drugs, and their high cytotoxic activities as anticancer compounds have recently received strong scientific interest[62].

Glycoalkaloids are plant secondary metabolites that are toxic to microorganisms, viruses, insects, animals, and also humans[63]. The glycoalkaloid content is toxic at levels above 20 mg/100 g of fresh tuber. Its concentration in commercially available potatoes is about 2–15 mg/100 g, but it sharply rises if the tubers are stressed and exposed to light[64, 65].

The glycoalkaloid content of potato tubers can vary considerably and is influenced by factors such as exposure to light, irradiation, mechanical injury, and conditions of storage. Unpeeled potatoes had higher glycoalkaloid content[66]. Peeling and blanching decreased the glycoalkaloids in dried potato granules[67]. The major glycoalkaloids present in potatoes are α- chaconine and α-solanine, which share the same aglycone, solanidine. Structurally, these compounds differ in the saccharide moiety in that α-solanine contains the tri-saccharide solatriose, whereas in α-chaconine the aglycone attached is chacotriose. Stepwise removal of a sugar moiety from the trisaccharides leads to the formation of β- and γ-glycoalkaloids and finally to solanidine, which displays a lower toxicity compared to the parent compounds[68]. Solanidine, the aglycone of the potato glycoalkaloids, is a hexacyclic tertiary amine, which can be released from the glycosylated parent compounds after either enzymatic or acid-catalyzed hydrolysis. Because of its steroid structure, solanidine represents an interesting educts for conversion to an intermediate for the synthesis of steroid hormones[27].

V. EFFECT OF POSTHARVEST STORAGE ON BIOACTIVE COMPOUNDS

Potatoes are stored at 4°C in cold storage in countries such as India[69]. Storage generally increases total phenols content in potatoes but little change or a decrease in bioactive compounds after storage have also been reported in some studies. So variable results have been reported, a few of them are discussed below:

Increase in total phenol content in irradiated potatoes after storage at 10, 20 and 30°C[70] and at 5 and 20°C[71] has been reported. Cold storage of the tubers over 5 months led to a significant increase in anthocyanidin concentrations. Total phenols content in four potato cultivars stored for

180 days at 4, 8, 12, 16 and 20°C were determined[72] and was found to increase after storage and the increase was higher at 4 and 16°C. Another study[73] showed increase in the total phenols, in potato tubers continued up to 271 days of storage at 6°C.

When total phenolics[32] were analyzed from 8 potato genotypes after 112 and 263 days of storage at 5°C, two genotypes showed sharp rise in total phenolics after storage, while four genotypes showed increase to a lesser extent and two genotypes showed little change. Stored tubers showed higher antioxidant activity than fresh tubers.

Potatoes of two cultivars grown at four locations in the North Indian plains did not show significant difference in total phenols content up to 115 days of storage at 4 and 10°C. No significant differences were observed in total phenolic content, chlorogenic acid, caffeic acid and vanillic acid after storage at 4 or 20°C for 110 days[74]. Total phenols in potato tubers decreased after 220 days of storage at 20°C[73]. Phenols content in the peels and flesh of three Indian potato varieties decreased after 90 days of storage at 2–4°C, 10–12°C and in heap storage, a traditional method of storage. The decrease in the peels was minimum at 2–4°C and maximum in heap storage[75].

The anthocyanin content of tubers was determined[50] in 14 cultivars/ clones immediately after harvest and after 135 days of storage at 4°C and 86% relative humidity, and no significant change in anthocyanin content of tubers was observed. The fact that cold storage had no significant effect on the anthocyanin content of potatoes indicated that there was no risk of degradation of these compounds during storage of potatoes over a longer period. Sugar accumulation is favoured by low temperature (4°C), which could be helpful in anthocyanin retention in stored potatoes. Degradation of anthocyanins is primarily caused by oxidation or cleavage of covalent bonds, which increases with increase in temperature during storage or processing. The carotenoid content per 100 g of tubers increased[61] in some of the studied varieties after a period of storage but at the same time decrease in carotenoid content per 100 g of tubers was observed in some of the varieties.

VI. EFFECT OF PROCESSING ON BIOACTIVE COMPONENTS

It is known that like storage, processing also changes the physical and chemical composition of foods thus, affecting their antioxidant activity[76]. Since potatoes are almost always consumed after processing (baked, chipped, fried, boiled or microwaved) and these processing techniques can affect the profile of endogenous nutrients in potatoes so it is important to understand the effect of such processing techniques on the activity and composition of bioactive compounds in potatoes.

Domestic cooking such as microwaving, boiling or frying can result in partial losses in the phenolic content[40]. Specifically, chlorogenic acid has also been shown to undergo degradation after home processing[29]. A 100% loss in the chlorogenic acid content of potatoes baked at 212°C for 45 minutes was reported[77], which suggested that chlorogenic acid is susceptible to heat. However, some studies have reported a reduction but not a complete loss of chlorogenic acid.

Studies on the effect of baking (oven heating) on the chlorogenic acid content[29] showed 90% to 100% retention of the chlorogenic acid and its isomer in baked potatoes. Other workers[74] reported that baking increased the total phenolic content and antioxidant activity in studied genotypes of potato. Samples showed higher concentration of chlorogenic acid, caffeic acid, *para*-coumaric acid and vanillic acid. Similarly, baking for 30 minutes increased the total phenolic content and chlorogenic acid content in three potato genotypes[33]. However, with increase in baking time to 45 minutes, the total phenolic content decreased. However, the phenolic content was still more than that of raw potatoes. This increase could be due to improvement in the extractability of phenolic compounds as cooking weakens the matrix, and inactivates enzymes that use phenolic compounds as substrate[3].

Raw purple potatoes were found to be a good source of anthocyanins (219 mg/kg FW) and the level of total phenolic compounds (TP), total anthocyanins (TA) and anti-oxidant activity (AOA) increased using different cooking techniques, with the exception of baking[78].

Retention of bioactive compounds in white-fleshed potatoes, yellow-fleshed potatoes, red-fleshed potatoes and purple fleshed potatoes after drum-drying and Refractance Window drying (RW) was studied[79]. They observed that total carotenoid and lutein concentration decreased with drum-drying and Refractance Window drying in different fleshed potatoes, whereas both types of drying did not show significant effect on total antioxidants in different fleshed potatoes. Microwaving caused a significant reduction (16-29%) in the anthocyanin content[80] of purple-fleshed cultivars. However, no alteration in the phenolic acid content was observed.

In the Piccolo cultivar, stir-frying for 18 minutes led to an increase in the total phenolic content[33]. The reports also claimed that the extractability of phenolic compounds improved after frying which led to an increase in the total phenolic content. This increase could be due to the presence of antioxidants of the frying oils.

Monro *et al.*[81] studied at potato genotype differences in nutritionally distinct starch fractions after cooking and cool storing. Their results suggest that the glycaemic impact of some potatoes may be substantially reduced by cool-storing after cooking.

VII. HEALTH BENEFITS OF BIOACTIVE COMPOUNDS

Polyphenols, together with carotenoids and vitamin C, significantly contribute to the antioxidant capacity of potatoes. Chlorogenic acid has strong antioxidant activity and potatoes are an excellent source of this. Chlorogenic acid is also well known for health promoting effects such as protection against degenerative diseases, cancer, heart disease, hypertension[82] and viral and bacterial diseases. It has been reported that chlorogenic acid blocks nitrosamine formation through competitive reaction with nitrite and to bind the carcinogen benzo(a) pyrene in a cellulose model system[17]. Chlorogenic acid has been found to be a strong and selective inhibitor of matrix metalloproteinase (MMP-9), an angiogenic enzyme responsible for tumor invasion and metastasis. Chlorogenic acid has been demonstrated to exhibit several desirable anticarcinogenic properties including inhibition of A549 human lung cancer cells[83].

The anti-proliferative and pro-apoptotic properties of potato bioactive compounds have been reported *in vitro* on human prostate cancer cell lines. The anticancer activity and mechanism of action of phenolic compounds on androgen-dependent (LNCaP) and androgen-independent (PC-3) prostate cancer cells was studied [84]. Their results suggested that α-chaconine and gallic acid were the most active components which showed effect against cancer cell lines.

When athymic nude mice received xenografts of human prostate cancer cells (PC-3) and were administered extracts of potato bioactives from either the white flesh *Solanum bulbocastanum* (PI243510) or CO112F2-2P/P (purple-flesh Colorado selection), neither potato extract provided a significant reduction in tumor growth nor reduced levels of the pro-angiogenic protein VEGF, but both potato extracts reduced MTA1 expression in lungs, and the extract from *S. bulbocastanum* reduced expression of metastasis associated protein 1 (MTA1) in tumors[85].

Potatoes with low glycemic index may prove beneficial for diabetic patients and result in a lower risk of type II diabetes. Moreover, chlorogenic acid has been shown to suppress the release of glucose in the blood[86]. In different studies, chlorogenic, caffeic and protocatechuic acids were the phenolic compounds that contributed most to the antioxidant activity, which may be explained by the *o*-dihydroxy structure that is common to all three components.

Anthocyanins function as antioxidants and are known to prevent diseases such as cardiovascular diseases, cancer and diabetes. Anthocyanidins such as delphinidin, pelargonidin, petunidin and malvidin inhibited[87] MCF-7 breast cancer cell cultures. Colored potato extracts and an anthocyanin rich fraction have been reported to suppress lymph-node carcinoma of the prostate and prostate cancer cell proliferation[84]. Polyphenols and anthocyanin rich purple potato flakes were found to play an important role in the protection against adverse effects related to oxidative damage in rats fed a high-cholesterol diet, and red potato flakes improved the antioxidant system by enhancing hepatic superoxide dismutase mRNA in rats.

A 95% ethanolic extracts of red-skinned cultivars had antioxidative activities similar to butylated hydroxyanisole (BHA)-butylated hydroxytoluene (BHT)-corn oil mixtures (2:2:6) in the carotene bleaching and linoleic acid oxidation assays[88]. Pure soybean oil treated with freeze-dried extracts from peels of six potato cultivars showed a reduction in the peroxide value during storage. Potato peel extract retarded the oxidation of soy bean oil and increased the induction time of oil per oxidation and potato peel extract was as good as BHA and BHT[89]. Consumption of unpeeled cooked potatoes improved the lipid metabolism and antioxidant status in cholesterol fed rats[90]. Ah-Hen *et al.*[26] determined the antioxidant activity (AA) and total phenolic content (TPC) of native Chilean potatoes. Total phenolic content and AA were compared with two commercial varieties of Chile, i.e. Shepody and Desiree. Antioxidant activity was higher in unpeeled potatoes, and was the highest in the unpeeled NG-6 or 'Bruja' native potato. The commercial var. Shepody showed pro-oxidant activity and had a relatively lower TPC. Results also indicated a higher concentration of total phenolics in the periderm of the colored native Chilean potatoes. Potato tubers as a source of antioxidants and minerals for the human diet were evaluated[8] and it was found that the hydrophilic antioxidant capacity and the total phenolic content were highly and positively correlated.

Apart from being promising intermediates in steroid drug synthesis, glycoalkaloids and their aglycones may also have beneficial properties, such as antiallergic, antipyretic, anti-inflammatory, hyperglycemic effects, and antibiotic activities against pathogenic bacteria, viruses, protozoa, and fungi[28]. Moreover, some evidence indicates that they may possess anti-carcinogenic activities and other health benefits when present at lower levels indicating a possible use in the phyto-pharmaceutical industry. Furthermore, in *vitro* anticarcinogenic effects of potato glycoalkaloids against a series of human cancer cells were studied[91]. The effects were concentration dependent in the range of 0.1-10 μg/ml. Besides the potential pharmacological applications, these compounds have an inhibitory effect on the growth of many insects and other phytopathogens and antifungal activity[92, 93]. These properties can be used to modify the host plant resistance in transgenic potatoes (*Solanum tuberosum* L.). α-solanine and α-chaconine have been shown to act synergistically in inhibiting snail feeding and find uses for pest control. In addition glycoalkaloids can act as a hatch agent which may be explored to control nematode infestation in potato crops.

VIII. PROCESSING AND PRODUCT DEVELOPMENT

The demand for processed potato products like chips and French fries is increasing continuously mainly due to improved living standard, increased urbanization, preference to fast foods and expanding tourist trade. Potato-processing industry is emerging as a fast growing industry particularly in the north-western India in Punjab and around Delhi[94]. Potato processing industry in India is mainly composed of three segments, *viz.*, potato snacks (including potato chips, *Alu Bhujia* etc.), potato flakes/ powder and French fries. However, many ready to eat products can be prepared by utilizing potato. Flours, granules, mashed products from potato are used in the preparation of soup powder, instant foods etc. Dehydrated potato products can also be used as binding materials for preparing curry based food items and for making *tikkia*, *pakoda*, cutlet, stuffed *parotha*, *kofta* etc. Poato flour based products are also for mass consumption and can be marketed easily[95].

(a) Potato Chips

Nowadays, fried products are very popular with people of all ages. Their preparation is easy, fast and economical. For these reasons, in recent decades, frying processes have been generalized not only in fast food establishments but also in restaurants, food industries, and homes, etc[96]. Potato chips, a popular salty snack have been the leading segment of potato processing industry in India. Potato chips are thin potato slices that are dehydrated by deep fat frying to moisture content of 0.02 kg/kg or less[97]. Potato chips contain oil content from 35% to 45% (wet basis) and give the product a unique texture-flavor combination[98]. Frying in hot oil at temperatures between 160°C and 180°C is characterized by high drying rates that are critical for ensuring favorable structural and textural properties of the final product[97]. When potatoes are thermally processed, the energy provided by cooking causes the amylose and amylopectin in the starch granules to disassociate, allowing water to penetrate into the starch granule. This process is called gelatinization and is important in the texture of cooked potatoes. As most of the water in a cooked potato is entrapped in the starch granule, it has a dry texture. Fat is especially compatible with enhancing the texture of cooked potatoes[99]. Amount of oil absorbed during frying and post-frying process affects the quality of final product. One of the most important quality parameters of fried potato products strictly related to consumer perception is the colour. Colour is critically appraised by consumers and often is the basis for their selection or rejection of chips. Potato chip colour is the result of the Maillard reaction, which depends on the content of reducing sugars and proteins, temperature and time of frying. Preparation of potato chips involves washing, peeling, slicing and frying (Fig. 2). Steps involved in preparation of potato chips are described below

Receiving and sorting of potato

Since the quality of final product is affected by raw material, so before proceeding to processing removal of rotted and damaged tubers, extraneous matter and other unwanted materials is carried out.

Receiving of potato
↓
Dry sorting
↓
Destoning
↓
Washing
↓
Peeling
↓
Washing, sorting and trimming
↓
Slicing
↓
Removal of water
↓
Frying
↓
Seasoning and flavouring
↓
Packaging

Fig. 2: Flow chart for potato chips production.

Destoning of potato

Destoning from potatoes is performed by use of destoner which removes stones, wood, or any other foreign matter that may have been dug when harvesting the potatoes. Stones may cause serious damage to the equipment and problems during the slicing operation. Potatoes are placed in a tube shape tanks, the stones sink lower down in the tube due to difference in density between stones and potatoes.

Washing

Potatoes are washed with high pressure sprays of water on tubers. This process removes sand, dirt and any other extraneous matter.

Peeling

For the processing of potatoes in any form, removal of peel is an important unit operation. Peeling may be accomplished by steam, abrasive or lye method depending on the product to be prepared. Steam peeling is one of the most popular among modern methods of peeling. Its widespread application is due to its high automation, precise control of time, temperature and pressure by electronic devices to minimize peeling loss and reduced environmental pollution as compared to chemical peeling[100]. Potatoes may be peeled with steam in steam pressure chambers. The dwell time in steam should only be long enough to heat through the peel. Immediately after heating they should be cooled quickly to stop penetration and the peel is then brushed away.

Another method of peeling is chemical peeling also known as lye peeling. In lye peeling of potato, the potatoes are propelled through the lye by helical conveyors. The lye solution may be between 82-105°C, with a caustic concentration between 5-15%. The dwell time may be as little as 10 s and as long as 1 min depending upon age of the tuber. Solute (NaOH) diffusion, energy transfer and chemical kinetics are different parameters involved in this process[101].

Abrasion peeling is used in particular, in potato chips plants where complete removal of skin is not essential. In abrasion peeling, machines are used to remove the peel. The main parts of the machine are a peeling drum and a water spraying unit. The peeling drum with protrusions on the inside surface rotates and detaches peel from potatoes by abrasion. The water spraying unit washes the potatoes and simultaneously peel is removed from the drum through the perforation along with the flow of water[102].

Washing, sorting and trimming

Before slicing the washing, sorting and trimming are important unit operations for production of good quality chips. Electronic sorters can also be used at the forefront of slicer to take out defective potatoes which will ultimately save energy, oil and increase efficiency of operation.

Slicing

Smooth surfaced slices of potato are desirable. Presence of starch, sugars and other substances may cause sticking of potato slices during frying; therefore, the sliced potatoes are washed.

Blanching

Sliced potatoes are blanched to obtain a more uniform colour after frying, to inactivate enzymes and to form a layer of gelatinized starch that limits oil absorption and improves texture. Extreme blanching conditions might, however, cause a significant loss in nutrients, product taste, colour and texture[103]. Blanching may be carried out in water or solutions such as sodium bisulphate, phosphoric acid, sodium citrate, calcium chloride or magnesium chloride. The slices are dipped in solution and scattered throughout the tank on mesh chain belts. The time may vary from 1 minute to longer depending upon the batch.

Drying/Frying

Partial drying of sliced potatoes is carried out to remove excess surface moisture which may lead to higher oil absorption and lengthen the process. Excess moisture can be removed with vibrating mesh belt, centrifugal extraction, heated air etc. Frying generates tasty products that have crispy crusts, tempting aromas and visual appeal. Type of oil, oil temperature, and duration of cooking greatly affects the final quality attributes of fried foods[104]. Frying process may be batch or continuous. The chips are fed into a kettle filled with heated oil and get fried for some time and forwarded to the area submerged in cooking oil to ensure complete cooking. In continuous frying system frying oil is continuously circulated and filtered to remove burnt particles.

Oil absorption of the food is affected by many factors including process conditions (temperature, time), pre-treatment of the food (such as dehydration methods), physico-chemical characteristics of food, oil origin, chemical composition of oil and others. Longer times and lower frying temperatures usually lead to higher final oil contents in fried potato products[105]. For example, par-fried or completely fried potatoes have fat concentrations of 3-5% and 16-18%, respectively [106].

The market trend for the consumption of healthier foods has forced the snacks industry to develop alternative technologies to traditional frying, maintaining their characteristic flavour and texture. Use of microwave frying, vacuum frying for food products may be considered as a new way of improving the quality of the fried foods.

Oztop *et al.*[105] studied the effects of microwaves on quality of fried potatoes (moisture content, oil absorption, color and hardness) and optimized the process by using *Taguchi Technique*. Microwave power level (400 W, 550 W and 700 W), frying time (2.0, 2.5, 3.0 min) and oil type (sunflower, corn and hazelnut oil) were the parameters used in the study. The optimum conditions were found as frying at 550 W microwave power levels, for 2.5 min in sunflower oil. At this condition, the oil content of fried potatoes was lower than that of conventionally fried ones.

The vacuum frying (VF) stands out among the studied processes to achieve quality improvement. In this operation, the food is immersed in oil at 40 mmHg in a closed system. The reduced pressure decreases the boiling temperature of water contained in the food, without affecting the characteristic texture of the fried food. The benefits of this technology are associated with frying at lower oil temperatures and reduced exposure to oxygen. Vacuum frying can decrease fat content by 43–57% by using reduced frying temperatures that remove less water.

Vacuum frying was tested as an alternative technique to develop low oil content potato chips. The effect of oil temperature (118, 132, 144°C) and vacuum pressure (16.66, 9.89 and 3.12 kPa) on the drying rate and oil absorption of potato chips and on the product quality attributes such as shrinkage, color, and texture was investigated. Furthermore, the characteristics of the vacuum-fried potato chips (3.12 kPa and 144°C) were compared to potato chips fried under atmospheric conditions (165°C). During vacuum frying, oil temperature and vacuum pressure had a significant effect on the drying rate and oil absorption rate of potato chips. Potato chips fried at lower vacuum pressure and higher temperature had less volume shrinkage. Color was not significantly affected by the oil temperature and vacuum pressure. Hardness values increased with increasing oil temperature and decreasing vacuum levels[107].

A new approach to reducing the fat content of fried potatoes is controlled dynamic radiant frying (CDRF). CDRF is an alternative to the immersion frying process, which can reduce oil content by 30 to 50% relative to traditional immersion-fried counterparts. CDRF uses high-temperature radiant emitters to generate high-flux infrared radiation to heat the product[108]. CDRF technology has been successfully used to produce fried-food products like potato cakes, and French fries[109]. It differs from other cooking methods because it mimics the heat flux experienced during immersion frying, supplying energy to the surface of the food to cause water evaporation, crust formation, and browning. Benefits of this technique are high heat flux generation, reduced fat and caloric content, accelerated product heating, removal of safety hazards associated with hot frying oil, and elimination of fry oil waste.

Inspection and seasoning

After frying the chips are inspected to remove burnt, discolored and defective chips and then seasoned to achieve the taste and flavor. Generally seasoning is done immediately after frying to have easy binding of seasoning particles to chips.

Packaging

Selection of packaging material for chips is quite sensitive as chips are receptive of moisture and exposure to light may lead to rancidity so ready chips are packaged into heat sealed packs of single or double walled depending upon the material from which they are made of.

(b) French Fries

French fries, sticks of deep fried potatoes, are probably the most typical example of fried foods world-wide. They have a crust with a thickness of the order of 0.5 mm and a water content of 2-4%. The formation of this thin layer is the result of the advancement of the dehydration front towards the center of the potato with simultaneous pore formation and shrinkage[110].

The preliminary steps in regular cut French fry production include washing of raw potatoes, peeling, sorting, and cutting into strips. After this, there are mainly two ways of producing deep-fat fried French fries: (a) the potato strips are immediately deep-fat fried and served, and (b) the potato strips are blanched, partially dehydrated, deep-fat par-fried, frozen, deep-fat finish-fried (Fig. 3) and served. In industrial production, the potato strips are generally blanched with water (60-85 °C) for more than 10 min mainly to inactivate enzymes and to obtain a uniform color, and then pre-dried with warm air to improve texture. The blanched potato strips are par-fried in hot oil (170–190 °C), cooled at room temperature, frozen, packaged and distributed. The process of finish-frying is usually accomplished in restaurants or at home and each of these steps is important for the final product quality and its oil content[111].

Receiving of potato
↓
Dry sorting
↓
Destoning
↓
Washing
↓
Peeling
↓
Trimming
↓
Cutting and grading
↓
Blanching
↓
Excess moisture removal from strips
↓
Par-frying
↓
De-oiling and cooling
↓
Freezing and Packaging
↓
Stoarge

Fig. 3: Flow chart for French fries production.

While use of heat in the preparation of potato dish is almost essential, this also initiates several chemical reactions between the endogenous chemicals such as reducing sugars, amino acids. Potato, when processed at elevated temperature, produces acrylamide. It is well established that acrylamide is formed, *via* the Maillard reaction between reducing sugars and asparagines, in potatoes during deep frying for chips and French fries at high temperature for a long period. Published data clearly establish potato products (French fries, chips) are the major dietary sources of acrylamide for adults, children in the developed countries where these products are consumed routinely in meals and/or as snack[112]. Acrylamide is a low molecular weight compound which has been classified by the International Agency for Research on Cancer (IARC) as a probable carcinogen[113]. Thus, the reduction of acrylamide in French fries and potato chip would go a long way in reducing its overall human exposure.

Thus, various efforts are being made to eliminate and/or reduce acrylamide content in fried products. Microwave pre-thawing of frozen strips reduced the acrylamide level of French fries by 10% (from 17.7 to 15.9 ng/g), 89% (from 72.1 to 8.0 ng/g), and 64% (from 50.5 to 18.4 ng/g) for frying at 170, 180, and 190°C, respectively, in comparison to the control samples. Quality attributes (texture, color, and oil content) of pre-treated strips were found to be comparable to those of the control[114]. Acrylamide formation could be inhibited by 30.3%, 53.3% and 89.3% when the sliced potato strips were soaked in 0.001, 0.01 and 0.1 M vanadyl sulphate solutions, respectively, for 60 min before frying. Moreover, 57.7%, 71.4% and 92.5% inhibition of acrylamide formation was observed when chips were soaked in the respective vanadyl sulphate solution before frying[115].

(c) Dehydrated Potato Products

Drying is one of the oldest methods of food preservation and an important factor in production technology. Perishable goods, such as raw vegetables or fruit, benefit from dehydration in many ways, primarily in substantially reducing losses resulting from raw material storage (up to 20%), lowering water activity of the material, slowing down many enzymatic reactions and microbial growth, which in write turn permits much longer storage than fresh material.

These products should all have high nutritional value, low content of toxic compounds and good organoleptic characteristics. Dehydrated potato products and intermediates obtained during the processing of potatoes into granules contain the same compounds as the raw material but in different quantities and proportions. Processing into refined products, including those dehydrated, usually uses potato varieties with a low tendency to accumulate anti-nutritional or toxic compounds[67].

During drying moisture evaporation will also lead to some structural changes such as shrinkage and development of surface porosity and roughness, which highly affect the mouthfeeling. When Potato slices were subjected to drying for 7 h at 62°C, it was observed that shrinkage varied linearly with respect to moisture content and reduction in radial dimension of potato slices was around 35%. However, Porosity undergoes rapid increase after attaining certain moisture content in final stages of drying[116].

Drying provides a whole range of different potato products. Potato granules, powder, shredded and sliced potato, potato flakes are some of the most important form of dehydrated potato products consumed world over[117].

(a) Potato flakes

Potato flakes are dehydrated potato products mostly processed by cooking, mashing and dehydration to produce a conveniently packaged and easy to reconstitute by adding hot water or milk making them desirable to many consumers[118]. They can be used as substitute for fresh mashed potatoes. Flakes can, however, be stored for longer duration on the shelf[119] compared to most fresh potato products and raw potatoes that can only be stored for up to 12 weeks in cooler highland areas[120]. Flakes can have different flavours such as onion, chilly, garlic, and butter and can be used as a thickener enhancing creamy frozen desserts, gravies and chocolate milk. The process of preparing flakes from potato involves many stages involving peeling, slicing, blanching, cooking and drum drying (Fig. 4). Potato flakes find wide range of uses such as thickening agent[121], extruded products[122], fabricated potato chips, snack pellets, battered breaded products, and potato powder[123].

Potatoes
↓
Washing
↓
Steam peeling
↓
Dry peel
↓
Slice cutting
↓
Slice washing (removal of starch)
↓
Blanching at 65-70°C for 20 to 25 min.
↓
Cooling at 25°C
↓
Cooking (30-45 min.)
↓
Mixing of additives
↓
Drum drying
↓
Inspection of dried flakes
↓
Magnet removal of metal particles, if any
↓
Milling of flakes from screen
↓
Equalization of moisture
↓
Packaging

Fig. 4: Flow chart for potato flakes.

(b) Potato granules

Potato granules, one of the important products prepared from dehydrated mashed potatoes, contain 6-7% moisture. They are reconstituted to a texture that is either dry and mealy or moist and creamy, according to individual preference. Potato granules come as a fine powder that is mixed with boiling water to yield a food similar in texture and taste to mashed potatoes. The production process involves peeling, cooking and drying potatoes, which are then reduced to granules of one or a few cells each (Fig. 5). Some chemical additives are used to prolong shelf life.

Pre-processing operations
↓
Slicing and washing
↓
Blanching
↓
Cooling
↓
Steaming
↓
Mashing, Mixing, sulphiting and other additives
↓
Conditioning
↓
Air lift drying
↓
Sieving
↓
Fluidized bed drying
↓
Final product

Fig. 5: Flow chart for potato granules.

(c) Potato flour

Potato flour can be made from potatoes by slicing, blanching, sulphitation, and drying in a hot air dehydrator (Fig. 6). Because it is rich in starch, potato flour can improve the functional properties of several food products. It can be used in a great variety of foods. It has its own distinctive flavour and texture but is sufficiently bland to be incorporated easily into traditional local preparations. Misra and Kulshrestha[124] studied effect of storage at

Raw potato
↓
Washing
↓
Peeling
↓
Cooking
↓
Mashing
↓
Chemical treatment
↓
Drying
↓
Dried product
↓
Grinding
↓
Potato flour
↓
Packaging

Fig. 6: Flow chart for potato flour.

room temperature and refrigerated conditions on nutritional value of potato flour made from OP-1, Kufri Sutlej and Kufri Ashoka varieties of India. Significant varietal differences occurred in all the nutritional parameters measured. However, storage of potato flour for 6 months did not result in any significant changes in the nutritional parameters, except that moisture increased and ascorbic acid decreased.

The dehydrated products have various advantages:

- The storage of dehydrated potato products is very easy in comparison to fresh potatoes.
- Dehydrated potato products have no specific storage requirements, making them perfect for use in any climate also without a need for refrigerated containers.
- Dehydrated potato products can easily be fortified with extra vitamins and minerals to further enhance their nutritional benefits
- Dehydrated potato products can be used by individuals with allergies to wheat and gluten intolerance.
- Using potato flakes in place of fresh potatoes is more hygienic in quick service restaurants, processing facilities and other places of use.
- Potato flakes assure consistent quality of the end product due to fixed quality parameters while in case of fresh potatoes quality cannot be consistent and would result in inconsistency of end products.
- There is defined period of availability of fresh potato although availability is extended due to storage in cold storages by another 6-7 months, there is a period of 3-4 months when availability of the product is not good and quality is also poor. The potato flakes offer the advantage of use round the year.

(d) Canned Potatoes

The canning technology has come to the forefront amongst all the other processing or preservation technologies. Different forms of potato such as whole, diced, sliced, strips, and julienne are canned. During processing potato should not disintegrate in good canning. Low specific gravity potatoes are preferred for canning. Disintegration of potatoes in the cans can be prevented by adding calcium chloride, which tends to firm the tissue. Process of canning involves washing, peeling, cutting, filling in cans, brining with $CaCl_2$, exhausting, sealing etc. (Fig. 7).

Rattan and Ramaswamy[125] optimized processing conditions for maximizing quality retention in canned potatoes during agitation thermal processing. Potato cubes were cut into 1.6 × 1.6 × 1.6 cm cubes and filled in No. 2 cans (307 × 409). Then the potato cubes were covered with 1% aqueous solution of carboxy-methyl cellulose and the processing times were adjusted to provide an equivalent lethality (F_o value) of 10 min with process

Graded Potatoes
↓
Washing
↓
Peeling
↓
Cutting
↓
Filling
↓
Brining(with $CaCl_2$)
↓
Exhausting
↓
Sealing
↓
Processing
↓
Cooling

Fig. 7: Flow chart for canned potatoes.

parameters retort temperature 115–125°C; rotation speed 20 rpm; can headspace 10 mm; and still and agitating modes (end-over-end, fixed and free axial). The study indicated that potato particles processed at higher temperatures and under better agitating conditions demonstrated better quality retention.

(e) Potato Starch

Potato starch is the third most important material used in food and non-food sectors with an estimated 6% share of global starch production. A significant fraction of the potato industry is involved in starch production. Potato starch is considered to be a great alternative to corn or wheat starch due to its gluten free composition. Potato starch is isolated from mashed potatoes and water and impurities are removed from it (Fig. 8). It is available in powder form as well as in liquid form. Potato starch has unique characteristics, such as a high phosphorus content, high viscosity, high swelling power, high paste clarity, low tendency to retrograde, as well as large granule size[126]. Alvani *et al.*[127] studied physico-chemical properties of starches characterized from 10 potato cultivars grown under the same conditions. The chemical composition revealed some differences amongst the starches with protein ranging from 0.30% to 0.34%, amylose 25.2% to 29.1% and phosphorus 52.6–66.2 mg/100 g. High performance size-exclusion chromatography (HPSEC) fractionation of isoamylose debranched amylopectin showed that the amylopectin molecules were less branched and consisted of more B1, but less A-chains, than cereal starches. Gelatinization onset (T_o), peak (T_p) and conclusion (T_c) temperatures of the native potato starches ranged from 58.7 to 62.5°C, 62.5 to 66.1°C and 68.7 to 72.3°C, respectively, whilst the gelatinization enthalpies ranged from 15.1 to 18.4 J/g. Granule diameter was reported from 20.6–30.9 μm.

Cleaning
↓
Mashing
↓
Starch Isolation
↓
Separation of Fruit Water by Hydro separators
↓
Removal of the Impurities
↓
Filtration
↓
Drying

Fig. 8: Flow chart for production of potato starch.

Esterification with acetic anhydride and adipic anhydride (E 1422 starch) caused the most pronounced changes in physicochemical parameters compared with native potato starch powder. In native and E 1422 starches, mean granule diameter changed from 23.1 to 25.4 μm, shape factor from 1.47 to 1.33, specific surface from 0.2115 to 0.1695 m^2/g, porosity from 41.31% to 45.29%, water absorption increased up to 50%, and flow behavior changed from easy to cohesive[128].

(f) By Product Utilization

In processing, 10–25% of the raw product is discharged as waste. Peels are the major waste of potato processing that is perishable and cause many management problems in terms of disposal and sanitation. Because the skins are an excellent source of fiber, minerals, vitamins, and antioxidants, they represent a nutritionally rich food ingredient that could potentially add value for the potato industry. At present, a large proportion of this waste is used as an animal feed and for the production of bio-fuels. Dry potato peels have been used as an excellent carbohydrate source in cattle feed. Using potato wastes instead of corn in cattle feed did not affect the metabolic state or milk status of the cattle. Typically, potato wastes are fed in a dry, dewatered form. The use of wet potato wastes in cattle feed has been investigated to reduce drying expenditures. Wet potato processing wastes can be introduced into cattle feed up to 20% without negative results. Potato wastes have also been evaluated as a potential source for the production of acetone, butanol, and ethanol by fermentation.

Potato peels provide a good source of dietary fiber, particularly when processed by lye peeling technique. Potato peels contain 40 g dietary fiber/100 g dry matter, depending on the variety of potato processed and the method of peeling. Application of extruded and unextruded potato peels as a source of dietary fiber in baked goods has been evaluated.

The addition of peels obtained from various processes (abrasion, steam, caustic, and manual peeling) to wheat flours at levels of 5, 10 and 15%

resulted in breads of acceptable sensory quality, but sensory evaluation of the products made with peels from caustic peeling revealed a musty aroma. In USA acceptable muffins were made with a 25% replacement potato peel for wheat flour. Potato peels were also found to prolong muffin shelf-life by controlling lipid oxidation. Extrusion cooking of potato peels affected the colour of baked goods, and some physical and chemical properties of the peels. Extrusion of abrasion potato peels also reduced the hydration capacity and increased non-starch polysaccharides[129]. When potato peels originating from steam peeling were extruded at barrel temperatures of 104°C and 143°C and various feed moistures, the application of high-temperature processing resulted in darker extrudates, probably caused by enzymatic and non-enzymatic browning reactions. Color was the parameter that was affected most by the extrusion process, as indicated by lower Hunter L values[130].

Potato peels have been reported to be as a major dietary source of phenolics and a number of antioxidants. The recovery of these valuable phenolics from the potato peel waste could improve the economics of the potato processing industries and also provide a cheaper and effective alternative to synthetic antioxidants which are commercially utilized in the food industries.

IX. CONCLUSION AND FUTURE PERSPECTIVE

It has now been established through researches in potato chemistry that there is a lot more in potatoes than starch. Potatoes contain enough phytochemicals to justify the claim of being health promoters, therefore, should form a substantial part of our daily diet. The existing level of processing and the extent of value addition are very low as compared to the developed countries. In India the food processing industry is ranked fifth in terms of production, consumption, export and expected growth. Potato processing is very important, in making India self reliant in sustaining the food production and nutrition. Value added processed products are opening up new market avenues in the national and international markets, and as a result, the farmers are finding it highly remunerative to grow processing potatoes even in areas traditionally unknown to potato cultivation. Growing urbanization, changing food habits and preference for ready-cooked snacks have made potatoes to travel from kitchen and dining tables to pouches and packets liked by everyone. The national and multinationals potato processing companies in India now prefer indigenous processing varieties over exotic ones. There is a big market for inexpensive dehydrated potato chips, cubes and other products which can be easily prepared at the cottage industry level and can provide employment to the rural youth and village women. As potatoes produce more energy per unit land and time, it is thus prudent to make concerted efforts to promote potatoes as a source for dietary energy in poor and poverty stricken countries.

REFERENCES

[1] FAO, http://www.fao.org/potato-2008 /en/potato/index.html (2008).
[2] Burlingame B., Mouille B., Charrondiere R., "Nutrients, bioactive non-nutrients and anti-nutrients in potatoes", *J Food Compos Anal.*, **22,** 494–502 (2009).
[3] Ezekiel R., Singh N., Sharma S., Kaur A., "Beneficial photochemicals in potato–A review", *Food Res Int.*, **50**, 487–496 (2013).
[4] National Horticulture Board, http://nhb.gov.in/area%20_production.html (2014).
[5] National Horticulture Board, http://nhb.gov.in/area pro/Indian%20Horticulture%202013. pdf. (2013).
[6] http://www.nsepp.in/national-seminar-on-emerging-poblems-of-potato-introduction.html
[7] Pandey S.K., Singh, J.P., Gopal J., "Potato varieties and cropping systems in India", *Potato J.*, **35**(3–4), 103–110 (2008).
[8] Andre C.M., Ghislain M., Bertin P., Oufir M., Mdel.H., Hoffmann L., Hauseman J.F., Larondelle Y.E., Evers D., "Andean potato cultivars (*Solanum tuberosum* L.) as source of antioxidant and mineral micronutrients", *J Agric Food Chem.*, **55**, 366–378 (2007).
[9] Negi P., Nath N., "Effect of partial dehydration on quality of canned potatoes", *Eur Food Res Tech.,* **215**(3), 231–234 (2002).
[10] Jimenez M.E., Rossi A.M., Samma´n N.C., "Phenotypic, agronomic, nutritional characteristics of seven varieties of Andean potatoes", *J Food Comp Anal.*, **22**, 613–616 (2009).
[11] Ritter E., Barandalla L., Lopez R., Galarreta D.J.I. "Exploitation of exotic, cultivated *Solanum* germplasm for breeding and commercial purposes", *Potato Res.*, **51**, 301–311 (2008).
[12] Liu R.H., "Health-Promoting Components of Fruits and Vegetables in the Diet", *Advances in Nutrition.*, **4**(3), 384S–392S (2013).
[13] Luis G., Rubio C., Weller G.D., Gutierrez A.J., Revert C., Hardisson A., "Comparative study of the mineral composition of several varieties of potatoes (*Solanum tuberosum* L.) from different countries cultivated in Canary Islands (Spain)", *Int J Food Science and Technology*., **46**(4), 774–780 (2011).
[14] Lisinska G., Leszczynski W., "Potato Science and Technology". Elsevier Applied Science Publishers Ltd., London, New York (1989).
[15] Toledo A., Burlingame B., "Biodiversity and nutrition: A common path toward global food security and sustainable development", *J Food Comp Ana1.*, **19**(6), 477–483 (2006).
[16] Vreugdenhil D., Bradshaw D., Gebhardt C., Govers F., Mackerron D.K.L., Taylor M.A., Ross H.A. (*eds.*), Potato Biology and Biotechnology: Advances and Perspectives. Elsevier, Amsterdam, the Netherlands, (2007).
[17] Friedman M., "Chemistry, biochemistry, and dietary role of potato polyphenols- A review", *J Agric Food Chem.*, **45,** 1523–1540 (1997).
[18] McCann S.E., Ambrosone C.B., Moysich K.B., Brasure J., Marshall J.R., Freudenheim J.L., Wilkinson G.S., Graham S., "Intakes of Selected Nutrients, Foods, and Phytochemicals and Prostate Cancer Risk in Western New York", *Nutr Cancer.*, **53**(1), 33–41 (2005).
[19] Shakya R., Navarre D.A., "Rapid screening of ascorbic acid, glycoalkaloids, and phenolics in potato using high-performance liquid chromatography", *J Agric Food Chem.*, **54**, 5253–5260 (2006).
[20] Manach C., Scalbert A., Morand C., Remesy C., Jimenez L., "Polyphenols: Food sources and bioavailability", *The Am J Clinical Nutr.*, **79**, 727–747 (2004).
[21] Shahidi F., Naczk M., "Phenolics in Food and Nutraceuticals", Boca Raton, FL: CRC Press (2003).

[22] Cieslik E., Greda A., Adamus W., "Contents of polyphenols in fruit and vegetables" *Food Chem.*, **94**, 135–142 (2006).

[23] Romero G.M., Roman A.D., Carretero S.A., Gutierrez F.A., "Analytical determination of antioxidants in tomato: Typical components of the Mediterranean diet", *J Sep Sci.*, **30**, 452–461 (2007).

[24] Chun O.K., Kim D.O., Smith N., Schroeder D., Han J.T., Lee C.Y., "Daily consumption of phenolics and total antioxidant capacity from fruit and vegetables in the American diet", *J Sci Food Agric.*, **85**, 1715–172 (2005).

[25] Helmja K., Vaher M., Gorbatsova J., Kaljurand M., "Characterization of bioactive compounds contained in vegetables of the *Solanaceae* family by capillary electrophoresis", *Proc Estonian Academy Sci Chem.*, **56**, 172–186 (2007).

[26] Hen A.K., Fuenzalida C., Hess S., Contreras Galvez V.A., Mondaca, L.R., "Antioxidant capacity and total phenolic compounds of twelve selected potato landrace clones grown in Southern Chile Chilean", *J Agric Res.*, **72**(1), 3–9 (2012).

[27] Schieber A., Saldana A.D.A., "Potato peels: A source of nutritionally and pharmacologically interesting compounds—a review", *Food.*, **3**(2), 23–29 (2009).

[28] Friedman M., "Potato glycoalkaloids and metabolites: Roles in the plant and in the diet," *J Agric Food Chem.*, **54**, 8655–8681 (2006).

[29] Im H.W., Suh B.S., Lee S.U., Kozukue N., Kameyama O.M., Levin C. E., Friedman M., "Analysis of phenolic compound by high-performance liquid chromatography and liquid chromatography/mass spectrometry in potato plant flowers, leaves, stems and tubers and in home-processed potatoes", *J Agric Food Chem.*, **56**, 3341–3349 (2008).

[30] Reddivari L., Hale A., Miller J., "Determination of phenolic content, composition and their contribution to antioxidant activity in specialty potato selections", *Am J Potato Res.,* **84**(4), 275–282 (2007).

[31] Reddivari L., Hale, A., Miller J., "Genotype, location, and year influence antioxidant activity, carotenoid content, phenolic content, and composition in specialty potatoes", *J Agric Food Chem.*, **55**(20), 8073– 8079 (2007).

[32] Stushnoff C., Holm D., Thompson M.D., Jiang W., Thompson H.J., Joyce N.I., Wilson P. "Antioxidant properties of cultivars and selections from the Colorado potato breeding program", *Am J Potato Res.*, **85**, 267–276 (2008).

[33] Navarre D., Shakya R., Holden J., Kumar S., "The Effect of different cooking methods on phenolics and vitamin C in developmentally young potato tubers", *Am J Potato Res.*, **87**(4), 350–359 (2010).

[34] Lewis C.E., Walker J.R.L., Lancaster J.E., Sutton K.H., "Determination of anthocyanins, flavonoids and phenolic acids in potatoes. I: Coloured cultivars of *Solanum tuberosum* L", *J Sci Food Agric.*, **77**, 45–57 (1998).

[35] Mattila P., Hellstrom J., "Phenolic acids in potatoes, vegetables, and some of their products", *J Food Comp Anal.*, **20**,152–160 (2007).

[36] Nara K., Miyoshi T., Honma T., Koga H., "Antioxidative activity of bound-form phenolics in potato peel", *Biosci Biotech Biochem.*, **70**, 1489–1491 (2006).

[37] Hye L.J., Kyung Y.Y., "Cultivar differences in phenolic contents/biological activities of color-fleshed potatoes and their relationships", *Hort Environ Biotech.*, **53**(2), 175–181 (2012).

[38] Pietta P.G., "Flavonoids as antioxidants", *J Nat Prod.,* **63**, 1035–1042 (2000).

[39] Lewis C.E., Walker J.R.L., Lancaster J.E., Conner A.J., "Light regulation of anthocyanin, flavonoid and phenolic acid biosynthesis in potato mini tubers in vitro", *Aust J Plant Physiol.,* **25**, 915–922 (1998).

[40] Tudela J.A., Cantos E., Espin J.C., Barberan F.A., Gil M.I., "Induction of antioxidant flavonol biosynthesis in fresh-cut potatoes. Effect of domestic cooking", *J Agric Food Chem.*, **50**, 5925–5931 (2002).

[41] Chu Y.H., Chang C.L., Hsu H.F., "Flavonoid content of several vegetables and their antioxidant activity", *J Sci Food Agric.*, **80**, 561–566 (2000).

[42] Wang L.S., Carmella S., Keyes R., Kuo C.T., Huang Y.W., Hecht S.S., Stoner G.D. Anthocyanins and Cancer Prevention. Nutraceuticals and Cancer. Sarkar F.H. (*ed.*), Springer Netherland, pp. 201–229 (2012).

[43] Elbe V.J.H., Schwartz S.J., Fennema O.R., Colorants: Food Chemistry, New York, Marcel Dekker, pp. 651"722 (1996).

[44] Lachman J., Hamouz K., Sulc M., Orsak M., Pivec V., Hejtmankova A., Dvorak P., Cepl J., "Cultivar differences of total anthocyanins and anthocyanidins in red and purple-fleshed potatoes and their relation to antioxidant activity", *Food Chem.*, **114**, 836–843 (2009).

[45] Eichhorn S., Winterhalter P., "Anthocyanins from pigmented potato (*Solanum tuberosum* L.) varieties", *Food Res Int.*, **38**, 943–948 (2005).

[46] Naito K., Umemura Y., Mori M., Sumida T., Okada T., Takamatsu N., Okawa Y., Hayashi K., Saito N., Honda T., "Acylated pelargonidin glycosides from red potato", *Phytochem.*, **47**, 109–112 (1998).

[47] Saona R.L.E., Giusti M.M., Wrolstad R.E., "Anthocyanin pigment composition of red-fleshed potatoes", *J Food Sci.*, **63**, 458– 465 (1998).

[48] Brown C.R., Culley D., Yang C., Durst R., Wrolstad R., "Variation of anthocyanin and carotenoid contents and associated antioxidant values in potato breeding lines", *J Amer Soc Hort Sci.*, **130**(2),174–180 (2005).

[49] Brown C.R., "Breeding for phytonutrient enhancement of potato", *Am J Potato Res.*, **85**, 298–307 (2008).

[50] Jansen G., Flamme W., "Coloured potatoes (*Solanum Tuberosum L*)-Anthocyanin content and tuber quality", *Genetic Resour Crop Evol.*, **53,** 1321–1331 (2006).

[51] Clark J., Tsang C., Clapham M.E., "Phenolic composition and antioxidant capacity of a novel variety of purple potato", *Proceed Nutr Soc*, **69,** OCE6: E510 (2010).

[52] Navarre D.A., Shakya R., Holden J., Crosslin J.M., "LC-MS analysis of phenolic compounds in tubers showing zebra chip symptoms", *Am J Potato Res.*, **86**, 88–95 (2009).

[53] Konings E.J., Roomans H.H., Dorant E., Goldbohm R.A., Saris W.H., Vanden B.P., "Folate intake of the Dutch population according to newly established liquid chromatography data for foods", *Am J Clinical Nutr.,* **73**, 765–776 (2001).

[54] Goyer A., Navarre D.A., "Determination of folate concentrations in diverse potato germplasm using a trienzyme extraction and a microbiological assay", *J Agric Food Chem.*, **55**, 3523–3528 (2007).

[55] Tanemura Y., Yoshino M., "Regulatory role of polyamine in the acid phosphatase from potato tubers", *Plant Physio Biochem.*, **44**, 43–48 (2006).

[56] Matsuda F., Morino K., Ano R., Kuzawa M., Wakasa K., Miyagawa H., "Metabolic flux analysis of the phenylpropanoid pathway in elicitor-treated potato tuber tissue", *Plant Cell Physio.*, **46**, 454–466 (2005).

[57] Della P.D., Pogson B.J., "Vitamin synthesis in plants: Tocopherols and carotenoids", *Annual Review Plant Bio.*, **57**, 711–738 (2006).

[58] Tan J.S., Wang J.J., Flood V., Rochtchina E., Smith W., Mitchell P., "Dietary antioxidants and the long-term incidence of age-related macular degeneration: The Blue Mountains Eye Study", *Ophtha.*, **115**, 334–341 (2008).

[59] Ducreux L.J.M., Morris W.L., Hedley P.E., Shepherd T., Davis H.V., Millam S., Taylor M.A., "Metabolic engineering of high carotenoid potato tubers containing significant levels of beta-carotene and lutein", *J Exper Bot.*, **56**, 81–89 (2005).

[60] Diretto G., Al-Babili S., Tavazza R., Papacchioli V., Beyer P., Giuliano G., "Metabolic engineering of potato carotenoid content through tuber-specific over expression of a bacterial mini-pathway", *PloS One*, **2,** 350 (2007).

[61] Murniece I., Kruma Z., Skrabule I., Vaivode A., "Carotenoids and phenols of organically and conventionally cultivated potato varieties", *Int J Chem Engg Appl.*, **4**(5), 342–348 (2013).

[62] Vaananen T., Ikonen T., Rokka V.M., Kuronen P., Serimaa R., Ollilainen V., "Influence of incorporated wild *Solanum* genomes on potato properties in terms of starch nanostructure and glycoalkaloid content", *J Agric Food Chem.*, **53**, 5313–5325 (2005).

[63] Camire M.E., Kubow S., Donnelly D.J. "Potatoes and human health. Critical reviews", *Food Sci Nutr.*, **49**, 823–840 (2009).

[64] Dao L., Friedman M., "Chlorophyll, chlorogenic acid, glycoalkaloids, and protease inhibitor content of fresh and green potatoes", *J Agric Food Chem.*, **42**, 633–639 (1994).

[65] Haddadin M.S., Humeid M.A., Qaroot F.A., Robinson R.K., "Effect of exposure to light on the solanine content of two varieties of potato (*Solanum tuberosum*) popular in Jordan", *Food Chem.*, **73**, 205–208 (2001).

[66] Tajner C.A., Rytel E., Kita A., Pêksa A., Hamouz K., "The influence of thermal process of coloured potatoes on the content of glycoalkaloids in the potato products", *Food Chem.,* **133**, 1117–1122 (2012).

[67] Elzbieta R., "The effect of industrial potato processing on the concentrations of glycoalkaloids and nitrates in potato granules", *Food Control.*, **28**, 380–384 (2012).

[68] Rayburn J.R., Bantle J.A., Friedman M., "Role of carbohydrate side-chains of potato glycoalkaloids in developmental toxicity", *J Agric Food Chem.,* **42**, 1511–1515 (1994).

[69] Gottschalk K., Ezekiel R., "Storage. *In*: Gopal J. and Khurana S.M.P. (*eds*.), Handbook of Potato", New York, Food Products Press, pp. 489–522 (2006).

[70] Thomas P., "Wound induced suberization and periderm development in potato tubers as affected by temperature and gamma-irradiation", *Potato Res.*, **25**, 155–164 (1982).

[71] Mondy N.L., Gosselin, B., "Effect of irradiation on discolouration, fenouls and lipids of potatoes", *J Food Sci.,* **54,** 982–984 (1989).

[72] Ezekiel R., Singh B., "Changes in contents of sugars, free amino acids and phenols in four varieties of potato tubers stored at five temperatures for 180 days", *J Food Sci Tech.,* **44**, 471–477 (2007).

[73] Ezekiel R., Paul V., Singh B., Peshin A., Shekhawat G.S., "Effect of low temperature, desprouting and gibberellic acid treatment on little tuber formation on potatoes during storage", *J Ind Potato Assoc.*, **27**, 13–23 (2000).

[74] Blessington T., Nzaramba M. N., Scheuring D.C., Hale A.L., Reddivari L., Miller J.C., "Cooking methods and storage treatments of potato: Effects on carotenoids, antioxidant activity, and phenolics", *Am J Potato Res.*, **87**, 479–491 (2010).

[75] Kumar D., Ezekiel R., "Changes in glycoalkaloids and phenolic contents in potato tubers stored under different conditions", *J Food Sci Tech.,* **46**, 480–483 (2009).

[76] Dewanto V., Wu X., Adom K., Liu R., "Thermal processing enhances the nutritional value of tomatoes by increasing total antioxidant activity", *J Agric Food Chem.*, **50**(10), 3010–3014 (2002).

[77] Dao L., Friedman M., "Chlorogenic acid content of fresh and processed potatoes determined by ultraviolet spectrophotometry", *J Agric Food Chem.*, **40**(11), 2152–2156 (1992).

[78] Lemos A., Aliyu M.M., Hungerford G., "Influence of cooking on the levels of bioactive compounds in Purple Majesty potato observed *via* chemical and spectroscopic means", *Food Chem.*, **173**, 462–467 (2015).

[79] Kaspar K.L, Park J.S., Mathison B.D., Brown C.R., Massimino S., Chew B.P. "Processing of pigmented-flesh potatoes (*Solanum tuberosum* L.) on the retention of bioactive compounds", *Int J Food Sci Tech.*, **47**(2), 376–382 (2012).

[80] Mulinacci N., Ieri F., Giaccherini C., Innocenti M., Andrenelli L., Canova G., Saracchi M., Casiraghi M.C., "Effect of cooking on the anthocyanins, phenolic acids, glycoalkaloids, and resistant starch content in two pigmented cultivars of *Solanum tuberosum", L. J Agric Food Chem.*, **56**, 11830–11837 (2008).

[81] Monro J., Mishra S., Blandford E., Anderson J., Genet R., "Potato genotype differences in nutritionally distinct starch fractions after cooking and cooking plus storing cool", *J Food Comp Anal.*, **22**, 539–45 (2009).

[82] Yamaguchi T., Chikama A., Mori K., Watanabe T., Shioya Y., Katsuragi Y., Tokimitsu I., "Hydroxyquinone-free coffee: A double-blind, randomized controlled dose-response study of blood pressure", *Nutr Meta Cardiovas Dis.,* **18**, 408–414 (2007).

[83] Feng R.T., Lu Y., Bowman L.L., Qian Y., Castranova V., Ding M, "Inhibition of activator protein-1, NF-kappaB, and MAPKs and induction of phase 2 detoxifying enzyme activity by chlorogenic acid", *J Bio Chem.*, **280**, 27888–27895 (2005).

[84] Reddivari L.H., Vanamala J., Safe S., Miller J.C., "The bioactive compounds α-chaconine and gallic acid in potato extracts decrease survival and induce apoptosis in LNCaP and PC3 prostate cancer", *Nutr Cancer.*, **62**, 601–610 (2010).

[85] Turner S.D., "Effects of bioactive compounds from different potato genotypes on prostate cancer development in athymic nude mice", M.Sc. Thesis, Texas A&M University, College Staion, TX (2012).

[86] Bassoli B.K., Cassolla P., Borba-Murad G.R., Constantin J., Pagadigorria S.C.L., Bazotte R.B., Silva D.R.S., Souza D.H.M., "Chlorogenic acid reduces the plasma glucose peak in the oral glucose tolerance test: Effects on hepatic glucose release and glycaemia", *Cell Biochem Func.*, **26**, 320–328 (2008).

[87] Zhang Y., Vareed S.K., Nair M.G., "Human tumor cell growth inhibition by nontoxic anthocyanidins, the pigments in fruits and vegetables", *Life Sci.*, **76**, 1465–1472 (2005).

[88] Onyeneho S.N., Hettiarachchy N.V., "Antioxidant activity. Fatty acids and phenolic acid compositions of potato peels", *J Sci Food Agric.,* **62**, 345–350 (1993).

[89] Samarin A.M., Poorazarang H., Hematyar N., Sarah A.E., "Phenolics in Potato Peels: Extraction and utilization as natural antioxidants", *World Appl Sci J.,* **18**(2), 191–195 (2012).

[90] Robert L., Narcy A., Rock E., Demigne C., Mazur A., Rémésy C., "Entire potato consumption improves lipid metabolism and antioxidant status in cholesterol-fed rat", *Eur J Nutr.,* **45**, 267–274 (2006).

[91] Friedman M., Lee K.R., Kim H.J., Lee I.S., Kozukue N., "Anticarcinogenic effects of glycoalkaloids from potatoes against human cervical, liver, lymphoma, and stomach cancer cells", *J Agric Food Chem.,* **53**(15), 6162–6169 (2005).

[92] Chataing B., Concepcion J.L., Lobaton R., Usubillaga A., "Inhibition of *Trypanosoma cruzi* growth *in vitro* by solanum alkaloids: A comparison with ketoconazole", *Planta Medica.,* **64,** 31–36 (1998).

[93] Hollister B., Dickens J.C., Perez F., Deahl K.L., "Differential neurosensory responses of adult Colorado potato beetle, *Leptinotarsa decemlineata*, to glycoalkaloids", *J Chem Eco.,* **27**, 1105–1118 (2001).

[94] Pandey S.K., Singh S.V., Sarkar D., "Potato (*Solanum tuberosum* L) for sustaining food and nutrition security in developing world. *Indian J Agric Sci.*, **75**, 3–18 (2005).

[95] Marwaha R. S., Pandey S. K., Kumar D., Singh S. V., Kumar P., "Potato processing scenario in India: Industrial constraints, future projections, challenges ahead and remedies – A review", *J Food Sci Technol.,* **47**(2), 137–156 (2010).

[96] Heredia A., Castelló M.L., Argüelles A., Andrés A., "Evolution of mechanical and optical properties of French fries obtained by hot air-frying", *LWT - Food Sci Tech.,* **57,** 755–760 (2014).

[97] Baumann B., Escher E., "Mass and heat transfer during deep fat frying of potato slices. I. Rate of drying and oil uptake", *LWT - Food Sci Tech.,* **28**, 395–403 (1995).

[98] Mellema M., "Mechanism and reduction of fat uptake in deep-fat fried foods", *Trends in Fd Sci Techn.*, **14**, 364–373 (2003).

[99] Decker EA., Ferruzzi M.G., "Innovations in food chemistry and processing to enhance the nutrient profile of the white potato in all forms", *Adv Nutr.* , **4**, 345S–350S (2013).

[100] Garrote R.L., Silva E.R., Bertone R.A., "Effect of thermal treatment on steam peeled potatoes", *J Food Engg.,* **45**(2), 67–76 (2000).

[101] Chavez M.S., Lunaa J.A., Garrote R.L., "A mathematical model to describe potato chemical (naoh) peeling. Energy and mass transfer model resolution", *J Food Engg.,* **32**(2), 209–223 (1997).

[102] Singh K.K., Shukla B.D., "Abrasive peeling of potatoes", *J Fd Engg.,* **26**, 431–442 (1995).

[103] Mestdagh F., Wilde T.D., Fraselle S., Govaert Y., Ooghe W., Degroodt J.M., Verhé R., Peteghem C.V., Meulenaer B.D., "Optimization of the blanching process to reduce acrylamide in fried potatoes", *LWT - Food Sci Tech.,* **41**(9), 1648–1654 (2008).

[104] Arias-Mendez A., Warning A., Datta A.K., Balsa-Canto E., "Quality and safety driven optimal operation of deep-fat frying of potato chips"., *J Food Engg.*, **19**(1), 125–134 (2013).

[105] Oztop M.H., Sahin S., Sumnu G., "Optimization of microwave frying of potato slices by using Taguchi technique", *J Food Engg.,* **79**(1), 83–91 (2007).

[106] USDA-NIFA., "USDA backs acrylamide reduction project for processed potatoes". http://www.foodnavigator-usa.com/Science/USDA-backs-acrylamidereduction-project-f. (2011).

[107] Garayo J., Moreira R., "Vacuum frying of potato chips", *J Food Engg.,* **55**(2), 181–191 (2002).

[108] Lloyd B.J., "Analysis of radiant heating to produce an alternative frying process", PhD Dissertation, North Carolina State University, Raleigh, NC (2003).

[109] Nelson III L.V, Jensen., Keener J.N.K.M., "Evaluation of Potato cakes and breaded chicken patties fried by the fry less 100K radiant system compared to oil immersion frying", *American Society of Agricultural and Biological Engineers Annual International Meeting.*, **3**, 2496–2504 (2010).

[110] Kalogianni P., Papastergiadis E., "Crust pore characteristics and their development during frying of French-fries", *J Food Engg.*, **120**, 175–182, (2014).

[111] Bingol G., Wang B., Zhang A., Pan Z., McHugh T.H., "Comparison of water and infrared blanching methods for processing performance and final product quality of French fries", *J Food Engg.,* **121**, 35–142 (2014).

[112] Zaheer K., Akhtar M.H., "Recent advances in potato production, usage, nutrition- A Review", doi: 10.1080/10408398.2012.724479 (2014).

[113] IARC, "Monographs on the evaluation of carcinogenic risks to human (Acrylamide)", *Int Agency Res Cancer.*, IARC, Lyon, France, **60**, 389–433 (1994).

[114] Tuta S., Palazoðlu K.T., Gökmen V., Effect of microwave pre-thawing of frozen potato strips on acrylamide level and quality of French fries. *J Food Engg.*, **97**(2), 261–266 (2010).

[115] Kalita D., Jayanty S.S., "Reduction of acrylamide formation by vanadium salt in potato French fries and chips", *Food Chem.,* **138**(1), 644–649 (2013).

[116] Aprajeeta J., Gopirajah R., Anandharamakrishnan C., "Shrinkage and porosity effects on heat and mass transfer during potato drying", *J Food Engg.,* **144**, 119–128 (2015).

[117] GAS. Potato flakes. http://www.scribd.com/doc/46426710/Potato-Flakes, (2011).

[118] Lamberti M., Geiselmann A., Conde-Petit B., Escher F., "Starch transformation and structure development in production and reconstitution of potato flakes", *LWT - Food Sci Tech.,* **37**, 417–427 (2004).

[119] Mardiah A., Huda N., Ahmad R., "A Study on the physicochemical properties, microstructure and sensory characteristics of fish flakes", *Am J Food Techn.*, **5**(6), 469–482 (2010).

[120] Abong G.O., Okoth M.W., Karuri E.G., Kabira J.N., Mathooko F.M., "Levels of reducing sugars in eight kenyan potato cultivars as influenced by stage of maturity and storage conditions", *J Anim Plant Sci.,* **2**(2), 76–84 (2009).

[121] Van Hecke, E., Nguyen P.U., Clausse D., Lanoiselle, J.L., "Flow behaviour of carrot puree: Modelling the influence of time, temperature and potato flakes addition", *Int J Fd Sci Tech.*, **47**, 177–185 (2012).

[122] Dushkova M.A., Menkov N.D., Toshkov N.G., "Functional characteristics of extruded blends of potato flakes and whey protein isolate", *Int J Food Engin.*, 7(6), 12 (2011).
[123] Kakade R.H., Das H., Ali S., "Performance evaluation of a double drum dryer for potato flake production", *J Food Sci Technol.*, **48**(4), 432–439 (2011).
[124] Misra A., Kulshrestha K., "Effect of storage on nutritional value of potato flour made from three potato varieties", *Plant Foods Human Nutr.,* **58**, 1–10 (2003).
[125] Rattan N.S., Ramaswamy H.S., "Quality optimization of canned potatoes during rotary autoclaving", *J Food Qual.,* **37**(3), 168–176 (2014).
[126] Zhou H., Wang C., Shi L., Chang T., Yang H., Cui M., "Effects of salts on physicochemical, micro structural and thermal properties of potato starch", *Food Chem.*, **156**(1), 137–143 (2014).
[127] Alvani K., Qi, X., Tester R.F., Snape C.E., "Physico-chemical properties of potato starches", *Food Chem.,* **125**(3), 958–965 (2011).
[128] Fornal J., Sadowska J., B³aszczak W., Jeliñski T., Stasiak M., Molenda M., Hajnos M., Influence of some chemical modifications on the characteristics of potato starch powders. *J Food Engg.,* **108**(4), 515–522 (2012).
[129] Camire M.E., Flint S.I., "Thermal processing effects on dietary fiber composition and hydration capacity in corn meal, oat meal, and potato peels", *Cereal Chem.*, **68**, 645–647 (1997).
[130] Arora A., Zhao J., Camire M.E., "Extruded potato peel functional properties affected by extrusion conditions", *J Food Sci.,* **58**, 335–337 (1993).

3

Developments in Processing, Quality Evaluation Techniques and by - Product Utilization of Tomato and Tomato Products

DATTATREYA M. KADAM[1*], MONIKA SHARMA[1] AND MANPREET KAUR GREWAL[1]

ABSTRACT

The tomato processing industry is an enormous industry worldwide as about 30% of it is processed into various products. Conventional methods for processing tomato such as concentration and drying are still in vogue. However, the energy demands of these methods are very high, and due to high processing temperatures, a significant loss in quality becomes inevitable. This has led to adoption of various alternative processing methods including osmotic concentration / osmotic distillation, aseptic packaging technology, high pressure processing, membrane technology, oscillating magnetic fields, ohmic heating, pulsed electric fields etc. Moreover a lot of solid waste is generated during tomato processing which contains a significant amount of fibrous matter, skin and seeds. These by products are rich in various nutritional components and can be exploited for recovery of biologically active components, natural antioxidants etc. and can be used in the preparation of dietary supplements, nutraceuticals, functional food ingredients, food additives and pharmaceuticals. Also, utilization of tomato by–products as food ingredients would lead to reduction in environmental pollution and development of high value compounds from cheaper raw material. In addition to by– product utilization, quality evaluation of whole and processed tomatoes is an important area of concern as the buying behaviours of the consumers is influenced by product quality. In order to achieve uniform product quality,

[1] ICAR–Central Institute of Post–Harvest Engineering and Technology (ICAR–CIPHET), Ludhiana–141004, Punjab, India.

**Corresponding author* E-mail: kadam1k@yahoo.com

it becomes extremely important to have efficient quality evaluation and control during the processing operations. Several instrumental methods both destructive and non–destructive methods have been adopted for quality evaluation of tomatoes such as traditional analytical methods, high–performance liquid chromatography, ultrasonic transducers, in–line process refractometers, consistometers, automatic firmness meter, atomic absorption spectrophotometry, Universal Testing Machine, continuous in– line, at– line and on– line sensors, electronic nose, Near–infrared reflectance spectroscopy (NIRS), Raman chemical imaging, hyperspectral reflectance imaging, laser–induced fluorescence spectroscopy, magnetic resonance imaging etc. This book chapter reviews the processing techniques, by–product utilization and quality evaluation methods of whole and processed tomato products.

Key words: Tomato, Harvesting, Handling, Processing, Techniques, Products, Quality, Waste utilizing

---

I. INTRODUCTION

Tomato is an important vegetable crop native to South America belonging to *Solanaceae* family. It is grown throughout the world for its fruit and is available round the year. Its high nutritive value and multiple uses in culinary preparations have made it the second most consumed vegetable in the world, after potato. As a processing crop, tomato ranks first among vegetables as approximately 30% is consumed as transformed products. Ripe tomato fruit is consumed fresh and utilized in the manufacture of range of fresh products such as puree, paste, powder, ketchup, sauce, soup and canned whole fruits.

Production

Tomato (S. *lycopersicon*) is the second most important vegetable crop next to potato. Global tomato production for the year 2010 was around 1.45 million tons, of which China dominated the world by contributing around 28.74 %, followed by USA (8.86%), India (8.22%), Turkey (6.90%), Egypt (5.87%), Italy (4.14%), Iran (3.61%), Spain (2.96%), Brazil (2.53%) and Mexico (2.06%)[1]. Continued increase in production is expected due to the purported health benefits associated with tomatoes in the diet.

Worldwide, tomato plants are grown in several environments. There are specific zones in which it is impossible to grow the crop in open fields because of extreme climate conditions (Netherlands, northern Europe), and there are zones in which it is possible only in specific seasons (México). Currently, the trend is toward protected production in controlled greenhouse environments because this technique is highly efficient and has the potential to increase food production to meet market demands. Most greenhouse production is in North America and Europe, where production systems are extremely intensive and can give very high yields (up to 700 tonnes/ha). Field production is much less intensive, and is the most common system in tropical and subtropical areas[2].

Chemical Composition

Tomatoes have a high nutritional value because they include proteins, carbohydrates, fiber, folic acid, tartaric acid, succinic acid and salicylic acid. A summary of the important nutrients is shown in Table 1. Tomatoes are rich in vitamins A, B1, B2, B6, C and E and in minerals such as potassium, magnesium, manganese, zinc, copper, sodium, iron and calcium. Tomatoes are rich in antioxidant lycopene, the pigment that gives them their characteristic red colour. Besides lycopene, tomatoes also contain the antioxidant glutathione which helps to cleanse the body of toxic products and prevents the accumulation of heavy metals[3]. Tomatoes contain a large quantity of water (93.5%), calcium (0.07%) and niacin, all of which are of great importance in the metabolic activities of humans[4].

Table 1: Nutritional composition per 100 g of fresh tomato[3]

Element	*Quantity*
Water (%)	93.5
Protein (g)	0.9
Fat (g)	0.1
Calories	23.0
Carbohydrates (g)	3.3
Fiber (g)	0.8
Phosphorus (mg)	19.0
Calcium (mg)	7.0
Iron (mg)	0.7
Vitamin A(UI)	1.100
Vitamin B1(mg)	0.05
Vitamin B2(mg)	0.02
Vitamin C(mg)	20.0
Niacin(mg)	0.6

Tomato Classification

Tomatoes are differentiated either according to their use for fresh consumption or industrial processing or according to their external shape. According to their end use, there are four types of tomatoes: milano, chonto, cherry and industrial[3]. The milano type tomato is generally used in salads and consumed in the green mature or red state. This type has significant commercial value and palatability. The chonto type is consumed when fresh and is used in the preparation of stews or pasta. The cherry type has very small fruits, grouped in clusters of 15 or more and consists of many different colours including yellow, red and orange. The cherry type is consumed fresh as a snack, in cocktails and as a garnish. The industrial type is characterized by large amounts of soluble solids that make it attractive for processing, primarily in the production of sauces, ketchup and pastas etc.

Horticultural Maturity Stages

Kader and Morris[5] suggested four main maturity stages according to internal appearance of equatorial section. These are as follows:

(i) M–1: Seeds are immature (white) and can be cut when the tomato is sliced; no gel in the locule.
(ii) M–2: Seeds mature (tan); gel formation in at least two locules.
(iii) M–3: Seeds pushed aside when tomato sliced; all locules have gel; internal colour is still green.
(iv) M–4: Appearance of red colour in gel and pericarp tissue.

Depending on the market and production area, tomatoes are harvested at stages of maturity ranging from physiological maturity (mature–green stage, M–3) through full–ripe. Immature tomatoes are available for certain regional dishes. It is difficult to accurately determine the completion of physiological maturity. Tomatoes harvested at the mature–green stage (M–3 or M–4) will ripen to high quality if handled properly[6]. Tomatoes harvested at the M–2 stage will ripen to moderate quality, while those harvested at M–1 stage will not ripen to acceptable levels of quality. Depending on the growing area and time of harvest, the percent of immature tomatoes (M–1) in lots of green–harvested tomatoes can range from 20 to 80%[7].

Tomato Physiology

Tomatoes are climacteric and show a pronounced increase in respiration during ripening. The intensity and duration of the climacteric varies among fruit species[8]. Respiration also varies with temperatures and atmospheric composition. Tomato fruit produces moderate amounts of ethylene at 1 to 10 $\mu L\ kg^{-1}\ h^{-1}$ at 20°C[9] and are sensitive to ethylene exposure. As little as 0.5 $\mu L\ L^{-1}$ ethylene is sufficient to trigger ripening and other associated metabolic processes[10]. For commercial ripening, green tomatoes should be held at 20 to 21°C with 90% RH and 50 $\mu L\ L^{-1}$ ethylene; this will promote uniform ripening. Upon reaching breaker stage, tomatoes produce sufficient ethylene and no longer require gassing. Highest quality tomatoes are those reaching the breaker stage within 3 days of ethylene exposure. These fruits were harvested at the mature green stage and will ripen with quality similar to tomatoes harvested at the breaker or later[6].

Physiological Disorders

Blotchy ripening is a physiological disorder characterized by the randomized development of green or green–yellowish areas on the surface of red tomato fruit. The development of this disorder is related to the availability of potassium and inorganic nitrogen in the soil system. Areas showing blotchy ripening have less organic acids, soluble solid content, and starch[11].

Sunburn is associated with excessive exposure to the sunlight. This elevates tissue temperature during fruit development which disrupts lycopene synthesis and resulting in the appearance of yellow areas in the affected tissues that remain during the ripening process. Blossom–end rot is a physiological disorder which is either due to poor uptake or translocation of calcium into the fruit. Symptoms begin in the green fruit as a small discolouration at the blossom end that increases in size and becomes dry and dark–brown. Occurrence increases dramatically when calcium levels in the soil system drop below 0.08%[11]. Eventually, secondary decay organisms colonize weakened tissues. Gray wall (blotchy ripening) is noticeable as fruit is uneven in colour, both inside and outside, with hardened patches of grayish or yellowish tissue[12]. It begins developing at the green stage and has been associated with marginal growing conditions such as cool weather, low light levels, poor nutrition, saturated (water and fertilizer) soils, tobacco mosaic virus, and bacteria; however, the cause is still undetermined. Gray wall can be a serious disorder in both field and greenhouse production systems.

Irregular ripening is characterized by the appearance of non–uniform ripening and white internal tissue. It has been associated with the feeding of sweet–potato whitefly (*Bemisia argentifolii*) on tomato fruit[13].

Internal bruising is recognized by the appearance of yellow to green locular gel in ripe tomatoes. It is caused by an impairment of normal ripening of the locular gel following a physical impact at the green or breaker stage of ripeness[14]. Fruit with internal bruising show significant reductions in Vitamin C content, consistency and total carotenoids[15]. Besides altering quality attributes, internal bruising also affects tomato fruit flavor[16]. Breaker–stage tomatoes are more sensitive to internal bruising than those handled at the green stage[17].

Postharvest Pathology

Tomatoes are susceptible to numerous fruit decays, from the field through postharvest handling. Postharvest decays often develop in wounds, bruised tissue and during fruit softening. Sound tomatoes can be inoculated by plant pathogens via cross–contamination from diseased fruits, dirty harvest containers and from poorly sanitized water handling systems and packing line components. Populations of decay pathogens can be adequately controlled through a regular sanitation program in the field and during handling, packing and ripening/storage operations.

Bacterial decays include soft rots (*Bacillus spp., Erwinia carotovora* ssp., *Pseudomonas* spp., and *Xanthomonas campestris*); lactic acid decay (bacterial sour rot) (*Lactobacillus* spp. and *Leuconostoc mesenteroides*)[18, 19].

Fungal decays include alternaria rot (black rot) (*Alternaria alternata*); Fusarium rot (*Fusarium* spp.); Gray Mold rot (*Botrytis cinerea*); Mucor rot (*Mucor mucedo*); Phoma rot (*Phoma ssp.*); Phomopsis rot (*Diaporthe* ssp.); Phytophthora rot (Buckeye rot) (*Phytophthora* spp.); Pleospora rot (*Pleospora herbarum; Stemphylium botryosum*,imp. stage); Rhizopus rot (*Rhizopus stolonifer*, R. oryzae); Ring rot (*Myrothecium roridum*); Sclerotium rot (*Sclerotium rolfsii*); Sour rot (*Geotrichum candidum*); Target spot (*Corynespora cassiicola*); Watery soft rot (*Sclerotinia minor, S. sclerotiorum*). Tomato spotted wilt virus induces a mottled colouration at red stage[11, 20].

II. TOMATO HARVESTING, HANDLING, GRADING AND PREPROCESSING

Tomato Quality Indices

Tomato fruit are berries with different attributes such as colour, ripening rate, firmness, size, shape and composition that determine its quality. Consumers consider a tomato to be of high quality if it has good colour and uniform ripening as well as acceptable firmness, shape, size, composition and taste; producers consider these qualities in addition to a high nutritional content and a long shelf life.

(i) Colour

Colour in tomato is the most important visible characteristic used to assess ripeness and postharvest life. It is also a major factor in the consumer's purchase decision. Mature fruits can be red, pink or yellow. The degree of ripeness is usually estimated visually by human graders by comparing the tomato colour to a classification chart. Based on the external colour, the USDA has established six ripening stages reflecting human ability to differentiate ripeness: ***green***, 100% green; ***breaker***, a noticeable *break* in colour with lesser than 10% of other than green colour; ***turning***, between 10 and 30% of surface, in the aggregate, of red(ish) colour; ***pink***, between 30 and 60% of red(ish) colour; ***light red***, between 60 and 90% and ***red***, more than 90% red (Fig. 1) Some European countries have established 10 or even more different colour stages, but an average consumer has trouble in finding differences between them. This manual practice of tomato maturity classification often results in errors due to human subjectivity, visual stress and fatigue[21]. As colour is an indication of tomato ripeness, a green to red gradient can also be used to assess the stage at which a tomato should be harvested and consumed.

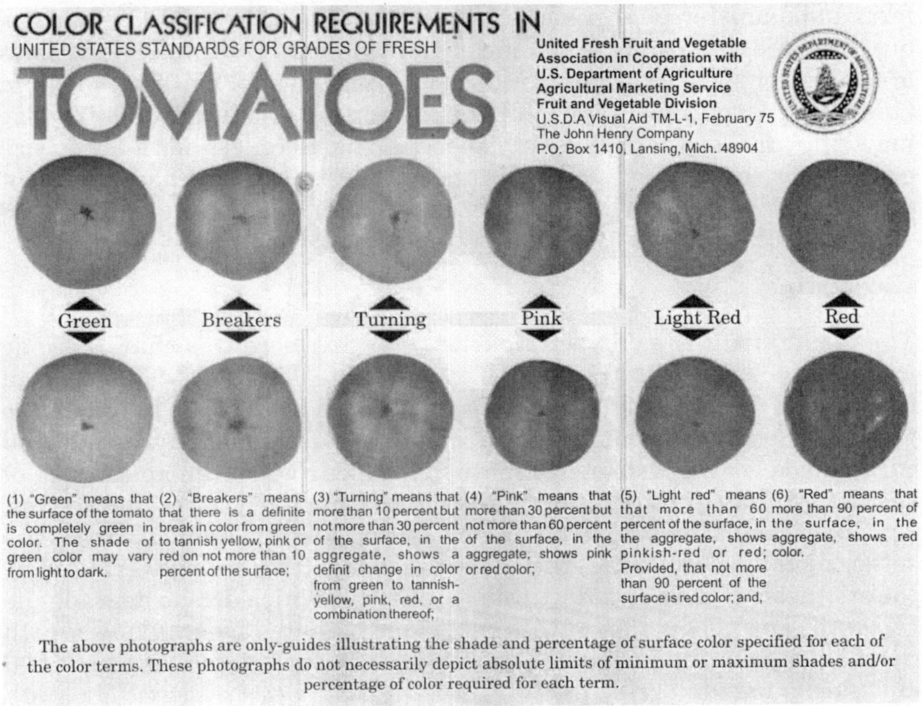

Fig. 1: USDA visual aid TM–L–1 chart for different ripening stages of tomato

(ii) Firmness

Other than visual appearance, the most important factor in tomato quality is firmness. Tomato quality is influenced by the hardness of the epidermis, the firmness and the internal structure of the fruit, which vary widely among cultivars. The production of the cell wall–solubilizing enzyme (polygalacturonase) during maturation plays a significant role in texture changes[3]. Generally, firmness refers to the force required to pierce the tomato with a standard probe. The most common technique for obtaining this is the Magness–Taylor (MT) firmness measurement, a destructive type of test method. Most of the non–destructive test (NDT) methods (*e.g.* acoustic firmness, laser air–puff and near IR) require expensive and cumbersome instrumentation[22].

(iii) Shape & size

The shape and size of tomato depends on the type, the growing treatment (fertilization) and the chemical modification of seeds to get a specific form or composition. The major attributes that determine tomato size are weight, form and diameter. The grades are mostly based on the condition and the quality of the fruits and not specifically on their size. However, on the basis of the size of the fruits three grades are formed: small (<100 g), medium (100–255 g) and large (> 255 g). Even if not a factor of grade quality, size may influence commercial quality expectations. Retailers normally do size grading for the local market and internal urban markets have different prices for size grades as against ungraded fruit.

Harvesting

The tomato fruit is a climacteric fruit, and therefore this characteristic must be considered at the time of harvesting. The most visible sign of organoleptic ripening in a tomato is the change from green to red due to the decomposition of chlorophyll and the synthesis of lycopene and carotenoids. The physiological maturity is clear when the apical part of the fruit begins to show an orange colouration while the rest of the fruit remains green. The second characteristic sign of tomato maturity is the softening that accompanies the change of colour. This change is caused by the synthesis of the polygalacturonase enzyme, which acts to degrade the cell wall. The production of this enzyme is initiated by ethylene, which helps to explain the importance of ethylene in the natural and artificial ripening of tomato[3].

In developing countries, most produce for internal rural and urban markets is harvested by hand. Larger commercial producers may find a degree of mechanization, an advantage, but the use of sophisticated harvesting machinery will be limited for the most part to agro-industrial production of cash crops for processing or export or both. In most circumstances, harvesting by hand, if done properly, will result in less damage to produce than will machine-harvesting. Manually tomato should be harvested from their natural break point by a "lift, twist and pull" series of movements which leaves the stalk attached to the fruit. Greenhouse-grown tomatoes are generally harvested at turning stage or later and packed according to ripeness and size (count) in single- or double-layer cartons, with or without bottom trays. Cluster tomatoes are usually picked when the least mature tomato begins to show red colour.

Grades, Sizes and Packaging

Fruit within a grade should have similar varietal characteristics, be uniformly mature, not overripe or soft, be clean, well developed, fairly well formed, fairly smooth, and have uniform colour. Fruit should be free from decay, freezing injury, sunscald, and damage due to bruising, discolouration, sunken scars, cuts and broken skins, puffiness, cat faces, other scars, radial or concentric growth cracks, and insect injury.

Tomatoes are rigorously sized with a 0.8 mm (1/32 in) overlap permitted between sizes, according to the dimensions mentioned in Table 2 below[23]:

Table 2: USDA dimensions for different sizes of tomato

Size	*Minimum diameter*[1]	*Maximum diameter*[2]
Small	5.40 cm (2 $^{4}/_{32}$ in)	5.79 cm (2 $^{9}/_{32}$ in)
Medium	5.72 cm (2 $^{8}/_{32}$ in)	6.43 cm (2 $^{17}/_{32}$ in)
Large	6.35 cm (2 $^{16}/_{32}$ in)	7.06 cm (2 $^{25}/_{32}$ in)
Extra Large	7.00 cm (2 $^{24}/_{32}$ in)	

USDA grades for field-grown tomatoes include U.S. No. 1, No. 2, U.S. Combination, and No. 3. Field-grown tomatoes are typically packed in lidded, 11.4 kg, 30 × 40 × 24 cm (W × L × H), that stack 10 cartons per layer on a 100 × 120 cm pallet. Grades for greenhouse tomatoes include U.S. No. 1, No. 2, and Unclassified[24]. Ripeness stages and defects are similar to field-grown tomatoes. These fruit are sized as follows: Small < 99.4 g; Medium 99.4 to 256 g); Large > 256 g. Bureau of Indian Standards has specified 4 grades *viz.*, Super A, Super, Fancy and Commercial for tomato crop.

For local markets, the fruits are packed in bamboo baskets or plastic crates. Plastic crates can be conveniently stacked one on the other and a contoured rim keeps the product safe and natural and allows sufficient air circulation. The packing should ensure careful handling *i.e.*, rigid enough to protect the fruits from being crushed. For exports, the fruits are packed in cardboard telescopic boxes with capacities of not more than 15 kg. Size graded tomatoes are pattern packed in layers to make best use of the box.

Pre–cooling Conditions

Following commercial packing, tomatoes are routinely palletized and cooled to 20°C for ripening or to 12°C for storage. While room–cooling is common, forced–air cooling is more uniform and produces a better quality fruit. Packed, palletized tomatoes with pulp temperature of 28°C actually increased by 2°C immediately after being stored at 20°C, and get cooled to 23°C only after 24 h using room–cooling[25]. However, with forced–air cooling, tomatoes are cooled to 20°C in 2.5 h and ripened more uniformly throughout the pallet than those room–cooled.

---

[1] Will not pass through a round opening of the designated diameter when tomato is placed with the greatest transverse diameter across the opening.

[2] Will pass through a round opening of the designated diameter in any position.

Optimum Storage Conditions

The main objective in storage after harvest is to control the rate of ripening to extend the marketing period. As the tomatoes are chilling sensitive, the recommended storage temperatures differ depending on the fruit maturity. A storage temperature of 13°C with 90–95% relative humidity is recommended for slow ripening. At this temperature, most varieties keep in good condition for 2–3 weeks and change colour very slowly. In cold storage, unripe tomatoes can be stored for 4 weeks at temperature of 8–10°C with 85–90% relative humidity. Fully ripe fruits are stored at 7°C with 90% relative humidity for 1 week.

Tomato fruit are chilling sensitive and the recommended storage temperature varies with the maturity stage. Mature–green fruit will ripen normally at 13 to 21°C (55 to 70°F). On the other hand, ripe tomato fruits can be stored at 10°C (50°F) without visible symptoms of chilling injury but with a negatively affected flavor and aroma[26]. Visual symptoms of chilling injury include pitting, non–uniform ripening and storage decays[8].

Optimal storage temperatures depend on the maturity stage of the tomatoes. Ideal conditions for ripening are 19 to 21°C with 90 to 95% RH. Storage >27 °C reduces intensity of red colour, while storage <13°C retards ripening and can lead to development of chilling injury, particularly in tomatoes at the mature–green stage. Red tomatoes can be stored at 7°C for a couple of days, although tomatoes stored at 10°C were rated lower in flavor and aroma than those held at 13°C[26].

Controlled Atmosphere (CA)

Tomatoes can be stored under CA to extend product quality. The exact combination of CO_2 and O_2 varies among maturity stages and cultivars; but a satisfactory CA is 3% O_2 + 2% CO_2[8]. Storage under CA delays quality loss as measured by several factors, such as lycopene synthesis and sugar and chlorophyll degradation[27, 28]. Storage in 3% O_2 + 97% N_2 extended postharvest–life of mature–green tomatoes for 6 weeks at 13°C without the development of off–flavors[29]. Storage under CA may reduce development of undesirable symptoms due to mechanical injury[30]. However, Moretti *et al.*[31] observed that CA storage did not alleviate development of internal bruising (disruption of locular gel ripening) following impacts of tomato fruits. Many researchers carried extensive studies on CA storage of tomatoes and found similar results with minor variation depending upon the geographical location, variety, climatic condition, seasons etc.

Retail Outlet Display Considerations

Tomatoes are normally displayed at the retail level in single–layer, corrugated cartons or in plastic clamshell containers at about 20°C.

Grading, sizing, and packing fruit adds value and convenience to the final product (Fig. 2). An important issue in the marketing of high–quality tomatoes is uniformity of size, grade, firmness and colour. Consumers tend to avoid packages of fruit with different colours, decay or external blemishes. Some distributors also offer tomatoes in bulk containers, giving the consumer the option to choose among different maturities, sizes and types of tomatoes. However, since tomato fruits are sensitive to compression stresses, care must be taken to avoid overloading the display.

Fig. 2: (A) Tomato grading, (B) Large–grade tomatoes packed in a single layer and (C) Pre–packed plastic boxes of cherry tomatoes.

Transport

Tomatoes are highly perishable in nature hence quick means of transportation is necessary. Tomatoes are transported by road through tractors, trucks and also by rail and air to distant markets. Village produce is transported to the nearby towns and city market only by road. Cool chain is essential during the transport of export quality commodity all the way from the farm to the customer. This helps in maintaining the temperature inside the box at the same low level as in the cold storage.

The various stages of the cool chain are:

(1) Cold store at the farm.

(2) Refrigerated truck from farm to the airport.

(3) Cold store at the airport.

(4) Building up of the pallet in a cold store at the airport.

(5) Loading the aircrafts directly from the cold store in a short time.

(6) Cargo aircraft maintains cold store temperature in hold.

(7) Off loading direct into a cold store in the receiving country.

(8) Refrigerated truck to the customers.

III. CONVENTIONAL PROCESSING TECHNIQUES AND PROCESSED PRODUCTS

Conventional Tomato Processing

Tomato as fresh–cut product

Despite efforts to commercialize fresh–cut tomatoes, such products are still only available in limited quantities to the food service industry, particularly fast–food restaurants and catering services. After processing, loss of the gel–like locule tissue, desiccation, water–soaking and the development of decay are the principle constraints challenging the worldwide fresh–cut industry. Crossing commercial varieties with mutants with delayed softening, and slicing less ripe tomatoes, *e.g.*, breaker stage, for subsequent application of ethylene are strategies being researched in different parts of the world to obtain a fresh–cut tomato with sufficient postharvest–life to be readily commercialized. Mature–green, sliced tomatoes ripened normally at 20°C[32]. Sliced, red tomatoes maintained good quality for 14 days when stored in modified atmosphere packaging (MAP) at 5°C[33].

Canning

Canned tomatoes are the most popular canned vegetable. The main steps in canning whole, sliced or diced tomatoes are:

(i) ***Grading:*** Tomatoes for canning whole, diced or sliced tomatoes are graded on the basis of colour, firmness, defects and size. Solids content is important only for processing into juice or paste.

(ii) ***Washing:*** Mechanically harvested tomatoes are usually thoroughly washed by high–pressure sprays or by strong–flowing streams of water while being passed along a moving belt or on agitating or revolving screens. Tomatoes are soaked for several minutes in tanks with paddles and agitators to loosen adhering dirt. After soaking the remaining dirt is washed using overhead sprays.

(iii) ***Sorting:*** The raw produce may need to be sorted for size and maturity. Most of the tomato processing plants use photoelectric sorters both before and after peeling. The initial sorter removes green tomatoes while the second sorter after peeling removes pink tomatoes. Green or pink tomatoes are ejected by pneumatic finger as the tomatoes fall between conveyor belts. Green tomatoes can be diverted to the juice line while pink tomatoes will reduce the colour score of juice. A few plants use hydrosorter, size sorter and texture sorter. Hydrosorters are used to remove immature fruits while doubling as washing step. Immature fruit floating in water tank of hydrosorter is either discarded or diverted to the juice line. Sorting for size is accomplished by passing the raw tomatoes through a series of moving screens with different mesh sizes or over differently spaced rollers. Size sorters remove excessively small tomatoes undesirable for canning and divert them to juice or crushed tomato line. Texture sorters remove over soft tomatoes which are again diverted to juice line. Separation into groups according to degree of ripeness or perfection of shape is done by hand.

(iv) ***Peeling and Coring:*** Formerly, tomatoes were initially scalded followed by hand peeling, but steam peeling and lye peeling have also become widely used. With steam peeling, the tomatoes are treated with steam to loosen the skin, which is then removed by mechanical means. In lye peeling, the fruit is immersed in a hot lye bath or sprayed with a boiling solution of 10 to 20 percent lye[34]. The excess lye is then drained and any lye that adheres to the tomatoes is removed with the peel by thorough washing. Coring is done by a water–powered device with a small turbine wheel. A special blade mounted on the turbine wheel spins and removes the tomato cores.

(v) ***Dicing and calcification:*** For canning diced tomatoes, peeled tomatoes are cut into 3/8", ½" and 1" cubes, inspecting to remove green or blemished dices and calcification by direct addition of calcium to the container or by conveying the dices through a calcium bath.

(vi) ***Filling:*** After peeling and coring, the tomatoes are conveyed by automatic runways, through washers, to the point of filling mostly by machine. Before being filled, the can or glass containers are cleaned by hot water, steam, or air blast. The containers are filled with the solid product and then usually topped with a light puree of tomato juice. Acidification of canned whole tomatoes with 0.1 to 0.2% of citric acid has been suggested as a means of increasing acidity to a safer and more desirable level. Because of the increased sourness of the acidified product, the addition of 2 to 3% sucrose is used to balance the taste. The addition of salt is important for palatability[34].

(vii) ***Exhausting:*** The objective of exhausting containers is to remove air so that the pressure inside the container following heat treatment and cooling will be less than atmospheric. The reduced internal pressure (vacuum) helps to keep the can ends drawn in, reduces strain on the containers during processing, and minimizes the level of oxygen remaining in the headspace. It also helps to extend the shelf life of food products and prevents bulging of the container at high altitudes. Vacuum in the can may be obtained by the use of heat or by mechanical means. The tomatoes may be preheated before filling and sealed hot. For products that cannot be preheated before filling, it may be necessary to pass the filled containers through a steam chamber or tunnel prior to the sealing machine to expel gases from the food and raise the temperature. Vacuum also may be produced mechanically by sealing containers in a chamber under a high vacuum.

(viii) ***Sealing:*** In sealing lids on metal cans, a double seam is created by interlocking the curl of the lid and flange of the can. Many closing machines are equipped to create vacuum in the headspace either mechanically or by steam–flow before lids are sealed.

(ix) ***Heat Sterilization:*** During processing, microorganisms that can cause spoilage are destroyed by heat. The temperature and processing time vary with the nature of the product and the size of the container. Acidic products, such as tomatoes, are readily preserved at 100°C (212°F). The containers holding these products are processed in atmospheric steam or hot–water cookers. The rotary continuous cookers, which operate at 100°C (212°F), have largely replaced retorts and open–still cookers for processing canned tomatoes. Some plants use hydrostatic cookers and others use continuous–pressure cookers.

(x) ***Cooling:*** After heat sterilization, containers are quickly cooled to prevent overcooking. Containers may be quick cooled by adding water to the cooker under air pressure or by conveying the containers from the cooker to a rotary cooler equipped with a cold–water spray.

(xi) ***Labelling and Casing:*** After the heat sterilization, cooling, and drying operations, the containers are ready for labelling. Labelling machines apply glue and labels in one high–speed operation. The labelled cans or jars are the packed into shipping cartons.

Concentrated Tomato Products: Paste, Juice Concentrate and Sauce

(i) ***Tomato paste:*** Good quality tomato paste is a red coloured high density homogenous mass, without foreign bodies (seeds, peel, etc.) and with an agreeable taste and smell close to those of fresh tomatoes. There are usually three types of tomato paste: 36, 30 and 24 which have refractometric extracts of respectively 34–38%, 28–32% and 24–26%. Paste of good quality must have a volatile acidity of maximum 0.15% as lactic acid. An 8% salt addition is accepted. The product with highest production volumes among concentrated products is tomato paste which is manufactured in a various range of concentrations even up to 44% refractometric extract. Tomato paste is the product obtained by removal of peel and seeds from tomatoes, followed by concentration of juice by evaporation under vacuum. In some cases, in order to prolong production period, it may be advisable to preserve crushed tomatoes with sulphur dioxide. Manufacturing steps fall into three successive categories:

(a) Tomatoes are first washed and sorted. Tomatoes are then crushed and seeds are separated with a centrifugal separator. Tomato pulp is pre–heated at 55–60°C and then passed to the sieving equipment's *i.e.*, pulper, refiner and super refiner with sieves of 1.5 mm, 0.8 mm and 0.4–0.5 mm respectively in order to give the smoothest possible consistency to the tomato paste.

(b) In continuous installations with three evaporation steps, the tomato paste is submitted in step/body I to pasteurization at 85–90°C for 15 min under vacuum of 330 mm Hg. for microbiological stability of finished product. In evaporating steps II and III, temperatures are around 42–46°C and vacuum at 680–700 mm Hg. Concentration occurs gradually and continuously in the three evaporating bodies. The advantages of continuous concentration are as follows:

  (i) Improved taste, colour, flavour, "shine" and consistency of tomato paste as the real concentration is performed in evaporating steps II and III at low temperatures (42–46°C)

  (ii) The whole concentration process time from the input of juice in body I until the output of paste from step III is of about 1 hour (for paste with 30–35% refractometric extract). Thus, production capacity is raised by about 30% as compared to discontinuous installations with the same evaporation surface.

  (iii) The steam consumption is reduced by 60% because heating in steps II and II is done with vapours resulting from juice evaporation in step I (double effect); water and electricity consumptions are also reduced by 30–40%.

(c) For tomato paste pasteurization, the paste coming out from concentration equipment is passed continuously and in a "forced" mode through a tubular pasteurizer from which it emerges at a temperature of 90–92°C. Sometimes it is possible to obtain a tomato paste with a concentration of 44% refractometric extract. For this purpose it is necessary to eliminate a part of cellulose from tomatoes using a separating turbine.

(d) Tomato paste storage and preservation is carried out after packing which is done usually in drums, metallic cans or glass jars; some modern equipment has been developed for packing in aluminium bags. As far as the concentration of tomato paste is concerned it is not possible to reduce water content down to 30% which corresponds to a water activity (a_w) of 0.70–0.75 (minimum limit of mould growing), it is necessary to take special measures like pasteurization, cold storage or salt addition. Salt is not a preservative in itself but contributes to the lowering of water activity. In drums, the preservation of tomato paste with minimum 30% refractometric extract is carried out in two ways: Either the hot paste (about 90°C) flows directly from pasteurization equipment into drums that have been previously steamed or the paste is cooled down to 30°C through a heat exchanger and is introduced into drums that have been previously steamed. For preservation purposes, it is possible to add 3–8% salt.

(ii) ***Concentrated tomato juice:*** Concentrated tomato juice is a homogenous mass, finely sieved product with 17–19 % refractometric extract without foreign bodies and any evidence of deterioration. A good quality product has a red colour, an agreeable and specific taste and smell. The installations, equipment and flow sheets for concentrated tomato juice are same as for the production of tomato paste. The final concentration is regulated between the above specified limits. The concentrated tomato juice is filled in receptacles (metal tinplate cans or glass bottles) and then pasteurized at 100°C during 15–25 minutes according to receptacle type. With modern production lines it should be possible to pass the concentrated tomato juice through a tubular pasteurizer and then pack aseptically and cool, without the need to pasteurize the receptacles.

(iii) ***Tomato sauces:*** Under the USA Code of Federal Regulation 7 CFR 52, 1991; tomato sauce is the concentrated product prepared from the liquid extract from mature, whole tomatoes or the residue from preparing such tomatoes for canning, or the residue from partial extraction of juice, or any combination of these ingredients, to which is added salt and spices and to which may be added one or more nutritive sweetening ingredients, a vinegar or vinegars, and onion, garlic, or other vegetable flavouring ingredients. The refractive index of the tomato sauce at 20°C is not <1.3461. These products are widespread in some countries and are used in order to spice some meals. Sauces can be obtained from fresh tomatoes or from concentrated products (tomato paste or concentrated tomato juice). Those from fresh tomatoes are of superior quality. Technological processing covers the following steps: concentrated juice processing, addition of flavour/taste ingredients (salt, sugar, vinegar, spices, etc.), boiling, fine sieving, filling of receptacles, closing and pasteurization (45 min at 85°C).

(iv) ***Tomato juice:*** The main characteristics of high quality tomato juice are natural red colour, taste and flavour of fresh tomatoes, uniformity (without pulp sedimentation), minimum total soluble solids 6%, and vitamin C content not less than 15 mg/100 ml. Juice microbiological stability is assured by a flash pasteurization which assures the destruction of natural micro-flora, while maintaining the initial properties. The modern technological flow-sheet covers the following main operations:

(a) Pre-washing is carried out by immersion in water, cold or heated up to 50°C (possibly with detergents to eliminate traces of pesticides). This operation is facilitated by bubbling of compressed air in the immersion vessel/equipment.

(b) Washing is performed with water sprays with a pressure of 15 atm or more.

(c) Sorting using rolling tables which enables the removal of non-standard tomatoes with green/yellow coloured parts, etc.

(d) Crushing in special equipment.

(e) Preheating at 55–60°C facilitates the extraction, dissolves pectic substances and contributes in maintaining vitamins and natural pigments content. In some modern installations, this step is carried out under vacuum at 630–680 mm Hg and in very short time.

(f) Extraction of juice and part of pulp (maximum 80%) is performed in tomato extractors with the care to avoid excessive air incorporation. In some installations, as an additional special care, a part of pulp is removed with continuous centrifugal separators.

(g) De-aeration under high vacuum of the juice brings about its boiling at 35–40°C.

(h) Homogenisation is done for mincing of pulp particles and is mandatory in order to avoid future potential product "separation" in two layers.

(i) Flash pasteurization at 130–150°C for 8–12 seconds followed by cooling at 90°C, which is also the filling temperature in receptacles (cans or bottles).

(j) Aseptic filling

(k) Closing of receptacles is followed by their inversion for about 5 to 7 minutes.

(l) Cooling has to be carried out intensely. Full cans do not need further pasteurization because the bacteria that have potentially contaminated the tomato juice during filling are easily destroyed at 90°C due to natural juice acidity. For bottles, it may be possible to avoid further sterilisation if the following conditions have been maintained: washing and sterilising of receptacles, cap sterilisation (with formic acid), filling and capping under aseptic conditions, in a space with UV lamps. As this is quite difficult to achieve, bottles are generally pasteurised using water baths.

(v) ***Dehydrated tomatoes:*** Fresh tomato used for manufacturing dehydrated products should be consistent in size, free from blemishes, red in colour but not overripe, firm with thick walled flesh, and most importantly free from bacteria rot or mold. The tomatoes are dehydrated using following steps: Tomatoes are washed using soak tanks or high pressure jets to remove the contaminants and initial microbial load. Usually hypochlorite is added in water in concentrations ranging from 10–2000 ppm to control microbial populations. Recommendations for water chlorination in tomato processing plants usually range from 100–300 ppm (well above the minimum lethal dose for spores) because chlorine is highly unstable in solution and is easily inactivated by the presence of soil.

After sorting and grading the tomatoes are washed in chlorinated water (50–200 ppm), rinsed, and transported to the cutting machine prior to the drying process. Hot air drying/Conventional drying using preheated air is most common and economical system used in the industry. In this method, preheated air at 60°C to 110°C at velocity of 0.5 to 2 m/s and varying relative humidity depending on the weather is blown across the product surface. Water extracted from the produce is then carried away by the air in the form of vapour. Drying times are usually very long (10 to 20 hours) and depend on the load. The method has low drying rates and thus poor product quality. Tomato process flow chart for different products discussed above is given in Fig. 3.

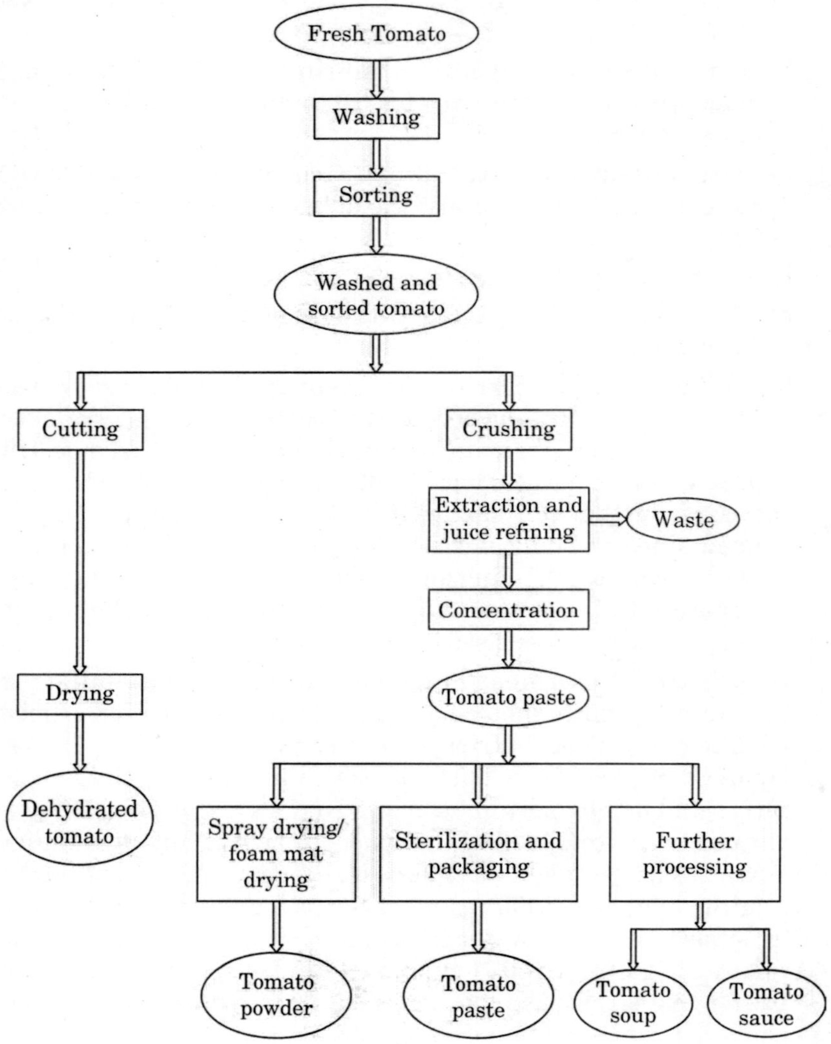

Fig. 3: Process flow chart for value added tomato products.

IV. ADVANCED PROCESSING TECHNIQUES/ RECENT DEVELOPMENTS

The increase in the consumer awareness of the health benefits of fruits and vegetables and the emerging need for convenience due to a fast-paced lifestyle is leading to an increasing demand of minimally processed fruits and vegetables. Over the last few years consumption of processed fruits and vegetables has increased and there is convincing epidemiological evidence that consumption of fruits and vegetables is in general beneficial to health[35]. Minimal processing has been defined as a combination of procedures such as washing, sorting, trimming, peeling and slicing or chopping, that do not affect the fresh-like quality of the food[36].

Tomatoes are often used as an ingredient in foods such as sauces and soups. Besides microbial safety, important quality aspects of processed tomatoes are colour, taste and consistency. Despite efforts to update the technology of the tomato processing factories, which have led to the introduction of cutting edge technology such as aseptic packaging technology, the conventional concentration method of vacuum vaporization still remains the dominant method for obtaining the concentrate from the raw juice. However, the energy demands of this method are very high, and deterioration in quality seems to be unavoidable, since the processed juice is exposed to temperatures of about 65–72°C for an extended period of time. Heat treatment is also the conventional method for reducing microbial population and to inactivate enzymes in fruit and vegetable processing. Heat application adversely affects the flavor, colour and nutritional value of the food in most cases. Thus, addition of preservatives, irradiation technology, centrifugation and ultra filtration etc. are being explored to reduce microbial load. Several physical methods for microbial and enzyme inactivation are under extensive study including the utilization of oscillating magnetic fields, high hydrostatic pressure, light pulses and electric field treatments *viz*. ohmic heating and pulsed electric fields.

In the food industry, electrical treatment may generally be used for non-thermal pasteurization of food and for intensifying juice yield in juice production, wine making, sugar production and tomato paste production[37].

Membrane Technology

There is great potential in using a combined low temperature and pressure ultra-filtration and direct osmosis process to produce tomato concentrates. A study was carried out by using a new tubular direct osmosis module to investigate the effect of clarifying tomato juice on the rate of direct osmotic concentration[38]. Under virtually the same operating conditions five respective clarifications of juice were tried including un-clarified tomato juice. A remarkable increase of permeation flux (over 100%) was observed shifting from un-clarified to the clarified tomato juice. Cldarification was carried out by passing the juice through 35 μm mesh and also by using membranes of molecular weight cut–off 200000, 100 000 and 25000 Daltons (Da), respectively, in order to obtain the juice ultra filtrates.

Attempts to replace the conventional evaporative concentration method with membrane technology have led to proposals of reverse osmosis as an alternative methodology. Further, attempts were made to combine Reverse Osmosis (RO) technology with Ultra filtration (UF) or Membrane Filtration (MF) or even centrifugation. Cross[39] patented a process which used UF pre–treatment of the juice and utilized hollow fiber RO in order to concentrate the clear UF permeate up to 30–60° Brix and remix it after this concentration with the UF retentate to obtain a highly concentrated juice. Osmosis is a natural phenomenon in which a solvent (usually water) passes through a semi-permeable barrier from the side with lower solute concentration to the higher solute concentration side. At equilibrium, the pressure difference between the two sides of the membrane is equal to the osmotic pressure of the solution. A novel tubular module was used to investigate the direct osmosis concentration process in the case of tomato juice. This module was constructed using an external stainless steel shroud accommodating, internally, a set of two identical RO membrane tubes having no support lengthwise and properly sealed at their ends. Higher osmotic medium concentrations yielded to higher osmotic permeation rates. Increasing the juice temperature was found to markedly increase the permeation flux. However, only a slight enhancement of flux was observed by increasing the juice flow rate. Moreover, higher juice concentrations up to approximately 12.0°Brix led to a lowering of the osmotic flux. Finally, as far as the membrane thickness was concerned, a strong trend was revealed for exponential increase of permeation by shifting towards lower membrane thicknesses[43].

During freezing, a part of the aqueous fraction freezes out and forms ice crystals that damage the integrity of the cellular compartments. Dehydration can be applied as a pretreatment before freezing, by removing a part of the water from the product in order to decrease the amount of crystals formed during the freezing process. Osmotic dehydration was reported as a pretreatment in freezing[40]. It has been reported that osmotic pretreatment reduces product water activity and enzyme activity with slight changes in product characteristics[41]. It decreases enzymatic browning and often retains or even improves food product colour. It improves texture characteristics of thawed fruits and vegetables and reduces structural collapse and drip loss during thawing[42].

During industrial processing of tomato, oxidation and/or trans–cis isomerization of the lycopene take place. The increase in the bioavailability of lycopene as a consequence of thermal processing is related to the biochemical properties of the *cis* isomers. However, Toor and Savage (2006)[44] pointed out the possibility of preserving the lycopene by means of applying hot air drying at 42°C with the aim of obtaining a product with 70% moisture content. Further during drying, the typical red colour of fresh tomato gradually changes towards more brownish hues or intense red depending on the mechanisms involved. This change occurs mainly as a result of the changes that lycopene undergoes, the oxidation of the ascorbic acid and enzymatic and non–enzymatic browning reactions[45]. Due to the changes suffered by the product as a consequence of being exposed to high temperatures and long processing times mentioned above, the food industry tends to use microwave energy in order to reduce drying times. Whilst traditional drying exposes the product to high temperatures and long processing times, the combination of hot air and microwaves could reduce degradation in the nutritional compounds[46]. Moreover, submitting the product to an osmotic dehydration prior to the drying process may help to preserve its organoleptic characteristics such as colour and nutritional properties, as well as to reduce the subsequent drying time. Among the different processing techniques, the osmotic dehydration presents clear advantages for preserving lycopene due to the low temperatures implied in the process as well as the protective effect of the surrounding osmotic solution to limit oxygen exposure. Heredia *et al.*[47] dehydrated cherry tomato by a combination of different techniques (osmotic dehydration, convective drying, and microwaves assisted air drying) in order to evaluate the effect of the process variables on the degradation and isomerization of lycopene, as well as on the optical properties. Obtained results showed that the osmotic pre-treatment limited the isomerization during the later stage of drying, whereas both the loss of total lycopene and the trans-cis isomerization, mainly to the 13-*cis* form, were favored by an increase in temperature and the microwave power.

High Intensity Pulsed Electric Field

High intensity Pulsed Electric Field (HIPEF) processing involves the application of short high power electrical pulses (typically 20–80 kV cm^{-1}) for short time periods (<1 sec) to fluid foods placed between 2 electrodes[48]. High-intensity pulsed electric field is a non-thermal technology extensively studied as an alternative to the traditional thermal treatment[49]. Typical systems for the treatment of pumpable fluids consist of a PEF generation unit, which is in turn composed of a high voltage generator and a pulse generator, a treatment chamber, a suitable product handling system and a set of monitoring and controlling devices[50]. High level of microbial destruction and few losses of flavor, colour, taste or nutrients have evinced the suitability of PEF technology to obtain high quality fresh-like foods. In comparison to the extensive research devoted to the destruction of microorganisms by HIPEF, the information available about the effect on enzymes is more limited. However, enzymes are less sensitive than microorganisms to HIPEF and their inhibition depends on the enzyme itself, the media where they are suspended and the processing parameter[51].

Min *et al.*[52] studied and compared the effects of commercial scale pulsed electric field (PEF) processing on the quality of tomato juice with those of thermal processing. Both thermally and PEF processed juices had shelf life of 112 days at 4°C. The lipoxygenase activity of thermally and PEF processed juices was 0 and 47%, respectively. Sensory evaluation indicated that flavor and overall acceptability of PEF processed juice were preferred to those of thermally processed juice ($p<0.05$). Aguiló-Aguayo *et al.*[53] looked at volatile compounds and flavour-related enzymes in tomato juice subjected to high–intensity pulsed electric field (HIPEF) processing compared with those of thermal processing (90°C for 30 or 60 s). Tomato juice treated by HIPEF showed lower residual lipoxygenase (LOX) activity than heated juice. Almost 50% of the hydroperoxide lyase remained active after HIPEF and heat treatment for 30 seconds, but was almost completely inactivated by 90°C for 60 seconds. HIPEF treatment was found suitable by the researchers in improving the organoleptic properties of tomato juice.

Ultra Sonication

Ultrasound is defined as sound waves with frequencies higher than 16 kHz. These waves can propagate in a liquid media as alternating compression. The application of power ultrasound in food enzyme inactivation has been explored in recent years. Generally, ultra sonication in combination with other treatment (s) is more effective in enhancing the inactivation efficacy. The concept of combination treatment has been further explored by introducing elevated static pressure in an ultrasound treatment chamber in a process called manothermosonication (MTS). Manothermosonication has been used to deactivate various enzymes such as lipoxygenase, peroxidase, lipase, pectinmethylesterase etc.

The inactivation effect of ultrasound is attributed mainly to a phenomenon called cavitation. If ultrasound has sufficient energy, a phenomenon known as cavitation occurs. Cavitation involves the formation, growth and rapid collapse of microscopic bubbles. Based on theoretical considerations, extremely high temperatures and pressures are momentarily delivered to the liquid media during the collapse of bubbles[54]. However, the ultimate reason for microbial inactivation *via* ultrasound is believed to be the mechanical damage caused by cavitation[55].

Pectin methylesterase (PME) is an endogenous pectic enzyme found primarily in tomato cell walls that de–esterifies the methyl group of pectin and converts it into low methoxy pectin or pectic acids[56]. To avoid quality losses, partial or complete inactivation of PME together with inactivation of polygalacturonase (PG) during tomato processing is required[57].

Inactivation of this enzyme is achieved by cold and hot break methods in processed tomato products. But due to certain disadvantages of these methods, different non thermal technologies, including power ultrasound, have been tested in recent years to deactivate tomato enzymes for improving the final product quality.

According to a study conducted by Raviyan *et al.*[54] thermosonication at 61°C, increased the inactivation of PME by 39 to 374 fold compared to the thermal inactivation test at the same temperature, while at 72°C, the increase was 36–84 fold depending upon cavitation intensity. In a study, the effect of sonication on the resistance of yeast (*Pichia fermentans*) in tomato juice was investigated and it was found that desired 5 log reductions (D5 value) and shape factors are achieved through the application of ultra–sonication, which correlate exponentially with amplitude level. Results presented in this study show that sonication alone is an effective process to achieve the desired level of yeast inactivation in tomato juice[58].

High Pressure Processing (HPP)

On industrial scale, high pressure processing (HPP) of food products is mainly applied for pasteurization purposes. High pressure treatment (100–1000 MPa) is an alternative to thermal processing of food. It is claimed that the process is clean and energy efficient compare to many conventional processes. Non covalent bonds are primarily affected by application of high pressure. Therefore, the flavor and nutritional quality of the food is usually unaffected, although microorganisms and enzymes may be inactivated. In recent years researchers focused mainly on inactivation of vegetative microorganisms by means of HPP. Relatively mild treatments (*e.g.,* 1–3 min, 500 MPa, approximately 25°C) resulted in a stable refrigerated tomato juice product[59]. Conventional sterilization has a negative effect on quality due to the intense heat treatment. As a novel technology for preservation, HP alone cannot inactivate bacterial spores; a combination with other techniques, such as temperature elevation is required. Meyer[60] has patented a HP sterilization process based on at least two–pressure pulses at a minimum initial temperature of 7–8°C. This HP process aims to inactivate vegetative micro–organisms and spores, without adversely affecting flavor and minimally changing texture and colour. HP processes based on this concept may be a useful alternative for preserving tomato based products.

The effect of high pressure treatment (200–600 MPa for 20 min) on the texture of cherry tomatoes and on the key softening enzymes (pectin methylesterase and polygalacturonase) was investigated by Tangwongchai *et al*[61]. According to them, polygalacturonase was more susceptible to pressure (compared to pectinmethylesterase), being almost totally inactivated after treatment at 500 MPa and pectinmethylesterase is barotolerant.

Alternative Current Heating Treatment

The heat produced by electricity could be utilized to kill microorganisms, and the death rates are comparable to that of conventional heating. Ohmic heating can also be used for microbial inactivation. Yildiz and Baysal[37] investigated the effects of alternative current (AC) treatment on the inhibition of *Aspergillus niger* and the inactivation of PME. Moreover, the change in total pectin content of chopped tomato was addressed. The main purpose of using AC heating treatment is to inactivate enzymes and microorganisms. The inactivation of pectin methylesterase (PME) and *Aspergillus niger*, and the changes in total pectin content by AC heating with different electric field strengths in the range of 36–108 V/cm for different treatment times in tomato fruit were investigated. PME activity decreased with an increase in treatment time at 108 V/cm. However, PME activity increased by AC treatment at 48 V/cm for 5 s and at 36 V/cm for 5, 10, 15, and 30 s compared to untreated control. The highest amount of total pectin was determined as 3.56% at 89°C by AC treatment at 36 V/cm for 80 s. The *A. niger* inactivation was increased with increase in electric field strength and increased temperature. The concentration of *A. niger* (cfu/g) was determined to be <10 at 108, 68, 48, and 36 V/cm electric field strengths for the treatment times of 6, 12, 20 and 30 s, respectively.

Ozone Treatment

The fresh fruit and vegetable industry is still working to increase the assortment of minimally processed vegetable products that meet consumers' needs for quick and convenient products that preserve their nutritional value, retain a natural and fresh colour, flavor and texture and contain fewer additives such as preservatives[62]. A treatment using a sanitizer is considered beneficial because it controls spoilage microorganisms. However, different washing and disinfection treatments may negatively affect the nutritional and sensory quality of the product[63]. Furthermore, the production of other disinfection byproducts such as chlorinated organic compounds (*e.g.*, trihalomethanes) which are potential carcinogens[64] has created the need to investigate the efficacy of nontraditional sanitizers and other alternative technologies such as ozone.

Das[65] employed gaseous ozone treatment in combination with CA storage to control *Salmonella enteritidis* in CA and MA packing of cherry tomatoes. Ozone is known to be effective sanitizer against all forms of microorganisms even at low concentration and at room temperature[66]. Zambre *et al.*[67] have conducted a study to assess the effect of ozone treatment on the ripening behavior and quality of tomatoes. According to them, Ozone treatment delayed both the development of red colour as well as rotting. Colour development and rotting followed a trend similar to the one which can be described by Hill's equation. Shelf life was enhanced by 12 days when treated tomatoes were stored at 15°C. The longer shelf life was mainly due to a reduction in surface microbial count. Further, enrichment of the storage atmosphere with ozone, which may be of value in reducing the microbial activity of sliced tomatoes, caused a transient increase in respiration, resulting in increased fructose, glucose, ascorbate and fumarate levels, both in intact and sliced fruit[68].

Irradiation

Irradiation of food products is achieved by exposing the product to a source of ionizing or non ionizing energy for the purpose of extending the products shelf life. The ionizing radiation source could be high energy electrons, X–rays (machine generated), or gamma rays (from Cobalt–60 or cesium–137), while the non–ionizing radiation is electromagnetic radiation mainly from ultraviolet rays (UV–A, UV–B and UV–C), visible light, microwaves and infrared[69]. Irradiation is not being widely used because of some misconceptions by consumers about its role in causing cancer. Scientific studies have established that 10 kGy and above does not produce any toxicological hazards or nutritional or microbiological problems in food[70].

While irradiation may prolong the shelf life of many fruits, one deleterious consequence is the development of abnormal textural changes that can seriously compromise quality and handling endurance. The softening of tomato fruit in response to irradiation is typically evident within hours following treatment, even for mature–green fruit[71]. Irradiation–induced softening was evident in mature–green and pink fruit within hours following irradiation doses of 0.7 to 2.2 kGy from gamma or X–ray sources, and differences between irradiated and control fruit persisted throughout post–irradiation storage. Fruit irradiated at the mature–green stage softened during post irradiation storage but exhibited an apparently irreversible suppression in polygalacturonase activity[72].

The use of UV–C radiation as a post harvest elicitor of beneficial responses in crops has become a key area of interest[73]. Mature–green tomato fruits (var. Capello) irradiated with doses of ultraviolet light (UVC, 254 nm) corresponding to 0 (control), UV – 3.7 kJ m^{-2} (hormic dose) and UV – 24.4 kJ m^{-2} (hyper dose) and stored at 16°C, under high relative humidity for a period of 28–35 days were evaluated for various attributes of senescence. It was observed that the development of colour and lycopene as well as the decline in chlorophyll retarded significantly (pd”0.001) during storage in response to the treatment with both the hormic and hyper doses. However the hyper dose impaired ripening and caused abnormal browning, manifested as sun–scalding of the fruit's surface. One possible mode of UV action is attributable to the significantly higher levels of total carotenoids found in irradiated fruits compared to controls indicative of a phytochemical antioxidant defense mechanism[73].

Barka *et al.*[74] also studied the effect of a hormic dose of UV–C (254 nm) on changes in fruit firmness and cell wall–degrading enzyme (CWDE) activity was determined using tomato fruit. It was found that irradiated tomatoes were firmer than control and consequently may explain how UV–C could delay the ripening and senescence process. It can thus be suggested that the CWDE are one of the targets of the UV–C, and by this action, irradiation contributed to a delay of the cell wall degradation and consequently retarded softening of the tomato fruit tissues.

Controlled or Modified Atmospheres

Controlled or modified atmospheres have been shown to delay ripening during storage. Modified atmospheres resulting from the enclosure of tomatoes in polyethylene or other forms of plastic packaging may also delay fruit ripening and prolong storage life. Passive and active modified atmosphere packages (MAP) with different concentrations of CO_2 and O_2 have been used to maintain the quality of fresh–cut tomato. Shelf life of tomato slices could be maintained for 10 days at 2°C under both active and passive MAP (7.5% O_2 and 0% CO_2)[75]. Odriozola–Serrano *et al.*[76] evaluated the feasibility of minimal processing and modified atmosphere packaging (5% O_2 + 5% CO_2) to preserve colour attributes and bioactive compounds of fresh–cut tomato from different cultivars (Rambo, Durinta, Bodar, Pitenza, Cencara and Bola) through storage under refrigeration. Neither the content of health–related compounds such as lycopene, vitamin C and phenolic compounds nor the antioxidant capacity changed significantly between whole and just processed fresh–cut tomatoes. Furthermore the initial colours of fresh–cut tomatoes as well as vitamin C were maintained for 3 weeks under cold storage. The antioxidant capacity was well correlated with vitamin C and phenolic content, whereas lycopene was directly related to colour measurements (a value indicating redness, L value indicating brightness and H representing hue angle). According to this study, minimal processing maintains the main antioxidant compounds and colour parameters of slices tomatoes for 21 days at 4°C, thus preserving their initial nutritional value. Srinivasa *et al.*[77] stored green, physiologically mature tomatoes in cartons covered with chitosan (biodegradable membrane) for 30 days at 27±1°C. Fruit showed better colour retention and firmness not only in relation to the control (fruit stored in air) but also in comparison with those enclosed in low density polyethylene (LDPE) film. The inclusion of an ethylene adsorbent (granular activated carbon) within polypropylene bags, in which tomatoes at the turning stage were stored at 8°C, reduced the level of ethylene within the storage atmosphere for 21 days. Colour change, softening and weight loss were reduced and quality on ripening was improved. Mondal *et al.*[78] reported that when tomatoes are enclosed in polyethylene and stored at 25°C they are more resistant to oxidative stress than unenclosed fruit stored at the same temperature. Low oxygen levels in modified or controlled atmospheres also inhibit polygalacturonase activity, thus reducing the rate of fruit softening[79].

Foam Mat Drying

Among the agricultural processing technologies, drying has a great importance as a post–harvest food preservation technology. Drying is the most ancient and most common type of food preservation[80]. Drying of agricultural products helps in reduction of moisture content, weight, and volume and minimizes packaging, storage, and transportation costs[81, 82,83]. In the recent years, foam mat drying technology has revamped and renewed attention for its added ability to process hard–to–dry materials to produce products of desired properties (*e.g.*, favourable rehydration, controlled density), retaining its volatiles that otherwise would be lost during the drying of non–foamed materials. Foam mat drying involves the incorporation of a foaming agent into liquid foods with subsequent whipping to form stiff foam[84,85]. Foam mat drying is carried out for the liquid or semi solid food by making them into foam with the addition of food foaming and stabilizing agents. The foam thus formed can be spread into a thin mat/sheet and dried by using hot air. Then, the dehydrated product is conditioned and converted into powder. Generally, drying rates are comparatively higher in foamed pulps because of increased surface area at the liquid–gas interface thus allowing rapid drying through internal moisture movement within the pulp. The dehydrated powder/flakes are superior to drum dried and spray dried products because of its honey comb structure and better reconstitution properties[84, 85]. In general, drying rate of foamed materials is faster than non–foamed materials and is greatly accelerated at the end. Many researchers have reported that the increased interfacial areas of foamed materials are the responsible factor for its induced drying rate. Advantages of the foam mat dried product are its higher stability against deteriorative microbial, chemical and biochemical reactions, reducing handling, packaging and transportation costs and negligible energy requirements during storage. Retention of colour, flavor, vitamins (particularly the ascorbic acid), nutritional and organoleptic characteristics and free from any scorched odor/taste, satisfactory shelf life in the product are the main attraction of the process technology. The powder developed using this method reconstitutes well and remains free flowing during storage[86].

V. QUALITY EVALUATION (DESTRUCTIVE AND NON-DESTRUCTIVE) OF FRESH AND PROCESSED TOMATO

Monitoring quality of tomatoes is very important at all levels of the industry, including the breeding programs, the agricultural production, and the post harvest management as well as processing. Several different parameters such as soluble solids, dry matter, acidity, contents of sugars and organic acids are often used to assess the quality of tomatoes. Sugars and organic acids, which are responsible for the sweetness and tartness and influence on tomato flavour, are the major factors affecting consumer acceptability. Soluble solids and dry matter are very important for the food industry, affecting processing yield of many tomato products such as pulp, paste and concentrate, which are evaporated to a definite percentage of solids[45,47]. In consideration of the role played by sugars and acids' contents in the definition of tomato quality, several analytical techniques have been developed to evaluate these parameters. However, traditional methods are not well adapted to on–line and require an expensive and time consuming sample preparation. Thus, the tomato production sector is interested in development of analytical method allowing simultaneous measurement of a large number of samples for routine quality control program[87].

Consistency of Tomato Products

Consistency is an important property and measure of quality in products made from tomato concentrate such as tomato sauce, tomato puree, ketchup, pizza sauce and spaghetti sauce. Concentrated tomato products are commonly produced in multi–effect evaporators where the degree of concentration and consistency are governed by the residence time in the evaporator.

Mohr[88] compared several instrumental methods for measuring the consistency of tomato juice. These methods were Bostwick consistometer, brookfield, efflux tube, and lawa texturometer system and connon–fensk viscometer. Bostwick consistometer and Ottawa Texture Measuring System (OTMS) were the most valid of them which showed a good correlation (with R–values equal to 0.91 and 0.82, respectively) with organoleptic results. Bostwick consistometer was recommended by Mohr for daily analysis of tomato products because of its simplicity and low cost[89]. Concentrated tomato products are commonly bought or sold based on their natural tomato soluble solids (NTSS) content. Therefore, commercial manufacturing operations primarily use in–line process refractometers to continuously monitor and control the NTSS content during evaporation. In–line absorption photometry has been used to predict consistency of concentrated tomato products [90]. The optical density of various concentrated tomato products was found to correlate highly with water insoluble solids (WIS) content and consistency as measured by the Bostwick consistometer.

Firmness of Tomato Fruit

Firmness of tomato fruit is one of the major factors contributing to the quality of fruit. It is directly related to fruit development, ripening after maturity and storage potential. It is also related to the likelihood of bruising when fruit are subjected to impacts during handling. The maturity of fruit is a qualitative measure, which is difficult to identify. Maturity grading of fruits in harvesting before delivery to the market is beneficial for indicating the optimal time for marketing or for optimization of storage management. The firmness of a fruit is an index of the mechanical, chemical and rheological properties of the fruit. It is negatively proportional to the maturity of the fruit and can hence be used as an alternative indicator to maturity in fruit grading and sorting. However, firmness is a loosely defined concept, evaluated by techniques that are mainly subjective such as judged by a sensory panel or by market experts by manual touch. Firmness can be determined more objectively with a range of different destructive or non–destructive measuring devices, such as the hand–operated penetrometer, Zwick densimeter and Durofel. There are some disadvantages associated with these hand–operated devices which include their user–dependent behavior, and they measure locally, whereas fruit firmness may vary from one location to another on the same fruit. "Firmness" in food materials corresponds to "stiffness" in engineering materials and is more conveniently described by the force–deformation relationship[81].

An alternative way to measure fruit firmness is determination of the force of impact of the fruit against a firm surface[92]. A third way to measure firmness is based on a correlation with the natural frequencies of the fruit[93]. Firmness has also been investigated by methods using sonic transmission. The measurements are conducted by exciting the fruit with an impact and measuring the acoustic response by means of a small microphone. The responses are analysed according to their resonant frequencies which are mathematically related to firmness[94, 95].

An effort has also been made by Lesage and Destain[94] to develop a mechanical sensor for determining the firmness of tomato fruit. It consists of a small plunger constrained to penetrate slightly into the fruits, by using an accurate lever mechanism. A highly significant correlation exists between firmness measurements performed with this device and the Stable Micro System (SMS), fitted with the same plunger diameter.

Another method for firmness determination was based on the non–destructive acoustic impulse–response technique introduced[96]. The natural frequencies of the intact fruit are obtained by recording the sound which is produced by hitting the fruit and then performing a Fourier transformation on the signal[97]. This technique was applied for tomato fruits by Schotte *et al.*[92] which provide a stiffness factor S, based on the first resonance frequency and the mass of the intact fruit. A logarithmic relation was found between objective stiffness measurements on tomatoes by the acoustic impulse–response technique and subjective firmness measurements, suggesting that subjective measurements distinguish firmness differences more easily in soft than in firm tomatoes. Accuracy and repeatability of the acoustic impulse–response technique and of subjective scores by market experts were equally high. However, the advantage of using S is that it is objective and no scaling problems between different experts occur. The ultrasonic method might be used as a nondestructive means of monitoring the firmness of tomatoes during various stages of storage[98].

A non–destructive method for assessing the maturity of tomatoes was developed using the mechanical properties of the fruit under the falling impact test. The levels of maturity were classified with cluster and discriminant analyses on the primitive impact measurements and their derivatives. The accuracy of classification was improved with linear discriminant analysis and the number of indices being processed was reduced with stepwise regression analysis. The accuracy of classification is 82.3% by the use of all nine indices and 79.2% by the three most dominant indices. The performance shows that falling impact together with linear discriminant analysis provides a promising non–destructive approach in assessing the maturity of tomatoes[99].

Flavour/Aroma of Tomato

Extensive research has been focused on the development of non–destructive techniques for measuring quality attributes of fruit. Aroma sensing is a particularly promising method of assessing fruit quality. Soluble solid content (SSC) and titrable acidity (TA), the main components responsible for tomato flavour[30] are usually measured by traditional analytical methods which are quite time consuming. Flores *et al.*[100] predicted the internal quality of intact tomato by Near Infrared Spectroscopy (NIRS) using soluble solid content (SSC) and titrable acidity (TA) as the two internal quality indices.

Aroma of tomatoes has been subject of much research but unfortunately, the two basic techniques used to assess aroma—sensory analysis and conventional gas chromatography–mass spectrometry (GC–MS) are generally too time consuming, complex and labour intensive for routine quality application. Sensing the aromatic volatiles emitted by fruit using electronic olfactory systems can be a strategy for predicting the stage of maturity. These systems are concerned with the exploitation of the information contained in the headspace of fruits. They have been studied in the recent past with conventional analytical chemistry equipment and the correlation between the state of over–ripening and the fruit aroma has also been found both in quantitative and qualitative terms. Some specific compounds have been identified as being responsible for the aroma of particular fruit. Two systems namely, electronic nose and headspace fingerprint mass spectrometry (HFMS) have been successfully applied for pre and postharvest aroma profiling of tomatoes[101].

The electronic nose offers a fast and non–destructive alternative to sense aroma and hence, may be advantageously used to predict the optimal harvest date. Commercially available electronic noses use an array of sensors combined with pattern recognition software. Unlike most chemical sensors, which are designed to detect specific gas molecules, the sensors in the E–nose are not specific to a particular volatile[102]. Gomez *et al.*[103] evaluated the capacity of electronic nose to monitor the change in volatile production of ripeness states for tomato, using a specific electronic nose device with 10 different metal oxide sensors (Portable E–nose (PEN2)). The electronic nose was able to detect a clearer difference in volatile profile of tomato and it was also possible to differentiate and to classify the different tomato maturity states, and this method was able to classify 100% of the total samples in each respective group.

An alternative to the E–nose is headspace fingerprint mass spectrometry (HFMS). This system uses a mass spectrometry detector in which all volatiles are fragmented in the source resulting in a complex mass spectrum that can be interpreted as a fingerprint of the aroma of the fruit. The HFMS approach offers important advantages over current solid–state sensors, including no sensitivity to water or alcohol, a linear response to vapour concentration, much less drift, and significantly improved reproducibility[104].

Chilling Injury in Tomatoes

Chilling injury is characterized by development of mealiness, accelerated softening relative to that of non–chilled fruit and non–uniform surface colour development, and usually becomes apparent only after later exposure to higher temperatures. Chilling injury in tomatoes affects their textural quality and is caused by exposure of the fruit to low temperatures between freezing point and 10–12°C. Gloss measurement has been suggested as a possible indicator of chilling injury[105]. Further, ultrasonic wave propagation has been used successfully to measure texture properties of various fruit[106,107]. Verlinden *et al.*[105] exploited ultrasonic waves for the determination of chilling injury in tomatoes. The ultrasonic transducers were fitted with beam focusing probes that were driven into the tomato flesh so that both ends of the transmitter and receiver probes were spaced 10 mm apart. The attenuation of 50 kHz ultrasonic waves increased when chilling injury developed, but only when the tomatoes were harvested at a mature stage.

Lycopene Determination by Chemical Imaging

The tomato ripening process is associated with the loss of chlorophyll and the accumulation of carotenoids (e.g. lutein, β–carotene, and lycopene) in the fruit. As the major carotenoid, lycopene contributes more than 90% to the total pigmentation in full–ripe tomatoes, and a 500–fold increase occurs for the lycopene content during the tomato ripening[108]. Although tomato surface colour is conventionally used as a maturity index, there is room for improvement by measuring internal fruit compositions. Lycopene content can provide complementary information to surface colour. Traditional methods for evaluating lycopene and other carotenoids in tomatoes, such as high–performance liquid chromatography (HPLC), require complicated procedures for sample preparation, processing, and analysis. Hence, they cannot satisfy the requirements of rapid and nondestructive measurement, despite the accurate results from the time–consuming process[109].

Near–infrared spectroscopy (NIR) has been used to assess carotenoid content[110], surface colour measurement to estimate internal lycopene content[111]. Since traditional analytical methods for assessing internal properties of tomatoes usually require destructive procedures to extract particular components, obtaining spatial–temporal information of the internal composition of a tomato can be an alternative approach. The recent development of a magnetic resonance imaging (MRI) technique provides a potential tool for detecting lycopene content[112] during tomato ripening. Raman chemical imaging, a new technique that shows promise for mapping constituents of interest in complex food matrices, has been exploited by Qin *et al.*[109] to detect and visualize internal lycopene distribution during postharvest ripening of tomatoes.

VI. UTILIZING WASTE PRODUCTS OF TOMATO PROCESSING

One of the most important problems of food industry is the management of waste and their conversion in products of higher value. The majority of tomatoes for processing are made into products such as juices, ketchup, sauce, paste, puree and powder. The solid waste, remaining after the juice/pulp extraction process, consists of skin, seeds, fibrous matter, trimmings, cores and cull tomato. Tomato seed and peel wastes from industrial processing, represents between 3% and 5% (in weight) of fresh product. These byproducts, as most of organic vegetable solid residues, are mainly used for livestock feed production—primarily cattle and sheep. These residues have a high moisture content, which leads to some storage difficulties. This waste is dumped, flushed into sewage, streams and rivers or to some extent used as fertilizers. However, using modern eco–compatible technologies, the use of food waste can be promoted to obtain biopolymers that can be re–used in the same sector of the raw materials. Nowadays, there is an increasing trend towards utilization of food processing byproducts as a source of functional components. The unnecessary waste of valuable commodities can be checked by processing into value added products.

Tomato seed protein has been found to have high lysine content; therefore, could substantially improve the protein quality of cereal products, which are low in lysine[113]. Tomato seed supplemented bread had improved loaf volume, texture and crumb quality, due to anti–staling properties[114]. In vitro digestibility tests have also been carried out for tomato seed protein and it was reported that the protein digestibility is 61.8% for tomato seed meal[115] as compared to 80% for casein. While, Sogi *et al.*[116] conducted some *in vivo* studies by feeding whole meal, deoiled meal and protein concentrate (PC) of tomato seeds to male albino rats for weight gain, along with standard and non–protein diets. Corrected protein efficiency ratio was 1.82 for whole meal, 1.93 for deoiled meal and 1.99 for PC as compared to 2.5 for casein. The net protein retention (NPR) was highest for casein (2.91), followed by whole meal (2.65) and deoiled meal (2.52) and least for PC (2.51). Tomato seeds contained high quality plant proteins that could be supplemented into various food products.

The typical waste from tomato processing is made up of the outer tegument of the berry (epicarp) and of seeds. The presence of polysaccharides is marked. From tomato production waste it is possible to re–use up to 20% of polysaccharides with gel and film–structured remarkable properties. These biopolymers can interact with others polymers as proteins, lipids and polysaccharides and modify their properties. Biopolymers obtained from tomato processing waste, because of their primary structure, play an interesting role in some biotechnological applications. The polysaccharide film–generating capacity allows the production of bio–films that can be used for solarization and mulching in agriculture. It is also possible that the polysaccharides extracted from tomato waste can serve as potential plant growth promoters[117].

Tomatoes and processed tomato products (juice, sauce, soup, spaghetti sauce etc.) constitute the major sources and accounts for more than 85% of all the dietary sources of lycopene. Among the agro–industrial byproducts, tomato wastes are the only byproducts that are rich in lycopene[118]. These byproducts from the tomato puree and sauce industry are commonly used in the development of lycopene rich products. The lycopene content varies from 0.85mg to 13.6 mg/ 100g in tomatoes. Although it has been used as a food colourant for many years, it has recently received attention with respect to its antioxidant activity and potential in preventing prostate cancer and cardiovascular diseases in humans[119].

Many methods have been used for carrying out on the extraction of lycopene from by–products especially tomato waste such as solvent extraction, high hydrostatic pressure processing, enzymatic treatment[120], supercritical fluid extraction. There are initiatives by food scientist to recycle the lycopene–rich by–products as food ingredients. Fortification with lycopene in dry fermented sausage was also done by adding dried tomato peel to the meat mixture during the sausage production[121]. The barley–tomato pomace blends have been processed into snacks using extrusion processing. Besides, enrichment of low quality edible oils such as refined olive oil, extra virgin olive oil and refined sunflower oil by lycopene from tomato peels or tomato puree was proven to induce thermal stability to these edible oils[119].

Tomato processing byproducts, together with unprocessed tomatoes were also analyzed for several bioactive phytochemicals, like sterols, tocopherols, carotenes, terpenes, total and simple polyphenols. Most of the phytochemicals determined exert antioxidant activities, tomato processing byproducts could be successfully utilized as functional ingredient for the formulation of antioxidant rich functional foods[122].

VII. CONCLUSIONS AND FUTURE PERSPECTIVES

Recent developments in agricultural technology have led to the demand for a new era of automated, non–invasive methods that leave the crop intact and do not interfere with its natural growth. The future trend in attribute analysis of fresh produce involves the development of portable, low–cost devices that take measurement none invasively from crops in the field to instantly determine quality characteristics. Post–harvest, the major research input in the near future is still likely to revolve around the role of ethylene in fruit ripening and the physiological basis of ripening control. As the heat application adversely affects the flavor, colour and nutritional value of processed products, the efforts have been made to update existing technology. The non–thermal technologies *viz.* oscillating magnetic fields, high hydrostatic pressure, light pulses and electric field treatments, addition of preservatives, irradiation technology, centrifugation and ultra filtration, ohmic heating etc. will be further extensively explored. Post–harvest phenomena, such as chilling injury will continue attracting attention. Although studies on the application of novel storage techniques are likely to continue, the degree to which currently available techniques may be further improved to provide a significant commercial impact is unclear.

Acknowledgment

The authors would like to thank all the information sources in the public domain used in the compilation of this chapter.

REFERENCES

[1] FAO 2010. www.faostat.fao.org.

[2] Rico–García E., Hernández H.F., Zarazúa S.G., Ruiz H.G., "Two new Methods for the Estimation of Leaf Area using Digital Photography", *Int J Agric Biol.*, **11**(4), 397–400 (2009).

[3] Jaramillo J., Rodriguez V., Guzman M., Zapata M., Rengifo T., "Technical manual: Good Agricultural Practices in the Production of tomato under protected conditions", FAO (2007).

[4] Olaniyi J., Akanbi W., Adejumo T., Akande O., "Growth, fruit yield and nutritional quality of tomato varieties", *Afr J Fd Sci.,* **4**(6), 398–402 (2010).

[5] Kader A.A., Morris L.L., "Correlating subjective and objective measurements of maturation and ripeness of tomatoes". *In*: Proc. 2nd Tomato Quality Wkshp. Veg. Crops Series 178. Univ Calif Davis (1976).

[6] Maul F., Sargent S.A., Balaban M.O., Baldwin E.A., Huber D.J., Sims C.A., "Aroma volatile profiles from ripe tomato fruit are influenced by physiological maturity at harvest: An application for electronic nose technology", *J Amer Soc Hort Sci.*, **123**(6), 1094–1101(1998).

[7] Sargent S.A., VanSickle J.J., "Premium–quality tomato program for Florida. Proc. Florida Tomato Inst", C.S. Vavrina (*ed.*), Publ. PRO 108. Univ. Florida, Inst. Fd & Agric. Sci., Gainesville. pp. 8–9 (1996).

[8] Wills R.B.H., McGlasson B., Graham D., Joyce D., "Postharvest: An introduction to the physiology and handling of fruits, vegetables and ornamentals". Univ New South Wales Press, p. 262 (1998).

[9] Kader A.A., "Postharvest biology and technology: An overview. *In*: Kader, A.A. (*ed.*) Postharv. Technol. Hort", Crops. 3rd ed. Pub. No. 3311. Oakland: Univ. Calif. pp. 39–47 (2002).

[10] Abeles F.B., Morgan P.W., Saltveit M.E., "Ethylene in Plant Biology", 2nd Ed. Acad. Press, p 414 (1992).

[11] Moretti C.L., Calbo A.G., Henz G.P., "Postharvest physiology and handling. *In*: Silva J.B.C. and Giordano L.B. (*eds.*), Processing tomatoes", Brasilia, Embrapa/SCT pp. 136–143 (In Portuguese) (2000).

[12] Jones B., Jones J., Stall R., Zitter T., "Compendium of tomato diseases". APS Press, St. Paul MN (1991).

[13] Khan S.H., Bullock R.C., Stoffella P.J., Powell C.A., Brecht J.K., McAuslane H.J., Yokomi R.K., "Possible involvement of altered gibberellin metabolism in the induction of tomato irregular ripening in dwarf cherry tomato by silverleaf whitefly", *J Plant Growth Reg.* **16**(4), 245–251 (1997).

[14] Mac Leod R.F., Kader A.A., Morris L.L., "Stimulation of ethylene and CO2 production of mature–green tomatoes by impact bruising", *Hort Science.,* **11**(6), 604–606 (1976).

[15] Moretti C.L., Sargent S.A., Huber D.L., Calbo A.G., Puschmann R., "Chemical composition and physical properties of pericarp, locule and placental tissues of tomatoes with internal bruising", *J Amer Soc Hort Sci.,* **123**(4), 656–660 (1998).

[16] Moretti C.L., Baldwin E.A., Sargent S.A., Huber D.H., "Internal bruising alters aroma volatile profiles in tomato fruit tissues", *Hort Science.,* **37**(2), 378–382 (2002).

[17] Sargent S.A., Brecht J.K., Zoellner J.J., "Sensitivity of tomatoes at mature–green and breaker ripeness stages to internal bruising", *J Amer Soc Hort Sci.,* **117**(1), 119–123 (1992).

[18] Conn K.E., Ogawa J.M., Manji B.T., Adaskaveg J.E., "Leuconostoc mesenteroides subsp. mesenteroides, the first report of a coccoid bacterium causing a plant disease", *Phytopathology,* **85**, 593–599 (1995).

[19] Bartz J.A., Hodge N.C., Perez Y., Concelmo D., "Two different bacteria cause a decay in tomatoes that is similar to sour rot", *Phytopathology,* **85**, 1123 (1995).

[20] Snowden A.L., "Colour atlas of postharvest diseases and disorders of fruits and vegetables". Vegetables Vol. 2, CRC Press, Boca Raton FL. (1992).

[21] De Grano A., Pabico J., "Automating the classification of tomato (lycopersicon esculentum) maturity using image analysis and neural networks", *Trans Nat Acad Scien Tech Phili.*, **29**(1), 131–132 (2007).

[22] Ranatunga C., Jayaweera H., Suraweera S., Ariyaratne T., "Effect of measurement of non–destructive firmness on tomato quality and comparison with destructive methods", *Proce Tech Sess.*, **25**, 29–35 (2009).

[23] USDA. U.S. Standards for Grades of Fresh Tomatoes. USDA, Agri Mktg Serv., Washington, DC (1991).

[24] USDA. U.S. Standards for Grades of Greenhouse Tomatoes. USDA, Agri Mktg Serv., Washington, DC (http://www.ams.usda.gov/standards/stanfrfv.htm) (1966).

[25] Brecht J.K., Tomato response to ethylene at high temperatures: are pressure ripening systems worth considering. Proc. Florida Tom. Inst., Vavrina C.S. (*ed.*), Publ. PRO 108. Univ. Florida, Inst. Fd & Agric. Sci., Gainesville, pp. 11–13 (1996).

[26] Maul F., Sargent S.A., Sims C.A., Baldwin E.A., Balaban M.O., Huber D.J., "Recommended commercial storage temperatures affect tomato flavor and aroma quality", *J Food Sci.*, **65**(7), 1228–1237 (2000).

[27] Goodenough P.W., Thomas T.H., "Comparative physiology of field grown tomatoes during ripening on the plant or retard ripening in controlled atmospheres", *Ann Appl Biol.*, **94**, 445–455 (1980).

[28] Nakhasi S., Schlimme D., Solomos T., "Storage potential of tomatoes harvested at the breaker stage using modified atmosphere storage", *J Food Sci.*, **56**(1), 55–59 (1991).

[29] Parsons C.S., Anderson R.E., Penney R.W., "Storage of mature–green tomatoes in controlled atmospheres", *J Amer Soc Hort Sci.*, **95**,791–794 (1970).

[30] Kader A.A., "Effects of postharvest handling procedures on tomato quality", *Acta Horticulturae,* **190**, 209–221 (1986).

[31] Moretti C.L., Sargent S.A., Huber D.J., Puschmann R., Fontes R.R., "Delayed ripening does not alleviate symptoms of internal bruising in tomato fruit", *Proc Fla State Hort Soc.,* **112**, 169–171 (1999).

[32] Mencarelli F., Saltveit M.E., "Ripening of mature–green tomato fruit slices", *J Amer Soc Hort Sci.,* **113**(5), 742–745 (1988).

[33] Hong J.H., Gross K.C., "Maintaining quality of fresh–cut tomato slices through modified atmosphere packaging and low temperature storage", *J Food Sci.*, **66**(7), 960–965 (2001).

[34] Hui Y.H., Ghazala S., Graham D.M., Murell K.D., Nip W., Handbook of vegetable Preservation and Processing. Chapter 6, 18, 23. Marcel Dekker Inc., New York, U.S.A. (2004).

[35] Duthie G.G., Duthie S.J., Kyle J.A.M., "Plant polyphenols in cancer and heart disease: Implications as nutritional antioxidants", *Nutr Res Rev.*, **13**, 79–106 (2000).

[36] Wiley R.C., Introduction to minimally processed refrigerated fruits and vegetables. *In*: Wiley. R.C. (*ed.*), Minimally processed refrigerated fruits and vegetables New York, Chapman & Hall. pp. 1–14 (1994).

[37] Y1ld1z H., Baysal T., "Effects of alternative current heating treatment on Aspergillus niger, pectin methylesterase and pectin content in tomato", *J Food Eng.,* **75**, 327–332 (2006).

[38] Petrotos K.B., Quantick P.C., Petropakis H., "Direct osmotic concentration of tomato juice in tubular membrane– module configuration. II. The effect of using clarified tomato juice on the process performance", *J of Membrane Sci.*, **160**, 171–177 (1999).

[39] Cross S., Achieving 60° Brix with membrane technology. Paper presented at Annual Meeting, Institute of Food Technologist, New Orleans, La., June pp. 19–22, (1988).

[40] Dermesonlouoglou E.K., Giannakourou M.C., Taoukis P., "Stability of dehydrofrozen tomatoes pretreated with alternative osmotic solutes", *J Food Eng.*, **78**, 272–280 (2007).

[41] Giraldo G., Talens P., Fito P., Chiralt A., "Influence of sucrose solution concentration on kinetics and yield during osmotic dehydration of mango", *J Food Eng.*, **58**, 33–43 (2003).

[42] Forni E., Torreggiani D., Crivelli G., Maestrelle A., Bertolo G., Santelli F., "Influence of osmosis time on the quality of dehydrofrozen kiwi fruit", *Acta Horticulturue*, **282**, 425–434 (1987).

[43] Petrotos K.B., Quantick P.C., Petropakis H., "A study of the direct osmotic concentration of tomato juice in tubular membrane– module configuration. I. The effect of certain basic process parameters on the process performance", *J of Memb Sci.*, **150**, 99–110 (1998).

[44] Toor R.K., Savage G.P., "Effect of semi–drying on the antioxidant components of tomatoes", *Food Chemistry*, **94**(1), 90–97 (2006).

[45] Gould W.V., Tomato production, processing and technology. Baltimore, USA: CTI Publications. (1992).

[46] Vadivambal R., Jayas D.S., "Changes in quality of microwave–treated agricultural products– A review", *Biosystems Engineering*, **98**, 1–16 (2007).

[47] Heredia A., Peinado I., Rosa E., Andrés A., "Effect of osmotic pre–treatment and microwave heating on lycopene degradation and isomerization in cherry tomato", *Food Chemistry*, **123**, 92–98 (2010).

[48] Senorans F.J., Ibanez E., Cifuentes A., "New trends in food processing", *Crit Rev Food Sci Nutr.*, **43**, 507–526 (2003).

[49] Dunn J., Pulsed electric field processing: An overview. *In*: Barbosa–Canovas G.V. and Zhang Q.H. (*eds.*), Pulsed electric fields in food processing: Fundamental aspects and applications. Lancaster, Pa., Technomic Publishing Co. pp. 1–30 (2001).

[50] Chipurura B., Muchuweti M., *Asian Journal of Clinical Nutrition*, **2**, 190–199 (2010).

[51] Aguiló–Aguayo I., Odriozola–Serrano I., Quintão–Teixeira L.J., Martín–Belloso O., "Inactivation of tomato juice peroxidase by high–intensity pulsed electric fields as affected by process conditions", *Food Chemistry*, **107**(2), 949–955 (2008).

[52] Min S., Jin Z.T., Zhang Q.H., "Commercial Scale Pulsed Electric Field Processing of Tomato Juice", *J Agric Food Chem.*, **51**(11), 3338–3344 (2003).

[53] Aguiló–Aguayo I., Soliva–Fortuny R., Martín–Belloso O., "Volatile compounds and changes in flavour–related enzymes during cold storage of high–intensity pulsed electric field– and heat–processed tomato juices", *J Sci Food Agric.* **90**(10), 1597–1604 (2010).

[54] Raviyan P., Zhang Z., Feng H., "Ultrasonication for tomato pectinmethylesterase inactivation: effect of cavitation intensity and temperature on inactivation", *J Food Eng.*, **70**, 189–196 (2005).

[55] Raso J., Pagan R., Condon S., Sala J.F., "Influence of the temperature and pressure on the lethality of ultrasound", *Appl Environ Microbiol.*, **64**, 465–471 (1998).

[56] Giner J., Gimeno V., Espachs A., Elez P., Barbosa–Ca´novas G.V., Martýn O., "Inhibition of tomato (Licopersicon esculentum Mill.) pectinmethylesterase by pulsed electric fields", *Innovative Food Science & Emerging Technology*, **1**, 57–67 (2000).

[57] Porretta S., Birzi A., Ghizzoni C., Vicini E., Effects of ultra–high hydrostatic pressure treatments on the quality of tomato juice. *Food Chemistry*, **52**, 35 –41 (1995).

[58] Adekunte A., Tiwari B.K., Scannell A., Cullen P.J., O'Donnell C., "Modelling of yeast inactivation in sonicated tomato juice", *Int J of Food Microbiology*, **137**, 116–120 (2010).

[59] Porretta S., Birzi A., Ghizzoni C., Vicini E., "Effects of ultra–high hydrostatic pressure treatments on the quality of tomato juice", *Food Chemistry*, **52**, 35 –41 (1995).

[60] Meyer R.S., US Patent 6017572. Ultra high–pressure, high temperature food preservation process. Tacoma, WA 98443, USA. (2000).

[61] Tangwongchai R., Ledward D.A., Ames J.M., "Effect of high–pressure treatment on the texture of cherry tomato", *Journal of Agricultural and Food Chemistry*, **48**, 1434–1441 (2000).

[62] Camelo A.L., Gomez P., Artez–Calero F., "Use of a and b colour parameters to assess the effect of some growth regulators on carotenoid biosynthesis during postharvest tomato ripening", *Acta Horticulture* **599**, 305–308 (2003).

[63] Laurila E., Ahvenainen R., Minimal processing of fresh fruits and vegetables. *In*: Jongen W. (*ed.*), Fruit and Vegetable Processing. Woodhead Publishing Limited/ CRC Press LLC, Cambridge, UK/Boca Raton, FL. (2002).

[64] Fawell J., "Risk assessment case study–chloroform and related substances", *Food and Chemical Toxicology*, **38**, S91–S95 (2000).

[65] Das G., Effect of CA storage. *Food Microbiology*, **23**(5), 430–438 (2006).

[66] Peter J.C., Ozone cure for some sanitation problems. *Food Technology,* **58**(4), 75–76, (2004).

[67] Zambre S.S., Venkatesh K.V., Shah N.G., "Tomato redness for assessing ozone treatment to extend the shelf life", *Journal of Food Engineering*, **96**(3), 463–468 (2010).

[68] Aguayo E., Escalona V.H., Artés F., "Effect of cyclic exposure to ozone gas on physicochemical, sensorial and microbial quality of whole and sliced tomatoes", *Postharvest Biology and Technology,* **39**, 169–177 (2006).

[69] Mahapatra A.K., Muthukumarappan K., Julson J.L., "Applications of ozone, bacteriocins and irradiation in food processing: A review", *Crit Rev Food Sci Nutr.*, **45 (6)**, 447–461 (2005).

[70] WHO, High–Dose Irradiation: Wholesomeness of food irradiated with doses above 10 kGy. World Health Organization, Geneva. (1999).

[71] Yasir M.S., Chacha K., Iwata T., "Effects of gamma irradiation on tissue firmness, some cell wall degrading enzymes and pectic substances of tomato fruit", *Bull Univ Osaka Prcf Scr.*, **39**, 9–20 (1987).

[72] Assi N.E., Huber D.J., Brecht J.K., "Irradiation–induced changes in tomato fruit and pericarp firmness, electrolyte efflux, and cell wall enzyme activity as influenced by ripening stage", *J Amer Soc Hort Sc.*, **122**(1), 100–106 (1997).

[73] Maharaj R., Arul J., Nadeau P., "UV–C Irradiation of Tomato and its Effects on Color and Pigments", *Advances in Environmental Biology*, **4**(2), 308–315 (2010).

[74] Barka E.A., Kalantari S., Makhlouf J., Arul J., "Impact of UV–C irradiation on the cell wall–degrading enzymes during ripening of tomato (Lycopersicon esculentum L.) fruit", *J Agric Food Chem.*, **48**(3), 667–671 (2000).

[75] Artés F., Conesa M.A., Hernández S., Gil M.I., "Keeping quality of fresh–cut tomato", *Postharv Biol Technol.,* **17**, 153–162 (1999).

[76] Odriozola–Serrano I., Soliva–Fortuny R., Martý´n–Belloso O., "Effect of minimal processing on bioactive compounds and color attributes of fresh–cut tomatoes", *LWT–Food Science and Technology*, **41**(2), 217–226 (2008).

[77] Srinivasa P.C., Prashanth K.V., Susheelamma N.S., Ravi R., Tharanathan R.N., "Storage studies of tomato and bell pepper using eco–friendly films", *Journal of the Science of Food and Agriculture*, **86**, 1216–1224 (2006).

[78] Mondal K., Sharma N.S., Malhotra S.P., Dhawan K., Singh R., "Oxidative stress and antioxidant systems in tomato fruits stored under normal and hypoxic conditions", *Physiology and Molecular Biology of Plants*, **12**, 145–150 (2006).

[79] Passam H.C., Karapanos I.C., Bebeli P.J., Savvas D., "A Review of Recent Research on Tomato Nutrition, Breeding and Post–Harvest Technology with Reference to Fruit Quality", *The European Journal of Plant Science and Biotechnology*, **1**(1), 1–21 (2007).

[80] Dall'Aglio G., Capri G., Palmieri L., *Industrie Conserve,* **61**, 147–151 (1986).

[81] Okos M.R., Narsimhan G., Singh R.K. Witnauer A.C., "Food dehydration", *In:* Handbook of Food Engineering, Heldman D.R. and Lund D.B. (*eds.*), Marcel Dekker: New York, (1992).

[82] Singh K.K., Development of a small capacity dryer for vegetable. *J Food Eng.*, **21**, 19–30 (1994).

[83] Kadam D.M., Samuel D.V.K., Chandra P., Sikarwar H.S., "Impact of processing treatments and packaging material on some properties of stored dehydrated cauliflower", *International Journal of Food Science and Technology*, **43**, 1–14 (2008).

[84] Kadam D.M., Balasubramanian S., "Foam mat drying of tomato juice", *Journal of Food Processing and Preservation.* **35**(4), 488–495 (2011).

[85] Kadam D.M., Wilson R.A., Kaur S., Manisha, "Influence of foam mat drying on quality of tomato powder", *International Journal of Food Properties,* **15**(1), 211–220 (2012).

[86] Mishra H.N., Tomato Powder http://www.iitkgp.ac.in/tech/tech_details.php?slno=153 (accessed May 1, 2016), (2009).

[87] Kamil M.M., Mohamed F.G., Shaheen M.S., "Fourier Transformer Infrared Spectroscopy for Quality Assurance of Tomato Products", *Journal of American Science*, **7**(6), 559–572 (2011).

[88] Mohr W.P., "Measuring consistency of tomato thin pulp and puree", *Canadian Journal of Plant Science*, **67**(3), 919–921 (1987).

[89] Mazaheri M.T., Tehrani, Ghandi A., "Modification of Bostwick method to determine tomato concentrate consistency", *J Food Eng.*, **79**, 1483–1486 (2007).

[90] Haley T.A., Smith R.S., Evaluation of in–line absorption photometry to predict consistency of concentrated tomato products. *Lebensm.–Wiss. U.–Technol.,* **36**, 159–164 (2003).

[91] Lesage P., Destain M.F., "Measurement of tomato firmness by using a non–destructive mechanical sensor", *Postharvest Biology and Technology*, **8**, 45–55 (1996).

[92] Schotte S., Belie N.D., Baerdemaeker J.D., "Acoustic impulse–response technique for evaulation and modelling of firmness of tomato fruit", *Postharvest Biology and Technology,* **17**, 105–115 (1999).

[93] Clark H.L., Mikelson, Fruit ripeness tester. U.S. Patent 2, 227, 037 (1942).

[94] De Baerdemaeker, J., Lemaitre, L., Meire, R., Quality detection by frequency spectrum analysis of the fruit impact force. *Trans Am Soc Agr Eng.,* **25**, 175–178 (1982).

[95] Affeldt H.A., Abbott J.A., Apple firmness and sensory quality using contact acoustic transmission. *In*: Dodd V.A. and Grace P.H. (*eds*.), Proceedings of the Eleventh International Congress on Agricultural Engineering, Dublin, pp. 2037–2045 (1989).

[96] Yamamoto H., Iwamoto M., Haginuma S., "Acoustic impulse response method for measuring natural frequency of intact fruits and preliminary applications to internal quality evaluation of apples and watermelons", *J Texture Stud.,* **11**, 117–136 (1980).

[97] Shmulevich I., Galili N., Rosenfeld D., Acoustic excitation for quality control of fruit. ASAE paper 936027 (1993).

[98] Mizrach A., "Nondestructive ultrasonic monitoring of tomato quality during shelf–life storage", *Postharvest Biology and Technology*, **46**, 271–274 (2007).

[99] Lein C.C., Chyung A., Ching–Hua T., "Non–destructive impact test for assessment of tomato maturity", *J Food Eng.,* **91**, 402–407 (2009).

[100] Flores K., Sánchez M.T., Pérez–Marín D., Guerrero J.E., Garrido–Varo A., "Feasibility in NIRS instruments for predicting internal quality in intact tomato", *J Food Eng.,* **91**, 311–318 (2009).

[101] Berna A.Z., Lammertyn J., Saevels S., Natale C.D, Nicola B.M., "Electronic nose systems to study shelf life and cultivar effect on tomato aroma profile", *Sens Actuators B Chem.*, **97**, 324–333, (2004).

[102] Marsili R., Combining mass spectrometry and multivariate analysis to make a reliable and versatile electronic nose. *In*: Marsili R. (*ed.*), Flavor, Fragance, and Odor Analysis. Rockford, Illinois, pp. 349–374 (2002).

[103] Gomez A.H., Hua G., Wang J., Pereira A.G., "Evaluation of tomato maturity by electronic nose", *Computers and Electronics in Agriculture,* **54**, 44–52 (2006).

[104] Berna A.Z., Buysens S., Di Natale C., Gr¨un I.U., Lammertyna J., Nicola B.M., "Relating sensory analysis with electronic nose and headspace fingerprint MS for tomato aroma profiling", *Postharvest Biology and Technology*, **36**, 143–155 (2005).

[105] Verlinden B.E., Smedt V.D., Nicola B.M., "Evaluation of ultrasonic wave propagation to measure chilling injury in tomatoes", *Postharvest Biology and Technology*, **32**, 109–113 (2004).

[106] Mizrach A., Galili N., Rosenhouse G., "Determination of fruit and vegetable properties by ultrasonic excitation", *Trans. ASAE*, **32**, 2053–2058 (1989).

[107] Mizrach A., Galili N., Gan–Mor S., Flitsanov U., Prigosin I., "Models of ultrasonic parameters to assess avocado properties and shelf life", *J Agric Eng Res.,* **65**, 261–267 (1996).

[108] Heuvelink E., Tomatoes. CABI Publishing, Wallingford, Oxfordshire, UK. (2005).

[109] Qin J., Chao K., Kim M.S., "Investigation of Raman chemical imaging for detection of lycopene changes in tomatoes during postharvest ripening", *J Food Eng.,* **107**(3–4), 277–288 (2011).

[110] Pedro A.M.K., Ferreira M.M.C., "Nondestructive determination of solids and carotenoids in tomato products by near–infrared spectroscopy and multivariate calibration", *Analytical Chemistry*, **77**(8), 2505–2511 (2005).

[111] Arias R., Lee T.C., Logendra L., Janes H., "Correlation of lycopene measured by HPLC with the LD, aD bD color readings of a hydroponic tomato and the relationship of maturity with color and lycopene content", *Journal of Agricultural and Food Chemistry,* **48**(5), 1697–1702 (2000).

[112] Cheng Y., Wang T., Chen J., Lin T., Spatial–temporal analyses of lycopene and sugar contents in tomatoes during ripening using chemical shift imaging. *Postharvest Biology and Technology,* **62**, 17–25 (2011).

[113] Brodowski D., Geisman J.R., "Protein content and amino acid composition of protein of seeds from tomatoes at various stages of ripeness", *Journal of Food Science*, **45**, 228–229, 235 (1980).

[114] Morrison W.R., Lipids in flour, dough and bread. Bakers Digest, **50**, 29–31 (1976).

[115] Rahma E.H., Moharram Y.G., Mostafa M.M., "Chemical characterization of tomato seed proteins (Var Pritchard)", *Egyptian Journal of Food Science*, **14**, 221–230 (1986).

[116] Sogi D.S., Bhatia R., Garg S.K., Bawa A.S., "Biological evaluation of tomato waste seed meals and protein concentrate",*Food Chemistry,* **89**, 53–56 (2005).

[117] Suzuki T., Yokotani K.T., Tsubura H., Yoshida S., Kusakabe I., Yamada K., Miki Y., Hasegawa K., "Plant growth– promoting oligosaccharides produced from tomato waste", *Biosource Technology*, **81**, 91–96 (2002).

[118] Krishna G., "Carotene and tocopherol in agro–industrial by–products and wastes of the tropics", *Agric Wastes*, **12**, 235–239 (1985).

[119] Chauhan K., Sharma S., Agarwal N., Chauhan B., "Lycopene of tomato fame: Its role in health and disease", *International Journal of Pharmaceutical Sciences Review and Research*, **10**(1), 99–115 (2011).

[120] Lavecchia R., Zuorro A., "Improved lycopene extraction from tomato peels using cell–wall degrading enzymes", *Eur Food Res Technol.,* **228**, 53–158 (2008).

[121] Calvo M.M., García M.L., Selgas M.D., "Dry fermented sausages enriched with lycopene from tomato peel", *Meat Sci.*, **80**, 167–172 (2008).

[122] Kalogeropoulos N., Chiou A., Pyriochou V., Peristeraki A., Karathanos V.T., "Bioactive phytochemicals in industrial tomatoes and their processing byproducts", *LWT – Food Science and Technology*, **49**(2), 213–216 (2012).

4

Lycopene: Ripening Behaviour, Extraction and Stability

DATTATREYA M. KADAM[1]* AND DEVINDER KAUR[2]

ABSTRACT

Tomato (Solanum lycopersicum) belongs to the family Solanaceae and is one of the world's major vegetables which comprises a significant part of the human diet. Tomatoes are the most widely used fruits, being consumed in the raw state and used for the preparation of a wide range of products such as juice, puree, paste, sauces and soups. The characteristics deep red colour of tomato and its products is due to pigment lycopene. Lycopene is a carotenoid and its concentration increases massively during fruit ripening. Lycopene content in tomatoes depends on variety, ripeness and environmental factors. Tomatoes and its products are rich source of lycopene, ascorbic acid and phenolic compounds. Lycopene is one of the strongest antioxidant and its singlet oxygen quenching ability is two times that of â-carotene and more than ten times that of á-tocopherol. In ripe tomato fruits lycopene occurs essentially in the all-trans configuration. Under different food processing conditions, lycopene undergoes degradation via isomerization and oxidation which results in decrease of biological properties of lycopene. As a polyene it can undergoes trans-cis isomerization induced by light, thermal energy or chemical reactions. The bioavailability of cis-isomers in food is higher than that of all-trans isomers and with the increase in temperature and processing time cis isomers also increases. Reisomerization from cis to trans form of lycopene takes place during storage. Oxidation of lycopene causes loss of colour and development of off flavours. Clinical evidences also showed the greater health benefits of lycopene therefore, preserving lycopene during processing and storage has drawn attention.

[1] ICAR-Central Institute of Post-Harvest Engineering and Technology (ICAR-CIPHET), Ludhiana – 141004, Punjab, India.
[2] Centre of Food Technology, University of Allahabad, Allahabad, U.P., India.
**Corresponding author* E-mail: kadam1k@yahoo.com

***Key words*:** Tomato, Fruit Ripening, Composition, Synthesis, Lycopene, Properties, Processing, Stability, Oxidation

I. INTRODUCTION

Tomato (*Solanum lycopersicum*) is one of the most important worldwide agricultural crops, is rich in a plethora of natural antioxidants. The tomato is native to South America. Genetic evidence shows the progenitors of tomatoes were herbaceous green plants with small green fruit and a center of diversity in the highlands of Peru, Bolivia, and Ecuador. *Solanum lycopersicum*, was transported to Mexico where it was grown and consumed by Mesoamerican civilizations[1].

Tomatoes are a member of the botanical family *Solanaceae,* which contains many potentially poisonous plants [nightshade, nicotianas (includes tobacco and petunias), jimson weed (belladonna), and mandrake], as well as edible plants (potatoes, capsicums, and egg plants)[2]. Commercial tomato production did not begin until after 1860 when tomatoes were finally accepted by consumers. Since 1890, tomato breeding has developed varieties adopted for use around the world. There are two types of tomato plant growth habits: determinate and indeterminate. Determinate tomato plants will grow to a genetically specified height and produce all of their fruiting flowers at one time. Indeterminate tomato plants continue to grow and produce fruiting flowers throughout the entire season. The fruit yield of tomato plant is determined by the weight and numbers of individual fruits. Fruit quality (size, shape, firmness, color, taste and solids content) and market demand, which varies from season to season, determine the market value of fruit.

Leading tomato producing countries include China, India, USA, Turkey, Egypt, Iran, Italy, Spain, Brazil and Mexico. Global tomato production for the year 2013 was around 163.4 million tonnes, of which China dominated the world by contributing around 28.74%, followed by India (9.22%), USA (8.92%), Turkey (7.12%), Egypt (5.97%), Italy (4.86%), Iran (4.11%), Spain (3.42%), Brazil (2.93%) and Mexico (2.59%)[3].

Continued increase in production is expected due to the purported health benefits associated with tomatoes in the diet. Tomatoes can be grown either in the field or under greenhouse conditions. Most greenhouse production is in North America and Europe, where production systems are extremely intensive and can give very high yields (up to 700 tonnes/ha). Field production is much less intensive, and is the most common system in tropical and subtropical areas. In order to satisfy the demand of growers, processors and consumers for a high health-promoting functional food, a large number of new tomato cultivars, with increased lycopene content have been developed by conventional plant breeding techniques[4, 5].

II. TOMATO FRUIT RIPENING

Fruit ripening is a complex, genetically programmed process begins when seeds are completely formed and the fruit reaches its final growth stage. Tomato ripening involves a series of physicochemical and organoleptic changes that lead to the development of a soft and edible ripe fruit with desirable quality attributes including acceptable eating texture, attractive color, pleasant aroma and flavor, and various health-promoting phytochemicals. It is an irreversible process involving series of biochemical, molecular and physical changes such as increased respiration, increased activity of cell wall-degrading enzymes, a transient increase in ethylene production, chlorophyll degradation, biosynthesis of carotenoids and flavour and aroma components are some of the major changes[6,7,8,9,10,11].

The tomato fruit has been classified as 'climacteric' in which respiration passes through a minimum followed by an increase to a maximum during ripening referred to as the climacteric peak, and then declines irreversibly. This respiratory peak is preceded by a climacteric-like rise in ethylene production[12,13,14]. The change in color during the ripening of tomato is due to the degradation of chlorophyll and dismantling of the photosynthetic apparatus followed by the accumulation of carotenoids such as β-carotene and lycopene[15, 16]. The characteristic pigment of red tomato fruit is due to deposition of lycopene, the predominant carotenoids found in tomato fruit, and β-carotene which are associated with the change from green to red as chloroplasts are transformed to chromoplasts. The lycopene content increased from 0.41 mg/kg fm (fresh matter) at the breaker stage to 9 mg/kg fm (pink stage), 22 mg/kg fm (light red stage) and 83 mg/kg fm at the red stage[17] (Fig. 1). At the breaker stage of tomato ripening *i.e.* when a red/orange coloration becomes apparent to the human eye, lycopene begins to accumulate and its concentration increases 50 fold in the tomato fruit[18].

Ethylene in minute amounts can trigger many events of cell metabolism including initiation of ripening and senescence, particularly in a climacteric fruit. The first step in ethylene biosynthesis by the enzyme (1-aminocyclopropane-1-carboxylate synthase, EC 4.4.114) ACC synthase is to convert S-adenosylmethionine (SAM) to 1-aminocyclopropane carboxylic acid (ACC)[19] (Fig. 2). The expression of multiple ACC synthase genes are activated, resulting in increased production of ACC at the onset of fruit ripening. Therefore, in most cases, the rate of ethylene biosynthesis is determined by the ACC synthase activity. ACC oxidase then oxidizes ACC to ethylene. Inhibition of ethylene biosynthesis by antisense RNA for ACC synthase[20] and ACC oxidase[21] was demonstrated first in the tomato fruit. Deamination of ACC to α-ketobutyrate by over-expressing ACC deaminase enzyme also inhibited ethylene formation and fruit ripening[22].

Degradation of cell walls starts by the action of several enzymes, the most important being polygalacturonase, starch hydrolysis, sugar

Fig. 1: Development of lycopene in tomato during the ripening stages.

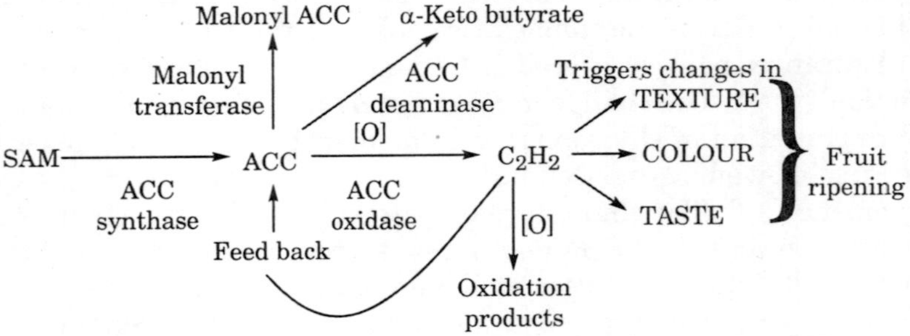

Fig. 2: Pathway for ethylene biosynthesis and metabolism[19]

accumulation, disappearance of organic acids and phenolic compounds, including tannins and results in a soft, juicy texture[23,24]. The noted drop in turgor pressure results from the accumulation of dissolved substances in the cells and water loss, leading to changes in the fruit textural properties, decay of cells walls and a temporary increase in the pressure of the chambers surrounding the seeds[25]. Such transformations are also accompanied by an increase in respiration level and the release of heat energy[26].

Texture is one of the main acceptability factors used by consumers to evaluate quality[27] and it is composed of several properties. These properties include firmness, characteristics of edible tissue, juiciness and crispness[28]. The texture of fruits and vegetables is affected by cellular organelles, biochemical constituents, water content or turgor and cell wall composition. Thus, any external factor affecting these traits can modify texture and can therefore lead to changes in final product quality[29]. Changes in turgor pressure and degradation of cell wall polysaccharides and starch determines the extent of fruit softening. Changes in texture occur due to changes in the chemistry of the primary cell wall compounds which occur during growth and development and are thought to involve genetically programmed changes in the cell wall structure and other physiological factors involved in the developmental plant function[30].

Fruit softening during ripening is brought about by a disassembly of the fruit cell wall, the major changes including depolymerization and solubilization of pectins, depolymerization of matrix glycans (hemicellulose), and loss of pectic galactose side chains and galacturonic acid, followed by solubilization of the remaining sugar residues and oligosaccharides[31]. The common components of the primary cell wall and middle lamella are pectins which contribute to the fruit texture. Their degradation during ripening seems to be responsible for tissue softening[32,33]. The conversion of protopectin into soluble polyuronides results into softening of fruit during ripening which are found loosely bound to the cell walls[34]. This phenomenon is attributed to textural softening during ripening[35]. The degradation of pectic substances and their inter-relationship is shown in Fig. 3. There is progressive loss of firmness during ripening is due to continuous solubilization of protopectin to form pectin and other products in the cell walls of the fruit[34,36,37]. During ripening of fruits the most noticeable change taking place in cell wall is depolymerization and de-esterification of pectic polysaccharides followed by solubilization of oilgosaccharides[33,38,39,40] and a rapid rise in the polygalacturonase (PG) activity[6,24,41,42] resulting in fruit softening. In general, cell wall modication and depolymerization during fruit ripening are variable among different fruits[40,43]. In case of tomato the depolymerization begins early to mid ripening[44]. Since the pectic polymers begin to acquire solubility only after PG has become active, it is believed that this enzyme is involved in the breakdown of the insoluble complex polysaccharides by reducing the length of the chains cross-linked by calcium[45]. PG and pectin methyl esterase PME activity increased remarkably in peach, tomato, and pear[15]. PG antisense constructs for various tomato lines have little effect on the fruit characteristics, *viz*, reduced susceptibility to cracking, and decay and other damages at the later stages of ripening[46].

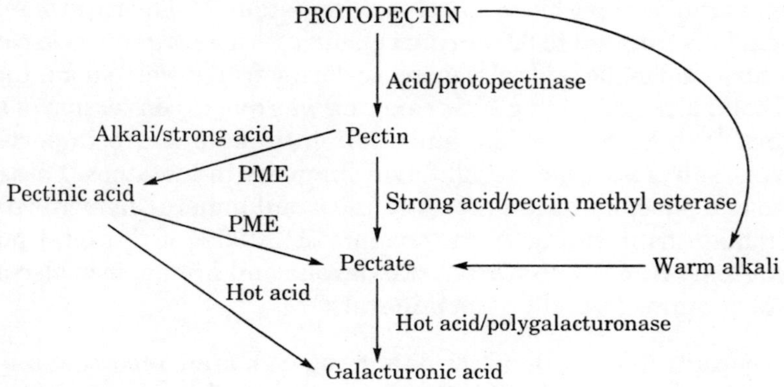

Fig. 3: Inter-relationship of pectic substances[19]

Examples of the best characterized ripening genes include those encoding the fruit specific polygalacturonase (PG) and pectinesterase (PE), two enzymes involved in cell wall degradation associated to fruit softening[47,48], as well as, phytoene synthase (PSY), responsible for the synthesis of lycopene, the red pigment characterizing ripe tomatoes[49]. However, fruit softening is not only a consequence of cell metabolism; biomechanical properties of fruit pericarp are also important which regulate fruit development[50,51].

During ripening the fruit softening is due to the action of carbohydrate hydrolases which results in degradation of cell wall polymers. Most of these enzymes are present in low levels and are constitutive throughout fruit development and ripening[41], but during ripening, generally all the hydrolases increase in activity, particularly cell wall hydrolases, showing a peak activity at climacteric stage. One novel approach to elucidate the role of enzymes in cell wall degradation and softening is to employ antisense RNA technology. This technology was one of the first molecular approaches used for delaying fruit ripening[52]. By specific suppression of PG gene expression with antisense RNA it has been possible to obtain firmer tomatoes with longer shelf-life[48]. Solid content in tomato can be increased by suppression of Pectin methyl esterase (PME)[53]. The genes coding for PG, PME, and other enzymes have been cloned in tomato[46] and in other fruits[52].

III. CHEMICAL COMPOSITION

The chemical composition of the tomato fruit depends on factors such as cultivar, maturity and the environmental conditions in which they are grown[54,55,56,57]. It has been shown that ripening processes and storage temperatures can severely affect the final nutrient composition of the fruit[58]. The dry matter content of the tomato fruit is about 5-7.5% of the total weight. Sugars, mainly glucose and fructose, represent 2-4% of the fresh weight while sucrose accounts for only 0.1-0.2%. Young tomato fruits undergo a transient period of starch accumulation[59]. The ratio of malic to citric acid falls from 1.3 to 0.6 during ripening[54]. The organic acid content, mainly citric and malic acids also increase during fruit development. Increase in total soluble solids during fruit ripening was due to conversion of starch to sugars[60,61]. Sugars, acids, and their interactions are important to sweetness, sourness, and overall flavor intensity in tomatoes. Potassium, nitrogen and phosphorus are the most important mineral elements present in the tomato fruit and account for more than 90% of the total mineral content. Apart from its characteristic flavour and aroma, it is also a good source of vitamins (A and C) and minerals[62, 63].

The concentration of organic acids and sugars has an important influence on the taste of ripening fruits. The characteristic flavour of edible fruits results from the aroma volatiles produced within the fruit during ripening and on maceration. The volatile profile of fruits can be determined by gas chromatography and mass spectroscopy[10]. Levels of ascorbic acid increase according to the maturity of the fruit, with the level highest in the periods of active growth; that is, during spring and early summer[64,65,66]. Fruit ripening at relatively high temperatures, whether on or off vine, along with relatively low light intensity levels probably leads to a decrease in the ascorbic acid content due to oxidation[67]. A seasonal increase has been observed in the ascorbic acid content of field-grown fruit from early to late summer[68]. Light intensity prior to harvesting can also affect ascorbic acid content. Transfer of fruit from shade to sun results in increase in ascorbic acid content[69,70]. However, ascorbate may also decrease with increased fruit solar exposure, probably due to high temperatures resulted in increased ascorbic acid degradation and the enzymes involved in ascorbate recycling (ascorbate peroxidase, dehydroascorbate reductase, and monodehydroascorbate reductase) showed stronger activity with increased light exposure[71]. The ascorbic acid content of tomatoes picked green and allowed to ripen at 22-24.5°C increased from 11 to 26 mg/100 g fw[67]. Salad tomatoes grown in field conditions contained ascorbic acid between 15-21 mg/100 g fw, while a range of industrial grades of tomatoes had a mean value of 19 mg/100 g fw[56]. There are large variations within tomato species (from 8 up to 119 mg/100 g fw in some wild species)[72]. Similar patterns have been observed in other studies[73, 74], but Abushita *et al.*[75] reported an increase during initial reddening of tomato but then a decrease in ascorbic acid. In postharvest-ripened fruit, ascorbic acid decreased and then returned to original levels, while in vine-ripened fruit it increased and then decreased to a similar or slightly lower level. Ascorbic acid content ranged from 8.4 to 32.4 mg/100 g in tomato pulp. Substantial amounts of ascorbic acid were also detected in skins (9.0-56.0 mg/100 g fw and 104-462 mg/100 g dry weight (DW)[76].

IV. SYNTHESIS OF CAROTENOIDS

Carotenoids are part of the isoprenoid pathway (Fig. 4), the same pathway from which spring over 22,000 primary and secondary metabolites, including menthol, camphor, rubber, gibberellic acid, abscisic acid, plastiquinones, spearmint, clove oil and sterols[77, 78]. All products resulting from this pathway are generally termed isoprenoids, terpenoids or terpenes. All isoprenoid units derive from acetate of acetyl CoA molecules *via* the mevalonic acid pathway. They are constructed with 5-carbon isoprene units. The isoprenoid pathway involves many intermediate compounds and enzymes. All carotenoids derive from gerynlgerynl pyrophosphate (GGPP), and are called tetra-terpenoids because all have a 40 carbon backbone. Since all carotenoids are produced in the same pathway, some carotenoids are precursors to others, thus, certain carotenoids predominate at the expense of others (Fig. 5). For example, tomatoes are high in lycopene, which is a precursor to carotene. In order for a tomato to have high lycopene, it must halt the pathway reactions before β-carotene is produced. Thus, red tomatoes have little β-carotene[79], whereas carrot is low in lycopene, as well as in every other carotenoid but it contain high amount of β-carotene.

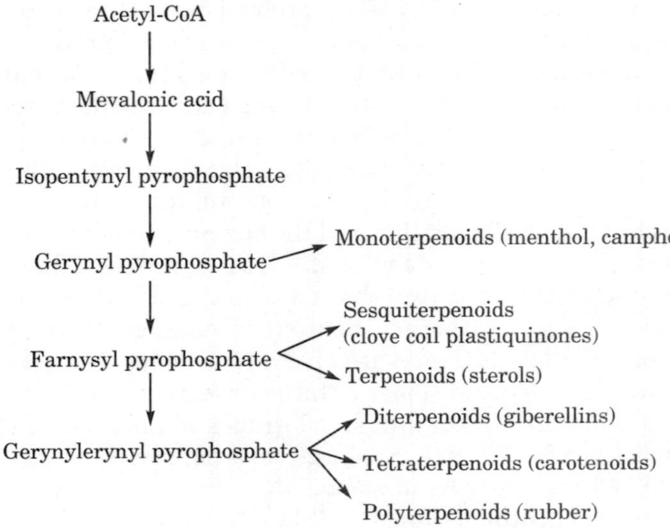

Fig. 4: Isoprenoid pathways with major classes of terpenoids.

The synthesis pathway of carotenes is well worked out (Fig. 5). The first carotenoid in the pathway is phytoene, a colorless compound resulting from the condensation of two GGPP molecules by phytoene synthase, an enzyme associated with the chromoplasts and stroma of chloroplasts. The next step involves dehydrogenation of phytoene into phytofluene and β-carotene by phytoene desaturase. Further dehydrogenations are catalyzed by β-carotene desaturase to yield neurosporene and lycopene. Lycopene cyclase acts on lycopene, an open-chain structure, to add aromatic rings on each end to form γ-carotene (one ring) and β-carotene (two rings). The mechanisms behind the pathway beyond β-carotene to the oxidized xanthophylls are not as clear. The oxygen moieties added to the carbon chain originate from molecular oxygen, and the hydroxylation reactions may be catalyzed by a cyt P450-dependent mixed-function oxidase. Neurosporene, lycopene and all the carotenoids are subject to cyclization reactions at one or both ends of the open chain structure. The enzymes involved in the cyclization reactions are not well known in all cases.

The genetic control of the carotenoid pathway is less clear. The carotenoid pathway is different in anoxygenic non-photosynthetic bacteria and fungi and oxygenic photosynthetic organisms[80]. The genes coding for phytoene synthase (CrtE (aka Psy)), phytoene desaturase (CrtB (aka Pds)) and lycopene cyclase (CrtL) have been sequenced and cloned from photosynthetic

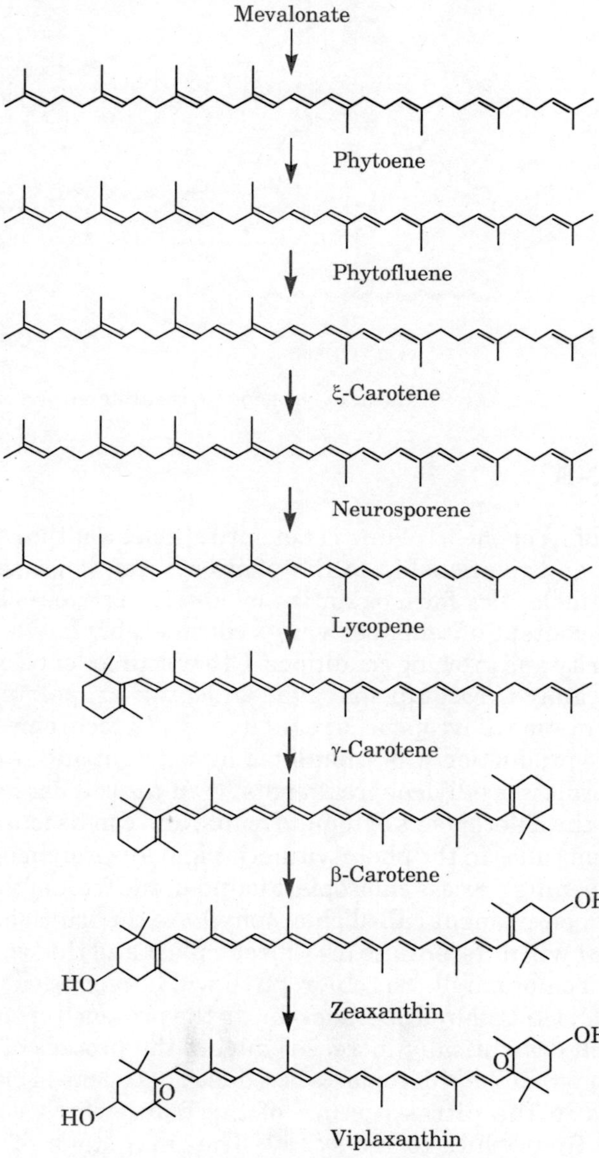

Fig. 5: Carotenoid biosynthesis pathway[77]

organisms[77,81]. The gene coding for lycopene cyclase was sequenced and cloned from cyanobacterium (*Synechococcus* sp strain PCC 7942) and named CrtL[82]. The same gene was cloned from tomato[83], also named CrtL. A gene called CrtI has been cloned from *Erwinia uredovora* which codes for bacterial phytoene desaturase, which desaturates phytoene to lycopene in one step, bypassing ξ-carotene desaturase. It is this CrtI gene that was over expressed in tomatoes to produce fruit with elevated levels of β-carotene[84,85] (Fig. 6).

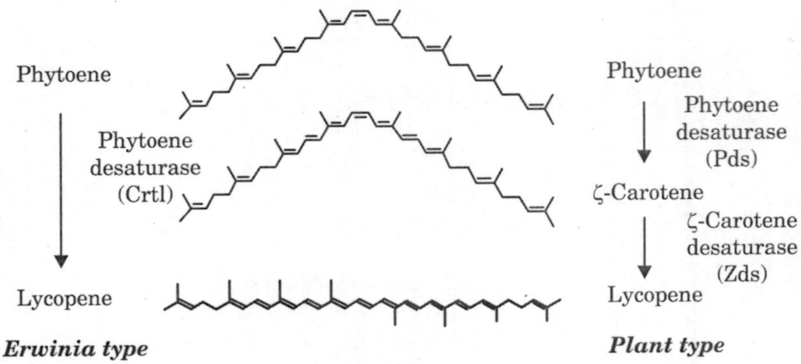

Fig. 6: Synthesis of lycopene (ψ, ψ-carotene)

V. LYCOPENE

The function of lycopene in plants is not entirely clear, although because of its proximity and relationship to the photo system, it probably protects photo system molecules form oxidation by absorbing excess blue light[86]. The lycopene content of tomatoes varies considerably between cultivars, stage of maturity and growing conditions. Other sources of lycopene include watermelon, guava, rosehip, papaya, and pink grapefruit[87,88,89,90,91] (Table 1). Formation of lycopene is regulated by phytochromes in the fruit, since lycopene production was stimulated by red light and reversed by far red light, regardless of ethylene treatments[92]. At the cellular level, lycopene is localized in the chloroplasts of tomato fruits, and can be found among the thylakoid membranes in the photosynthetic pigment-protein complex[93, 94]. During the ripening process chloroplast found in the fruit of the tomato are high in the green pigment called chlorophyll. As the fruit begins to ripen the chloroplast will differentiate into chromoplast and this will trigger the degradation of chlorophyll. The chlorophyll will be degraded into colorless tetrapyroles[95]. Chlorophyll degrades due to the presence of chlorophyllase activity, which dramatically increases during the process of ripening[94]. When chlorophyll in the chloroplasts is reduced, lycopene is biosynthesized with changes in the ultrastructure of the fruit, which results in the color change from white to red[96,97,98]. The final stage of chromoplast development is the formation of lycopene crystals that occupy a large portion of chromoplasts and appear as volumnous red sheets in the chromoplasts[99,100,101,102,103,104].

The red color of the tomato is due to its major carotene, lycopene, which is present at levels up to 90% of the total carotenoids[105]. A range of other carotenoids is commonly reported including β-carotene, δ-carotene, γ-carotene and neurosporene[88]. Lycopene epoxide, an oxidation product of lycopene, was reported as the most predominant carotenoid in tomatoes[56, 106]. The common red tomato also contains the colourless

Table 1: Approximate lycopene content of various foods (mg/100 gm wet weight)[87, 88, 89, 90].

Food	*Lycopene content (mg/100 gm wet basis)*
Tomatoes, fresh	0.88–20
Tomatoes, cooked	3.70
Tomato sauce	5.46–6.20
Tomato paste	5.40–150.00
Tomato soup (condensed)	7.99
Tomato powder (drum/spray dried)	112.63–126.49
Tomato juice	5.00–11.60
Sun-dried tomato in oil	46.50
Pizza sauce (canned)	12.71
Ketchup	9.90–13.44
Apricot	<0.01
Apricot, canned	0.06
Pumpkin	0.38–0.46
Apricot, dried	0.86
Rosehip puree	0.68–0.71
Grapefruit	0.35–3.36
Guava, fresh	5.40
Guava, juice	3.34
Watermelon, fresh	2.30–7.20
Papaya, fresh	2.00–5.30

precursor's phytoene and phytofluene. Some tomato strains are orange because they do not synthesize lycopene or because other carotenes, such as β-carotene, predominate[107].

The optimum temperature for lycopene synthesis in tomato was 24°C[108] and severely impaired above 30°C or below 10°C[109,110,111]. Tomatoes matured in sunlight produced more red color (lycopene) than those which were shaded[112]. Tomatoes held at constant 22°C developed significantly more lycopene when exposed to any one of three types of flourescent light relative to dark controls[113]. Lycopene concentration in tomatoes was higher in summer (from June to August) and lower in winter (from October to March)[114]. Tomato fruits grown in the greenhouse either in summer or winter are lower in lycopene content than fruits produced outdoors during summer, and fruits picked green and ripened in storage are substantially lower in lycopene than vine-ripened fruits[105]. High temperatures (38ºC) inhibited lycopene production while low temperatures inhibited both fruit ripening and lycopene production[115]. The rates of synthesis of lycopene and β-carotene at favourable temperatures (22-25ºC) can be increased by illuminating tomato plants during the ripening of the fruit. It is difficult to know which parameters caused these results, but they may include the level of intercepted light, and the high temperatures that occur under protected growth conditions[68].

The lycopene content of fresh tomatoes can vary up to 18 mg/100 g fw, but most values for typical red tomatoes are between 5-8 mg/100g fw[68, 116]. Some studies showed that lycopene content in whole tomato fruit was more than 9.27 mg/100g. Some deep-red varieties contain more than 15 mg/100g, whereas the yellow varieties contain only about 0.5 mg /100g[116, 117]. Field-grown tomatoes appear to contain higher levels of lycopene, ranging from 5.2 to 23.6 mg/100g fw[56, 118, 119], than greenhouse-grown tomatoes, which are reported to contain between 0.1 and 10.8 mg/100g fw[120]. The lycopene content of vine-ripened tomatoes was 33% higher than when ripened off the vine[121] and lycopene values for fresh tomatoes range from 8.8 to 42.0 mg/100g fw[122].

The lycopene content of the tomatoes varies considerable with the stages of maturity and growing conditions[18,123,124,125,126]. A number of researchers have shown that significant differences in lycopene content occur between cultivars. It has also been noted that cultivars with smaller fruits, such as the cherry tomatoes, appear to have higher concentrations of antioxidants than larger tomatoes[120,128, 129]. It has been hypothesized that this is due to the fact that they possess a larger surface area to weight ratio, which is important as lycopene reportedly concentrates in the pulp close to the skin of the fruit. A wild relative of the domestic tomato, *Lycopersicon pimpernellifolium*, produces tiny, currant-like fruit that are said to contain 40 times more lycopene than regular tomatoes, and is therefore a prime candidate for use in the breeding of new hybrids. Researchers have also isolated a crimson gene and found higher lycopene content in varieties carrying this gene than in those without it[57]. A genetically modified cultivar purportedly contained twice the lycopene level of conventional cultivars[130]. Gomez *et al.*[118] grew 15 different cultivars (standard red type) under the same conditions and fruit harvested at optimum ripeness had lycopene levels that varied from 5.04 to 13.46 mg/100g fw. Tomato skin contains 1.2 mg lycopene/100g (wet basis), while whole mature tomato contains only 3.4 mg lycopene/100g (wet basis). Therefore, the concentration of lycopene in tomato skin is about three times higher than in whole mature tomatoes and skin and pericarp of tomato fruits are rich in lycopene[127,131,132]. Researchers reported that tomato skin is a rich source of lycopene, as they contained about five times more lycopene (53.9 mg/ 100 g) than the whole tomato pulp (11 mg/100 g). This indicates that most of the lycopene is found attached to the insoluble fiber portion of the tomatoes[133].

Traditional analytical methods for assessing internal properties of tomatoes usually require destructive procedures to extract particular components and then analyze the same. The MRI technique provides a non-destructive means to assess both the qualitative and the quantitative properties of materials containing protons. Because MRI can accurately display internal characteristics without destroying the sample, it is often used in the assessment of biological materials and in medical diagnosis. Many researchers have used MRI to evaluate internal quality and physiological changes of fruit and vegetables[134,135,136,137,138]. Chemical shift imaging (CSI) and high performance liquid chromatography (HPLC) was used to analyze the internal changes of sugar and lycopene contents. Bruker 7T MRI system using the gradient-echo sequence for scanning was used for imaging experiments on tomatoes during the ripening. The CSI results indicate that there was no consistent trend in the variation of sugar signal intensity during ripening while the lycopene content significantly increased in the outer pericarp and columella of the tomatoes at the red ripe stage, as compared to the mature green stage. The result of sugar and lycopene signal intensities was further confirmed by HPLC analyses[139]. Carotenoids exhibit two strong Raman peaks in two separate spectral regions, 1100-1200 and 1400-1600/ cm, due to the stretching vibrations of the carbon-carbon bonds in the polyene chain[140]. Bench-top point-scan Raman chemical imaging system has been used to detect and visualize internal lycopene distribution during postharvest ripening of tomatoes. Hyperspectral Raman images were acquired from fruit cross-sections in the wavenumber range of 200-2500/ cm with a spatial resolution of 1 mm. At different ripening stages Raman chemical images were created to visualize the spatial distribution of the lycopene. The methods and findings of this study form a basis for developing a Raman-based nondestructive method to detect internal lycopene as a maturity indicator for tomatoes[141].

VI. LYCOPENE ISOLATION

Lycopene (ψ, ψ-carotene) (Fig. 7) is one of over 600 carotenoids found in nature. It accumulates in relatively few tissues, and responsible for a characteristic red pigmentation of the ripe tomato, watermelon, pink grapefruit, rose hip etc.[87,88,89,90] (Table 1). Carotenoids are not water soluble owing to its non polar nature, but dissolve in many organic solvents. Because of conjugated diene double bonds, some carotenoids also act as chromophores, refracting light in different wavelengths to produce various colors in the red-orange-yellow spectrum. β-carotene gives carrots a yellow/orange color, while lycopene confers red color to tomatoes. Lycopene is lipid soluble, and usually extracted with organic solvents such as chloroform, hexane, acetone, benzene, petroleum ether, or carbon disulfide, prior to chemical analysis for quantitative determination. In cases where solvent extraction may be slow and incomplete, the efficient mechanical grinding of the material can be used to facilitate extraction. Exposure of extracted lycopene to light should be avoided and only gold, yellow[142] or red light should be used. Antioxidants such as butylated hydroxy toluene (BHT) should be used in extraction and separation solvents to control oxidation and isomerization reaction of lycopene[90].

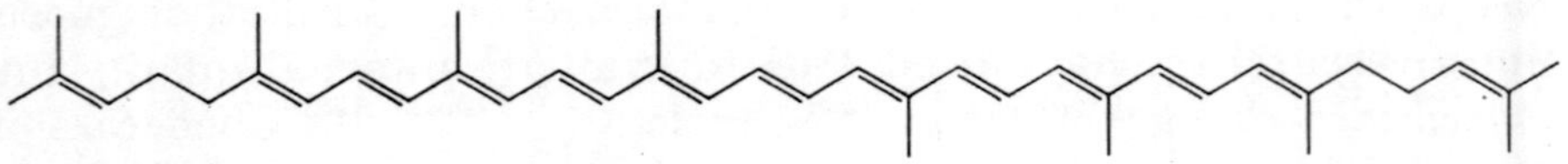

Fig. 7: Molecular structure of lycopene (ψ, ψ-carotene)

Different methods may be used for the analysis of carotenoids in food products: HPLC, spectrophotometry or color evaluation[143]. A traditionally spectrophotometric measurement has been used for the determination of lycopene concentration in tomatoes. The use of hexane and acetone is common for the extraction of lycopene from tomato tissue and then measurement of the OD at 460 to 470 nm[144,145]. A pure lycopene sample is necessary for the preparation of calibration curves. Although spectrophotometry or colorimetry can be used to rapidly assess the lycopene content of products, but for reliable analysis a highly versatile, sensitive and selective method such as HPLC is required. HPLC analysis of carotenoids is usually done with C_{18} or C_{30} RP-columns, operated with isocratic or gradient elution with a wide variety of mixtures of different organic solvents as mobile phases, using UV–*vis* (g ≈ 450 nm) or photodiode array or MS detection[146,147,148,149,150]. Variations in the properties of silica packing material in terms of particle size, porosity, carbon load, end-capping technique, and polymerization can greatly affect the sensitivity and selectivity and selectivity of lycopene analysis[151]. Heating of column is sometimes used to improve separation and standardization conditions of pigment[143]. C_{30} stationary phase is employed to achieve superior selectivity of lycopene isomers compared to conventional C_{18} reversed-phase and silica normal-phase columns[151,152]. A polymerically synthesized C_{30} column provides excellent separation of all-*trans* lycopene isomers from the *cis* counterpart[152, 153]. Identification and structure elucidation of isomeric carotenoids have been facilitated with the aid of high-resolution NMR spectroscopy. 15 (E/Z)-isomeric forms of lycopene can be characterized with H- and C-NMR, UV/VIS, mass, and IR spectroscopy[154].

Different systems can be used for the extraction of carotenoids from the samples: like liquid-liquid extraction, solid phase extraction or supercritical fluid extraction[146, 148,155]. For quantitative determination of lycopene various organic solvents such as ethyl acetate, ethanol, acetone, petroleum ether, hexane, benzene, chloroform, etc, can be used for its extraction[131,133,147, 149,156]. AOAC[157] recommends methanol/tetrahydrofuran (50:50 v/v) for extracting the carotenoids. Extraction of lycopene and β-carotene has also been done with methanol (100%), ethyl ether (100%), tetrahydrofuran (100%), methanol/tetrahydrofuran (92:8 v/v), methanol/ tetrahydrofuran (50:50 v/v) and hexane/acetone/ethanol (50:25:25 v/v/v)[158]. A mixture of hexane with acetone and ethanol or methanol has been used for lycopene extraction because other components such as diethyl ether and tetrahydrofuran might contain peroxides that react with carotenoids[156,158,159,160]. The stability of lycopene extracts obtained with hexane/acetone or hexane/ethanol was higher than that of extracts obtained with other organic solvents such as chloroform, methanol or dichloromethane[161].

Supercritical fluid extraction is an advanced separation technique based on the enhanced solvating power of gases above their critical point. One of the most frequently used supercritical fluids is carbon dioxide. Use of supercritical fluid technology for industrial production of lycopene starting from very economical and natural sources (*e.g.* Surplus of tomato production and tomato processing by-products) provides an opportunity for functional food development and lycopene products in the global market. Recent studies have described a lycopene extraction process based on supercritical CO_2, which enables the extraction of over 60% of lycopene from tomato waste. Lycopene and β-carotene from the pulp and skins of ripe tomatoes was isolated by supercritical carbon dioxide extraction. The extractions were conducted at pressures (2500-4000 psi) and temperatures (40-80ºC). Lycopene (65 %) and β-carotene (35%) were obtained at 4000 psi and 80ºC. Lycopene and β-carotene showed a different solubility in the supercritical fluid depending on process parameters. With a proper choice of operative parameters, it has been possible to obtain a product that contained 87% lycopene and 13% β-carotene[162]. Rozzi *et al.*[155] also extracted lycopene from tomato by product using supercritical fluid extraction at temperatures (32-86°C) and pressures (2000-7000 psi). The percentage of lycopene extracted increased with elevated temperature (86°C) and pressure (34.47 psi), after which the amount of lycopene extracted decreased. These conditions resulted in the extraction of 61% of the lycopene. Gomez-Prieto *et al.*[163] proposed a procedure for the supercritical fluid extraction of all-*trans* lycopene from freeze dried tomato pulp and skin using carbon dioxide at pressure ranging from 32.5 to 53.2 MPa at temperature of 40°C and minimizes the risk of degradation *via* isomerization and oxidation of lycopene. The highest total lycopene recovery (42%) was achieved at 53.2 MPa, resulting in 88% all-*trans*-lycopene and 12% *cis*-lycopene. Similar type of study was done by Ciurlia *et al.*[164] by using innovative SC-CO_2 procedure for lycopene extraction from dried tomatoes (pulp and skins without seeds). The pigment was extracted in hazelnut oil at temperature of 60°C, pressure of 400 bar, flow rate of 10 kg CO_2/h. The extraction yields were: 72.5% for lycopene, 80% for hazelnut oil. The addition of centrifugation step separates the lycopene into its isomeric forms: *all-trans* lycopene was collected as precipitate, *cis*-lycopene remained in solution in the oil phase. After centrifugation, the obtained lycopene was five-fold more concentrate than the starting solution (4.12% vs. 0.74%, w/w) and the extraction yield was 72.5%.

Higher lycopene yield from tomatoes can be obtained with minor degradation and isomerization by using ultrasonic assisted extraction (UAE) with response surface methodology (RSM). The yield of *trans* lycopene was

75.93% compared to optimized conventional method of extraction and the purity of total-lycopene obtained was 98.27 ± 0.52% with β-carotene constituted 1.73 ± 0.52% of the extract[165].

The instability of lycopene during processes of extraction, handling, and elimination of organic solvents makes the preparation of a sample for analysis an extremely delicate task, often requiring successive and complex procedures to ensure that all the carotenoids are extracted. Besides, not all the analytical methods available for carotenoid analysis in food products are suitable for lycopene-rich foods due to the low solubility of lycopene in some of the solvents employed as in the case of methanol and due to the fact that the use of other solvents may interfere with the mobile phases applied for carotenoid separation[143]. There are many HPLC methods that can be applied to the determination of carotenoids. However, this kind of compounds needs a very careful and tedious manipulation due to their chemical labiality. Therefore the development of new methodologies of extraction/separation is of relevance and the necessity for a reliable and rapid analysis method for lycopene in vegetable products has been recognized[166].

VII. PHYSICAL AND CHEMICAL PROPERTIES OF LYCOPENE

Lycopene, an acyclic isomer of β-carotene, is a 40-carbon ($C_{40}H_{56}$) aliphatic, highly unsaturated molecule containing 11 conjugated arranged in a linear array, making it longer than any other carotenoid and two unconjugated double bonds[167] (Fig. 7). Two central methyl groups are in the 1, 6 position, while the remaining methyl groups are in the 1, 5 position relative to each other. A series of conjugated double bonds constitutes a chromatophore of variable length. The long chromophore in the polyene chain accounts for the red color of lycopene and also for its powerful antioxidant activity[159]. Since lycopene lacks the β-ionone ring structure, it is devoid of provitamin-A activity. Lycopene can be readily cyclised to β-carotene by the plant enzyme lycopene-β-cyclase (lcy-b)[168, 169].

The antioxidant activities of lycopene and other carotenoids are related to their abilities to quench singlet oxygen (O_2) and to trap peroxyl radicals ($ROO^{\cdot}$)[170, 171]. It is generally believed that an increased number of conjugated double bonds and the opening of the β-ionone ring increase the quenching ability towards singlet oxygen[172]. Lycopene has been reported to deactivate an array of free radicals, such as hydrogen peroxide, nitrogen dioxide, thyol, and sulphonyl[173,174,175]. The quenching constant of lycopene was found to be double than that of β-carotene and 100 folds that of α-tocopherol[172, 176, 177, 178, 179] (Table 2).

In ripe tomato fruits, lycopene exists as elongated, needlelike crystals having molecular weight of 536.85 Da and melting point of 172-175°C.

Table 2: Comparison of the antioxidant activities (quenching of singlet oxygen) of different antioxodants[172, 176, 177, 178, 179]

Carotenoid	*Rate constant for quenching of singlet oxygen Kq × 10^9 ($mol^{-1}s^{-1}$)*
Lycopene	31
γ –Carotene	25
Astaxanthin	24
Canthaxanthin	21
α-Carotene	19
β-Carotene	14
Bixin	14
Zeaxanthin	10
Lutein	8
α-tocopherol	0.3

Lycopene is more soluble in chloroform, benzene, acetone and other organic solvents than in water[167,180]. Because of the highly conjugated nature of lycopene, it is particularly subject to oxidative degradation and isomerization. Chemical and physical factors known to degrade lycopene including exposure to light, oxygen, elevated temperature, extremes in pH, and active surfaces and metallic ions such as Cu^{2+}, Fe^{3+} catalyze its oxidation[181, 182, 183].

In nature, lycopene exists in *all-trans* form and seven of these bonds can isomerizes from the *trans*-form to the *mono* or *poly-cis* form under the influence of heat, light, or certain chemical reactions. As a result of the 11 conjugated carbon-carbon double bonds in its backbone, lycopene can theoretically be arranged in 2048 different geometrical configurations. Although a large number of geometrical isomers are theoretically possible for *all-trans* lycopene, but only certain ethylenic groups of a lycopene molecule could participate in *cis-trans* isomerization because of steric hindrance[184,185] (Fig. 8). Interconversion of isomers is thought to take place with exposure to thermo energy, absorption of light, or by involvement in specific chemical reactions. *Cis* isomers of lycopene have chemical and physical characteristics distinctly different from their *all-trans* counterparts. Some of the differences observed as a result of a *trans*-to-*cis* isomerization reaction include lower melting points, decreased color intensity, a shift in the lambda max, smaller extinction coefficients, and the appearance of a new maximum in the ultraviolet spectrum[186]. Stereoisomeric forms of lycopene were described with special reference to the properties of light absorption in relation to their molecular structures[187].

A is the predominant structure found in plants (95%). Structures B, C, E and G are found in human serum, making up approximately half the total lycopene content. Structure H is the naturally occurring form of lycopene in fresh Tangerine-type tomatoes[188]

all-*trans* lycopene
15-*cis* lycopene
13-*cis* lycopene
11-*cis* lycopene
9-*cis* lycopene
7-*cis* lycopene
5-*cis* lycopene
7,9,9′,7′-*cis* lycopene

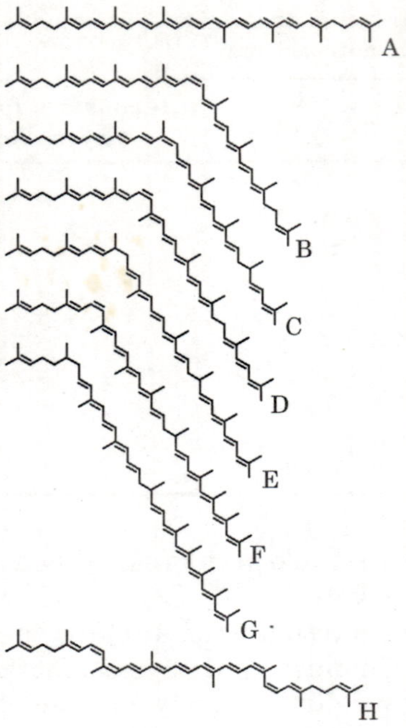

Fig. 8: Common lycopene isomers structure

VIII. EFFECT OF PROCESSING ON STABILITY OF LYCOPENE

Various physical and chemical factors such as time, temperature, reaction medium, environmental conditions etc. affect the stability of lycopene in food matrix. On the other hand, the three most important factors which affect the chemical stability of lycopene are heat, light and oxygen. Stability of the lycopene is also affected by the packaging material[189]. Processing of tomatoes with processing techniques either hot/cold break or super cold break methods resulted in 9-28% loss of total lycopene[90,190]. Both high pressure and thermal treatment have harmful effect on stability of carotenoids[191]. Heat concentration of tomato pulp resulted in up to 57% loss of lycopene[192]. Spray drying had more detrimental effect on lycopene content as compare to 6-8% loss of lycopene in air, vacuum or osmotic vacuum drying of tomato pulp. 8.07 to 20.93% of lycopene loss was reported in spray drying of tomato pulp. This loss was due to high air inlet temperature and increased compressed air flow rate[193]. Lycopene generally undergo two type of changes during processing and storage; isomerization from *all-trans* to *mono* or *poly-cis* forms initially followed by auto-oxidation of unsaturated bonds[194]. Lycopene in fresh tomatoes exists predominately in the *all-trans* form configuration. Carotenoids undergo trans-*cis* isomerization during thermal processing of food (Table 3). The *cis*-isomers increase with temperature and processing time. As lycopene is highly unsaturated, it is susceptible to oxidation. During the thermal processing it has been observed that isomerization of lycopene from *all-trans* to *mono* or *poly-cis* form occurs due to changes in the 7 of conjugated double bonds. On storage no further *trans-cis* isomerization takes place instead *cis-trans* reversion are observed mainly, followed by the depletion of *cis* isomers and then by *all-trans* auto-oxidation[195].

Table 3: Lycopene isomers in various thermally processed tomatoes products[90]

Sample	*Total lycopene (mg/100 g, dry basis)*	*Cis isomers (%)*
Peeled tomato	149.89	5.37
Tomato juice (hot-break)	161.23	5.98
Tomato juice (retorted)	180.10	3.56
Tomato (whole, retorted)	183.49	3.67
Tomato paste (concentrated)	174.79	5.07
Tomato paste (retorted)	189.26	4.07
Tomato soup (retorted)	136.76	4.34
Tomato sauce (retorted)	73.33	5.13

The greater percentage of lycopene loss as compared to the gain in the *cis*-isomer form suggests that oxidation of lycopene was the main mechanism for lycopene degradation during both heating and illumination treatments[196]. Lycopene is also susceptible to oxidation during the commercial frozen storage. Degradation of total lycopene in the frozen tissues is due to the oxygen and formation of ice crystals[197]. The decrease in lycopene content during spray drying was due to an actual degradation of lycopene, rather than to a progressive conversion from the all-*trans*-lycopene to a less strongly colored, less intensely absorbing *cis* form. It was previously suggested that the first stage of lycopene degradation during drying and storage of tomato powders is the reversible isomerization of all-*trans*-lycopene to less colored, more oxidizable *cis* isomers. *Cis-trans* reisomerization may take place due to the return of molecule from a high energy state to stable or ground state as introduction of *cis*-bonds twists and contracts the molecule making it energy richer and more unstable. Autooxidation of lycopene is irreversible and will lead to fragmentation of the molecule, producing acetone, methylheptenone, laevulinic aldehyde, and probably glycoxal also, which causes apparent color loss, and typically hay or glass-like odours evolve[198]. Photosensitized oxidation with 500 W incandescent of purified lycopene from tomato puree produced 2-methyl-2-hepten-6-one and apo-62-lycopenal during 40 hrs of exposure[199], and oxidation with hydrogen peroxide and *m*-chloroperbenzoic acid produced lycopene-1, 2-epoxide and lycopene-5, 6-epoxide[174, 200]. Autooxidation of lycopene lead to the formation of number of cleavage products like acycloretinal apo-142-lycopenal, apo-122-lycopenal, apo-102-lycopenal, apo-82-lycopenal, and apo-62-lycopenal etc. These products are formed by the cleavage of 11 conjugated double bonds of lycopene[201]. The stability of lycopene is about 0.2 g/L at room temperature[202]. Determination of the degree of lycopene isomerization during processing would provide a measure of the potential health benefits of tomato based foods. Degradation of lycopene due to isomerization and oxidation affects the bioavailability and decreases its functionality for health benefits. Studies dealing with extraction and purification of lycopene warn of its liability and advise a range of precautions from low light to gold light to freezing temperatures to prevent degradation during extraction or storage[203,204,205].

(a) Effect of Heat

The degradation kinetics and isomerization of lycopene in water and oil based tomato model systems at thermal (80-140°C for 1-4 hours) and light irradiation treatments (1-6 days) were studied. An increase in temperature or irradiation intensity and in exposure time results in increased degradation and isomerization of lycopene. The stability of lycopene in oil based tomato products favored at temperature of 80 and 100°C where as increased isomerization of lycopene and resulting in degradation of total lycopene and *cis*-isomers in both water- and oil-based tomato products were favored at high heating temperatures (120°C and 140°C). However, the levels of degradation of total lycopene content and *cis*-isomers were greater in water-based samples than in oil-based model systems under different treatments[196]. Heat and light both promoted lycopene isomerization of the all-*trans* form to the *cis*-isomers and further oxidation of *cis*-isomers. Thermal and light irradiation results in lycopene degradation *via* isomerization initially and followed by auto-oxidation of unsaturated bonds. Cherry tomatoes were dehydrated by a combination of different dehydrating techniques (osmotic dehydration, convective drying, and microwaves assisted air drying) to assess the effect of the process variables on the degradation and isomerization of lycopene. With convective and microwave drying the loss of total lycopene and the *trans-cis* isomerization, mainly to the 13-cis form, were favored, where as osmotic pre-treatment limited the isomerization during the later stage of drying[206].

When subjected to different processing treatments like baking (177°C and 218°C, microwave cooking (1000 W), and frying (145°C and 165°C) significant decrease in lycopene content of tomato was observed as lycopene is not stable when exposed to cooking temperatures above 100°C. Higher temperature encountered during baking and frying causes greater lycopene losses where as microwave cooking resulted in lower lycopene losses due to shorter exposure and lower temperature treatment. Formation of hydroperoxide free radicals at higher frying temperature also accelerate the degradation of lycopene, which was clear from the very high losses observed in lycopene during frying as compared to microwave cooking and baking of tomato slurry[207]. Lycopene starts degrading at a temperature as low as 25°C through oxidatioin[208]. Isomerization occurs at a very low rate at the low temperatures and as the temperature increased above 75°C the isomerization of lycopene also increased[208,209]. Thermal stability and isomerization of lycopene in oleoresins from tomato were studied at 25, 50, 75 and 100°C in the dark. As the storage temperature increased from 20 to 100°C, the degradation of total lycopene in oleoresin increased. At lower temperature lycopene may degrade mainly through isomerization without oxidation. Isomerization of lycopene in tomato oleoresins increased at 75 and 100°C[208]. The stability of lycopene during heating (50, 100, 150 °C) and illumination (2,000 to 3,000 lux) at 25°C was studied for varied lengths of time. At 50°C no significant change in all-*trans* isomer was reported during 12 hours of heating, however, the content begins to decline thereafter. At the higher temperature all-*trans* isomers decreased and the level of formation of mono-*cis* isomers was reported high during the initial phase of heating and then depletion of mono-*cis* isomers and isomerization to di-*cis* isomers increased with increased heating time which is then followed by degradation of the molecule. Degradation was favored at higher temperature instead of isomerization where as, during the illumination for 32 hours isomerization was mainly favoured[210] (Fig. 9). Encapsulation of lycopene by spray drying was resulted in increase in lycopene purity from 96 - 98%. The increase in purity of lycopene was due to reduction of 0.7 % cis isomers (9-*cis* & 13-*cis*) and 0.3 % of lycopene 1, 2-epoxide. Small amount of 15-*cis*-lycopene was only detected before spray-drying[211] (Fig. 10).

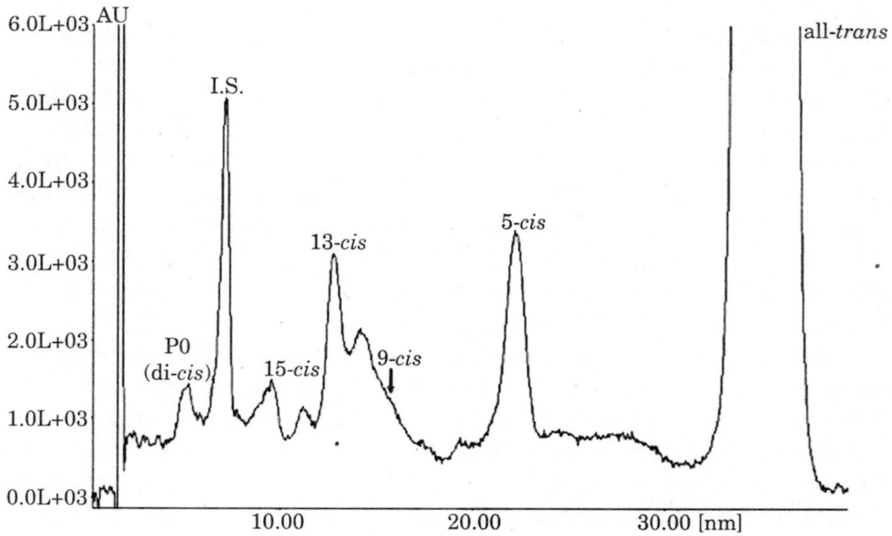

Fig. 9: HPLC chromatogram of lycopene and its *cis* isomers after illumination for 32 hour using C30 column eluted with 1-butanol-acetonitrile–methylene chloride (30:70:10, v/v/v) as mobile phase with a flow rate of 2.0 ml/ min at 476 nm[210].

Sterilization and pasteurization of tomato juice resulted in 35 and 50% decrease of lycopene[212]. Less than 10% loss of lycopene has been reported during the production of semidried tomato at 48°C[213]. However, some studies found that lycopene is increased during thermal processing of tomato. An

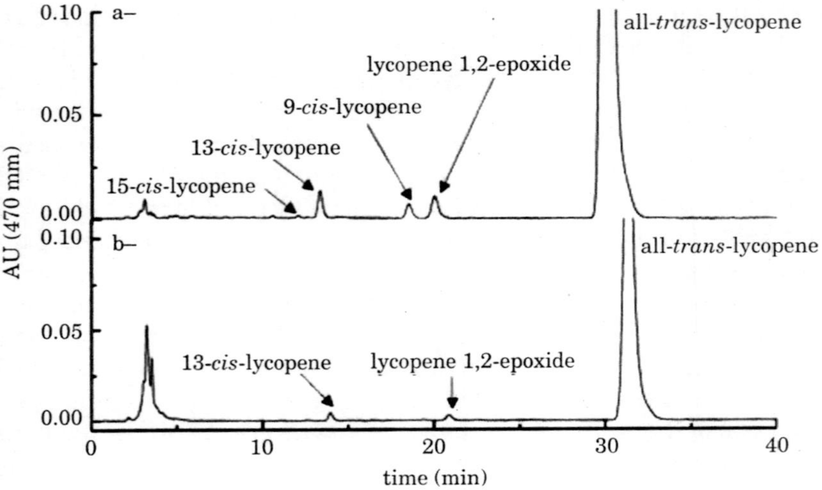

Fig. 10: Chromatogram obtained by HPLC of lycopene before (a) and after (b) spray drying encapsulation. Chromatographic conditions: YMC C_{30} column (3 µm, 250 × 4.6 mm) with MeOH (0.1 TEA)/TBME (1:1) as mobile phase at 1 ml/min and column temperature set at 33°C[211].

increase of *trans*-lycopene was found during 30 minutes heating at temperature of 88°C[214]. Shi and Maguer[159] reported an increase of total *cis*-isomers of lycopene from 0 to about 1.75 mg/kg of tomato puree after heat treatment at 90°C for 2 hours. The microwave cooking resulted in relatively higher percentage of retained lycopene because the tomato slurry reached to its boiling point after only 60 seconds of microwave cooking and only 35% of lycopene was decreased. For baking the percentage of retained lycopene for the samples baked at 177°C for 15, 30 and 45 mins, were 64%, 65.6% and 37.3% compared to the 51.5 %, 41.3% and 21.1% for the samples baked at 218°C. The studies conducted on the degradation of lycopene shows that various factors for *e.g.*, enzymes, heating temperature, food additives etc. affect the stability of lycopene in the raw and processed products. Heating tomato juice[215] and preparation of a sauce from canned tomatoes[216] was shown to increase the level of *cis*-isomers of lycopene. However, common heat treatments during food preparation did not significantly alter carotenoid distribution in green vegetables[217]. Thermal treatment enhanced the formation of *cis*-lycopene from all-*trans* lycopene in tomato matrices as reported by several authors[90, 159, 218]. Lycopene undergo degradation *via* isomerization and oxidation under different heating exposures but at the same time, heating can enhance the extractability of lycopene from the tomato matrix. Cole and Kapur[198] studied the kinetics of lycopene degradation by the effect of oxygen, temperature and light and concluded that the oxidative degradation of lycopene at 50°C led to the formation of oxidation products. There was significant loss of lycopene in serum-free tomato pulp at 100°C in the presence of oxygen, with or without light. The intensity of illumination and temperature were found to be in direct correlation with lycopene degradation in the presence of oxygen[219].

(b) Effect of Light and Oxygen

The tomato puree was exposed to temperature range of 60-120°C, 1-6 hours and to visible light for 1-6 days to observe the effect on lycopene content. When exposed to light no significant difference was found for total and all-*trans* lycopene. Isomerization of lycopene is favored at 60 and 80°C where as lycopene extraction was more favorable at the higher temperature range. The *cis*-isomer lycopene reduction was statistically significant and the percent changes were 30% lower than the untreated samples. Therefore, *cis*-isomer of lycopene is less stable than all-*trans* lycopene under the light exposure[220]. Lycopene standard was heated at 50-150ºC and illuminated (2,000 to 3,000 lux) at 25ºC for varied lengths of time. The content of *all-trans*-lycopene was found to decrease with increasing illumination time after a 144 hour exposure to light it amounted to a 94% loss. The result indicated that the isomerization reaction was favored at the beginning of the processing and then the degradation dominated after prolonged heating at 150°C. All the *mono-cis* isomers of lycopene showed inconsistent changes. Comparing the results, apparently with increasing temperature and heating time, degradation dominated over isomerization and degradation of lycopene and its *cis* isomers, during illumination, might proceed simultaneously[210]. Degradation of total amount of lycopene during illumination at 25°C for 144 hour follows the first order reaction. Tomato powder was subjected to three treatments: light exposure at room temperature, 6°C, and 45°C in the dark. The amount of *cis* isomers as a percentage of the total lycopene increased to the 14-18 % range, regardless of the storage conditions[221]. Shi *et al.*[222] studied the effect of irradiation on canola oil at 2010 (Outdoor), 900, 650 and 140 (Indoor) $\mu mol/m^2 s^{-1}$ for 1-6 days and found that as the intensity of light irradiation increased there was significant decrease in the amount of total lycopene as well as in *trans* lycopene, whereas level of *cis* isomers found to be increased. Results also showed that irradiation caused more loss in lycopene than heating treatments at 25°C, 100°C and 180°C and the rate of *trans* isomer degradation was greater then the *cis*-isomers formation in both the experiments. Lycopene extraction should be done in gold, yellow and red light and its exposure should be avoided to visible light[142].

As discussed earlier heat, light and oxygen are the three potent factors, leads to the degradation of carotenoids and loss of their biological activity which is very important from health point of view. Tomato juice powder packed in air-packed packaging material retained the maximum lycopene content as compared to CO_2^-, N_2^-, or vacuum-packed samples[223]. But later on it was found that vacuum and dark storage combination gave the lowest lycopene loss in tomato pulp[133]. The effect of oxygen on the degradation of carotenoids in an aqueous system was studied by Henry *et al.*[183]. There was 90% loss of lycopene after 2 hrs whereas other carotenoids did not show any comparable loss even 7 hours of exposure. Ax *et al.*[224] studied the stability of lycopene by dissolving it in the oil phase of oil-in water emulsions, under oxygen saturation or oxygen-free conditions. At 25°C, about 25% of lycopene was degraded within 30 hours in the oxygen- removed emulsions, whereas a lycopene loss of about 80% was found for oxygen saturation. Lycopene destabilization was about three times higher in the presence of oxygen than under inert conditions. Removal of oxygen from water by flushing the system with nitrogen did not produce encouraging results. However, complete exclusion of oxygen using enzymes such as glucose oxidase led to significant improvement in lycopene stability[225]. Nitrogen or argon head space can also be provided to improve the stability of lycopene by keeping the exposure to atmospheric oxygen to a minimum[188].

(c) Effect of Processing

Lycopene is important not only because of the color it imparts but also because of the recognized nutritive and health benefits associated with its presence. Degradation of lycopene not only affects the sensory quality of the products but also their nutritive value. Processing dehydrates the product, resulting in higher concentration of lycopene. Minimizing the loss of lycopene throughout the processing and during storage has always been important. One study determined that the isomeric configuration of lycopene was not significantly different from 95% *trans* in raw tomatoes, paste or juice[226]. A different study concluded that heating tomato products changed *trans* lycopene to *cis* lycopene, a more readily oxidized and bioavailable form, and thus more beneficial[218]. Lycopene is more stable in native tomato fruit tissues and matrices because of its all-*trans* configuration than in isolated or purified form[227]. The degradation of lycopene in tomato based products during processing are due to isomerization and oxidation. Isomerization converts all-*trans* isomers to mono, di or poly-*cis*-isomers due to additional energy input and results in an unstable, energy-rich molecule. Determination of the degree of lycopene isomerization during processing would provide a measure of the potential health benefits of tomato-based foods. Heat induces isomerization of the all-*trans* to *cis* forms. The *cis*-isomers increase with temperature and processing time. In general, dehydrated and powdered tomatoes have poor lycopene stability unless carefully processed and promptly placed in a hermetically sealed and inert atmosphere for storage. A significant increase in the *cis*-isomers with a simultaneous decrease in the all-*trans* isomers can be observed in the dehydrated tomato samples using the different dehydration methods. Frozen foods and heat-sterilized foods exhibit excellent lycopene stability throughout their normal temperature storage shelf life[159,228]. Researchers showed that cooked tomato or tomato products have more bioavailable lycopene than the raw fruit[226, 229]. This may be attributed to its release from the plant tissue matrix, weakening of lycopene-protein complexes resulting from heat-induced cellular disruption and heat-induced *trans*-to *cis*-isomerization.

Tomato products of varying moisture and fat content, container type and thermal treatments during food preparation and industrial food processing operations do not result in significant losses of lycopene or a shift in the distribution of *cis*-lycopene isomers[90]. Dehydration of tomatoes at a mild temperature results in the conversion of *trans* to *cis*-isomers without cause significant losses in total lycopene content. Kinetics of lycopene degradation during spray drying of tomato pulp follows a first-order reaction with a reaction rate constant dependent on product moisture content, in addition to temperature[230]. *Cis*-lycopene isomers, while present at low levels in raw tomatoes, may have increased during spray-drying of the powders[221]. *Cis* isomers were not detected in the fresh tomato samples, but a significant increase was found in the *cis*-isomers with simultaneous decrease in the all-*trans* isomers during the osmotic-vacuum- drying, vacuum-drying, and air-drying dehydration to produce dehydrated tomato samples[74]. Dehydration of tomatoes at mild temperature does not usually causes significant losses in total lycopene content, but the conversion of trans to *cis*-isomers always occurs in the dehydrated products[90]. Muratore *et al.*[231] studied the effect of partial dehydration at 40, 60 and 80°C for different lengths of time on lycopene content. The highest value of lycopene was found in the dehydrated sample at 80°C (76.4 mg/ 100 g dry matter), while the lowest value was found at 40°C (58.5 mg/ 100 g d.m.), confirming that in these samples lycopene was damaged by the length of the drying process. Shi *et al.*[74] analyzed the degradation of lycopene in tomatoes using different dehydration methods: vacuum-drying (55°C, 4-8 h), conventional air-drying (95ºC, 6-10 h), and osmotic-vacuum-drying (first dehydrated with an osmotic treatment at 25°C for 4 h, followed by vacuum-drying at 55°C for 4-8 h). During osmotic treatment, lycopene content remained essential constant[74]. After osmotic-vacuum-drying, total lycopene retention in tomatoes was greater than that using vacuum-drying. Maximum loss of lycopene was observed in conventional air-dried tomato samples, which was attributed to the influence of heat and oxygen. Tomato pulp was air-dried to two final moisture levels: dried pulp (DP) (moisture 9%, water activity 0.35) and intermediate-moisture pulp (IMP) (moisture 23%, water activity 0.7) and stored in airtight clear glass bottles in the dark for 5 months at 4, 20, and 37°C. All-*trans*-lycopene degradation was more significant in IMP and was inversely related to storage temperature, showing about 28, 38, and 75% loss after 5 months of storage at 37, 20, and 4°C, respectively. The result implied that IMP was more sensitive than DP toward lycopene degradation reactions[232].

Thermal processing enhances the nutritional value of small tomatoes (cv Corbarini) by increasing significantly the content of carotenoids without significant changes in the total flavonoids. The improvement of nutritional value in terms of total carotenoids was due to the presence of the peels which were not removed during the canning procedure[233]. The feasibility of minimal processing and modied atmosphere packaging (5%O_2+5%CO_2) under refrigeration were studied to preserve the color attributes and bioactive compounds of fresh-cut tomato from different cultivars. Neither the content of bioactive compounds (lycopene, vitamin C and phenolic compounds) nor the antioxidant capacity changed signicantly between whole and just-processed fresh-cut tomatoes. Minimal processing thus preserves the initial nutritional value by maintaining the main antioxidant compounds and color parameters of slices tomatoes for 21 days at 4°C[234].

Heating tomatoes at 80°C for 10 h had no effect on lycopene, and heating at 100°C only reduced lycopene 10% in the same amount of time[235]. No change in lycopene concentration was observed as fresh tomatoes were processed into paste in a commercial processing facility[56]. Processes like cooking, cooling and canning do not usually cause large changes in total lycopene content, but most lycopene can be converted from the all-*trans* form into the *cis* isomer. Lycopene is stable during heating and industrial treatment, and it also improves bioavailability of lycopene[218, 226]. One study demonstrated no change in lycopene content after tomatoes were cooked at 100°C for 16 min[57]. In fact, more benefits are realized from lycopene than processed tomato products.

(d) Changes in Lycopene During Storage

During storage light and oxygen have similar effect on lycopene as during processing and temperature plays crucial role on stability of lycopene during storage. The long chromophore in the polyene chain accounts for the red color of lycopene[159] and increases its affinity for singlet oxygen and radical scavenging capacity beyond the other carotenoids[221, 236]. These chemical properties are also responsible for the spontaneous isomerization and oxidation of lycopene outside tomato tissue matrices, resulting in the degradation of lycopene during tomato processing[159]. Minimizing the loss of lycopene throughout the production process and in storage is important. Color, total lycopene and all-trans lycopene in foam-mat-dried tomatoes were better retained at 20°C then at –10°C, 2°C and 37°C. Isomerization of *all*-trans to *cis* isomer is favored during the drying temperatures and reisomerization from *cis* to all *trans* isomers is favored at 20°C[195]. When high amount of *cis* isomers are present, lycopene oxidation during storage occurred faster because *cis* isomers are more readily oxidized then *trans* isomers. Akanbi and Oludemi[63] studied the stability of lycopene during processing and in-package storage conditions of tomato cultivars. Dried slices showed better retention of lycopene than in dried pulp. Comparison of the k-data (reaction rate constant) at common reference temperatures reveals that, lycopene degraded by three to six times faster at 40°C than at 29°C during storage. Sharma and Maguer[159] reported the kinetics of lycopene degradation in tomato pulp solids to be a pseudo first order reaction. Pureed watermelon had less ice crystal formation and lost less lycopene then small chunks held for a year at –20°C. 4-6%, lycopene was degraded during an initial freeze-thaw experiment. 30-40% lycopene loss was observed over a year's storage at –20°C and a loss of 5-10% over the same period at –80°C. Thus, it can be concluded that lower temperature is helpful to prevent lycopene loses[197]. Frozen tomatoes also lost less lycopene when kept at lower temperature. After 12 months storage at –20°C, the losses of carotenoids reached 36%, and of total lycopene 48% and at –30°C, the losses was less then at –20°C. The gradual reduction of carotenoids was accompanied by a change of colour from red to red with a yellow shade and therefore the colour of fruits seemed increasingly lighter[237,238]. However, the occurrence of enzymes in frozen tomato, especially of lipase, was certainly responsible for the deterioration of organoleptic traits. Tomato powder stored at 45°C for 6 weeks encountered lycopene loss up to 30-60% due to *trans*-cis isomerization. Gassy off odors developed during the storage of dried tomato pulp. These oxidative changes are due to the auto-oxidation of lycopene to ketone and aldehyde. Rate of loss of pigment dependent on temperature and light conditions under storage[221]. Cold and frozen conditions also negatively affected the stability of lycopene in tomato powder and pulp. 30% loss of lycopene in tomato powders were reported when stored at –10, 2 and 6°C in the dark[195, 221]. Similar types of the finding were reported for the frozen tomato pulp under different frozen storage conditions[239].

IX. LIPID OXIDATION

Antioxidant when present at low concentrations compared to those of an oxidizable substrate significantly delays or prevents oxidation[240]. Pro-oxidant can be defined as a component, such as a metal ion, able to lower the activation energy for the initiation of lipid oxidation[241]. Lipid oxidation can be inhibited first, by antioxidants reacting with ROO. which further stops propagation and inhibit the formation of ROOH. Also oxidation is inhibited by reaction of antioxidants with alkoxyl radicals (RO) which decreases the decomposition of hydroperoxides and the formation of aldehydes. In the evaluation of antioxidants, varied results can be obtained by methods which measure products at different stages of lipid oxidation. Peroxide value (PV) and conjugated dienes (CD) are the most common methods used to measure lipid hydro peroxides. The aldehydes can be measured by anisidine value (AnV) analysis and thiobarbituric acid (TBA) test.

The antioxidant/pro-oxidant action of carotenoids on lipid oxidation has been of interest in food lipids as well as in biological membrane lipids. Since the major constituents of biological membranes are lipid and protein, lipid oxidation can damage membrane lipids[240]. It is known that carotenoids may act as antioxidants by quenching singlet oxygen or by reacting with free radicals. Moreover, the antioxidant/pro-oxidant properties of carotenoids are affected by the concentration of carotenoid, oxygen partial pressure and the nature of the environment. The mechanisms of reactions between carotenoids and radical species may involve radical addition, hydrogen abstraction and electron transfer, but the precise antioxidant/pro-oxidant mechanisms remain unclear[242, 243, 244].

Extensive investigation has been focused on lipid oxidation, which is of importance leading to rancidity. Although carotenoids have not been considered as food antioxidants[245], some authors have proposed antioxidant role for carotenoids in food lipids. β-carotene has been the prototype for examining antioxidant action of carotenoids in different *in vitro* lipid models. Antioxidant activity of carotenoids in organic solutions is related to oxygen concentration, the chemical structure of carotenoids, and the presence of other antioxidants[246]. In different experimental conditions, the discrepant data has shown carotenoids act as both antioxidant[247, 248, 259, 250] and pro-oxidant in lipids in light[251, 252, 253] and dark[254, 255]. Henry *et al.*[183] immersed lycopene in sunflower oil prior to heating, which might enhance the stability of lycopene because of the presence of antioxidants in the oil[183]. Carotenoids such as β-carotene, lutein, and lycopene act as antioxidants in lipid phases by trapping free radicals or physically quenching singlet oxygen[215].

X. CONCLUSION

Lycopene is one of the strongest antioxidant and its singlet oxygen quenching ability is higher than other carotenoids and tocopherols. As a powerful antioxidant lycopene helps neutralize harmful free radicals. The increasing awareness of the health benefits of lycopene resulted in its increased uses as a neutraceutical and in functional foods. Details of tomato fruit ripening, chemical composition, synthesis of carotenoids and lycopene isolation were studied to better understanding. Various physical and chemical factors such as time, temperature, reaction medium, environmental conditions etc affect the stability of lycopene in food matrix. During cooking, industrial food processing and storage lycopene isomerize from all-*trans* to *cis*-isomers would lead to a change in lycopene stability. On the other hand the three most important factors which affect the chemical stability of lycopene are heat, light and oxygen which can easily degrade lycopene, *via* isomerization or auto-oxidation that affect its stability, bioactivity and bioavailability and hence its health benefits which is matter of concern. Processing also plays an important role on stability of lycopene and lipid oxidation. Stability of the lycopene is also affected by the packaging material. Further research is required to establish the effect of processing and storage conditions on bioactivity, bioavailability, and stability of lycopene isomers and on emerging technologies and new approaches in food packaging, such as, MAP and active packages, in modifying isomers of lycopene.

Acknowledgment

The authors would like to thank all the information sources in the public domain used in the compilation of this chapter.

REFERENCES

[1] Gould W.V., "Tomato production, processing and technology", Baltimore: CTI Publications (1992).

[2] Jenkins J.A., "The origin of the cultivated tomato", *Eco Botany.*, **2**, 379–392 (1948).

[3] FAO., Food and Agriculture Organization, United Nation, Rome. www.fao.org (2010).

[4] Armendariz R., Macua J.I., Lahoz I., Gamica J., Bozal, J.M., "Lycopene content in commercial tomato cultivars for paste in Navarra", *Acta Horti (ISHS).*, **724**, 259–262 (2006).

[5] Cantore V., Boari F., Vanadia S., Pace B., Depalma E., Leo L., Zacheo G., "Evaluation of yield and qualitative parameters of high lycopene tomato cultivars", *Acta Horticulturae (ISHS)*, **789**, 173–180 (2008).

[6] Brady C.J., McGlasson W.B., Speirs J., Nevins D.J., Jones R.A., Allen R., Liss Inc., "*The biochemistry of fruit ripening, Tomato Biotech, New York*, 279–288 (1987).

[7] Gillaspy G., Ben-David H., Gruissem W., "Fruits: A developmental perspective", *Plant Cell.*, **5**, 1439–1451(1993).

[8] Giovannoni J.J., "Molecular biology of fruit maturation and ripening", *Ann Rev Plant Physio Plant Mole Biolo.;* **52**, 725–749 (2001).

[9] Rondeau-Mouro C., Crepeau M. J., Lahaye M., "Application of CP-MAS and liquid-like solid-state NMR experiments for the study of the ripening-associated cell wall changes in tomato", *Int J Biol Macromole.*, **31**, 235–244 (2003).

[10] Alexander L., Grierson D., "Ethylene biosynthesis and action in tomato: A model for climacteric fruit ripening", *J Exp Bot.*, **53**, 2039–2055 (2002).

[11] Seymour G.B., Manning K., Eriksson E.M., Popovich A.H., King G.J., "Genetic identification and genomic organization of factors affecting fruit texture", *J Exp Bot.,* **53**, 2065–2071 (2002).

[12] Picton, S., Barton, S.L., Bouzayen M., Hamilton A.J., Grierson D., "Altered fruit ripening and leaf senescence in tomatoes expressing an antisense ethylene-forming enzyme transgene", *The Plant J.*, **3**, 469–481 (1993).

[13] Lelievre J.M., Latche A., Jones B., Bouzayen M., Pech J.C., "Ethylene and fruit ripening", *Physio Plant.,* **101**, 727–739 (1997).

[14] Hoeberichs F.A., Van Der Plas H.W., Woltering E.J., "Ethylene perception is required for the expression of tomato ripening-related genes and associated physiological changes even at advanced stages of ripening", *Post Harvest Biol Tech.*, **26**, 125–133 (2002).

[15] Tucker G.A., Grierson D., Davies D., "Fruit ripening", *The Biochemistry of Plants*, Academic Press New York, **12**, 265–319 (1987).

[16] Lizada C., Seymour G.B., Taylor J.E., Tucker G.A., "*Mango -Biochemistry of Fruit Ripening*", Chapman and Hall, London. pp. 255–271, (1993).

[17] Meredith F.I., Purcell A.E., "Changes in the concentration of carotenes of ripening homestead tomatoes", *Proceed Am Soc Horti Sci.*, **89,** 544–550 (1966).

[18] Fraser P.D., Truesdale M.R., Bird C.R., Schuch W., Bramley P.M., "Carotenoid biosynthesis during tomato fruit development", *Plant Physio.*, **105**, 405–413 (1994).

[19] Prasanna V., Prabha T.N., Tharanathan R.N., "Fruit ripening phenomena: An overview", *CRC Crit Rev Fd Sci Nutr.*, **47**, 1 (2007).

[20] Oeller P.W., Min-Wong L., Taylor L.P., Pike D.A., Theologis A., "Reversible inhibition of tomato fruit senescence by antisense RNA", *Sci.*, **254,** 437–439 (1991).

[21] Hamilton A.J., Lycett G.W., Grierson D., "Antisense gene that inhibits synthesis of the hormone ethylene in transgenic plants", *Nature,* **346**, 284–287 (1990).

[22] Klee H.J., Hayford M.B., Kretzmer K.A., Barry G.F., Kishore G.M., "Control of ethylene synthesis by expression of a bacterial enzyme in transgenic tomato plants", *The Plant Cell.,* **3**, 1187–1193 (1991).

[23] Ahrens M.J., Huber D.J., "Physiology and firmness determination of ripening tomato fruit", *Physiologia Plantarum,* **78**, 8–14 (1990).

[24] Fisher R.L., Bennett A.B., "Role of cell wall hydrolases in fruit ripening", *Annual Review of Plant Physiol Plant Mole Bio.,* **42**, 675–703 (1991).

[25] Almeida D.P., Hube D.J., "Transient increase in locular pressure and occlusion of endocarpic apertures in ripening tomato fruit", *J Plant Physiol.,* **158**, 199–203 (2001).

[26] Saltveit M.E., "Postharvest biology and handling", *In*: Tomatoes (*ed.*), Heuvelink E. CAB Press, Wallingford, CT, USA (2005).

[27] Afoakwa E.O., Sefa-Dedeh S., "Changes in cell wall constituents and mechanical properties during post-harvest hardening of trifoliate yam *Dioscorea dumetorum* (Kunth) pax tubers", *Fd Res Int.,* **35**, 429–434 (2002).

[28] Harker F.R., Redgwell R.J., Hallett I.C., Murray S.H., "Texture of fresh fruit", *Horticultural Review,* **20**, 121–224 (1997).

[29] Bourne M.C., "Texture evaluation of horticultural crops", *Hort Sci.,* **15**, 51–57 (1980).

[30] Abbott J.A., Conway W.S., Sams C.E., "Potsharvest calcium infiltration affects textural attributes of apples", *J Am Horti Sci.,* **114**, 932–936 (1989).

[31] Brummell D.A., Harpster M.H. Cell wall metabolism in fruit softening and quality and its manipulation in transgenic plants. *Plant Mole Bio.,* **47**, 311–340 (2001).

[32] Poovaiah B.W., Nukuya A., "Polygalacturonase and cellulase enzymes in the normal Rutgers and Mutant rin tomato fruits and their relationship to the respiratory climacteric", *Plant Physiology,* **64**, 534–53 (1979).

[33] Seymour G.B., Harding S.E., Taylor A.J., Hobson G.E., Tucker, G.A., "Polyuronide solubilization during ripening of normal and mutant tomato fruit", *Phytochem.,* **26**, 1871–1875 (1987).

[34] John M.A., Dey P.M., "Postharvest changes in fruit cell wall", *Advances Food Res.,* **30**,139–193 (1986).

[35] Doreyappa Gowda I.N., Huddar A.G., "Studies on ripening changes in mango (*Mangifera indica* L.) fruits", *J Food Sci Tech.,* **38**, 135–137 (2001).

[36] Grierson D., Tucker G.A., Robertson N.G., Friend J., Rhodes M.J.C., "The molecular biology of ripening", *Recent Advances in the Biochemistry of Fruits and Vegetables,* Academic Press, London, pp. 149–160 (1981).

[37] Sakai T., Sakamoto T., Hallaert J., Vandamme E.J., "Pectin, pectinase and protopectinase: Production, properties and application", *Adv in App Microbio.,* **39**, 213–294 (1993).

[38] Bartley I.M., Knee M., "The chemistry of textural changes in fruit during storage", *Food Chem.,* **9**, 47–58 (1982).

[39] Brummell D.A., Cin V.D., Crisosto C.H., Labavitch J.M., "Cell wall metabolism during maturation, ripening and senescence of peach fruit", *J Exper Bot.,* **55**, 2029–2039 (2004).

[40] Brummell D.A., "Cell wall disassembly in ripening fruit", *Func Plant Bio.,* **33**, 103–119 (2006).

[41] Tucker G.A., "Introduction", *In*: Seymour, G.B., Taylor, J.E., Tucker G.A., "Biochemistry of Fruit Ripening", Chapman and Hall, London, pp. 1–51 (1993).

[42] Pressey R., "Changes in polygalacturonase isoenzymes and converter in tomatoes during ripening", *Hort Sci.,* **21**, 1183–1185 (1986).

[43] Toivonen P.M.A., Brummell D.A., "Biochemical bases of appearance and texture changes in fresh-cut fruit and vegetables", *Postharvest Biol Technol.,* **48**, 1 (2008).

[44] Huber D.J., O'Donoghue E.M., "Polyuronides in avocado (*Persea americana*) and tomato (*Lycopersicon esculentum*) fruits exhibit markedly different patterns of molecular weight downshifts during ripening", *Plant Physio.,* **102**, 473–480 (1993).

[45] Wong D.W.S., "Pectic enzymes", *In: Food Enzymes. Structure and Mechanisms*", Chapman and Hall, New York, pp. 212–236 (1995).

[46] Gray J., Picton S., Shabbeer J., Schuch W., Grierson D., "Molecular biology of fruit ripening and its manipulation with antisense genes", *Plant Mole Bio.,* **19,** 69–87 (1992).

[47] Hall L., Tucker G., Smith C., Watson C., Seymour G., "Antisense inhibition of pectin esterase gene expression in transgenic tomatoes", *The Plant J.,* **3,** 121–129 (1993).

[48] Smith C.J.S., Watson, C.F., Ray J., Bird C.R., Morris P.C., Schuch W., Grierson D., "Antisense RNA inhibition of polygalacturonase gene expression in transgenic tomatoes", *Nature.,* **334,** 724–726 (1988).

[49] Fray R.G., Grierson D., "Identification and genetic analysis of normal and mutant phytoene synthase genes of tomato by sequencing, complementation, and co-suppression", *Plant Mole Bio.,* **22**, 589–602 (1993).

[50] Thompson D.S., Davies W.J., Ho L.C., "Regulation of tomato fruit growth by epidermal cell wall enzymes", *Plant Cell Env.,* **21**, 589–599 (1998).

[51] Saladie M., Matas A.J., Isaacson T., Jenks M.A., Goodwin S.M., "A reevaluation of the key factors that influence tomato fruit softening and integrity". *Plant Physio,* **144**, 1012–1028 (2007).

[52] Bansal K.C., Chadha, K.L., Ravindran, P.N., Sahijram L., "Biotechnology for improving product quality in horticultural crops", *Biotechnology in Horticultural and Plantation Crops*", Malhotra Publishing House, New Delhi, India, pp. 201–218 (2000).

[53] Tieman D.M., Harriman R.W., Ramamohan G., Handa A.K., "An antisense pectinmethylesterase gene alters pectin chemistry and soluble solids in tomato fruit", *The Plant Cell.,* **4**, 667–679 (1992).

[54] Davies J.N., Hobson G.E., "The constituents of tomato fruit- the influence of environment, nutrition, and genotype", *CRC Criti Rev Food Sci Nutr.,* **15**, 205–280 (1981).

[55] Giovanelli U., Lavelli V., Pen C., Nobili S., "Variation in antioxidant components of tomato during vine and post-harvest ripening", *J Sci Food Agri.*, **79**, 1583–1588 (1999).

[56] Abushita A.A., Daood H.G., Biacs P.A., "Change in carotenoids and antioxidant vitamins in tomato as a function of varietal and technological factors", *J Agri Food Chem.*, **48,** 2075–2081 (2000).

[57] Thompson K.A., Marshall M.R., Sims C.A., Wei C.I., Sargent S.A., Scott J.W., "Cultivar, maturity and heat treatment on lycopene content in tomatoes", *J Food Sci.*, **65,** 791–795 (2000).

[58] Madhavi D.L., Salunkhe D.K., Kadam S.S., "Handbook of vegetable science and technology: Production, composition, storage and processing Tomato", New York, Marcel Dekker, pp. 171–201 (1998).

[59] Wang F., Sanz A., Brenner M.L., Smith A.G., "Sucrose synthase, starch accumulation, and tomato fruit sink strength", *Plant Physio.*, **101**, 321–327 (1993).

[60] Biale J.B. "The post-harvest biochemistry of tropical and subtropical fruits", *Adv Food Res.*, **10**, 293–354 (1960).

[61] Seymour G.B., Taylor J.E., Tucker G.A., "*Biochemistry of fruit ripening*", London, Chapman and Hall (1993).

[62] Kaur C., Khurdiya D.S., Pal R.K., Kapoor H.C., "Effect of microwave heating and conventional processing on the nutritional qualities of tomato juice", *J Food Sci Tech.*, **36**, 331–333 (1999).

[63] Akanbi C.T., Oludemi F.O., "Effect of processing and packaging on the lycopene content of tomato products", *Int J Food Prop.*, **7**, 139–152 (2004).

[64] Fox B.A., Cameron A.G., "*Food Science, Nutrition and Health* (6th *Edn.*)", London, Edward Arnold Press, pp 236–242 (1995).

[65] Cano A., Acosta M., Arnao M.B., "Hydrophilic and lipophilic antioxidant activity changes during on-vine ripening of tomatoes (*Lycopersicon esculentum* Mill.)", *Postharvest Bio Tech.*, **28**, 59–65 (2003).

[66] Gautier H., Diakou-Verdin V., Bénard C., "How does tomato quality (sugar, acid, and nutritional quality) vary with ripening stage, temperature, and irradiance?", *J Agri Food Chem.*, **56**, 1241–1250 (2008).

[67] Murneek A.E., Maharg L., Wittwer S.H., "Ascorbic acid (Vitamin C) content of tomatoes and apples", *Univ Missouri Agri Exp Station Res Bull.*, **568,** 3–24 (1954).

[68] Dumas Y., Dadomo M., Lucca D.G., Grolier P., "Effects of environmental factors and agricultural techniques on antioxidant content of tomatoes", *J Sci Food Agri.*, **83**, 369–382 (2003).

[69] Hamner K.C., Bernstein L., Maynard, L. A., "Effects of light intensity, day length, temperature, and other environment factors on the ascorbic acid content of tomatoes", *J Nutr.*, **29**, 85–97 (1945).

[70] Somers G.F., Kelly W.C., Hamner K.C., "Inuence of nitrate supply upon the ascorbic content of tomatoes", *Am J Bot.*, **38**, 472–475 (1951).

[71] Torres R., Faini F., Modak B., Urbina F., Labbe C., Guerrero J., "Antioxidant activity of coumarins and flavonols from the resinous exudate of Haplopappus", *Phytochem.*, **67**, 984–987 (2006).

[72] Stevens M.A., Rick C.M., Atherton J.G., Rudich J., "Genetics and breeding: Fruit quality", *The tomato crop, a scientific basis for improvement*", London: Chapman and Hall Ltd, pp. 84–96 (1986).

[73] Venter F., "Solar radiation and vitamin content of tomato fruits", *Acta Horticulturae*, **58**, 121–127 (1977).

[74] Shi J., Maguer M.L., Liptay A., Wang S.L., "Chemical composition of tomatoes as affected by maturity and fertilization practices", *J Food Qual.*, **22**, 147–156 (1999).

[75] Abushita A.A., Hebshi E.A., Daood H.G., Biacs P.A. "Determination of antioxidant vitamins in tomatoes", *Food Chem.*, **60**(2), 207–212 (1997).

[76] George B., Kaur C., Khurdiya D.S., Kapoor H.C., "Antioxidants in tomato (*Lycopersium esculentum*) as a function of genotype", *Food Chem.,* **84**, 45–51 (2004).

[77] Dey P.M., Harborne J.B., "*Plant Biochemistry*", Academic Press. San Diego, CA, USA pp. 417–437 (1997).

[78] Salisbury F.B., Ross C.W., "*Plant Physiology*", 4 Ed. Wadsworth Publishing Co. pp. 315–324 (1992).

[79] Chalukova M., Manuelyan H., Kalloo, Verlag S., Berlin, "*Breeding for carotenoid pigments in tomato-Monographs on theoretical and applied genetics*", **14**, 179–195 (1991).

[80] Armstrong G.A., "Eubacteria show their true colors: Genetics of carotenoid pigment biosynthesis from microbes to plants", *J Bacteriology*, **176**(16), 4795–4802 (1994).

[81] Miura Y., Kondo K., Shimada H., Saito T., Nakamura K., Misawa N., "Production lycopene by food yeast, *Candida utilis* that does not naturally synthesize carotenoid", *Biotech Bioengin.*, **58**(2/3), 306–308 (1998).

[82] Cunningham F.X., Sun M.D., Chamovitz Z., Hirschberg D., Gantt E., "Molecular structure and enzymatic function of lycopene cyclase from the *Cyanobacterium synechococcus* sp strain PCC7942", *Plant Cell*, **6**(8), 1107–1121 (1994).

[83] Pecker I., Gabbay R., Cunningham F.X., Hirschberg J., "Cloning and characterization of the cDNA for lycopene beta-cyclase from tomato reveals decrease in its expression during fruit ripening", *Plant Molecular Biology*, 30(4), 807–819 (1996).

[84] Pfander H., Traber B., Birkhauser B., "*Carotenoids Synthesis*", Vol 2, p. 319 (1996).

[85] Romer S., Frasier P.D., Kiano J.W., Shipton C.A., Misawa N., Schuch W., Bramley P. M., "Elevation of the provitamin-A content of transgenic tomato plants", *Nat Biotech.*, **18**, 666 (2000).

[86] Hopkins W.G., "*Introduction to Plant Physiology*", New York: John Wiley & Sons, (2nd Ed.) pp.134–135 (1999).

[87] Gross J., "Pigments in Fruits", Academic Press, London (1987).
[88] Gross J., "Pigments in Vegetables", Van Nordstrand Reinhold, New York (1991).
[89] Mangels A.R., Holden J.M., Beecher G.R., Forman M.R., Lanza E., "Carotenoid content of fruits and vegetables: An evaluation of analytic data", *J Am Dietetic Assoc.*, **93**, 284–296 (1993).
[90] Nguyen M.L., Schwartz S.J., "Lycopene stability during food processing", *Proceed Soc Exp Bio Med.*, **218**, 101–105 (1998).
[91] Rao A.V., Agarwal S., "Role of lycopene as antioxidant carotenoid in the prevention of chronic diseases: A review", *Nutri Res.*, **19**, 305–323 (1999).
[92] Alba R., Cordonnier-Pratt M.M., Pratt L.H., "Fruit-localized phytochromes regulate lycopene accumulation independently of ethylene production in tomato", *Plant Physio.*, **123**, 363–70 (2000).
[93] Bouvier F., Backhaus R.A., Camara B., "Induction and control of chromoplast specific carotenoid genes by oxidative stress", *J Bio Chem.*, **273**, 30651–30659 (1998).
[94] Akhtar M.S., Goldschmidt E.E., John I., Rodoni S., Matile P., Grieson D., "Altered patterns of senescence and ripening in *gf*, a stay-green mutant of tomato (*Lycopersicon esculentum* Mill.)", *J Exp Bot.*, **50**, 1115–1122 (1999).
[95] Pruzinska A., Tanner G., Anders I., Roca M., Hortensteiner S., "Chlorophyll breakdown pheophorbide an oxygenase is a rieske-type iron-sulfur protein, encoded by the accelerated cell death 1 gene", *Proceed Nat Acad Sci USA.*, **100**, 15259–15264 (2003).
[96] Harris W.M., "Chromoplasts of tomato fruits. III. The high-delta tomato", *Bot Gazette.*, **131**, 163–166 (1970).
[97] Khudairi A.K., "The ripening of tomatoes", *Am Sci.*, **60**, 696–707 (1972).
[98] Matienco B.T., Yedalty E.M., "*Ultrastructure of carotenoidoplasts*", New York, Academic Press (1973).
[99] Ellis G.H., Hammer K.C., "The carotene content of tomatoes as influenced by various factors", *J Nut.*, **25**, 539–553 (1943).
[100] Hobson G., Davies J., Hulme A., "*The tomato-The biochemistry of fruits and their products*", Norwich, U.K, Academic Press, pp. 453–457 (1971).
[101] Laval-Martin D., "La maturation du fruit de tomato cerise: Mise en evidence, parcryodecapage, de levolution des chloroplasts en duextypes de chromoplates" , *Protoplasma*, **82**, 33–59 (1974).
[102] Kirk J.T.O., Tilney-Basset R.A.E., "*The plastids: Their chemistry, structure, growth and inheritance* (2nd Ed.)", Amsterdam, Elsevier (1978).
[103] Burton G.W., Ingold K.U., "β-Carotene: An usual type of lipid antioxidant", *Sci.*, **244**, 569–557 (1984).
[104] Lee M.T., Chen B.H., "Stability of lycopene during heating and illumination in a model system", *Food Chem.*, **78**, 425–432 (2002).
[105] Gould W.V., "*Tomato production, processing and technology*", Baltimore: CTI Publications (1992).
[106] Khachik F., Carvalho L., Bernstein P.S., Muir G.J., Zhao D.Y., Katz N.B., "Chemistry, distribution, and metabolism of tomato carotenoids and their impact on human health", *Exp Bio Med.*, **227**, 845–851 (2002).
[107] Rymal K.S., Nakayama T.O.M., "Major carotenoids of the seeds of three cultivars of the tomato", *J Agri Food Chem.*, **22**, 715–717 (1974).
[108] Vogele A.C., "Effect of environmental factors upon the color of the tomato and watermelon", *Plant Physio.*, **12**, 929–955 (1937).
[109] Goodwin T.W., Jamikorn M., "Biosynthesis of carotenes in ripening tomatoes", *Nature*, **170**, 104–105 (1952).
[110] Tomes M.L., "Temperature inhibition of carotene biosynthesis in tomato", *Bot Gazette.*, **124**, 180–185 (1963).
[111] Mustafa A.A., "Effect of low and high temperature on tomato fruit pigments when harvested at mature-green stage", *Emi J Agri Sci.*, **1**, 125–133 (1989).

[112] Duggar B.M., "Lycopersicon: The red pigment of the tomato and the effects of conditions on its development", *Washington Uni Study*, **1**, 22–45 (1913).
[113] Shewfelt A.L., Halpin J. E., "The effect of light quality on the rate of tomato color development", *Am Soc Horti Sci"*, **91**, 561–565 (1967).
[114] Heinonen M.I., Ollilainen V., Linkola E.K., Varo P.T., Koivistoinen P.E., "Carotenoids in Finnish foods, vegetables, fruits, and berries", *J Agri Food Chem.*, **37**, 655–659 (1989).
[115] Lurie S., Handros A., Felix A., Shapiro R., "Reversible inhibition of tomato fruit gene expression at high temperature", *Plant Physio.,* **110**, 1207–1214 (1996).
[116] Hart D.J., Scott K.J., "Development and evaluation of an HPLC method for the analysis of carotenoids in foods, and the measurement of the carotenoid content of vegetables and fruits commonly consumed in the UK", *Food Chem.*, **54**, 101–111 (1995).
[117] Tonucci L.H., Holden J.M., Beecher G.R., Khachik F., Davis C.S., "Mulokozi G. Carotenoid content of thermally processed tomato-based food products", *J Agri Food Chem.*, **43**, 579–586 (1995).
[118] Gomez R., Costa J., Amo M., Alvarruiz A., Picazo M., Pardo J.E., "Physicochemical and colorimetric evaluation of local varieties of tomato grown in Spain", *J Sci Food Agri.*, **8**, 1101–1105 (2001).
[119] Takeoka G.R., Dao L., Flessa S., Gillespie D. M., Jewell W. T., Huebner B., Bertow D., Ebeler S., "Processing effects on lycopene content and antioxidant activity of tomatoes", *J Agri Food Chem.*, **49**, 3713–3717 (2001).
[120] Leonardi C., Ambrosino P., Esposito F., Fogliano V., "Antioxidative activity and carotenoid and tomatine contents in different typologies of fresh consumption tomatoes", *J Agri Food Chem.*, **48,** 4723–4727 (2000).
[121] Arias R., Lee T.C., Specca D., Janes H., "Quality comparison of hydroponics tomatoes (*Lycopersicon esculentum*) ripened on and off vine", *J Food Sci.*, **65**, 545–548 (2000).
[122] Bramley P.M., "Is lycopene beneficial to health?", *Phytochem.,* **54**, 223–236 (2000).
[123] Fraser P.D., Bramley P.M., "The biosynthesis and nutritional uses of carotenoids", *Progress in Lipid Res.*, **43,** 228–265 (2004).
[124] Sahlin E., Savage G.P., Lister C.E., "Investigation of the antioxidant properties of tomatoes after processing", *J Food Compo Anal.,* **17**, 635–647 (2004).
[125] Kaur D., Sharma R., Wani A.A., Gill B.S., Sogi D.S., "Physicochemical changes in seven tomato (*Lycopersicon esculentum*) cultivars during ripening", *Int J Food Prop.*, **9**, 747–757 (2006).
[126] Ilahy R., Hdider C., Lenucci M.S., Tlili I., Dalessandro G., "Antioxidant activity and bioactive compound changes during fruit ripening of high-lycopene tomato cultivars", *J Food Compos Anal.,* **24**, 588–595 (2011).
[127] Toor R.K., Savage G.P., "Antioxidant activity in different fractions of tomatoes", *Food Res Int.*, **38**, 487–494 (2005).
[128] Orlowski M., Grzeszczuk M., Gorecka K., "Evaluation of the fruit biological value of small fruity tomato cherry type grown in unheated high plastic tunnel", *Veg Crops Res Bull.*, **54**, 21–24 (2002).
[129] Raffo A., Leonardi C., Fogliano V., Ambrosino P., Salucci M., Gennaro L., Bugianesi R., Giuffrida F., Quaglia, G., "Nutritional value of cherry tomato harvested at different ripening stages", *J Agri Food Chem.*, **50**, 6550–6556 (2002).
[130] Manzano N.T., "Farmacy tomato", *Agroanalysis.*, **21,** 36–38 (2001).
[131] Al-Wandawi H., Rahman M.A., Al-Shaikhly K., "Tomatoes processing wastes as essential raw material sources", *J Agric Food Chem.,* **33**, 804 (1985).
[132] D'Souza M., Singha S., Ingle M., "Lycopene concentration of tomato fruit can be estimated from chromaticity values", *HortSci.*, **27**(5), 465–466 (1992).
[133] Sharma S.K., Maguer M.L., "Kinetics of lycopene degradation in tomato pulp solids under different processing and storage conditions", *Food Res Int.*, **29**, 309–315 (1996).

[134] Chen P., McCarthy M.J., Kauten R., "NMR for internal quality evaluation of fruits and vegetables", *Trans Am Soc Agri Engin.*, **32**, 1747–1753 (1989).

[135] Chen P., McCarthy M.J., Kim S.M., Zion B., "Development of a high-speed NMR technique for sensing maturity of avocados", *Trans Am Soc Agri Engin.*, **39**, 2205–2209 (1996).

[136] Ishida N., Ogawa H., Koizumi M., Kano H., "Ontogenetic changes of the water status and accumulated soluble compounds in growing cherry fruits studied by NMR imaging", *Magnetic Resonance Chem.*, **35,** S22–S28 (1997).

[137] Clark C.J., Drummond L.N., MacFall J.S., "Quantitative NMR imaging of kiwifruit (*Actinidia deliciosa*) during growth and ripening", *J Sci Food Agri.*, **78**, 349–358 (1998).

[138] Koizumi M., Naito S., Haishi T., Utsuzawa S., Ishida N., Kano H., "Thawing of frozen vegetable observed by a small dedicated MRI for food research", *Magn Reson Imag.*, **24**, 1111–1119 (2006).

[139] Cheng Y.C., Wang T.T., Chen J.H., Lin T.T. "Spatial-temporal analyses of lycopene and sugar contents in tomatoes during ripening using chemical shift imaging", *Postharvest Bio Tech.*, **62**, 17–25 (2011).

[140] Withnall R., Chowdhry B.Z., Silver J., Edwards H.G.M., De Oliveira, L.F.C., "Raman spectra of carotenoids in natural products", *Spectrochimica Acta Part A: Molecular and Biomolecular Spectroscopy.*, **59**, 2207–2212 (2003).

[141] Qin J., Chao K., Kim M.S., "Investigation of Raman chemical imaging for detection of lycopene changes in tomatoes during postharvest ripening", *J Food Engin.*, **107**, 277–288 (2011).

[142] Landers G.M., Olson J.A., "Absence of isomerization of retinyl palmitate, retinol, and retinal in chlorinated and unchlorinated solvents under gold light", *J Assoc Offic Anal Chem.*, **69**, 50–55 (1986).

[143] Schoefs B., "Chlorophyll and carotenoid analysis in food products. Properties of the pigments and methods of analysis", *Trends Food Sci Tech.*, **13**, 361–371 (2002).

[144] Mencarelli F., Saltveit M.E., "Ripening of mature-green tomatoes fruit slices", *J Am Soc Horti Sci.*, **113**, 742–745 (1988)

[145] Tan B., Soderstrom D.N. "Qualitative aspects of UV-VIS spectrophotometry of carotene and lycopene", *J Chem Edu.*, **22,** 21–31 (1988).

[146] Gomez P.M.S., Caja M.M., Maria S.G., "Solubility in supercritical carbon dioxide of the predominant carotenes of tomato skin", *J Am Oil Chem Soc.*, **79**, 897–902 (2002).

[147] Pichini S., Zuccaro P., Pellegrini M., Carlo D.S., Bacosi A., Palmi I., Tossini G., Pacifici R., "Determination of fat-soluble nutrients in serum by liquid chromatography and multi-wavelength detection", *J Liq Chrom Tech.*, **25**, 781–786 (2002).

[148] Tzouganaki Z.D., Alta-P.J., Koupparis M.A., "Development and validation of liquid chromatographic method for the determination of lycopene in plasma", *Analy Chim Acta.*, **467**, 115–123 (2002).

[149] Lin C.H., Chen B.H., "Determination of carotenoids in tomato juice by liquid chromatography", *J Chrom A.*, **1012**, 103–109 (2003).

[150] Burns J., Fraser P.D., Bramley P.M., "Identification and quantification of carotenoids, tocopherols and chlorophylls in commonly consumed fruits and vegetables", *Phytochem.*, **62**, 939–947 (2003).

[151] Sander L.C., Sharpless K.E., Craft N.E., Wise S.A., "Development of engineered stationary phases for the separation of carotenoid isomers", *Anal Chem.*, **66,** 1667–1674 (1994).

[152] Emenhiser C., Simunovic N., Sander L.C., Schwartz, S.J., "Separation of geometrical carotenoid isomers in biological extracts using a polymeric C_{30} column in reversed-phase liquid chromatography", *J Agri Fd Chem.*, **44**, 3887–3893(1996).

[153] Rouseff R., Raley L., Hofsommer H.J., "Application of diode array detection with a C_{30} reversed phase column for the separation and identification of saponified orange juice carotenoids", *J Agri Food Chem.*, **44**, 2176–2181 (1996).

[154] Hengartner U., Bernhard K., Meyer K., "Synthesis, isolation, and NMR spectroscopic characterization of fourteen (Z)-isomers of lycopene and of some acetylenic didehydro- and tetradehydro lycopenes", *Helvetica Chimica Acta.*, **75**, 1848–1865 (1992).

[155] Rozzi N.L., Singh R.K., Vierling R.A., Watkings B.A., "Supercritical fluid extraction of lycopene for tomato processing byproducts", *J Agri Food Chem.*, **50**, 2638–2643 (2002).

[156] Sadler G., Davis J., Dezman D., "Rapid extraction of lycopene and β-carotene from reconstituted tomato paste", *J Food Sci.*, **55**, 1460–1461 (1990).

[157] AOAC (Association of Official Analytical Chemists), Peer Verified Methods Program. AOAC, Manual on Policies and Procedures. AOAC: Arlington, (1993).

[158] Barba A.I.O., Hurtado M.C., Mata M.C.S., Ruiz V.F., Tejada M.L.S., "Application of a UV-vis detection-HPLC method for a rapid determination of lycopene and β-carotene in vegetables", *Food Chem.*, **95**, 328–336 (2006).

[159] Shi J., Manguer M.L., "Lycopene in tomatoes: Chemical and physical properties affected by food Processing", *CRC Crit Rev Food Sci Nutr.*, **40**, 1–42 (2000).

[160] Van den Berg H., Faulks R., Fernando-Granado H., Hirschberg J., Olmedilla B., Sandmann G., Southon S., Stahl W., "The potential for the improvement of carotenoid levels in foods and the likely systemic effects", *J Sci Food Agri.*, **80**, 880–912 (2000).

[161] Tomes M.L., "Temperature inhibition of carotene biosynthesis in tomato", *Botani Gaz.*, **124**, 180–185 (1963).

[162] Cadoni E., De Giorgi M.R., Medda E., Poma G., "Supercritical CO_2 extraction of lycopene and α-carotene from ripe tomatoes", *Dyes and Pigments.*, **44**, 27–32 (2000).

[163] Gomez-Prieto M.S., Caja M.M., Herraiz M., Maria S.G., "Supercritical fluid extraction of *all-trans*-lycopene from tomato", *J Agri Food Chem.*, **51**, 3–7 (2003).

[164] Ciurlia L., Bleve M., Rescio L., "Supercritical carbon dioxide co-extraction of tomatoes (*Lycopersicum esculentum* L.) and hazelnuts (*Corylus avellana* L.): A new procedure in obtaining a source of natural lycopene", *J Supercrit Fluids,* **49**, 338–344 (2009).

[165] Eh A.L., Teoh S.G., "Novel modified ultrasonication technique for the extraction of lycopene from tomatoes", *Ultrasonics Sonochem.*, **19**, 151–159 (2012).

166] Bicanic D., Fogliano V., Luterotti S., Swarts J., Piani G., Graziani G., "Quantification of lycopene in tomato products: Comparing the performances of a newly proposed direct photo-thermal method and high-performance liquid chromatography", *J Sci Food Agri.*, **85**, 1149–1153 (2005).

[167] Davies B.H., Goodwin T.W., "Carotenoids", *Chemistry and Biochemistry of Plant Pigments* (2nd Ed.)", New York, Academic Press, pp. 38–165 (1976).

[168] Clinton S. K., "Lycopene: Chemistry, biology, and implications for human health and disease", *Nut Rev.*, **56**, 35–51 (1998).

[169] Khachik F., Carvalho L., Bernstein P.S., Muir G.J., Zhao D.Y., Katz N.B., "Chemistry, distribution, and metabolism of tomato carotenoids and their impact on human health", *Exp Bio Med.*, **227**, 845–851 (2002).

[170] Foote C.S., Denny R.W., "Chemistry of singlet oxygen VII. Quenching by β-carotene", *J Am Chemi Soc.*, **40**, 6233–6235 (1968).

[171] Burton G.W., Ingold K.U., "β-Carotene: An usual type of lipid antioxidant", *Sci.*, **244**, 569–573 (1984).

[172] Di Mascio P., Murphy M.C., Sies H., "Antioxidant defense systems, the role of carotenoid, tocopherol and thiols", *Am J Clini Nut.*, **53,** 194–200 (1991).

[173] Bohm F., Tinkler J.H., Truscott T.G., "Carotenoids protect against cell membrane damage by nitrogen dioxide radical", *National Med.*, **1**, 98–99 (1995).

[174] Lu Y., Etoh H., Watanabe N., "A new carotenoid, hydrogen peroxide oxidation products from lycopene", *Biosci Biotech Biochem.*, **59**, 2153–2155 (1995).

[175] Mortensen A., Skibsted L.H., Sampson J., Rice-Evans C., Everett S.A., "Comparative mechanisms and rates of free radical scavenging by carotenoid antioxidants", *FEBS Letters*, **418**, 91–97 (1997).

[176] Di Mascio P., Kaiser S., Sies H., "Lycopene as the most efficient biological carotenoid singlet oxygen quencher", *Archives of Biochem and Biophy.*, **274**, 532–538 (1989).
[177] Conn P.F., Lambert C., Land E.J., Schalch W., "Truscott T.G., "Carotene-oxygen radical interactions", *Free Radical Research Communications*, **16**, 401–408 (1992).
[178] Conn P.E., Schalch W., Truscott T.G., "The singlet oxygen and carotenoid interaction", *J Photochem Photobiology B-Biology*, **11**, 41–47 (1991).
[179] Miller N.J., Sampson J., Candeiar L.P., Bramley P.M., Rice-Evans C.A., "Antioxidant activities of carotenes and xanthophylls", *FEBS Letters*, **384**(3), 240–6 (1996).
[180] Moss G.P., Weedon B.C.L., Goodwin T. W., "*Chemistry of the carotenoids.* (2nd Ed.), *Chemistry and biochemistry of plant pigments*", New York, Academic Press, pp. 149–224 (1976).
[181] Crouzet J., Kanasawud P., "Formation of volatile compounds by thermal degradation of carotenoids", *Methods in Enzymology*, **213**, 54–62 (1992).
[182] Scita G., "Stability of β-carotene under different laboratory condition", *Methods in Enzymology*, **213**, 175–185 (1992).
[183] Henry, L.K., Catignani, G.L., Schwartz, S.J., "Oxidative degradation kinetics of lycopene, lutein, 9-*cis* and all-*trans* β-carotene", *J Am Oil Chemists Soc.*, **75**, 823–829 (1998).
[184] Pauling L., "Recent work on the configuration and electronic structure of molecules with some applications to natural products: Isomerism and the structure of carotenoids", *Fortschritte der Chemie Organischer Naturstoffe*, **3**, 227–229 (1939).
[185] Zechmeister L., LeRosen F.W., Went F.W., Pauling L., "Prolycopene, a naturally occurring stereoisomer of lycopene", *Proceed Nat Academy Sci.*, USA, **27**, 468–474 (1941).
[186] Zechmeister L., Polgar A., "*Cis-trans* isomerization and spectral characteristics of carotenoids and some related compounds", *J Am Chem Soc.*, **65**, 1522–1528 (1943).
[187] Zechmeister L., "*Cis-trans isomeric carotenoids, vitamins A and arylpolyenes*", New York, Academic Press (1962).
[188] Nguyen M.L., Schwartz S.J., "Lycopene: Chemical and biological properties" *Food Tech.*, **53**, 38–45 (1999).
[189] Lee H.S., Coates G.A., "Characterization of color fade during frozen storage of red grapefruit juice concentrates", *J Agric Food Chem.*, **50**, 3988–3991 (2002).
[190] Re R., Bramley P.M., Rice-Evans C., "Effects of food processing on flavonoids and lycopene status in a Mediterranean tomato variety", *Free Radical Res.*, **36**, 803–810 (2002).
[191] De Ancos B., Gonzalez E., Cano M.P., "Effect of high pressure treatment on the carotenoid composition and the radical scavenging activity of persimmon fruit puree", *J Agri Food Chem.*, **48**, 3542–3548 (2000).
[192] Noble A.C., "Investigation of the color changes in heat concentrated tomato pulp", *J Agri Food Chem.*, **23**, 48–49 (1975).
[193] Goula A.M., Adamopoulos K.G., "Spray drying of tomato pulp in dehumidified air: II. The effect on powder properties", *J Food Engin.*, **66**, 35–42 (2005).
[194] Boskovic M.A., "Fate of lycopene in dehydrated tomato products: Carotenoid isomerization in food system", *J Food Sci.*, **44**, 84–86 (1979).
[195] Lovric T., Sablek Z., Boskovic M., "*Cis-trans* isomerisation of lycopene and color stability of foam-mat dried tomato powder during storage", *J Sci Food Agri.*, **21**, 641–647 (1970).
[196] Chen J., Shi J., Xue S. J., Ma Y., "Comparison of lycopene stability in water- and oil-based food model systems under thermal and light-irradiation treatments", *LWT-Food Sci Tech.*, **42**, 740–747 (2009).
[197] Fish W.W., Davis A.R., "The effect of frozen storage conditions on lycopene stability in watermelon tissue", *J Agri Food Chem.*, **51**, 3582–3585 (2003).
[198] Cole E.R., Kapur N.S., "The stability of lycopene I. Degradation by oxygen", *J Sci Food Agri.*, **8**, 360–365 (1957).

[199] Ukai N., Lu Y., Etoh H., Tagi A., Ina K., Oshima S., Ojima F., Sakamoto H., Ishiguro Y., "Photosensitized oxygenation of lycopene", *Biosci Biotech Biochem.*, **58**, 1718–1719 (1994).

[200] Ben-Aziz A., Britton G., Goodwin T.W., "Carotene epoxides of *Lycopersicon esculentum*", *Phytochem.*, **12**, 2759–2764 (1973).

[201] Kim S., Nara E., Kobayashi H., Terao J., Nagao A., "Formation of cleavage products by autoxidation of lycopene", *Lipids*, **36**(2), 191–200 (2001).

[202] Borel P., Grolier P., Armand M., Partier A., Lafont H., Lairon D., Braesco A., "Carotenoids in biological emulsions: Solubility, surface-to-core distribution, and release from lipid droplets", *J Lipid Res.*, **37**, 250–261 (1996).

[203] Thurnham D., Smith E., Flora P., "Concurrent liquid-chromatographic assay of retinol, α-tocopherol, β-carotene, α-carotene, lycopene and β-cryptoxanthin with tocopherol acetate as internal standard", *Clinical Chem.*, **34**, 377–381 (1988).

[204] Barua A.B., Furr H.C., "Extraction and analysis by high performance liquid chromatography of carotenoids in human serum", *Methods in Enzymology*, **213**, 273–281 (1992).

[205] Scott K.J., "Observations on some of the problems associated with the analysis of carotenoids in foods by HPLC", *Food Chem.*, **45**, 357–364 (1992).

[206] Heredia A., Peinado I., Rosa E., Andrés A., "Effect of osmotic pre-treatment and microwave heating on lycopene degradation and isomerization in cherry tomato" *Food Chem.*, **123**, 92–98 (2010).

[207] Mayeaux M., Xu Z., King, J. M., Prinyawiwatkul W. "Effects of cooking conditions on the lycopene content in tomatoes", *J Food Sci.,* **71***,* 461–464 (2006).

[208] Hackett M.M., Lee J.H., Francis D., Schwartz S.J., "Thermal stability and isomerization of lycppene in tomato oleoresins from different varieties", *J Food Sci.,* **69***,* 536–541 (2004).

[209] Mayer-Meibach E., Schuchmann H.P., Regier M., Behsnilian D., "Thermal processing of carrots: Lycopene stability and isomerization with regard to antioxidant potential", *Food Res Int.,* **38***,* 1103–1108 (2005).

[210] Lee M.T., Chen B.H., "Stability of lycopene during heating and illumination in a model system", *Food Chem.,* **78***,* 425–432 (2002).

[211] Nunes I. L., Mercadante A.Z., "Encapsulation of lycopene using spray-drying and molecular inclusion processes", *Brazilian Archives Bio Tech.*, **50**(5), 893–900 (2007).

[212] Seybold C., Frohlich K., Bitsch R., Otto K., Bohm V., "Changes in contents of carotenoids and vitamin-E during tomato processing", *J Agri Food Chem.,* **52***,* 7005–7010 (2004).

[213] Toor R.K., Savage G.P., "Effect of semi-drying on the antioxidant components of tomatoes", *Food Chem.*, **94**(1), 90–97 (2006).

[214] Dewanto V., Wu X., Adom K.K., Liu R.H., "Thermal processing enhances the nutritional value of tomatoes by increasing total antioxidant activity", *J Agri Food Chem.,* **50***,* 3010–3014 (2002).

[215] Sies H., Stahl W., Sundquist A.R., "Antioxidant functions of vitamins- Vitamins E and C, β-carotene, and other carotenoids", *Annals New York Acad Sci.*, **368**, 7–19 (1992).

[216] Schierle J., Bretzel W., Buhler J., Faccin N., Hess D., Steiner K., Schuep W., "Content and isomeric ratio of lycopene in food and human blood plasma", *Food Chem.*, **59**, 459–465 (1997).

[217] Khachik F., Goli M.B., Beecher G.R., Holden J., Lusby W.R., Tenorio M.D., Barrera M.R.J., "Effect of food preparation on qualitative and quantitative distribution of major carotenoid constituents of tomatoes and several green vegetables", *J Agric Food Chem.*, **40**, 390–398 (1992).

[218] Stahl W., Sies H., "Uptake of lycopene and its geometrical isomers is greater from heat-processed than from unprocessed tomato juice in humans", *J Nutr.*, **122**, 2161–2166 (1992).

[219] Cole E.R., Kapur N.S., "The stability of lycopene II. Oxidation during heating of tomato pulp", *J Sci Food Agri.*, **8**, 366–368 (1957).

[220] Shi J., Dai Y., Kakuda Y., Mittal G., Xue S.J., "Effect of heating and exposure to light on the stability of lycopene in tomato puree", *Food Control.*, **19**, 514–520 (2008).

[221] Anguelova T., Warthesen J., "Lycopene stability in tomato powders", *J Food Sci.*, **65**, 67–70 (2000).

[222] Shi J., Wu Y., Bryan M., Maguer M.L., "Oxidation and isomerization of lycopene under thermal treatment and light irradiation in food processing", *Nutra Food.*, **7**, 179–183 (2002).

[223] Wong F.F., Bohart G.S. "Observation on the color of vacuum-dried tomato juice powder during storage", *Food Tech.*, **5**, 293–296 (1957).

[224] Ax K., Mayer-Miebach E., Link B., Schuchmann H., Schubert H., "Stability of lycopene in oil-in-water emulsions", *Engin Life Sci.*, **4**, 199–201 (2003).

[225] Ribeiro H.S., Ax K., Schubert H., "Stability of lycopene emulsions in food systems", *J Food Sci.*, **68**, 2730–2734 (2003).

[226] Gartner C., Stahl W., Sies H., "Lycopene is more bioavailable from tomato paste than from fresh tomatoes", *Am J Clin Nutr.*, **66**, 116–122 (1997).

[227] Simpson K.L., Lee T.C., Rodriguez D.B., Chichester C.O., Goodwin T.W. "Metabolism in senescent and stored tissues- Chemistry and biotechnology of plant pigments", New York: Academic Press, 2nd Ed, Vol. 1, pp. 779–842 (1976).

[228] Xianquan S., Shi J., Kakuda Y., Yueming J., "Stability of lycopene during food processing and storage", *J Med Food.*, **8**, 413–422 (2005).

[229] Zimmerman M., "Antioxidant powerhouse", *Taste for Life.*, **8**, 1–3 (2002).

[230] Goula A.M., Adamopoulos K.G., Chatzitakis P.C., Nikas V.A., "Prediction of lycopene degradation during a drying process of tomato pulp", *J Food Engin.*, **74**, 37–46 (2006).

[231] Muratore G., Rizzo V., Licciardello F., Maccarone E., "Partial dehydration of cherry tomato at different temperature and nutritional quality of the products", *Food Chem.*, **111**, 887–891 (2008).

[232] Giovanelli G., Paradiso A., "Stability of dried and intermediate moisture tomato pulp during storage", *J Agri Food Chem.*, **50**, 7277–7281 (2002).

[233] Pernice R., Parisi M., Giordano I., Pentangelo A., Graziani G., Gallo M., Ritieni A., "Antioxidants prole of small tomato fruits: Effect of irrigation and industrial process", *Scientia Horti.*, **126**, 156–163 (2010).

[234] Odriozola S.I., Soliva F.R., Martýn B.O., "Effect of minimal processing on bioactive compounds and color attributes of fresh-cut tomatoes", *Lebensmittel-Wissenschaft und Technologie*, **41**, 217–226 (2008).

[235] Zanoni B., Peri C., Nani R., Lavelli V., "Oxidative heat damage of tomato halves as affected by drying", *Food Res Intern.*, **31**, 395–401(1999).

[236] Henry L.K., Puspitosan N.N.L., Jaren G.M., Van Breeman R.B., Catignani G.L., Schwartz S.J., "Effects of ozone and oxygen on the degradation of carotenoids in an aqueous model system", *J Agri Food Chem.*, **48**, 5008–5012 (2000).

[237] Van den Berg H., Faulks R., Fernando G.H., Hirschberg J., Olmedilla B., Sandmann G., Southon S., Stahl W., "The potential for the improvement of carotenoid levels in foods and the likely systemic effects", *J Sci Food Agri.*, **80**, 880–912 (2000).

[238] Lisiewska Z., Kmiecik W., "Effect of storage period and temperature on the chemical composition and organoleptic quality of frozen tomato cubes", *Food Chem.*, **70**, 167–173 (2000).

[239] Biacs P., Wissgott U., "Investigation of color changes of some tomato products during frozen storage", *Mole Nutr Food Res.*, **41**(5), 306–310 (2006).

[240] Halliwell B., Gutteridge J.M.C., "The definition and measurement of antioxidants in biological systems", *Free Rad Biol Med.*, **18**, 125–26 (1995).

[241] Labuza T., "Kinetics of lipid oxidation in foods", *Crit Rev Food Tech.*, **10**, 355–405 (1971).

[242] Tsuchihashi H., Kigoshi M., Iwatsuku M., Niki E., "Action of β-carotene as an antioxidant against lipid peroxidation", *Archives Biochem Biophysics.*, **323**, 137–147 (1995).

[243] Liebler D.C., McClure T.D., "Antioxidant reactions of β-carotene: Identification of carotenoid-radical adducts", *Chem Res Tox.*, **9**, 8–11 (1996).

[244] Edge R., Mc Garvey D.J., Truscott T.G., "The carotenoids as antioxidants - A review", *J Photochem Photobio B-Bio.,* **41,** 189–200 (1997).

[245] Loliger J., Aruoma O.I., Halliwell B., "*The use of antioxidants in foods - Free radicals and food additives*", London: Taylor & Francis, pp. 121–150 (1991).

[246] Krinsky N.I., "Actions of carotenoids in biological systems", *Ann Rev Nutr.*, **13**, 561–587 (1993).

[247] Kiritsakis A., Dugan L.R., "Studies in photo-oxidation of olive oil", *J Am Oil Chem Soc.*, **62**, 892–896 (1985).

[248] Fakourelis N., Lee E.C., Min D.B., "Effects of chlorophyll and β-carotene on the oxidation stability of olive oil", *J Food Sci.*, **52**, 234–235 (1987).

[249] Lee E.C., Min D.B., "Quenching mechanism of β-carotene on the chlorophyll sensitized photo-oxidation of soybean oil", *J Food Sci.*, **53**, 1894–1895 (1988).

[250] Jung M.Y., Min D.B., "Effects of quenching mechanisms of carotenoids on the photosensitized oxidation of soybean oil", *J Am Oil Chem Soc.*, **68**, 653–658 (1991).

[251] Terao J., Yamauchi R., Murakami H., Matsushita S., "Inhibitory effects of tocopherols and b-carotene on singlet oxygen-initiated photo-oxidation of methyl linoleate and soybean oil", *J Food Sci Process Preserv.*, **4**, 79–93 (1980).

[252. Faria J.A.F., Mukai M.K., "Use of a gas chromatographic reactor to study lipid photo-oxidation", *J Am Oil Chem Soc.*, **60**, 77–81 (1983).

[253] Haila H., Heinonen M., "Action of α-carotene on purified rapeseed oil during light storage", *S.*, **27**, 573–577 (1994).

[254] Warner K., Frankel E.N., "Effects of β-carotene on light stability of soybean oil", *J Am Oil Chem Soc.,* **64**, 213–218 (1987).

[255] Suzuki T., Usuki R., Kaneda T., "The role of carotenoids in the oxidative deterioration of edible oils", *J Japan Oil Chem Soc.*, **38**, 486–491 (1989).

5

Nutritive Value and Processing of Green Leafy Vegetables

VINITA PURANIK[1], DEVINDER KAUR[1*], PINKI SAINI[1] AND NEELAM YADAV[1]

ABSTRACT

Green Leafy Vegetables (GLV) is treasure of nutrients and micronutrients that forms the second important category of vegetables. GLV like spinach, fenugreek, amaranth, chenopodium and mustard is discussed in this chapter. They are the rich sources of essential minerals, vitamins, dietary fibre and mainly antioxidants. GLV constitutes ascorbic acid, tocopherols, carotenoids and phenolic compounds as sources of antioxidants. GLV are found to be highly perishable due to high moisture content and non-availability of sufficient storage, transport and proper processing facilities at the production that causes mechanical injury, moisture loss, decay and aging. For the reduction of losses, proper packaging, transport and handling practices must be used. GLV processing steps include washing, trimming, refrigeration, blanching, freezing, canning, drying, packaging, use of preservatives, irradiation, storage and ultimately cooking or heating. Utilization of existing technologies and the development of new technologies for tracking product from the field to retail establishments must be emphasized. This would facilitate rapid and precise recalls, reducing the number of human illnesses and limiting monetary loss to an entire industry.

***Key words*:** Green Leafy Vegetables, Processing, Antioxidants Preservatives

I. INTRODUCTION

India is a country with diverse agro-climatic conditions that favours the cultivation and availability of wide array of fruits and vegetables. India is

[1] Centre of Food Technology, University of Allahabad, Allahabad - 211006, India
**Corresponding author* E-mail: devi_sonu@yahoo.com

the second largest producer of vegetables next to China, contributing about 10 per cent of the world production[1]. Being a major component of total agricultural produce of the country, horticulture products including fruits and vegetables encompass a large variety of crops. Green leafy vegetables (GLV), a rich source of nutrients in general and micronutrients in particular, are available at low cost throughout the year. They seem to be capable of delivering health benefits besides fulfilling physiological needs. Routine consumption of fruits and vegetables provides significant health benefits to human body[2]. Epidemiological data as well as *in vitro* studies suggest that foods containing phytochemicals with antioxidation potential have strong protective effects against major disease risks including cancer[3,4], cardiovascular diseases[5,12], type II diabetes[13], dementia[14] and macular degeneration[15]. In addition to the well-established benefits of essential vitamins and minerals found in high quantities in a wide range of fruit and vegetables, they also provide a good source of fiber to the diet and a diverse array of non-essential nutrients. The protective action of green leafy vegetables has been attributed to the presence of antioxidants, especially antioxidant vitamins including ascorbic acid, α-tocopherol and β-carotene[16,19].

The most important quality attributes contributing to the marketability of fresh and minimally processed produce include appearance, colour, texture, flavour, nutritional value, and microbial safety. These quality attributes are determined by plant variety, stage of maturity or ripening, and pre- and post-harvest conditions; which can change rapidly during post-harvest storage. There are numerous pre-harvest factors affecting the post-harvest quality and life of fruits and vegetables, including harvest maturity, cultivar, climate, and soil in which the produce grew, chemicals applied, and water status[20]. Within each commodity there is a range of genotypic variation in composition, quality, and post-harvest life[21]. The soil type and its fertility also affect the chemical composition of a produce. Fruits and vegetables undergo many physiological changes during post-harvest storage, including tissue softening, increase in sugar level, and decrease in organic acid levels, degradation of chlorophyll accompanied by the synthesis of anthocyanins or carotenoids upon maturation, production and losses of volatile flavour compounds, decrease in phenolic and amino acid contents, and breakdown of cell materials due to respiration[22]. Fruits and vegetables are susceptible to spoilage by microorganisms. Post-harvest spoilage was estimated to be about one fourth of the harvested produce[23, 24]. The survival and growth of pathogenic microorganisms are the major food safety concerns in fresh and minimally processed fruits and vegetables due to possible contamination during preparation and increased nutrients and cellular fluid on the fresh-cut surface. Implementation of appropriate post-harvest sanitation operation procedures during preparation, processing and maintenance of product in refrigerated and sanitized condition is essential for controlling microbial growth, thus providing good quality and safe products.

II. COMPOSITION

Nutrient composition of vegetables is very complex and difficult to assess. Presence of plant metabolites are strongly affected by genetic and environmental factors as well as transportation and storage conditions. Various growth factors such as light, temperature, humidity, type of soil, application of fertilizers, damage caused by microorganisms and insects, stress induced by UV radiation, heavy metals, and pesticides alter the metabolite composition of plants[25]. Before vegetables reach the retail market they have been handled by plant growers, transporters, packagers, storehouse operators, distributors, and/or processors. The physicochemical changes that occur in vegetables during these phases can lead to loss of potentially beneficial components[26].

GLVs contain around 2 to 7 per cent protein[27], *i.e.*, 30 to 3200 mg/100g, 0.09 to 40 mg/100g of iron and 120 to 14190 μg/100g of carotene[28]. They are found to be rich in carotenoids, iron, calcium, ascorbic acid, riboflavin, folic acid and appreciable amounts of other minerals[29, 30]. It has been observed that 95 percent of the total β-carotene available from all sources in the country, is derived from fruits and vegetables and, out of this, 52 percent is contributed by green leafy vegetables. An enhanced intake of β-carotene rich foods in daily diet may be preferred to the synthetic vitamin A dosage approach and can be one of the strategies for improving nutritional status in the country[31]. India, having a variety of natural surroundings and varying climates and seasons, has a number of species of edible leafy vegetables such as spinach, amaranth, bengal gram leaves, fenugreek, cauliflower leaves, mint and coriander.

Six green leafy vegetables *viz.* spinach, amaranth, bengal gram, cauliflower, mint, and coriander moisture content of the leaves varies from 75.1 percent to 95.4 percent and protein from 9.83 percent to 30.9 percent. Ascorbic acid content in leafy vegetables varies from 4.9 to 110.4 mg/100g[32] higher than those reported by some workers in coriander (88.6 mg/100 g) and spinach (25.4 mg/100 g)[33]; and in amaranth leaves (91.0 mg/100 g)[34]. Differences in leaf maturity and in the ascorbic acid assay employed are the reasons that accounts for the differences in ascorbic acid content of leafy vegetables[35, 36].

Among fresh leafy vegetables, spinach is an important source of nutrients ranking 2nd behind kale in total carotenoids and folate[37, 38] and 3rd in total antioxidant capacity behind only garlic and kale[39]. A single 30 g serving of fresh spinach containing 58 μg of folate and 2015 IU of vitamin A is equivalent to 29% and 20% of the daily value for each respective vitamin[40]. Adequate intake of folate is an important factor in the prevention of neural tube defects such as spina bifida and anencephaly, coronary artery disease, and colorectal cancer[41]. Carotenoids in the diet are essential for normal growth, reproduction and resistance to infection, and deficiencies have been linked to blindness and increased risk of several types of cancers[42]. Fenugreek (*Trigonella foenum graecum*) is an annual herb that belongs to the family *Leguminosae*, widely grown in India, Pakistan, Egypt, and Middle Eastern countries[43]. It is rich source of calcium, iron, α-carotene and other vitamins[44]. Fenugreek seed is widely used as a galactagogue by nursing mothers to increase breast milk supply[45]. The seeds of fenugreek contain lysine and L-tryptophan rich proteins, mucilaginous fiber and other rare chemical constituents such as saponins, coumarin, fenugreekine, nicotinic acid, sapogenins, phytic acid, scopoletin and trigonelline, which are accounted for many of its valuable therapeutic effects, may inhibit cholesterol absorption and also help lower sugar levels[46, 48]. Therefore, fenugreek seeds are used as a traditional remedy for the treatment of diabetes and hyper-cholesterolemia in Indian and Chinese medicines[49,50].

Vegetables are highly seasonal and perishable and are available in bulk during a particular season of the year. Large quantities of vegetables get spoiled due to the glut in the market because of non-availability of sufficient storage, transportation and proper processing facilities at the production point[51]. Enhancing utilization and increasing their availability in the off season may be done using suitable preservation techniques that are sustainable at the household level. The food industry uses a variety of preservation or processing methods to extend the shelf life of fruits and vegetables so that they can be consumed round the year and transported safely to consumers all over the world. Primarily, food preservation aims to create a microbiologically safe product, but processors also strive to produce the highest-quality food. Depending on the method of processing used there may be variations in change in color, texture, flavor and nutritional quality.

III. BLANCHING

Peroxidases are the enzymes present in plant tissues that cause oxidation of many compounds in the presence of oxygen during storage[52]. When these peroxidases react with hydrogen peroxide they form phototoxic free radicals which in turn cause loss of vitamins and develop cold damage symptoms[53,54]. A common technique employed to arrest enzyme activity and associated changes before processing is blanching[54,55]. Blanching consists of mild heat treatment of vegetables in different heating systems like steam, hot water and microwave for varying time periods. It is influenced by various factors like blanching media, temperature, time, physical and physiological characteristics of the vegetables, average size of the pieces and uniformity of heat distribution and penetration[56, 57]. Although blanching leads to some of the favourable effects like inactivation of enzymes, expelling air trapped in the intracellular regions, reduces any initial infections but it also causes loss of nutritional quality. As blanching is a prerequisite for preserving green leafy vegetables, it is very much essential to establish the conditions of blanching to enhance the utility of this nutritious economical source.

Several researchers have studied the effect of different processing methods on the level of carotenoids. Substantial losses of β-carotene have been reported during processing in different vegetables[58], but little or no change[59] or an increase in carotenoid levels as a result of processing[60] are also noticed depending on procedure. Higher losses of ascorbic acid of amaranth in water blanching followed by sun drying than steam blanching and cabinet drying have been reported[61]. Due to high z-value of spinach chlorophyll, a high temperature and short time treatment can increase chlorophyll retention[62] but improvement in colour of processed spinach has been observed by blanching at 65 ± 3C for up to 45 min[63].

Researches reports indicate that microwave blanched vegetables retain nutrients better when compared to conventional steam or water blanching due to the reduction of leaching losses during blanching. Because of its thermal resistance, peroxidase is generally used as an indicator enzyme to evaluate the adequacy of blanching[64]. Several studies on microwave blanching, either independently or in combination with steam or water heating have been reported[65,70].

When compared to conventional blanching, microwave blanching has the potential of reducing the loss of valuable nutrients. The kinetics of peroxidase inactivation indicates that microwave blanching is comparable to water blanching with higher reaction rate in the case of water blanching. The temperature and absorbed power levels during microwave blanching are influenced by the vegetable itself, its quantity, shape, location in the oven, and the microwave power applied[71].

IV. DRYING

Due to highly perishable in nature and low bulk density, a vegetable drying is considered to be the most suitable and widely used method of preservation. Hence, dehydration of leafy vegetables by reducing the moisture content below a critical level at which enzyme activity and growth of microorganisms is hindered adversely; the vegetables can be preserved for fairly long period. Dehydration is one of the traditional methods of preservation of vegetables, which converts the vegetable in to light weight, easily transportable and storable product. Advantage of this method is that the vegetable can be easily converted in to fresh-like form by rehydrating it and can be used throughout the year. In addition to increasing variety in the menu, reducing wastage, labour and storage space, dehydrated vegetables are simple to use and have longer shelf life than fresh vegetables[72]. The quality of the dehydrated product in terms of rehydration ratio, colour and flavour retention depends on the pretreatments applied and method of drying.

The sun and solar drying techniques of fruits and vegetables are dates back to many centuries. The poor quality and product contamination led to the development of alternate drying technologies[73]. The most applicable method of drying includes freeze, vacuum, osmotic, cabinet or tray, fluidized bed, spouted bed, ohmic, microwave and combination thereof[74]. Except for freeze drying, applying heat during drying through conduction, convection and radiation are the basic techniques used to force water to vapourise, while forced air is applied to encourage the removal of vapour. A large number of food and biomaterials are dehydrated in a variety of units with diverse processing conditions. The choice of drying method depends on various factors such as the type of product, availability of dryer, cost of dehydration and final quality of desiccated product. Energy consumption and quality of dried products are other critical parameters in the selection of a drying process.

Drying curry leaves in oven at 50°C was reported to be completed in 5.02 hours while drying in sun (35 ± 2°C) and shade (25 ± 2°C) required 3.16 and 7.20 days, respectively[75]. Among the three drying methods, oven drying was superior with emerald green colored curry leaf powder. A study was conducted to dehydrate the selected green leafy vegetables (fenugreek leaves, mustard leaves, *chenopodium leaves* and spinach) by mechanical and sun drying methods. The dehydration kinetics revealed that the moisture content decreased very rapidly during the first hour of drying with sun drying requiring about eight hours and mechanical drying about four hours to reach the desired moisture level of 9-11 per cent[76]. A study on improvement in rehydration and shelf life stability of hot air dried and sun dried cabbage by pre-drying treatment was carried out. Both treated (soaking in solutions containing 3% salt, 6% sugar and a combination of both at 0-4°C for 12-16 hours) and untreated (only blanched) cabbage was soaked at room temperature in 0.2 per cent KMS solution for 10 minutes and dried in hot air oven, solar cabinet drier and under direct sun. Addition of sugar increased the drying time of cabbage by one hour in hot air oven (8 hours), four hours in direct sun (13 hours) and three hours in solar cabinet drier (12 hours) apparently due to the humectant effect of sugar[77]. It is reported that, sun drying (35-40°C) of amaranth, curry leaves, *gogu* and mint required 14, 10, 24 and 21 hours, respectively, while in cabinet drier (60-70°C) the same leafy vegetables could be dried in 2.5, 1, 3 and 2.5 hours respectively[78].

Fenugreek required 4.5, 3.5, 3.0 and 2.5 hours for drying in a forced circulation drier at 40, 45, 50 and 60°C respectively whereas coriander leaves could be dried in 3.5, 3.0 and 2.5 hours at respective temperatures[79]. The time required for sun drying of cabbage, coriander leaves and *palak* has been reported to be 4.5, 3.0 and 3.0 hours respectively, while in solar cabinet drier the vegetables could be dried in 14.0, 8.0 and 8.0 hours respectively[80]. The vegetables blanched in osmotic solution of salt (5%) + KMS (0.1%)

maintained at room temperature (cold) or at 60°C (hot) produced dehydrated product with 7.5:1, 5.2:1 and 7.4:1 dehydration ratio, respectively after drying in electric tray drier. The osmo-dehydration treatment caused a weight loss of 12.8 and 18.8 per cent in cold and hot brining. It is also reported that the total time of 5, 3 and 6 hours was required for cold and hot brined and unbrined *methi* to be dried in a cabinet drier at 60°C[81]. The effect of different modes of drying on moisture content and drying time of selected leafy vegetables has been studied. The cabinet drying was a faster drying mode requiring less time for drying the vegetables than sun and shade drying. Shade drying requires maximum time for drying the samples. The time required for drying cabbage, fenugreek and spinach was 5.30, 2.30 and 3.30 hours, respectively in cabinet drier, 8.30, 5.15 and 6.00 hours, respectively under sun, and 32, 27 and 29 hours, respectively under shade. The moisture content of cabinet tray dried cabbage, fenugreek and spinach (8.05, 7.50 and 8.33% respectively) was significantly less than sun (9.13, 9.16 and 9.53% respectively) and shade dried samples (9.60, 9.25 and 10.26% respectively[82]. The effect of drying conditions on quality of dehydrated leafy vegetables was studied. Drumstick leaves took seven hours for drying while others took six hours under the same drying conditions. Among the driers, cabinet dryer was superior for dehydration of leafy vegetables as it reduced maximum moisture (2.5%) in both curry leaves and *methi*. The moisture content was higher in drumstick leaves (5.5%) and amaranth (5.2%) after dehydration[83].

V. ADVANCED PROCESSING TECHNIQUES

Advanced processing includes relatively newer technologies which may or may not be in commercial practice. They include high-pressure processing and use of various electric methods such as microwave, pulsed electric fields and between electric fields, ohmic processing. One tremendous advantage of these advanced methods is the uniform application of pressure or electric fields to the product as a whole, rather than needing to rely on heat or freezing temperature penetration from the external surface to the inside of the container. During pressurization there is some heating of the material, but this is generally less if temperature was the only means of preservation. Electric field processing generates heat locally, which also minimizes the amount of heat required. Advanced processes therefore minimize the temperature (and hence enhance the quality) gradient in the product and shorten the process time required. Heating through the use of electric fields and/or high pressure differs from conventional thermal processing because it is able to uniformly penetrate several centimetres into the food. Heat is generated quickly and evenly throughout the mass, and steam generated heats adjacent areas by conduction.

Ionizing irradiation is effective in inactivating pathogens both on the surface and inside of fresh produce[84, 87]. After irradiation at 0.7 kGy, *E. coli* O157:H7 and *Salmonella*, inoculated on the surface of spinach, were reduced by 4 logs[88]. Furthermore, *E. coli* O157:H7, internalized in leaves of spinach and lettuce, was inactivated by irradiation although higher doses were required compared to when the bacterium was on the surface[84]. Recently, the Food and Drug Administration (FDA) allowed the use of ionizing radiation for the control of food borne pathogens and extension of shelf-life in fresh iceberg lettuce and fresh spinach up to a maximum absorbed dose of 4.0 kGy[89]. There have been many studies related to the effect of irradiation on quality of lettuce[90, 92]. It was found that vitamin C in spinach was reduced by 25% after 1 kGy irradiation. The appearance of spinach was not significantly affected by 1 kGy irradiation[85]. A similar study showed that the chlorophyll content and overall visual quality were not affected by irradiation at doses up to 1 kGy. Firmness was reduced by irradiation after 7 days of storage at 4°C[86]. Another study was conducted to investigate the effects of irradiation at 1 kGy, a dose that potentially inactivates *E. coli* O157:H7 by 5 logs, on the quality of 13 common fresh-cut vegetables: iceberg, romaine, green and red leaf lettuce, spinach, tomato, cilantro, parsley, green onion, carrot, broccoli, red cabbage, and celery. The appearance of irradiated samples was similar to the non-irradiated ones except that irradiated carrots, celery, cilantro, and green onions had higher appearance scores than corresponding non-irradiated vegetables. There was no difference in the instrumental texture between irradiated samples and non-irradiated ones. The aroma of several irradiated vegetables was significantly better than controls after 14 days storage, because these control samples decayed or senesced. The 1 kGy irradiation did not affect vitamin C content of most vegetables; however, irradiated green and red leaf lettuce had 24% to 53% lower vitamin C contents than the controls. The results suggested that most fresh-cut fruits and vegetables tested can tolerate up to 1 kGy irradiation without significant losses in any of the quality attributes[85].

Electron beam irradiation has the potential to ensure the safety and quality of whole and fresh-cut fruits and vegetables[93, 96]. The main assets of this technology include its high lethality, low quality detrimental effects, unmatched penetration into the product core, cold application, capabilities to be applied in a conveyor belt, and tested safety of the intervention technology[97]. A preliminary study on the reduction of *E. coli* O157:H7 on baby spinach by electron beam irradiation using a 10-MeV linear accelerator (LINAC) showed that doses as low as 0.9 kGy reduced the population of different strains of *E. coli* O157:H7 by 5 to 6 log cycles. At doses of 1.2 kGy or higher, no pathogens were detected in any of the samples. The respiration rate of irradiated spinaches was not different from non-irradiated controls except at 3 kGy. Sensory characteristics such as crispiness or color were not affected by irradiation up to 2 kGy[98] however, whenever e-beams are used to irradiate a food product, a compromise must be achieved between the quality of the treatment (that is, decontamination, surface pasteurization) and the radiation dose received by the food.

VI. PACKAGING

Shelf life and quality of fresh-cut produce may be greatly reduced due to high rates of respiration. Traditional packaging and handling techniques reduce the shelf life and sensory quality of fresh leafy vegetables. Temperature control and atmospheric modifications help to maintain produce quality by reducing respiration rate and enhance shelf life by minimizing the adverse effects of cutting.

Among the available methods to control or moderate the degradative processes, packaging technology is vitally important. In particular, the use of permeable polymeric films could retard microbial growth by affecting the concentration of gases within the package[99,100]. However, improperly modified atmospheric conditions may adversely affect product quality. Therefore, the selection of suitable packaging materials is critical to obtain the desirable environment to extend the shelf life of fresh-cut vegetables.

Modified atmosphere packaging (MAP) along with low temperature storage extends the shelf life of fresh produce packaged in polymeric film packages through interaction of the natural process of respiration of produce with the restricted gas exchange across the polymeric film package to control the in-pack O_2 and CO_2. Appropriate steady-state in-pack gas concentrations have been reported to delay senescence, maintain physicochemical constituents and extend shelf life of different crops[101,103]. Selection of an appropriate packaging film and the in-pack weight are the two important variables which influence the various physicochemical constituents of the packaged produce through creation of different in-pack gas compositions of O_2 and CO_2. Increase in the produce weight, for the same package size, reduces package free volume and causes more intensive and faster changes in the in-pack atmosphere.

In general, an inverse relationship has been shown between respiration rates of fruits and vegetables and their postharvest shelf-life. Reduced O_2 and high CO_2 levels have also been proved to effectively control enzymatic browning, firmness and decay of fresh-cut fruits and vegetables. Besides, the proliferation of aerobic spoilage microorganisms can be substantially delayed with reduced O_2 levels. However, under certain conditions, the growth of some anaerobic psychrotrophic pathogens might be allowed or even stimulated. An adequate O_2 concentration inside packages is required to limit aerobic respiration without triggering anaerobic processes, as too low O_2 and D or excessive CO_2 concentrations may induce the production of undesirable metabolites and the unleashing of several physiological disorders[104]. In the last few years, the use of elevated O_2 atmospheres (70 kPa O_2) has been proposed as an alternative to low O_2 atmospheres to inhibit the growth of naturally occurring spoilage microorganisms, prevent undesired anoxic respirative processes and maintain the fresh-like quality of fresh-cut produce[105, 107].

VII. EFFECT OF PROCESSING ON NUTRIENTS

Ascorbic Acid

Ascorbic acid is water soluble and sensitive to oxygen, light, heat and oxidizing agents. During the various processing treatments ascorbic acid is most difficult to preserve. Various factors *viz.* pH, trace metals, enzymes, oxygen, temperature, low relative humidity, extended storage influenced the oxidation of ascorbic acid. In aqueous solution ascorbic acid is easily oxidized in presence of oxygen, heavy metals, alkaline pH and high temperature. DHA can be reduced to ascorbic acid with use of reducing agents and this can be irreversibly oxidized to diketogluconic acid, which has no vitamin C activity[108]. Ascorbic acid oxidase is also responsible for the enzymatic degradation of ascorbic acid. In the presence of molecular oxygen ascorbate oxidase oxidizes ascorbic acid to DHA[109]. The activity of ascorbic acid oxidase is prevented during the blanching of GLV whereas phenolase, cytochrome oxidase and peroxidase, are indirectly responsible for ascorbic acid degradation. Conventional food processing treatments results in significant decrease in vitamin C content. Various preliminary preparations like cutting, slicing etc. and heating in water make vitamin C more susceptible to leeching. Blanching is initial step in processing of vegetables in which vegetables are boiled or steamed for few minutes for the inactivation of enzymes, which other wise would cause the development of off flavors and loss of pigments and ascorbic acid. 20-70% of vitamin C is lost during the blanching of green leafy vegetables. This leeching loss is due to surface to volume ratio of green leafy vegetables are high as compare to other vegetables[110, 119]. GLV results in 9-67% losses of vitamin C during cooking[84, 88]. During processing the loss of vitamin C is due to the fact that it is soluble in water and is sensitive to high processing temperatures.

Ascorbic acid, the most water-soluble and heat labile vitamin was found to be lost in large quantities on dehydration. The exposure of GLV to direct sunlight resulted in higher vitamin C loss and this may be due to the most accessible atmospheric constituents like oxygen, light, temperature. Vitamin C content of the dehydrated leaves was less than the fresh sample. This was the only nutrient, which reduced after dehydration as it is oxidized rapidly on exposure to heat and air. The maximum amount of vitamin C was in shadow dried sample as in this technique the leaves were not exposed to direct heat and air. A considerable variation was seen in retention of vitamin C on dehydration of different leaves. Almost all vitamin C was destroyed in leaves of *Chenopodium album* whereas *Centella asiatica* retained 14% of its initial vitamin C content. Among the other GLV, the retention of vitamin C was in the range of 4–8%. Conversely Kaur *et al,*[120] have reported the retention of vitamin C in mustard, mint and spinach to be as high as 13–38%.

The pre-freezing operations results in 19.1-51.5% loss in vitamin C content of various vegetables. 60-70% vitamin C loss has been observed in storage of frozen green leafy vegetables[121, 122]. The process is dependent on the freezing rate, the heat transfer coefficient and amount of heat removed from the food product. Depending on the vegetable type, pre-freezing operations have a major impact on the vitamin C contents and this influence persists in the frozen storage[123]. It was reported that during frozen storage, in order to minimize the vitamin C loss, proper inactivation of the oxidases and prevention of storage temperature fluctuations are the key issues to be considered[124, 125].

Oxalic Acid

Studies reported that storage of fresh leaves with or without polyethylene bags for 24 and 48 h in refrigeration, storage at 30ºC in polyethylene bags or drying in the sun or in an oven did not cause any significant differences in the oxalic acid contents as oxalic acid contents were found to be similar in sun and oven dried leaves. Blanching significantly lowers the oxalic acid content in leaves as compared to the fresh leaves and this decrease was found to be higher in the leaves blanched for a longer time period. Non significant differences in oxalic acid content were observed in the fenugreek and *chenopodium* leaves blanched for 5, 10 and 15 min. Significant differences were observed in the oxalic acid contents of spinach and amaranth leaves blanched for 10 and 15 min. Cooking seemed to decrease oxalic acid content in the leaves to a greater extent than drying; open pan and pressure cooking had similar effects. Blanching involves interaction of leaves with hot water, which may rupture the cell wall, and soluble oxalic acids may leach out in the blanching medium. This leaching out of oxalic acid may account for the loss in the blanched leaves. Similarly some workers also have reported the loss of oxalate upon boiling and frying[126].

Phytic Acid

Drying (sun and oven) and storage (at 30°C in polyethylene bags) conditions did not change the phytic acid content of the amaranth, *chenopodium*, fenugreek and spinach leaves significantly[126] while some workers have reported lower phytic acid content in amaranth leaves[127] whereas higher phytate content in amaranth and fenugreek by others[128,130]. Studies reported that the Phytic acid content of the leaves blanched for 5, 10 and 15 min were ranged from 127.33 to 231.33, 126.0 to 230.07 and 125.87 to 229.56 mg/100 g dry weight, respectively and Phytic acid content of these leafy vegetables cooked in an open pan and a pressure cooker varied from 129.47 to 234.36 and 129.43 to 234.32 mg/100 g dry weight, respectively. Cooking did not significantly change the phytic acid content of the leaves. During cooking endogenous phytases are inactivated by heat[131]. They are, therefore, unavailable to breakdown phytate which can then only be degraded by high temperature processing[132]. Blanching, however, reduced the level of this antinutrient; the longer the period of blanching the greater the lowering effect. Blanching involves interaction of leaves with hot water, which may rupture the cell wall, and soluble phytic acid may leach out in the blanching medium which may account for losses from the blanched leaves. Similarly other workers have also reported that the level of phytic acid was significantly reduced by conventional blanching of green leafy vegetables[133].

Polyphenols

Drying (sun and oven) and storage (at 30°C in polyethylene bags) methods used in the study did not significantly alter the polyphenol content of the leafy vegetables[126]. Higher polyphenol content in fenugreek leaves and lower in amaranth leaves have been reported earlier[128], while higher tannins in amaranth leaves have been reported by others[130]. Blanching could significantly lower polyphenol level. Cooking on the other hand did not significantly decrease the polyphenol contents. It was observed that the longer the period of blanching, the greater the lowering effect. Lower polyphenol contents in the blanched leaves might have been due to leaching of this antinutrient into the blanching medium.

Under MAP, the pigments and phenolic content were largely influenced by the dynamics of in-pack gaseous atmosphere inside the polymeric film packages as Kaur *et al*.,[134] reported in the study that LDPE film-packaged spinach samples had higher phenolic content than PP film-packaged samples and low in-pack O_2 (1–2 kPa) along with the buildup of CO_2 (8–10 kPa) seems to have enhanced the retention of antioxidant components, *i.e.*, β-carotene and ascorbic acid, in LDPE packages which helped in chlorophyll retention and also prevented formation of oxidation compounds of phenols that otherwise caused browning of cut surfaces. Thus, O_2 and CO_2 permeability of packaging film or in-pack weight of produce might be such that equilibrated O_2 partial pressures remain near to 1–2 kPa so as to affect a beneficial increase in the phenolic content along with concomitant retention of chlorophyll. There was decrease in total phenolics in vegetable spinach (42% loss) which was microwaved at 800 W for drying[135]. Phenolic content was found to be more degraded after boiling than microwaving[135].

Antioxidants

Different cooking methods (boiling, frying and microwave cooking) appreciably affect the antioxidant activity of the green leafy vegetables. On measurement of total phenolic contents (TPC), reducing power and percentage inhibition in linoleic acid system, TPC of vegetables decreases by boiling, frying and microwave cooking, reducing power increases as a result of frying, however, boiling and microwave cooking does not affect reducing power. Inhibition of per oxidation increases by boiling and frying, whereas, in contrast, it decreases by microwave cooking. So, all the cooking methods affect the antioxidant properties of the vegetables; however, microwave treatment exhibits more deleterious effects when compared with those of other treatments. Thus, an appropriate method might be sought for the processing of such vegetables to retain their antioxidant components at maximum level[136].

The spinach is found to be more tolerant to irradiation than many other leafy greens such as lettuce, probably due to its higher antioxidant capacity and greater ascorbic acid content. Ascorbic acid serves as the primary water soluble antioxidant in plant tissues and preferentially interacts with reactive species such as those radicals generated from radiolysis of water, thus protecting cell constituents from oxidative damage. As visual quality of spinach after storage is correlated with initial ascorbic acid content, so, spinach with higher ascorbic acid content has longer storage life. Irradiation at doses up to 2 kGy does not significantly affect the liking of cut spinach or overall antioxidant capacity of spinach even though irradiation promotes the loss of ascorbic acid during cold storage[137].

On storage of modified atmosphere packaged fresh-cut spinach leaves for 4 days at 15°C and 75% relative humidity with differential in-pack gas atmospheres generated through different packaging films and variable in-pack weight shows retention of chlorophyll, β-carotene and ascorbic acid better in low-density polyethylene (LDPE) packages than in polypropylene (PP) packages. LDPE film-packaged samples have higher phenolic content than PP film-packaged samples. Low in-pack O_2 (1–2 kPa) along with the buildup of CO_2 (8–10 kPa) enhances the retention of antioxidant components, *i.e.*, β-carotene and ascorbic acid, in LDPE packages. This helps in chlorophyll retention and also prevents formation of oxidation compounds of phenols that otherwise causes browning of cut surfaces. Under MAP, the pigments and phenolic content are largely influenced by the dynamics of in-pack gaseous atmosphere inside the polymeric film packages. The chlorophyll, ascorbic acid and β-carotene retention is considerably higher in LDPE packages than in PP packages at the end of 4 days of storage. Phenolic content increases in LDPE packages owing to very low O_2 partial pressures as compared to PP packages where it decreases[138].

Conclusion

Green Leafy Vegetables (GLV), a rich source of nutrients in general and micronutrients in particular, are available at low cost throughout the year. They seem to be capable of delivering health benefits besides fulfilling physiological needs. Routine consumption of fruits and vegetables provides significant health benefits to human body. Post-harvest spoilage was estimated to be about one fourth of all produce harvested. Implementation of appropriate post-harvest sanitation operation procedures during preparation and processing and maintenance of product in refrigerated and sanitized condition is essential for controlling microbial growth, thus providing high-quality and safe products. Depending on the method of processing used there may be variations in change in color, texture, flavor and nutritional quality. It was found that oxalic acid, phytic acid and polyphenol contents of the leaves were the same even after processes like drying and storage. Blanching and cooking resulted in significant reduction in oxalic acid; only blanching resulted in significant reduction in phytic acid and polyphenol contents. The reduction in these antinutrients might have asserted a beneficial effect on availability of minerals like iron, calcium and zinc, as antinutrients are shown to have hindering effects on mineral bioavailability.

REFERENCES

[1] Singh U., Sagar V.R., "Influence of Packaging and storage temperature on the quality of dehydrated selected leafy vegetables", *J Food Sci Tech.*, **45**(5), 450–453 (2008).

[2] Steinmetz K.A., Potter J.D., "Vegetable, fruit and cancer prevention - A review", *J of Am Diet Asso.*, **2**, 325–351(1996).

[3] World Cancer Research Fund/American Institute for Cancer Research (2007). Food, Nutrition, Physical Activity, and the Prevention of Cancer: A Global Perspective. AICR, Washington DC (2007).

[4] Appel L.J., Moore T.J., Obarzanek E., Vollmer W.M., Svetkey L.T., Sacks F.M., "A clinical trial of the effects of dietary patterns on blood pressure", *N Engl J Med.*, **336**,1117–1124 (2011).

[5] Steinberg D., "Antioxidants and atherosclerosis: A current assessment", *Circulation.*, **84**, 1420–1425 (1991).

[6] Block G., Patterson B., Subar A., "Fruits vegetables and cancer prevention: A review of the epidemiological evidence", *Nutr and Cancer.*, **18**, 1–29 (1992).

[7] Ames B.N., Shigenga M.K., Hagen T.M., "Oxidants, antioxidants and degenerative diseases of aging", *Prod of National Aca of Sci.*, USA, **90**, 7915–7922 (1993).

[8] Hertog M. G., Feskens E. J. M., Hollman P.C.H., Katan M.B., Kromhout D., "Dietary antioxidant flavonoids and risk of coronary heart disease: The Zutphen elderly study", *Lancet*, **342**, 1007–1011 (1993).

[9] Byers T., Guerrero N., "Epidemiological evidence of vitamin C and vitamin E in cancer prevention", *Am J of Clin Nutr.*, **62**(Suppl. 6), 1385S–1392S (1995).

[10] Knekt P., Jarvinen R., Seppanen R., Heliovaara M., Teppo L., Aroma A., "Dietary flavonoids and the risk of lung cancer and other malignant neoplasm", *Am J of Epidem.*, **146**, 223–230 (1997).

[11] Elliott J.G., "Application of antioxidant vitamins in foods and beverages", *Food Technol.*, **53**, 46–48 (1999).

[12] Kaur C., Kapoor H.C., "Antioxidants in fruits and vegetables – the millennium's health", *Int J Food Sci and Technol.*, **36**, 703–725 (2001).

[13] Carter P., Gray L.J., Troughton J., Khunti K., Davis M.J., "Fruit and vegetable intake and incidence of type 2 diabetes mellitus: Systematic review and meta-analysis". *Br Med J.,* **341**, c4229 (2010).

[14] Hughes T.F., Andel R., Small B.J., Borenstein A.R., Mortimer J.A., "Midlife fruit and vegetable consumption and risk of dementia in later life in Swedish twins", *Am J Geriatr Psychiat.,* **18**, 413–420 (2010).

[15] Cho E., Seddon J.M., Rosner B., Willett W.C., Hankinson S.E., "Prospective study of intake of fruits, vegetables, vitamins, and carotenoids and risk of age-related maculopathy", *Arch Ophthalmol.,* **122**, 883–892 (2004).

[16] Gey K.F., Puska P., Jordan P., Moser U.K., "Inverse correlation between plasma vitamin E and mortality from ischemic heart disease in cross cultural epidemiology". *Am J of Clini Nutr.*, **53**, 326–334 (1991).

[17] Willet W.C., "Micronutrients and cancer risk", *Am J of Clini Nutr.*, **59**, 162–165 (1994).

[18] Kalt W., Kushad M.M., "The role of oxidative stress and antioxidants in plant and human health: Introduction to the colloquium", *Hort Sci.*, **35**, 572–574 (2000).

[19] Prior R.L., Cao G., "Antioxidant phytochemicals in fruits and vegetables. Diet and health implications", *Hort Sci.,* **35**, 588–592 (2000).

[20] Thompson A.K., "Preharvest factors on postharvest life", *In: Fruit and vegetables. Ames, Iowa*: Blackwell Publishing Ltd. pp. 1–8 (2003a).

[21] Kader A.A., "Quality parameters of fresh-cut fruit and vegetable products", *In: Fresh-cut fruits and vegetables*. Boca Raton, Fla.: CRC Press LLC. pp. 11–20 (2002).

[22] Sharma R.M., Singh R.R., "Harvesting, postharvest handling and physiology of fruits and vegetables", *In: Postharvest technology of fruits and vegetables* Vol. 1. Handling, processing, fermentation and waste management. Tagore Garden, New Delhi: Indus Publishing Co. pp. 94–147 (2000).

[23] Salunkhe D.K., "Developments in technology of storage and handling of fresh fruits and vegetables", *Cri Rev Food Technol.,* April, pp. 15–54 (1974).

[24] Salunkhe D.K., Kadam S.S., "Handbook of vegetable science and technology: Production, composition, storage, and processing", New York: Marcel Dekker (1998).

[25] Orcutt D.M., Nilsen E.T., "The physiology of plants under stress soil and biotic factors", New York: John Wiley and Sons Inc. p. 680 (2000).

[26] MacEvilly C., Peltola K., "The effect of agronomy storage processing and cooking on bioactive substances in food", *In:* Goldberg G. (*ed.*), *Plants: Diet and health.* Oxford, U.K., Blackwell Publishing Ltd. pp. 226–39 (2003).

[27] Narasinga Rao B.S, Deosthale Y.G., Pant K.C., "Nutritive Value of Indian Foods", National Institute of Nutrition. Indian Council of Medical Research, Hyderabad, India (1989).

[28] Gopalan C., Rama Sastri B.V., Balasubramanian S.C., "Nutritive value of Indian foods", National Institute of Nutrition. Indian Council of Medical Research, Hyderabad, India, (2004).

[29] Devadas R.P., Saroja S., "Availability of Fe and β-carotene from amaranthus to children", *In*: Emmaus P. (*ed.), Proceeding of the 2nd Amaranthus Conference, Rodala Press Inc.*, pp. 15–21 (1980).

[30] Prakash D., Pal M., "Nutritional and antinutritional comparison of vegetable and grain Amaranthus leaves", *J Sci Food Agric.,* **57**, 573–585 (1991).

[31] Gopalan C., "New dimensions of old problem", *In: Nutrition in Development Transition in South-East Asia,* New Delhi: World Health Organisation, Regional Office for South-East Asia, pp. 34–48 (1992).

[32] Singh G., Kawatra A., Sehgal S., "Nutritional composition of selected green leafy vegetables, herbs and carrots", *Plant Foods for Human Nutr.,* **56**, 359–364 (2001).

[33] Tapadia S.B., Arya A.B., Rohinidevi P., "Vitamin C contents of processed vegetables", *J Food Sci Technol.*, **32**, 513–515 (1995).

[34] Yadav S.K., Sehgal S., "Effect of home processing on ascorbic acid and β-carotene content of spinach (*Spinacia oleracia*) and Amaranthus (*Amaranthus tricolos*) leaves" *Plant Food Hum Nutr.*, **47**, 125–131 (1995).

[35] Oke O.L., *Exp Agric.*, **4**, 345 (1968).
[36] Stepanova E.N., Grigoreva M.P., Smirnova E.V., "Vopa Pitan" 32, 57; *In*: *Nutr Abstr Rev.*, 1974, **44**, 241. No. 2220 (1973).
[37] Holden J.M., Eldridge A.L., Beecher G.R., Buzzard I.M., Bhagwat S.A., Davis C.S., Douglass L.W., Gebhardt S.E., Haytowitz D.B., Schakel S., "Carotenoid content of U.S. foods: An update of the database", *J Food Comp Anal.*, **12**,169–96 (1999).
[38] U.S. Dept. of Agriculture, Agricultural Research Service (2003). http:// www.nal.usda.gov/ fnic/foodcomp.
[39] Cao G., Sofic E., Prior R.L., "Antioxidant capacity of tea and common vegetables", *J Agric Food Chem.*, **44**, 3426–31 (1996).
[40] National Academy of Sciences (1989). Recommended dietary allowances. 10th ed. Washington, D.C., Subcommittee on the 10th Edition of the RDAs. Food and Nutrition Board. Commission on Life Sciences, Natl. Research Council (1989)
[41] Herbert V., "Folic acid", *In:* Shils M., Olson J., Shike M. and Ross A.C. (*eds.*), *Nutrition in health and disease*. Baltimore: Williams & Wilkins (1999).
[42] Tee E.S., "Carotenoids and retinoids in human nutrition", CRC *Cr Rev Food Sci.*, **31**,103–163 (1992).
[43] Alarcon-Aguilara F.J., Roman-Ramos R., Perez-Gutierrez S., Aguilar-Contreras A., C. Contreras-Weber C., Flores-Saenz J. L., "Study of the anti-hyperglycemic effect of plants used as antidiabetics", *J Ethnopharmacol.*, **61**,101 (1998).
[44] Sharma R.D., Sarkar A., Hazra D. K., "Use of fenugreek seed powder in the management of non-insulin dependent diabetes mellitus", *Nutr Rep..*, **16**(8), 1331–1339 (1996).
[45] Fleiss P., "Herbal remedies for the breastfeeding mother", *Mothering* Summer, **68** (1988).
[46] Billaud C., "Fenugreek: Composition, nutritional value and physiological properties" *Sciences-des-ailments.*, **21**, 3–26 (2001).
[47] Sauvaire Y., Ribes G., Baccou J.C., Loubatieres-Mariani M.M., "Implication of steroid saponins and sapogenins in the hypocholesterolemic effect of fenugreek", *Lipids Mar.*, **26**(3), 191–197 (1991).
[48] Ribes G., Sauvaire Y., Costa C.D., "Antidiabetic effects of subfractions from fenugreek seeds in diabetic dogs", *Proc Soc Exp Biol Med.*, **182**(2), 159–166 (1986).
[49] Basch E., Ulbricht C., Kuo G., Szapary P. Smith M., "Therapeutic applications of fenugreek", *Altern Med Rev.*, **8**(1), 20–27 (2003).
[50] Miraldi E., Ferri S., Mostaghimi V., "Botanical drugs and preparations in the traditional medicine of West Azerbaijan (Iran)", *J Ethnopharmacol.*, **75**(2–3), 77–87 (2001).
[51] Pande V.K., Sonune A.V., Philip S.K., "Solar drying of coriander and methi", *J Food Sci Tech.*, **37**(2), 110–113 (2000).
[52] Murcia M.A., Lopez Ayerra B., Martinenz-Tome M., Vera A.M., Garcia-Carmona F., "Evolution of ascorbic acid and peroxidase during industrial processing of broccoli", *J Sci Food agric.*, **80**(13), 1882–1886 (2000).
[53] Brewer M.S., Begum S., Bozeman A., "Microwave and conventional blanching methods on chemical, sensory and color characteristics of frozen broccoli", *J Food Qual.*, **18**(6), 479–493 (1995).
[54] Ihl M., Monsalves M., Bifani V., "Chlorophyllase in activation as a measure of blanching efficacy and color retention of artichokes (*Cynara scolymus L.*)", *Lebensm Wiss Technol*, **31**, 50–56 (1998).
[55] Gaiser M., Rathjen A., Spiess W.E.L., "Nitrate extraction during blanching of spinach". *Lebensm Wiss Technol*, **30**, 432–435 (1996).
[56] Muftagil N., "The peroxidase enzyme activity of some vegetables and its resitance to heat", *J Sci Food Agric.*, **36**, 877–880 (1985).
[57] Masure M.P., Campbell H., "Rapid estimation of peroxidase in vegetable extracts – an index of blanching for frozen vegetables", Fruit Prod *J Am Food Manufact.*, **23**, 36–374 (1944).

[58] Speak A.J., Speak S.S., Sshreurs W.H.P., "Total carotenoid and b-carotene contents of Thai vegetables and the effect of processing. *Food Chemistry*, **27**, 245–257 (1988).

[59] Khachik F., Goli B.M., Beecher G.R., Holden J., Lushy W.R., Tenorio M.D., Barrera M.R., "Effect of food preparation on qualitative and quantitative distribution of major carotenoid constituents of tomatoes and several green vegetables", *Journal of Agri and Food Chem.*, **40**, 390–398 (1992).

[60] Grando F., Olmedilla B., Blanco I., Rojashidalgo, "Carotenoid composition in raw and cooked Spanish vegetables", *J of Agric and Food Chem.*, **40**, 2135–2140 (1992).

[61] Onayemi O., Badify G.I.O., "Effect of blanching and drying methods on the nutritional and sensory quality of leafy vegetables", *Plant Food for Human Nutr.*, **37**, 291–298 (1987).

[62] Gupte S.M., Elbisi H.M., Francis F.J., "Kinetics of thermal degradation of chlorophyll in spinach puree", *J of Food Sci.*, **29**, 379–382 (1964).

[63] Loef H.W., Thung S.B., Uber den Einub von chlorophyllase auf die Farbe von spinat von wabrend und nach der verwertung. *Zeitschrift fur ebensmittel ntersuchung Forschung.*, **126**, 401–406 (1965).

[64] Günes B., Bayindirh A., "Peroxidase and lipoxygenase inactivation during blanching of green beans, green peas, and carrots", *Lebensm-Wiss und-Technol.*, **26**, 406–410 (1993).

[65] Collins J.L., McCarty I.E., "Comparison of microwave energy with boiling water for blanching of whole potatoes", *Food Technol.*, **23**(3), 63–66 (1969).

[66] Dietrich W.C., Huxsoll C.C., Guadagni D.G., "Comparison of microwave, conventional, and combination blanching of brussel sprouts for frozen storage", *Food Technol.*, **24**(5),105–109 (1970).

[67] Huxsoll C.C., Dietrich W.C., Morgan J.A.I., "Comparison of microwave with steam or water blanching of corn-on-the-cob", *Food Technol.*, **24**(3), 84–90 (1970).

[68] Chen S.C., Collins J.L., McCarty I.E., Johnston M.R., "Blanching of white potatoes by microwave energy followed by boiling water", *J Food Sci.*, **36**,742–743 (1971).

[69] Avisse C., Varoquaux P., "Microwave blanching of peaches", *J Microwave Power.*, **12**(1), 73–75 (1977).

[70] Ramaswamy H.S., Fakhouri M.O., "Microwave blanching: Effect on peroxidase activity, texture and quality of frozen vegetables", *J Food Sci Technol.*, **35**, 216–222 (1988).

[71] Ramesh M.N., Wolf W., Tevini D., Bognar A., "Microwave Blanching of Vegetables", *J of Food Science.*, **67**(1), 390–398 (2002).

[72] Chauhan S.K, Sharma C.R., "Development of instant dehydrated saag", *Beverage and Food World*, **20**(4), 25–26 (1993).

[73] Bezyma L.A., Kutovoy V.A., "Vaccum drying and Hybrid technologies", *Stewart Pst Harvest Rev.*, **4,** 6–13 (2005).

[74] George S.D., Cenkowski S., Muir W.E., "A review of drying technologies for the preservation of nutritional compounds in waxy skinned fruit", North Central ASAE/CSAE Conf. Winnipeg, Manitoba, Canada, 24-25 September, MB04-104. (2004).

[75] Madalageri B.B., Mahadev, Hiremath S.M., "Dehydration methods, Oil extraction and flavor components- Detection in curry leaf (*Murraya koenigii* Spreng) and detection of flavor components", *Karnataka J Agric Sci..,* **9**(2), 284–288 (1996).

[76] Singh H., Bawa A.S., Ahmed J., "Dehydration characteristics of some green leafy vegetables", *Indian Food Packer*, **51**(2), 5–13 (1997).

[77] Gupta D.K., Baburao N., Jayaraman K.S., "Improvement in rehydration and shelf life stability of hot air dried and sun dried cabbage by pre-drying treatment", *Indian Food Packer*, **53**(6), 51–60 (1999).

[78] Lakshmi B., Vimala V., "Nutritive value of dehydrated green leafy vegetable powders", *J Food Sci Tech.*, **37**(5), 465–471 (2000).

[79] Pande V.K., Sonune A.V., Philip S.K., "Solar drying of coriander and methi", *J Food Sci Tech.*, **37**(2), 110–113 (2000).

[80] Unde P.A., Padian S.M., Chaudhari R.S., "Osmotic dehydration of leafy vegetables", *Beverage and Food World*, **26**(1), 26–27 (2000).

[81] Birar R.M., Ferande V.R., Unde P.A., "Effect of pretreatments on quality of dehydrated fenugreek (*methi*)", *Beverage and Food World*, **28**(12), 23–24 (2001).

[82] Bhosle B.S., Arya A.B., "Effect of different modes of drying time of selected vegetables", *The Indian J Nutr Diet.*, **41**(7), 293–298 (2004).

[83] Singh U., Sagar V.R., Behera T.K. , Suresh Kumar P., "Effect of drying conditions on the quality of dehydrated selected leafy vegetables", *J. Food Sci Tech..*, **43**(6), 579–582 (2006).

[84] Niemira B.A., "Relative efficacy of sodium hypochlorite wash versus irradiation to inactivate *Escherichia coli* O157:H7 internalized in leaves of Romaine lettuce and baby spinach". *J Food Prot.*, **71**(11), 2526–2532 (2007).

[85] Fan X., Niemira B.A., Prakash A., "Irradiation of fresh and fresh-cut fruits and vegetables", *Food Technol.*, **3,** 36–43 (2008).

[86] Gomes C., Da Silva P., Moreira R.G., Castell-Perez E., Ellis E.A., Pendleton M., "Understanding *E. coli* internalization in lettuce leaves for optimization of irradiation treatment", *Intern J Food Microbiol.*, **135**(3), 238–247 (2009).

[87] Mahmoud B.S.M., "Effects of X-ray radiation on *Escherichia coli* O157:H7, *Listeria monocytogenes*, *Salmonella enterica* and *Shigella flexneri* inoculated on shredded iceberg lettuce", *Food Microbial,* **27**,109–114 (2009).

[88] Neal J.A., Cabrera-Diaz E.M., Crquez-Gonza´alez M., Maxim J.E., Castillo A., "Reduction of *Escherichia coli* O157:H7 and *Salmonella* on baby spinach, using electron beam radiation", *J Food Protect.,* **7,** 2415–2420 (2008).

[89] FDA 2008, "Irradiation in the production, processing and handling of food", Fed Regist., **73**(164), 49593–603 (2008). http://www.fda.gov/OHRMS/DOCKETS/98fr/cf00135.pdf

[90] Prakash A., Guner A.R., Caporaso F., Foley D.M., "Effects of low-dose gamma irradiation on the shelf life and quality characteristics of cut romaine lettuce packaged under modified atmosphere", *J Food Sci.,* **65**(3), 549–553 (2000).

[91] Fan X., Sokorai K.J.B., "Sensorial and chemical quality of gamma irradiated fresh cut iceberg lettuce in modified atmosphere packages", *J Food Protect.,* **65**(11), 1760–1765 (2002).

[92] Fan X., Toivonen P.M.A., Rajkowski K.T., Sokorai K.J.B., "Warm water treatment in combination with modified atmosphere packaging reduced undesirable effects of irradiation on the quality of fresh-cut iceberg lettuce", *J Agric Food Chem.*, **51**, 1231–1236 (2003).

[93] Han J., Gomes-Feitosa C.L., Castell-Perez M.E., Moreira R.G., Silva P.F., "Quality of packaged romaine lettuce hearts exposed to low-dose electron beam irradiation", *Lebens Wiss Technol.*, **37**(7), 705–715 (2004).

[94] Moreno M., Castell-Perez M.E., Gomes C., Da Silva P., Moreira R.G., "The effects of electron beam irradiation on physical, textural and microstructural properties of Tommy Atkins mangoes (*Mangifera indica* L.)", *J Food Sci.,* **71**(2), E80–E86 (2006).

[95] Moreno M., Castell-Perez M.E., Gomes C., Da Silva P., Kim J., Moreira R.G., "Optimizing electron beam irradiation of "Tommy Atkins" mangoes (*Mangifera indica* L.)", *J Food Process Eng.,* **30**(4), 436–457. (2007).

[96] Moreno M.A., Castell-Perez M.E., Gomes C., Da Silva P., Moreira R.G., "Quality of electron beam irradiation of blueberries (*Vaccinium corymbosum* L.) at medium dose levels (1.0 to 3.2 kGy)", *LebensWiss Technol / Food Sci Technol.*, **40,** 1123–1132. (2007).

[97] Miller R.B., "Electronic irradiation of foods: An introduction to the technology", New York, Springer, (2005).

[98] Castillo A., Personal communication. Dept. of Animal Science, Texas A & M Univ., College Station, Tex. (2006).

[99] Schlimme D.V., Rooney M.L., "Packaging of minimally processed fruits and vegetables". *In*: Wiley R.C. (*ed.*), "Minimally processed refrigerated fruits and vegetables", New York, Chapman & Hall, Inc., pp. 135–182 (1994).

[100] Babic I., Watada A.E., "Microbial populations of fresh-cut spinach leaves affected by controlled atmospheres", *Postharvest Biol Technol.*, **9**, 187–93. (1996).

[101] Pentima J.H.D., Rios J.J., Clemente A., Olias J.M., "Biogenesis of off-odour in broccoli storage under low-oxygen atmosphere", *J. Agric. Food Chem.*, **43**, 1310–1313 (1995).

[102] Saito M., Rai D. R., Mastuda A.R., "Effect of modified atmosphere packaging on glutathione and ascorbic acid content of asparagus spears". *J Food Process Preserv.*, **24**, 243–251 (2000).

[103] Rai D. R., Paul S., "Packaging requirements of highly respiring produce under modified atmosphere packaging", *J Food Sci Technol.*, **44,** 10–15 (2007).

[104] Soliva-Fortuny R.C., Martýn-Belloso O., "New advances in extending the shelf-life of fresh-cut fruits: A review". *Trends in Food Science and Technology*, **14**, 341–353 (2003).

[105] Amanatidou A., Smid E.J. and Gorris L.G.M., "Effect of elevated oxygen and carbon dioxide on the surface growth of vegetable-associated microorganisms". *Journal of Applied Microbiology*, **86**, 429–438 (1999).

[106] Jacxsens L., Devlieghere F., Van der Steen C., Debevere J., "Effect of high oxygen atmosphere packaging on microbial growth and sensorial qualities of fresh-cut produce", *International Journal of Food Microbiology*, **71**, 197–210 (2001).

[107] Van der Steen C., Jacxsens L., Devlieghere F., Debevere J., "Combining high oxygen atmospheres with low oxygen modified atmosphere packaging to improve the keeping quality of strawberries and raspberries"', *Postharvest Biology and Technology*, **26**, 49–58 (2002).

[108] Tolbert B.N., Dowing N., Carlson R. W., "Chemistry and Metabolism of ascorbic acid and ascorbate sulfate", *Ann Ny Acad Sci.*, **258**, 48–69 (1975).

[109] Saari N.B., Fujita S., Niyazoe R. and Okugawa N., "Distribution of ascorbate oxidase activities in the fruits of family cucurbitacae and some of their properties", *Journal of Food Biochemistry,* **19**, 321–327 (1995).

[110] Selman J.D., "Vitamin retention during blanching of vegetables", *Food Chem.*, **49**, 137–149 (2001).

[111] Lisiewska Z., Kmiecik W., "Effect of level of nitrogen fertilizer, processing conditions and period of storage for frozen broccoli and cauliflower on vitamin C retention", *Food Chem.*, **57**, 267–270 (1996).

[112] Negi P.S. and Roy S.K., "Effect of blanching and drying methods on β-carotene, ascorbic acid and chlorophyll retention of leafy vegetables", *Lebens Wiss Technol.*, **33**, 295–298 (2000).

[113] Favell D.J., "A comparison of the vitamin C content of fresh and frozen vegetables", *Food Chem.*, **62**, 59–64 (1998).

[114] Prochaska L.J., Nguyen X.T., Donat N., Piekutowski W.V., "Effects of food processing on the thermodynamic and nutritive value of foods: Literature and database survey", *Medical Hypotheses*, **54**(2), 254–262 (2000).

[115] Krehl W. A., Winters R.W., "Effect of cooking methods on retention of vitamins and minerals in vegetables", *Journal of the American Dietitians Association*, **26**, 966–972 (1950).

[116] Eheart M.S., Gott C., "Chlorophyll, ascorbic acid and pH changes in green vegetables cooked by stir-fry, microwave and conventional methods and comparison of chlorophyll methods". *Food Technology,* **19**(5), 185–188 (1965).

[117] Warthesen J.J., Vickers Z.M., Whitney-West S., Wolf I.D., "Cookery methods for vegetables: Influence on sensory quality, nutrient retention, and energy consumption", *Home Economics Research Journal,* **13,** 61–79 (1984).

[118] Johnson J.M., Driskell J., "Vitamin retention and sensory evaluation of selected foods prepared by boiling, boiling-in-bag, and steaming methods", *Home Econ Res J.*, **16**, 136–142 (1987).

[119] Schnepf M., Driskell J., "Sensory attributes and nutrient retention in selected vegetables prepared by conventional and microwave methods", *Journal of Food Quality*, **17**, 87–99 (1994).

[120] Kaur A., Kaur D., Oberoi D.P.S., Gill B.S., Sogi D. S., "Effect of dehydration on physicochemical properties of mustard, mint and spinach", *Journal of Food Processing and Preservation*, **32**, 103–116 (2008).

[121] Erickson M., "Lipid oxidation: Flavor and nutritional quality deterioration in frozen foods". *Quality in frozen food* by Erickson M. and Hung Y-C. (*eds.*), pp. 141–173. (1997).

[122] Lisiewska Z., Kmiecik W., "Effect of freezing and storage on quality factors in Hamburg and leafy parsley", *Food Chem.*, **60**, 633–637 (1997).

[123] Tosun B.N., Yücecan S., "Influence of commercial freezing and storage on vitamin C content of some vegetables", *International Journal of Food Science and Technology*, **43**, 316–321 (2008).

[124] Wu Y., Perry A.K., Klein B.P., "Vitamin C and ß-carotene in fresh and frozen green beans and broccoli in a simulated system", *J Food Qual.*, **15**, 87–96 (1992).

[125] Howard, L.A., Wong, A.D., Perry, A.K., Klein, B.P., "Beta-carotene and ascorbic acid retention in fresh and processed vegetables", *J Food Sci.*, **64**, 929–936 (1999).

[126] Yadav S.K., Sehgal S., "Effect of domestic processing and cooking on selected antinutrient contents of some green leafy vegetables", *Plant Foods for Human Nutrition*, **58**, 1–11 (2003).

[127] Bawa S.F., Yadav S.P., "Protein and mineral contents of green leafy vegetables consumed by Sokoto population", *J Sci Food Agric.*, **37**, 504–506 (1986).

[128] Gupta K., Barat G.K., Wagle D.S., Chawala H.K.L., "Nutrient contents and antinutritional factors in conventional and non conventional leafy vegetables", *Food Chem.*, **31**(2), 105–116 (1989).

[129] Aleotor V.A., Adeogun O.A., "Nutrient and antinutrient components of some tropical leafy vegetables", *Food Chem.*, **53**(4), 375–379 (1995).

[130] Wesch-Ebeling P., Maiti P., Garcia-Diaz G., Gonzalez D.I., Sosa-Alvarado F., "Contributions to the botany and nutritional value of some wild amaranth species (Amaranthaceae) of Nuevo Leon, Mexico", *Economic Botany*, **49**(40), 423–430 (1995).

[131] Ryden P., Selvedran R.R., "Phytic acid: Properties and determination", *In*: Macrae R. Robinson R.K. and Sadler M.J. (*eds.*), *Encyclopaedia of Food Science, Food Technology and Nutrition*. London: Academic Press, pp. 3582–3587 (1993).

[132] Plaami S., Kumpulainen J., "Inositol phosphate content of some cereal based foods", *J Food Comp Anal.*, **8**, 324–335 (1995).

[133] Mosha T.C., Gaga H.E., Pace R.D., Laswai H.S., Mtebe K., "Effect of blanching on the content of antinutritional factors in selected vegetables", *Plant Food Hum Nutr.*, **47**, 361–367 (1995).

[134] Kaur P., Rai D.R., Paul S., "Quality changes in fresh-cut spinach (*Spinacia oleracea*) under modified atmospheres", *Journal of Food Quality*, **34**, 10–18 (2011).

[135] Diane M., Barrett, Beate Lloyd, "Advanced preservation methods and nutrient retention in fruits and vegetables", Wiley Online Library: (wileyonlinelibrary.com) DOI 10.1002/jsfa.4718 (2011).

[136] Bushra S., Farooq A., Shahid I., "Effect of different cooking methods on the antioxidant activity of some vegetables from Pakistan", *International Journal of Food Science and Technology*, **43**, 560–567 (2008).

[137] Xuetong F., Kimberly J.B., "Changes in quality, liking, and purchase intent of irradiated fresh-cut spinach during storage", *Journal of Food Science*, **76**(6), S363–S368 (2011).

[138] Kaur P., Rai D.R., Paul S., "Quality changes in fresh-cut spinach (*Spinacia oleracea*) under modified atmospheres with perforations", *Journal of Food Quality*, **34**(1), 10–18 (2011).

6

Processing, Value Addition and Quality Characteristics of Beetroot

MRIDULA D.[1*], S.K. TYAGI, DILIP JAIN, SHEETAL BHADWAL, AND R.K. GUPTA

ABSTRACT

Beetroot (Beta vulgaris) is botanically classified as an herbaceous biennial from Chenopodiaceae family and has several varieties with bulb colors ranging from yellow to red. Deep red-colored beet roots are the most popular for human consumption, both cooked and raw as salad or juice. Beetroots are rich in valuable, active compounds such as carotenoids, glycine betaine, saponins, betacyanines, folates, betanin, polyphenols and flavonoids. They have antimicrobial and antiviral effects and also can inhibit the cell proliferation of human tumor cells. The use of betalains as food colorant is approved by European Union and betalains are labelled as E-162. Dried beetroots can be consumed directly in the form of chips as a substitute of traditional snacks, extruded products and also as a component of instant food and health mixes.

Key words: Beetroot, Processing, Colour pigments, Betanin, Stability.

I. INTRODUCTION

Beetroot is a nutritious vegetable and is an ideal component of a healthy diet of any individual. This vegetable has been considered for blood formation, heart health and also for digestion. Beetroot is now been advocated a cancer preventive vegetable which also boost up our immune system. From the

[1] ICAR-Central Institute of Post-Harvest Engineering & Technology, Ludhiana-141004, Punjab, India
ICAR-Central Arid Zone Research Institute, Jodhpur-321003, Rajasthan, India
**Corresponding author* E-mail: mridulads4@gmail.com

middle ages, this precious vegetable was used for treatment of many conditions, particularly problems related to blood as well as digestion. Beetroot now is still championed as a cure for many health problems.

Beetroot, being grown since Assyrian Times, was highly esteemed by the ancient Greeks and was used in offerings to Apollo. There were many Roman recipes for beetroot, which they regarded more highly than greatly revered cabbage. It appeared in fourteenth century English recipes and was first described as the beetroot in Germany in 1558, though it was a rarity at that time in northern Europe[1]. Beetroot (*Beta vulgaris*) is botanically classified as an herbaceous biennial from Chenopodiaceae family. It has several varieties with bulb colours ranging from yellow to red. Deep red-colored beetroots are popular for human consumption and being utilized in cooked as well as in the form of juice and salad. Nowadays, consumers are showing interest in the use of natural food colors because of adverse effect of synthetic dyes on health of consumers. But in food processing industry, carotenoids and anthocyanins are often preferred than betalains, though betalains are water-soluble pigments and stable between pH 3 and 7[2]. Fresh beet/beet powder or its extracted pigments are used to improve the red color of tomato pastes, sauces, soups, desserts, jams, jellies, ice creams, sweets and breakfast cereals[3, 4]. Addition of beetroot in any form not only add colour to that product but also contributes to health and wellbeing of consumers due to the presence of antioxidants in it.

II. BEETROOT-A CROP

There are three subspecies of beetroot *viz. Beta vulgaris* subsp. *Maritima* (plant smaller to 80 cm tall, root not swollen), *Beta vulgaris* spsp. *Vulgaris* (plant larger to 2 m tall, with a rounded fleshy taproot and) *Beta vulgaris* subsp. *Cicla,* also known as Chard, Swiss Chard, Silverbeet, perpetual Spinach or mangold. Different varieties of beetroot may be grouped according to their shape. They are flat, short-top shaped, deep oblate to round, globular to oval, half long and long. The Indian Agricultural Research Institute (ICAR), New Delhi, India had recommended *Crimson Globe* and *Datroit dark red* varieties of beetroot for Indian farmers[5]. In case of *Crimson Globe*, the roots are globular to flattened globe, medium dark red with light redcening, tender and finely grained. The top is small with glossy dark green tinged with maroon leaves.

Datroit dark red: The roots are preferably round with deep red skin, smooth, uniform and attractive with small collar and small tap root.

Eastern and Central Europe has traditionally been associated with the production of beetroot. Russia, Poland, France, Italy and the USA are now considered as the major beetroot producing countries, where beetroot is mainly utilized for processing purposes with little for fresh consumption.

Beetroot crop is suitable for a wide range of soils, but sandy to loamy soils are considered good for beetroot production with pH 6 to 8. Optimum temperature for beetroot production is 15°C to 18°C, with maximum temperature as 24°C. Though it is winter season crop but it can also be grown at slightly higher temperature with adequate soil moisture. Beetroot crop required about 75 to 90 days in summer, and 105 days to 120 days or longer in winter to reach to full maturity.

Beetroots are usually harvested after two months of sowing but it may require three or more months for full maturity. Its final size will depend on the cultivar and crop spacing and not solely on to the degree of maturity. Beetroot should be harvested carefully generally by lifting them in upward position, because bruised roots may bleed and diminished the quality of produce. Beetroot is generally harvested with its leaves and preferably with manual operation. Leaves are usually not being utilized for human consumption but there high minerals content makes them useful for compost heap. Commercially, beetroot are harvested using mechanical harvesting. The first mechanical harvester for cultivated beets was introduced in 1930s in USA[6].

Beetroot in US market is sold after proper grading them according to their size, which fall into three different grades; grade 1 being below 5 cm (up to 2.5/8 inches) in diameter, grade 2 being around 5-7 cm (2-2.5/8 inches) and grade 3 being around 7.5-10 cm (3-4 inches)[6]. Grade 2 beets are consistently available season long, but grades 1 and 3 have greater seasonal variation. For beetroot sold fresh, the USDA (United States Department of Agriculture) set a maximum permissible weight loss of 41% from harvesting to point of sale, although this is 7% if roots are sold with tops as the leaves soon wilt due to moisture loss[6].

III. CHEMICAL COMPOSITION OF BEETROOT

Beetroot is a nutritious vegetable and considered good to add micronutrients in a healthy meal. In healthy diets, it is always advised to include some amount of most important vegetables like carrot, beetroot, and spinach due to the role played by plant pigments in disease prevention. Beetroot is sweet in taste thus a rich source of carbohydrates hence one should be cautious while adding it in the diet of diabetics. It is a good source of protein, dietary fibre and has significant amount of minerals and vitamins. It is practically has no fat and free from cholesterol that makes beetroot comparatively low in energy value. Beetroot (raw, peeled, but not grated) contains 87.1% moisture, 7.6% carbohydrate, 1.7% protein and 0.1% fat and 154 kJ (36 kcal) energy[4]. Beetroot, boiled for 45 min contains 82.4% moisture, 9.5% carbohydrate, 2.3% protein and 0.1% fat and 195 kJ (46 kcal) of energy. Pickling boiled beetroot decreases the carbohydrate and protein content than raw beetroot, with a corresponding decrease in energy value. Acetic acid or vinegar will contribute to the energy value of pickled beetroot, however, if it is eaten along with vinegar. Raw and boiled beetroot provides about 0.27% and 0.37% total nitrogen, respectively. Beetroot has highest sugar content amongst the vegetable group, which make this valuable vegetable unsuitable for diabetics.

Beetroot is a good source of micronutrients *i.e.* minerals and vitamins. Raw beetroot contains calcium, magnesium, phosphorus, iron, sodium, potassium, coppers, zinc to the tune of 18.3, 9.0, 55, 1.19, 59.8, 43.0, 0.29, 0.91 mg/100 g, respectively. It also contains a significant amount of vitamins particularly thiamin 0.04 mg, riboflavin 0.09 mg, niacin 0.40 mg and vitamin C 10 mg/100 g[7]. Folate *i.e.* folic acid occurs in higher levels in beetroot than in most other vegetables. There is 150 mg and 110 mg of folate per 100 g in raw and boiled beetroot, respectively.

The color of red beetroot is due to the presence of a purple pigment betacynin (betanin, isobetanin, probetanin and neobetanin) and a yellow–orange pigment betaxanthin known collectively as betalins[8,9]. The pigments are present in cell vacuoles. The pigment is quite stable in acidic conditions. This is the reason of pickling the beetroot to preserve it for a longer duration. The structural formula of betanine ($C_{24}H_{26}N_2O_{13}$) is given hereunder (Fig. 1).

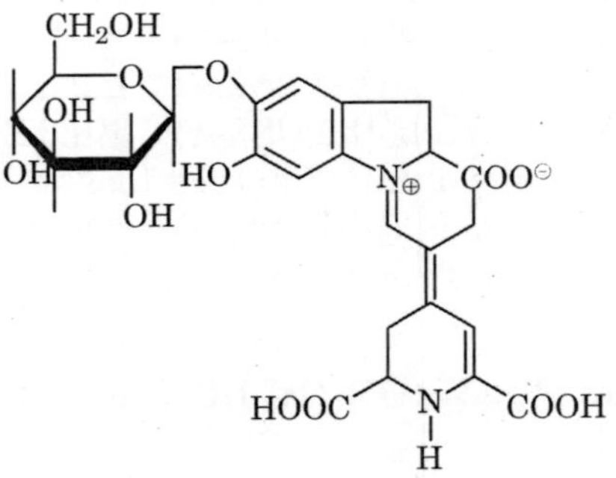

Fig. 1: General process used for beetroot canning.

IV. HEALTH BENEFITS OF BEETROOT

Beetroot (*Beta vulgaris*) is a useful vegetable crop due to its medicinal properties. Beetroot has been considered as a laxative and useful to treat bad breath, coughs and headaches. It is rich in Betaine (betacyanin), which plays an important role in lowering potentially toxic levels of homocysteine, a homologue of the naturally occurring amino acid cysteine that can be harmful to blood vessels thereby contributing to the development of heart disease, stroke, and peripheral vascular disease. Betaine functions in conjunction with S-adenosylmethionine, folic acid, and vitamins B_6 and B_{12}

to carry out this function[10]. Beetroot is sweetest amongst all vegetables, contain more sugar even higher than carrot or sweet corn. The sugar content in beetroot is about 8-10% but in the sugar beet cultivar, it may be up to 20%. The red pigment of the beetroot is also a very strong hydrogen acceptor. Researches had been carried out to study the mechanism of action of beetroot on cancer cells in their electrochemical tests with anticancer substances[11]. It was found that the combination of beetroot with dextarotary lactic acid was capable of activating the cellular respiration of highly virulent cancer cells by 500-1200% and the respiration of cancer cells was completely normalized[12].

Betanin is the most prevalent betalain in red beets. Red beet is one of the most potent vegetables with respect to antioxidant activity[13, 14]. The ingestion of a single dose of red beet juice showed an increase of antioxidant compounds including betalains in urinary excretion[15]. Betalains and other phenolic compounds present in red beet decreased oxidative damage of lipids and improved antioxidant status in humans[2]. Antioxidant activity in red beet is associated with the scavenging of free radicals and consequently in the prevention of diseases like cancer, cardiovascular diseases[16]. Betanin has indicated an inhibitory effect on a wide range of oxidative reactions *e.g.* lipid peroxidation and the decomposition of heme in the blood in laboratory trials. Betaxanthins, which are yellow and orange betalain pigments also shown antiradical activity, but lesser than betanin[6]. The regular consumption of red beet products may provide protection against certain oxidative stress-related disorders in humans[17].

Antioxidant activity of natural pigments such as betalains has attracted the attention of researchers in recent years[18]. Betalains are plant pigments present in red beetroot (*Beta vulgaris* var. *rubra*) and some other foods such as amaranthus, rose etc. which are being used in processed meat products, candies and yogurt etc.[19]. Betanin (betanidin 5-O-β-D-glucoside, red) is the most abundant betalain component, making up 75–95% of total betalains[20]. The betanin molecule includes phenolic and cyclic amine groups, which are good electron donors that endow betanin with exceptionally high free-radical scavenging ability[21b]. Various studies indicated that betanin and betalains possess antioxidant and anti-inflammatory activities[22, 23]. Red beetroots are one of the 10 most potent antioxidant vegetables due to the presence of betalains[24], which are reported to protect mice during gamma irradiation[25]. Betanin also plays inhibitory role in lipid peroxidation and heme decomposition *in vitro* even in small concentrations[17] and attenuates carbon tetrachloride-induced liver injury in fish (*Cyprinus carpio* L.)[21a]. In cell culture and animal studies observed that beetroot juice had significant tumour-inhibiting and antimutagenic effects[26, 27].

V. PROCESSING AND PRESERVATION OF BEETROOT

Storage of Minimally Processed Beetroot

Beetroot store well and are acquiescent to long-distance transport. Beetroots are sensitive to heat, light, high temperature and air. Traditionally, this crop is stored on farms or in covered heaps. However, when the crop is grown on large scale for commercial purposes, beetroots are now stored under controlled environmental conditions. Before storing beetroot, it is important to either twisted off leaves or cut 5 to 8 cm above the crown. Beetroot should be stored in frost-proof environment. Under controlled environmental conditions, beetroot can be stored for about five to six months. Beetroots for fresh marketing may be topped, graded, washed and packaged in polyethylene bags for convenient handling. Beetroots (without leaves) comprised of around 88% moisture content. High humidity is important to minimize the moisture loss as well as wilting and withering of roots. Best temperature for storage of beetroot is near freezing point. The freezing point for topped beetroot, with leaves removed is –0.9°C while for bunched beetroot, it is –0.4°C. Below these temperatures, freezing of the roots/leaves damage will occur[6]. Storage of beetroot at 6-7°C extend the storage period compared to 0-1°C but reduces the quality[28]. Elevated levels of carbon dioxide during storage may also be detrimental to beetroot storage. High levels of carbon dioxide (30-70%) increases respiration rate of beetroot that leads to a higher metabolic rate[29]. Therefore, storage of beetroot with some degree of air circulation prevents carbon dioxide from building up and enables beetroot to be stored for longer time. The best storage conditions for beetroot are the cool temperatures (0 to 5°C) and 90 to 95% relative humidity[30]. Beet roots, stored for up to 30 weeks at 4°C, showed very little change in the ratio of reducing sugars to soluble or insoluble sugars[31].

Minimal processing comprises grading, washing, peeling and cutting to produce a product that is fresh and convenient to prepare and consume[32]. Minimally processed fruits and vegetables are more perishable than the whole one, due to physical stresses undergone during peeling and cutting operations. Cut vegetables when stored at high temperature showed increased respiration rate and enzymatic activities. Another study revealed that 0°C temperature for storage of sliced peeled beet root (2 mm-thick disc) while 10 and 15°C temperatures, drastically reduced the quality of minimally processed beet roots during storage. Respiratory activity and mass loss were higher in the storage at 15°C. No significant differences in color, total soluble solids, betacyanin and betaxanthin were observed among various storage temperatures[33]. Beetroot contain an antioxidant compound named Betalaine, which give purple-red color[17]. With minimal processing, more specifically the washing step and rinsing, these compounds lose beet, making the product whitish due to surface discoloration[34]. The treatment with doses up to 4.0 kGy of gamma radiation to maintain the colour quality of minimally processed red beetroot was recommended[35].

Dehydration

Dehydrated beetroot can be easily utilized to manufacture vegetable noodles, soups, snacks and fast food. In a general process of beetroot drying involved washing, peeling, size reduction, hydrothermal treatment and drying to reduce the moisture content. Convective drying in hot air is still the most popular method applied to reduce the moisture content of fruits and vegetables[36], including beetroots[37, 38]. Vacuum-microwave treated beetroot cubes as well as freeze-drying ones exhibited lower compressive strength, better rehydration potential and higher antioxidant activity than those dehydrated in convection. Study further indicated that increasing the microwave wattage and decreasing the time of convective pre-drying improved the quality of beetroot cubes dried by the combined method[39]. The best drying temperatures required for sequential reduction in temperature (120 to 50°C) resulting in good color retention. This also reduced batch time to 4 h compared to 6 h batch of conventional isothermal drying at 50°C[40]. The beetroot samples obtained from the Microwave convection drying system for drying of beet root cube (10 mm thickness) to 7% moisture content showed lower final moisture content than those obtained from the other system[41].

Spray drying of beetroot extract may also be carried out with added maltodextrin or other binders as prerequisites. Spray dried powder is less brightly colored due to maltodextrin and thermal degradation due to exposure at higher temperature. In addition, the powder's hygroscopic nature is another limitation that may lead to operational inconvenience such as lump formation while mixing, blending, rehydration, etc[40]. In a study, on spray drying of betalain dye from red beet roots using malto dextrine, gum acacia and soluble starch using spray drier temperature between 150°–165°C, it was observed that with decrease in percentage of carriers in beetroot juice, the percent yield of betalains increased. This was observed by optical density and betalain content estimation. The shelf life studies of this spray dried betalain dye powder, over 180 days showed that the dye is quite stable in the temperature range of - 4°C to 20°C[3].

Osmotic dehydration is described as the partial dehydration of fruits through the process of osmosis, which involves immersion of fruit in hypertonic solution at given temperature and for specific period of time[42]. Dehydration of osmotically pretreated slices of beetroot with hot air followed by microwave finish drying retained the quality of beetroot, so it is considered the worth method for the preservation[43]. The optimized osmotic dehydration of beetroot candy and recommended osmotic solution concentration, temperature and process duration as 60°Brix, 55°C and 180 min, respectively for osmotic dehydration of beetroot[2].

Pickling and Canning

Beetroot may be preserved by different methods such as pickling and canning. Pickling of beetroot may be done in malt vinegar. Processing of beetroot involves washing, dicing, steaming to soften and sterilize them followed by mechanical skinning. The beetroot may be then vacuum-packed in a preserving solution such as vinegar (Fig. 2).

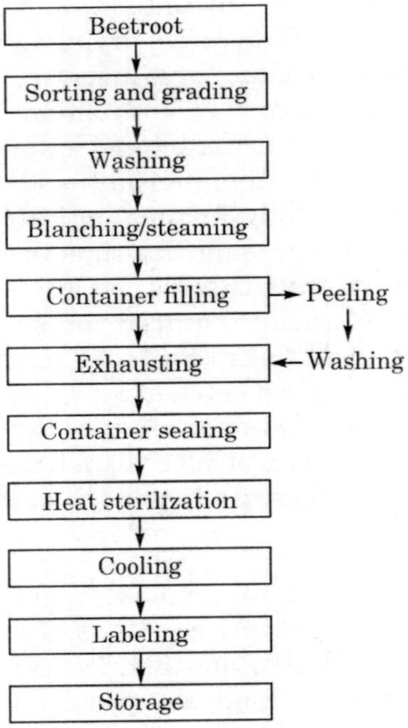

Fig. 2: General process used for beetroot canning H_2O

Beetroot Juice

Beetroot is commonly consumed in fresh form as salad. It may be used in food industry to produce juice and concentrated juices which may be natural colorants (betalains) as additives in processed food products. Lactic acid fermentation, one of the methods of natural preservation may have a preservative effect on beetroot juice. There are three main types of vegetable juice fermentation: spontaneous fermentation caused by the natural micro flora of a particular raw material; fermentation directed by the addition of a starter culture; and controlled fermentation involving the introduction of starter cultures to a thermally-preserved (by pasteurization) raw material[44]. The effect of Lactic acid fermentation on quality of two beetroot cultivars namely Chrobry and Czerwona Kula and found that lactic acid fermentation influenced both red and yellow pigments. In both of these cultivars, the dominant red-violet pigment components were betanin, isobetanin, while vulgaxanthin I was the major yellow pigment. Betanin was the main red-violet component in juices of these cultivars before fermentation. After fermentation also in this study, betanidin and isobetanidin were observed. Huge amounts of betanidin were found especially in fermented Chrobry juice. It was found that fermented juice from Chrobry cultivar contained higher betanidin than betanin levels, while in juice from Czerwona Kula cv., betanin level was much higher than betanidin level[45].

A method for preparing a solution of freeze-dried juice from beetroot was developed and optimum heat processing regimes were established along with the requirements for the storage of the juice. The solution, stabilized with citric acid, heated twice for 20 min to 90°C with 24 h incubation at 20-22°C after the first heating. The product can then be kept at less than or equal to 8°C for 2-3 months. The juice is used for colouring ice cream without affecting its microbiological condition[46].

Gradations in colour changes of beetroot juice were studied and reported that the red colour of beetroot juice is due to the pigment betaine[47]. The stability of colour of beetroot Juice may be improved by addition of 0.1% ascorbic acid and 0.1% citric acid during crushing of the raw material. The study indicated that the amount of colouring matter reduced during blanching. Heating above 121°C resulted in complete betaine decomposition. Mg, Cu, Zn and Fe contents cause a reduction of betaine content in juice heated to 100°C in 30 min. The method for beetroot processing, which prevents betaine decomposition, involved beetroot crushing using equipment with a coarse surface, addition of ascorbic and citric acid, pressing for 20 min, heating of the juice to 60°C and filling into hermetically sealed containers prior to sterilization.

VI. BEETROOT PIGMENTS

There are four main classes of plant pigments, which are chlorophylls, carotenoids, flavonoids and betalains. Chlorophylls and carotenoids are found in the organelles of cells and of water insoluble nature. Organelles namely nucleus, chloroplasts and mitochondria present in cell perform a particular function. When colour pigments are present in organelles, they are called chromoplasts. Chlorophylls, present in chloroplasts are blue-green or yellow-green pigments, presents in plant. Chlorophylls, the primary light-trapping pigments plays an important role in photosynthesis. The main role of chlorophylls is to absorb the light energy and convert it to the chemical energy. Carotenoids, present in plant products also trap light energy and prevent the degradation of chlorophyll molecules. Carotenoid pigments mainly present in yellow, orange, red and also in brown coloured plant materials *i.e.* fruits and vegetables. Carotenoids may be divided into two categories: carotenes and xanthophylls. Flavonoids and betalains are water soluble pigments found in cell vacuoles.

The purple-red color of beetroot is due to the presence of betalain pigments. These pigments are similar to anthocyanins and flavonoids, which were wrongly termed in the old literature as anthocyanins containing nitrogen. Betalains are water-soluble pigments, may be divided into two subgroups: red betacyanin and yellow betaxanthin, which provide beet roots with their typical color[48]. Besides, these pigments often exhibit antioxidant activity and potential health benefits of vegetables[17].

Betalains are a group of pigments that are yellow, orange, pink, red and purple in colour. Unlike the other plant pigment, betalains have a limited food distribution. Red colouration in most of plants is due to the presence of carotenoids and flavonoids. Betalain pigments were first isolated from the red roots of beetroot (*Beta vulgaris*); hence named after the plant genus *Beta*. Betanin was first discovered in around 1920, while a crystalline form of betanin dye was produced in the 1960s. Betanin normally occurs at much higher levels in the roots of red beetroot than other betacyanin pigments. Like all betacyanins, betanin is metabolically derived from a molecule called 3,4-dihydroxyphenylalanine (L-DOPA). It is derived from two L-DOPA molecules. Firstly one L-DOPA molecule undergoes a change to form betalamic acid. The second L-DOPA molecule is changed to cyclo-DOPA glucoside (CDG) that condenses with betalamic acid to produce betanidin. A change in structure, involving the addition of glucose, converts betanidin to betanin[6].

Apart from betanin, the yellow betaxanthin pigments vulgaxanthin-I and vulgaxanthin-II are the next most important pigments in beetroot. The characteristic red-purple colour of beetroot is mainly due to variable level of different betalain pigments, particularly the relative concentrations of betanin and the yellow betaxanthin pigments. Distinct light and dark rings are often observed when beetroot is cut transversely, mainly due to variation in the amounts of pigment in the vascular system and storage tissue of beetroot. Due to higher levels of pigment, the vascular system appears darker, while the storage tissue appears as lighter bands. The betalain pigments are distributed mainly towards the outer parts of the beetroot and also contribute to the redness in leaves, stalks and flowers.

Red beetroot contains a large concentration of betanin, 300-600 mg/kg, and lower concentrations of iso-betanin, betanidin, and betaxanthins[17]. The prickly pear (*Opuntia ficus indica*) contains about 50 mg/kg of betanin and 26 mg/kg of indicaxanthin[49]. The bioavailability of betalains is at least as high as flavonoids, which are well-accepted natural antioxidants. Betalains, as natural antioxidants, may provide protection against oxidative stress-related disorders[17, 50].

Food Colouring Applications of Beetroot

The use of naturally derived colours was expanding, due to public concerns about synthetic food dyes and improvements in their stability during processing and storage[51]. Beetroot juice is used for food colouring purposes in the home for many years. Betalains consist of two sub-classes: betacyanins (red-violet pigments) and betaxanthins (yellow-orange pigments[16, 66]. The use of betalains as food colorant is approved by European Union and betalains are labeled as E-162[2]. Betalain pigments are available to the food industry either as juice concentrate or in the form of powder. During commercial extraction, beetroots are first crushed and the coloured juice is collected and volume is reduced by way of concentrating it. Beetroot juice is usually concentrated under vacuum until it remains with 60-65% of total solids. Freeze drying is an appropriate processing technique to produce beetroot powder with 0.3-1.0% pigment[6].

Beetroot pigment concentration in juice and its powder can be enhanced through fermentation technology. The fermentable solids present in beetroot juice can be removed in a biofermentor, using yeasts such as *Candida utilis* and *Saccharomyces cerevisiae*, to yield a product with higher pigment content[52]. Beetroot powders prepared following fermentation of beetroot juice contain more betacyanin than from unfermented beet juice.

Betalain pigments have several food applications *viz.* dry mixes, confectioneries, desserts, dairy and meat products. The concentration of pure pigment required to obtain desired hue is relatively small, rarely exceeding 50 mg/1kg, calculated as betanin[16]. According to Codex Alimentarius commissio, betalain application is limited to only good manufacturing practice. Beetroot red extracted from beetroots are commercialized in European Union and USA as food colorant[53, 54]. Betanin ($C_{24}H_{27}N_2O_{13}$) makes up 75-95% of the total coloring matter found in the beetroot, therefore it is used as a natural food colour additive[24]. Betanin pigment, like other betacyanins, is highly susceptible to both pH and temperature[55]. Being powerful antioxidant, betanin pigment may provide protection and reduce risk of cardiovascular disease and cancer[56]. Betalains have attracted additional interest because of their antioxidative, anti-inflammatory and anticarcinogenic properties. The main source of commercially produced betalains is red beetroot, but alternative sources are found in plants from the Amaranthaceae and Cactaceae families[57].

Betalain stability is important for its application in food industry. Betalain extracts need to be use with great precautions because they are sensitive to processing conditions, particularly storage temperature, processing heat, pH, light, moisture and oxygen. These processing and storage conditions have interactive effects, and the added pigments can quickly go off under unfavourable conditions. For example, the red pigment betanin may degrade on exposure to bright light, hot and humid air, oxygen and high temperatures to a light brown colour.

Betalain colouration is stable in the pH range 3.5 to 7.0 *i.e.* acid to neutral. The optimum pH for both betacyanin and betaxanthin pigments is in slightly acidic pH range 5.0-6.0. pH of red beetroot extracts above 7.0 changes its colour from red towards blue. The beetroot tissues exposed to higher or alkaline pH (7.5-8.5) may get discoloured. Beetroot, after size reduction retains its purple-red colour well in acidic solutions such as malt vinegar or acetic acid. The heating of betalains may result in the discolouration of pigment. Heating of red betanin pigment for longer duration can change them to light brown, particularly if high temperature is accompanied by higher pH *i.e.* alkaline pH. High light intensities also accelerate the pigment discolouration. Moisture content of the product and powder is also a determining factor for betalains stability. The stability of betalains is usually higher in low moisture products, which have been stored properly. Higher moisture content is favourable for degradation of pigment. Exposure to oxygen also increases the discolouration of pigment in food products during storage. Betalains react with the oxygen in the air, but discolouration is partially reversible if oxygen levels are subsequently lowered[6].

Fermentation of vegetables having sugar content is an important method to removes sugars, which may interfere during processing. Fermentation also offers greater recoveries, stabilize betanin content and ensure its retention during storage. The fermentation of beetroot by *Lactobacillus plantarum* and *Lactobacillus delbrueckii* stabilized the betanin content, reduced the losses due to refrigerated storage and during the processing of beetroot[58].This study indicated that refrigerated storage reduced 50% of the betanin content in 10 days, while steaming reduced 40%. Lower pH during heat treatment showed better betanin retention compared with higher pH. This study further showed that fermentation of beetroot by either *L. plantarum* or *L. delbrueckii* at 4°C was able to preserve betanin up to 10-15 mg/100 mL beetroot extract up to 32 days of refrigerated storage and ensured a safe, good quality product.

Betalain pigments are widely used in food products despite having lower stability than synthetic food colourings. The betalain pigments are suitable for use in food products with a short shelf-life that have been produced with minimum heat treatment and packaged in a dry state under reduced levels of oxygen, light and humidity[16]. The discolouration tendency of betalain pigments under certain environmental conditions showed the possibility for being used them in the detection of food oxidation. The inclusion of beetroot juice at the maximum of 4 per cent in the preparation of herbal ice cream without much alteration is its acceptability[59].

VII. UTILIZATION OF BEETROOT FOR WINE PRODUCTION

Wine is quite expensive product than all other alcoholic beverages. Preparation of wine from a cheaper source could change its economic value. The market price of sweetish starchy food like potatoes, sweet potato, banana, and beetroot is much lower, when compared to that of grapes and apple. Hence, wine produced from cheaper fruits and vegetables can be a cost effective solution. The sweet potato with beetroot combination resulted in better ethanol concentration of 7-9.5% w/v in white wine than mango[60]. The optimum pH and the inoculums level of yeast best suited for the maximum production of wine from beetroot (*Beta vulgaris* L) was investigated by Bangale[61]. This study showed the highest alcohol content (9.2%) for the wine sample having pH of 4 and 10% yeast (*Saccharomyces cerevisiae*) level.

VIII. VALUE ADDED READY-TO-EAT PRODUCTS UTILIZING BEETROOT

Beetroot is usually consumed in the form of salads, juices, powder, and pickles. However, beetroot supplemented food products are hardly available in the market. In order to enhance the functional properties of food products, incorporation of dried beetroot have been tried in a number of preparations. In a study on development of beetroot incorporated extrudates, it was observed that physio-chemical quality of extrudates was significantly affected by dried beetroot supplementation and level of legumes (defatted soy flour or chickpea splits)[63]. Bulk density of extrudates increased while expansion ratio decreased with increasing proportion of dried beetroot in feed material. The increase of dried beetroot in feed material resulted in increased the water absorption index. Incorporation of beetroot resulted in the darker extrudates that ultimately affected the sensory acceptability for appearance and colour. Extrudates, prepared from dried beetroot and DSF, were observed to have higher protein (12.53 to 13.05%), ash (1.20 to 1.73%) and crude fibre (1.04 to 1.47%) content compared to extrudates with chickpea splits. Mean sensory score, more than 7 for extrudates having dried beetroot up to 6% in feed material, indicated commercial scope for production of rice and DSF or chickpea splits based extruded snack food supplemented with dried beetroot[63]. The commercial scope for manufacturing of good quality vegetarian biscuits with dried beetroot powder up to 10% in flour blends along with ascorbic acid is observed. In this study, it was observed that expansion in diameter and thickness of the developed biscuits decreased while hardness and breaking strength increased with increased levels of beetroot powder in biscuit samples but these quality parameters did not deteriorate the sensory acceptability[62]. The beetroot powder up to 2.5% can be successfully incorporated in health drink mixes without affecting the sensory quality[64].

QUALITY EVALUATION OF BEETROOT PRODUCTS

Extraction of betalains

Freeze dried sample (0.1 g) dissolved in 10 ml of 50% ethanol, agitated for 10 s and the homogenate centrifuged at 6000 rpm for 10 min. The supernatant is required to be collected as it is after centrifugation and the same is to be repeated for 2 more times to ensure maximum extraction of betalains. The supernatant was further used for determination of betalains.

Determination of betalain compounds with spectrophotometric analyses

The content of betaxanthins and betacyanins in the extracts is to be determined spectrophotometrically at 538 nm and 480 nm with a UV–Vis spectrometer, respectively[65]. The absorbance reading obtained is used to calculate the betalain concentration for each sample. The betalain content (BC) may be calculated as

BC (mg/L) = [(A × DF × MW × 1000)/(e × l)], where A is the absorption, DF the dilution factor and l the pathlength (1 cm) of the cuvette. For quantification of betacyanins and betaxanthins, the molecular weights (MW) and molar extinction coefficients (e) (MW = 550 g/mol; e = 60,000 L/mol cm) and (MW = 308 g/mol; e = 48,000 L/mol cm in H_2O) were applied.

Determination of betalain compounds with HPLC analyses

HPLC analyses is to be performed on HPLC, equipped with P680 HPLC pump, ASI-100 automated sample injector. The analytical column Merck Lichrocart 250 × 4 RP-18e (5 μm) was operated at 30°C. Solvents were 0.2% (v/v) formic acid in water (A) and acetonitrile (B). At a flow rate of 1 mL/min, simultaneous monitoring was performed at 538 nm and 476 nm for betalains.

Antioxidant activity of extracts

The antioxidant activity of the extracted betalain samples is to be determined by DPPH method. The DPPH assay is utilized with some modifications[67]. The stock reagent solution (1×10^{-3} M) is to be prepared by dissolving 22 mg of DPPH in 50 mL of methanol and stored at –20°C until use. The working solution (6×10^{-5} M) may be prepared by mixing 6 mL of stock solution with100 mL of methanol to obtain an absorbance value of 0.8 ± 0.02 at 515 nm, as measured using a spectrophotometer. Extracts each of 0.1 ml can be vortexed for 30 s with 3.9 ml of DPPH solution and left to react for 30 min, after which the absorbance at 515 nm can be recorded. A control with no added extract is also to be analyzed. The DPPH solution with no added extract is required to be analyzed as control.

Scavenging activity was calculated as follows:

"DPPH radical" scavenging activity (%) = $[(A_{control} - A_{sample})/A_{control}] \times 100$ where A is the absorbance at 515 nm

IX. CONCLUSION

Fresh produce of beetroot not only involves the appropriate postharvest processing and proper storage to increase the shelf life but also the application of novel processing techniques to increase its usage in health foods. Beetroots are rich source of various macro and micro nutrients that requires evolving exclusive processing methods for beetroot processing and value addition so that minimum damage occurs to the beetroot nutrients and phytofactors present in them with their maximum shelf life. Many techniques such as canning, osmotic dehydration have been used previously to increase their availability throughout the year. They are a rich source of betalains which has a potential of becoming a major natural food colourant in future. Thus, there is a huge scope for future research to improve the processing quality through novel techniques as well as to increase the extraction and availability of betalains. Research should also be directed towards the enhanced use of beetroot and their co-products in various food combinations for development of ready-to-eat and instant food preparations.

X. REFERENCES

[1] Bigg M., "Complete book of vegetables". August Books Ltd, 12 Ravensbury Terrace, London, SW 18 4RL. (1997).

[2] Singh B., Hathan B.S., "Optimization of osmotically dehydrated beetroot candy using response surface methodology". *International Journal of Food and Nutritional Sciences*, **2**(1), 15–21 (2013).

[3] Koul V.K., Jain M.P., Koul S., Sharma V.K., Tikoo C.L., Jain S.M., "Spray drying of beet root juice using different carriers". *Indian J Chem Technol.*, **9**(5), 442–445 (2002).

[4] McCance R.A., Widdowson E.M., "The Composition of Foods", 5th Ed. Cambridge, UK: The Royal Society of Chemistry (1995).

[5] Barooah S., "Vegetable Growing in India", Kalyani Publisheres, Ludhiana p. 25 (1993).

[6] Nottingham S., "Beetroot" (E-book), http://www.stephennottingham.co.uk beetroot.htm (2004).

[7] Gopalan C., Sastri B.V.R., Balasubramanian S.C., "Nutritive Value of Indian Foods", National Institute of Nutrition, Indian Council of Medical Research, Hyderabad, India (2007).

[8] Alard D., Wray V., Grotjahn L., Reznik H., Strack D., "Neobetanin: Isolation and identification from Beta vulgaris", P*hytochemistry,* **24**, 2383–2385 (1985).

[9] Kaur C., Kapoor H.C., "Anti-oxidant activity and total phenolic content of some Asian vegetables", *International Journal of Food Science and Technology,* **37,** 153 161 (2002).

[10] A.D.A.M., Inc., ed., "Betaine", University of Maryland Medical Center, http://www.umm.edu/altmed/ConsSupplements/Betainecs.html (2002).

[11] Jurasunas S., "Oxygen, Mitochondria and Cancer", Lisbon–2006. http://www.sergejurasunas.com/accessed on 10.4.2015, (2006).

[12] Seeger P.G., "Studies on Ehrlich's tumor ascites carcinoma of mice. IV. Respiratory intensity of tumor ascites cells in Todt's biometer after addition of various substances", *Dtsch Gesundheitsw.*, **12**(41), 1265–1268, (1957).

[13] Vinson J.A., Hao Y., Su X., Zubik L., "Phenol antioxidant quantity and quality in foods: Vegetables", *Journal of Agricultural and Food Chemistry*, **46**, 3630–3634 (1998).

[14] Žitòanová I., Ranostajová S., Sobotová H., Demelová D., Pecháò I., Ïuraèková Z., "Antioxidative activity of selected fruits and vegetables", *Biologia.*, **61**, 279–284. (2006).

[15] Netzel, M., Stintzing, F.C., Quaas, D., Strass, G., Carle, R., Bitsch, R. (2005). Renal excretion of antioxidative constituents from red beet in humans," *Food Research International*, 38, pp. 1051–1058.

[16] Delgado-Vargas F., Jiménez A.R., Parades-López O., "Natural pigments: Carotenoids, anthocyanins, and betalains- characteristics, biosynthesis, processing, and stability", *Critical Reviews in Food Science and Nutrition*, **40**(3), 173–289 (2000).

[17] Kanner J., Harel S., Granit R., "Betalains—a new class of dietary cationized antioxidants", *Journal of Agricultural and Food Chemistry*, **49**(11), 5178–5185 (2001).

[18] Sreekanth D., Arunasree M.K., Roy K.R., Reddy T.C., Reddy G.V., Reddanna P., "Betanin a betacyanin pigment purified from fruits of *Opuntia ficus-indica* induces apoptosis in human chronic myeloid leukemia Cell line-K562", *Phytomedicine*, **14**(11), 739–746 (2007).

[19] Nemzer B., Pietrzkowski Z., Spórna A., Stalica P., Thresher W., Michalowski T., Wybraniec S., "Betalainic and nutritional profiles of pigment-enriched red beet root (*Beta vulgaris* L.) dried extracts", *Food Chem.*, **127**, 42–53 (2011).

[20] Gliszczyñska-Œwig³o A., Szymusiak H., Malinowska P., "Betanin, the main pigment of red beet: Molecular origin of its exceptionally high free radical-scavenging activity", *Food Additives & Contaminants*, **23**(11), 1079–1087 (2006).

[21a] Han J., Gao C., Yang S., Wang J., Tan D., "Betanin attenuates carbon tetrachloride (CCl_4)-induced liver injury in common carp (*Cyprinus carpio* L.)", *Fish Physiology and Biochemistry*, **40**(3), 865–874 (2014).

[21b] Han J, Ma D, Zhang M, Yang X, and Tan D (2015). Natural antioxidant betanin protects rats from paraquat-induced acute lung injury interstitial pneumonia. *BioMed Research International*, http://dx.doi.org/10.1155/2015/608174.

[22] Livrea M.A., Tesoriere L., "Lipoperoxyl radical scavenging and antioxidative effects of red beet pigments," in *Red Beet Biotechnology: Food and Pharmaceutical Applications,* Neelwarne B. (*ed.*), pp. 105–124, Springer, New York, NY, USA (2012).

[23] Winkler C., Wirleitner B., Schroecksnadel K., "*In vitro* effects of beet root juice on stimulated and unstimulated peripheral blood mononuclear cells", *American Journal of Biochemistry and Biotechnology*, **1**, 180–185 (2005).

[24] Azeredo H.M.C., Santos A.N., Souza A.C.R., Kenya C.B., Mendes., Andrade M.I.R., "Betacyanin stability during processing and storage of a microencapsulated red beetroot extract", *Am J Food Tech.,* **2**, 307–312 (2007).

[25] Lu X., Wang Y., Zhang Z., "Radioprotective activity of betalains from red beets in mice exposed to gamma irradiation", *European Journal of Pharmacology*, **615**(1–3), 223–227 (2009).

[26] Edenharder R., Kurz P., John K., Burgard S., Seeger K., "*In vitro* effect of vegetable and fruit juices on the mutagenicity of 2-amino-3-methylimidazole[4,5-f]quinoline,2-amino-3,8-dimethylimide-3-ol[4,5f] quinoxaline", *Food Chem Toxicol.*, **32**, 443–459 (1994).

[27] Kapadia G.J., Tokuda H., Konoshima T., Nishino H., "Chemoprevention of lung and skin cancer by *Beta vulgaris* (beet) root extract", *Cancer Letters*, **100**, 211–214 (1996).

[28] Schouten S.P., Van Schaik A.C.R, "The storage of red beets", *Proc Symp Postharvest Handling of vegetables, Acta Hortic J Hort Sci.,* **56**, 97 (1981).

[29] Ryall A.L., Lipton W.J., "Handling transportation and storage of fruits & vegetables, Vol. I", Vegetables and Melons, AVI Pub. Co., Westport, CT (1972).

[30] Kays S.J., "Postharvest Physiology of Perishable Plant Products", Athens, Georgia, USA: Exon Press, pp. 347–356(1997).

[31] Rutherford P.P., "Carbohydrates changes in stored vegetables with special reference to red beet and parsnip", *Annals of Applied Biology*, **85**, 440–444 (1977).

[32] Burns J.L., "Lightly processed fruits and vegetables: Introduction to the Colloquium", *Hort Science.,* **30**, 14–17 (1995).

[33] Vitti M.C.D., Yamamoto L.K., Sasaki F.F., Aguila J.S.del., Kluge R.A., Jacomino A.P., "Quality of minimally processed beet roots stored in different temperatures" **48**(4), 503–510 (2005).

[34] Kluge R.A., Costa C.A., Vitti M.C.D., Ongarelli M.D., Jacomino A.P., Moretti C.L., "Armazenamento refrigerado de beterraba minimamente processada em diferentes tipos de corte", *Ciência Rural*, **36(1**), 263–270 (2006).

[35] Nunes T.C.F., Rogovschi V.D., Fabbri A.D.T., Sagretti J.M.A., Hirashima F.K., Sabato S.F., "Colorimetric evaluation of irradiated red beetroots", *International Nuclear Atlantic Conference* –INAC 2013 Recife, PE, Brazil, November 24-29, 2013 (2013).

[36] Lewicki P.P., "Design of hot air drying for better foods", *Trends in Food Science & Technology*, **17**, 153–163 (2006).

[37] Kaminski W., Tomezak E., Skorupska., "Estimation of the effect of shape and temperature on drying kinetics using MLP", *Drying Technology*, **22,** 191–200 (2004).

[38] Shynkaryk M., Lebovka N., Vorobiev E., "Pulsed electric fields and temperature effects on drying and rehydration of red beetroots", *Drying Technology*, **26**, 695–704 (2008).

[39] Figiel A., "Drying kinetics and quality of beetroot dehydrated by combination of convective and vacuum microwave methods", *Journal of Food Engineering,* **98**, 461–470 (2010).

[40] Gokhale S.V., Lele S.S., "Dehydration of red beet root (*Beta vulgaris*) by hot air drying: Process optimization and mathematical modeling', *Food Sci Biotechnol.,* **20**(4), 955–964 (2011).

[41] Singh S., Gaikwad K., Omre P.K., Kumbhar B.K., "Microwave convection drying characteristics of beet root (*Beta vulgaris L.*) using modeling equations for drying", *J Food Process Technol.*, **4**(9), 2–4 (2013) doi: 10.4172/2157-7110.1000263.

[42] Ponting J.D., "Osmotic dehydration of fruits- recent modify: Cations and applications", *Proc Biochem.*, **8**, 18–20 (1973).

[43] Kaur K., Singh A.K., "Drying kinetics and quality characteristics of beet root slices under hot air followed by microwave finish drying". *African Journal of Agricultural Research*, **9**(12), 1036–1044 (2014).

[44] Karovièová J., Kohajdová Z., "Lactic acid fermented vegetable juices", *Horticultural Sci. (Prague)*, **30,** 152–158 (2003).

[45] Czyzowska A., Klewicka E., Libudzisz Z., "The influence of lactic acid fermentation process of red beet juice on the stability of biologically active colorants", *European Food Research and Technology*, **223**(1), 110–116 (2006).

[46] Derbinova E.S., "Freeze-dried beetroot juice used in the manufacture of ice cream", *Kholodil' naya Tekhnika*, **11**, 34–36 (1979).

[47] Fan Yung, A.F., Khotivari A.V., "Changes in betaine during the production of beetroot juice", *Izvestiya Vysshikh Uchebnykh Zavedenii, Pishchevaya Tekhnologiya,* **6**, 152–153 (1975).

[48] Fenena, O.R. (1995). Quimica de los alimentos. 2. (ed.) Zaragoza: Acribia, p. 585.

[49] Butera D.L., Tesoriere F., Di Gaudio A., Bongiorno M., Allegra A.M., Pintaudi R., Kohen., Livrea M.A., "Antioxidant activities of sicilian prickly pear (*Opuntia ficus indica*) fruit extracts and reducing properties of its betalains: Betanin and indicaxanthin", *J Agri Food Chem.,* **50**, 6895–6901 (2002).

[50] Tesoriere L.D., Butera M., Allegra M., Fazzari., Livrea M.A., "Distribution of betalain pigments in red blood cells after consumption of cactus pear fruits and increased

resistance of the cells to *ex vivo* induced oxidative hemolysis in humans", *J Agri Food Chem.*, **53**, 1266–1270 (2005).

[51] Downham A., Collins P., "Colouring our foods in the last and next millennium", *International Journal of Food Science and Technology*, **35**, 2–22 (2000).

[52] Adams J.P., von Elbe J.H., Amundsen C.H., "Production of a betacyanine concentrate by fermentation of red beet juice with *Candida utilis*", *Journal of Food Science*, **41**(1), 78–81 (1976).

[53] Codex Alimentarius Commission (2004). ALINORM 05 D 28 D 27 - Report of the twenty-second session of the codex committee on processed fruits and vegetables. Available at: http://www. codexalimentarius.net/download/report/624/al28_27e.pdf (2004).

[54] Castellar M.R., Obo´n J.M., Alacid M., Ferna´ndez- Lo´pez J.A., "Color properties and stability of betacyanins from Opuntia fruits", *J Agr Food Chem.*, **51**, 2772–2776 (2003).

[55] Pedren˜o M.A., Escribano J., "Correlation between antiradical activity and stability of betanine from *Beta vulgaris L.* roots under different pH, temperature and light conditions", *J Sci Food Agri.*, **81**, 627–631 (2001).

[56] Rakin M., Vukasinovic M., Siler-Marinkovic S., Maksimovic M., "Contribution of lactic acid fermentation to improved nutritive quality vegetable juices enriched with brewer's yeast autolysate", *Food Chem.*, **100**, 599–602 (2007).

[57] Pavokovi D., Krsnik-Rasol M., "Complex biochemistry and biotechnological production of betalains. Biotechnological production of betalains", *Food Tech Biotech.*, **49**, 145–155 (2011).

[58] Jagannath A., Kumar M., Raju P.S., "Fermentative stabilization of betanin content in beetroot and its loss during processing and refrigerated storage". *Journal of Food Processing and Preservation*, **39**(6), 606–613 doi: 10.1111/jfpp.12267 (2015).

[59] Manoharan A., Ramasamy D., Dhanalashmi B., Gnanalashmi K.S., Thyagarajan D., "Organoleptic evaluation of beetroot juice as natural color for strawberry flavor ice cream", *Indian Journal of Medicine and Healthcare*, **1**(1), 34–37 (2012).

[60] Karthik R.G., Ezhilarasan V., Yazhini K.A., Sridhar S., Chinnathambi V., "Comparative evaluation of white wine production from different carbohydrate rich substrates using air-lift bio-reactor", *International Journal of Pharma and Bio Sciences.*, **3**(1), 392–404 (2012)

[61] Bangale S.R., Rajsekaran P., Wilson I., "Impact of pH and yeast on the preparation of wine from beetroot (*Beta vulgaris L.*)", *Adv Bio Tech.*, **11**(11), 47–48 (2012).

[62] Mridula D., Kumar M.N., Jain D., Tyagi S.K., Rai D.R., "Effect of addition of beetroot powder on physical properties and acceptability of defatted soya flour fortified biscuits", *Indian Journal of Nutrition and Dietetics*, **46**(2), 59–69 (2009).

[63] Mridula D., Jain D., Singh K.K., Patil R.T., Gupta M.K., "Physico-chemical and sensory quality of extruded snack foods developed from rice and defatted soy flour/ chickpea splits supplemented with dried beetroot", *Journal of Agriculture Engineering*, **48**(4), 17–23 (2011).

[64] Mridula D., Singh K.K., Patil R.T., "Value added health mix from extrudates, beetroot and carrot powder", *Indian Journal of Nutrition and Dietetics*, **47**, 453–462 (2010).

[65] Stintzing F.C., Schiebe A., Carle R., "Evaluation of colour properties and chemical quality parameters of cactus juices", *European Food Research and Technology*, **216** (4), 303–311 (2003).

[66] Stintzing F.C., Carle R., "Functional properties of anthocyanins and betalains in plants, food and in human nutrition", *Trends in Food Science and Technolgy.*, **15**, 19–38 (2004).

[67] Lee S.C., Kim J.H., Jeong S.M., Kim D.R., Ha J.U., Nam K.C., "Effect of far infrared radiation on the antioxidant activity of rice hulls", *Journal of Agricultural and Food Chemistry*, **51**, 4400–4403 (2003).

7

Processing of Root Crops and Functionality of Their Biocomponents

TANUSHREE MAITY[1]*, ALOK SAXENA[2] AND P.S. RAJU[1]

ABSTRACT

Sweet potato and cassava are the two root crops which can tolerate harsh environmental conditions and can grow in specific regions where other crops fail to grow. Both these crops are considered as cash crops for farmers. These crops are high yielding and drought tolerant with wide adaptability to various climates and farming systems. One of the challenges faced by most of the developing countries is the lack of storage facilities, resulting in significant amount of the loss of root crops due to improper storage. In order to reduce such losses, value addition to these root crops in the form of other useful products is necessary. Moreover, cassava roots have a short shelf-life of only one to three days due to post-harvest physiological deterioration which can cause significant economic losses. The color of sweet potato varies from white to deep orange, whereas, cassava is a white root. Sweet potato could be considered as an excellent novel source of natural health-promoting compounds such as carotenoids and anthocyanins, for the functional food market. Sweet potatoes are consumed fresh or after cooking / baking. They can also be dehydrated or canned. Cassava roots are processed into various traditional forms in different regions of the world. Both roots can be dried and milled into flours. Sweet potato and cassava flours can be used in various baked products. Starch from sweet potato and cassava is a raw material for manufacturing sugar, alcohol, citric acid etc. Starch processing from sweet potato and cassava can create new economic

[1] Fruits and Vegetables Technology Discipline, Defence Food Research Laboratory, Mysore 570011, Karnataka, India

[2] Amity Institute of Food Technology, Amity University Uttar Pradesh, Noida 201313, Uttar Pradesh, India

**Corresponding author* E-mail: tanumft@gmail.com

and employment activities for farmers and rural households and can add nutritional value to food systems. The functional properties of the components present in these crops could increase utilization by consumers and the food industry.

***Key words*:** Sweet potato, Utilization, Starch, Carotenoids, Anthocyanins, Processing, Phytochemicals, Cassava, Composition, Cyanogenic glycosides, Products

I. INTRODUCTION

Root and tuber crops, such as cassava, sweet potato and yam, taro are important staple foods for millions of people and are grown in the tropical and semitropical regions of the world. Sweet potatoes and cassava are root vegetables which are grown for their underground food reserves. Both the crops have excellent shelf life and good transportability and handling characteristics. In developing countries, sweet potato and cassava crops offer very high potential for income generation through small-scale processing because they are efficient producers of carbohydrates. Due to their high starch content, root and tuber crops play a significant role in the manufacturing and are also used as ingredients of fabricated foods around the world. These crops are important calorie sources. Approximately one-third of the worldwide population depends on them[1]. However, research and development work in tuber and root crops have been neglected even though these staple crops are clearly important in the tropical countries as an important part of the food industry and as animal feed[2]. It is often believed that there is an inverse relationship between root crop consumption and the standard of living. However, in recent years there has been an increasing political and scientific awareness of the importance of increasing root crop production in many countries[2]. Sweet potato (*Ipomea batatas*), which has been a traditional starch source in Africa and Asia, is mainly consumed in baked, boiled or steamed forms[3]. It contains abundant nutritional substances and could become an important food resource to solve the problem of food shortages caused by desertification and population growth[4]. Cassava root (*Manihot esculenta* Crantz) plays certain important roles in African development such as famine-reserve crop, rural staple food, cash crop for both rural and urban households and raw material for feed and chemical industries[5]. In Asia, consumption of fresh cassava roots and exportation to the European Union for use in animal feed are important and its use for bio-fuel production is also increasing[6, 7]. Both the root crops such as sweet potato and cassava possess multi-dimensional uses in food, feed and chemical industries. The chapter deals with the advancements in processing of these root crops and utilization of bio-components from them.

II. SWEET POTATO

A. Origin, Botany and Production

Sweet potato (*Ipomoea batatas* Lam) is a tuberous rooted tender perennial herbaceous dicotyledonous species of the family Convolvulaceae. It is an important vegetable crop of tropical and sub-tropical countries including Africa, India, China, Japan, pacific islands and southern United States. It is a native of tropical America. In southern parts of United States it is a standard food article than the white potatoes[8]. The origin of the sweet potato is proposed to be in between the Yucatan Peninsula of Mexico and the Orinoco River in Venezuela[9]. Sweet potato crop is known as a food security crop because it can be harvested little by little over several months. It has been emphasized that the sweet potato roots and leaves (greens or tips) can support more people per unit hectare than any other food[10]. Sweet potato crop is high yielding and drought tolerant, with wide adaptability to various climates and farming systems[11, 12]. A single sweet potato plant may produce 40–50 roots ranging in length from a few to 30 cm, and weighing between 100 and 1000 g[13]. Sweet potato is more or less sweet flavored vegetable. The flesh of sweet potato is either soft or firm. The edible storage organ of the sweet potato is a true root. Sweet potato is considered as more drought resistant than other vegetable crops and requires long and warm growing season. The stems are usually prostate and slender. Sweet potato crop does well in almost all types of soils except heavy clay. The ideal pH of soil for sweet potato crop is slightly to moderately acidic (pH 5.2-6.7). The harvested crop can store well for six months at 13 to 16°C under 85 to 90% relative humidity. Storage of fresh sweet potatoes requires nicely controlled temperature (13-15°C)[14] and relative humidity (85-95%)[15]. Sweet potatoes are harvested as soon as they reach marketable size regardless of the stage of maturity. However it should be harvested at well mature stage when intended for storage. Sweet potato varieties exist in many colors of skin and flesh, ranging from pure white through cream, yellow, orange, to a very deep purple, although white and yellow-orange fleshed varieties are the most common[16] (Fig. 1). Presently, hundreds of sweet potato cultivars are grown throughout the world and are unique to the growing regions. The varieties marketed in the USA include CA Oriental type, Yellow Jersey, Japanese type, One beauregard, White type, Okinawan type, and Beauregard White type. Indian Agriculture Research Institute, New Delhi released three improved varieties of sweet potato namely, Pusa *Safed*, Pusa *Sunehari* and Pusa *Lal*. Tamilnadu Agricultural University also developed certain other varieties of sweet potato such as Co 3, Co CIP 1, *Sree Nandini, Sree Vardhini, Kiran, Sree Bhadra, Sree Rethna, Gouri* and *Sankar*. Central Tuber Crop Research Institute, Trivandrum, India identified open pollinated progenies OP 1, 2, 34, 217 and 219, selection S-43, 73, 107 and 162 genotypes of sweet potato. Pentaploids (2N =75) of sweet potato have been developed and few among them have shown tuber development comparable with cultivated hexaploid sweet potato[17].

Fig. 1: Sweet potato with different flesh colors

The world production of sweet potato was estimated to be 108004174 MT in 2012[18]. Asia and Africa regions account for 95% of the world's production[19]. The leading countries in sweet potato production are China, Nigeria, Tanzania, Uganda, Indonesia, Vietnam, United States, Ethiopia, India and Rwanda (Table 1). In India the production was reported to be 1072800 MT in 2012. In India, the major sweet potato producing states are in the states of Odisha, Uttar Pradesh and West Bengal.

Table 1: Major sweet potato producing country [18]

Country	*Production (MT)*
China	77375000
Nigeria	3400000
United Republic of Tanzania	3018175
Uganda	2650000
Indonesia	2483467
Vietnam	1422501
USA	1201203
Ethiopia	1185050
Madagascar	1144000
India	1072800

B. Sweet Potato: Composition and Nutritional Value

The nutrient composition of sweet potato varies greatly according to genetic and environmental factors. Different genotypes and environmental conditions result in different quality, nutritional and health values and processing suitability of genotypes grown in different localities[20]. Sweet potato contains many nutrients including protein, carbohydrates, minerals, carotenoids, dietary fiber and vitamins. The crude protein content of sweet potato generally ranges from 1.3 % to 10 % on dry weight basis[21]. However, several researchers have reported different protein content in different varieties or cultivars of sweet potato. Two sweet potato cultivars, Koganesengan and Beniazuma, were reported with 2.1% and 1.3% protein respectively[19]. Nigerian varieties of sweet potato were reported with 1.4% and 9.4% protein contents[22]. Hydroponically grown sweet potatoes were reported with protein contents ranging from 1.2% to 1.8% (on fresh weight basis)[23, 24]. The protein contents ranges from 3.0% to 7.2% on dwb in 16 cultivars of sweet potato grown in Sri Lanka[25]. Cambie and Ferguson[26] reported 1.7% protein content for sweet potato while Gichuhi *et al.*[27] reported 4.5%, 4.7%, and 9.0% protein (dwb) for cultivars J6/66, Beauregard and TU-82-155, respectively. The important major storage proteins in sweet potato, which account for more than 80% of the total protein, are Sporamins A and B[27, 28]. Sporamins A and B are proteinase inhibitors with antioxidant activity and reported to have some anti-carcinogenic properties[29, 30]. Among various minerals present in sweet potato calcium has been found in highest concentration followed by magnesium, iron, copper, zinc, and manganese in two Egyptian sweet potato cultivars[31]. However, Ekpenyong[32] reported that phosphorous is found in highest concentration in sweet potatoes. Olaofe and Sanni[33] reported potassium (3617 mg/100 g), magnesium (580 mg/100 g) and calcium (112 mg/100 g) in sweet potato. Oboh *et al.*[22] analyzed 3.8 to 5.9% crude dietary fiber in 49 varieties of sweet potatoes. The total dietary fiber content of orange fleshed sweet potato, purple-fleshed sweet potatoes and yellow/ white-fleshed sweet potato ranges from 2.0 to 3.2, 2.3 to 3.9, 2.3 to 3.3 g/100 g[34]. Total dietary fiber content of two kinds of sweet potato roots were reported as 3.4 and 2.3 g/100 g by Ishida *et al.*[19]. Sweet potato is rich source of antioxidants such as polyphenols. The following phenolic compounds have been found in sweet potatoes: neochlorogenic acid (3-CQA), cryptochlorogenic acid (4-CQA), chlorogenic acid (5-CQA), and three isochlorogenic acid isomers–isochlorogenic acid A (3, 5-diCQA), isochlorogenic acid B (3, 4-diCQA) and isochlorogenic acid C (4, 5-diCQA)[35, 36]. The phenolic content in American sweet potato has been reported to be from 0.14 to 0.51 mg chlorogenic acid equivalent/g fresh weight after peeling the storage root. Sweet potatoes contain high levels of chlorogenic acid derivatives, a family of esters formed from cinnamic and quinic acids. Chlorogenic acid and iso-chlorogenic acid are main components in polyphenols of American sweet potato[37, 38].

Sweet potato peels (SPP) can be utilized as valuable source for biological active compounds with antioxidant activity. SPP are reported to have almost three times more antioxidant activity than other plant tissues. Sweet potato peels had also received attention for food and feed application as they possess high levels of phenolics[39, 40]. SPP extracts has been reported to be beneficial against Alzheimer's disease as it reduces the level of lipid peroxidation and increases catalase activities in biochemical studies using the brain tissue of mice. SPP possess potent wound healing activity possibly due to an underlying antioxidant mechanism[41, 42]. Extraction of phenolic compounds has been found significantly affected by solvent: solid ratio (positively) and peel cut depth (negatively)[43].

Sweet potato leaf

Yield of sweet potato leaves is much higher than those of green leafy vegetables and can be harvested several times from the crop during the year[44]. In Africa and Japan, the leaves of the sweet potato are eaten, and the protein content has been reported to be as high as 27.0% protein on dry weight basis[11]. In some parts in Nigeria, sweet potato leaves are acceptable as soup ingredients in terms of flavor, appearance, palatability, softness, and acceptability[45]. Pace *et al.*[46] analyzed 4.0–6.0% protein, 8.0–12.0% carbohydrate, 60 mg/100 g calcium and 80 mg/100 g phosphorus in sweet potato leaves. The leaves of the sweet potato are dark green and have nutritive values comparable to common dark green leafy vegetables. Leaves contain a large amount of protein with high amino acid score. Sweet potato leaves are rich in soluble dietary fiber and mineral content, particularly iron. Vitamin content of leaves such as carotene, vitamin B_2, C and E are high in comparison with other leafy vegetables. Leaves contain high amounts of protein (3.8 and 3.7 g/100 g), total dietary fiber (5.9 and 6.9 g/100 g), and ash (1.9 and 1.5 g/100 g)[19]. Tewe *et al.*[45] reported that the protein content of sweet potato leaves was 18.4%, while fiber content was between 3.3% and 6.0%. A new sweet potato cultivar (Suioh) has been developed for utilization as vegetable greens[36]. Sweet potato leaves contain at least 15 biologically active anthocyanins, which are beneficial to human health and may also be useful as natural food colorants[47]. The crude protein, crude fiber, crude fat, carbohydrate and ash contents of 40 Chinese cultivars of sweet potato ranged between 16.69–31.08, 9.15–14.26, 2.08–5.28, 42.03–61.36 and 7.39–14.66 g/100 g dwb, respectively. The average crude fat content was 3.69 g/100 g dwb. The major minerals found in these cultivars were reported to be potassium, phosphorus, calcium, magnesium and sodium with average values of 1625.1, 1248.2, 744.9, 405.2 and 159.98 mg/100 g dwb, respectively[48]. Sweet potato leaves contains active polysaccharides and flavonoids compounds. Sweet potato leaves also contain chlorogenic acid. Optimum parameters for chlorogenic acid extraction were reported to be alcohol concentration (30%), liquid material ratio (1:20), pH value (5.0), extract times (four), distillation (90 minutes). The content of chlorogenic acid extracted from sweet potato leaves by the optimum extracting technology has been reported to be 3.52%[49].

C. Utilization of Sweet Potato

Sweet potato is among the most under-exploited of the developing world's major crops[50]. In the developing world, sweet potato consumption varies by countries, regions and income. In developing countries, where limited transport infrastructure exists, processing the sweet potato into value-added products provides an alternative to the difficulties associated with storage and transport of the raw sweet potato roots[24]. Sweet potatoes are consumed after cooking. They can either be canned or dehydrated. For canning both yellow and white sweet potatoes are used but yellow sweet potatoes are preferred because of attractive color, flavor and cooking qualities. McConell and Gottschall[51] recommended freshly dug sweet potatoes for canning purpose. The dehydrated sweet potatoes are excellent for home consumption and for making pies. During Second World War large quantities of sweet potatoes were dried to make pies[52]. For dehydration, raw sweet potatoes are peeled in abrasive peelers or lye peelers in 15-20% NaOH lye solution. Further they are trimmed, sliced or cut and the cut material is steam blanched for 7 minutes and dehydrated on trays at 71.1°C to a moisture percentage of less than 7.0%. Sulfiting of sweet potatoes prolongs the keeping quality. Processing of sweet potato into a powdered form (flour) not only prolongs the shelf life but also improves the nutritional value of bread made from sweet potato powder, which acts as a substitute for wheat flour[53]. The unique purple color, flavor and supplemented nutrients as well as the thickness of purple sweet potato flour make it an ideal additive for addition in soups, sauces, baby foods and bakery products[54]. The aid of maltodextrin and mixture of maltodextrin-cyclodextrin as carriers in spray drying technique has been used to make flour from purple sweet potato with better fluidity and storage properties (lower moisture content and higher *Tg* values) as compared to flours without carrier[55].

Peters and Wheatley[56] identified sweet potato flour (SPF) as a promising product to increase rural income with potential markets in bakery, noodle, and various snacks in an in-depth diagnostic assessment of the feasibility of establishing a household-level SPF processing industry in East Java. Their results suggested that SPF could be marketed as a low-cost alternative to imported wheat flour. SPF can serve as a source of carbohydrate, β-carotene, minerals, can add natural sweetness, color and flavor to processed products, and can prevent food allergies[57]. SPF can serve as an alternative for individuals diagnosed with celiac disease or with allergies to the gluten in wheat. SPF can also be used directly or as a raw material for processing into other products. In India, dried sweet potatoes grounded into flour are used to supplement flours in bakery products, *chapatis*, and puddings, while in the Philippines, dried sweet potato chips are pounded into flour for use in gruel[58]. Dawkins and Lu[59] reported 13-18% yield in flour prepared from steam blanched, microwave blanched and unblanched sweet potatoes. A flour yield ranging from 17% to 38% for different sweet potato cv. was also reported by Gakonyo[60]. Suismono[61] reported 24% fresh root: flour conversion rate, while Martin[62] reported a 17–38% range. Average sweet potato root: flour conversion rates ranging from 23% to 27% were reported by Peters *et al.*[63]. Van Hal[57] estimated that SPF could contribute 0-100%, 20%, 20-40%, 17%, and 10% of the daily nutrient needs for β-carotene, thiamin, iron, vitamin C, and protein, respectively. The quality characteristics of SPF include high starch content, white color, low acidity, crude fiber, and ash contents[64]. It is important to understand the functional, physical, and chemical characteristics of SPF if its suitability for use in value -added products has to be determined. A variety of products such as doughnuts, biscuits, cookies, fried sweet potato cakes, extruded sweet potato chips, ice creams, porridge, brownies, pies, breakfast foods, and weaning foods have been made from SPF[65,66]. SPF has also been utilized for efficient conversion to ethanol. For ethanol production, SPF is prepared using abrasive peeling, hand peeling, lye peeling, and drum drying. The SPF is then treated with pectinase, and then with culture filtrate containing α-amylase and glucoamylase of Endomycopsis fibuligera. The inoculated samples are fermented for 3 days. A wine like product with ethanol up to 8.6% (w/v), desirable aroma and color can be processed in this way[67].

D. Sweet Potato Starch

Physico-chemical characterization

Sweet potato is a good raw material for the starch industry as the roots contain 80-90% carbohydrate, mainly in the form of starch[68, 69]. Sweet potato is an important starch source in China, Vietnam, Korea, Taiwan, and the Philippines[70, 71]. Starch is the major carbohydrate in roots and ranges from 73.7% to 84.9% of the root on dry weight basis[72]. Earlier, an average total starch content of 61.5% dwb was reported in sweet potato[73]. However, Miller *et al.*[74] reported 81.0% starch content on dwb in sweet potato. Quality of the fresh sweet potato root and the amount of extractable starch are considered to be important for the production of sweet potato starch (SPS) because of their influence on final starch quality and process efficiency. SPS production is also influenced by the chemical composition of the starch including ash, protein, fiber etc. and functional properties of the starch itself. The process of starch isolation from the sweet potato using a modified technology with efficient drying has been reported by Bovell-Benjamin *et al.*[23] and is shown in Fig. 2. Industries based on SPS extraction has been developed in several regions in China[75] and may account for more than two million tones annually, but the statistics are not quite precise[76]. Starches from different sweet potato cultivar exhibit a variety of nutritive, physical and sensory properties, such as higher starch content and more extractable starch[77]. Sweet potato varieties with high dry matter and starch contents yield maximum starch. Processing (milling, sifting), type of cultivar and chemical or enzymatic modifications can interfere with the functional properties of SPS. SPS exhibits wide variation in granule size (3–40 μm) and amylose content (15–30%). It has a gelatinization temperature between 61 and 70°C, and gives both A and C type of X-ray diffraction pattern[78]. Bovell-Benjamin *et al.*[79] reported granule sizes of SPS in the range of 9.0–34.2, 2.6–15.8, and 5.3–21.8 mm in three different varieties of sweet potato, namely, J6/66, Beauregard, and TU-82-155, respectively. They reported 4.4-6.0% moisture, 1.1-1.3% protein, 0.6-0.7% ash and 0.1-0.3% fat in starches isolated from these varieties. Woolfe[10] reported that SPS granules are found in oval, round, or polygonal shapes and mean granule size ranges from 12.3 to 21.5 mm. Kays[80] also reported a range from 1 to 30 mm for SPS granules. Srichuwong *et al.*[81] reported 24.2-24.4% amylose content (% w/w dry starch), onset temperature of gelatinization 49.5-70.9°C and peak temperature of gelatinization 57.7-76.9°C in starches isolated from two varieties (K159 and DCY) of sweet potato. DSC study has shown gelatinization peak temperature of 77.6°C of sweet potato starch[82]. SPS exhibits a higher viscosity profile and paste clarity compared to corn starch[83]. But has low viscosity values on pasting compared to potato starch (Wheatley and Bofu[76]. SPS is not sufficiently white in color. However, Miller *et al.* [74] reported L* color values of SPS comparable to corn starch.

Sweet potato
↓
Washing and peeling
↓
Homogenization with water (sweet potato: water; 1:2)
↓
Centrifugation for 5 minutes at 3000 rpm
↓
Decantation
↓
Repeat homogenization, centrifugation and decantation
↓
Moist starch
↓
Dehydrate at 70°C for 12 h
↓
Grinding & milling
↓
Sweet potato starch (SPS)

Fig. 2: Isolation of SPS from sweet potato

Chemical modifications are necessary to expand the range of applications of native starches in the food industry. Sweet potato starches were modified by hydroxypropylation to evaluate the effect of molar substitution from 0.042 to 0.153 on the rheological properties, thermal properties, freeze-thaw stability, paste clarity, and gel strength of hydroxypropylated sweet potato starches (HPSPS). Hydroxypropylation increased the swelling power and solubility values of the native SPS. Increase in molar substitution also increased these parameters. The transition temperatures and enthalpy (ΔH) of gelatinization of HPSPS were lower than those of native SPS, and significantly decreased with an increase in molar substitution[84]. The rheological properties of HPSPS pastes such as consistency index, apparent viscosity, Casson yield stress, complex viscosity, and dynamic moduli values decreased with an increase in molar substitution, while their flow behavior index and tan d values increased. The paste clarity of HPSPS paste was more pronounced with increasing molar substitution of hydroxypropyl groups[84].

Uses of sweet potato starch (SPS)

SPS is widely used in a variety of food and industrial applications in Asia[85]. For use in various food products SPS should have a smooth texture, which is consistently soft and flexible at low temperature and must be able to retain its thickening power at high temperature[68]. It is used as an ingredient in processed sweet potato products such as noodles, soups, sauces, snacks, and breads[86]. SPS can be used in the preparation of noodles and other wheat-based foods[87]. Starch quality is an important factor, and starch from sweet potato is preferred over starch from other crops such as corn and cassava for making noodles[88]. It has also been used as dough conditioners in bread manufacturing, stabilizers in ice cream, and thickeners in soups and sauces. It can be enzymatically processed into glucose syrup, which has a wide range of applications in the food and pharmaceutical industries[89]. About 5% of total SPS production is used for making distilled spirit '*shochu*' in Japan as well as lactic acid, butanol, acetone, vinegar, and yeast[10]. SPS is used for production of maltose and glucose syrups, which are used in the manufacture of baby foods in some developed countries. Bovell-Benjamin *et al.*[90] processed SPS to make syrup (Fig. 3). For syrup preparation, the SPS was rehydrated, heated to 102°C, treated with thermostable bacterial α amylase at 90°C for 5 h, cooled, and further treated with glucoamylase at 62.5°C for 12 hours.

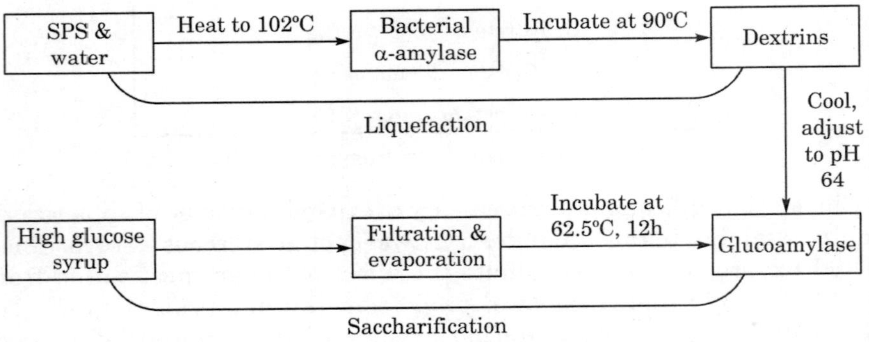

Fig. 3: Enzymatic hydrolysis of sweet potato starch into glucose syrup[90]

Recently, sweet potato has been utilized as substitute as energy resource other than corn grains for bio-ethanol production. Cultivation of sweet potato requires lower input costs and could also provide higher ethanol yield per cultivation area than corn grains[91]. Ethanol productivity largely depends on liquefaction temperature. Sweet potato mashes containing greater than 0.3 kg dm^{-3} total solids were prepared using enzymatic viscosity reduction, and utilized for liquefaction and simultaneous saccharification and fermentation (SSF). In a study, liquefaction at 60°C was found to be sufficient to efficiently accelerate SSF process and ethanol yields of 15.1–15.4% (v/v) during 48-72 h[81]. Sweet potato starch is also reported as a raw material for preparing citric acid. Yuguo *et al.*[92] reported citric acid production from mash dried sweet potato with dregs by *Aspergillus niger* in an external-loop airlift bioreactor. Betiku and Adesina[93] showed that SPS hydrolyzate could serve as sole carbon source for citric acid production. They employed *Aspergillus niger* for the production of citric acid from SPS hydrolyzate by optimizing concentrations of SPS hydrolyzate, $(NH_4)_2$ HPO_4, KH_2 PO_4, fermentation time and pH. In their study, citric acid concentration was taken as response. A statistical model predicted the highest citric acid concentration of 83.01 g L^{-1}, at the optimal condition of SPS hydrolyzate of 153.77 g L^{-1}, $(NH_4)_2HPO_4$ of 3.55 g L^{-1}, KH_2PO_4 of 2.58 g L^{-1}, time of 8 days and pH of 6. Starch can be enzymatically processed to obtain other products such as starch hydrolysates by amylases and further converted into trehalose through combination of amylases and trehalose synthase[94]. For industrial production of trehalose, the high thermostability of the enzyme is the most important property for reducing the microbial contamination and lower the production cost. A new recombinant *Picrophilus torridus* TSase (PTTS) has been developed with catalytic ability for the conversion of maltose to trehalose by intramolecular transglucosylation. The trehalose yield from sweet potato starch was found to be 1.3-fold higher with application of mutant of PTTS than the wild type at 50°C for 4 h[95].

E. Carotenoids in Sweet Potatoes

Carotenoids are one of the most valuable food constituents in promoting human health. As natural pigments, they give the appealing yellow, orange, or red color to many fruits and vegetables. There are a variety of flesh colors, such as white, yellow, orange and purple, in sweet potato root. Sweet potato contains abundant β-carotene, the precursor of vitamin A, particularly in the orange-flesh and yellow-flesh sweet potatoes[96, 97]. The intensity of the yellow or orange flesh color of the sweet potato is directly correlated to the carotenoid content [98]. The flesh colors and β-carotene content in sweet potato roots vary widely because they are affected by genetic variety, maturity, growing conditions, postharvest storage and season[99, 100]. Orange fleshed varieties are associated with sensory attributes, such as pumpkin flavor, watery texture and orange color, while yellow and white varieties are associated with the descriptive terms, creamy color, starchiness, hard texture, coarse texture, yellow color, fibrous texture and sweet taste[101]. All-trans- β-carotene is the major carotenoid present in sweet potatoes. β-carotene is exclusively present in considerable amounts in the orange-fleshed varieties of sweet potato. The β-carotene concentration of sweet potato varies considerably in different varieties and cultivars (Table 2). The white-fleshed sweet potatoes are the most commonly consumed in Africa, Asia, and Latin America. Aside from the remarkable varietal differences, the carotenoid content is also influenced by root age and production site[96]. Indian sweet potatoes are white fleshed with negligible carotenes, however, a new yellow-fleshed variety, '*kiran*' showed significant amounts of β carotene with an average value of 1087 mg/100 g[102]. Increasing commercial interest in carotenoid production is due to their natural origin, null toxicity, high versatility, providing both lipo- and hydro-soluble colorants with colors ranging from yellow to red, provitamin A activity, role in lipid oxidation and anti-carcinogenic properties[103]. Enzyme extracted pigments remain in their natural state, bound with proteins through covalent bonding or weak interactions, depending on their structure.

Table 2: β -Carotene content in various sweet potatoes varieties/cultivars

Place	*Cultivar/varieties*	*β -Carotene content (μg/g fresh root)/*	*No. of cultivar varieties*	*Refere-nce(s)*
Hawaii	Purple-fleshed varieties	10–50	4	[34]
	Yellow/white-fleshed varieties	10–60	7	
	Orange-fleshed varieties	67–131	7	
Kenya	White cultivars	Less than 1	6	[96]
	Cream cultivars	1–10	3	
	Pale orange to orange cultivars	10–80	7	
	White/ yellow cultivars	Less than 1	9	[104]
	Cream/yellow cultivars	20–110	7	
	Light yellow fleshed	1–10	7	
	Light to deep orange fleshed	21–63	6	
	Yellow & orange fleshed	12–109	6	[105]
Uganda	Orange fleshed	68–125		[106]
Brazil	White fleshed	0.6	8	[107]
Taiwan	Orange fleshed	2.13–3.76[a]	1	[108]
	Yellow fleshed	0.41–0.83[b]	1	
India	Yellow fleshed	10.8	1	[102]
United States	Orange fleshed	11.8–167	10	[109]
	Purple fleshed	5.4–56.6	6	
	Yellow fleshed	1.5–2.3	2	
	White fleshed	0.2	1	

a; orange-fleshed sweet potato (Tainung 66) harvested at different times
b; yellow-fleshed sweet potato (Tainung 57) harvested at different times

Almeida and Penteado[107] identified luteochrome (191 μg/100g), β-carotene 5, 6, 5' 6 diepoxide (114 μg/100g), β-carotene (61 μg/100g), β -carotene 5,6-monoepoxide (25 μg/100g), ζ-carotene (22 μg/100g), α-zeacarotene (19 μg/100g), and neurosporene (13 μg/100g) in raw white fleshed sweet potatoes. Maoka *et al.* [110] reported a total carotenoid content of 3.1 mg/100 g in freeze-dried yellow-fleshed sweet potato variety 'Benimasari'. They found β-carotene (10% of the total carotenoid), β -carotene-5,8-epoxide (6.5%), β-carotene-5, 8, 50, 80-diepoxide (40.5%), β cryptoxanthin-50, 80-epoxide (10.5%), β-cryptoxanthin-5, 8; 50, 80-diepoxide (5.5%), auroxanthin (2.2%), and neochrome (4.5%) using high-performance liquid chromatography after high pressure liquid chromatographic separation technique. Fig. 4 shows a number of bio-components identified in sweet potato.

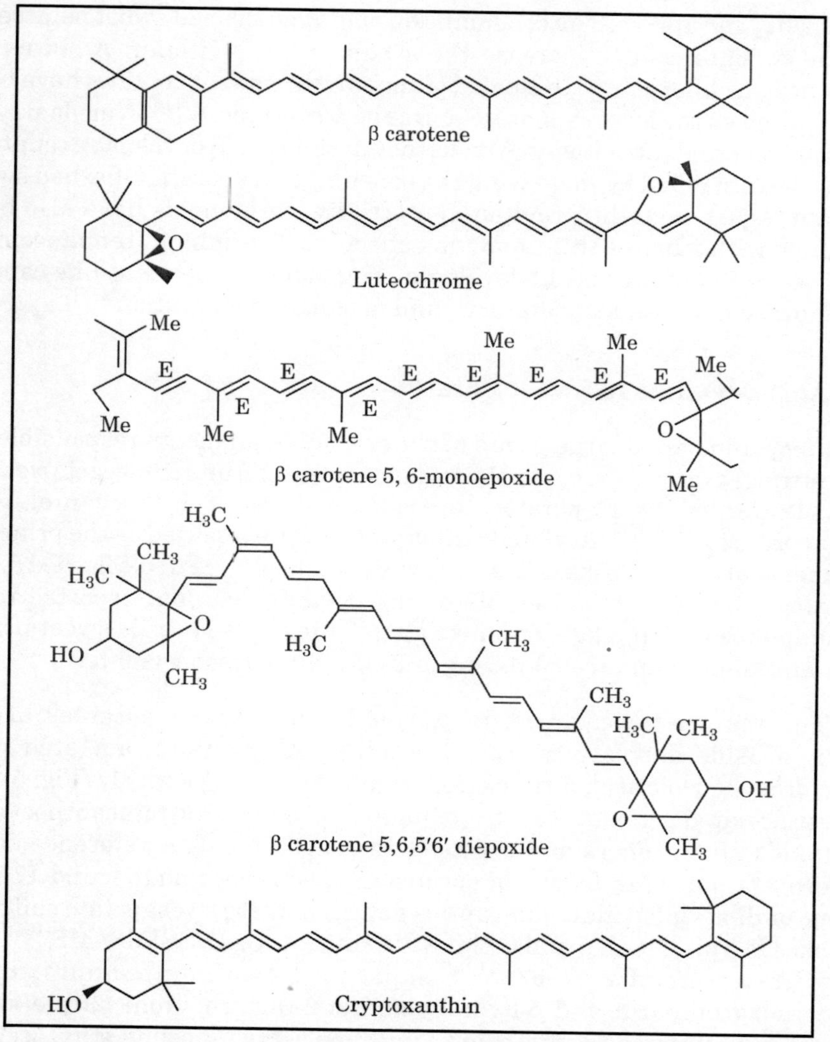

Fig. 4: Various compounds (carotenoids) identified in sweet potato[107, 110]

β-Carotene is cleaved in the intestinal mucosa by carotene dioxygenase, yielding retinaldehyde, which is reduced to retinol (vitamin A). Literature says that pro-vitamin A from orange fleshed sweet potato appears to be more bio-available than that obtained from dark green leafy vegetables[111, 112]. Bureau and Bushway[113] reported highest pro-vitamin A contents in sweet potato (882.97 to 1712.42 retinol equivalents/l00 g) among 22 fruits and vegetables from the United States, analyzed using a HPLC technique. Regular consumption of yellow or orange fleshed sweet potatoes containing about 3 mg/100 g β -carotene on a fresh weight basis provide the recommended daily amount of vitamin A for children less than 5 years old[114]. Sweet potato is reported to eliminate symptoms of vitamin A deficiency in rats[115]. The consumption of orange fleshed sweet potatoes is found to significantly increase the serum retinol (vitamin A status) in Indonesian children[112]. In Kenya, orange-fleshed sweet potatoes have been recognized as the least expensive source of provitamin A[116]. Van Jaarsveld *et al.*[97] reported that vitamin A deficiency in children in developing countries could be controlled by increasing the consumption of orange-fleshed sweet potato. Apart from the provitamin a activity, carotenoids have also been credited with other health-promoting effects such as immuno enhancement and reduction of the risk of developing degenerative diseases like cancer, cardiovascular diseases, cataracts, and macular degeneration[117].

F. Anthocyanins in Sweet Potato

Anthocyanins are a large group of water soluble pigments responsible for the attractive orange, red, purple, and blue colors of fruits and vegetables[118]. Purple fleshed sweet potatoes are good sources of anthocyanins with aromatic acids[119, 120]. Acylated anthocyanins were reported as the principal pigments in purple flesh cultivars of sweet potatoes[121]. Purple fleshed sweet potatoes contain large amounts of anthocyanins content reported to be around 802-1747 mg kg^{-1} fresh weight[122]. Japanese purple sweet potato contains approximately 0.4-0.6 mg anthocyanin/g fresh weight.

The main anthocyanins in purple-fleshed sweet potatoes are 3, 5-diglucoside derivatives from cyanidin and peonidin, acylated with p-hydroxybenzoic acid, ferulic acid, or caffeic acid, respectively (Fig. 5)[123]. These acylated pigments constitute more than 98% of the total anthocyanin content in both tubers and shoots of sweet potato[124]. A reference-based interpretation of MS fragment spectra has been reported to found 17 non, mono or di-acylated individual anthocyanins in three sweet potato cultivars (Stokes Purple, NC 415 and Okinawa) using LC-DAD/ESIMS/ MS[125]. For the first time, He *et al.*,[126] separated two components, 6, 7-dimethoxycoumarin and 5-hydroxymethyl-2-furfural from purple sweet potato extracts by successive sample injection using n-hexane-ethyl acetate-methanol–water (1:2:1:1, v/v/v/v) as the solvent system. However, thirteen anthocyanins were identified in the purple-fleshed sweet potato cultivar 'Jihei No. 1'. The main anthocyanins in cv. 'Jihei' No. 1 were 3-sophoroside-5-glucoside derivatives from cyanidin and peonidin, acylated with p-hydroxybenzoic acid, ferulic acid, or caffeic acid. A unique anthocyanin, delphinidin-3, 5-diglucoside was also found[127]. Two mono-acylated forms, cyanidin 3-caffeoyl sophoroside-5-glucoside and peonidin 3-caffeoyl sophoroside-5-glucoside, were isolated and identified using FAB-MS and NMR techniques from flesh of purple sweet potato cv. 'Yamagawa murasaki'[128]. Six anthocyanins di-acylated with p-hydroxybenzoic acid, caffeic acid and ferulic acid were subsequently identified by Terahara *et al.*[129] followed by the identification of two non-acylated forms, cyanidin 3-sophoroside-5-glucoside and peonidin 3-sophoroside- 5 glucoside[130].

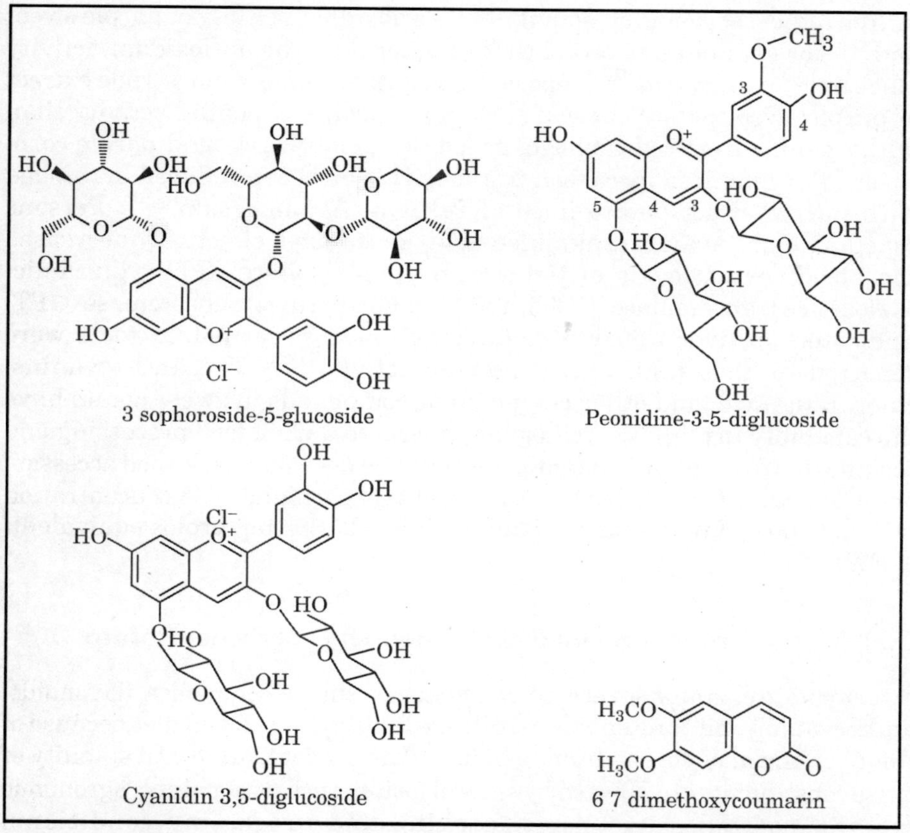

Fig. 5: Various compounds (anthocyanins) identified in purple sweet potato.

There is an increasing demand for natural food colorants because of their natural origin and lack of toxicity[131, 103]. Purple-fleshed sweet potatoes contain large quantities of anthocyanins, which have excellent biological properties and potential value as natural colours for foods[120]. The anthocyanins extracted from purple fleshed sweet potatoes are being used as a source of natural colorant in various food applications. Cevallos-Casals and Cisneros-Zevallos[40] reported purple sweet potato colorant as having much higher color retention and more stability than commercial red grape colorant. Anthocyanins extracted from fermented purple sweet potato were reported to be more pure than chemistry extraction. More than 80.7% of the anthocyanins were found to be acylated anthocyanins and cyanidin and peonidin were the major anthocyanidins. Extracted anthocyanins from fermented purple sweet potato were found to be more stable at low pH values exploring the use of sweet potato anthocyanins as natural colourants in food industry[132].

The anthocyanins of purple sweet potatoes not only show great potential as natural colorants but also provide various nutraceutical properties such as free radical scavenging activity[133]. Antioxidant activity of purple sweet potato was evaluated to be 3.2 times higher than the antioxidant activity blueberry[40]. Kano *et al.*[134] reported that anthocyanins from a crude extract of purple sweet potato showed stronger radical scavenging activity than anthocyanins from red cabbage, grape skin, elderberry, and purple corn. Liver injury induced by carbon tetrachloride has been reduced in rats fed with purple sweet potato juice of cultivar 'Ayamurasaki'[120]. Persons consuming cv. Ayamurasaki juice daily for 44 days effectively decreased the blood serum levels of l-guanosine triphosphate (GTP), glutamic-oxaloactive transaminase (GOT), and glutamic-pyruvic transaminase (GPT) (indicators of liver injury)[120]. Purple-fleshed sweet potato roots were reported to have high antimutagenic activity[135]. The anthocyanins, chlorogenic acid, and other polyphenolic compounds in sweet potato have the capability to inhibit carcinogens generated during food processing and cooking due to their antioxidant potential. The deep purple fleshed accession of sweet potato 'Guincho' has beeen found to show highest TA concentration (418 mg/100 g, FW) and antioxidant activity (17,305 mg Trolox equivalent/g, FW)[136].

G. Effect of Processing on Phyto-Chemicals of Sweet Potato

Sweet potato contains several phytochemicals such as phenolics, flavonoids, anthocyanins and carotenoids, which are highly desirable in diet because of their beneficial effects on human health. The concentration and stability of these constituents are affected by several factors such as genotype, agronomic factors, postharvest storage, cooking and processing conditions. Phytochemical content was evaluated in four genotypes of sweet potato, namely, NCPUR06-020, Covington, Yellow Covington, and NC07-847, immediately after harvest and after curing/storage for 4 or 8 months. Covington genotype was reported to have the highest carotenoid content at harvest time (281.9 μg/g DM), followed by NC07-847 (26.2 μg/g DM), which increased by 25% and 50%, respectively, after 8 months of storage. Curing/storage for up to 8 months did not significantly affect total phenolic content in Covington, Yellow Covington and NC07-847 genotypes, however, for purple fleshed genotype-NCPUR06-020, total phenolic content declined due to anthocyanin degradation during storage[137]. Enzyme assisted extraction of pigments from sweet potato has higher stability and can be use as coloring agents in food systems for commercial production. Blanched sweet potato samples gave the highest color retention after 120 days of storage at 4°C. However, reasonable retention rates were found during storage at 25°C in light and dark[103]. Hagenimana *et al.*[104] found that drying sweet potato slices in a forced air oven at 60°C for 12 h reduced the total carotenoid content by 30%. Further, storage of the dried chips in opaque paper bags at ambient conditions for 11 months resulted in a 10% loss. Several research groups have reported variable carotenoid losses in the production of sweet potato flour, which are affected by the variety utilized, the size and shape of the pieces submitted to drying, and the drying method and condition. In a study of spray drying of purple sweet potato, various carrier agents were assessed for their ability to retain phyto-chemicals in the final flour. The flour with malto-dextrin showed higher retention rate of anthocyanins, flavonoids and total phenolics than flours without and with other carriers in spray dried purple sweet potato flour[138]. Sung Puag *et al.*[139] demonstrated that boiling yellow fleshed sweet potato resulted in loss (43%) of β-carotene. Van Jaarsveld *et al.*[140] achieved better β-carotene retention in sweet potato with oven drying compared to sun drying, with the latter provoking greater degradation than drying in the shade. Lower retention in open sun drying was due to the destructive effect of direct sunlight, greater and more prolonged exposure to oxygen and the uncontrolled environmental conditions. Drying thicker slices retained more β-carotene. β-carotene was found to be 121-136 μg/g in boiled mashed orange sweet potato variety 'Resisto' of different sizes. The raw root contained 153-167 μg/g β-carotene. Carotenoid is retained better when the sweet potatoes were boiled intact covered with water in a pot with the lid on at the shortest possible time, i.e. cooking to just doneness[141]. Kidmose *et al.*[105] observed that drying of chips in shade resulted in a significant reduction of all-trans-β -carotene (approximately 21%), which was further reduced when flour was produced from the dried chips. The concentrations of total phenols and antioxidant activity determined in boiled purple fleshed sweet potatoes were higher than in raw roots[136]. Bechoff *et al.*[142] reported that all trans β-carotene losses in flour made from dried chips varied between 16 and 34% under different drying treatments. Hot air cross-flow drying retained significantly more provitamin A than open sun drying, but drying in a green house solar dryer and sun drying did not result in significantly different provitamin A retention. The shape of the sweet potato pieces (chip or crimped slice) influenced retention during sun drying, with the crimped slices retaining more provitamin A. Bengtsson *et al.*[143] found that drying slices of orange fleshed sweet potato at 57°C in a forced air oven for 10 h reduced all-trans- β-carotene content by 12%. The retention of all-trans-β-carotene was 78% in sweet potato boiled in water for 20 min. Retention was 77% in sweet potato steamed for 30 min, whereas deep frying of sweet potato for 10 min resulted in retention levels of 78%. Drying sweet potato slices at 57°C in a forced air oven for 10 h reduced the all-trans-β-carotene content by 12%. Solar drying and open-air sun drying of slices to moisture content of 10% resulted in all-trans- β -carotene losses of 9% and 16%, respectively. The high content of all-trans- β -carotene in orange fleshed sweet potato varieties and the moderately low losses due to degradation and isomerization during traditional cooking renders sweet potato a suitable food source of provitamin A.

III. CASSAVA

A. Origin, Botany and Production

Cassava (*Manihot esculenta* Crantz) is a tropical perennial shrub and belongs to the family Euphorbiaceae[144]. It is originated in South America and subsequently has been distributed to tropical and subtropical regions of Africa and Asia. Cassava is the chief source of dietary food energy for the majority of the people living in the lowland tropics and sub-humid tropics of West and Central Africa[145]. Cassava provides greater proportion of energy for low-income households than any other food item. Cassava root is easily propagated. It requires 9 to 24 months for growth. It can be stored under the ground for long even after maturity and it need not to be harvested all at the same time. This ensures a constant food supply to farmers for most of the year without storage problems. Cassava plays a vital role in alleviating famine conditions by providing a sustained food supply when other crops fail. The crop is an important source of carbohydrate for humans and animals, having higher energy density (610kJ/100g fresh weight).

The estimate total world cassava production in 2012 was 262 million tones according to FAO[18], which is an increase of 40% since 2000. Africa represents the continent with the largest cassava production of more than 50% of the annual world production (149 million tonnes). Nigeria is the largest producer of cassava with an annual production of 54 million tonnes in 2012 followed by Indonesia with production of 23 million tonnes of cassava.

B. Composition and Nutritional Value of Cassava

Cassava roots and cassava leaves are used for human consumption and animal feed. The edible part of the root comprises about 85% of the total weight of the root[144]. The cassava root contains mainly carbohydrate, 64-72% of which is starch. Cassava starch granules are mainly composed of two polysaccharides, amylose (20%) and amylopectin (80%)[146]. Cassava starch is pure white in colour and is free from other components and is simple to extract from roots[147]. There is a large variation in sucrose content between cassava genotypes. Sweet varieties have been reported to contain about 17% sucrose and small quantities of fructose and dextrose[148]. Cassava roots are low in protein and fat. The lipid content of cassava is only 0.5%. Cassava is reasonably rich in calcium (16–35 mg/100 g) and vitamin C (15–45 mg/100 g)[149,150]. Investigations showed that cassava contained 9.2–12.3% moisture, 1.2–1.8% crude protein, 0.1–0.8% crude lipid, 1.5–3.5% crude fibre, 1.3–2.8% ash, 80.1–86.3% carbohydrate, 1406–1465 kJ 100 g^{-1} DM and 95–135 mg g^{-1} of phytic acid in 5 cassava genotypes[151]. Chemical composition of cassava roots on dry weight basis is presented in Table 3.

Table 3: Chemical composition of cassava roots [152]

Nutrient	*Dry weight basis (%)*
Dry matter	35.00
Starch	30.21
Crude protein	1.10
Fat	0.47
Crude fibre	1.10
Ash	0.70
Calcium	0.10
Phosphorus	0.15

Padonou *et al.*[153] analyzed 13 sweet and 7 bitter varieties of cassava flour. The protein, lipid and fibre content in these varieties ranged from 1.66 to 4.08, 0.42 to 0.65 and 2.63 to 4.92 %, respectively. A great variation in the water content of fresh roots has been observed. It ranged from 60.3% to 87.1% (wb) without any significant difference between sweet and bitter cultivars. Cassava root is deficient in most of the essential amino acids. It has less than the recommended minimum limit of all essential amino acids, except tryptophan. Therefore, cassava roots should be eaten along with other crops rich in essential amino acids to supplement the deficiency[154]. It has been reported that cassava can produce 250×10^3 cal/ha/day compared to 176×10^3 for rice, 110×10^3 for wheat, 200×10^3 for maize, and 114×10^3 for sorghum[149].

Cassava leaves are much richer in protein than the roots, although the leaf contains a lower proportion of methionine than the root protein. The levels of all other essential amino acids in leaf protein exceed the FAO's recommended reference patterns[149,154]. Cassava leaves are also consumed in many African countries[155].

C. Cyanogenic Glycosides in Cassava

The major constraint in cassava roots consumption as human food is the presence of toxic compounds known as cyanogenic glycoside. Cassava contains 4 to 5 cyanogenic glycosides[156]. The hydrogen cyanogen (HCN) in the human body is metabolised to thiocyanate[157]. It is believed that cyanogens protect the plant from attacks by some worms, arthropods and mammals because of that many farmers in Southern Africa prefer to grow the variety with high level of cyanogens, known, as bitter varieties[158, 159]. The illness related to cassava consumption is due to prolong consumption of cyanogens present in cassava. These illnesses include tropical ataxic neuropathy, epidemic spastic paraparesis, endemic goitre and cretinism[160, 161]. Residual cassava cyanogens, when ingested, are hydrolysed in the human digestive system. It is assumed that intestinal microbes hydrolyse the cyanogens[162].

The sweet varieties of cassava are believed to contain much lower HCN content than the bitter varieties. The cyanogenic potential in sweet varieties has been reported to be 8.0 to 20.0 ppm and in bitter varieties from 9.5 to 15.0 ppm. Bitter varieties are associated with high concentrations of cyanogenic glycosides (> 100 mg/kg fresh weight)[163, 164]. Sweet varieties have a high concentration of free sugars but it does not always follow that they have low concentrations of cyanogenic glycoside[165, 166]. However, bitter taste and high level of cyanogens can also be related to environmental stress conditions, such as drought, low soil fertility and pest attack[167]. The safe level of cyanogens in cassava flour has been set by WHO as 10 ppm[168]. Age, variety, and environmental conditions influence the occurrence and concentration of HCN in various parts of the cassava plant and at different stages of development, respectively. The concentrations of cyanogens vary in different varieties, between tissues in the same plant and even between compartments of the same tissue[163, 167, 169, 170]. The main cyanogens present in cassava are linamarin and lotaustralin in ratio 97:7[166] (Fig. 6). The cyanogenic glycoside linamarin is enzymatically hydrolysed to glucose, acetone and hydrogen cyanide during cell rupture[171]. Cassava tissues also contain enzyme linamarase, which can hydrolyse cyanogenic glycoside but the enzyme is not located in the same cell compartments as the cyanogenic glycosides[172, 162]. Cyanogenic glycosides are located inside vacuoles and the enzyme linamarase in the cell wall[171]. Disruption of cassava tissues initiates the hydrolysis of cyanogenic glycosides by the enzyme to produce acetone cyanohydrin from linamarin and 2-butanone cyanohydrin from lotaustralin[171]. These cyanohydrins are unstable and decompose spontaneously to the corresponding ketones and hydrogen cyanide (HCN) at pH values above 5 and temperatures above 30°C[173].

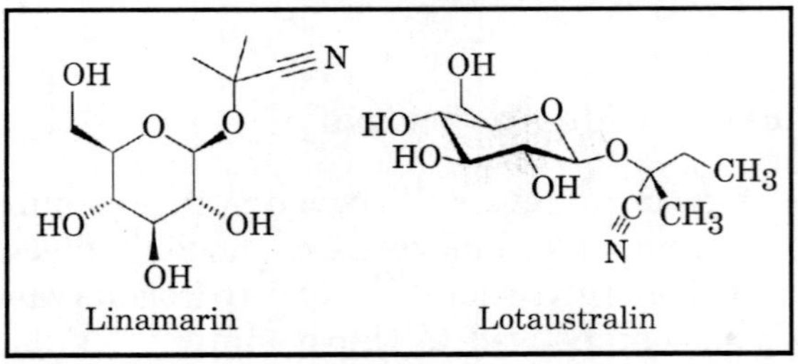

Fig. 6: Major cyanogens present in Cassava roots

Various methods have been used to determine cyanogens in cassava. The most common analysis involves: (i) extraction of cyanogens from cassava, (ii) hydrolysis of cyanogenic glycosides to cyanide and (iii) analysis of cyanogenic glycosides. Extraction of cyanogens from the plant material is carried out using dilute acid to stop endogenous linamarase activity and to stabilise the cyanohydrins. Cyanogens can be hydrolysed to cyanide by acid hydrolysis, by autolysis or by enzymatic hydrolysis. Both acid and enzymatic hydrolysis methods convert linamarin to cyanohydrins, which further decompose to cyanide and acetone at alkaline pH levels. Various methods have been developed to determine endogenous cyanide liberated from cyanogenic glycosides. In the titration method, endogenous cyanide is precipitated with $AgNO_3$ after steam distillation of hydrogen cyanide from autolysed cassava samples. Another method, alkaline picrate method is a semi-quantitative method in which cyanide reacts with alkaline picrate paper, and the change in color is matched against a color chart. Alternatively, the resulting chromophore can also be dissolved from the picrate paper for more accurate quantitative determination using spectrophotometer[174-177]. Tivana *et al.*[178] demonstrated straightforward applicability of the corrin based chemosensor, aquacyanocobyrinic acid (ACCA), for the instantaneous detection and rapid quantification of endogenous cyanide in fresh and processed cassava roots. Hydrolytically liberated endogenous cyanide from cyanogenic glycosides reacts with ACCA to form dicyanocobyrinic acid (DCCA), accompanied by a change of colour from orange to violet. The method was successfully tested on various cassava samples containing between 6 and 200 mg equiv. HCN/kg as verified with isonicotinate/1, 3-dimethylbarbiturate as an independent method. The affinity of ACCA sensor to cyanide has been reported to be very high, coordination occurs fast and the colorimetric response can be instantaneously monitored with spectrophotometric methods.

The cassava roots need to be processed if these negative aspects are to be overcome, and experienced cassava producers are aware of different ways and methods of processing cassava[179]. The cyanogenic glycoside levels can be reduced to acceptable levels that would not impose hazard to the consumer when the cassava roots are adequately processed. Processing techniques were found to be highly effective in substantially reducing mean total cyanogen contents (197.3–951.5 mg HCN equivalent/kg) to low levels (1.1–27.5 mg HCN equivalent/kg) in cassava and cassava based products. The processing techniques which are more effective in reducing the cyanogen content are those that maximize root tissue disintegration to allow endogenous linamarase to react with linamarin and favourable conditions for the break down of cyanohydrins. The degree of root size reduction, the pH values of the fermented and softened mash influence the break down of cyanohydrins. Soaking and mashing/ dewatering, which are accompanied with fermentation of the roots, are prerequisites in developing the required organoleptic characteristics of the cassava foods. Different processing techniques are effective in reducing total cyanogen contents of a wide range of cassava varieties to low levels. Many studies confirmed the reduction in total cyanogens during cassava fermentation and soaking processes used in the production of a few foods[174, 180]. The cyanogens in cassava are hydrolyzed into volatile free cyanide by allowing contact between the cyanogenic substances localized in the vacuoles of the cells with hydrolyzing enzymes in the cell walls. This can be achieved by damaging the cells mechanically or by fermentation. Studies carried out in Nampula, Mozambique showed that heap fermentation of cassava roots may reduce toxic compounds by up to 96%. Another process used for the removal of toxic compound in cassava roots is shredding of roots, which is also an efficient detoxification method (98% reduction). Foodstuffs prepared from cassava roots had lower cyanogen contents (2–31 mg HCN/kg) than those prepared as cassava root slices or from cassava flour (28–88 mg HCN/kg)[149]. The effectiveness of processing techniques is essential if cassava foods are to be produced on a large scale and in case the techniques have to be modified and/or improved without causing adverse effects to health. It has been reported that the need to breed low cyanogen containing cassava varieties for the production of different cassava foods is not necessary as processing techniques are capable of effectively reducing cyanogenic glucosides in cassava acid-fermented foods to levels safe for human consumption[181].

D. Post-Harvest Changes in Cassava

A major limitation of cassava is the rapid post-harvest physiological deterioration which often begins within 24 h of harvest[182]. Within 2-3 days of harvesting, the roots show blue to black vascular streaking and are unmarketable affecting the economic value of the crop. The root deteriorates in air at ambient temperature within 3–4 days[183, 184]. This post-harvest physiological deterioration is followed by microbial deterioration within 5–7 days that is caused largely by pathogens[185, 186]. The physiological deterioration has been reported to be an oxidative burst which initiates within 15 min of the root being injured[187] followed by altered gene expression. The genetic change is predicted to play role in reactive oxygen species turnover, cell wall repair, programmed cell death, ion, water or metabolite transport, signal transduction or perception, stress response, metabolism and biosynthesis, activation of protein synthesis[188] and the accumulation of secondary metabolites (hydroxycoumarins)[189,190]. Scopoletin and its glucoside scopolin are the hydroxycoumarins which by oxidation and polymerisation result in blue/black discolouration. During post-harvest deterioration of cassava root phenylalanine ammonia-lyase activity has been also observed[189, 191]. Isamah *et al.*[192] reported that lipid peroxidation on the very 1st day of storage of cassava roots and the sustained decrease in the activities of superoxide dismutase and catalase contributed to the post-harvest deterioration of cassava root tubers. ATPases and peroxidases have an important role in the post-harvest deterioration of cassava tuber. In a study it is found that ATPase activities peaked within 24 h of harvest of the roots and thereafter progressively decreased. Calcium-ATPases rapidly pump excess calcium into organelles, vesicles, and in plant cells, out into the cell wall. Pectic acid in plant cell walls is mainly in the form of calcium pectate which confers a high level of plasticity in the cell wall. The decrease in calcium pump activities after 24 h in storage lead to destabilization of cell membranes, since calcium is not available to form cross-linkages with pectic substances required for cell wall rigidity. This has been found to induce cell wall softening[193, 194]. Peroxidase activity has been found significantly greater after 24 hours of storage of cassava roots than at zero time. However it decreased subsequently. The initial increase in peroxidase activity might be a defence mechanism in response to various biotic and abiotic stresses[195]. Decrease in peroxidase activity thereafter has been attributed to the biological phenomenon of lignification. The presence of peroxidases predisposes cells to lignification. Therefore, the decrease in peroxidase activity after the 24 h of harvest may be due to the softening of cassava root after harvest[191, 196].

E. Processing of Cassava

Cassava products are used in various forms for human consumption, livestock feed, and manufacturing of industrial products. Cassava is used in the preparation of several household foods and derivatives, such as paste, biscuits, bread sagos and sauce. Cassava processing improves palatability, increases shelf-life, facilitates transport and, most importantly, detoxifies cassava roots by removing cyanogens[6,172,179]. A large proportion of the root is processed into a variety of foods using various methods, including drying, roasting, frying and steaming. However, in many African countries, cassava is still mainly consumed just boiled or pounded into *fufu*[197, 198]. A wide range of traditional cassava products such as *gari,* starch, *lafun, abacha,* etc. are produced for human consumption[199, 200]. Eagleston *et al.*[201] reported that the whole cassava plant, boiled cassava root, cassava root meal, chips and pellets could be used in compounding livestock feed. The roots could be dried, ground and fed to ruminants and it could be used as substitutes for maize in poultry feed. Cassava starch, cassava flour, cassava juice and fermented cassava are now used in industries[200, 202, 203]. Terry *et al.*[202] noted that considerable attention has been given to cassava as a raw material for ethanol production with particular example in Brazil due to rapid escalation of energy cost. Cassava must be used immediately because of its rapid physiological and microbial deterioration after harvest. Cassava is used mainly by the farmers themselves as a subsistence crop because the fresh cassava roots have a short shelf-life (1 to 2 days) after harvest. Therefore, the manufacture of cassava-based industrial products may be of interest as a potential way of boosting the economic value of the crop. The various products based on cassava root are discussed in the subsequent section.

Boiling of the roots

Boiling of cassava roots is performed in almost all countries where cassava is used as a staple food. The cell structure during heating remains more or less intact until the cell membranes are destroyed around 60-70 °C. Boiled cassava is one of the most common forms of consumption in both rural and urban areas. It is prepared by peeling, cutting and boiling fresh cassava roots. The enzymatic detoxification using boiling is limited. As consequences of this inefficiency, boiling alone can only be used for sweet varieties where the cyanogen content is less than 100 mg HCNeq/kg fresh weight[145, 179]. Boiled roots are usually eaten directly, fried or stewed with some spice to improve the taste. Cultivars are thus selected and cultivated by farmers based on consumer demand concerning quality, which is mostly governed by the final texture (particularly the friability or mealiness) and the taste of the boiled roots[153]. The physicochemical characteristics of starch do not appear to be directly linked to mealiness (friability) and ease of cooking[153, 204, 205].

Age of plant and environmental conditions during growing may affect texture (friability) and taste of boiled cassava roots. Concerning texture, friability or mealiness is universally cited by consumers as the most important quality attribute of boiled cassava[198, 204, 206, 207]. Cassava roots are consumed in Benin in several forms, including boiled in water, roasted, fried or after conversion into intermediate products, such as gari, flour (from dried chips), tapioca or starch. The main quality attributes of boiled cassava are whiteness, sweetness and friability, all of which should be high[153].

Sun dried cassava products

Sun-drying is employed in most of the common processed cassava products in Africa[6]. The process involves peeling the cassava roots, chipping (or grating) and spreading on dried grass or on the roof of the houses for sun drying. The efficiency of removal of cyanogenic glycosides in this process depends somehow on the rate of moisture loss. Fast drying results in lower cyanogen removal, while slower rates of drying result in a higher reduction. More important is the degree of disruption of cellular tissues of the cassava roots. The disruption of the tissue and the cells is creating contact between cyanogenic glycosides and hydrolytic enzymes[163]. Processes that promote high disruption of cassava roots such as grating and crushing before sun drying remove much of the cyanogenic glycoside[208]. Dried roots are then milled to make a sticky porridge known as *karakata* in Mozambique.

Soaking in water and sun drying

Soaking the cassava roots in water and sun drying is a common practice in Nigeria, Democratic Republic of Congo, Tanzania, Zambia and Malawi[6, 179]. Peeled cassava roots are soaked in water for 3-5 days, followed by sun drying. The process is reported to be the best for cyanogen removal[6, 208]. However, Nyirenda *et al.*[179] reported that in Zambia, pre-sundried roots are soaked in water for 3 to 14 days. Soaking in water leads to a type of fermentation which disrupts the cellular tissues of roots, which has combined effect of allowing cyanogenic glycosides to come in contact with linamarase, a toxic principle[209].

Heap fermentation and sun drying

In heap fermentation the cassava roots are dried prior to fermentation. Heap fermentation of cassava root products is common in Tanzania[210], Uganda and Mozambique [211]. The process involves peeling of cassava roots, sun drying for 1 to 3 days, heaping and covering, fermentation, scraping off the mould mycelia, crushing into crumbs, sun drying, pounding and sieving into flour. The heap fermentation has been reported to be dominated by mould mycelium growth of *Neurospora sitophila*, *Geotrichum candidum* and *Rhizopus oryzae*. The growth of moulds softens the roots by enzymatic degradation of cell wall of cassava tissues. The degrading of tissue structures in the cassava roots enables contact between cyanogenic glycosides and linamarase. Heap fermentation of cassava roots followed by sun drying is capable of reducing the cyanogen levels up to 95%[211, 212].

Grating and fermentation prior to roasting

Another processing method of cassava roots involves grating and fermentation prior to roasting which is used in most of the West African countries. The grated cassava roots are allowed to ferment in sacks for 1 to 7 days, which encourages lactic acid fermentation. After 3 days the pH decreases from 6.0 to 4.0. Grating is important for the disruption of cellular tissues of cassava roots. The process allows the reduction of roots to particle size of between 0.1 to 1 mm. The reduction of cassava tissues maximizes the hydrolysis of cyanogenic glycosides as they come in contact with linamarase. The combined process of grating (shredding) and fermentation of cassava roots is important in the processing of roots for the development of many products such as roasted granules (*gari*), steamed granules (*Iattieke*) and some of the fermented pastes (*agbelina* and *placali* from Ghana and Cote d'Ivoire, respectively)[6, 209, 213].

Cassava chips and crackers

The post-harvest storage of cassava roots is difficult due to its highly perishable nature and ease of microbial contamination. To minimize the post-harvest losses, cassava requires immediate processing after harvest to ensure stability during storage. One of the simplest ways is to process the cassava roots into dried cassava chips[214]. Cassava chips are derived cassava products traditionally produced by farmers in sub-Saharan Africa following fermentation, and drying of fresh roots of cassava and are widely consumed. After production, cassava chips can be stored for more than two months. However, it may be contaminated by a wide array of harmful microbes. Fungi are common contaminants of stored cassava chips[215]. Vitrac *et al.*[216] described a two stage frying process of thin slices of fresh cassava for the development of fried cassava chips. The process involves inner vaporization of high available water, which can be transported inside the material, at temperature around the boiling point and inner vaporization of "bound" water at higher temperature depending on water desorption equilibrium.

Cassava crackers are popular snack foods in Southeast Asian countries. In the manufacturing process starch is mixed with hot water and partially gelatinized to form dough, which is subsequently shaped and steamed or boiled to complete gelatinization. The cooked dough is then sliced and dried to a half finish product with about 12% moisture content. The half finish product is dipped in hot oil whereby it expands to a porous, low density and crispy product[217]. Saeleaw and Schleining[218] showed that with increasing frying time and temperature, the moisture content, the linear expansion, the number of force and sound peaks significantly increased while the density and peak force decreased during deep fat frying of cassava crackers.

Cassava flour

Cassava flour can be used for other purposes such as starch production, glucose, adhesives, fuel alcohol, animal feed and other industrial uses. The flour is usually made into hot water dough (*fufu*) and eaten with soup. Cassava flour is made from the roots after peeling, washing and chipping into 2–3 cm sizes. The chips are then dried in a solar dryer at 50-55°C for 3 days, milled, screened through a mesh sieve and packaged in plastic container. Aryee *et al.*[219] analyzed flour from 31 varieties of cassava and reported starch content 67.92% to 88.11%. Amylose content of cassava flour varied from 10.9% to 44.3%. Gelatinization temperature was found to be in range of 66.8 to 70.4°C with peak temperatures varying between 73.1°C and 84.5°C. The cyanogenic potential ranged from 0.58 to 20.0 mg HCN per 100 g of dry weight. However, Padonou *et al.*[153] reported the amylose content of 18.2 to 22.6% on starch basis in cassava. The mean values of the enthalpy change and onset temperature of starch gelatinization has been reported to be 11.5 J g^{-1} db and 65.7°C, respectively. Lafun (cassava flour) is sundried, fermented whole roots that are milled and sieved. Cassava flour is not only used in variety of products such as spaghetti and macaroni, but also, has been identified to be useful in the manufacture of cassava beer in the brewery industry[203].

Ethanol production

The potential of cassava as a feedstock for ethanol production has been researched by several workers[220, 221, 222]. Cassava has short cooking time and low gelatinization temperature. Hence, it is a suitable feedstock for fuel ethanol production. Conventional production of ethanol is from raw materials containing fermentable sugars which could be directly fermented or from starchy materials which could be hydrolyzed to sugars and then fermented. Lignocellulosic biomass is also considered as a most promising feedstock for the production of ethanol, due to its versatility and availability at low cost[223, 224, 225]. The yield of ethanol depends largely on the starch content in the raw material. The process of cassava-based ethanol production includes three phases: cassava production, ethanol conversion, and ethanol distribution and blending. Farmers start cultivating cassava in the beginning of the rainy season and harvest after 7-10 months. After the harvest, cassava is sliced and dried in the sun before delivery to ethanol plants in the form of dried chips. The conversion ratio of fresh root to dried chips is 2.4 $kg\ kg^{-1}$. The conventional process for bioethanol production from cassava chips involves three steps *viz.*, liquefaction, saccharification and fermentation, the former two reactions being performed by alpha-amylase and glucoamylase, respectively and the latter by yeast. In the process of fuel ethanol production, cassava chips are smashed to disintegrate the cellular tissue, then the mixture of pretreated cassava dry chips, water and special amylase are kept under certain temperature and are liquidized and subsequently saccharificated, to convert starch into monomeric sugars. After that sugars are fermented into ethanol by yeast. After fermentation, low concentration ethanol is distillated[226]. Besides ethanol, by products include dried distillers grains which are sold for animal feed production, biogas used as a supplemental energy, and CO_2 collected for sale. The ethanol is delivered to blending stations. At the blending stations, the tank blending process uses pumping machines to deliver gasoline and ethanol into one tank and to perform recirculation within storage tank. Blends are then transported to gas stations for domestic consumption[227, 228].

A low temperature process for fuel ethanol production from cassava chips has been developed involving highly efficient enzymes such as Spezyme and starch hydrolyzing enzymes, which could co-act with yeast[229, 230]. Shanavas *et al.*[231] optimized process conditions for the conversion of cassava starch to bioethanol using novel eco-friendly enzymes Spezyme and Stargen, for initial thinning of starch slurry, followed by the simultaneous saccharification and fermentation processes. The liquefying enzyme Spezyme was reported to be active at 90°C and pH 5.5 on 10% (w/v) starch slurry at levels of 20.0 mg (280 Amylase Activity Units) for 30 min. However, Stargen enzyme at levels of 100 mg (45.6 Granular Starch Hydrolyzing Units) was found to be sufficient to completely hydrolyze 10% (w/v) starch slurry at room temperature (30°C). Ethanol yield and fermentation efficiency were found to be 533 g/kg and 94.0%, respectively, in the Stargen and yeast process with 10% (w/v) starch for 48 h.

Cassava starch

Starch is the main component of the cassava flour, ranges from 80 to 86 g/ 100 g dwb[205]. Cassava starch is also industrially modified to provide products with physical and chemical properties for specific applications, including the preparation of baby food, jelly, thickening agents, gravies, custard powders, glucose, and confectioneries. For instance, cassava starch is used in making products such as biscuits, bread and derivatives such as sagos and sauce. Cassava starch has also been industrially modified to provide products with physical and chemical properties for specific applications, including the preparation of jelly, thickening agents, gravies, custard powders, baby food, glucose and confectioneries[200]. Cassava starch has been extensively used to produce films[232, 233, 234, 235].

F. Traditional Products from Cassava

Baton de manioc production

Baton de manioc is a ready-to-eat cassava based product which is eaten in African sub-continent, especially in Cameroon. It is prepared by peeling, cutting and splitting the cassava roots in transverse chunks (1–4 cm). The central fibers are removed, the chunks washed twice and soaked for 3 days in water contained in a large plastic drums. During this soaking period the roots are allowed to ferment and soften. They are removed from the soak water, mashed, dewatered and more fibers removed. It is placed in polyethene sacks and left weighted with stones for an hour to remove excess water. The mash is then ground between a stone and a wooden slab to yield a fine paste that is kneaded into long cylindrical shapes (25–30 cm). Kneaded mash is wrapped in large leaves (*Zingiber officinale*), bound with raffia ropes and steam cooked for 1–1.5 h to obtain a ready-to-eat food [181]. Fig. 7 represents the steps involved in the preparation of Baton de manioc, fufu and gari.

Fufu production

Fufu is a fermented wet-paste from cassava and it is ranked next to gari as an indigenous food of most Nigerians[236]. It is consumed in the eastern and western parts of Nigeria as well as some other parts of West and Central Africa[237, 238]. Traditionally it is produced and marketed as a wet, pasty food product. It is also made into porridge in boiling water before consumption. The cassava roots are peeled and cut into chunks of size 5–9 cm and soaked for 2 days. The fermented and softened roots are finely chopped, manually mashed, squeezed to remove excess water and evenly distributed on concrete floors or roofing sheets. During steeping, fermentation occurs resulting in a decrease in pH, softening of the roots and reduction of the cyanogenic potential of the roots. The fibres are removed by sieving. The fibre produced

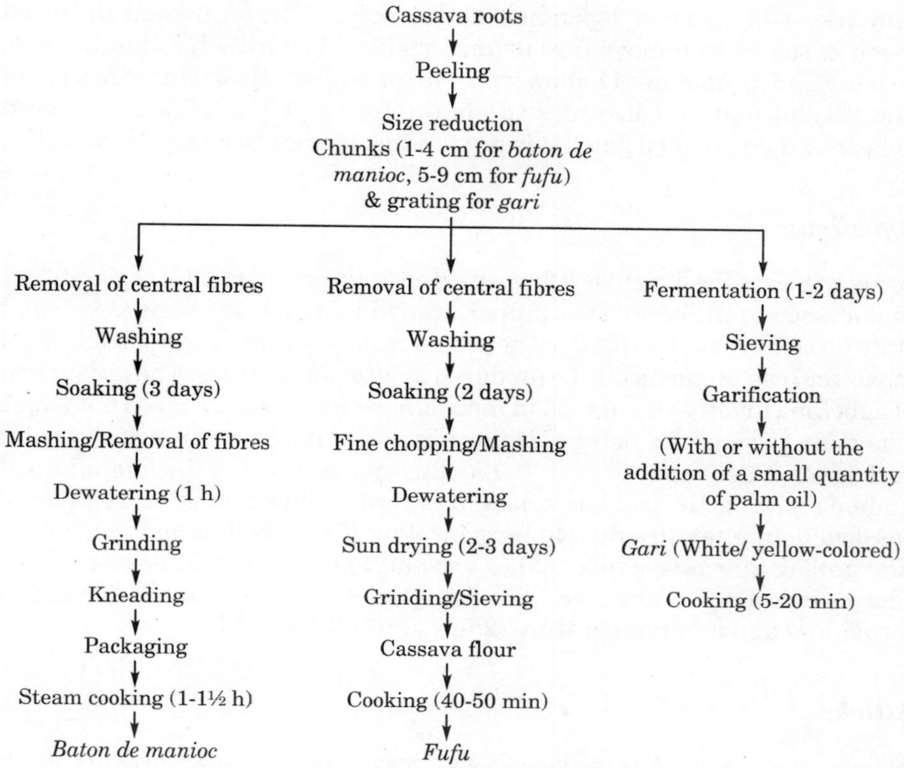

Fig. 7: Various steps in preparation of *Baton de manioc, fufu* and *gari*

is used for animal feed, either in its wet form or after sun drying. In the traditional method, the sieved mass is allowed to sediment in a large container for about 24 h. After sedimentation, the water is decanted and the clean sediment is dewatered overnight in cotton bags, using heavy stones. The fufu is collected and sold to consumers in its wet form in small units packaged in plastic or polypropylene bags or in cooked form. However, alternatively, they are dried under the sun for 2–3 days and then ground into fine cassava flour. For preparation of fufu, the flour is sieved and stirred continuously in an adequate quantity of boiling water for 40–50 min until the required texture of a gelatinized paste is obtained[181, 239].

Gari

Gari can be eaten with cold water poured over it or mixed with hot water to form a paste and eaten with vegetables. Gari preparation involves peeling, washing and grating of cassava roots. The resulting mash is thoroughly mixed, packed in polyethene sacks and dewatered by placing the sacks between a set of thick and long wooded poles arranged beneath and on top such that the ends are strongly fastened together with ropes. Dewatering may take a day or more depending on the variety. The fermented and dried mash is sieved to remove fibrous materials and "garified" (simultaneous cooking and drying) in a shallow wok with or without the addition of a small quantity of palm oil in order to obtain either yellow or white colored gelatinized grits called gari. This is a product similar to Eba in Nigeria[181].

Agbelima

Agbelima (cassava dough) is a fermented cassava product and is of significant importance to different food preparations in Ghana. It is based on solid state fermentation. It is used in the preparation of a wide range of traditional meals and can economically be produced in larger quantities. The production of agbelima involves the use of an inoculum which is locally called 'kudeme'. The main purpose for using this inoculum is to improve the texture, color and flavor of agbelima[240, 241, 242]. Bacillus spp. dominates the microflora of kudeme with lactic acid bacteria and is responsible for the breakdown of cassava dough texture during fermentation[243]. Agbelima is prepared by first peeling the cassava tubers, then washing and grating the peeled cassava. The grated cassava tubers are then inoculated with kudeme and allowed to ferment at ambient temperature (25-27°C) for 3 days[244].

Attieke

Attieke is a steamed granular cassava meal, couscous-like product, with slightly sour taste and whitish colour[245]. It is consumed two to three times a day with meat, fish or vegetables. This food is processed by fermented grated cassava, dewatering the mash, followed by sieving, granulating, sun drying and steaming of the granular product. The popularity of attieke to urban dwellers in recent years has been associated with its cheapness, lower bulk (as compared to other cassava product) and its characteristic of ready to eat food. Assanvo *et al.*[246] reported the consumption of attieke between 28,000 and 34,000 tons per year, the equivalent of 40,000–50,000 tons of fresh Cassava. It was originally prepared and consumed exclusively in a restrict ethno cultural setting among the ethnic groups (Adjoukrou, Alladjan, Ebrie, Avikam, and Ahizi), living in the Laguna area in the south of Côte d'Ivoire[246]. The preparation methods, such as fermentation time vary according the locations or within ethnic groups and these are expected to affect the quality of attieke samples. In addition, differences in the traditional starter used for attieke fermentation are the basis of different organoleptic qualities of attieke sold in the country[180, 247, 248].

IV. CONCLUSION

The sweet potato could be considered as an excellent novel source of natural health-promoting compounds, such as β -carotene and anthocyanins, for the functional food market. Cassava is a staple food crop cultivated in several developing countries, largely by small farmers. It is a source of subsistence and of cash income for poor farmers as well as a source of rural employment. Sweet potato and Cassava crops have immense potential and have a major role to play in human nutrition, food security, and poverty alleviation in developing countries. The processing of sweet potato and cassava into flour not only prolongs the shelf life but also increases the intake of many nutrients into specialty bread made from their flour, which acts as a substitute for wheat flour. Starch processing from sweet potato and cassava can create new economic and employment activities for farmers and rural households, and can add nutritional value to food systems. Repositioning sweet potato and cassava production and their potential for value-added products will contribute substantially to utilizing its benefits and many uses in human food systems. Multidisciplinary, integrated research and development activities aimed at improving production, post-harvest storage and processing technologies, and quality of the sweet potato and cassava based value-added products are critical issues, which should be addressed globally.

REFERENCES

[1] Howeler R.H., Ezumah H.C., Midmore D.J., "Tillage systems for root and tuber crops in the tropics", *Soil & Tillage Research*, **27,** 211–240 (1993).

[2] Hahn S.K., John C.G., Isoba, Ikotun T., "Resistance breeding in root and tuber crops", The international institute of tropical agriculture (IITA), Ibadan, Nigeria. Crop Production, **8,** 147–168 (1989).

[3] Chun S.Y., Yoo B., "Steady and dynamic shear rheological properties of sweet potato flour dispersions", *European Food Res Technol.,* **223**, 313–319 (2006).

[4] Rumbaoa R.G.O., Cornago D.F., Geronimo I.M., "Phenolic content and antioxidant capacity of Philippine sweet potato (*Ipomoea batatas*) varieties", *Food Chem.*, **113**, 1133–1138 (2009).

[5] Nweke F.I., Spencer D.S.C., Lynam J.K., *"The cassava transformation: Africa's best-kept secret"*, Michigan University Press, USA, (2002).

[6] Westby A., "Cassava utilization, storage and small-scale processing", *In:* Hillocks R.J., Thresh J.M. and Bellotti A.C. (*eds.*), *Cassava biology, production and utilization*. CABI Publishing, Wallingford, UK, pp. 281–300 (2002).

[7] Jansson C., Westerbergh A., Zhang J., Hu X., Sun, C., "Cassava, a potential biofuel crop in (the) People's Republic of China". *App Energy,* **86**, *Supplement* 1, S95–S99 (2009).

[8] Thompson H.C., Kelly W.C., *"Sweet potato",* Tata McGraw-Hill Pub. Comp. Ltd. N.Y. p. 405 (1957)

[9] Austin D.F., "The taxonomy, evolution and genetic diversity of sweet potatoes and related wild species". *In*: *"Exploration, Maintenance, and Utilization of Sweet potato Genetic Resources"* Gregory P. (*ed.*), pp. 27–60. CIP, Lima, Peru, (1988).

[10] Woolfe J.A., *"Sweet potato: An untapped food resource"*. Cambridge, UK: Cambridge University Press, p. 643 (1992).

[11] Diop A., "Storage and processing of roots and tubers in the tropics". In *Food and Agriculture Organization of the United Nations, Agro-Industries and Post-Harvest Management Service*. Calverley D.J.B. (*ed.*), pp. 38–50. Agricultural Support Systems Division. Food and Agriculture Organization, Rome, Italy, (1998).

[12] Jiang X., Jianjun H., Wang Y., "Sweet potato processing and product research and development at the Sichuan Academy of Agricultural Sciences". In *Sweet potato Post-Harvest Research and Development in China* (Fuglie K.O. and Hermann M. *eds.*), Proceedings of an International Workshop held in Chengdu, Sichuan, PR China, November 7–8, 2001, International Potato Center (CIP), Bogor, Indonesia (2004).

[13] CIAD, SAAS, SAHVRI. *Social and economic assessment of the small scale sweet potato base pig production system in Suining Municipality, Sichuan Province.* Center for Integrated Agricultural Development (CIAD), Sichuan Academy of agricultural Sciences (SAAS) and Sichuan Animal Husbandry and Veterinary Research Institute (SAHVRI); CIAD, Beijing Agricultural University, Beijing, China, p. 54 (1996).

[14] Rees D., Van Oirschot Q.E.A., Amour R., Rwiza E., Kapinga R., Carey T., "Cultivar variation in keeping quality of sweet potatoes", *Postharvest Biol Technol.,* **28**, 313–325 (2003).

[15] Mortley D.G., Bonsi C.K., Loretan P.A., Hill W. A., Morris C.E., "Relative humidity influences yield, edible biomass, and linear growth rate of sweet potato", *Hort Science.,* **29,** 609–610 (1994).

[16] Onwueme I.C., "*The Tropical Root Crops: Yam, Cassava, Sweet potato and Cocoyam*". Wiley, New York (1978).

[17] Salunkhe D.K., Desai B.B., "*Sweet potato in Postharvest biotechnology of vegetables*", CRC Press, Florida, **1**, 111 (2000).

[18] FAO. Agricultural statistics. Rome: Food and Agricultural Organization of the United Nations (http://www.faostat.fao.org) (2013).

[19] Ishida H., Suzuno H., Sugiyama N., Innami S., Tadokoro T., Maekawa A., "Nutritive evaluation on chemical components of leaves, stalks and stems of sweet potatoes (*Ipomoea batatas poir*)", *Food Chemistry*, **68,** 359–367 (2000).

[20] Guoquan L., "Effect of Genotype and Environment Factors on Quality Characters of Sweet Potato", Beijing: China Meteorological Press (2003).

[21] Purcell A.E., Walter Jr. W.M., Giesbrecht F.G., "Root, hill, and field variance in protein content of North Carolina sweet potatoes", *J Agric Food Chem.*, **26**, 362–364 (1978).

[22] Oboh S., Ologhobo A., Tewe O., "Some aspects of the biochemistry and nutritional value of the sweet potato (*Ipomea batatas*)", *Food Chem.*, **31**, 9–18 (1989).

[23] Bovell-Benjamin A.C., Bromfield E., Pace R, Wang B., "Technical Paper 2001-01-2278". Society for Automative Engineers, Warrendale, PA. (2001).

[24] Dansby M.Y., BovellBenjamin A.C., "Production and proximate composition of a hydroponic sweet potato flour during extended storage", *J Food Process Preserv.,* **27**, 153–164 (2003).

[25] Ravindran V., Ravindran G., Sivakanesan R., Rajaguru S.B., "Biochemical and nutritional assessment of tubers from 16 cultivars of sweet potato (*Ipomoea batatas* L.)", *J Agric Food Chem.,* **43**, 2646–2651 (1995).

[26] Cambie R.C., Ferguson L.R., "Potential functional foods in the traditional Maori diet", *Mutat Res.*, pp. 523–524, 109–117 (2003).

[27] Gichuhi P., Bromfield E., Alvarez N., Dean D., Biswas M., Bovell-Benjamin A.C., "Technical Paper No. 04ICES-291", Society for Automotive Engineers, Warrendale, PA (2004).

[28] Scott G.K., Symes C.W., "Isolation, characterisation and cell growth-regulatory properties of kumara (sweet potato) trypsin inhibitors", *Biochem Mol Biol Int.,* **38**, 333–344 (1996).

[29] Maeshima M., Sasaki T., Asahi T., "Characterization of major proteins in sweet potato tuberous roots", *Phyto-chemistry*, **24**, 1899–1902 (1985).

[30] Hou W.C., Chen H.J., Lin Y.H., "Dioscorins the major tuber storage proteins of yam (*Dioscorea batatas Decne*), with dehydroascorbate reductase and monodehydroascorbate reductase activities", *Plant Science,* **149**, 151–156 (1999).

[31] Makki H.M., Abdel-Rahman A.Y., Khalil M.K.M., Mohamed M.S., "Chemical composition of Egyptian sweet potatoes", *Food Chem.*, **20**, 39–44 (1986).
[32] Ekpenyong T.E., "Composition of some tropical tuberous foods", *Food Chem.*, **15**, 31–36 (1984).
[33] Olaofe O., Sanni C.O., "Mineral contents of agricultural products". *Food Chem.*, **30**, 73–77 (1988).
[34] Huang A.S., Tanudjaja L., Lum D., "Content of alpha- beta-carotene, and dietary fiber in 18 sweet potato varieties grown in Hawaii", *J Food Comp Anal.*, **12**, 147–151 (1999).
[35] Jung J.K., Lee S.U., Kozukue N., Levin C.E., Friedman M., "Distribution of phenolic compounds and antioxidative activities in parts of sweet potato (*Ipomoea batata L.*) plants and in home processed roots", *J Food Comp Anal.*, **24**, 29–37 (2011).
[36] Ishiguro K., Yahara S., Yoshimoto M., "Changes in polyphenolic content and radical-scavenging activity of sweet potato (*Ipomoea batatas L.*) during storage at optimal and low temperatures", *J Agric Food Chem.*, **55**, 10773–10778 (2007).
[37] Walter Jr W.M., Purcell A.E., "Evaluation of several methods for analysis of sweet potato phenolics", *J Agric Food Chem.*, **27**, 942–946 (1979).
[38] Clifford M.N., Johnston K.L., Knight S., Kuhnert N., "Hierarchical scheme for LC-MS n identification of chlorogenic acids", *J Agric Food Chem.*, **52**, 2900–2911 (2003).
[39] Zhu F., Cai Y.Z., Yang X., Ke J., Corke H., "Anthocyanins, hydroxycinnamic acid derivatives, and antioxidant activity in roots of different Chinese purple-fleshed sweet potato genotypes", *J Agric Food Chem.*, **58**, 7588–7596 (2010).
[40] Cevallos-Casals B.A., Cisneros-Zevallos L.A., "Bioactive and functional properties of purple sweet potato (*Ipomoea batatas* (*L.*) Lam)", *Acta Horticulturae.*, **583**, 195–203, (2002).
[41] Kim H.W., Kim J.B., Cho S.M., Chung M.N., Lee Y.M., Chu S.M., Kim J.H., "Anthocyanin changes in the Korean purple-fleshed sweet potato, Shinzami, as affected by steaming and baking", *Food Chem.*, **130**, 966–972 (2012).
[42] Panda V., Sonkamble M., Patil S., "Wound healing activity of Ipomoea batatas tubers (sweet potato)", *Func Foods Health Disease.*, **10**, 403–415 (2011).
[43] Anastácio A., Carvalho I.S., "Phenolics extraction from sweet potato peels: Key factors screening through a Placket–Burman design", *Industrial Crops and Products*, **43**, 99–105 (2013).
[44] Frankow-Lindberg B.E., Lindberg J.E., "Effect of harvesting interval and defoliation on yield and chemical composition of leaves, stems and tubers of sweet potato (*Ipomoea batatas L.*(Lam.)) plant parts", *Field Crops Res.*, **82**(1), 49–58 (2003).
[45] Tewe O.O., Ojeniyi F.E., Abu O.A., *"Sweet potato Production, Utilization and Marketing in Nigeria"*, Social Sciences Department, International Potato Center (CIP), Lima, Peru (2003).
[46] Pace R.D., Phillips B.W., Adeyeye S.O., Gilliard F.M., *"Eat smart menus and recipes with sweet potato greens"*, Publication No. 110–196 (1996). Tuskegee University, Tuskegee, AL (1996).
[47] Islam S., "Sweet potato (*Ipomoea batatas L.*) leaf: Its potential effect on human health and nutrition", *J Food Sci.*, **71**, R13–R21 (2006).
[48] Sun H., Mu T., Xi L., Zhang M., Chen J., "Sweet potato (*Ipomoea batatas L.*) leaves as nutritional and functional foods", *Food Chem.*, **156**, 380–389 (2014).
[49] Li G., Yu S., Yang X.S., Chen Q.F., "Study on Extraction Technology for Chlorogenic Acid from Sweet Potato Leaves by Orthogonal Design", *Procedia Environmental Sciences.*, **8**, 403–407 (2011).
[50] Walker T.S., Crissman C.C., *"Case studies of the economic impact of CIP related technologies"*, 92-9060-181-7. Lima, Peru: International Potato Center (1996).
[51] McConnell E.R., Gottschall P.B., "Effects of canning on new and familiar sweet potato varieties with particular emphasis on breakdown and firmness", *Food technol.*, **11**, 209 (1957).

[52] Cruess W.V., "*Commercial fruits ad vegetables products*", McGrawHill, NY, **4**, 619 (1958).

[53] Hathorn C.S., Biswas M.A., Gichuhi P.N., Bovell-Benjamin A.C., "Comparison of chemical, physical, micro-structural, and microbial properties of breads supplemented with sweet potato flour and high-gluten dough enhancers", *LWT - Food Sci Technol.*, **41**, 803–815 (2008).

[54] Ahmed M., Akter M.S., Lee J.C., Eun J.B., "Encapsulation by spray drying of bioactive components, physicochemical and morphological properties from purple sweet potato". *LWT-Food Sci Technol.*, **43,** 1307–1312 (2010).

[55] Peng Z., Li J., Guan Y., Zhao G., "Effect of carriers on physicochemical properties, antioxidant activities and biological components of spray-dried purple sweet potato flours", *LWT - Food Sci Technol.*, **51**, 348–355 (2013).

[56] Peters D., Wheatley C., "Small scale agro-enterprises provide opportunities for income generation: Sweet potato flour in East Java, Indonesia", *Q J Int Agric.*, **36**, 331–352 (1997).

[57] Van Hal M., "Quality of sweet potato flour during processing and storage", *Food Rev. Internat.*, **16**, 1–37 (2000).

[58] Nair G.M., Ravindran C.S., Moorthy S.N., Ghosh S.P., "*International Sweet potato Symposium*", Baybay, Philippines, May 20–26, (Mackay K.T., Palomar M.K. and Sanico R.T. *eds.*), p. 301. SAEMEO-SEARCA, Philippines (1987).

[59] Dawkins N.L., Lu J.Y., "Physicochemical properties and acceptability of flour prepared from microwave blanched sweet potatoes", *J. Food Process Pres.*, **15,** 115–124 (1991).

[60] Gakonyo N., "Sweet potato post-harvest system. *In*: *Product Development for Root and Tuber Crops*", Scott G., Fergusun P. and Herere J. (*eds.*), p. 317. CIP and IITA, Lima, Peru, and Ibadan, Nigeria (1993).

[61] Suismono I., "*Sweet potato processing for flour and noodles*", M.Sc. Thesis, Bogor Agricultural University (IPB), Bogor, Indonesia, p. 137 (1995).

[62] Martin F.W., "Techniques and problems in small scale production of flour from sweet potato", *J Agric Univ Puerto Rico.*, **68**, 423–432 (1984).

[63] Peters D., Wheatley C., Heriyanto C., Antarlina S.S., "*Participatory process improvement for small scale sweet potato flour production in East Java, Indonesia*", International Potato Center (CIP), Bogur, Indonesia http://www.eseap.cipotato.org/MF-ESEAP/Fl-Library/KNGTRIAL (2005).

[64] Antarlina S., UPWARD "*Proceedings of the Inaugural Planning Workshop on the User's Perspective with Agricultural Research and Development*", Baguio City, Philippines, April 3–5, 1990, Los Banos, Laguna, Philippines, UPWARD, CIP Reg. OYce Region VII, Philippines p. 68 (1990).

[65] Fuglie K.O., Hermann M. (*eds.*), "Sweetpotato post-harvest research and development in China". *In*: *Proceedings of an International Workshop held in Chengdu, Sichuan*, PR China, November 7–8, 2001. International Potato Center (CIP), Bogor, Indonesia, (2004).

[66] Greene J.L., Bromfield E., Dean D., Abdalla M., Bovell-Benjamin A.C., Technical Paper No. 2003–01–2617. *Society for Automotive Engineers* (SAE), Warrendale, PA (2003).

[67] Reddy V.S.O., Basappa S.C., "Preparation of sweet potato flour and its fermentation to ethanol", *J Food Sci Technol.*, **34**, 108–112 (1997).

[68] Garcia A.M., Walter Jr W.M., "Physicochemical characterization of starch from Peruvian sweet potato selections ASA Alimentos SA, Lima, Peru", *Starch/Starke,* **50,** 331–337 (1998).

[69] Lu G.Q., Sheng J.L., "Application of near infrared reflectance spectroscopy (NIRS) in sweet potato quality breeding", *Scientia Agricultura Sinica,* **23**, 76–81 (1990).

[70] Collado L.S., Mabesa R.C., Corke H., "Genetic variation in the physical properties of sweet potato starch", *J Agric Food Chem.*, **47**, 4195–4201 (1999).

[71] Marter A.D., Timmins W.H., "Small-scale processing of sweet potato in Sichuan Province, People's Republic of China", *J Trop Sci.,* **32**, 241–250 (1992).

[72] Sriburi P., Hill S.E., Mitchell J.R., "Effects of L-ascorbic acid on the conversion of cassava starch", *Food Hydrocoll.,* **13**, 177–183 (1999).

[73] Brabet C., Reynoso D., Dufour D., Mestres C., Arredondo J., Scott G., "Starch content and properties of 106 sweet potato clones from the world germplasm collection held at CIP, Peru", CIP Program Report 1997–1998, CIP, Lima, Peru.

[74] Miller S.A., Dean D., Ganguli S., Abdalla M., Bovell-Benjamin A.C., *"Physicochemical and viscometric properties of a sweet potato syrup",* Paper presented at ARD 2003. Association of Research Directors, Inc., 13th Biennial Research Symposium, March 29–April 2, Atlanta, GA. Ministry of Agriculture, Animal Industries and Fisheries 1992. Kampala, Uganda (2003).

[75] Li W.G., Wu X., Cai H., Du R., "Sweet potato in China", *In*: *Product Development for Root and Tuber Crops, Volume 1—Asia*. Scott G.J., Wiersema S. and Ferguson P.I. (*eds.*), pp. 41–50, Proceedings of the International Workshop, April 22–May 1, 1991, ViSCA, Baybay, Leyte, Philippines. CIP, Lima, Peru (1991).

[76] Wheatley C.C., Bofu S., "Sweet potato starch in China: Current status and future prospects", *In*: *Cassava, Starch and Starch Derivatives*. Howeler R.H., Oates C.G. and O'Brien G.M. (*eds.*), pp. 201–205. Proceedings of the International Symposium held in Nanning, Guangxi, China, November 11–15, 1996, A CIAT Publication, CIAT Regional Cassava Program for Asia (2000).

[77] Madhusudhan B., Gowda L.R., Tharanathan R.N., "Studies on sweet potatoes—III. Distribution of unit chains of branched and unbranched molecules of starch", *Food Chem.,* **57**, 201–204 (1996).

[78] Moorthy S.N., "Tropical sources of starch", *In*: Eliasson A. (*ed.*), *Starch in food,* pp. 321–360 Boca Raton: CRC Press (2004).

[79] Bovell-Benjamin A.C., Gichuhi P., Abdalla M., Biswas M., Bromfield E., Alvarea M., Kpomblekou A.K., Dean D., Technical Paper 2004-01-2527. Society for Automotive Engineers, Warrendale, PA.

[80] Kays S.J., "The chemical composition of the sweet potato". *In*: *Sweetpotato Technology for the 21st Century*. Hill W.A., Bonsi C.K. and Loretan P.A. (*eds.*), pp. 201–262 Tuskegee University, Tuskegee, AL (1992).

[81] Srichuwong S., Orikasa T., Matsuki J., Shiina T., Kobayashi T., Tokuyasu K., "Sweet potato having a low temperature-gelatinizing starch as a promising feedstock for bioethanol production", *Biomass and Bioenergy,* **39,** 120–127 (2012).

[82] Zaidul I.S.M., Norulaini N.N., Omar A.M., Yamauchi H., Noda T., "RVA analysis of mixtures of wheat flour and potato, sweet potato, yam, and cassava starches", *Carbohydrate Polymers*, **69**, 784–791 (2007).

[83] Seog H.M., Park Y.K., Nam Y.J., Shin D.H., Kim J.P., "Physicochemical properties of several sweet potato starches", *J Korean Agric Chem Soc.,* **30,** 179–185 (1987).

[84] Lee H.L., Yoo B., "Effect of hydroxypropylation on physical and rheological properties of sweet potato starch", *LWT - Food Sci Technol.,* **44,** 765–770 (2011).

[85] Tian S.J., Rickard J.E., Blanshard J.M.V., "Physicochemical properties of sweet potato starch", *J Sci Food Agric.*, **57,** 459–491 (1991).

[86] Cho S.A., Yoo B., "Comparison of the effect of sugars on the viscoelastic properties of sweet potato starch pastes", *International Journal of Food Science and Technology*, **45**, 410–414 (2010).

[87] Chen Z., Schols H.A., Voragen A.G.J., "Starch granule size strongly determines starch noodle processing and noodle quality", *J Food Sci.,* **68**, 1584–1589 (2003).

[88] Fuglie K.O., Oates C.G., "Starch Markets in Asia", Paper presented in CIP-Indonesia Research Review Workshop, Bogor, Indonesia, March 26–27, (2002).

[89] Sarikaya E., Higasa T., Adachi M., Mikami B., "Comparison of degradation abilities of á-and â-amylases on raw starch granules", *Process Biochem.,* **35**, 711–715 (2000).

[90] Bovell-Benjamin A.C., Yousif-Ibrahim S., Gichuhi P., Bromfield E., Technical Paper 2005-01-3113. Society for Automative Engineers, Warrendale, PA (2005).

[91] Ziska L.H., Runion G.B., Tomecek M., Prior S.A., Torbet H.A., Sicher R., "An evaluation of cassava, sweet potato and field corn as potential carbohydrate sources for bioethanol production in Alabama and Maryland", *Biomass Bioenergy,* **33**, 1503–1508 (2009).

[92] Yuguo Z., Zhao W., Xiaolong C., "Citric acid production from the mash of dried sweet potato with its dregs by Aspergillus niger in an external-loop airlift bioreactor", *Process Biochem.,* **35**(3), 237–242 (1999).

[93] Betiku E., Adesina O.A., "Statistical approach to the optimization of citric acid production using filamentous fungus Aspergillus niger grown on sweet potato starch hydrolyzate", *Biomass and Bioenergy,* **55**, 350–354 (2013).

[94] Akoh C.C., Chang S.W., Lee G.C., Shaw J.F., "Biocatalysis for the production of industrial products and functional foods from rice and other agricultural produce", *J Agric Food Chem.,* **56**, 10445–10451 (2008).

[95] Chou H.H., Chang S.W., Lee G.C., Chen Y.S., Yeh T., Akoh C.C., Shaw J.F., "Site-directed mutagenesis improves the thermostability of a recombinant Picrophilus torridus trehalose synthase and efficiency for the production of trehalose from sweet potato starch", *Food Chem.*, **119,** 1017–1022 (2010).

[96] Kosambo L.M., Carey E.E., Misra A.K., Wilkes J., Hagenlmana V., "Influence of age, farming site, and boiling on pro-vitamin A content in sweet potato (*Ipomoea batats* (L.) Lam.) storage roots", *Journal of Food Composition and Analysis,* **11**(4), 305–321 (1998)

[97] Van Jaarsveld P.J., Faber M., Tanumihardjo S.A., Nestel P., Lombard C.J., Benade A.J.S., "β-Carotene orange-fleshed sweet potato improves the vitamin A status of primary school children assessed with the modified-relative-dose-response test", *Am J Clin Nutr.*, **81,** 1080–1087 (2005).

[98] Ameny M.A., Wilson P.W., "Relationship between Hunter Color Values and βCarotene Contents in White Fleshed African Sweet potatoes (*Ipomoea batatas Lam*)", *J Sci Food Agric.,* **73**, 301–306 (1997).

[99] Hart D., Scott K.J., "Development and evaluation of an HPLC method for the analysis of carotenoids in foods and measurement of the carotenoid content of vegetables and fruits commonly consumed in the UK", *Food Chem.,* **54,** 101–111 (1995).

[100] Hulshof P.J., Xu C., van de Bovenkamp P., Muhilal S., West C.E., "Application of a validated method for the determination of provitamin A carotenoids in Indonesian foods of different maturity and origin", *J Agric Food Chem.,* **45**, 1174–1179 (1999).

[101] Tomlins K., Owori C., Bechoff A., Menya G., Westby A., "Relationship among the carotenoid content, dry matter content and sensory attributes of sweet potato", *Food Chem.,* **131,** 14–21 (2012).

[102] Bhaskarachary K., Rao D.S., Deosthale Y.G., Reddy V., "Carotene content of some common and less familiar foods of plant origin", *Food Chem.*, **54**, 189–193 (1995).

[103] Çinar I., "Effects of cellulase and pectinase concentrations on the colour yield of enzyme extracted plant carotenoids", *Process Biochem.*, **40**, 945–949 (2005).

[104] Hagenimana V., Carey E.E., Gichuki S.T., Oyunga M.A., Imungi J.K., "Carotenoid contents in fresh, dried and processed sweet potato products", *Ecology of Food and Nutrition,* **37**, 455–473 (1998).

[105] Kidmose U., Christensen L.P., Agili, S.M., Thilsted S.H., "Effect of home preparation practices on the content of provitamin A carotenoids in coloured sweet potato varieties (*Ipomoea batatas* Lam.) from Kenya", *Innov Food Sci Emerg Technol*, **8,** 399–406 (2007).

[106] Namutebi A., Natabirwa H., Lemaga B., Kapinga R., Matovu M., Tumwegamire S., Nsumba J., Ocom J., "Long-term storage of sweet potato by small-scale farmers through improved post harvest technologies", *Uganda J Agric Sci.*, **9**, 922–930 (2004).

[107] Almeida L.B., Penteado M.V., "Carotenoids and pro-vitamin A value of white fleshed Brazilian sweet potatoes (*Ipomoea batatas* Lam.). *J Food Comp Analys.,* **1**, 341–352 (1988).

[108] Liu S.C., Lin J.T., Yang, D.J., "Determination of cis-and trans-α-and β-carotenoids in Taiwanese sweet potatoes (*Ipomoea batatas* (L.) Lam.) harvested at various times. *Food Chem.*, **116,** 605–610 (2009).

[109] Teow C.C., Truong V.D., McFeeters R.F., Thompson R.L., Pecota K.V., Yencho G.C., "Antioxidant activities, phenolic and β-carotene contents of sweet potato genotypes with varying flesh colours" *Food Chem.*, **103,** 829–838 (2007).

[110] Maoka T., Akimoto N., Ishiguro K., Yoshinaga M., Yoshimoto M., "Carotenoids with a 5, 6-dihydro-5, 6-dihydroxy-β-end group, from yellow sweet potato "Benimasari", *Ipomoea batatas* Lam", *Phyto-chem.,* **68,** 1740–1745 (2007).

[111] De Pee S., Bloem M.W., Gorstein J., Sari M., Yip R., Shrimpton R., "Reappraisal of the role of vegetables in the vitamin A status of mothers in Central Java, Indonesia", *Am J Clin Nutr.,* **68**, 1068–1074 (1998).

[112] Jalal F.N.M.C., Nesheim M.C., Agus Z., Sanjur D., Habicht J.P., "Serum retinol concentrations in children are affected by food sources of beta-carotene, fat intake, and anthelmintic drug treatment", *Am J Clin Nutr.,* **68**, 623–629 (1998).

[113] Bureau J.L., Bushway R.J., "HPLC determination of carotenoids in fruits and vegetables in the United States", *J Food Sci.,* **51**(1), 128–130 (1986).

[114] Tsou S.C.S., Hong T.L., "The nutrition and utilization of sweet potato". *In*: *Sweet potato Technology for the Twenty-First Century*. Hill W.A., Bonsi C.K. and Loretan P.A. (*eds.*), pp. 359–366. Tuskegee University Press, Tuskegee, AL (1992).

[115] Steenbock H., "White corn vs. yellow corn and a probable relation between the fat-soluble vitamin and yellow plant pigments", *Science,* **50**, 352–353 (1919).

[116] Low J., Kinyae P., Gichuki M.A., Taná P., *"Combating vitamin A deficiency through the use of sweet potato",* Results from Phase I of an action research project in South Nyanza, Kenya. International Potato Center, Lima, Peru (1997).

[117] Krinsky N.I., Johnson E.J., "Carotenoid actions and their relation to health and disease", *Molecular Aspects of Medicine*, **26**, 459–516 (2005).

[118] Plata N., Konczak-Islam I., Jayram S., McClelland K., Woolford T., Franks P., "Effect of methyl jasmonate and p-coumaric acid on anthocyanin composition in a sweet potato cell suspension culture". *Biochem Eng J.,* **14**, 171–177 (2003).

[119] Yang J., Gadi R.L., "Effects of steaming and dehydration on anthocyanins, antioxidant activity, total phenols and color characteristics of purple-fleshed sweet potatoes (*Ipomoea batatas*)", *American Journal of Food Technology*, **3**, 224–234 (2008).

[120] Suda I., Oki T., Masuda M., Kobayashi M., Nishiba Y., Furuta S., "Physiological functionality of purple-fleshed sweet potatoes containing anthocyanins and their utilization in foods", *JARQ,* **37**, 167–173 (2003).

[121] Yoshinaga M., Yamakawa O., Nakatani M., "Genotypic diversity of anthocyanin content and composition in purple-fleshed sweet potato (*Ipomoea batatas* (L.) Lam)", *Breed Sci.,* **49**, 43–47 (1999).

[122] Steed L.E., Truong V.D., "Anthocyanin Content, Antioxidant Activity, and Selected Physical Properties of Flowable Purple Fleshed Sweet potato Purees", *J Food Sci.*, **73**, 215–221 (2008).

[123] Kim H.W., Kim J.B., Cho S.M., Chung M.N., Lee Y.M., Chu S.M., Kim J.H. (2012). Anthocyanin changes in the Korean purple-fleshed sweet potato, Shinzami, as affected by steaming and baking. *Food Chemistry,* **130**, 966–972 (2012).

[124] Fossen T., Andersen O.M., "Anthocyanins from tubers and shoots of the purple potato, *Solanum tuberosum*", *J Hort Sci Biotech.*, **75**, 360–363 (2000).

[125] Truong V.D., Deighton N., Thompson R.T., McFeeters R.F., Dean L.O., Pecota K.V., Yencho G.C., "Characterization of anthocyanins and anthocyanidins in purple-fleshed sweet potatoes by HPLC-DAD/ESI-MS/MS", *J Agric Food Chem.,* **58**, 404–410 (2010).

[126] He K., Ye X., Li X., Chen H., Yuan L., Deng Y., Li X., "Separation of two constituents from purple sweet potato by combination of silica gel column and high-speed counter-current chromatography", *J Chromatography* B., **881–882,** 49–54 (2012).

[127] Li J., Li X. D. Zhang Y., Zheng Z.D., Qu Z.Y., Liu M., Qu L., "Identification and thermal stability of purple-fleshed sweet potato anthocyanins in aqueous solutions with various pH values and fruit juices", *Food Chem.*, **136,** 1429–1434 (2013).

[128] Goda Y., Shimizu T., Kato Y., Nakamura M., Maitani T., Yamada Terahara N., Yamaguchi M., "Two acylated anthocyanins from purple sweet potato", *Phytochem.*, **44**, 183 (1997).

[129] Terahara N., Konczak I., Ono H., Yoshimoto M., Yamakawa O., "Characterization of acylated anthocyanins in callus induced from storage root of purple-fleshed sweet potato, *Ipomoea batatas* L", *J Biomed Biotechnol.*, **5**, 279–286 (2004).

[130] Saigusa N., Terahara N., Ohba R., "Evaluation of DPPH-radical-scavenging activity and antimutagenicity and analysis of anthocyanins in an alcoholic fermented beverage produced from cooked or raw purple-fleshed sweet potato (*Ipomoea batatas* cv. Ayamurasaki) roots", *Food Sci Technol Res.*, **11**(4), 390–394 (2005).

[131] Ben-Amotz A., Fishier R., "Analysis of carotenoids with emphasis on 9-cis β-carotene in vegetables and fruits commonly consumed in Israel", *Food Chem.*, **62**, 515–520 (1998).

[132] Fan G., Han Y., Gu Z., Gu F., "Composition and colour stability of anthocyanins extracted from fermented purple sweet potato culture", *LWT.*, **41,** 1412–1416 (2008).

[133] Oki T., Masuda M., Furuta S., Nishiba Y., Terahara N., Suda I., "Involvement of anthocyanins and other phenolic compounds in radical scavenging activity of purple fleshed sweet potato cultivars", *J Food Sci.*, **67**, 1752–1756 (2002).

[134] Kano M., Takayanagi T., Harada K., Makino K., Ishikawa F., "Antioxidative activity of anthocyanins from purple sweet potato, *Ipomoera batatas* cultivar Ayamurasaki", *Biosci Biotechnol Biochem.,* **69**(5), 979–988 (2005).

[135] Yoshimoto M., Okuno S., Yoshinaga M., Yamakawa O., Yamaguchi M., Yamada J., "Antimutagenicity of sweet potato (*Ipomoea batatas*) roots", *Biosci Biotechnol Biochem.,* **63**(3), 537–541 (1999).

[136] Burgos G., Amoros W., Muñoa L., Sosa P., Cayhualla E., Sanchez C., Bonierbale M., "Total phenolic, total anthocyanin and phenolic acid concentrations and antioxidant activity of purple-fleshed potatoes as affected by boiling", *J Food Compos Analys.*, **30,** 6–12 (2013).

[137] Grace M.H., Yousef G.G., Gustafson S.J., Truong V.D., Yencho G.C., Lila M.A., "Phyto-chemical changes in phenolics, anthocyanins, ascorbic acid, and carotenoids associated with sweet potato storage and impacts on bioactive properties", *Food Chem.,* **145**, 717–724 (2014).

[138] Peng Z., Li J., Guan Y., Zhao G., "Effect of carriers on physicochemical properties, antioxidant activities and biological components of spray-dried purple sweet potato flours", *LWT - Food Sci Technol.,* **51,** 348–355 (2013).

[139] Sungpuag P., Tangchitpianvit S., Chittchang U., Wasantwisut E., "Retinol and beta carotene content of indigenous raw and home-prepared foods in Northeast Thailand", *Food Chem.*, **64,** 163–167 (1999).

[140] Van Jaarsveld P.J., Marais D.W., Harmse E., Laurie S.M., Nestel P., Rodriguez-Amaya D.B., "Beta- Carotene Content of Sun-Dried and Oven-Dried Chips of Orange-Fleshed Sweet Potato", Lima Peru. Paper presented at the XXII IVACG Meeting, (2004).

[141] Van Jaarsveld P.J., Harmse E., Nestel P., Rodriguez-Amaya D.B., "Retention of β-carotene in boiled, mashed orange-fleshed sweet potato", *J Food Compos Analys.*, **19**(4), 321–329 (2006).

[142] Bechoff A., Dufour D., Dhuique-Mayer C., Marouzé C., Reynes M., Westby A., "Effect of hot air, solar and sun drying treatments on provitamin A retention in orange-fleshed sweet potato", *J Food Engg.,* **92**, 164–171 (2009).

[143] Bengtsson A., Namutebi A., Alminger M.L., Svanberg U., "Effects of various traditional processing methods on the all-trans-β-carotene content of orange-fleshed sweet potato", *J Food Compos Analys.*, **21**, 134–143 (2008).

[144] Alves A.A.C., "Cassava botany and physiology", *In*: Hillocks R.J., Thresh J.M. and Bellotti A.C. (*eds.*), *Cassava biology, production and utilization*. CABI Publishing, Wallingford, UK, pp. 67–89 (2002).

[145] Tsegai D., Kormawa P.C., "Determinants of Urban Household Demand for Cassava Products in Kaduna, Northern Nigeria", *In*: Conference of International Research for Development, Witzenhause, 9-10 October (2002).

[146] RodriguezSandoval E., FernandezQuintero A., Cuvelier G., Relkin P., BelloPerez L.A., "Starch retrogradation in cassava flour from cooked parenchyma", Starch/ *Starke*, **60,** 174–180 (2008).

[147] Moorthy S.N., "Physicochemical and functional properties of tropical tuber starches: A review", *Starch-Starke*, **54**, 559–592 (2002).

[148] Hendershott C.H., "A literature review and research recommendations on cassava *(Manihot esculenta, Crantz")"*, University of Georgia, Georgia. p. 336 (1972).

[149] Okigbo B.N., "Nutritional implications of projects giving high priority to the production of staples of low nutritive quality. In the case for cassava (*Manihot esculenta, Crantz*) in the humid tropics of West Africa", *Food Nutri Bull.*, **2**, 1–10 (1980).

[150] Charles A.L., Chang Y.H., Ko W.C., Sriroth K., Huang T.C., "Some physical and chemical properties of starch isolates of cassava genotypes", *Starch / Starke*, **56**, 413–418 (2004).

[151] Charles A.L., Sriroth K., Huang T.C., "Proximate composition, mineral contents, hydrogen cyanide and phytic acid of 5 cassava genotypes", *Food Chem.*, **92**, 615–620 (2005).

[152] Buitrago J.A., "La yucca em alimentacion animal, Publicacion 85, CIAT, Cali, Colombia. (1990).

[153] Padonou W., Mestres C., Nago M.C., "The quality of boiled cassava roots: Instrumental characterization and relationship with physicochemical properties and sensorial properties" *Food Chem.*, **89**(2), 261–270 (2005).

[154] Eggum B.O. "The protein quality of cassava leaves", *British J Nutr.,* **24**, 761–768 (1970).

[155] Ufuan A.A., Ajayi O.A., Bokanga M., Maziya-Dixon B., "The use of cassava leaves as food in Africa", *Ecology Food Nutr.*, **44**, 423–435 (2005).

[156] McMahon J.M., White W.L., Sayre R.T., "Review Article Cyanogenesis in cassava (*Manihot esculenta* Crantz)", *J Exp Bot.,* **46**, 731–741 (1995).

[157] McMahon T.F., Birnbaum L.S., "Age-related changes in toxicity and biotransformation of potassium cyanide in male C57BL6N mice", *Toxicol Appl Pharmacol.*, **105**, 305–314 (1990).

[158] Bellotti A.C., Riis L., "Cassava cyanogenic potential and resistence to pests and diseases", *In*: Bokanga M., Essers A.J.A., Poulter N., Rosling H. and Tewe O. (*eds.*), Proceedings of the international workshop on cassava safety, March 1-4, 1994, Ibadan, Nigeria, *Acta Horticulturae*, **375**, 141–151 (1994).

[159] Vetter J., "Plant cyanogenic glycosides", *Toxicon.,* **38**(1), 11–36 (2000).

[160] Delange F., Ekpechi L.O., Rosling H., "Cassava cyanogenesis and iodine deficiency disorders", *In*: Bokanga M., Essers A.J.A., Poulter N., Rosling H. and Tewe O. (*eds.*), Proceedings of the 38 international workshop on cassava safety, March 1-4, 1994, Ibadan, Nigeria, *Acta Horticulturae.*, **375**, 289–293 (1994).

[161] Rosling H., "Cassava toxicity and food security–A review of health effects of cyanide exposure from cassava and of ways to prevent such effects", Uppsala: International Child Health Unit, Uppsala University (1988).

[162] Teles F.F.F., "Toxicidade crônica da mandioca na Africa e América Latina", *Revista Brasileira de mandioca* **1**, CNPMF, Bahia, Brasil, 107–116 (1995).

[163] Nambisan B., Sundaresan S., "Effect of processing on the cyanoglucoside content of cassava", *J Sci Food Agric.*, **36**, 1197–1203 (1985).

[164] Chiwona-Karltun L., Brimer L., Kalenga Saka J.D., Mhone A.R., Mkumbira J., Johansson L., Bokanga M., Mahungu N.M., Rosling H., "Bitter taste in cassava roots correlates with cyanogenic glucoside levels". *Journal of the Science of Food and Agriculture*, **84**, 581–590 (2004).

[165] Borges M., de F., Fukuda W.M.G., "Teor de cianeto em raízes frescas e processadas de mandioca (*Manihot esculenta* Crantz) de mesa", *Revista Brasileira de Mandioca, Cruz das Almas,* **8**(2), 71–76 (1989).

[166] King N.L., Bradbury J.H., "Bitterness of cassava: Identification of a new apiosyl glucoside and other compounds that affect its bitter taste", *J Sci Food Agric.*, **68**, 223–230 (1995).

[167] Bruijn G.H., "Etude du caractère cyanogénetique du manioc", *Meded Landb Hogesch Wageningen* **71**, Wageningen, The Netherlands (1971).

[168] FAO/WHO., *Joint FAO / WHO food standards program.* Rome: Codex Alimentarius Commission XII, Supplement 4, FAO (1991).

[169] Wheatley C.C., Chuzel G., "Cassava: The nature of the tuber and use as raw material", *In*: Macrare R., Robinson R.K. and Sadler M.J. (*eds.*), Encyclopedia of Food Science, Food Technology and Nutrition. Academic Press, San Diego, California, pp. 734–743 (1993).

[170] Burns A.E., Gleadow R.M., Zacarias A.M., Cuambe C.E., Miller R.E., Cavagnaro T.R., "Variations in the chemical composition of cassava (*Manihot esculenta* Crantz) leaves and roots as affected by genotypic and environmental variation", *J Agric Food Chem.,* **60,** 4946–4956 (2012).

[171] Conn E.E., Cyanogenesis–a personal perspective", *In*: Bokanga M., Essers A.J.A., Poulter N., Rosling H. and Tewe O. (*eds.*), Proceedings of the international workshop on cassava safet. Ibadan, Nigeria, *Acta Horticulturae.*, **375,** 31–43 (1994).

[172] Nweke F.I.' "Cassava distribuition in Africa", Study of cassava in Africa, IITA, Ibadan, Nigeria, work paper no. 12 (1994).

[173] White W.L.B., McMahon J.M., Sayre R.T., "Regulation of cyanogenesis in cassava", *In*: Bokanga M., Essers A.J.A., Poulter N., Rosling H. and Tewe O. (*eds.*), Proceedings of the international workshop on cassava safety, March 1-4, 1994, Ibadan, Nigeria, *Acta Horticulturae.*, **375**, 69–77 (1994).

[174] Brien G.M., Taylor A.J., Poulter N.H., "Improved enzymic assay for cyanogens in fresh and processed cassava", *J Sci Food Agric.,* **56**, 277–289 (1991).

[175] Bradbury J.H., Egan S.V., Lynch M.J., "Analysis of cyanide in cassava using acid hydrolysis of cyanogenic glucosides", *J Sci Food Agric.*, **55**, 277–290 (1991).

[176] Bradbury M.G., Egan S.V., Bradbury J.H., "Picrate paper kits for determination of total cyanogens in cassava roots and all forms of cyanogens in cassava products", *J Sci Food Agric.*, **79**, 593–601 (1999).

[177] Bradbury J.H., "Development of a sensitive picrate method to determine total cyanide and acetone cyanohydrin contents of gari from cassava", *Food Chem.*, **113**(4), 1329–1333 (2009).

[178] Tivana L.D., Francisco J.D.C., Zelder F., Bergenståhl B., Dejmek P., "Straight forward rapid spectrophotometric quantification of total cyanogenic glycosides in fresh and processed cassava products", *Food Chem.,* **158,** 20–27 (2014).

[179] Nyirenda D.B., Chiwona-Karltun L., Chitundu M., Haggblade S., Brimer L., "Chemical safety of cassava products in regions adopting cassava production and processing–Experience from Southern Africa", *Food Chem Toxicol.*, **49**, 607–612 (2011).

[180] Westby A., "Importance of fermentation in cassava processing", *In*: Ofori F. and Hahn S.K. (*eds.*), Proceedings of ninth symposium of the international society for tropical root crops, Accra, Ghana, ISTRC, pp. 249"255 (1991).

[181] Agbor-Egbe T., Mbome I.L., "The effects of processing techniques in reducing cyanogen levels during the production of some Cameroonian cassava foods", *J Food Compos Analys.,* **19**, 354–363 (2006).

[182] Asaoka M., Blanshard J.M.V., Rickard J.E., "The effect of preharvest pruning on the quality of cassava starch", *Annals of Applied Biol.,* **122**, 337–344 (1993).

[183] Rickard J.E., Coursey, D.G., "Cassava storage. I Storage of fresh cassava roots. *Trop Sci.*, **23,** 1–32 (1981).

[184] Eka O.U., "Roots and tubers", *In*: Eka O.U. and Osagie A.U. (*eds*.), Nutritional quality of plant foods. Post-harvest Research unit, Department of Biochemistry, University of Benin, Benin-city, Nigeria, pp. 1–31 (1999).

[185] Kojima M., Iwatsuki N., Data E.S., Villegas C.D.V., Uritani I., "Changes of cyanide content and linamarase activity in wounded cassava roots", *Plant Physiol.,* **72**, 186–189 (1983).

[186] O'Brien G.M., Mayer J.E., Chavez A.L., Orozco O., Wheatley C.C., Wehnam J.E., "Biochemical studies of cassava root post-harvest physiological deterioration", Proceedings of the 6th Symposium on ISTRC-AB, pp. 279–283 (1995).

[187] Reilly K., Góomez-Váasquez R., Buschmann H., Tohme J., Beeching J.R., "Oxidative stress responses during cassava post-harvest physiological deterioration", *Plant Mol Biol.,* **56,** 625–641 (2004).

[188] Reilly K., Bernal D., Cortés D.F., Gómez-Vásquez R., Tohme J., Beeching J.R., "Towards identifying the full set of genes expressed during cassava post-harvest physiological deterioration", *Plant Mol Biol.,* **64,** 187–203 (2007).

[189] Tanaka Y., Data E.S., Hirose S., Taniguchi T., Uritani I., "Biochemical changes in secondary metabolites in wounded and deteriorated cassava roots. *Agric Biol Chem.,* **47,** 693–700 (1983).

[190] Buschmann H., Rodriguez M.X., Tohme J., Beeching J.R., "Accumulation of hydroxycoumarins during post-harvest deterioration of tuberous roots of cassava *Manihot esculenta* Crantz", *Ann Bot.,* **86**, 1153–1160 (2000).

[191] Uritani I., "Biochemistry on postharvest metabolism and deterioration of some tropical tuberous crops", *Botanical Bulletin of Academia Sinica.,* **40**, 177–183 (1999).

[192] Isamah G.K., Asagba S.O., Ekakitie A.O., "Lipid peroxidation, activities of superoxide dismutase and catalase during post-harvest deterioration of cassava (*Manihot esculenta* Crantz) root tubers", *Int Biodeterioration & Biodegradation.*, **52** 129–133 (2003).

[193] Trewavas A., "Le calcium, c'est la vie: Calcium makes waves", *Plant Physiol.*, **120,** 1–6 (1999).

[194] Isamah G.K., "ATPase, peroxidase and lipoxygenase activity during post-harvest deterioration of cassava (*Manihot esculenta* Crantz) root tubers", *Int Biodeterioration & Biodegradation.,* **54,** 319–323 (2004).

[195] Kwak S.S., Kim S.K., Park I.H., Liu J.R., "Enhancement of peroxidase activity by stress-related chemicals in sweet potato", *Phytochem.,* **43**, 565–568 (1996).

[196] Converso D.A., Fernández M.E., Peroxidase isozymes from wheat germ: Purification and properties. *Phytochem.*, **40**, 1341–1345 (1995).

[197] Nago C.M., Hounhouigan D.J., "La transformation alimentaire traditionnelle des racines et tubercules au Bénin" (97p). CERNA, FSA/UAC, Bénin, (1998).

[198] Ngeve J.M., "Cassava root yields and culinary qualities as affected by harvest age and test environment", *J Sci Food Agric.,* **83**(4), 249–257 (2003).

[199] Kormawa P., Akoroda M.O., "Cassava Supply Chain Arrangement for Industrial Utilization in Nigeria", Ibadan (2003).

[200] Ene L.S.O., "Prospects for Processing and Utilization of Root and Tuber Crops", In National Root Crops Promotion of Root Crop-Based Industries. pp. 7–11 (1992).

[201] Eagleston G.B., Okoye M.B., Jeon V.M., "Traditional African Methods of Cassava Processing and Utilisation and Research Needs", *In*: Tropical Root Crops Production of Roots Crops-Based Industries, pp. 4–6 (1992).

[202] Terry E.R., Akoroda M.O., Arene O.B., "Tropical Roof Crops and the African Food Crisis" in Proceeding of the third Triennial symposium of the International Society for Tropical Crops-Africa Branch Held in Owerri, Nigeria, 17-23 August, 1986 (1983).

[203] Olomu J.M., "Monogastric Animal Nutrition. Principles and Practices", Benin. Ajachem Publications. pp. 130–133 (1995).
[204] Asaoka M., Blanshard J.M.V., Rickard J.E., "Seasonal effects on the physicochemical properties of starch from four cultivars of cassava", *Starch / Starke*, **43**, 455–459 (1991).
[205] Charoenkul N., Uttapap D., Pathipanawat W., Takeda Y., "Molecular structure of starches from cassava varieties having different cooked root textures", *Starch–Starke,* **58**(9), 443–452 (2006).
[206] Eggleston G., Asiedu R., "Effects of boiling on the texture of cassava clones: A comparison of compressive strength, intercellular adhesion and physicochemical composition of the tuberous roots" *Tropical Science*, **34**(3), 259–273 (1994).
[207] Favaro S.P., Beléia A., Junior N.D.S.F., Waldron K.W., "The roles of cell wall polymers and intracellular components in the thermal softening of cassava roots", *Food Chem.*, **108**(1), 220–227 (2008).
[208] Cardoso A.P., Mirione E., Ernesto M., Massaza F., Cliff J., Haque M.R., Bradbury J.H., "Processing of cassava roots to remove cyanogens", *J Food Compos Analys.*, **18**, 451–460 (2005).
[209] Westby A., Choo B.K., "Cyanogen reduction during lactic fermentation of cassava", *In*: Bokanga M., Essers A.J.A., Poulter N., Rosling H. and Tewe O. (*eds*.), Proceedings of the international 44 workshop on cassava safety, March 1-4, 1994, Ibadan, Nigeria, *Acta Horticulturae*, **375**, 209–215 (1994).
[210] Ndunguru G.T., Thomson M., Waida T.D.R., Rwiza E., Jeremiah S., Westby A., "Relationship between quality and economic value of dried cassava products in Mwanza, Tanzania", *In*: Akoroda M.O. and Terri J. (*eds*.) *Food Security and Crop Diversification in SADC Countries. The Role of Cassava and Sweet potato.* SARRNET, Malawi, pp. 408–414 (1999).
[211] Essers A.A., Ebong C., van der Grift R.M., Nout M.R., Otim-Nape W., Rosling H., "Reducing cassava toxicity by heap-fermentation in Uganda", *Int J Food Sci Nutri.*, **46**, 125–136 (1995).
[212] Esser A.J.A., Jurgens C.M., "Contribution of selected fungi to the reduction of cyanogen levels during solid-substrate fermentation of cassava", *Int J Food Microbiol.,* **26**, 251–257 (1995).
[213] Obilie E.M., Tano-Debrah K., Amoa-Awua W.K., "Souring and breakdown of cyanogenic glucosides during the processing of cassava into akyeke". *Int J Food Microbiol.*, **93**, 115–121 (2004).
[214] Bricas N., Vernier P., Ategbo E., Hounhouigan J., Mitchikpe E., N'Kpenu K.E., "Le développement de la filière cossettes d'igname en Afrique de l'Ouest (Development of yam chip sub-sector in West Africa)", *Les Cahiers de la Recherche-Développement,* **44**, 100–114 (1997).
[215] Essono G., Ayodele M., Akoa A., Foko J., Filtenborg O., Olembo S., "Aflatoxin-producing Aspergillus spp. and aflatoxin levels in stored cassava chips as affected by processing practices", *Food Control.,* **20**, 648–654 (2009).
[216] Vitrac O., Dufour D., Trystram G., Raoult-Wack A.L., "Characterization of heat and mass transfer during deep-fat frying and its effect on cassava chip quality", *J Food Engg.,* **53**, 161–176 (2002).
[217] Tongdang T., Meenun M., Chainui J., "Effect of sago starch addition and steaming time on making cassava cracker (Keropok)", *Starch– Stärke*, **60**(10), 568–576 (2008).
[218] Saeleaw M., Schleining G., "Effect of frying parameters on crispiness and sound emission of cassava crackers", *J Food Engg.,* **103,** 229–236 (2011).
[219] Aryee F.N.A., Oduro I., Ellis W.O., Afuakwa J.J., "The physicochemical properties of flour samples from the roots of 31 varieties of cassava", *Food Control.*, **17,** 916–922 (2006).
[220] Shetty J.K., Chotani G., Gang D., Bates D., "Cassava as an alternative feedstock in the production of renewable transportation fuel", *Int Sugar J.,* **109**, 3–11 (2007).

[221] Abraham T.E., Krishnaswamy C., Ramakrishna S.V., "Effect of hydrolysis conditions of cassava on the oligosaccharide profile and alcohol fermentation", *Starch / Starke*, **39**, 237–240 (1986).

[222] Vijayagopal K., Balagopal C., "Saccharification of cassava starch with barley malt for ethanol production", *J Root Crops.*, **4**, 9–11 (1978).

[223] Lee J., "Biological conversion of lignocellulosic biomass to ethanol", *J Biotechnol.*, **56**, 1–24 (1997).

[224] Lin T., Wang L., Chen Y., Tian C., Pan X., Tang J., Li J., "Sources and preservation of sedimentary organic matter in the Southern Bohai Sea and the Yellow Sea: Evidence from lipid biomarkers", *Appl Microbiol Biotechnol.*, **69,** 627–462 (2006).

[225] Lynd L.R., Weimer P.J., Van Zyl W.H., Pretorius I.S., "Microbial cellulose utilization: Fundamentals and biotechnology", *Microbiol Mol Biol Rev.*, **66**, 506–577 (2002).

[226] Yang H., Chen L., Yan Z., Wang H., "Emergy analysis of cassava-based fuel ethanol in China", *Biomass and Bioenergy.*, **35**, 581–589 (2011).

[227] Le L.T., van Ierland E.C., Zhu X., Wesseler J., "Energy and greenhouse gas balances of cassava-based ethanol", *Biomass and Bioenergy*, **51,** 125–135 (2013).

[228] Leng R., Wang C., Zhang C., Dai D., Pu G., "Life cycle inventory and energy analysis of cassava-based fuel ethanol in China", *J Clean Prod.*, **16**(3), 374–384 (2008).

[229] Duan G., Shetty J.K., "Novel enzyme for tapioca fermentation to make alcohol without high temperature cooking", Bangkok: *Pacific Ethanol and Biofuel.,* (2004).

[230] Shetty J.K., Lantero O.J., Dunn-Coleman N., "Technological advances in ethanol production", *Int Sugar J.*, **107,** 1283 (2005).

[231] Shanavas S., Padmaja G., Moorthy S.N., Sajeev M.S., Sheriff J.T., "Process optimization for bioethanol production from cassava starch using novel eco-friendly enzymes", *Biomass and Bioenergy,* **35,** 901–909 (2011).

[232] Chen C.H., Lai L.S., "Mechanical and water vapor barrier properties of tapioca starch/decolorized hsian-tsao leaf gum films in the presence of plasticizer", *Food Hydrocoll.*, **22,** 1584–1595 (2008).

[233] Kaisangsri N., Kerdchoechuen O., Laohakunjit N., "Biodegradable foam tray from cassava starch blended with natural fiber and chitosan", *Industrial Crops and Products,* **37,** 542–546 (2012).

[234] Kechichian V., Ditchfield C., Veiga-Santos P., Tadini C.C., "Natural antimicrobial ingredients incorporated in biodegradable films based on cassava starch", *LWT-Food Sci Technol.,* **43,** 1088–1094 (2010).

[235] Pelissari F.M., Yamashita F., Garcia M.A., Martino M.N., Zaritzky N.E., Grossmann M.V.E., "Constrained mixture design applied to the development of cassava starch–chitosan blown films", *J Food Engg.*, **108,** 262–267 (2012).

[236] Sanni L.O., Akingbala J.O., Oguntunde A.O., Bainbridge Z.A., Graffham A.J., Westby A., "Processing of fufu from cassava in Nigeria: Problems and prospects for development", *Science Technology and Development*, **16**(1), 58–71 (1999).

[237] Longe O.G., "Effect of processing on the chemical composition and energy value of cassava", *Nutrition Reports International*, **21**(6), 819–828 (1980).

[238] Okafor N., Ijioma B., Oyolu C., "Studies on the microbiology of cassava retting for foofoo production", *J Appl Bacterial.,* **56,** 1–13 (1984).

[239] Obadina A.O., Oyewole O.B., Sanni L.O., Tomlins K.I., Westby A., "Identification of hazards and critical control points (CCP) for cassava fufu processing in South-West Nigeria", *Food Control.,* **19**, 22–26 (2008).

[240] Budu A.S., "Process and product characteristics of fermented cassava (Manihot esculenta, Crantz) dough, agbelima", M. Phil. Thesis, submitted to the Department of Nutrition and Food Science, University of Ghana (1990).

[241] Al-Hassan R.M., "Cassava as a food security crop in Ghana. Characterisation of village level processing 'and marketing", *In*: Root, tuber crows and plantain development in Ghana, Proceedings of the *1st National Workshop on Root, Tuber Crops and Plantain* held at U.S.T. Kumasi, Ghana, by the Crops Research Institute (CSIR), Kumasi, Ghana, August 14-15, pp. 1–18 (1991).

[242] Sefa-Dedeh S., "Effects of particle size on some physicochemical characteristics of agbelima (cassava dough) and corn dough", *Trop Sci.*, **29**, 21–32 (1989).

[243] AmoaAwua W.K.A., Jakobsen M., "The role of Bacillus species in the fermentation of cassava", *J Applied Bacteriology.,* **19**, 250–256 (1995).

[244] Ellis W.O., Dziedzoave N.T., Boakye K., Simpson B.K., Smith J.P., "Effect of cassava variety and processing methods on the performance of 'kudeme'in agbelima production", *Food Control.*, **8**(4), 9–204 (1997).

[245] Kouadio N.A., Mosso K., Kouakou K., Agbo S.F., "Etude comparative des méthodes traditionnelles de la préparation de l'attiéké dans le Sud de la Cote d'Ivoire", Cahiers de la recherche scientifique et technique, **108**, 703–706 (1991).

[246] Assanvo J.B., Agbo G.N., Behi Y.E.N., Coulin P., Farah Z., "Microflora of traditional starter made from cassava for "attiéké" production in Dabou (Côte d'Ivoire)", *Food Control*, **17**, 37–41 (2006).

[247] Desmazeaud M., "Les bactéries lactiques dans l'alimentation humaine: Utilisation et innocuité", *Cahiers agricultures*, **5**(5), 331–342 (1996).

[248] Akely P.M.T., Azouma O.Y., "Mechanical pressing and semolina preparation from fermented cassava paste during "attiéké"(yucca flour) processing", *J Food Engg.*, **101**(4), 343–348 (2010).

8

Cole Crops: Pre and Post Harvest Processing Technology

Jagbir Rehal[1]* and Gagan Jyot Kaur[1]

ABSTRACT

Cole crops are an important group of vegetables grown in cool season and primarily include cabbage, cauliflower, broccoli, Brussel's sprouts, etc. These are valued crops owing to their carotenoids, polyphenols and glucosinolates. This chapter elaborates on the nutritional profile, maturity indices and physiological disorders of cole crops besides emphasizing on their post-harvest handling, packaging and storage. Their quality indices, processed products and waste utilization are also discussed in detail.

Key words: Cole, Disorder, Processing, Waste utilization

I. INTRODUCTION

Brassica vegetables belong to the cruciferae family and include broccoli, cauliflower, cabbage, Chinese cabbage, Brussels sprout and others. All cole crops (cabbage, cauliflower, knol-knol, Brussel's sprouts, etc) have originated from a single ancestor, *Brassica oleraceae var. sylvestris*[1]. Cauliflower (*Brassica oleraceae var. botrytis*) is one of the major cruciferous vegetables and was introduced in India in 1822[1]. The initial introductions were "Cornish" types from England followed by the European types. The Indian cauliflower or the tropical types are a result of intercrossing among these types. The name cauliflower comes from the Latin words '*caulis*' meaning 'stalk' and '*floris*' meaning 'flower'. As suggested by its name, cauliflower is actually a flower. The edible part of the plant is the head of underdeveloped,

[1] Department of Food Science and Technology Department of Processing and Food Engineering Punjab Agricultural University, Ludhiana, Punjab, India
**Corresponding author* E-mail: jagbir@pau.edu

tender flower stems and buds. Cauliflower leaves are more elongated than those of cabbage and broccoli. Leaf color is often light green. Winter or late-season types have curds that consist of functional flower buds, similar to broccoli. Purple cauliflower is a winter variety of broccoli. If harvested before frost, the heads taste like broccoli; after frost, the heads taste like cauliflower. Purple cauliflower turns green when cooked.

Cabbage (*Brassica oleracea var. capitata, tuba, and sabauda*) is thought to be an early ancestor of the wild *brassica* species. The English name cabbage derives from the French word 'caboche' meaning 'head'. Cabbage is herbaceous flowering plant with leaves forming a compact head chrematistics. After the rosette stage, new cabbage leaves develop with shorter petioles and the leaves begin to cup inwards to form heads. Heads may be pointed or conical, oblong, round, or flattened. Leaf texture may be smooth or crinkled (Savoy). Leaf color may be green, red, or purple.

Brussels sprouts (*Brassica oleracea var. gemmifera*) are believed to derive from Savoy cabbage. Sprouts form in leaf axils of an elongated stem. Sprouts start forming in the lowest leaf axil and progress upward.

Broccoli (*Brassica oleracea var. italica*) is believed to be the first of cole crops to evolve from wild cabbage. The head is comprised of functional flower buds. There are two types of broccoli: Sprouting and heading. Sprouting broccoli forms small shoots in the leaf axils over a long period of time, while heading broccoli is a relatively recent introduction that produces one large central head.

All *brassica* vegetables especially cauliflower is generally used as cooked vegetable either singly or mixed with potato, carrot, and peas. In raw form, it is also mixed with green salad or its pieces are dipped into sauces. It is also used in the preparation of pickle or mixed pickle with other vegetables[2].

Production

Cole crops are biennials but are generally grown as annuals. The cruciferous crops may be classified as cool-season plants, and make best growth under cool moist conditions. Cauliflower is sensitive to its growing environment and requires cool temperatures, plenty of moisture and good fertility to produce quality heads.

The total production of cabbage and other brassicas in world is 68,840,531 tonnes with China being the top producer with 32,351,001 followed by India with a production of 7,949,000 tonnes[3]. Russian Federation (5,27,880 tonnes) occupies the third place in the world , followed by Republic of Korea (4,56,307 tonnes) and Japan (2,05,666 tonnes). In India, the area under cabbage and cauliflower are 265.4 and 320.6 thousand hectares with a production of 5,887.8 and 5,796.6 thousand tonnes respectively[4]. In Punjab, the area under cabbage and cauliflower are 3.34 and 6.02 thousand hectares with a production of 73.23 and 138.26 thousand tonnes respectively[5].

Chemical Composition

The consumption of Brassica vegetables remains related to human health and reduces the risk of certain cancers and cardiovascular diseases[6]. They are one amongst the most valued vegetables as far as the content of carotenoids, polyphenol compounds and glucosinolates is concerned[7]. Tables 1 and 2 summarizes the nutritive composition of the major brassica crops[8,9,10]. Glucobrassicin is one of the main glucosinolates (GLS) present in brassicas which are β-thioglucoside N-hydroxysulfates almost exclusively found in Brassicaceae plants and are the substrate part of the myrosinase based defense system[11]. The content of glucosinolates is influenced not only by species, botanic and trade variety or conditions of cultivation but also by their storage conditions after harvest and processing process parameters as well as the way meals are finally prepared[12]. Lipid-soluble antioxidants (carotenoids and vitamin E) are responsible for up to 20% of the brassica total antioxidant activity[13]. The health promoting potential of these components is the focus of intensive research. They exhibit antiproliferative effects on a wide variety of tumor cell lines[14,15,16].

Maturity Indices

Maturity indices are the guidelines used to determine the maximum/ minimum maturity and quality of horticultural crops. Through analyzing physiological, horticultural and economic quality, brassicas can be harvested at optimum maturity leaving the quality of the product at its best.

The harvest maturity of cauliflower is based on size and compactness of the head or curd. Mature heads are at least 15 cm (6 inches) in diameter. Protruding florets, browning and spreading of curds create 'ricy' appearance and are signs of over maturity (Fig. 1). The size of the curds varies greatly and can go up to a diameter of 30 cm and is dependent on the time of sowing as early cauliflower is generally smaller than late cauliflower.

Fig. 1: 'Riceyness' of cauliflower

Table 1: Mineral and vitamin content (100 g edible part) [8]

Crop	*Ca (mg)*	*P (mg)*	*Fe (mg)*	*Na (mg)*	*K (mg)*
Broccoli (raw)	103	78	1.1	15	382
Cabbage raw (White)	49	29	0.4	20	233
Cabbage raw (Red)	2	35	0.8	26	268
Cauliflower	21	64	0.7	9	380
Crop	***Vitamin (IU)***	***Thiamine (mg)***	***Riboflavin (mg)***	***Niacin (mg)***	***Ascorbic acid (mg)***
Broccoli (raw)	2500	0.1	0.23	0.9	113
Cabbage raw (White)	130	0.05	0.05	0.3	47
Cabbage raw (Red)	40	0.09	0.06	0.4	61
Cauliflower	50	0.17	0.05	–	43

Table 2: Composition of *brassica* vegetables in per 100 g of edible raw material [9, 10]

Vegetable	*Water (g)*	*Kilo calories*	*Fat (g)*	*Protein (g)*
Broccoli	89–91	28	0.3	3.1–4.0
Brussels sprouts	84–89	16–58	0.5	2.4–4.4
Cabbage	86–93	8–36	0.7	1.4–3.3
Cauliflower	87–92	11–34	0.3	1.8–3.4
Vegetable	***Total carbohydrate (g)***	***Sugar (g)***	***Fiber(g)***	
Broccoli	5.3	0.4–2.0	1.3–3.0	
Brussels sprouts	8.7	3.6–4.0	1.3–4.6	
Cabbage	5.4	2.7–3.8	0.6–3.4	
Cauliflower	5.2	2.4–2.6	0.8–2.4	

The harvest maturity of broccoli depends on head diameter and its compactness where all the beads (florets) should be closed. The heads must be harvested before the opening of the buds when the bud clusters are still compact and have not turned yellow. The lateral heads growth is encouraged by removing the main head and as their development is uneven, their harvesting also varies. The heads are harvested when they are about 5-25 cm in diameter with about 15 cm of stem attached. Parts of the foliage of the harvested shoots are removed.

Brussels sprouts should be harvested when the buds are firm. Over maturity is indicated by splitting of outer leaves. Harvest maturity is based on sprout size and compactness. Sprouts should be 2.5 cm (1 in) or more in diameter but not more than 7 cm (2.75 in) in length. Stem elongation, resulting in space between older leaves is also a sign of over-maturity.

Cabbage head should be compact, with relatively heavy head with turgid green leaves. If these are harvested too early, weight loss occurs during storage due to wilting and if the heads are harvested too late, they are whiter and may have internal damage caused by frost.

Heading-type cabbage may be harvested as small as 10 cm (4 in) in diameter and continued until 15 cm to 25 cm (6 in to 10 in). Harvest maturity for heading-types is also based on head compactness and firmness to the touch (Fig. 2). A firm or compact head is mature. A compact head can be only slightly compressed with moderate hand pressure. A very loose head is immature and should not be harvested (Fig. 3)[17].

Fig. 2: Head compact of cabbage

Fig. 3: Loose, immature cabbage head

II. PHYSIOLOGICAL DISORDERS

Various disorders are observed in cole crops. The physiological disorders[18] are discussed below:

Buttoning: The buttoning means production of small unmarketable button like curds. Plants that develop buttons are small and have small leaves that do not cover the developing head[19, 20]. This is due to sudden check in the vegetative growth. The reasons attributed are over aged seedlings, limited nitrogen supply, planting early varieties in late season and change in temperature.

Ricyness: There is elongation of peduncles of individual flower buds that result in a granular appearance of the curd[19]. Premature initiation of floral bud is characterized by richens. Such curds are poor in quality for marketing. Ricyness results due to unfavorable high temperature during curd development.

"Leafy curds" or Leafiness: This condition is characterized by green leaves found between sections of the curd and is thought to be caused by relatively high temperatures and delayed harvesting[19, 21].

Browning: This is caused due to deficiency of boron in soil. This disorder is characterized by pith water soaked symptoms on leaves, which turn brown with time[19, 22]. Affected curds are bitter to the taste. Application of boron containing amendment can control browning.

Blindness: Blind plants are those without terminal buds. It results in no head formation or multiple small heads form. The leaves which develop are large, thick leather and dark green. Blindness is supposed to be due to temperature or injury/damage to the terminal growing point due to cultural operation, insect and pest disease.

Whiptail: This is caused due to molybdenum deficiency. Young plants turn white particularly along the leaf margins and also become cupped, turn whiter and finally die. Application of molybdenum @ 1 kg/ha is used for controlling the malady[23]. Leaf blades fail to develop properly and only midrib develops.

III. HARVESTING AND POST-HARVEST HANDLING

It is recommended that harvesting can be conducted during the coolest time of the day, preferably in the morning when the head is most turgid. Cole crops like cabbage, cauliflower are manually harvested by bending the head to one side and cutting it with a sharp knife or small machete. Care should be taken to sharpen the knife often to reduce drudgery and damage to the crop. The snapping or twisting the stalk may result in damaged head and jagged appearance of stems which extend too far out from the base of the head and are also more susceptible to decay. The stalk should be cut flat and as close to the head as possible, yet long enough to retain two to four wrapper leaves. Great care should be exercised in handling the curds as they bruise easily, especially when turgid, and such bruises may show up as unsightly, discolored patches a day or two later at the market. For this reason, retain sufficient leaves, with tips cut off some way above the heads, to entirely protect the curds on all sides. Extra leaves act as cushions during handling and may be desired in certain markets but yellowed, damaged, or diseased wrapper leaves should be removed. Heads with insect damage and other defects should be discarded.

Broccoli is a much more perishable product than the other cole crops. Heads are picked, usually about a hand's breadth under the heads, as soon as they are well-sized and well-colored, but before there are any signs of loosening. As with cauliflower, heads continue to expand and loosen after being cut and quality deteriorates rapidly, especially when high temperatures occur. For Brussels sprouts harvesting commences when the lower sprouts on the stem are ready, and progresses as those higher up the stem achieve market maturity. This is a long-standing crop, which reaches maturity after about four months; picking may extend for a further three months or so, depending on temperatures. The sprouts ready for harvesting are generally a lighter green colour than less mature sprouts. As with cabbages, sprouts may crack if left too long before harvesting. The leaves below the sprouts to be harvested are first broken off. The sprouts are then picked by pushing them sideways, usually with the heel of the hand. Any pieces of leaf-stalk or stem remaining attached to picked sprouts are removed before packing them in suitable containers for marketing. The harvested produce should be put in baskets or well-ventilated picking containers and taken out of the field immediately after harvest. Otherwise they will heat up due to tissue respiration and start to wilt. It is essential that heads not harvested be left undamaged because fields may be harvested as many as three times for maximum yield.

Quality Indices

Cauliflower

1. Globular, intact, firm, clean, white-cream in color
2. Medium sized curds between 15-25 cm diameters are preferred while fewer than 10 cm are unacceptable.
3. Crown of well trimmed turgid green leaves
4. Free of bracts and absence of 'riceyness'
5. Free of pests and not affected by disease
6. Free of mechanical damage, decay, browning or yellowing
7. Curds in the same carton must have uniform size and weight

Cabbage

1. The head should be intact, firm, compact and free of cracks and relatively heavy
2. Free of adhering soil, insects, moulds
3. Should be devoid of discoloration of outer leaves
4. The stem should be cut off directly below the lowest leaf.

Broccoli

1. Dark or bright green closed florets without budding yellow flowers.
2. The head should be compact and firm
3. Free from diseases
4. The stalk should be cleanly cut and the attached leaves should be green and fresh.

Brussels sprouts

1. Should be firm, bright green, without yellowing
2. The inner leaves are light yellow, fairly tightly arranged and without large air pockets between them
3. Free of soil, insect, pests and diseases
4. Free from mechanical damage

Size and Grades

Brussels sprouts: There are two grades, U.S. No. 1 and U.S. No. 2. They are based on size, external appearance, decay, and damage[24]. To meet these grades, sprouts should be more than 2.5 cm (1 in) in diameter and less than 7 cm (2.75 in) in length. U.S. No. 1 consists of Brussels sprouts which are well colored, firm, not withered or burst, which are free from soft decay and seed stems, and free from damage caused by discoloration, dirt or other foreign materials, freezing, diseases, insects, or mechanical or other means. Unless otherwise specified, the diameter of each Brussels sprout shall be not less than one inch, and the length shall be not more than 2.75 inches. U.S. No. 2 consists of Brussels sprouts which are fairly well colored, fairly firm, not withered or burst, which are free from soft decay and seedstems, and free from damage caused by insects, and free from serious damage caused by discoloration, dirt or other foreign material, freezing, disease or mechanical or other means. Unless otherwise specified, the diameter of each Brussels sprout shall be not less than one inch, and the length shall be not more than 2.75 inches.

Cabbage: Grades include U.S. No. 1 and U.S. Commercial, based on defects (physical and decay), excessive wrapper leaves and off-size heads[25]. Size classification is optional. For the pointed-head type (oxheart) cabbage: Small <0.7 kg (1.5 lb); Medium is 0.7 to 1.4 kg (1.5 to 3 lb); and Large > 1.4 kg (3 lb). For domestic and Danish-types (green) cabbage: Small < 0.9 kg (2 lb); Medium is 0.9 to 2.3 kg (2 to 5 lb); and Large > 2.3 kg (5 lb). U.S. No. 1 shall consist of heads of cabbage of one variety or similar varietal characteristics, which are of reasonable solidity, and are not withered, puffy, or burst and which are free from soft rot, seedstems, and from damage caused by discoloration, freezing, disease, insects or mechanical or other means. Stems shall be cut so that they do not extend more than one-half inch beyond the point of attachment of the outermost leaves. U.S. Commercial shall consist of heads of cabbage which meet the requirements for U.S. No. 1 grade except for the increased tolerance for defects specified below, and except that the heads shall be reasonably firm.

Cauliflower: U.S. No. 1 consists of heads of cauliflower which meet the following requirements:

(a) basic requirements for curds: (1) clean; (2) compact; (3) color-white, creamy white, or cream; (4) size-not less than 4 inches in diameter.

(b) basic requirements for jacket leaves: (1) fresh; (2) closely trimmed, unless otherwise specified; (3) fairly clean.

(c) free from: (1) soft or wet decay.

(d) free from damage caused by: (1) bruising; (2) cuts; (3) discoloration; (4) enlarged bracts; (5) fuzziness; (6) hollow stem; (7) insects; (8) mold; (9) riciness; (10) wilting; and, (11) other means.

(e) free from serious damage by any cause

"U.S. Commercial" consists of heads of cauliflower which meet the requirements for the U.S.

No. 1 grade except for the increased tolerances for defects specified[26].

Broccoli: There are three grades, U.S. Fancy, U.S. No. 1 and U.S. No. 2. They are based on external appearance, damage, trimming and stalk diameter[27]. "U.S. Fancy" consists of Italian sprouting broccoli of the same type. The broccoli shall be free from decay, damage caused by over maturity, discoloration of bud clusters or leaves, freezing, wilting, dirt or other foreign materials, diseases, insects, mechanical or other means. The bud clusters shall be generally compact. Bunched broccoli shall be neatly and fairly evenly cut off at the base, and closely trimmed unless otherwise specified as "well trimmed." "U.S. No. 1" consists of Italian sprouting broccoli of the same type. It should meet the quality grades of US Fancy. The bunched broccoli shall be neatly and fairly evenly cut off at the base, and well trimmed, unless otherwise specified as "closely trimmed," "fairly well trimmed," or "leafy." Besides meeting the quality traits of U.S. Fancy the bunched broccoli shall be fairly well trimmed unless otherwise specified as "closely trimmed," or "well trimmed," or "leafy."

Precooling

Brassica vegetables are precooled within two to six hours of harvesting. Vacuum cooling works well for field packed heads for removing field heat. It is more effective if cauliflower is wetted before cooling[28]. The most common way is by forced air cooling which draws cool air through the produce packed in crates, pallets or packages and lowers the temperature of the vegetables rapidly. Other method is hydrocooling which employs chilled chlorinated water (100-300 ppm at 0.5°C) for cooling the produce. With broccoli, ice crushed or as flakes is sometimes used by growers to remove field heat quickly but there is a danger the ice may injure the florets so temperature management is more critical (Fig. 4). For broccoli hydrocooling followed by contact ice cooling using crushed ice is the best solution[29]. According to Tan[30] this method is suitable for both local markets as well as export by air freight.

Fig. 4: Contact ice cooling of broccoli

Packaging

Brussels sprouts are packaged loose in 11 kg (25 lb) cartons or in 3.6 to 4.6 kg (8 to 10 lb) flats or cartons containing 12 plastic containers of sprouts[31]. Plastic liners are often used in cartons with loose sprouts to reduce moisture loss. In addition, polyethylene bags are sometimes used in place of the plastic containers for consumer units. Cauliflower heads are normally well-trimmed and packed in a single layer with 12 to 24 heads in a carton weighing 11.4 to 13.6 kg (25 to 30 lb). Cauliflower may also be packed in 27 kg (60 lb) wire-bound crates or 23 kg (50 lb) cartons/crates[27]. Cabbage heads are shipped in sacks, wax-coated corrugated cardboard cartons and wire-bound crates of various sizes up to 22.7 kg (50 lb). Some cabbage is shipped in heavy fiberboard bulk pallet bins holding 227 to 455 kg (500 to 1,000 lb)[32]. It is important that each container should have produce of similar head size. Cartons and crates provide considerably more protection to the produce and are easier to stack and load. In the fresh market, broccoli is normally bunched, with two to three heads bound together. These are packed in a bunch of 14 to 18, weighing approximately 10 kg (21 lb), in waxed cardboard boxes[31]. Larger heads may have the stem trimmed off to produce "crowns" and these are packed loose in 9 kg (20 lb) boxes. Individual florets are also cut and packed in 2.5 to 5 kg (5.5 and 11 lb) film bags for hotel, restaurant and institutional use.

Storage

Storage at low temperatures slow down the physiological and biochemical processes associated with deterioration and decay. Low temperatures also reduce water loss through transpiration and delay the growth of micro-organisms which cause rot. An increase in the temperature of 10°C can increase the rate of deterioration and decay by two to three times[33]. Vitamin C is most sensitive to destroy the product is subjected to unfavorable handling and storage conditions. Losses are enhanced by extended storage, higher temperatures, low relative humidity, physical damage and chilling injuries. Temperature management is the most important tool to extend shelf-life and indirectly delay losses in nutrients such as vitamin C[34].

Cauliflower has high respiration rate and therefore it is highly perishable[35]. Lipid peroxidation caused by oxygen catabolites results in formation of degradation products which decrease the overall quality of vegetable during senescence in terms of flavor and nutritive value. Depending on the degree of curd maturity and storing conditions the storage life for cauliflower ranges from 2-10 weeks[36]. The compact and disease free heads with white or creamy white color should be selected for storage. Cauliflower is highly sensitive to exogenous ethylene. Discoloration of the curd and accelerated yellowing and detachment of wrapper leaf stalks will result from low levels of ethylene during distribution and short-term storage.

Do not mix loads such as apples, melons and tomatoes with cauliflower[37] as these commodities are known to produce considerable ethylene which can have undesirable effect like yellowing, decay and brown spotting of cauliflower. Mertens[38] as well as Ekman and Golding[39] have found that cauliflower stored at 0-1°C maintain good quality for 4 weeks. Storage life can be increased up to 5 weeks under the control or modified atmosphere. Weight loss of cauliflower during storage can be significantly reduced by control atmosphere containing 2% O_2 +2% CO_2[39] or 3% O_2 and 5% CO_2[40,41]. Controlled or modified atmospheres offer moderate to little benefit to cauliflower. Injury from low O_2 (< 2%) or elevated CO_2 (> 5%) may not be visual and will only be evident after cooking. When the curds become grayish, extremely soft, and emit strong off-odor.

Broccoli is highly perishable and is cooled immediately after harvest by hydrocooling or packing in ice. Low temperature is extremely important to achieve adequate shelf life in broccoli. Exposure to ethylene will accelerate floret yellowing (*e.g.* 2 ppm at 10°C, reduces shelf-life by 50%). The shelf life of broccoli benefits with 1-2% O_2 and 5-10% CO_2 at 0-5°C extreme gas levels and/temperature fluctuations result in offensive off-odors and off-flavors[42]. Fresh broccoli deteriorates quickly once harvested, mainly due to relatively high respiration and susceptibility to wilting when stored at room temperature[43]. Broccoli stored at 20°C showed intense yellowing caused by sepal chlorophyll degradation after two or three days of storage[44].

The storage duration of cabbage varies dramatically with cultivar *i.e.*, early cabbage which is harvested in summer for 3-4 weeks and late or winter cabbage (harvested in fall) for 3-7 months. It benefits with 2-3% O_2 and 3-6% CO_2 at 0-5°C and may be held in CA for up to 10 months. Extreme gas levels may result in off-odors and flavors. Storage at 0°C (32°F) with >95% RH is required to optimize cabbage storage life. Early crop round cabbage can be stored 3-6 weeks, while late crop cultivars can be stored for up to 6 months. For the latter, storage at –0.5°C (31°F) is sometimes recommended. Chinese cabbage can be stored from 2 to 6 months, depending on cultivar, at 0° to 2.5°C (32° to 36°F). Deterioration of cabbage during storage is associated with stem or seed stalk growth (bolting), root growth, internal breakdown, leaf abscission, discoloration, decay and black speck. Long-term storage usually results in extensive trimming of heads to remove deteriorated leaves. Some benefit to shelf-life can be obtained with low O_2 (2.5-5%) and high CO_2 (2.5-6%) atmospheres at temperatures of 0-5°C (32-41°F). CA storage will maintain color and flavor of cabbage, retard root and stem growth, and reduce leaf abscission. Atmospheric oxygen below 2.5% for round cabbage and 1% for Chinese cabbage will cause fermentation and CO_2 atmospheres >10% will cause internal discoloration[45].

Brussels sprouts are moderately perishable and can be stored for 3-5 weeks at temperature near the optimum 0°C with more than 95% relative humidity. Although they have considerable wax on their leaves they become flaccid due to water loss if high relative humidity is not maintained. They are sensitive to ethylene exposure which causes leaf yellowing and leaf abscission in the produce. Care should be taken to prevent freeze damage as it may result in small dark translucent areas and after thawing the entire bud becomes dark, soft and translucent[46]. The optimum conditions for increasing the shelf life of major brassica crops are compiled in Table 3.

Table 3: Shelf life of *Brassica* vegetables under different storage conditions[33]

Name cole crop	*Temp (°C)*	*Relative humidity (%)*	*Shelf life*
Cauliflower	0	90 – 95	3 weeks
	4	80 – 90	2 weeks
	20	60 – 70	2 – 3 days
Cabbage	0	90 – 95	4 – 16 weeks
	4	80 – 90	2 – 6 weeks
	20	60 – 70	<1 week
Brussels sprouts	0	90 – 95	2 – 3 weeks
	4	80 – 90	About 1 week
	20	60 – 70	2 – 3 days
Broccoli, air storage without plastic liner	0	90 – 95	1 – 2 weeks
Broccoli, polystyrene box with ice	0	90 – 95	3 – 4 weeks
Broccoli, waxed carton with plastic liner	0	90 – 95	4 weeks
Broccoli	4	80 – 90	4 – 7 days
Broccoli	20	60 – 70	1 – 2 days

IV. PROCESSING

Brassica vegetables are normally subjected to food processing, such as brief thermal treatments before technological use, and are cooked before human consumption. It is common culinary practice to slice, shred or dice vegetables before cooking which may decrease glucosinolate content by exposing them to enzyme myrosinase. During the initial steps of preparation of vegetables for processing tissues will disrupt and the GLS will come into contact with myrosinase leading to production of reactive compounds[11].The enzyme can be denatured by wet-thermal treatment but processing can result in reduction of constituents by leaching or due to thermal destruction[47]. Glucosinolates undergo thermal degradation above 100°C[48]. Microwave cooking has been used to inactivate myrosinase with assumed negligible loss of glucosinolates[49]. Steaming had minimal effects on chlorophyll, soluble proteins and sugars and vitamin C as well as glucosinolates[50]. Roy *et al,*[51] evaluated the effect of steaming on the nutritional quality of broccoli by assessing the total antioxidant activity of raw and steamed broccoli. Steaming significantly increased the extractability of these compounds in broccoli and its antioxidant potential. The effect of culinary processing and freezing process does not decrease the amino acid content drastically apart from tyrosine[52].

Microwave and steam cooking have been reported not to change the levels of GLS in a major way[53]. The total anthocyanin content remains almost unchanged (95.4% of total content in the raw sample) after microwave treatment whereas after the cooking process it was found to be only 19.8% of that of raw sample[54].

Drying is the most convenient and feasible method for processing. Dried cauliflower can be stored for up to 6 months at room temperature with good sensorial attributes. It was reported that for the optimum utilization of the drying conditions the bed thickness of the cauliflower can be raised from 4 cm to 8 cm without adversely affecting the quality of dehydrated cauliflower[55]. Blanching is necessary as a pre-treatment to freezing procedure, mainly in order to inactivate degradative enzymes but it causes significant reductions in the levels of phenolics and other phytochemicals and antioxidant potential[56]. High-temperature and short-time drying processes after blanching broccoli florets maximized the antioxidant activity[57].Combined method of heat treatment combined with a UV-C treatment on the antioxidant activity of broccoli stored at 20°C showed a significant increase in the antioxidant activity[58].

Cauliflower is quick frozen in such a way that the range of temperature of maximum crystallization is passed quickly and the product temperature has reached –180°C (0°F) at the thermal centre after thermal stabilization. It should be prepared from fresh, clean, sound heads of cauliflower which are washed and properly blanched to ensure stability of color and flavor.

Effect of shredding of brassica vegetables on the glucosinolate content was studied by Song and Thornalley[59] who reported that shredding into large pieces or coarse shredding (10–20 g florets, quartering of Brussels sprouts and 4 × 4 cm cabbage leaf sectors), resulted in less loss of total glucosinolate analyte content. The loss of total glucosinolate analyte content after boiling for 30 min was: broccoli 77%, Brussels sprouts 58%, cauliflower 75% and green cabbage 65% whereas steaming for 0–20 min, microwave cooking for 0–3 min and stir-frying for 0–5 min gave no significant loss of total glucosinolate analyte contents over these cooking periods.

Blanching by steaming of broccoli prior to freezing and refrigeration at 4°C were the best preservation processes for maintaining high level of glucosinolates in broccoli[60]. A large proportion of its production is commercialized fresh and the remaining production is destined for processing as freezing or fermentation. Sauerkraut is prepared from sound, well matured heads of cabbage which have been properly trimmed and cut to which salt (2 to 3%) is added and which is cured by natural fermentation. The finished product contains not less than 1.5% of acid expressed as lactic acid. The finished product is heated in hot brine to above 74°C and placed in cans, glass jars, or plastic bags. The filled containers are then conveyed to an exhaust box and to a container sealer. The sealed containers are water cooled to about 38°C and then stored in a cool place. The growth and fermentation pattern of sauerkraut are affected by variety of cabbage, temperature and salt content[61].

Although in recent years there has been interest in an intermediate form of commercialized cauliflower as a minimally processed product[2, 62]. It is reported that minimal processing of cauliflower irradiated with 0.5 kGy and 1.0 kGy doses, maintained its sensorial attributes[63].

V. WASTE UTILIZATION

Nowadays there is considerable emphasis on the recovery, recycling and upgrading of wastes. This is particularly valid for the food processing industry in which wastes, effluents, residues, and byproducts can be recovered and can often be upgraded to higher value and useful products[64]. Cauliflower has a very high waste index *i.e.* ratio of non-edible to edible portion after harvesting[65] which is an excellent source of protein (16.1%), cellulose (16%) and hemicellulose (8%)[66]. Disposal of the non-edible portion of cauliflower (cauliflower waste), which contributes to about 45–60% of the total weight of the vegetable, remains a crucial problem[67]. Its leaves which are generally thrown away as waste are also rich source of iron and β-carotene and thus can be utilized in various value added products[68]. The nutritional status of the powder obtained by drying the cauliflower leaves is tabulated in Table 4. The incorporation of cauliflower by-products into ready-to-eat snacks increased dietary fiber in the finished product by over 100%, increased protein content and Water Absorption Index (WAI)[13]. It was found that addition of cauliflower significantly affected expansion indices, bulk density, color and total cell area whereas acceptability score showed that cauliflower by-products could be added up to 10%. The upper stem, stalks and leaf midribs of cauliflower are processed to get cauliflower byproducts flour. The utilization of this white cauliflower by-products flour is donein production of meat products and in food fortification as a good and available inexpensive source of protein, minerals, antioxidant compounds especially phenolic compounds and crude fiber to improve the nutritional, physico-chemical and sensory quality criteria with lowering the cost of products[70]. Incorporation of cauliflower leaf powder in noodles up to 10 per cent along with roasted wheat flour not only improves the texture, taste and overall acceptability but also improves the nutritive value of these products without adding much to the cost of the product and cauliflower leaves, which are generally thrown away can be utilized in a better way thus reducing wastage[71].

The leaves of cauliflower/broccoli are available only for a short period but these can be dried or stored for use during lean season[72]. The woody stem parts of broccoli spears are considered unfit for consumption and are discarded. This portion accounts for nearly one third of the weight of broccoli[73]. This can be utilized by skinning, blanching and addition of taste enhancers as ingredient in soups and salads.

Table 4: Nutrient composition of dry cauliflower leaf powder [69]

Components	*Amount*
Moisture, %	2.52
Protein, %	29.98
Ascorbic acid, mg/100 g	58.27
Beta-carotene, mg/100 g	45.58
Total iron, mg /100 g	66.78
Ionizable iron, mg/100 g	2.62
Ionizable iron, % of total Fe	3.93
In vitro iron, % of total Fe	2.33
Copper, mg/100 g	1.52
Manganese, mg/100 g	5.63
Zinc, mg/100 g	4.68

VI. CONCLUSION

Owing to the highly susceptible nature of cole crops a large percent of these vegetables gets wasted due to inadequate post harvest handling which cuts down the export quantities and increases spoilage. Infrastructure facilities should be there to facilitate grading, packing and transportation of the crop. There is a need to develop minimally processed cole crops with high sensory quality, microbiological safety and nutritional value. The processing methods need to be developed further so that they become more environment friendly, mitigate climate change, are energy saving and cost effective. Even though there appears to be the prevalence of several problems in the pre and post harvest aspects of these crops, there is still a vast scope and demand for cole crops. Brussels sprout and Broccoli are yet to be recognized in India and scope to exploit their potential is bright.

VII. REFERENCES

[1] Thanki N., Joshi P., Joshi H., "Effect of household processing on reduction of reduction of pesticide residues in Cauliflower *(Brassica oleraceae var. botrytis)*", *Eur J Exptl Biol.,* **2**(5), 1639–1645 (2012).

[2] Sharma S.R., Singh P.K., Chable V., Tripathi S.K., "A review of hybrid cauliflower development" *J New Seeds.*, **6**(2), 151–193 (2005).

[3] FAO 2011.www.faostat3.fao.org.

[4] Package of practices for cultivation of vegetables. Punjab Agricultural University, Ludhiana, pp. 27–30.

[5] www.indiastat.com

[6] Francisco M., Velasco P., Moreno D.A., Garcýa-Viguera C., Cartea M.A., "Cooking methods of *Brassica rapa* affect the preservation of glucosinolates, phenolics and vitamin C", *Food Res. Int.*, **43,** 1455–1463 (2010).

[7] Lipecki J., Libik A., "Some components of vegetables andfruit with high biological value", *Folia Hort., Suppl.*, **1**, 16–21 (2003).

[8] Watt B.K., Merrill A.L., "Composition of foods: Raw, processed, prepared," Agriculture Handbook No. 8 (Science and Education Administration, USDA, Washington, D.C., (1975).

[9] U.S. Department of Agriculture, Agricultural Research Service. USDA Nutrient Database for Standard Reference, Release 14, Nutrient Data Laboratory Home page, http://www.nal.usda,gov/fnic/foodcomp, (2001).

[10] Duckworth R.B., Fruits and Vegetables, London: Pergamon Press Limited. ISBN 0080119735, Appendix A, p. 280 (1966).

[11] Bones A.M., Rossiter J.T., "The myrosinase-glucosinolate system, its organisation and biochemistry", *Physiologia Plantarum.*, **97,** 194–208(1996).

[12] Sosinska E., Obiedzinski M.W., "Effect of Processing on the content of glucobrassicin and its degradation products in broccoli and cauliflower," *Food Control,* **22**, 1348–1356 (2011).

[13] Stojceska V., Ainsworth P., Plunkett A., Ibanoglu E., Ibanoglu S., "Cauliflower by-products as a new source of dietary fibre, antioxidants and proteins in cereal based ready-to-eat expanded snacks", *J Food Engg.,* **87**, 554–563 (2008).

[14] Hertog M.G.L., Kromhout D., Aravanis C., Blackburn H., Buzina R., Fidanza F., Giampaoli S., Jansen A., Menotti A., Nedeljkovic S., Pekkarinen M., Simic F.S., Toshima H., Feskens E.J.M., Hollman P.C.H., Katan M.B., "Flavonoid intake and long-term risk of coronary heart disease and cancer in the seven countries study", *Arch Int Med.,***155**(4), 381–386 (1995).

[15] Boivin D., Lamy S., Lord-Dufour S., Jackson J., Beaulieu E., Cote M., "Antiproliferative and antioxidant activities of common vegetables: A comparative study", *Food Chem.*, **112**, 374–380 (2009).

[16] Paramithiotis S., Hondrodimou O.L., Drosinos E.H., "Development of the microbial community during spontaneous cauliflower fermentation," *Food Res Int.,* **43**, 1098–1103 (2010).

[17] National Agricultural Research Institute, PH Bulletin No. 25 Cabbage: Postharvest Care and Market Preparation, May 2004.

[18] http://www.agriinfo.in.

[19] Norman J.C., Tropical Vegetable Crops. Arthur H. Stockwell LTD, Elms Court, p. 252. (1992).

[20] Fritz V.A., Rosen C.J., Grabowski M.A., Hutchison W.D., Becker R.L., Tong C.B.S., Wright J.A., Nennich T.T., "Growing broccoli, cabbage and cauliflower in Minnesota", University of Minnesota Extension Bulletin, pp. 1–15 (2009).

[21] Loughton A., "Production and handling of broccoli", Factsheet ISSN 1198–712X (2009).

[22] Verma P., "Physiological Disorders of Vegetable Crops", Alfa Beta Technical Solutions. Jaipur-India, p. 170 (2009).

[23] Masarirambi M.T., Oseni T.O., Shongwe V.D., Mhazo N., "Physiological disorders of Brassicas/Cole crops found in Swaziland: A review", *African J Plant Sci.*, **5**(1), 8–14 (2011).

[24] USDA.1954. Agricultural Marketing Service, Washington, DC.

[25] USDA.1945. Agricultural Marketing Service, Washington, DC.

[26] USDA.1968. Agricultural Marketing Service, Washington, DC.

[27] USDA.2006. Agricultural Marketing Service, Washington, DC.

[28] Stewart J.K., Barger W.R., "Effects of cooling method on the quality of asparagus and cauliflower", *Proc Amer Soc Hort Sci.*, **78**, 295–301(1961).

[29] Ryall A.L., Lipton W.J., "Handling, Transportation and Storage of Fruit and Vegetables". Vol.1. Vegetables and Melon. AVI Publ. Co, Westport, CT. (1971).

[30] Tan S.C., "Post-Harvest Handling of Brassica Vegetables. Dep. Agriculture Western Australia", www.agric.wa.goof.au. (2005).

[31] Boyette M.D., Sanders D.C., Rutledge G.A., "Packaging requirements for fresh fruits and vegetables", http://www5.bae.ncsu.edu/programs/extension/publicat/postharv/ag-414-8/index.html. (1996).

[32] Boyette M.D., Sanders D.C., Estes E.A., "AG-413-5 Postharvest cooling and handling of cabbage and leafy greens" in "Postharvest Commodity Series" at http://www.bae.ncsu.edu/programs/extension/publicat/postharv/ag-413-5 (1999).

[33] Tan S.C., Haynes Y.S., Phillips, "Post-harvest handling of Brassica vegetables Farmnote No. 44/94",http://www.agric.wa.gov.au(2007).

[34] Mohamed Mahroop Raja M., Raja A., Mohamed Imran M., Habeeb Rahman A., "Quality aspects of cauliflower during storage" *Intl Food Res J.,* **18**, 427–431 (2011).

[35] Tamer C.E., "Quality evaluation of canned cauliflower pickles prepared with different ingredients", *African J Agric Res.*, **7**(10), 1550–1555 (2012).

[36] Kader A.A., "Postharvest technology of horticultural crops", Publication 3311, University of California, sec. ed., pp. 267–270 (1992).

[37] Anonymous., Cauliflower: Recommendations for Maintaining Postharvest Quality http://postharvest.ucdavis.edu

[38] Mertens H., "Ethylene and respiratory metabolism of cauliflower (*Brassica oleracea L.* convar. *botrytis*) in controlled atmosphere storage", *Acta Hort.,* **258,** 493–501 (1989).

[39] Ekman J.H., Golding J.B., "Preliminary evaluation of storage technologies for broccoli, cauliflower and head lettuces", *Acta Hort.,* **712**(1), 201–208(2006).

[40] Adamicki F., Elkner K., "Effect of temperature and controlled atmosphere on storage and quality of cauliflowers", *Biuletyn Warzywniczy,* **XXVIII**, pp. 197–224 (1985).

[41] Hodges D.M., Munro K.D., Forney C.F., McRae K.B., "Glucosinolate and free sugar content in cauliflower (*Brassica oleracea* var. *botrytis* cv. Fremont) during controlled atmosphere storage", *Postharvest Biol Technol.,* **40**(2), 123–132 (2006).

[42] De Ell J., Postharvest handling and storage of cole crops. http://www.omafra.gov.on.ca. 2010

[43] Gillies S.L., Toivonen P.M.A., "Cooling method influences the postharvest quality of broccoli", *Hort Sci.*, **30**, 313–315 (1995).

[44] Wang C.Y., "Effect of aminoethoxy analog of rhizobitoxine and sodium benzoate on senescence of broccoli", *Hort Sci.,* **12**, 54–56 (1977).

[45] *Cantwell M.,* Suslow *T.,* Cabbage (Round & Chinese types): Recommendations for maintaining postharvest quality http://postharvest.ucdavis.edu/pfvegetable/Cabbage/ . (2013).

[46] Cantwell M., Suslow T., Brussels sprouts: Recommendations for maintaining post harvest quality. http://postharvest.ucdavis.edu. (2013).

[47] Rungapamestry V., Duncan A.J., Fuller Z., Ratcliffe B., "Effect of cooking brassica vegetables on the subsequent hydrolysis and metabolic fate of glucosinolates" *Proc Nutr Soc.*, 66(1), 69–81 (2007).

[48] Oerlemans K., Barrett D.M., Bosch Suades C., Verkerk R., Dekker M, "Thermal degradation of glucosinolates in red cabbage", *Food Chem.*, **95,** 19–29 (2006).

[49] Beltramo E., Berrone, E., Buttiglieri, S. and Porta, M., "Thiamine and benfotiamine prevent increased apoptosis in endothelial cells and pericytes cultured in high glucose". *Diabetes – Metabol Res Rev.*, **20,** 330–336 (2004).

[50] Yuan G., Sun B., Yuan J., Wang Q., "Effects of different cooking methods on health-promoting compounds of broccoli", *J Zhejiang University Science B.*, **10**(8), 580–588 (2009).

[51] Roy M.K., Juneja L.R, Isobe S., Tsushida T., "Steam processed broccoli (*Brassica oleracea*) has higher antioxidant activity in chemical and cellular assay systems", Food Chem., **114**, 263–269 (2009).

[52] Slupski J., Lisiewska Z., Kmiecik W., Gebezynski P., Sobezynska L., "The effect of processing on the amino acid content in green cauliflower", *Agric Food Sci.,* **19**, 136–143 (2010).

[53] Rungapamestry V., Duncan A.J., Fuller Z., Ratcliffe B., "Changes in glucosinolate concentrations, myrosinase activity, and production of metabolites of glucosinolates in cabbage (*Brassica oleracea var.capitata*) cooked for different durations", *J Agric Food Chem.,* **54**(20), 7628–7634 (2006).

[54] Scalzo R.L., Genna A., Branca F., Chedin M., Chassaigne H., "Anthocyanin composition of cauliflower (*Brassica oleracea* L. var. botrytis) and cabbage (*B. oleracea L. var. capitata)* and its stability in relation to thermal treatments", *Food Chem.,* **107,** 136–144 (2008).
[55] Gupta M.K., Sehgal V.K., Kadam D.M., Singh A.K., Yadav Y.K., "Effect of bed thickness on cauliflower drying", *American J Biochem.*, **2**(5), 56–61(2012).
[56] Volden J., Borge G.I.A., Hansen M., Wicklund T., Bengtsson G.B., "Processing (blanching, boiling, steaming) effects on the content of glucosinolates and antioxidant-related parameters in cauliower (*Brassica oleracea* L. ssp. *botrytis*), *LWT - Food Sci Technol.,* **42**, 63–73(2009).
[57] Mrkiæ V., Cocci E., Dalla Rosa M., Sacchetti G., "Influence of drying conditions on bioactive compounds and antioxidant activity of dried broccoli (*Brassica oleracea*, L.)", *J Sci Food Agric.*, **86**, 1559–1566 (2006).
[58] Lemoine, M.L., Chaves, A.R., Marti-nez, G.A., "Influence of combined hot air and UV-C treatment on the antioxidant system of minimally processed broccoli (*Brassica oleracea* L. var. *Italica*)", *LWT – Food Sci Technol.,* **43**(9), 1313–1319 (2010).
[59] Song L., Thornalley P.J., "Effect of storage, processing and cooking on glucosinolate content of Brassica vegetables", *Food Chemical Toxicol.,* **45,** 216–224 (2007).
[60] Rodrigues A.S., Rosa E.A.S., "Effect of post-harvest treatments on the level of glucosinolates in broccoli", *J Sci Food Agric.,* **79**, 1028–1032 (1999).
[61] Hui Y.H., Ghazala S., Graham Dee M., Murrell K.D., Nip Wai-Kit (*eds.*), Handbook of Vegetable Preservation and Processing. CRC Press (2003).
[62] Sanz S., Olarte C., Echavarri J.F., Fernando A., "Evaluation of different varieties of cauliflower for minimal processing", *J Sci Food Agric.*, **87**, 266–273 (2007).
[63] Nunes T.C.F., Rogovschi V.D., Santillo A.G., Michels R., Pitombo R.N.M., Villavicencio A.L.C.H., International Nuclear Atlantic Conference - INAC 2009, Rio de Janeiro,RJ, Brazil, September27 to October 2, 2009, Associação Brasileira De Energia Nuclear – Aben ISBN: 978-85-99141-03-8 (2009).
[64] Laufenberg G., Kunz B., Nystroem M., "Transformation of vegetable waste into value added products: (A) the upgrading concept, (B) practical implementations", *Bioresource Technol*, **87**, 167–198 (2003).
[65] Kulkarni M., Motey R., Lele S.S., "Biotechnology in agriculture, industry and environment", *In*: Proceedings of the international conference of SAARC countries, organized by icrobiologists society at Karad, India, December 28–30. pp. 24–31 (2001).
[66] Wadhwa M., Kaushal S., Bakshi M.P.S., "Nutritive evaluation of vegetable wastes as complete feed for goat bucks", *Small Ruminant Research,* **64**, 279–284 (2006).
[67] Oberoi H.S., Kalra K.L., Uppal D.S., Tyagi S.K., "Effect of different drying methods of cauliflower waste on drying time, colour retention and glucoamylase production by *Aspergillus niger* NCIM 1054", *Inter J Food Sci Technol.*, **42**, 228–234 (2007).
[68] Kowsalya S., Sangeetha M., "Acceptability and nutrient profile of cauliflower leaves (*Brassica oleraceae var Botrytis*), *Ind J Nutr Dietetics.*, **36**, 332–338 (1999).
[69] Kocchar A., www.assocham.org/events/recent/event_638/Dr_Anita_Kochhar.ppt
[70] Abul-Fadl M.M., "Nutritional and chemical evaluation of white cauliflower by-products flour and the effect of its addition on beef sausage quality", *J App Sci Res.*, **8**(2), 693–704 (2012).
[71] Wani T.A., Sood M., Kaul R.K., "Nutritional and sensory properties of roasted wheat noodles supplemented with cauliflower leaf powder", *Annals Food Sci Technol.,* **12**(2), 102–107 (2011).
[72] Singh G., Asha K., Sehgal S., "Development and nutritional evaluation of products prepared from dried powder of cauliflower leaves", *J Food Sci Technol*, **42**(2), 137–139 (2005).
[73] Salunke D.K., Kadam S.S., Handbook of Vegetable Science and Technology: Production, Compostion, Storage, and Processing. CRC Press. (1998).

9

Brassica – Bioactive Compounds and Extraction

GAGAN JYOT KAUR[1]* AND JAGBIR REHAL[2]

ABSTRACT

Family of Brassicacene vegetables constitutes a variety of crops such as broccoli, turnip, cauliflower, kale, etc. These vegetables posses both antioxidant and anti carcinogenic properties. A large variation in antioxidants content in Brassicaceae family is recorded which is mainly due to variation in the variety, maturity at harvest, crop condition, genotype, soil state, post harvest condition etc. These bioactive compounds are important for health benefits, their isolation, extraction and utilization has attracted much needed attention. Various clinical studies have proved the health claims of these compounds. This chapter provides a detailed information regarding classification of various bioactive compounds, their extraction and the factors influencing their content.

Key words: Brassica, Bioactive Compounds, Brassicacene, Clinical studies extraction

I. INTRODUCTION

A plant based diet composed of vegetables and fresh fruits have become one of the most important recommendations of health organizations. The nutritional campaign of "five per day" is supported by various organizations in several countries. It has been predicted that the consumption pattern of fruits and vegetables is inversely related to the stroke incidence, mortality, ischemic heart mortality and cardiovascular disease (CVD) mortality. *Brassica oleracea* is known for its bioactive compounds which include three classes

[1] Department of Processing and Food Engineering, College of Agricultural Enginerring PAU, Ludhiana – 141004, India.
[2] Department of Food Science and Technology Punjab Agricultural University, Ludhiana - 141004, India
*Corresponding author E-mail: engg-gagan@pau.edu

of photosynthetic lipid soluble compounds, carotenoids, chlorophyll and tocopherols. These lipid soluble compounds are associated with good health responsible for lowering risks of human diseases[1]. Moreover the plant based phenols, flavanoids, isoflavones, terpenes, GLS have anti carcinogenic and anti-oxidant properties[2,3,4,5]. Family of *Brassicaceae* vegetables constitutes a variety of crops such as broccoli, turnip, cauliflower, kale, etc. These vegetables posses both antioxidant and anti carcinogenic properties[6,7,8] and their consumption has reported beneficial effect on human metabolism by preventing the chances of CVD cancer. The variation in antioxidants content in *Brassicaceae* family is due to many factors like variety, maturity at harvest, crop condition, genotype, soil state and post harvest conditions[9,13]. Other factors affecting the retention of an antioxidant in the processed food are industrial processing, blanching, cooking, steaming, microwave cooking, cutting etc. Since these bioactive compounds are important for health benefits, their isolation, extraction and utilization has attracted much needed attention.

II. CLASSIFICATION

The family of *Brassicaceae* (also known as *Cruciferae*) is a larger group, having 3500 species grouped in 350 genera including several types of edible plants. Petals of plants from this family have a distinctive cross form arrangement, which is the origin of the initial term "*Cruciferae*"[14]. The genus *Brassica* is the most important one within the tribe *Brassicaceae*.

Three of the *Brassica* species are derived from three ancestral genomes, denoted by the letters AA, BB, or CC as shown in Fig. 1. Alone, each of these diploid genomes produces a common *Brassica* species where *n* denotes the number of chromosomes in each genome, and is the number found in the pollen or ovule.

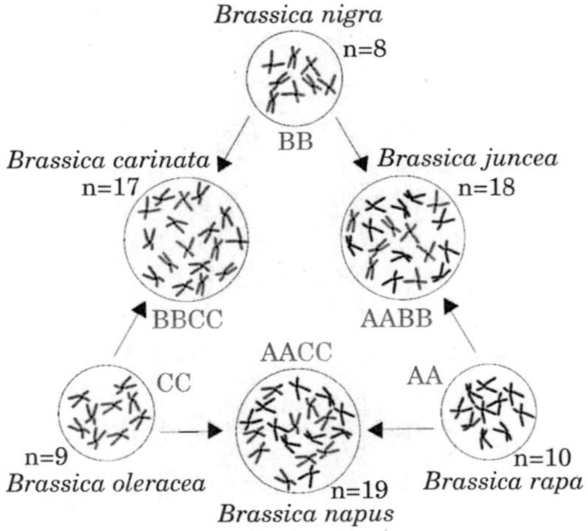

Fig. 1: Genetic configuration of *brassica* family

1. *Brassica nigra* (l) Koch (2n = 16)
2. *Brassica oleracea* L (2n = 18)
3. *Brassica rapa* L (2n = 20)

Brassica (*nigra, rapa and oleracea*) are genetically distinct. Their interbreeding allowed the creation of three new species of tetraploid *Brassica* [known as allotetraploid (containing four genomes from 2 species) or amphidiploids (containing one diploid genome from two different *Brassica* species)]. Molecular studies have named three species as Paleopolyploids[14].

1. *Brassica carinata* L (2n = 34)
2. *Brassica juncea* L (2n = 36)
3. *Brassica napus* L (2n = 38)

The major characterization of the genus is done into oilseed, forage, condiment and vegetable crops by using their buds, inflorescences, leaves, roots, seeds and stems. *Brassica oleracea* species is further divided into various groups like *Acepphala* (Kale and collards), *Capitata capitata* (Savoy cabbage), *Costata* (Tronchuda cabbage), *Gemmifera* (Brussel sprouts) *Botrytis botrytis* (cauliflower) *Botrytis italic* (Broccoli) *Gonglodes* (kohlrabi) and *Albogabra* (Chinese kale). Similarly *Brassica rapa* is classified as *Chinensis* (Pak Choy, Bak Choy), *Narinosa* (Chinese flat cabbage, wrutacai), *Nipposinica* (Mibuna, Mizuna), *Oleifera* (Turnip rape, rapeseed), pekinensis (Chinese cabbage), parachinensis (Choysum), *Rapa* (turnip greens and tops), *Ruvo* (Broccoleto), *Trilochulari* (Yellow sarson) *Brassica* napus and juncea as sub grouped into pabulara napo *Brassica* (leaf rape, nabicol swede, rutabaga) and *Brassica* Junce (Rugosa (mustard green), *Capitata* (head mustard), crispifolia (cut leaf mustard) respectively[15]. The same species may be utilized for several uses according to different forms or types. The common feature of all these crops is high content of bioactive compounds which are medically proven to prevent number life threatening ailments.

III. BIOACTIVE COMPOUNDS

Bioactive compounds are extra-nutritional constituents that typically occur in small quantities in plants and plant products. They are very important as they induce a variety of physiological functions including acting as direct or indirect antioxidant regulating enzymes, controlling apoptosis and the cell cycle. Cruciferous vegetables are good source of many health promoting and potentially protective phytochemicals including folic acid, phenolics, carotenoids, selenium, flavanoids, GLS (and its breakdown products), vitamin C etc and dietary essential minerals (Ca, Mg, Na, K, Fe, Zn etc). Since they vary widely in chemical structure and function they are grouped accordingly[16,17,18,19].

Glucosinolates (GLS)

More than 130 GLS have been characterized out of which 30 are present in *Brassica* family[20,21] which can be aliphatic, aromatic or idolyl. GSLs invariably occur in anionic nature[22, 23]. The type and concentration of GLS have been found to vary between the species and the cultivar[24, 25, 26, 27]. Different tissues of the same plant vary both quantitatively and qualitatively in their glucosinolate (GLS) content. The difference in the GLS can be due to factors like plant age, fertilization, environmental growth etc.[27, 28, 29]. *Brassica* family contains GLS the metabolic breakdown products of which are potent modulators in xenobiotic metabolizing enzymes that protect DNA from damage[28]. They are a class of plant secondary metabolite •-D thioglucoside, *N*-hydroxysulfates with a side chain and sulphur linked •-D-glucopyranosyl. Presence of glucose moiety and ionic forms makes GLS hydrophilic and non-volatile compounds[22]. The predominant GLS are listed in Table 1[20]. GLS and or their breakdown products are known for fungicidal, bactericidal, nematocidal properties and have attracted intense research due to chemo protective attribute[20]. Along with it these compounds have the potential for controlling soil borne pathogens[30]. The structure of the GLS consisting of a β-D glucose moiety linked to a sulphated thiohydroxiate is shown in Fig. 2.

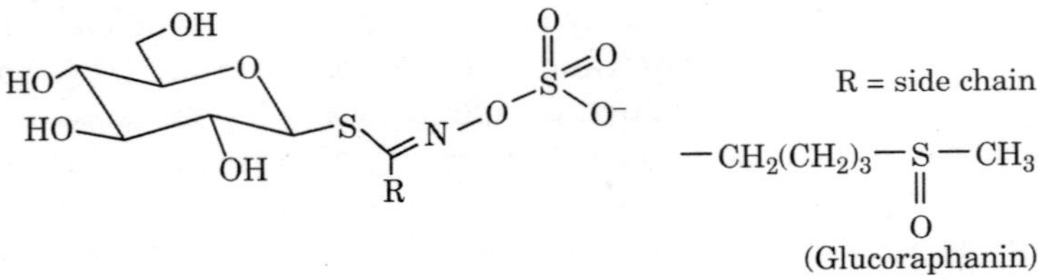

Fig. 2: Basic glucosinolate structure

Eight most significant glucosinolates have been identified so far from the *Brassica* genus. Ugolini *et al.*[31] reported sinigrin as a major component in Indian *Brassica* species. Glucosinolates, the thioglucosides, present in brassica vegetables are thought to contribute to the development/synthesis of this compound. Dietary glucosinolates have been reported to block formation of endogenous or exogenous carcinogens for preventing initiation of carcinogenesis[32].

IV. BRASSICA VEGETABLE AS A SOURCE OF ANTIOXIDANTS

An antioxidant can be defined both as a substance that opposes oxidation or inhibits reaction promoted by oxygen, and an organic substance capable of counter acting the damaging effect of oxidation in animal tissues. A dietary antioxidant has the capacity to scavenge ROS/RNS (reactive oxygen/ nitrogen species) to stop radical chain reaction or it can inhibit the formation of reactive oxidants. The antioxidants present in the *Brassica* family can be broadly classified into two categories

Table 1: Predominant glucosinolates identified in *Brassica* spp.

Chemical class	***Chemical name***	***Trivial name***
Sulphur-containing GLS	4-Mehylsulphinyl-3-butenyl	Glucoraphenin
	4-(Methylsulphinyl)butyl	Glucoraphanin
	5-(Methylsulphinyl)pentyl	Glucoalyssina
	3-(Methylsulphinyl)propyl	Glucoiberina
	4-(Methylthio)butyl	Glucoerucin
	5-(Methylthio)pentyl	Glucoberteroin
	3-(Methylthio)propyl	Glucoiberverin
Branched aliphatic GLS	1-Methylethyl	Glucoputranjivin, isopropyl
	2-Methylethyl	
	1-Methylpropyl	Glucochlearin, glucojiabutin, *sec*-butyl, 2-butyl
Aliphatic olefins	3-Butenyl	Gluconapina
	2(*R*)-2-Hydroxy-3-butenyl	Progoitrina
	2-Hydroxy-4-pentenyl	Gluco]napoleiferin
	4-Pentenyl	Glucobrassicanapina
	1-Pentenyl	
	2-Propenyl Allyl,	Sinigrin
Aromatic aryl GLS	4-Hydroxybenzyl	[Gluco]sinalbin
	2-Phenylethyl	Gluconasturtiina, phenethyl
Aromatic indole GLS	Indol-3-ylmethyl	Glucobrassicina
	4-Hydroxy-3-indolylmethyl	4-Hydroxy Glucobrassicina
	1-Methoxyindol-3-ylmethyl	Neoglucobrassicina
	4-Methoxyindol-3-ylmethyl	4-Methoxyglucobrassicina

- Water soluble antioxidants
  - Phenolic Compounds (Flavanoids)
  - Hydroxycinnamic acid
  - Phytic acid
  - Brassinosteroids
  - Vitamin C
- Lipid soluble antioxidants (water insoluble antioxidant)
  - Carotenoids
  - Vitamin E

Water Soluble Antioxidants

(a) Phenolic compounds

Phenolics range from simple, low molecular-weight, single aromatic-ringed compounds to large and complex tannins and derived polyphenols[33, 34]. The term polyphenol (secondary metabolites) includes more than 8000 compounds with great structural diversity (each having one aromatic ring with one or more hydroxyl groups). They can be divided into 10 different classes depending on the chemical structure. Table 2 shows the main families of phenolic compounds and their basic structure[35]. Most of the phenolic compounds are associated with mono or polysaccharides (glycosides) or functional derivatives (ester or methyl ester).

Phenolic acids present in fruits and vegetables are in the free esterified and insoluble forms. They are derivatives of benzoic and cinnamic acid[36, 37]. Phenols are able to scavenge reactive oxygen species due to their electron donating properties or hydrogen donating radical scavengers[38]. The most widespread and diverse group of polyphenols in *Brassica* species are flavanoids (mainly flavanols, anthocyanins) and hydroxycinnamic acids.

Table 2: Classification of families of phenolic compounds

Carbon numbers	*Class*	*Basic structure*
C_6	Simple Phenols	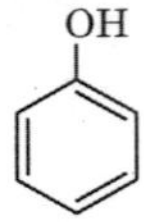
	Benzoquinones	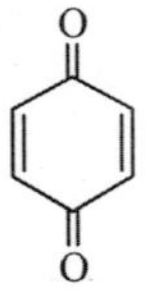
C_6-C_1	Benzoic acid	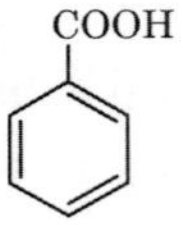
C_6-C_2	Acetophenones	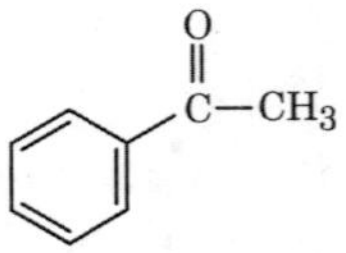
	Phenylacetic acid	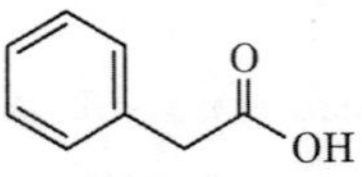
C_6-C_3	Cinnamic acid	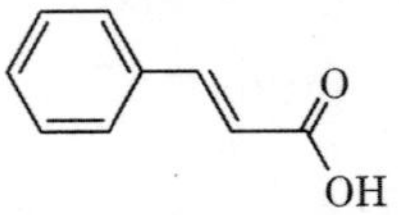
	Phenylopropene	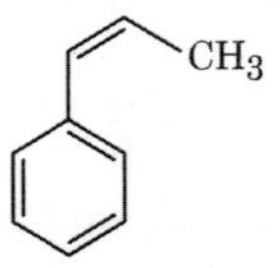
	Coumarins	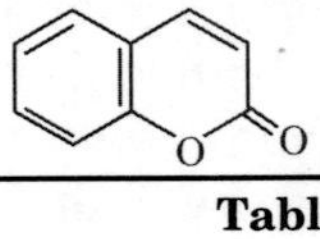

Table 2: (*Contd...*)

Table 2: (*Contd...*)

Carbon numbers	***Class***	***Basic structure***
	Chromones	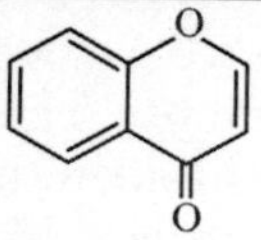
C_6-C_4	Naphthoquinones	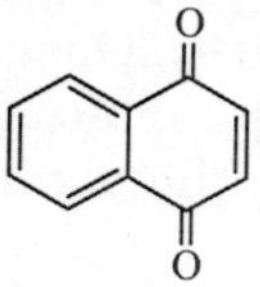
C_6-C_1-C_6	Xanthones	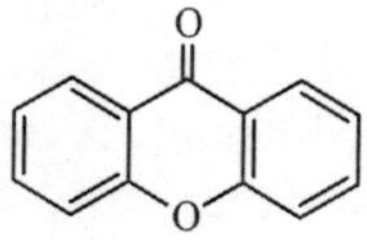
C_6-C_2-C_6	Stilbenes	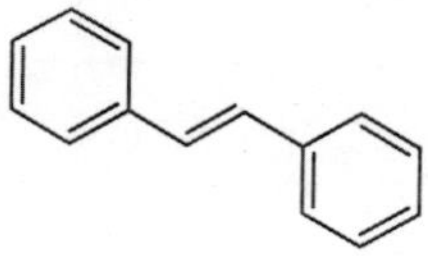
	Anthraquinones	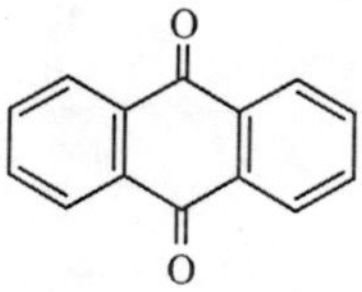
C_6-C_3-C_6	Flavonoids	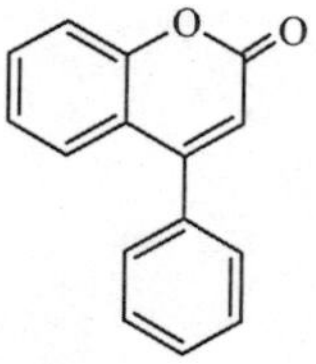
$(C_6$-$C_{3)2}$	Lignans,neolignans	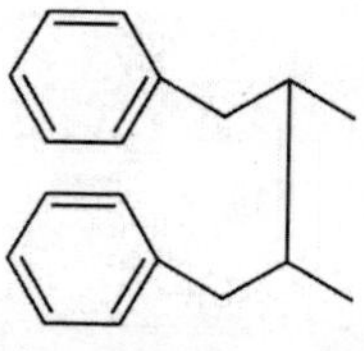
$(C_6$-$C_{1)n}$	Hydrolysable tannins	Heterogeneous polymer composed of phenolic acids and simple sugar
$(C_6$-$C_{3)n}$	Lignins	Highly crosslinked aromatic polymer

Flavanoids: are a group of poly-phenolic compounds, which can be subdivided into 13 classes: chalcones, dihydrochalcone, auron, flavones, flavonols, dihydroflavonol, flavanones, flavanols (catechins), flavandioles or leucoanthocyanidins, anthocyanidins (its glycoside is called anthocyanin), isoflavonas, flavonoids, and condensed tannins or proanthocyanidins[39,40].

These are the most abundant phenolics, present in high concentration in epidermis of leaves, fruits[41, 34] and often in vacuoles. Natural flavonoids are polyphenolic compounds comprising fifteen carbon with 2 aromatic rings connected by a 3 carbon bridge C_6-C_3-C_6.They usually occur as glycosides-glucose, rhamnose, rhamno glucosides and can be divided into subclasses according to the structure of C_6-C_3-C_6 flavone skeleton. They have reported bioactive properties like free radical scavenging, inhibition of hydrolytic and oxidative enzymes[15]. Common flavanoids are kaempferol, quercetin, Isorhamnetin (aglucones) found as O-glucosides (Fig. 3). Within colored flavanoids, anthocyanins (group of plant pigments) are the most important group of plant pigments having anti-oxidant and other beneficial properties[42, 43, 44]. The chemical structure of the anthocyanin determines the stability, colour intensity and potential biological activity. The most common anthocyanins are pelargonidin, cyaniding, delphinidin, peoidin, petunidin and malvidin, with cyanidin as the most common in *Bassicaceae* family[45, 46, 47].

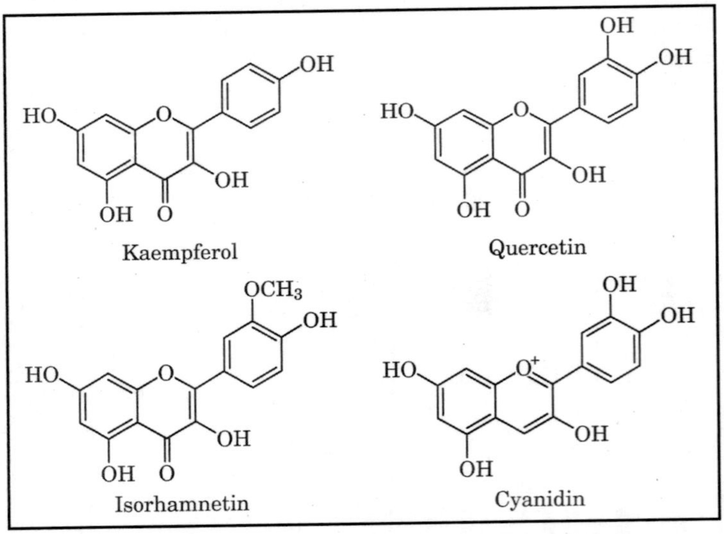

Fig. 3: Structural configuration of common flavanoids

(b) Hydroxycinnamic acids

Hydroxycinnamic acids are non flavanoid phenolics characterized by C_6-C_3 structure. They occur freely or as components of plants polymers. The most common are *p*-coumaric, sinapic and ferulic acids (Fig. 4), often found in conjugation with sugar or other hydroxycinnamic acids[48, 49,50, 51].

(c) Phytic acid

Inositol hexaphosphate (Ins P6) is the storage form of phosphorus in many crop plants (mustard)[52, 53]. It contains six phosphate groups thereby; it can

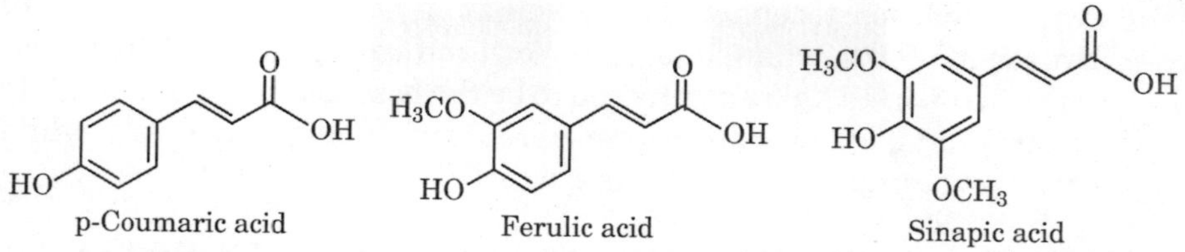

Fig. 4: Structural configuration of common hydroxyl-cinnamic acids

act as metal chelator (calcium, iron and nickel) to inhibit the generation of highly reactive oxygen species such as hydroxyl free radicals[54, 55, 56].

(d) Brassinosteroids

Brassinosteroids are naturally occurring poly-hydroxy steroids having 5-alpha-cholestan skeleton. About 42 brassinosteroid configurates have been characterized. They are classified as C27, C28 or C29 brassinosteroids on the basis of substitution pattern of side chain[57, 58]. Structural variation is due to the kind and orientation of functional group on skeleton. C28 steroids have a 24 alpha-methyl group in the side chain and C27 brassinosteroids with no substitution at 24.

(e) Vitamin C

Ascorbic acid is able to scavenge the superoxide and hydroxyl radicals. Vitamin C also has the ability to regenerate α-tocopherol[59]. The content of vitamin C in *Brassica* vegetables varies significantly among and within species, which could be due to climatic conditions[60] and nitrogen fixation[11]. The degradation product of vitamin C namely dehydroascorbic (DHA) and ascorbigens act as a radical scavenger and as a donor/acceptor in electron transport at the plasma membrane. They are capable of performing many biological activities in the human body.

Lipid Soluble Anti-oxidants:

(a) ***Carotenoids***: Out of 16 carotenoids identified in the Brassica; lutein and b-carotene, are the most abundant[61]. Carotenoids (carotenes and xanthophylls) are yellow, orange and red pigments present in many fruits and vegetables. Several of them are precursors of vitamin A (*i.e.*, b-carotene, g-carotene and b-cryptoxanthin) and due to conjugated double bonds; they are both radical scavengers and quenchers of singlet oxygen, particularly under low oxygen pressure[62].

(b) ***Vitamin E*** belongs to a group of lipid-soluble antioxidants whose effect is mainly due to α-tocopherol. The pre-dominant reaction responsible for tocopherol antioxidant activity is the donation of hydrogen atoms, where a tocopheroxyl radical is formed[63]. Kurilich *et al.*[62] reported kale and broccoli to be the best sources of lipid soluble antioxidants, followed by brussel sprouts with moderate levels, while cauliflower and cabbage are characterized by their low levels.

V. VARIATION OF ANTIOXIDANT POTENTIAL IN *BRASSICA* VEGETABLES

The antioxidant potential of *Brassica* vegetables has been measured in the most economically important crops like broccoli, kale, Brussels, sprouts, cabbage etc. Biosynthesis and concentration of phenolic compounds in plants depends on genetic and environmental factors. Phenolic contents are affected by biotic stresses (insect attack and pathogen infection) and abiotic stresses (light, temperature, nutrient supplies, water availability, growing conditions and UV radiation) besides storage conditions, post-harvest treatments and the estimation methods[64, 65].

(i) Influence of Plant Part and Genotype

The antioxidant capacity of the plant depends on the edible portion of the plant (root, stem, leaves of inflorescence). In Turnip the roots showed significantly lower anti oxidant capacity in comparison to flower buds, leaves and stems[66], the internal leaves of Tronchuda cabbage has low antioxidant potential than external leaves (which is not consumed)[67], kale seeds have higher antioxidant potential but lower phenolic content in comparison to kale leaves[68]. The maturity stage of the plant is also the deciding factor of the antioxidant activity. Juvenile cabbage reported higher content of flavanols in comparison to mature cabbage[69], high antioxidant activity in white and Chinese cabbage in the first 8-12 weeks which then gradually decreases[70]. Podsedek *et al.*[63] found red cabbage and Brussel sprouts have higher antioxidant potential than savoy white cabbage. The antioxidant potential in curly kale was at least 10-fold higher than that in cauliflower and white cabbage[71]. The radish seeds have higher antioxidant values in comparison to broccoli seeds[72], among kale, broccoli, brussels sprouts and green and white cauliflower, kale showed the highest antioxidant potential and cauliflower showed the lowest[73]. Red cabbage has higher anti-scavenging activity among Chinese cabbage, cauliflower, Brussels and cabbage [74].

(ii) Influence of Environment

Environmental factors like irrigation, fertilization and harvesting time affect the antioxidant activity of *Brassica* crops. An increase in the antioxidant capacity with nutritional soil supplements in green cabbage[69] and sulphur fertilization in two *B. rapa* crop widely grown in Southern Italy[75] has been recorded. Li *et al.*[76] reported a decrease in total phenolic content on increased nitrogen fertilization but an increase in phenolic content with increased sulphur fertilization in leaf mustard[76]. Similar results were recorded for broccoli[77] and 'friarello' (local *B. rapa* crop in southern Italy). Young *et al.*[78] found that pak choi samples grown organically had higher levels of total phenolics than conventional samples due to high instance of insect attack, which may be a biotic stress factor contributing to higher levels of phenolic agent.

(iii) Influence of Post-harvest Handling, Processing and Storage

The most commonly available commercial products prepared from *Brassica* vegetables are washed, pre-cut or fresh cut vegetables, fermented and frozen vegetables. Unit operations such as cutting and slicing may cause a rapid enzymatic degradation of several naturally occurring antioxidants as a result of the cellular disruption, which allows the contact of substrates with the enzymes. During vegetable cooking, antioxidants breakdown and leach out in cooking water which may influence the antioxidant activity of the vegetables[63]. The GLS levels are not affected after cutting but washed and chopped vegetables showed optimal conditions for myrosinase (thioglucoside glucohydrolase)[79, 80, 81, 82], which breaks GLS to isothiocyanates and/or nitriles in plant material. Storage at 4° C upto 5 months was possible for white and red cabbage without any affect on the levels of GLS[79].

Cooking (thermal treatment) is an essential primary process for preservation leading to increase in shelf life. The variation in cooking parameters significantly affects the textural, nutritional and antioxidant activity of the produce. Boiling, steaming, microwave, pressure cooking, freezing are the most commonly employed processing techniques. Lin and Chang[83] did not found any profound effect of thermal treatment on antioxidant but Pellegri *et at.*[84] and Roy *et al.*[85] both reported that steaming increased antioxidant activity while microwave had detrimental effect on the same. Better retention of carotenoids[86] and increased antioxidant activity[87] during boiling in comparison to steaming and microwave cooking respectively were recorded. Microwave cooking affected the total phenolic content significantly but not the total anti-oxidant activity of fresh broccoli[88]. Frying retained almost all the carbohydrates, vitamins and minerals in comparison to boiling and steaming where the reported loss in the mineral content was about 25-50%[89]. Steamed broccoli contains high amount of phenol compounds, better retention of vitamin C, $B_{6,}$ Mg and Ca in comparison to other cooking processes[90].

Reduction in anti-oxidant activity during blanching followed by boiling while higher anti-oxidant activity was reported in steaming followed by boiling and microwave[72, 91, 92]. Similar trend of higher reduction in the antioxidants during boiling followed by blanching and steaming is reported[93]. During boiling the polyphenols were reduced thus affecting the anti-oxidant activity[94]. It was observed that polyphenol content remained constant during freezing.

In vegetables phenolics are present both in free and conjugated forms and flavanoids are mostly reported in conjugated form. Aglycones are present as a result of food processing. The effect of thermal processing on phenolics, flavonoids or total antioxidant activities is different for different products and still needs further research to establish complete effect. Some antioxidant compounds like ascorbic acid and carotenoids are very sensitive to heat and storage and are lost during different vegetable processing treatments[95]. However, flavonoids and some phenolic compounds are quite stable at high temperature and over long period of storage[96].

The loss of dietary antioxidants is caused by boiling[97], the type of cooking[98], cooking time and amount of water. Steam cooking reported a loss of 20-30% in flavanoids for turnip tops[97]. On the contrary the phenolic compounds (water soluble) were recovered in the water after cooking[99]. Steam-cooking of broccoli increased flavonoid and phenolic content when compared to fresh broccoli due to retention of phenolic compounds in the edible part and inactivation of oxidative enzymes[96] whereas cooking in water reported the opposite effect. Microwave and pressure cooking reported less detrimental effect on phenol content in comparison to boiling for several vegetables, including cauliflower[100].

Lipid degradation and peroxidation are usually correlated with tissue deterioration after harvesting[101, 102]. Degradation of chlorophyll and membranes causes an increase in the production of free radicals. Senescence plays an important role, and leads to the formation of hydrogen peroxide (reduced oxygen)[103]. This can be delayed by refrigeration, modification of the surrounding atmosphere or heat treatments. Cold storage and modified atmospheres minimizes the decay of antioxidant and preserves high level of natural antioxidants (phenols and vitamin C) after the harvest. The use of modified atmosphere package reduces the respiration rate, maintaining sensory attributes and increased shelf life. The low oxygen atmosphere did not affect the nutritional quality, maintaining the initial phenol content, antioxidant potential and vitamin C content during cold storage[104, 105]. Increase in polyphenol content stored under different modified atmosphere packaging at low temperature could be due to biosynthesis of polyphenols for plant protection in the first days after harvest[106].

Heat treatments conserved tissue integrity thereby maintaining the antioxidant activity when compared with the raw produce[101]. Blanching caused significant reduction in phenols and antioxidants while deep freezing resulted in diminishing levels of antioxidant potential[93] while, pre-blanching induced an increase in radical scavenging activity[105]. High temperature short time (HTST) treatment followed by blanching of broccoli maximized anti-oxidant activity due to increased release of compounds from the matrix[107]. In another study carried out by Lemoine *et al.*[101] broccoli showed an increase in antioxidant activity due to combination of heat and UV treatment. Slicing lead to higher loss of antioxidants during hydrothermal treatments and degraded significantly during drying[108]. Low oxygen, controlled atmosphere, modified atmosphere and MAP with Argon (Ar) instead of Nitrogen improved the storage life of fresh fruits and vegetables. The storage of Broccoli in oriented polypropylene (OPP) with 14% oxygen, 10.5% carbon dioxide resulted in most of off odors and heat treatment of the broccoli increased the content of aroma compounds[109]. Cold storage, modified storage atmosphere conditions, refrigeration, freezing and thermal treatments are respective areas in which the studies have been conducted but still elaborate research needs to be done.

VI. CLINICAL STUDIES

The regular onsumption of *Brassica* vegetables have recorded lower incidence of different types of cancer and diseases. A high intake of cruciferous vegetables is associated with a reduced risk of cancer, particularly lung, gastrointestinal tract and prostate cancer[110]. Researchers have reported anticancer effects of dietary bioactives from broccoli, anti-proliferative effects of sulphoraphane in human breast cancer[111, 112, 113]; reduced risk of cancer *via* decreased damage to DNA[114, 115], effects on the regulation of intestinal cell growth and death in colon cancer[116], cancer-protective effect of high selenium broccoli[5], protective effect in prostatic tumors[117, 118, 119, 120], such as sulphoraphane induced apoptosis in prostate cancer cells, are highly promising chemopreventive agent[121]. Isothiocyanates are the breakdown products of glucosinolates, found to reduce the carcinogenic activity enhancing the detoxification, bioactives (isothiocyanates) from broccoli are effective in curing bladder carcinoma cells[122, 123]. The clearance of carcinogens can be done by enhancing the activity of phase II enzymes and inhibition of Phase I enzymes[124, 125, 126, 127]. Other degradation products of GLS like Benzyl-p-hydroxybenzyl and 2 hydroxy but 3-enyl glucosinolates[124, 125], Sulforaphane[20, 128, 129, 130, 131], Indole-3-carbinol (I3C)[132, 133, 134, 135, 136] have been reported for anti carcinogenic, antioxidant and anti-atherogenic activities, with increase in phase II enzymes though antioxidant response element (ARE) on antioxidant capacity and on cellular oxidative stress[137], as well as cholesterol lowering effects [138], protective effects on cardiovascular diseases[139] and *Helicobacter pylori* infections[140]. Carotenoids such as carotene present in broccoli prevent several diseases related to oxidative stress. Quercetin (flavanol) displayed the ability to prevent the oxidation of LDL by scavenging free radicals and chelating transition metal ion[15]. Phenolic compounds, especially flavonoids, perform different biological activities, but the most important are antioxidant activity, capillary protective effect and the inhibitory effect elicited in various stages of tumor[141]. They may protect against cancer development through several biological mechanisms.

VII. EXTRACTION OF GLS

The determination of GLS in the plants is complex, due to the presence of myrosinase enzyme. The processing of plant part (cutting, chopping, grinding) in presence of water, hydrolyses the parent compound to different compounds (glucose, nitrile, isothiocyanate and sulphur) as shown in the the Fig. 5.

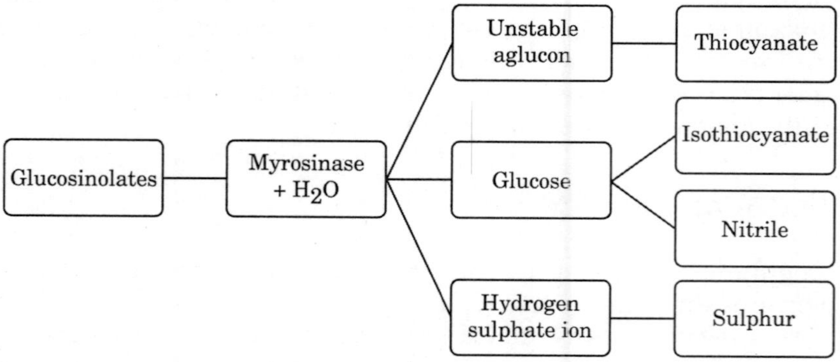

Fig. 5: Hydrolysis of GLS into various compounds in presence of enzyme

Analytical approach for determination of GLS can be done by assessing the total GLS/Individual GLS/breakdown products. For determination of intact Glucosinolates inhibition of myrosinase activity is essential, which is obtained by freeze drying/IQF (dipping/spraying the sample in liquid nitrogen) and then using aqueous methanol for extraction in combination with high temperature[79,142]. Another method can be employed by for determining the quantity of glucose released after the treatment with enzymes, followed by selective clean up to eliminate free glucose and other interfering compounds[81].

Gas liquid Chromatography is the traditional method for the determination of GLS, a number of modifications along with new procedures have been developed like reverse phase HPLC(High Pressure Liquid Chromatography). It is commonly used for analyzing various intact glucosinolates (glucoiberin, glucocheirolin, progoitrin, sinigrin, epiprogoitrin, glucoraphenin, sinalbin, gluconapin, glucosibarin, glucotropaeolin, glucoerucin, gluconasturtiin) in plants. The samples are extracted using an organic solvent, followed by filtration and evaporation. Interferences in the organic sample extracts can be removed by using SPE (Solid phase extraction). In the HPLC analysis, the twelve intact glucosinolate standards are baseline separated as shown in Fig. 6. Glucosinolate mixture standard solution is spiked into the sample extracts for the identification of the glucosinolates in the sample extracts. Comparison of the peak areas of the corresponding retention times in the chromatogram of original sample extract with the spiked one, the peak areas of the spiked one are increased and the respective GLS present are recorded. The most suitable technique is to use the HPLC analysis of the enzymatically desulfated glucosinolates. Desulfated glucosinolate gives a better separation. However, desulfated glucosinolates are often subject to the difficulties in interpreting results of the individual glucosinolates due to concerns over the effect of pH value, time, and enzyme concentration on desulfation products. Therefore, the direct analysis of the intact glucosinolates is needed for more specific and accurate determination, for better interpretation of analytical results, for the reduction of analytical time.

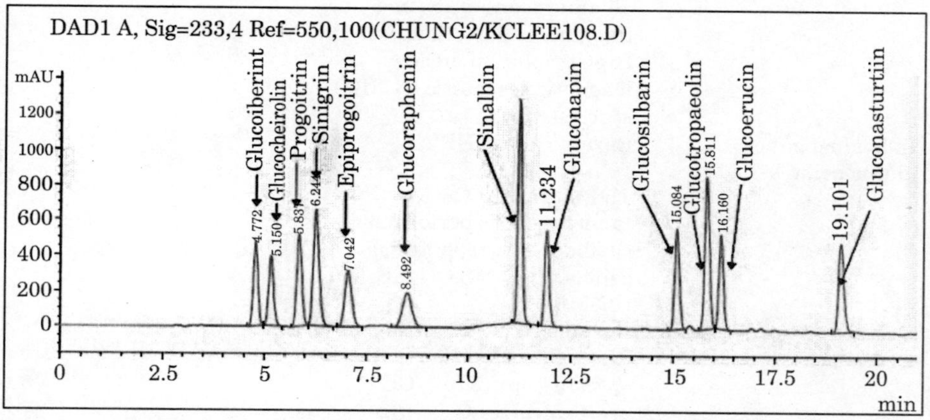

Fig. 6: HPLC analysis of twelve intact glucosinolate standards

A number of researchers have worked in relation to the extraction and estimation of GLS. Different methods like HPLC-MS, Thermospray LC with tandem MS, High Performance capillary electrophoresis, Capillary GC-MS, GC-MS, GC-MS-MS, X-ray fluorescence spectroscopy (XRF) etc are used for the determination. Table 3 gives the details of the various other methods used for extraction of GLS.

VIII. EXTRACTION OF PHENOLS

Extraction of phenolic compounds in plant materials is influenced by their chemical nature, the extraction method employed, sample particle size, storage time and conditions, as well as presence of interfering substances. The chemical nature of plant phenolics vary from simple to highly polymerized substances that include varying proportions of phenolic acids, phenylpropanoids, anthocyanins and tannins, among others. They may also exist as complexes with carbohydrates, proteins and other plant components; some high-molecular weight phenolics and their complexes may be quite insoluble. Therefore, phenolic extracts of plant materials are always a mixture of different classes of phenolics that are soluble in the solvent system used. Additional steps may be required for the removal of unwanted

Table 3: Commonly used methods for analysis of glucosinolates and their breakdown products

Compound	***Method***	***Reference(s)***
Total Glucosinolates	Palladium chloride and thymol assays	[79,111,140, 142, 143, 144, 145]
	Glucose and sulphate release enzyme assays	[79,142, 143, 144,146, 147, 148, 149, 150, 151]
	ELISA	[79,152, 153]
	Near infrared reflectance (NIR) spectroscopy, alkaline degradation and thioglucose detection	[79,146,147,150,151,154]
	High resolution nuclear magnetic resonance (NMR) spectroscopy	[23,144, 155, 156, 157, 158]
Individual intact glucosinolates	Reverse phase HPLC	[79, 142, 144, 150, 151, 158, 159]
	Thermospray LC with tandems, High performance capillary electrophoresis, capillary GC-MS,GC-MS, GC-MS-MS	[144,159, 160, 161]
Desulpho-glucosinolates	Reverse HPLC	[48, 79, 144, 153, 159, 162, 163, 164, 165, 166]
Degradation products	X-ray fluorescence spectroscopy (XRF), GC or GC-MS,HPLC	[79, 111, 144, 148,158, 167, 168]

phenolics and non-phenolic substances such as waxes, fats, terpenes and chlorophylls.

Different types of extraction methods are used for extraction of polyphenols. The extraction process is majorly influenced by temperature, polyphenolic structure, pressure, sample characteristics etc. Most commonly used extraction method is LLE (Liquid Liquid Extraction). Traditional methods are being replaced by new methods which are more sensitive, selective, rapid and environment friendly like SFE (Super-critical Fluid Extraction), Solid Phase Extraction (SFE), Pressurized Liquid Extraction (PLE), Microwave Assisted Extraction (MAE)/Microwave Assisted Process (MAP) and Ultra Assisted Extraction (UAE). The details of various methods are tabulated in Table 4[32].

Solid phase extraction (SPE) techniques and fractionation based on acidity are commonly used to remove unwanted phenolics and non-phenolic substances[175]. Solubility of phenolic compounds is governed by the type of solvent (polarity) used, degree of polymerization of phenolics, as well as interaction of phenolics with other food constituents and formation of insoluble complexes. Therefore, there is no uniform or completely satisfactory procedure that is suitable for extraction of all phenolics or a

Table 4: Methods for extraction of phenols

Method	***Cost***	***Temperature***	***Time***	***Medium***	***Limitation***
LLE	Cheap & Easy	Room Temp	Long duration resulting in degradation	Organic solvents	Takes very long time
SPE-Cost is more in comparison to LLE					
PLE	Less	Increase temperature (40-200ºC) Increase pressure (500-3000psi)	Shorter duration	Organic solvents	Need high pressure.
MAE/MAP	More than LLE		Reduced duration time	Flammable solvents	Rise in temperature due to flammable solvents
SFE (Latest Technique)	Expensive	Low temperature (Sub-critical conditions)		Need high % of organic modifier	Cannot be used for extraction of phenolic compounds due to low polarity of CO_2

specific class of phenolic substances in plant materials. Methanol, ethanol, acetone, water, ethyl acetate and, to a lesser extent, propanol, dimethylformamide, and their combinations are frequently used for the extraction of phenolics[169].

IX. CONCLUSION

Brassica family is one of the most important families in terms of bioactive compounds having antioxidant (free radical scavenging) properties. It contains a wide variety of water and lipid soluble antioxidants, finding utility in various medical treatments. Since the composition varies widely between the species the isolation of the individual bioactive compound and its extraction cannot be standardized. A number of methods have been reported for the extraction like LLE, SPE, HPLC, Reverse phase HPLC-MS, Thermospray LC with tandem MS, High Performance capillary electrophoresis, Capillary GC-MS, GC-MS, GC-MS-MS, X-ray fluorescence spectroscopy(XRF) but still further research is required.

X. REFERENCES

[1] Machlin L.J., "Critical assessment of the epidemiological data concerning the impact of antioxidant nutrients on cancer and cardiovascular disease", *Crit Rev Food Sci Nutr.,* **35,** 41–50 (1995).

[2] Craig W.J., "Phytochemicals: Guardians of our health", *J Am Diet Assoc.,* **97(suppl.)**, 199–204 (1997).

[3] Potter J.P., "Food nutrition and the prevention of cancer: A global perspective", Washington, DC: World Cancer Research Fund. (1997).

[4] Rhodes M.J.C., "Physiologically-active compounds in plant foods: An overview", *Proc Nutr Soc.,* **55**, 371–84 (1996).

[5] Zhang Y., Talalay P., Cho C.G., Posner G.H., "A major inducer of anticarcinogenic protective enzymes from broccoli: Isolation and elucidation of structure", *Proc Natl Acad Sci.,* **89,** 2399–403 (1992).

[6] Cohen J., Kristal R., Stanford J., "Fruit and vegetable intakes and prostate cancer", *J Nat Can Inst.,* **9,** 61–68 (2000).

[7] Chu Y.F., Sun J., Wu X., Liu R.H., "Antioxidant and antiproliferative activities of common vegetables", *J Agric Food Chem.*, **50**, 6910–6916 (2002).

[8] Verhoeven D.T., Verhagen H., Goldbohm R.A., van den Brandt P.A., van Poppel G., "A review of mechanisms underlying anticarcinogenecity by *Brassica* vegetables", *Chemico-Bio Inter.,* **103**, 79–129 (1997).

[9] Jeffery E.H., Brown A.F., Kurilich A.C., Keek A.S., Matusheski N., Klein B.P., "Variation in content of bioactive components in broccoli. Study review", *J Food Comp Anal.,* **16**, 323–330 (2003).

[10] Kurilich A.C., Jeffery E.H., Juvik J.A., Wallig M.A., Klein B.P., "Antioxidant capacity of different broccoli (*Brassica* oleracea) genotypes using the oxygen radical absorbance capacity (ORAC) assay", *J Agri Food Chem.,* **50**, 5053–5057 (2002).

[11] Lisiewska Z., Kmiecik W., "Effects of level of nitrogen fertilizer, processing conditions and period of storage of frozen broccoli and cauliflower on vitamin C retention", *Food Chem.*, **57**(2), 267–270 (1996).

[12] Vallejo F., Tomas-Barberan F.A., Garcia-Viguera C., "Potential bioactive compounds in health promotion from broccoli cultivars grown in Spain", *J Sci Food Agri.,* **82**, 1293–1297 (2002).

[13] Vander den Berg H., Faulks R., Granado H.F., Hirschberg J., Olmedilla B., Sandmann G., Southon S., Stahl W., "The potential for the improvement of carotenoid levels in foods and the likely systemic effects", *J Sci Food Agri.*, **80**(7), 880–912 (2000).

[14] Cartea M., Lema M., Francisco M., Velasco P., "Basic information on vegetable *Brassica* crops". *In*: Sadowski J. and Kole C. (*eds.*), Genetics, Genomics and Breeding of Vegetable *Brassicas*, Science Publishers, Enfield, New Hampshire, pp. 1–33 (2011).

[15] Cartea M.E., Francisco M., Soengas P., Velasco P., "Phenolic compounds in *Brassica* Vegetables", *Molecules,* **16**, 251–280 (2011).

[16] Howard L.A., Jeffery E.H., Wallig M.A., Klein B.P., "Retention of phytochemicals in fresh and processed broccoli", *J Food Sci.,* **62**, 1098–1100 (1997).

[17] Femina A., Robertson J.A., Waldron K.W., "Cauliflower (*Brassica oleracea* L.), globe antichoke (*Cynara scolymous* L.) and Chicory witloof (*Cichorium intybus*) processing by-products as source of dietary fibre", *J Sci Food and Agri.,* 77, 511–518 (1998).

[18] Kushad M.K., Brown A.F., Kurillicn A.C., Juvik J.A., Klein B.P., Wallig M.A., Jeffery E.H., "Variation in GLS in vegetable crops of *Brassica oleracea*", *J Agril Food Chem.,* **47,** 1541–1548 (1999).

[19] Conaway C.C., Getachun S.M., Liebes L.L., Pusateri D.J., Tophan D.K.W., Botero-Omary M., Chung F.L., "Disposition of GLS and sulphoraphanes in human after ingestion of steam and fresh broccoli", *Nutr Cancer.,* **38**, 168–178 (2001).

[20] Fahey J.W., Zalcmann A.T., Talalay P., "The chemical diversity and distribution of GLS and isothiocyanates among plants". *Phytochem.,* **56**(1), 5–51(2001).

[21] Sørensen J.C., "Biorefined oilseed products–high quality products with added value". PhD Thesis. Chemistry Department, KVL, Denmark (2001).

[22] Larsen P.O., "GLS". *In*: The biochemistry of plants. Stumf P.K. and Conn E.E. (*eds.*), vol. 7. New York, Academic Press, pp. 501–525(1981).

[23] Prestera T., Fahey J.W., Holtzclaw W.D., Abeygunawardana C., Kalchinski J.L., Talalay P., "Comprehensive chromatographic and spectroscopic methods for the separation and identification of intact GLS", *Anal Biochem.*, **239,** 168–179 (1996).

[24] Bradshaw J.E., Heaney R.K., Macfarlane Smith W.H., Gowers S., Gemmell D.J., Fenwick G.R., "The glucosinolate content of some fodder *Brassicas*", *J Sci of Food Agri.*, **35,** 977–981 (1984).

[25] Kirkegaard J.A., Sarwar M., "Biofumigation potential of *Brassicas*. I. Variation in glucosinolate profiles of diverse field-grown *Brassicas*", *Plant and Soil.*, **201**, 71–89 (1998).

[26] Rangkadilok N., Nicolas M.E., Bennett R.N., Premier R.R., Eagling D.R., Taylor P.W.J., "Developmental changes of sinigrin and glucoraphanin in three *Brassica* species (*Brassica nigra*, *Brassica juncea* and *Brassica oleracea* var. italica)", *Scientia Horticulturae.*, **96**, 11–26 (2002).

[27] Sang J.P., Minchinton I.R., Johnstone P.K., Truscott R.J.W., "Glucosinolate profiles in the seed, root and leaf tissue of cabbage, mustard, rapeseed, radish and swede", *Canadian Journal of Plant Sci.,* **64**, 77–93 (1984).

[28] Clossais-Besnard N., Larher F., "Physiological role of GLS in *Brassica napus*. Concentration and distribution pattern of GLS among plant organs during a complete life cycle" *J Sci of Food and Agri.,* **56**, 25–38 (1991).

[29] Sarwar M., Kirkegaard J.A., "Biofumigation potential of Brassicas. II. Effect of environment and ontogeny of glucosinolate production and implications for screening", *Plant and Soil.*, **201**, 91–101(1998).

[30] Moreno D.A., Carvajal M., L´opez-Berenguer C., Garc´ýa-Viguera C., "Chemical and biological characterisation of nutraceutical compounds of broccoli", *J Pharm Biomed Anal.*, **41,** 1508–1522 (2006).

[31] Ugolini L., De Nicola G., Palmieri S., "Use of reverse micelles for the simultaneous extraction of oil, proteins and GLS from cruciferous oilseeds", *J Agric Food Chem.*, **56**, 1595–1601(2008).

[32] Vig A.P., Rampal G., Singh T.S., Arora S., "Bioprotective effects of GLS- A review", *LWT Food Sci Technol.*, **42**, 1561–1572 (2009).

[33] Crozier A, Jaganath I.B., Clifford M.N., "Dietary phenolics: Chemistry, bioavailability and effects on health", *Nat Prod Reps.,* **26**, 1001–1043 (2009).

[34] Pereira D.M., Valentao P., Pereira J.A., Andrade P.B., "Phenolics: From chemistry to biology", *Molecules,* **14**, 2202–2211 (2009).

[35] Salas P.G., Aranzazu M.S., Antonio S.C., Alberto F.G., "Phenolic compound extraction systems for fruits and vegetable Samples", *Molecules*, **15,** 8813–8826 (2010).

[36] Aherne S.A., O'Brien N.M., "Dietary flavonols: Chemistry, food content, and metabolism", *Nutr.,* **18**, 75–81 (2002).

[37] Robards K., Prenzler P.D., Tucker G., Swatsitang P., Glover W., "Phenolic compounds and their role in oxidative processes in fruits", *Food Chem.,* **66**, 401–436 (1999).

[38] Rice-Evans C.A., Miller N.J., Paganga G., "Structure-antioxidant activity relationships of flavonoids and phenolic acids", *Free Radical Bio Med.,* **20**, 933–956 (1996).

[39] Escarpa A., Gonzalez M.C., "An overview of analytical chemistry of phenolic compounds in foods", *Crit Rev Anal Chem.,* **75**, 57–139 (2008).

[40] Wu X., Prior R.L., "Identification and characterization of anthocyanins by high-performance liquid chromatography-electrospray ionization-tandem mass spectrometry in common foods in the United States: Vegetables, nuts, and grains", *J Agr Food Chem.,* **53**, 3101–3113 (2005).

[41] Crozier A., Jaganath I.B., Clifford M.N., "Phenols, polyphenols and tannins: An overview", *In* Plant Secondary Metabolites: Occurrence, structure and role in the

human diet, Crozier A., Clifford M. and Ashihara H. (*eds.*), Blackwell, Oxford, UK, pp. 1–24 (2006).

[42] Sadilova E., Stintzing F.C., Carle R., Anthocyanins, colour and antioxidant properties of eggplant (*Solanum melongena* L.) and violet pepper (*Capsicum annuum* L.) peel extracts. *Z. Naturforsch.,* **61**, 527–535 (2006).

[43] Mc Dougall G.J., Fyffe S., Dobson P., Stewart D., "Anthocyanins from red cabbage-stability to simulated gastrointestinal digestion", *Phytochem.*, **68**, 1285–1294 (2007).

[44] Moreno D.A., Perez-Balibrea S., Ferreres F., Gil-Izquierdo A., Garcia-Viguera C., "Acylated anthocyanins in broccoli sprouts", *Food Chem.,* **123**, 358–363 (2010).

[45] Tatsuzawa F., Saito N., Shinoda K., Shigihara A., Honda T., "Acylated cyanidin 3-sambubioside-5-glucosides in three garden plants of the *Cruciferae", Phytochem.,* **67**, 1287–1295 (2006).

[46] Lo Scalzo R., Genna A., Branca F., Chedin M., Chassaigne H., "Anthocyanin composition of cauliflower (*Brassica oleracea* L. var. *botrytis*) and cabbage (*B. oleracea* L. var. *capitata*) and its stability in relation to thermal treatments", *Food Chem.,* **107**, 136–144 (2008).

[47] Vallejo F., Tomas-Barberan F.A., Ferreres F., "Characterisation of flavonols in broccoli (*Brassica oleracea* L. var. italica) by liquid chromatography-UV diode-array detection-electrospray ionisation måss spectrometry", *J. Chromatogr* A., **1054**, 181–193 (2004).

[48] Olsen H., Aaby K., Borge G.I.A., "Characterization and quantification of flavonoids and hydroxycinnamic acids in curly kale (*Brassica oleracea* L. convar. *acephala* var. *sabellica*) by HPLC-DAD-ESI-MSn", *J Agric Food Chem.,* **57,** 2816–2825 (2009).

[49] Lin L.Z., Harnly J.M., "Phenolic Component Profiles of Mustard Greens, Yu Choy, and 15 Other *Brassica* Vegetables", *J Agri Food Chem.,* **58**, 6850–6857 (2010).

[50] Price K.R., Casuscelli F., Colquhoun I.J., Rhodes M.J.C., "Hydroxycinnamic acid esters from broccoli florets", *Phytochem.,* **45**, 1683–1687(1997).

[51] Raboy V., "Seeds for a better future: Low phytic grains help to overcome malnutrition and reduce pollution", *Trends in Plant Sci.*, **6,** 458–462 (2001).

[52] Naczk M.R., Amarowicz R., Sullivan A. and Shahidi F., "Current research developments on poly phenolics of rapeseed/canola: A review", *Food Chem.,* **62**, 489–502 (1998).

[53] Jabri E., Carr M.B., Hausinger R.P., Karplus P.A., "The crystal structure of urease from *Klebsiella aerogenes", Science,* **268**, 998–1004 (1995).

[54] Rimbach G., Pallanf J., "Phytic acid inhibits free radical formation in vitro but does not affect liver oxidant or anti oxidant status in growing rats", *J Nutr.,***128**, 1950–1955 (1998).

[55] Grases F., Simonet B.M., Prietro R.M., March J.G., "Phytate levels in different rat tissues: Influence of dietary phytate", *British J Nutr.,* **86**, 225–231 (2001).

[56] Fujioka S., Sakurai A., "Biosynthesis and metabolism of brassinostroids", *Physiologia Plantarum,* **100,** 710–715 (1977).

[57] Yokota T., "The structure, biosynthesis and function of brassinosteroids", *Trends Plant Sci.,* **2**, 137–143 (1997).

[58] Davey M.W., Van Montagu M.I.D., Sanmartin M., Kanellis A., Smirnoff N., Benzie I.J.J., Strain J.J., Favell D., Fletcher J., "Plant L-ascorbic acid: Chemistry, function, metabolism, bioavailability and effects of processing", *J Sci Food Agri.,* **80**, 825–860 (2000).

[59] Howard L.A., Wong A.D., Perry A.K., Klein B.P., "β-carotene and ascorbic acid retention in fresh and processed vegetables", *J Food Sci.*, **64**(5), 929–936 (1999).

[60] Wills R.B.H., Rangga A., "Determination of carotenoids in Chinese vegetables", *Food Chem.,* **56**, 451–455 (1996).

[61] Kurilich A.C., Tsau G.J., Brown A., Howard L., Klein B.P., Jeffery E.H., Kushad M., Wallig M.A., Juvik J.A., "Carotene, tocopherol, and ascorbate con- tents in subspecies of *Brassica oleracea", J Agril Food Chem.,* **47**, 1576–1581 (1999).

[62] Podsedek A., "Natural antioxidants and antioxidant capacity of *Brassica* vegetables: A review", *LWT –Food Sci Technol.,* **40**, 1–11 (2007).

[63] Gawlik-Dziki U., "Effect of hydrothermal treatment on the antioxidant properties of broccoli (*Brassica oleracea* var. *botrytis italica*) florets", *Food Chem.,* **109**, 393–401(2008).

[64] De Pascale S., Maggio A., Pernice R., Fogliano V., Barbieri G., "Sulphur fertilization may improve the nutritional value of *Brassica rapa* L. subsp *sylvestris", Eur J Agron.,* **26**, 418–424 (2007).

[65] Fernandes F, Valentao P., Sousa C., Pereira J.A., Seabra R.M., Andrade P.B., "Chemical and antioxidative assessment of dietary turnip (*Brassica rapa* var. *rapa* L.)". *Food Chem.,* **105**, 1003–1010 (2007).

[66] Ferreres F., Sousa C., Vrchovska V., Valentao P., Pereira J.A., Seabra R.M., Andrade P.B., "Chemical composition and antioxidant activity of tron- chuda cabbage internal leaves". *Eur Food Res Tech.,* **222**, 88–98 (2006).

[67] Ferreres F., Fernandes F., Sousa C., Valentao P., Pereira J.A., Andrade P.B., "Metabolic and bioactivity insights into *Brassica oleracea* var. *ace- phala", J Agric Food Chem.,* **57**, 8884–8892 (2009).

[68] Kim D.O., Padilla-Zakour O.I., Griffiths P.D., "Flavonoids and antioxidant capacity of various cabbage genotypes at juvenile stage", *J Food Sci.,* **69**, C685–C689 (2004).

[69] Samec D., Piljac-Zagarac J., Bogovic M., Habjanic K., Gruz J., "Antioxidant potency of white (*Brassica oleracea* L. var. *capitata*) and Chinese (*Brassica rapa* L. var. *pekinensis* (Lour.)) cabbage: The influence of development stage, cultivar choice and seed selection", *Sci Hortic.,* **128**, 78–83 (2011).

[70] Nilsson J., Olsson K., Engqvist G., Ekvall J., Olsson M., Nyman M., Akesson B., "Variation in the content of GLS, hydroxycinnamic acids, carotenoids, total antioxidant capacity and low-molecular-weight carbohydrates in *Brassica* vegetables", *J Sci Food Agr.,* **86**, 528–538 (2006).

[71] Ciska E., Drabin N.,Narwojsz A., Honke J., "Stability of glucosinolates and glucosinolate degradation products during storage of boiled white cabbage", Food Chem., **203**, 340–347 (2016).

[72] Sikora E., Cieslik E., Leszczynska T., Filipiak-Florkiewicz A., Pisulewski P.M., "The antioxidant activity of selected cruciferous vegetables subjected to aquathermal processing", *Food Chem.,* **107**, 55–59(2008).

[73] Jagdish S., Upadhyay A.K., Singh S., Rai M., "Total phenolics content and free radical scavenging activity of *Brassica* vegetables", *J Fd Sci Technol.*, 595–597(2009).

[74] Stefania De P., Albino M., Rita P., Vincenzo F., Giancarlo B., "Sulphur fertilization may improve the nutritional value of *BrassicaRPMO* rapa L. subsp.*sylvestris*", *Eur J Agron.,* **26**, 418–424 (2007).

[75] Li J., Zhu Z., Gerendas J., "Effects of nitrogen and sulfur on total phenolics and antioxidant activity in two genotypes of leaf mustard", *J Plant Nutr.,* **31**, 1642–1655 (2008).

[76] Vallejo F., Tomas-Barberan F.A., Garcia-Viguera C., "Effect of climatic and sulphur fertilization conditions, on phenolic compounds and vitamin C, in the inflorescences of eight broccoli cultivars", *Eur Food Res Technol.,* **216**, 395–401 (2003).

[77] Young J.E., Zhao X., Carey E.E., Welti R., Yang S.S., Wang W.Q., "Phytochemical phenolics in organically grown vegetables", *Mol Nutr Food Res.,* **49**, 1136–1142 (2005).

[78] Mithen R.F., Dekker M., Verkerk R., Rabot S., Johnson I.T., "The nutritional significance, biosynthesis and bioavailability of glucosinolates in human foods". *J Sci Food Agric*., **80**, 967–984(2000).

[79] Goodrich R.M., Anderson J.L., Stoewsand G.S., "Glucosinolate changes in blanched broccoli and Brussels sprouts", *J Food Proc Preserv.,* **13,** 275–280 (1989).

[80] Wathelet J.P., Mabon N., Foucart M., Marlier M., "Influence of blanching on the quality of brussels-sprouts (*Brassica-oleracea* L-CV *gemmifera*)", *Sci Aliments* **16,** 393–402 (1996).

[81] Lin C.H., Chang C.Y., "Textural change and antioxidant properties of broc- coli under different cooking treatments", *Food Chem.*, **90**, 9–15 (2005).

[82] Pellegrini N., Chiavaro E., Gardana C., Mazzeo T., Contino D., Gallo M., Riso P., Fogliano V., Porrini M., "Effect of different cooking methods on color, phytochemical concentration, and antioxidant capacity of raw and frozen *Brassica* vegetables", *J Agril Food Chem.*, **58**, 4310–4321(2010).

[83] Roy M.K., Juneja L.R., Isobe S., Tsushida T., "Steam processed broccoli (*Brassica oleracea*) has higher antioxidant activity in chemical and cellular assay systems", *Food Chem.*, **114**, 263–269 (2009).

[84] Miglio C., Chiavaro E., Visconti A., Fogliano V., Pellegrini N., "Effects of different cooking methods on nutritional and physicochemical characteristics of selected vegetables", *J Agril Food Chem.*, **56**, 139–147 (2008).

[85] Ng Z.X., Chai J.W., Kuppusamy U.R., "Customized cooking method improves total antioxidant activity in selected vegetables", *Intl J Fd Sci Nutr.*, **62**, 158–163 (2011).

[86] Turkmen N., Sari F., Velioglu Y.S., "The effect of cooking methods on total phenolics and antioxidant activity of selected green vegetables" *Food Chem.*, **93**, 713–718 (2005).

[87] Bognar A., "Comparative study of frying to other cooking technique influence on the nutritive value", *Grasas Aceites.*, **49**, 250–260 (1998).

[88] Schnepf M., Driskell J., "Sensory attributes and nutrient retention in selected vegetables prepared by conventional and microwave methods", *J Food Qual.*, **17**, 87–99 (1994).

[89] Volden J., Borge G.I.A., Bengtsson G.B., Hansen M., Thygesen I.E., Wicklund T., "Effect of thermal treatment on GLS and antioxidant-related parameters in red cabbage (*Brassica oleracea* L. ssp *capitata* f. *rubra*)", *Food Chem.*, **109**, 595–605(2008).

[90] Wachtel-Galor S., Wong K.W., Benzie I.F.F., "The effect of cooking on *Brassica* vegetables", *Food Chem.*, **110**, 706–710 (2008).

[91] Volden J., Borge G.I.A., Hansen M., Wicklund T., Bengtsson G.B., " Processing (blanching, boiling, steaming) effects on the content of GLS and antioxidant-related parameters in cauliflower (*Brassica oleracea* L. ssp *botrytis*)", *LWT - Food Sci Technol.*, **42,** 63–73(2009)

[92] Gawlik-Dziki U., "Effect of hydrothermal treatment on the antioxidant properties of broccoli (*Brassica oleracea* var. *botrytis italica*) florets", *Food Chem.*, **109**, 393–401 (2008).

[93] Zhang D.L., Hamauzu Y., "Phenolics, ascorbic acid, carotenoids and antioxidant activity of broccoli and their changes during conventional and microwave cooking", *Food Chem.*, **88**, 503–509 (2004).

[94] Vallejo F., Tomas-Barberan F.A., Garcia-Viguera C., "Phenolic compound contents in edible parts of broccoli inflorescences after domestic cooking", *J Sci Food Agri.*, **83**, 1511–1516 (2003).

[95] Francisco M., Velasco P., Moreno D.A., Garcia-Viguera C., Cartea M.E., "Cooking methods of *Brassica rapa* affect the preservation of GLS, phenolics and vitamin C", *Food Res Int.*, **43**, 1455–1463 (2010).

[96] Sultana B., Anwar F., Iqbal S., "Effect of different cooking methods on the antioxidant activity of some vegetables from Pakistan", *Intl J Food Sci Technol.*, **43**, 560–567(2008).

[97] Gliszczynska-Swiglo A., Ciska E., Pawlak-Lemanska K., Chmielewski J., Borkowski T., Tyrakowska B., "Changes in the content of health-promoting compounds and antioxidant activity of broccoli after domestic processing", *Food Addit Contam.*, **23**, 1088–1098 (2006).

[98] Natella F., Belelli F., Ramberti A., Scaccini C., "Microwave and traditional cooking methods: Effect of cooking on antioxidant capacity and phenolic compounds content of seven vegetables", *J Food Biochem.*, **34**, 796–810 (2010).

[99] Costa M.L., Civello P.M., Chaves A.R., Martínez G.A., "Effect of hot air treatments on senescence and quality parameters of harvested broccoli (*Brassica oleracea* L var *italica*) heads", *J Sci Food and Agri.,* **85**, 1154–1160 (2005).

[100] Serrano M., Martínez-Romero D., Guillen F., Castillo S., Valero D., " Maintenance of broccoli quality and functional properties during cold storage as affected by modified atmosphere packaging", *Postharvest Biol Tech.,* **39**, 61–68 (2006).

[101] Lemoine M.L., Chaves A.R., Martinez G.A., "Influence of combined hot air and UV-C treatment on the antioxidant system of minimally processed broccoli (*Brassica oleracea* L. var. *italica*)", *LWT – Food Sci Technol.,* **43**, 1313–1319 (2010).

[102] Cefola M., Amodio M.L., Cornacchia R., Rinaldi R., Vanadia S., Colelli G., " Effect of atmosphere composition on the quality of ready-to-use broccoli raab (*Brassica rapa* L.)", *J Sci Food Agri.,* **90**, 789–797 (2010).

[103] Starzynska A., Leja M., Mareczek A., "Physiological changes in the antioxidant system of broccoli flower buds senescing during short-term storage, related to temperature and packaging", *Plant Sci.,* **165**, 1387–1395 (2003).

[104] Hagen S.F., Borge G.I.A., Solhaug K.A., Bengtsson G.B., "Effect of cold storage and harvest date on bioactive compounds in curly kale (*Brassica oleracea* L. var. *acephala*)", *Postharvest Biol Tech.,* **51**, 36–42 (2009).

[105] Viña S.Z., Olivera D.F., Marani C.M., Ferreyra R.M., Mugridge A., Chaves A.R., Mascheroni R.H., "Quality of Brussels sprouts (*Brassica oleracea* L. *gemmifera* DC) as affected by blanching method", *J Food Engg.,* **80**, 218–225 (2007).

[106] Mrkic V., Cocci E., Dalla Rosa M., Sacchetti G., "Effect of drying conditions on bioactive compounds and antioxidant activity of broccoli (*Brassica oleracea* L.)", *J Sci Food Agri.,* **86**, 1559–1566 (2006).

[107] Tanongkankit Y., Chiewchan N., Devahastin S., "Effect of processing on antioxidants and their activity in dietary fiber powder from cabbage outer leaves", *Dry Technol.,* **28**, 1063–1071 (2010).

[108] Jacobson A., Nielsen T., Sjoholm I., "Influence of temperature, modified atmosphere packaging, and heat treatment on aroma compounds in broccoli", *J Agric Food Chem.*, **52**, 1607–1614(2004).

[109] Kristal A.R., Lampe J.W., "Brassica vegetables and prostate cancer risk: A review of the epidemiological evidence", *Nutr Cancer.,* 42, 1–9 (2002).

[110] Jackson S.J.T., Singletary K.W., "Sulforaphane inhibits human MCF-7 mammary cancer cell mitotic progression and tubulin polymerization", *J Nutr.*, **134**, 2229–2236 (2004).

[111] Brandi G., Schiavano G.F., Zaffaroni N., De Marco C., Paiardini M. , Cervasi B., Magnani M., "Mechanisms of action and antiproliferative properties of *Brassica oleracea* juice in human breast cancer cell lines", *J Nutr.*, **135**, 1503–1509 (2005).

[112] Rose P., Whiteman M., Huang S.H., Halliwell B., Ong C.N., "betaPhenylethyl isothiocyanate-mediated apoptosis in hepatoma HepG2 cells", *Toxicol Appl Pharmacol.*, **66,** 437–438 (2005).

[113] Gill C.I.R., Haldar S., Porter S., Matthews S., Sullivan S., Coulter J., McGlynn H., Rowland I., "The effect of cruciferous and leguminous sprouts on genotoxicity, *in vitro* and *in vivo*", *Cancer Epidemiol Biomarkers Prev.,* **13**, 1199–1205 (2004).

[114] Wiseman A., "D. dietary anticancer isothiocyanates (ITC) in Brassica raise the reduced-glutathione barrier to DNA-damage in the colon?" *Trends Food Sci Technol.*, **16**, 215–216 (2005).

[115] Parnaud G., Li P., Cassar G., Rouimi P., Tulliez J., Combaret L., Gamet-Payrastre L., "Mechanism of sulforaphane-induced cell cycle arrest and apoptosis in human colon cancer cells", *Nutr Cancer.,* **48**, 198–206 (2004).

[116] Finley J.W., Davis C.D., Feng Y., "Selenium from high selenium broccoli protects rats from colon cancer", *J. Nutr.*, **130**, 2384–2389 (2000).

[117] Zhang Y., Talalay P., Cho C., Posner G.H., "A major inducer of anticarcinogenic protective enzymes from broccoli: Isolation and elucidation of structure", *Proc. Natl. Acad. Sci.*, **89**, 2399–2403 (1992).

[118] Giovannucci E., Rimm E.B., Liu Y., Stampfer M.J., Willett W.C., "A prospective study of cruciferous vegetables and prostate cancer". *Cancer Epidemiol Biomarkers Prev.,* **12**, 1403–1409 (2003).

[119] Canene-Adams K., Campbell J.K., Zaripheh S., Jeffery E.H., Erdman Jr. J.W., "The tomato as a functional food", *J Nutr.,* **135**, 1226–1230 (2005).

[120] Canene-Adams K., Clinton S.K., Erdman J.W., "Effects of dietary lycopene, tomato, and broccoli powders on the growth of Dunning R3327-H prostatic adenocarcinoma tumors *FASEB J.,* **19** , A79 (2005).

[121] Singh S.V., Srivastava S.K., Choi S., Lew K.L., Antosiewicz J., Xiao D., Zeng Y., Watkins S.C., Johnson C.S., Trump D.L., Lee Y.J., Xiao H., Herman-Antosiewicz A., "Sulforaphane-induced cell death in human prostate cancer cells is initiated by reactive oxygen species", *J Biol Chem.,* **280**, 19911–19924 (2005).

[122] Munday R., Munday C.M., "Selective induction of phase II enzymes in the urinary bladder of rats by allyl isothiocyanate, a compound derived from Brassica vegetables", *Nutr Cancer.,* **44**, 52–59 (2002).

[123] Tang L., Zhang Y., Dietary isothiocyanates inhibit the growth of human bladder carcinoma cells. *J Nutr.,* **134**, 2004–2010 (2004).

[124] Tawfiq N., Heaney R.K., Plumb J.A., Fenewick G.R., Musk S.S.R., Williamson G., "Dietary GLS as blocking agents against carcinogenesis: GLS breakdown products assessed by induction of quinine reductase activity in murinehepa1C1c 7 cells", *Carcinogenesis.,* **16**, 1191–1194 (1995).

[125] Fahey J.W., Zhang Y.S., Talalay P. "Broccoli sprouts: An exceptionally rich source of inducers of enzymes that protects against chemical carcinogens", *P Natl Acad Sci.,* USA **94**, 10367–10372(1997).

[126] Smith T.J., Guo Z., Thomas P.E., Chung F.L., Morse M.A., Elkind, K., Yang, C.S., "Metabolism of 4-(methylnitrosoamino)-1-(3, pyridyl)-1-butanone in mouse lung microsomes and its inhibition by isothiocyanates" *Cancer Res.,* **50**, 6817–6822 (1990).

[127] Conaway C.C., Jiao D., Chung F.L., "Inhibition of rat liver cytochrome p 450 isozymes by isothiocyanates and their conjugates: A structure activity relationship study" *Carcinogenesis.,* **17**, 2423–2427 (1996).

[128] Zhang Y., Kensler T.W., Cho C.G., Posner G.H., Talalay P., "Anti carcinogenic activities of sulforaphane and structurally related synthetic norbornyl isothiocyanate", *P Natl Acad Sci.,* USA **91**, 3147–3150 (1994).

[129] Zhang Y., Talalay P., "Anticarcinogenic activities of organic isothiocyanate: Chemistry and mechanisms", *Cancer Res.,* **54,** 1976–1981 (1994).

[130] Talalay P., Fahey J.W., Holtzclaw W.D., Prestera T., Zhang Y., "Chemo protection against cancer by phase 2 enzyme induction", *Toxicol Lett.,* **82**, 173–179 (1995).

[131] Khor T.O., Hu R., Shen G., Jeong W.S., Hebbar V., Chen C., Xu C., Nair S., Reddy B., Chada K., Kong A.N.T., "Pharmacogenomics of cancer chemo preventive isothiocyanate compound sulforaphane in the intestinal polyps of Apc Min/+ mice", *Biopharm Drug Dispos.,* **27**, 407–420 (2006).

[132] Jongen W.M.F., "GLS in *Brassica*: Occurrence and significance as cancer modulating agents", *P Nutr Soc.,* **55**, 433–446 (1996).

[133] Aggarwal B.B., Ichikawa H., "Molecular targets and anti cancer potential of indole-3-carbinol and its derivative, *Cell Cycle.,* **4**, 1201–1215 (2005).

[134] Ashok B.T., Chen Y., Liu X., Bradlow H.L., Mittelman A., Tiwari R.K., "Abrogation of strogen mediated cellular and biochemical effects by indole -3-carbinol", *Nutr Cancer.,* **41,** 180–187 (2001).

[135] Ashok B.T, Chen Y., Liu X., Garikapaty V.P., Seplovitz R., Tschorn J., Roy K., Mittelman A., Tiwari R.K., "Multiple molecular targets of indole-3-carbinol: A chemo preventive anti estrogen in breast cancer", *Eur J Cancer Prev.,* **11,** 86–93 (2002).

[136] Yuan F., Chend D.Z., Liu K., Sepkovic D.W., Bradlow H.L., Auborn K., "Anti estrogenic activities of indole-3-carbinol in cervical cell: Implication for prevention of cervical cancer", *Anticancer Res.,* **19,** 1673–1680 (1999).

[137] Eberhardt M.V., Kobira K., Keck A.S., Finley J.W., Juvik J.A., Jeffery E.H., " Correlations of chemical composition, antioxidant capacity and cellular oxidative stress in extracts from broccoli cultivars [abstract]", *FASEB J.*, **19**, A1475 (2005) .

[138] Suido H., Takeuchi A., Makino T., Tanaka T., "Serum cholesterol lowering effects of a broccoli and cabbage mixture in rats", *Proceedings of the XIIIth International Sympoium on athrosclerosis*, Kyoto, Japan, September 28–October 2, p. 238 (2003).

[139] Sesso H.D., Buring J.E., Gaziano J.M., Liu S., "Flavonoid intake and the risk of cardiovascular disease in women", *Am J Clin Nutr.,* **77**, 1400–1408 (2003).

[140] Galan M.V., Kishan A.A., Silverman A.L., "Oral broccoli sprouts for the treatment of *Helicobacter pylori infection*: A preliminary report", *Dig Dis Sci.,* **49,** 1088–90 (2004).

[141] Yokozawa T., Kim H.Y., Cho E.J., Choi J.S., Chung H.Y., "Antioxidant effects of isorhamnetin 3,7-di-O-beta-D-glucopyranoside isolated from mustard leaf (*Brassica juncea*) in rats with streptozotocin-induced diabetes", *J Agr Food Chem.,* **50**, 5490–5495 (2002).

[142] Heaney R.K., Fenwick G.R., "Methods in plant biochemistry", Waterman P.G. (*ed.*), Academic Press, London, pp. 531–550 (1993).

[143] Heaney R.K., Spinks E.A., Fenwick G.R., "Improved method for the determination of the total glucosinolate content of rapeseed by determination of enzymically released glucose", *Analyst.,* **113**, 1515–1518 (1988) .

[144] Kiddle, G., Bennett, R.N., Botting, N.P., Davidson, N.E., Robertson, A.A.B., Wallsgrove, R.M., "High-performance liquid chromatographic separation of natural and synthetic desulphoglucosinolates and their chemical validation by UV, NMR and chemical ionisation-MS methods", *Phytochem Anal.*, **12**, 226–242 (2001).

[145] Wang Q., Grubb C.D., Abel S., "Direct analysis of single leaf disks for chemopreventive glucosinolates", *Phytochem Anal.,* **13**, 152–157 (2002).

[146] Ettlinger M.G., Lundeen A.J., "The mustard oil of *Limnanthes douglasii* seed, m-methoxybenzyl isothiocyananthe", *J Am Chem Soc.,* **78**, 1952–1954 (1956).

[147] *Ettlinger M.G., Lundeen A.J.,* "First synthesis of a mustard oil glucoside: The enzymatic Lossen rearrangement", *J Am Chem Soc.*, **79**, 1764–1765 (1957).

[148] Daxenbichler M.E., Spencer G.F., Carlson D.G., Rose G.B., Brinker A.M., Powell R.G., "Glucosinolate composition of seeds from 297 species of wild plants", *Phytochem.,* **30**, 2623–2638 (1991).

[149] Schnug E., "Fine methode zur schnellen und einfachen bestimmung des gesamtglucosinolatgehaltes in gruenmasse und samen von kruziferen durch die quantitative analyse enzymatisch freisetzbaren sulfates", *Fat Sci Technol.*, **89**, 438–442 (1987).

[150] Underhill E.W. Kirkland D.F., "Gas chromatography of trimethylsilyl derivatives of glucosinolates", *J Chromatogr* A., **57**, 47–54 (1971).

[151] Thies W., "Quantitative gas liquid chromatography of glucosinolates on a microliter scale", *Fette Seifen Anstrichm.,* **78**, 231–234 (1976).

[152] Doorn van H.E., Holst van G.J., Kruk van der G.C., Ruijs N.C.M.E. Raaijmakers, Postma E., "Quantitative determination of the glucosinolates sinigrin and progoitrin by specific antibody ELISA assays in Brussels sprouts", *J Agric Food Chem.,* **46**, 793–800 (1998).

[153] Doorn van H.E., "Development of vegetables with improved consumer quality: A case study in Brussels sprouts", PhD Thesis, Wageningen Agricultural University, (1999).

[154] Thies W., "Isolation of sinigrin and glucotropaeolin from cruciferous seeds", *FETT Wissenschaft Technologie-Fat Science Technol.,* **90**, 311–314 (1988).

[155] Fenwick G.R., Eagles J., Self R., "The Fast Atom Bombardment Mass Spectra of Glucosinolates and Glucosinolate Mixtures", *Organic Mass Spectrosc.*, **17**, 544–546 (1982).

[156] West L.G., Meyer K.A., Balch B.A., Rossi F.J., Schultz M.R., Haas G.W., Glucoraphanin and 4-hydroxyglucobrassicin contents in seeds of 59 cultivars of broccoli, raab, kohlrabi, radish, cauliflower, brussels sprouts, kale, and cabbage. *J Agric Food Chem.,* **52**, 916–926 (2004).

[157] Font R., Río-Celestino M. Del, Cartea E., and Bailo A. De Haro., "Quantification of glucosinolates in leaves of leaf rape (*Brassica napusssp. pabularia*) by Near-infrared spectroscopy", *Phytochem.,* **66**, 175–185 (2005).

[158] Zhang Y., Wade K.L., Presteda T., Talalay P., "Quantitative determination of isothiocyanates, dithiocarbamates, carbon disulfide, and related thiocarbonyl compounds by cyclocondensation with 1,2-benzenedithiol", *Anal Biochem.,* **239**,160–167 (1996).

[159] Bennett R.N., Mellon F.A., Kroon P.A., "Screening crucifer seeds as sources of specific intact glucosinolates using ion-pair high-performance liquid chromatography negative ion electrospray mass spectrometry", *J Agric Food Chem.,* **52**, 428–438 (2004).

[160] Karcher A., Rassi El Z., "Capillary electrophoresis of glucosinolates and their degradation products", *Electrophoresis.,* **20**, 3181–3189 (1999).

[161] Tian Q., Rosselot R.A., Schwartz S.J., "Quantitative determination of intact glucosinolates in broccoli, broccoli sprouts, Brussels sprouts, and cauliflower by high-performance liquid chromatography-electrospray ionization-tandem mass spectrometry", *Anal Biochem.,* **343**, 93–99 (2005).

[162] Christensen B.W., Kjaer A., Ogaard-Madsen J., Olsen C.E., OlsenO., Sorensen H., "Mass-spectrometric characteristics of some pertrimethyl-silylated desulfoglucosinolates", *Tetrahedron.,* **38**, 353–357 (1982).

[163] West L., Tsui I., Haas G., "Single column approach for the liquid chromatographic separation of polar and non-polar glucosinolates from broccoli sprouts and seeds", *J. Chromatogr* A., **966**, 227–232 (2002).

[164] Helboe P., Olson O., Sorensen H., "Separation of glucocinolates by HPLC", *J Chromatogr.*, **197**, 199–205 (1980).

[165] Olsen O., Sùrensen H., "Isolation of glucosinolates and the identification of (a-L-rhamnopyranosyloxy)benzylglucosinolate from *Reseda odorata*", *Phytochem.,* **18**, 1547–1553 (1979).

[166] Truscott, R.J.W. Minchinton I.R., Burke D.G., Sang J.P., "A novel methoxyindole glucosinolate", *Biophys Res Commun.*, **107**, 1258–1264 (1982).

[167] Bertelli D., Plessi M., Braghiroli D., Monzani A., "Separation by solid phase extraction and quantification by reverse phase HPLC of sulforaphane in broccoli", *Food Chem.,* **63**, 417–21 (1998).

[168] Chiang W.C.K., Pusateri D.J., Leitz R.E.A., "Gas chromatography/mass spectrometry method for the determination of sulforaphane and sulforaphane nitrile in broccoli", *J Agric Food Chem.,* **46**, 1018–1021 (1998).

[169] Antolovich M., Prenzler P.D., Robards K., Ryan D., "Sample preparation in the determination of phenolic compounds in fruits", *Anal.,* **125**, 989 (2000).

10

Processing of Coriander: A Nutraceutical Perspective

RAMACHANDRA C.T.[1]*, SHIVABASAPPA K.[2] AND DATTATREYA M. KADAM[3]

ABSTRACT

Processing of coriander leaves, and seeds, physical properties of leaves and seeds. Different drying methods for leaves and seeds: solar drying, fixed bed drying, microwave drying and thin layer convective drying. Drying behaviour, drying kinetics and mathematical modeling of coriander. Effect of drying on quality parameters of leaves and seeds. Pre-treatments for drying and extraction: blanching and high hydrostatic pressure pre-treatment. Grinding of seeds: ambient grinding and cryogenic grinding. Extraction techniques-Essential oil extraction, extraction methods-Hydro-distillation method, steam distillation, super critical fluid extraction and microwave assisted hydro-distillation. Value added products: coriander puree, coriander chutney, powder and essential oil. Process flow charts and production processes. Packaging and storage of value added products: importance, requirements, different packaging materials, storage conditions, modified atmospheric packaging and changes during storage.

***Key words*:** Coriander processing, Nutraceutical, Bio-active compound, Drying, Grinding

I. INTRODUCTION

Coriander (*Coriandrum sativum* L.), a medicinal and culinary plant from the *Umbelliferae* (also known as *Apiaceae*) family, is one of the important

[1] Department of Processing and Food Engineering, University of Agricultural Sciences, Raichur, India.
[2] Department of Agricultural Engineering, Agriculture College, Hassan University of Agricultural Sciences, Bangalore, India.
[3] ICAR-Central Institute of Post-Harvest Engineering and Technology, Ludhiana, Punjab, India.
**Corresponding author* E-mail: ramachandract@gmail.com

spice crops[1]. The name "coriander" is derived from the Greek word "kopis" which means "bug". Coriander commonly called as cilantro/*dhania* is an important spice crop having a prime position in flavouring food. It is annual herb popularly grown and prized for its culinary and medicinal use. Coriander leaves and seeds are used for various purposes. The green leaves are used in foods as flavour, while powder grounded from the seed is used in spice products. It is commonly grown in India, Pakistan, Bangladesh, Russia, Central Europe, Morocco, and China[2]. India is the world's largest coriander producer, where the seeds and leaves of this spice are widely used for various food applications and essential oil extraction[3].

India is the biggest producer, consumer and exporter of coriander in the world with an annual production averaging around 3 lakh tonnes. Rajasthan (54%) and Madhya Pradesh (17%) are the two largest producing states in the country contributing over two-thirds to the country's total production in 2006-07. The other producers are Gujarat (6.9%), Assam (6.6%), Andhra Pradesh (3.5%), Karnataka (3.3%), Orissa (3.2%) and Tamil Nadu (2%). The production from Rajasthan and Madhya Pradesh are reported to be 1.55 and 0.50 lakh tonnes respectively. Coriander for seed cultivation is grown as rabi crop with sowing undertaken during October-November and new crop arrivals seen in February–March.

It is native to North Africa and Southwest Asia. Coriander seeds are used as whole seeds or powdered form for food seasoning and flavouring. The raw coriander seed that arrives in spot markets are generally cleaned and graded into various grades. Both whole seed and splits are used in preparing powders and oils. In India, most of the seeds are converted into powder and hardly any quantity goes for oil extraction. Seeds contain about 0.1-1.5% of oils. The crushed seed results in 97-98% of powder and the rest is wastage. The major buyers of coriander seeds are spice powder making industries located in the southern states of India like Tamil Nadu, Andhra Pradesh and around Delhi. India consumes about 75-80 lakh bags (3.0-3.2 lakh tonnes) annually and the southern states are major consumers with more than 50% of the total domestic consumption. Demand from spice-processing industries gains momentum from April and extends till June, which is also the peak arrival season (Karvy, Seasonal report on coriander seeds, 2008).

II. BRIEF DESCRIPTION OF THE CROP

Coriander is an annual herb and, according to the climatic conditions, is cultivated as a summer or winter annual crop. At flowering, the glabrous plant can reach heights between 20 cm and 140 cm. The germination is epigeal and the plant has a tap root. The stem is more or less erect and sympodial, monochasial-branched, sometimes with several side branches at the basal node. The coriandrum plant at the flowering stage has shown in Fig. 1.

Fig. 1: Coriandrum in flowering stage

Each branch finishes with an inflorescence. The colour of ribbed stem is green and sometimes turns to red or violet during the flowering period. The stem of the adult plant is hollow, and its basal parts can reach a diameter of up to 2 cm. The leaves alternate, and the first ones are often gathered in a rosette. The plant is diversifolious and the blade shape of the basal leaves is usually either undivided with three lobes, or tripinnatifid, while the leaves of the nodes following are to a higher degree pinnatifid. The higher the leaves are inserted, the more pinnate they are. Thus, the upper leaves are deeply incised with narrow lanceolate or even filiform-shaped blades. The lower leaves are stalked, while the petiole of the upper leaves is reduced to a small, nearly amplexicaul leaf sheath. The leaves are green or light green and their underside often shiny waxy. During the flowering period the leaves sometimes turn red or violet. They wither before the first fruits are ripe starting from the basal leaves. The inflorescence is a compound umbel. Sometimes there are one or two linear bracts. The umbel has two to eight primary rays, which are of different length, in such a way that the umbellets are located at the same level. Two, three or more bracteols carry the umbellets with five to twenty secondary rays.

Flowering starts with the primary umbel. In every umbel the peripheral umbellets, and in every umbellet the peripheral flowers are the first ones to flower. These flowers are protandrous. The central flowers of the umbellets are staminiferous or sometimes sterile. Coriander has an inferior ovary and the five calyx teeth surrounding the stylopodium are still visible in the ripe fruit. The five calyx teeth are of different length, as the petals in peripherally situated flowers. The flowers have five petals. The peripheral flowers of every umbellet are asymmetric, as the petals towards the outside of the umbellets are lengthened. The central flowers are circular, with small inflexed petals. The colour of the petals is pale pink or sometimes white.

Common Varieties of the Coriander

Annual/Perennial plant: It blooms and must be replanted yearly.

Part used: Fruit, whole plant

Habitats: Waste places and arable land, often by the sides of rivers.

Indian Varieties: J-16, J-214, K-45 and New Pusa,'CO-1', 'GAU-1', 'GAU-2', 'GAU-21', Karan, Rcr-14, RD-41, Sadhna (CS-4), Swathi (CS- 6), UD-20 and UD-21, Gujarat Coriander-1, Gujarat Coriander-2, RCr-41, UD-2, Rajendra Swati (RD-44), CS-287, CO-1, CO-2, CO-3, Sadhana, Swati and Sindhu.

Common Indian names (Vernacular names)

Bengali: *dhane, dhania*; Gujarati: *kothmiri, konphir, libdhane*; Kannada: *kothambri, kothmiri bija*; Kashmiri: *daaniwal, kothambalari*; Malayalam: *kothumpkalari bija, kothumpalari*; Marathi: *dhana, kothimber*; Oriya: *dhania*; Punjabi: *dhania*; Tamil: *kothamali*; Telugu: *dhaniyalu*

Chemical Composition

There exist very different uses of coriander and these are based on different parts of the plant. The traditional uses of the plant, which are based on the primary products, *i.e.*, the fruits and the green herb, are two-fold: medicinal and culinary. The chemical composition of coriander seed is given in Table 1.

Table 1: Chemical composition of coriander[4]

Sl. no.	*Components*	*Content (%)*
1.	Water	11.37
2.	Crude protein	11.49
3.	Fat	19.15
4.	Crude fibre	28.43
5.	Starch	10.53
6.	Pentosans	10.29
7.	Sugars	1.90
8.	Mineral constituents	4.98
9.	Essential oil	0.84

The essential oils and fatty oils of the coriander fruits are both used in industry, either separately or combination. After extraction of the essential oil, the fatty oil is obtained from the extraction residues either by pressing or by solvent extraction.

The essential oils and fatty acids of the fruits

The uses of coriander fruits are related to their chemical composition. The most important constituents are the essential oil and the fatty acids. The composition of the essential oil is shown in Table 2 and the composition of the fatty acid in Table 3.

Table 2: Essential oils in matured coriander[4]

Sl. no.	*Main components*	*Total essential oil (%)*
1.	Linalool	60–80
2.	Alpha-pinene	10.50
3.	Gamma-terpinene	09.00
4.	Geranylacetate	0.1–4.7
5.	Camphor	0.9–4.9
6.	Geraniol	1.2–4.6
7.	Terpinen-4-ol	3.0
8.	Alpha-terpineol	0.5
9.	Gamma-terpinene	1–8
10.	Limonene	0.5–4
11.	Alpha-pinene	0.2–8.5
12.	Camphene	1.4
13.	Myrcene	0.2–2
14.	Ketones	7–9
15.	Linalyl acetate	0-2.7

Table 3: Fatty acids in matured coriander[4]

Sl. no.	*Main components*	*Fatty acids (%)*
1.	Petroselinic acid C18:1	68.8
2.	Linoleic acid C18:2	16.6
3.	Oleic acid C18:1	7.5
4.	Palmitic acid C16:1	3.8

Minor components: Stearic acid, Vacceinic acid and Myristica acid

Maturity indices

Green colour turn to straw coloured. If harvesting is delayed– seeds shattered and fruits start to split. Plants are cut or pulled followed by piled in shade for drying to avoid grain shattering and loss of essential oil. After 2-3 days of shade drying, the grain is threshed, winnowed and dried. The moisture content is reduced from 2 to 6 per cent. The cleaned, dried produce is stored in gunny bags lined with white polythene.

Different Grades in Coriander

Coriander is classified into different grades based on quality of the produce. In general, coriander has been classified into grades like Badami, Eagle, Scooter, single parrot, double parrot, green medium, green extra and green special. After the harvesting of coriander, it is dried in sun; the excess drying leads to brown colour and fetch a low price. These brown colour seeds are called the Badami grade. The finer quality is called green and it is traded at a premium to all other grades. However, the Badami grade has the highest market share, constituting 50% of the total produce. The approximate price difference between these two largest grades is Rs. 2 per kg and this difference has been narrowing in the last few years due to higher prices (Karvy, Seasonal report on coriander seeds, 2008).

Pathology

Coriander is affected by some specialized and some general pathogens. An important disease is bacterial blight, caused by *Pseudomonas syringae* pv. *coriandricola*, which causes necroses and reduced number of fruits in the umbels and leaf spots. Since the application of antibiotic chemicals should be avoided, and the disease is systemic and thus spread by seeds. The disease has already been described by Palamarja and Chotina[5] who noted that fruits harvested from affected plants will redistribute the bacteriose. Pollinating insects carry the bacteria and spread the disease, which can then become systemic. Awareness of the possibility of seed transmission is therefore needed when coriander fruits are transferred. The bacterium can persist in the field on volunteer plants, which can be other species of cultivated plants or weeds. Heat treatment of the ripe fruits before sowing (for 6 days = 144 hours at 65°C with dry heat) and other disinfection treatments have been developed[6].

Other bacterial diseases affecting coriander caused by *Erwinia* and *Xanthomonas* species, which are not of great importance. Several fungal diseases affect coriander[7]. In Europe, *Ramularia coriandri* is of particular importance, and resistance breeding of this fungus has been carried out in the former Soviet Union[8] *Fusarium oxysporum* causes coriander wilt[9], and resistance screenings of this disease were done by Prakasam *et al.*[10]. Stem galls are caused by *Protomyces macrosporus*[11]. Grain mould diseases caused by *Helminthosporium* spp., *Fusarium* spp., *Curvularia* spp. and *Alternaria* spp. were studied by Rajan *et al.*[12]. The symptoms of several fungal diseases of the green plant are described by Palamarja and Chotina[4]. Two virus diseases are mentioned for coriander. Alfalfa mosaic virus was isolated from coriander in Yemen[13] and the transfer and infection of celery mosaic virus in Brazil has been described[14].

A specific insect pest of coriander is *Systole coriandri*, whose larva damages the fruits and survives in them. Seeds of the affected fruits must not be used for sowing. A detailed account of this pest is given by Nieto *et al.*[15], who describe the biology of several mites, moths and beetles. Serious losses of fruits caused by the chalcid fly *Systole albipennis* in Ethiopia, and reports the possibility of their biological control by the parasites *Tetrastichus* spp. and *Liondontomerus* spp[16]. Smallfield *et al.*[17] screened coriander for resistance to this pest in India. The aphids *Myzus persicae* and *Hyadaphis coriandri* have been mentioned by Araujo[18] and Jain and Yadav[19] for coriander.

III. USES OF THE CORIANDER

Gripe water: At times, coriander has been used instead of dill to make gripe water[20].

Liqueurs: Coriander is used in the preparation of the liqueurs Chartreuse and Benedictine, and was added in the early days because of its digestive properties (http://hubpages.com/hub/Condiments-as-Herbal-Medicines).

Anorexia: Coriander acts as a stimulant for the appetite and because of this is sometimes used in the treatment of anorexia nervosa. http://www.ciao.co.uk/sr/qbenedictine+ liqueur

Indigestion: Coriander Infusion - To make an infusion of coriander pour a cup of boiling water onto 1 teaspoon of bruised *coriander seeds* and leave to infuse for five minutes. Drink before meals in order to relieve indigestion and flatulence.

Acne: *Coriander extract* in a lotion along with turmeric, has been successfully used as a treatment for acne. http://www.liveandfeel.com/medicinalplants /coriander.html

Colic: A breastfeeding mother can drink coriander infusion to help relieve colic in the baby, since the anti-spasmodic properties absorb easily into breast milk. http://www.acne-reatmentblog.com/acnetreatment/ dealing-with-dry-facial-skin.html

Epistaxis: Nasal drops of green coriander act as a haemostat and thus stop bleeding[21].

Period pain: Coriander infusion is useful for relieving period pain and was once employed to help during child birth.

Conjunctivitis: Either juice or decoction of green coriander is put in eyes[21].

Gargle: Fresh juice of leaves is used as a gargle in sore throat and stomatitis[21].

Culinary and Medicinal Use of Coriander:

- Coriander seeds and seed powder taste equally delicious with either meat or sweet puddings.
- Coriander leaves make a delightful addition to salads and can be added to curry or stews.
- Chopped coriander leaves are often used as garnish on culinary dishes.

- Coriander tea is recommended for patients suffering from kidney problems. For preparing coriander tea, coriander seeds are allowed to boil in water for a few minutes and the decoction is taken as a medicine. Coriander tea also helps to cure mouth ulcers and swellings. Regular intake of coriander tea also helps to lower the blood cholesterol levels. In case of Diarrhoea, coriander seeds is soaked in water overnight and then taken along with buttermilk early in the morning.
- The juice of coriander is also used as an Ayurvedic medicine for treating nausea, and morning sickness. It is also used in the treatment of colitis and some of the liver disorders. Coriander seeds also help to reduce acid peptic disease and it is also used as Ayurvedic medicine in the treatment of dysentery. Coriander seeds also help to reduce body fever by inducing urination. Coriander is used along with other herbs such as dry ginger, helps to relieve respiratory tract infections and cough.
- Some of the Ayurvedic medicines also use coriander in the treatment of Typhoid fevers and in the treatment of menorrhagia, which is profuse bleeding during menses. The decoction of coriander seeds is used for this treatment. A mild decoction of coriander seed is also used as eyewash to reduce irritation and burning sensation.
- Coriander leaves along with Asafoetida, rock salt and cumin seed are mixed with buttermilk and taken after meals, helps to improve digestion and prevents abdominal distension.
- Coriander is used as Digestive, Astringent, Liver stimulant, Anthelmintic, Diuretric, Dyspeptic, Anti pyretic, Anti-inflammatory, Stomatitis, haemostat, Conjuctivitis, Conjunctivitis Headache, Tonic for majjadhatu, Appetizer. The fruit extract of coriander inhibit mycelial growth of Pythium *aphanidermatum*. The essential oil of Coriander exhibit strong antifungal activity at very low concentration. The plant can have a narcotic effect if it is eaten in very large quantities[22].

IV. MEDICINAL AND PHARMACOLOGICAL PROPERTIES OF CORIANDER

The antioxidant properties have been attributed to flavonoids, phenolics, and diterpens, which have the ability to scavenge free radicals, donate hydrogen atoms or electron, or chelate metal cations[23]. Some herbs have antioxidant activity in vegetable oils and lard, but others are effective in emulsion, water, sausage, or minced meat. For example, rosemary and sage are effective antioxidants in lard[24], but their activity is relatively low in oil-in-water emulsion as compared to clove.

Herbs and spices can be natural inhibitors of microbial growth and spoilage of foods. Their inhibitory properties have been ascribed to the naturally present phenols and phenolics, coumarins, terpenoids, and alkaloids[25,26]. The phenolics break down bacterial cell membranes, resulting in leakages of intracellular components or destroying bacterial enzyme system. It is reported that the gram-positive bacteria are more susceptible to phenolics than gram-negative bacteria as the phospholipid membranes of the later are impermeable to lipophilic compounds providing barrier for antibacterial action[27, 28].

One to two teaspoons of coriander juice, added to refreshing buttermilk, is incredibly beneficial in curing digestive disorders such as indigestion, nausea, dysentery, hepatitis and ulcerative colitis. Such gleaming benefits of coriander are also helpful in curing typhoid fever (The Ayurvedic Pharmacopoeia of India). *Coriandrum sativum* has antimicrobial potential and were active against *Pseudomonas aeruginosa* selectively, while oil was ineffective against *Proteus vulgaris*[29].

The aqueous infusion and decoction of coriander did not show any antimicrobial activity against urinary pathogens as well as against *Candida albicans*[30,31]. Coriander seed extract could minimize the drug induced oxidative stress and protecting the system against its toxicity[32].

The methanolic crude extracts of *Coriandrum sativum* was screened for their free radical scavenging properties using ascorbic acid as standard antioxidant. The aerial parts of *Coriandrum sativum* showed lower inhibitory concentration values (4.1-7.9 mg/ml), efficiency concentration values (178-321 mg/mg DPPH) and higher values of free radical scavenging activity (0.31-0.51) as compared with their seeds[33, 34, 35]. *Coriander* can be useful in the control of diabetes. Treatment with coriander reduced the level of hyperglycaemia during the development of streptozotocin diabetes[36,37]. Coriander is also used externally to treat ulcers and rheumatism[38].

V. PROCESSING OF CORIANDER

Pre-treatments

1. *Blanching*

Blanching is done for the coriander leaves with distilled water at 90°C, where, after reaching the desired temperature, the samples are to be immersed for 30 seconds and then immediately taken out and cooled in tap water[39]. Blanching can increase drying rate due to the elimination of the cellular membrane resistance to water diffusion, and by decreasing the cellular wall resistance to water flow[40].

2. *Gamma radiation*

The gamma radiations are being used to rescue the coriander from attack of micro-organisms. Different dosage levels are standardized for different group of food materials. The dosage level recommended for the coriander is ranging from 5 to 12.5 kGy. The 10 kGy of dosage will destroy all the micro-organisms in pre-packaged samples and 5 kGy is sufficient for preventing from mold attack and found that there is no effect on spice quality[41].

Drying

For safe storage, the moisture content of the food must be reduced to a level below that at which micro-organisms can grow. Coriander is a seasoning herb, being widely used in several countries in the world. The great preference for coriander leaves in food preparation generates the waste of its stems and roots waste up to about 40% of the bunch of coriander sold in the markets, thus losing a niche in the production of dehydrated seasonings, consequently, an increase in income for farmers. However, an important observation about the coriander is the moisture content, which favours both its microbial and nutritional degradation, requiring immediate processing in order to increase its shelf-life and provide the final product with quality.

The herbs and spices contain about 75–80% moisture; drying is commonly employed to lower the moisture content to less than 15%. In addition, there is a substantial reduction in weight, reducing packaging, storage, and transportation costs. However, drying of herbs and spices can cause some changes in aroma and appearance but the effect of a particular drying on the volatile compounds is not easily predictable and would depend on the stability of aroma compounds and the spice source.

Drying is one of the oldest methods of food preservation and represents a very important aspect of food processing. Drying of food products is aimed at longer storage periods, lower packaging requirements and shipping weights[42,43,44].

Drying is used to safeguard the material to prevent deterioration and loss in commercial value. The technique benefits in the transformation of the product, adding value to it and originating a new option in the market[39]. Drying has the advantage of facilitating the conservation of foods, prolonging shelf-life and reducing the volume of the product, which facilitates and reduces costs of transportation, promoting physico-chemical stability and adding economical value to the final product[45].

The use of a drying technique without any prior knowledge and parameter control, such as temperature, air velocity and initial moisture that controls this process, can lead to complete nutritional and sensorial degradation of agricultural products submitted to the technique[46].

The foremost intent of the researchers in dehydration of the herbs is to examine the effect of convective drying parameters on its retention of heat sensitive properties of the dried material and its rehydration efficiency. Various studies have concluded that the temperature is the predominant factor affecting drying rate and quality. A few drying methods have been discussed here.

Sun drying

Open sun drying of the coriander leaves done by spreading the material on a black polyethylene sheet and also covering it with a similar sheet.

Solar drying

Domestic solar dryer (Fig. 2) is of natural circulation type, small in size and portable. There are three trays (55.5 cm × 15.3 cm × 3 cm) and air entered from the bottom and passes through all the trays before leaving from the top. The material used for construction is 4 mm thick transparent glass, mild steel sheet and angle iron. There is a provision for covering the trays with shading plates (56 cm × 15.8 cm × 2 cm) and drying is carried out both with covered and open trays. The overall size of dryer is 64.4 cm × 27.2 cm × 54.4 cm and it is inclined at 45° with the horizontal.

Fig. 2: Solar dryer

Forced air convection dryer

Forced air convection drying was done in a wooden, cuboid-shaped dryer (Fig. 3) (60 cm × 100 cm × 125 cm) with a 30 cm exhaust fan installed on one side and its top surface. Forced air convection dryer having three slots (26.67 cm × 33 cm) to carry three perforated trays. The material to be dried is spread in the trays and the dryer is to be placed in the sun. When operated, the fans pull the heated air through the trays and the hallow cabinet below the trays thus enhancing the air movement rate.

Thin layer convection dryer

Advances in drying technology of the herbs have led to the development of new processes for drying which retain a higher percentage of the original

Fig. 3: Forced air convection dryer

flavour, colour and nutritional value of the dried product. The dried herbs should be good enough in reconstitution properties in minutes by adding cold/hot water. Fig. 4. shows the schematic diagram of the thin layer drying unit. It consists of an air conditioning unit, fan, air straightener, sample tray, data acquisition system and instruments for measuring temperature, mass, relative humidity and airflow. The air was conditioned to the required temperature and humidity using the air conditioner and was forced to the drying chamber using the fan. Type T thermocouples were used to measure the temperatures to an accuracy of ± 0.01°C. The mass, relative humidity and airflow were measured to accuracies of 0.01 g, 1% and 0.1 m/s, respectively.

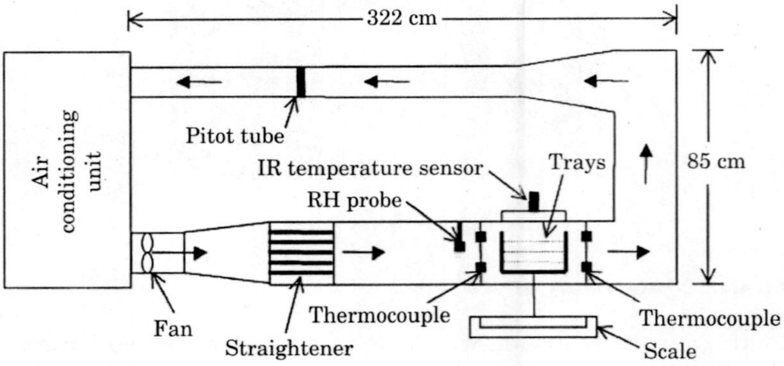

Fig 4: Schematic diagram of the thin layer (convection) dryer

Studies of thin layer (convection) drying of herbs and other food materials are enormous in number[47,48,49,50,51,52,53]. Three hundred minutes was required to dry coriander leaves to a final moisture content of 4.17% db using a convective dryer[54].

Microwave drying

Fig. 5 shows a combined microwave-convective dryer, used in drying the coriander samples. The microwave unit employs an inverter technology for continuous control of the microwave power output. The internal chamber dimensions were 242 mm (H) × 412 mm (W) × 426 mm (D). The microwave dryer has 10 different power levels with rated power ranging from 220 to 1000 W. It was operated at a frequency of 2450 MHz. The desired drying temperature was unable to be preset; instead, an appropriate power setting was chosen (Power level 3 corresponding to measured power of 295 W) which maintained the average product temperature as close as possible to 50°C. The microwave power output was determined according to the standard procedure prescribed by Schiffmann (1987). Temperature was logged using 4 fiber-optic probes connected to a 4-channel UMI Signal Controller (FISO Technologies Inc., Quebec, QB). The fiber-optic probe had a temperature measurement range of –50 to 250°C with an accuracy of ± 0.01°C.

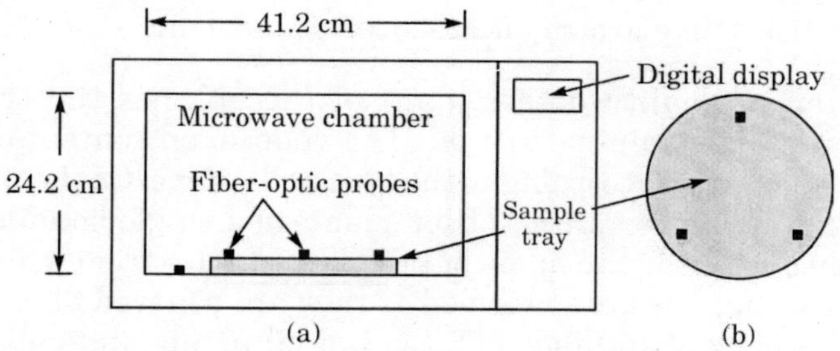

Fig 5: Microwave unit used in the dehydration of coriander foliage (a), and the placement of fiber-optic probes on the material tray (b).

The effect of microwave heating on the extraction of volatiles from the herbs and spices has been studied[55, 56]. Ahmed *et al.*[57] studied the drying characteristics of blanched coriander leaves. Air drying and its effects on the characteristics of coriander have been studied by researchers[54, 57, 58]. The entire coriander leaf drying process took place in the falling rate period, and there was no constant rate period. It was also noted that drying temperatures between 40-50°C are suitable for good quality finished products[58].

Extraction of Essential Oil

Plant oils and extracts have been used for a wide variety of purposes for many thousands of years. These purposes vary from the use of rosewood and cedar wood in perfumery, to flavoring drinks with lime, fennel or juniper berry oil, and the application of lemongrass oil for the preservation of stored food crops. In particular, the antimicrobial activity of plant oils and extracts has formed the basis of many applications, including raw and processed food preservation, pharmaceuticals, alternative medicine and natural therapies.

The spice extracts such as essential oils and oleoresins from leaves and flower tops of various herbal spices can be recovered using steam distillation, water and steam distillation, supercritical carbon dioxide extraction, and solvent extraction using low-boiling organic solvents. In steam distillation, the plant material is exposed to high temperatures, leading to the degradation of important components of essential oils, while extraction with organic solvents leaves residues of the solvent in spice extracts. Extraction using compressed carbon dioxide gas or supercritical fluid is an effective method as the extracts are free of solvent and there is no thermal and hydrolytic degradation of important components[59, 60, 61].

An essential oil is a concentrated, hydrophobic liquid containing volatile aromatic compounds from plants. Essential oils are also known as volatile, ethereal oils or aetherolea, or simply as the "oil of" the plant from which they were extracted, such as oil of clove. Oil is "essential" in the sense that it carries a distinctive aroma, or essence, of the plant.

The primary quality determinant of the spice is the content and composition of its steam-volatile oil. The volatile oil content of the spice can vary considerably according to the type and sourcezQs that are isolated by a physical method or process from plants of a single botanical species. The oils normally bear the name of the plant species from which they are derived. Essential oils are so termed as they are believed to represent the very essence of odor and flavor. Essential oil plants and culinary herbs include a broad range of plant species that are used for their aromatic value as flavorings in foods and beverages and as fragrances in pharmaceutical and industrial products. Essential oils derive from aromatic plants of many genera distributed worldwide. The aromagram of coriander oil is given in Fig. 6 and the structure of the major compound identified in the essential oil of coriander is given in Fig. 7.

Essential oils are a great benefit to help protect our bodies and homes from this onslaught of pathogens. Immune system needs support and these essential oils can give the required endorsement[45]. The major volatile compounds of essential oil of coriander and their GC–O odour description is given in Table 4.

Methods of extraction

The method of producing essential oils from plant materials is given in Fig. 8 which includes the studies on extraction of essential oil from coriander with different processing methods.

Maceration

Maceration actually creates more of "infused oil" rather than an essential oil. Plant matter is soaked in vegetable oil, heated and strained at which point it can be used for massage. This method is not desirable because it changes the composition of oil.

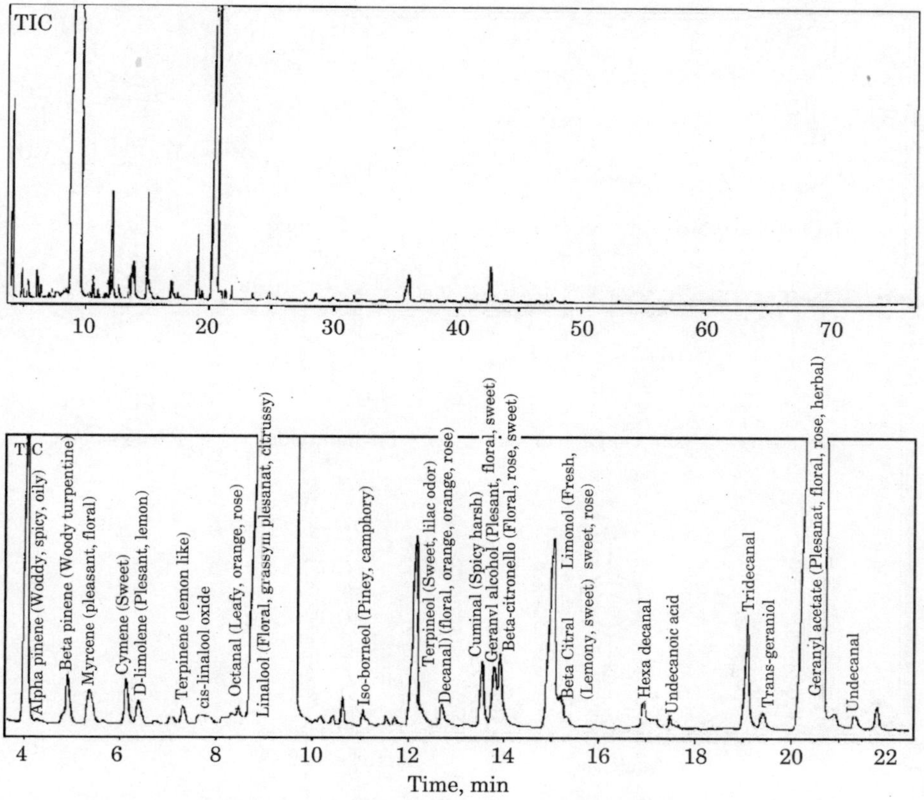

Fig. 6: Aromagram of coriander oil[63].

Hydrodistillation

In order to isolate essential oils by hydrodistillation (Fig. 9), the plant material is packed in a still and a sufficient quantity of water is added and brought to a boil; alternatively, live steam is injected into the plant charge. Due to the influence of hot water and steam, the essential oil is freed from the oil glands in the plant tissue. The vapour mixture of water and oil is condensed by indirect cooling with water. From the condenser, distillate flows into a separator, where oil separates automatically from the distillate water[65].

Disadvantages of water distillation

- Oil components like esters are sensitive to hydrolysis while others like acyclic monoterpene hydrocarbons and aldehydes are susceptible to polymerization (since the pH of water is often reduced during distillation, hydrolytic reactions are facilitated). Oxygenated components such as phenols have a tendency to dissolve in the still water, so their complete removal by distillation is not possible.

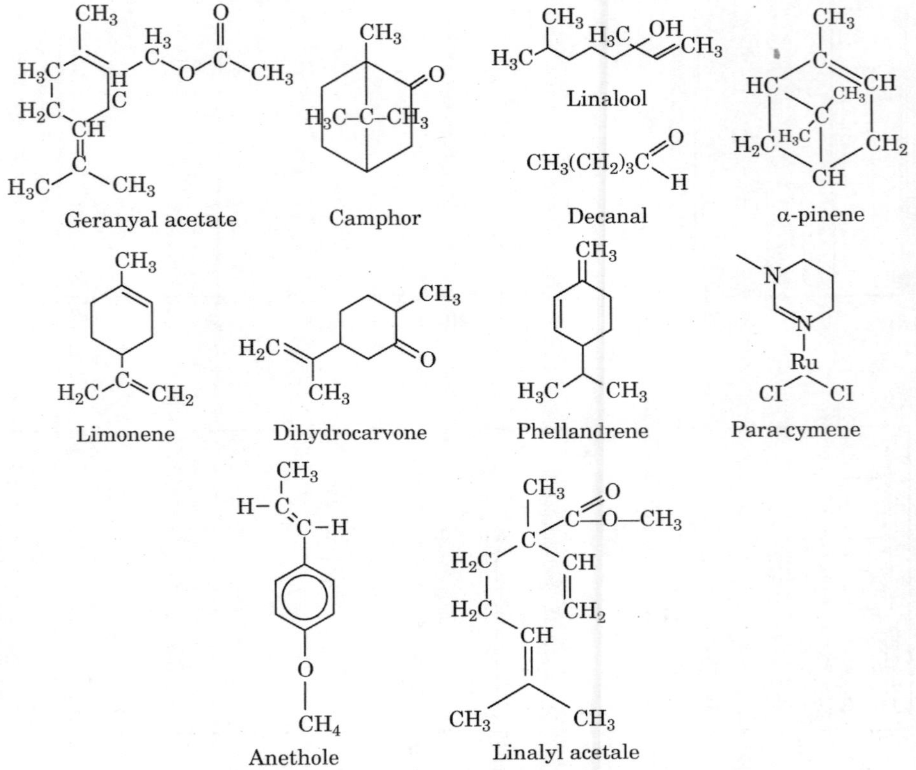

Fig. 7: Structures of the major compound identified in the essential oil of coriander (*Coriandrum sativum L.*) in previous investigations[64]

Table 4: Major volatile compounds of essential oil of *Coriandrum sativum* L. and their GC–O odour description[4]

Compound	*Concentration range (%)*	*Odour descriptors*
α-Pinene	2.36–23.23	Woody, spicy, oily
Linalool	57.52–75.14	Floral, grassy, pleasant, citrusy
Terpineol	0.08–5.37	Sweet, lilac odour
Cuminal	0.13–0.54	Spicy, harsh
Citronellol	0.65–1.67	Strong floral, rose, sweet like
Geraniol	0.45–2.35	Fresh, sweet, rose like
Geranyl acetate	8.95–24.51	Pleasant, floral rose, herbal

- As water distillation tends to be a small operation (operated by one or two persons), it takes a long time to accumulate much oil, so good quality oil is often mixed with bad quality oil.
- The distillation process is treated as an art by local distillers, who rarely try to optimize both oil yield and quality.
- Water distillation is a slower process than either water and steam distillation or direct steam distillation[65].

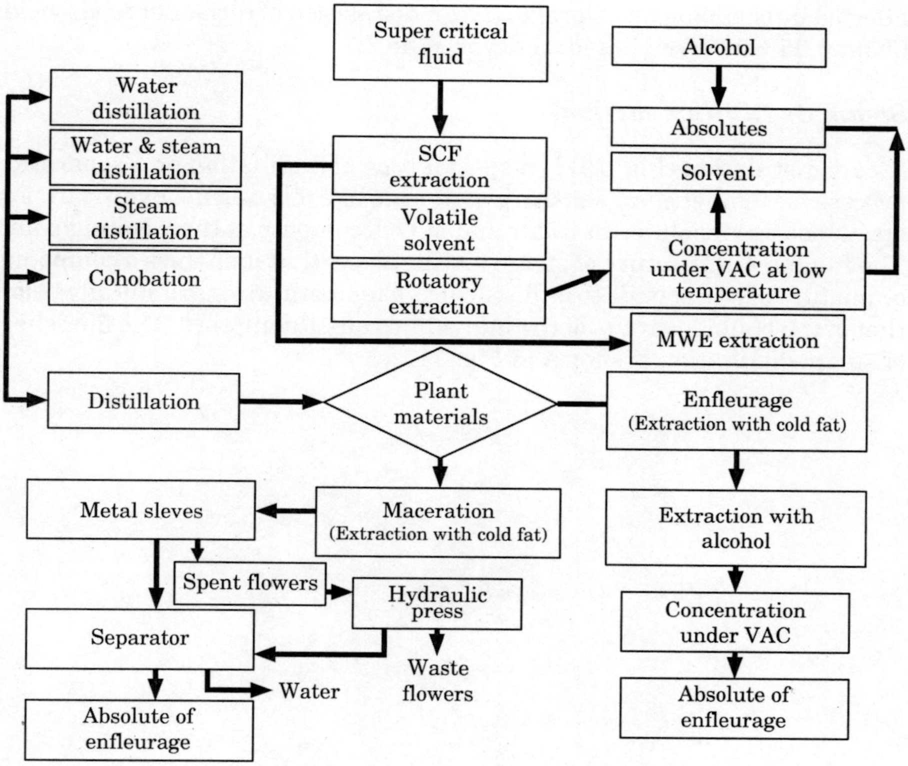

Fig. 8: Methods of producing essential oils from plant materials

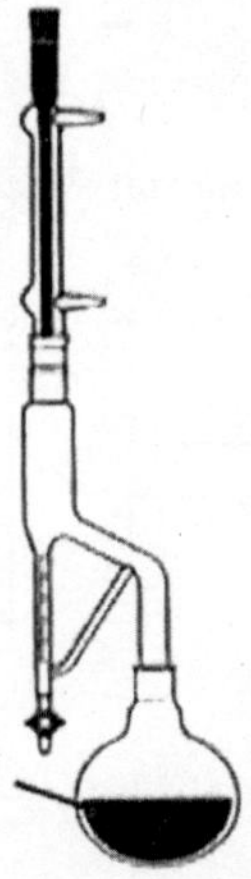

Fig. 9: Clevenger-type hydro-distillation apparatus

Chopping and immediate distillation gave an average increase in oil yield of 10.4 l/h, compared with unchopped herb[66]. However, oil yields of chopped herb fell after storage for more than 4 h. The oil composition was also affected by chopping and storage. Hydro-distillation of coriander seeds yields about 0.35% (w/w) oil based on dry weight[67].

Steam distillation method

Steam distillation (Fig. 10) is a special type of distillation or a separation process for temperature sensitive materials like oils, resins, hydrocarbons, etc. which are insoluble in water and may decompose at their boiling point. The fundamental nature of steam distillation is that it enables a compound or mixture of compounds to be distilled at a temperature substantially below that of the boiling point(s) of the individual constituent(s)[65]. The flow sheet of steam distillation is shown in Fig. 11.

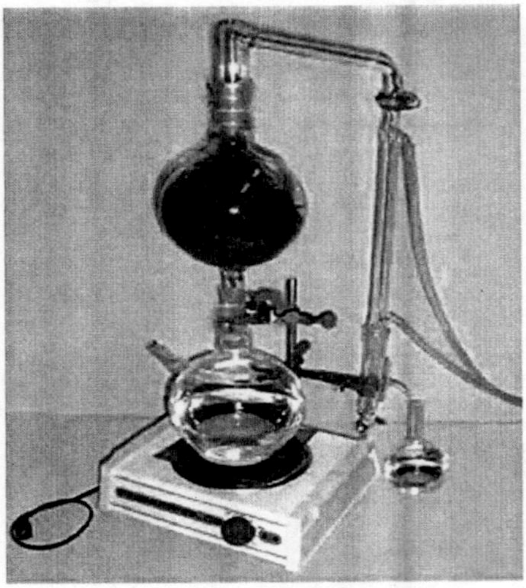

Fig. 10: Essential oil steam distiller

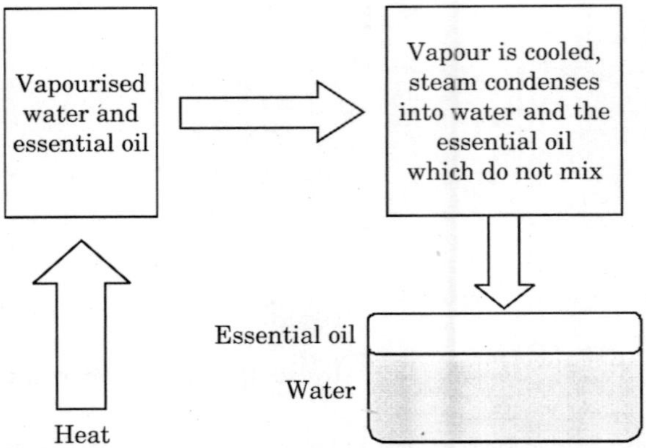

Fig. 11: Flow sheet of steam distillation

Essential oils contain substances with boiling points up to 200°C or higher temperatures. In the presence of steam or boiling water, however, these substances are volatilized at a temperature close to 100°C, at atmospheric pressure.

Solvent-extraction

The volatile isolates obtained commercially from aromatic plants are essential oils, concretes, absolutes, pomades and resinoids. Essential oils are isolated from plant material by distillation whereas other volatile isolates are obtained by solvent extraction.

The enriched solvent from the extractor is pumped into an evaporator for solvent recovery and the solvent content is reduced to about one-tenth of the original volume. The recovered solvent is pumped to the solvent tanks for further use. The concentrated material from the evaporator is pumped into a vacuum evaporator where the solvent is removed carefully under high vacuum and the recovered solvent is reused. The resultant concrete has an odour similar to but stronger than the raw material from which it was extracted[65].

In the solvent extraction method of essential oils recovery, an extracting unit is loaded with perforated trays of essential oil plant material and repeatedly washed with the solvent. A hydrocarbon solvent is used for extraction. All the extractable material from the plant is dissolved in the solvent. This includes highly volatile aroma molecules as well as non-aroma waxes and pigments. The extract is distilled to recover the solvent for future use. The remaining waxy mass is known as the concrete. The concentrated concretes are further processed to remove the waxy materials which dilute the pure essential oil. To prepare the absolute from the concrete, the waxy concrete is warmed and stirred with alcohol (ethanol). During the heating and stirring process, the concrete breaks up into minute globules. Since the aroma molecules are more soluble in alcohol than the waxes, an efficient separation of the two results. This is not considered as the best method for extraction as the solvents can leave a small amount of residue behind which could cause allergies and affect the immune system.

Most essential oil production is accomplished by extraction, using volatile solvents such as petroleum ether and hexane. The chief advantage of extraction over distillation is that uniform temperature (usually 50°C) can be maintained during the process, As a result, extracted oils have a more natural odour that is unmatched by distilled oils, which may have undergone chemical alteration by the high temperature. This feature is of considerable importance to the perfume industry; however, the established distillation method is of lower cost than the extraction process.

Super critical fluid extraction (SFE)

Supercritical fluid extraction (SFE) is a non conventional sample preparation method with intend of reduced use of organic solvents and increased sample throughput. The factors to consider include temperature, pressure, sample volume, analyte collection, modifier (co-solvent) addition, flow and pressure control, and restrictors. Generally, cylindrical extraction vessels are used for SFE and their performance is good beyond any doubt. The collection of the extracted analyte following SFE is another important step: significant analyte loss can occur during this step, leading the analyst to believe that the actual efficiency was poor.

There are many advantages to the use of CO_2 as the extracting fluid. In addition to its favourable physical properties, carbon dioxide is inexpensive, safe and abundant. The extraction procedure possesses distinct advantages

1. The extraction of constituents at low temperature, which strictly avoids damage from heat and some organic solvents,
2. No solvent residues
3. Environmental friendly extraction procedure.

The largest area of growth in the development of SFE has been the rapid expansion of its applications. SFE finds extensive application in the extraction of foods and fragrances, essential oils, and natural products[65].

Supercritical CO_2 extraction (SCO_2) involves heating of carbon dioxide to 31°C and pumped through the plant material at around 73 atm. pressure, as under these conditions; carbon dioxide behaves like a 'dense fog' or vapour. With release of the pressure in either process, the carbon dioxide escapes in its gaseous form, leaving the essential oil behind. The pictorial view of the super critical fluid extraction plant is shown in Fig. 12.

Optimum extraction conditions for coriander seeds were studied in the pressure range from 80 to 120 bar and for temperatures between 40 and 55°C. CO_2 densities were not tested since such densities produce low extraction selectivity; *i.e.*, the co-extraction of higher molecular weight compounds. To assess the optimum oil composition, GC-MS analysis of the extract was performed at each extraction condition. Based upon various studies, optimum extraction conditions for the extraction and fractionation of essential oil from coriander seeds are pressure 90 bar and temperature 50°C. Linalool is the principal component extracted. At 90 bar and 50°C, the linalool content in the essential oil extracted by SFE is greater than 72%. The yield obtained from hydrodistillation was 1.62% while from SFE at the optimum conditions was 1.16%[68].

Extraction of ripe fruits of *C. sativum* by steam distillation and by supercritical fluid extraction (SFE), using CO_2, was compared. The

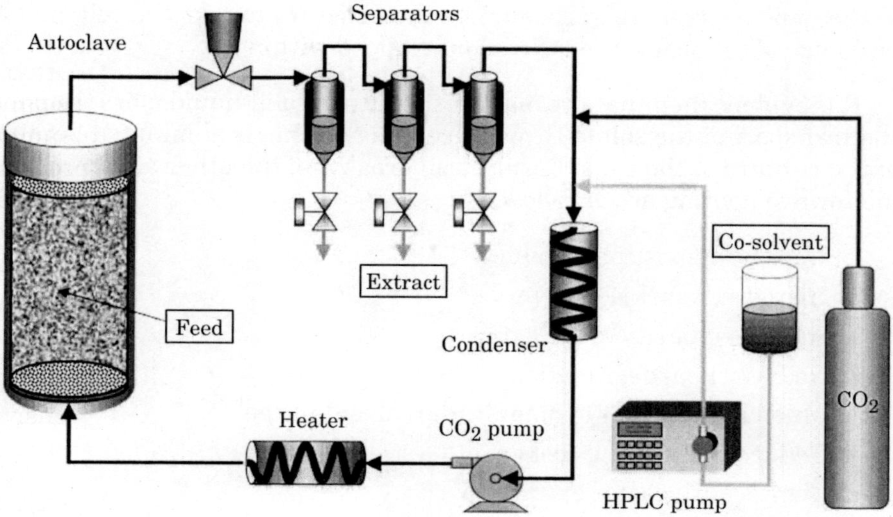

Fig. 12: Pictorial view of the super critical fluid extraction plant.

percentage composition of the 40 identified compounds was compared with the composition of commercial coriander oil extracted by hydrodistillation. The oil obtained by SFE showed some quantitative and qualitative differences, giving a superior aroma compared with that of the oil obtained by hydro-distillation[69].

Microwave-assisted extraction

Microwave radiation interacts with dipoles of polar and polarisable materials. The coupled forces of electric and magnetic components change directions rapidly (2450 MHz). Polar molecules try to orient in the changing field direction and hence get heated. In non-polar solvents without polarisable groups, the heating is poor (dielectric absorption only because of atomic and electronic polarizations). This thermal effect is practically instantaneous at the molecular level but limited to a small area and depth near the surface of the material. The rest of the material is heated by conduction. Thus, large particles or agglomerates of small particles cannot be heated uniformly, due to a major drawback of microwave which is heating. It may be possible to use high power sources to increase the depth of penetration. However, microwave radiation exhibits an exponential decay once inside a microwave-absorbing solid.

In microwave-assisted extraction (Fig. 13): 1) the heat of the microwave irradiation is directly transferred to the solid without absorption by the microwave-transparent solvent; 2) the intense heating in step 1 causes instantaneous heating of the residual microwave-absorbing moisture in the solid; 3) the heated moisture evaporates, creating a high vapour pressure;

4) the vapour pressure generated by the moisture breaks the cell; and 5) breakage of cell walls releases the oil trapped within it.

It is evident then that the main resistance to solid-liquid mass transfer, the transport of the solute through cell membrane, is eliminated because of the rupture of the cells. Besides cell breakage, the other advantages of microwave heating are as follows:

1. Improved "existing" products
2. Increased marker recovery
3. Increased purity of the extract
4. Reduced heat degradation
5. Much lower (order of magnitude) solvent usage[65]
6. Reduced processing costs
7. Significantly faster extraction
8. Much lower energy usage
9. Potential for "new" products

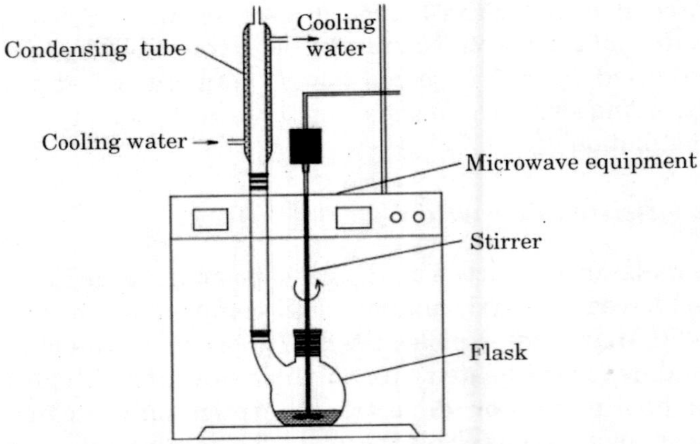

Fig. 13: Schematic diagram of microwave-assisted extraction

Value Added Products

This section deals with the value added products prepared form coriander *i.e.*, coriander puree and oil.

1. *Coriander puree*

Coriander leaves were separated from stems, washed in running tap water and blanched for 2 min in hot water at 90°C. The blanching time was based on preliminary trials that resulted in complete inactivation of peroxidase enzyme in coriander leaves. Coriander leaves were cooled immediately in chilled water, drained and comminuted in a laboratory size grinder fitted with a 14 mesh (1.19 mm aperture size) screen to obtain a puree of uniform size. The puree was stored at 0°C, immediately, until further use[70]. The process flow chart for preparation of coriander puree is given in Fig. 14.

Fresh coriander leaves
↓
Washing
↓
Blanching
↓
Cooling
↓
Grinding
↓
Storage

Fig. 14: Process flow chart for preparation of coriander puree

2. *Coriander Oil*

Seeds of the coriander plants are harvested from healthy, well-grown, two-year-old plants. Freshly harvested seeds (300 g) were grounded. The grounded seeds were subjected to hydrodistillation using a modified Clevenger-type glass apparatus for 4 hours for isolation of oils. The yield of an essential oil obtained from the seeds is 0.42%. The oil samples were stored at 0°C in airtight containers after drying them over anhydrous sodium sulphate and filtered before going to GC-MS analyses for chemical components[71]. The process flow chart for preparation of coriander oil is given in Fig. 15.

Coriander seeds
↓
Grinding
↓
Hydro-distillation
↓
Packaging
↓
Storage

Fig. 15: Process flow chart for preparation of coriander oil

VI. PACKAGING AND STORAGE OF CORIANDER

In order to maintain the quality of the spices during handling, transportation, storage and distribution, the packaging material to be used is to be selected with care, keeping in mind the functional as well as the marketing requirements. The packaging requirements for spices like coriander powder etc, in general are listed below:

- To protect the product from spillage and spoilage.
- To provide protection against atmospheric factors such as light, heat, humidity and oxygen. The selected packaging materials should have high water vapour and oxygen barriers.
- The packaging material should have a high barrier property to prevent aroma/flavour losses and ingress of external odour.
- The volatile oil present in the spice product has a tendency to react with the inner/ contact layer of the packaging material, at times leading to a greasy and messy package with smudging of the printed matter. The packaging material should therefore be grease and oil resistant and compatible with the product.

The flexible pouches are generally laminates. Some of the commonly used laminates are:

- Polyester/metallised polyester/LDPE
- BOPP/LDPE
- BOPP/metallised polyester/LDPE
- Polyester/Al foil/LDPE

Polyester and BOPP based laminates are generally more popular for spice packaging due to certain advantageous characteristics of each of these two films. Polyester used for lamination is generally 10 or 12 μ thick. The film is highly transparent with excellent clarity, gloss and printability thus enhancing the sales appeal. The film has very low moisture and gas permeability and, therefore, ensures prolonged shelf-life of the contents with aroma, flavour and taste retention. The very high mechanical strength (tear, puncture, burst and flex) minimises damage to the contents during handling and transportation. The film is free from additives and, therefore, does not impart any odour or taint to the sensitive spice product that is packed.

Pre-cooling Conditions

With the exception of basil, herbs should be cooled to just above 0°C as soon as possible after harvest. Vacuum-cooling is recommended[72]. Basil should be cooled to no lower than 12°C.

Storage Conditions

Chervil, coriander, dill and savoury should be stored at 0°C and 95 to 100% RH. Postharvest life ranges from 1 week for chervil to 2 weeks for coriander and dill, and up to 3 weeks for savoury[73]. Basil should be stored at 12°C and 95 to 100% RH. At this temperature and RH, quality can be maintained for 2 weeks

A 5 to 10% O_2 + 4 to 6% CO_2 CA is only moderately beneficial for fresh herbs[74]. However, MAP lengthens the shelf-life of coriander[75].

Use of water sprinklers is acceptable. Other herbs should be displayed in refrigerated units.

Chervil, coriander, dill and savoury are not sensitive to chilling temperatures and should be stored as cold as possible without freezing.

Annual herbs produce very little ethylene, but are highly susceptible to ethylene exposure[76]. Symptoms of ethylene damage include yellowing and leaf abscission[77].

VII. REFERENCES

[1] Gupta K., Thakral K.K., Arora S.K., Wagle D.S., "Studies on growth, structural carbohydrates and phytate in coriander (*Coriandrum sativum*) during seed development", *J Sci Food Agric.*, **54**, 43–46 (1991).

[2] Bhuiyan M.N.I., Begum J., Sultana M., "Chemical composition of leaf and seed essential oil of *Coriandrum sativum* L. from Bangladesh", *Bangladesh J Pharmacol.*, **4**, 150–153 (2009).

[3] Coskuner Y., Karababa E., "Physical properties of coriander seeds (*Coriandrum sativum* L.)", *J Food Eng.*, **80**, 408–416 (2007).

[4] Diederichsen A., "Promoting the conservation and use of underutilized and neglected crops: Coriander (*Coriandrum sativum L.*)", *1st edition, International Plant Genetic Resources Institute IPGRI, Italy*, (1996).

[5] Palamarja N.S., Chotina A.A., "Koriandr. Gosudarstvennoe izdatel'stvo", *sel'skochozjaistvennoj literatury, Moskva*, (eds.1953)

[6] Toben H.M., Beerentrup H.M., Rudolph K., Von der Schulenburg H.G., "Koriander als Petroselinsäurelieferant für die industrielle Verwendung", *Stand der Züchtung und Bekämpfungsmöglichkeiten der, Bakteriellen Doldenwelke, Vortr. Pflanzenzücht.* **30**, 173–183 (1994).

[7] Toben H.M., "Die durch *Pseudomonas syringae* pv. *coriandricola* hervorgerufene Doldenwelke an Koriander, Charakterisierung des Erregers und Strategien zur Bekämpfung", *Cuvillier Verlag Göttingen.*, (1995).

[8] Singh G., Baswana K.S., "Screening of coriander germplasm against chalcid fly (*Systole albipennis*)" *Ann. Appl. Biol. Supplement: Tests of Agrochemicals and Cultivars*, **104**(5), 114–115 (1984).

[9] Sultana S., Ripa F.A., Hamid K., "Comparative antioxidant activity study of some commonly used spices in Bangladesh, Pak", *J Biol Sci.*, **13**, 340–343 (2010).

[10] Prakasam V., Vedamuthu P.G.B., Khader M.A., Jeyarajan R., "Screening coriander lines for wilt resistance", *S Indian Hortic.*, **35**, 258–259 (1987).

[11] Das A.M., "Studies on stem gall disease of coriander incited by *Protomyces macrospores*", pp. 80–81 *in Abstracts of papers presented at the second international symposium on plant pathology, 27 January - 3 February 1971. (Indian Agricultural Research Institute ed.), New Delhi* (1971).

[12] Rajan F.S., Vedamuthu P.G.B., Khader M.K., Jeyarajan R., "Screening coriander lines against grain mould disease", *S Indian Hortic.*, **38**, 168–169 (1990).

[13] Alhubaishi A.A., Walkey D.G.A., Webb M.J.W., Bolland C.J., Cook A.A., "A survey of horticultural plant virus diseases in the Yemen Arab Republic", *FAO Plant Prot Bull.*, **35**,135–143 (1987).

[14] Oliveira M.L., Kitajima E.W., "Biological properties of celery yellow mosaic virus (in Port.)", *Fitopatol Bras.*, **6**, 35–46 (1981).

[15] Nieto A., Castro M.A., Alzamora S.M., "Kinetics of moisture transfer during air-drying of blanched and/ or osmotically dehydrated mango", *Journal of Food Engineering*, **50**, 175–185 (2001).

[16] Jansen P.C.M., "*Coriandrum sativum* L.", *in* Spices, condiments and medicinal plants in Ethiopia, their taxonomy and agricultural significance. (College of Agriculture, Addis Ababa University, Ethiopia and the Agricultural University, Wageningen, The Netherlands, *eds.*), *Centre for Agricultural Publishing and Documentation, Wageningen*, pp. 56–67 (1981).

[17] Smallfield B.M., Perry N.B., Beauregard D.A., Foster L.M., Dodds K.G., "Effects of post harvest treatments on yield and composition of coriander herb oil", *Journal of Agricultural and Food Chemistry,* **42**(2), 354–359 (1994).

[18] Araujo de C.Z.D., "Occurrence of *Myzus persicae* (Sulzer, 1776) (Homoptera, Aphididae) on a crop of coriander *Coriandrum sativum* (Umbelliferae) (in Port.)", *Anais Soc Entomol Bras.*, **15**, 173–174 (1986).

[19] Jain P.C., Yadav C.P.S., "Relative susceptibility of coriander varieties to coriander aphid (*Hyadaphis coriandri* Das)", *Indian J Entomol.*, **50**, 535–536 (1988).

[20] Hiromi I., Kohara K., Li X.X., Kobayashi F., Yasuyoshi H., "Identification of (E,E)-2,4-Undecadienal from Coriander (*Coriandrum sativum* L.) as a highly effective deodorant compound against the offensive odor of porcine large intestine", *Journal of Agricultural and Food Chemistry,* **58**, 11014–11017 (2010).

[21] Kasra M., "Importance of Coriander (*Coriandrum sativum* L.) Between the Medicinal and Aromatic Plants", *Advances in Environmental Biology*, **4**, 433–436 (2010).

[22] Verma A., Pandeya S.N., Sanjay K., Yadav S., Styawan , S. Pankaj S., "A Review on *Coriandrum sativum* (Linn.): An Ayurvedic Medicinal Herb of Happiness", *J of Adv in Pharm and Healthcare Res.,* **1**(3), 28–48 (2011).

[23] Pietta P., Simonetti P., Mauri P., "Antioxidant activity of selected medicinal plants", *J Agric Food Chem.,* **46**(11), 4487–4490 (1998).

[24] Palitzsch A., Schulte H., Metzl F., Baas H., "Effect of natural spices, spice extracts, essential oils, extraction residues, and synthetic antioxidants on the decomposition of pork fat and model lipids. I. Effect of natural spices and spice extracts on pork fat", *Fleischwirtschaft.,* **49**, 1349–1354 (1969).

[25] Cowan M.M., "Plant products as microbial agents", *Clin Microbiol Rev.,* **12**(4), 564–582 (1999).

[26] Bergonzelli G.E., Donnicola D., Porta N., Theulaz I.E.C., "Essential oils as components of a diet-based approach to management of *Helicobacter* infection", *Antimicrob Agents Chemother.,* **47**(10), 3240–3246 (2003).

[27] Shiffmann R.F., "Microwave and dielectric drying. "*Handbook of Industrial Drying*", Mujumdar A.S. and Marcel Dekker A.S. (*eds.*), New York, (1987).

[28] Farag R.S., Daw Z.Y., Abo-raya S.H., "Influenc of some spice essential oils on *Aspergillus parasiticus* growth and production of aflatoxin in a synthetic medium", *J Food Sci.,* **54**, 54–74 (1989).

[29] Shelef L.A., Naglik O.A., Bogen D.W., "Sensitivity of some common food borne bacteria to the spice sage, rosemary and allspice", *J Food Sci.,* **45**, 1042–1044 (1980).

[30] Sadrollah R., Rehmanian M., Jahanbin R., Mohajeri F., Rezaei M.R., Solaimani B., "Diurnal changes in essential oil content of coriander (*Coriandrum sativum L.*) aerial parts from Iran", *Res J Biol Sci.*, **4**(3), 277–281 (2009).

[31] Nazia M., Chaudhry A., Perween T., "Bactericidal activity of black pepper, bay leaf, aniseed and coriander against oral isolates, Pak", *J Pharm Sci.*, **19**, 214–218,

[32] Naveen S., Farhath K., "Antioxidant potential of coriander seed extract and its amelioration of liver antioxidant enzymes by CCl_4 induced toxicity in rats, *IJPBA*, **1** (2010).

[33] Leelamma S., Chithra V., "*Coriandrum sativum* changes the levels of lipid peroxides and activity of antioxidant enzymes in experimental animal", *Indian J Biochem Biophys.*, **36**, 59–61 (1999).

[34] Shaw M., Meda V., Tabil Jr. L., Leduc P., "Development and trends in drying of herbs and specialty crops in Western Canada." *CSAE / SCGR* 2005 Meeting, CSAE/SCGR Paper No. 05-030 (2005).

[35] Mishra A., Prakash D., Monika B., "The aerial parts of *Coriandrum sativum*, spinach and fenugreek have higher values of anti-radical power compared with their seeds", *Int J Food Sci Nutr.*, **56**, 473–481 (2005).

[36] Ebtesam A., Suhaimi A.L., "Effect of *Coriandrum sativum* a common herbal medicine on endocrine and reproductive organ structure and function", *The International Journal of Alternative Medicine*, **7** (2009).

[37] Day C., Bailey C.J., Flatt P.R., Flatt S.K.S., "Certain traditional plant treatments for diabetes, namely agrimony, alfalfa, coriander, eucalyptus and juniper, can retard the development of streptozotocin diabetes in mice", *Diabetologia.*, **33**, 462–464 (1990).

[38] Hegi G., (*ed.*), "Illustrierte Flora von Mitteleuropa", *Vol. 5-2. J. F. Lehmanns Verlag, München.* pp. 1071–1074 (1926).

[39] Silva A.S., Almeida F., de A.C., Lima E.E., Silva F.L.H., Gomes J.P., "Drying kinetics of coriander (*Coriandrum sativum*) leaf and stem", *Cienc. Tecnol. Aliment,* **6**(1), 13–19 (2008).

[40] Parichat P., Rattanachaikunsopon P., "Potential of coriander (*Coriandrum sativum*) oil as a natural antimicrobial compound in controlling *Campylobacter jejuni* in raw meat", *Biosci Biotechnol Biochem.*, **74**, 31–35 (2010).

[41] Munasiri M.A., Parte M.N., Ghaneekar A.S., "Sterilization of ground prepackaged Indian spices by gamma irradiation", *J Food Sci.*, **52**, 823–824 (1987).

[42] Okos M.R., Narsimhan G., Singh R.K., Weitnauer A.C., "Food Dehydration" *In*: Heldman D.R. and Lund D.B. (*eds.*), *Handbook of Food Engineering,* New York, Marcel Dekker, (1992).

[43] Kadam D.M., Samuel D.V.K., Parsad R., "Optimisation of pre-treatments of solar dehydrated cauliflower", *J. Food Eng.,* **77**(3), 659–664 doi:10.1016/ j.jfoodeng.2005.07.027. Special Section: CHISA 2004 (pp. 379–471) (2006).

[44] Kadam D.M., Goyal R.K., Singh K.K., Manoj K.G., "Thin layer convective drying of mint leaves", *J Med Plant Res.*, **5**(2), 164–170 (2011).

[45] Akpinar E.K., Bicer Y., Yildiz C., "Thin layer drying of red pepper", *J Food Engg.,* **59**, 99–104 (2003).

[46] Verma A., Pandeya S.N., Sanjay K., Yadav S., Styawan , S. Pankaj S., "A Review on *Coriandrum sativum* (Linn.): An Ayurvedic Medicinal Herb of Happiness", *J of Adv in Pharm and Healthcare Res.,* **1**(3), 28–48 (2011).

[47] Li Y, Morey R.V., "Thin-layer drying rates and quality of cultivated American ginseng." *Transactions of the ASAE*, **30**(3), 842–847 (1987).

[48] Temple S.J., Temple C.M., van Boxtel A.J.B., Clifford M.N., "The effect of drying on black tea quality", *Journal of the Science of Food and Agriculture*, **81**(8), 764–772 (2001).

[49] Erenturk S., Gulaboglu M.S., Gultekin S., "The thin-layer drying characteristics of rosehip", *Biosystems Engineering*, **89**(2), 159–166 (2004).

[50] Demir V., Gunhan T., Yagcioglu A.K., Degirmencioglu A., "Mathematical modelling and the determination of some quality parameters of air-dried bay leaves", *Biosystems Engineering*, **88**(3), 325–335 (2004).

[51] Sabahat Perween, "Antimicrobial activities of *Emblica officinalis* and *Coriandrum sativum* against gram positive bacteria and *Candida albicans*", *Pak J Bot.*, **39**, 913–917 (2007).

[52] Akpinar E.K., Bicer Y., Cetinkaya F., "Modelling of thin layer drying of parsley leaves in a convective dryer and under open sun." *J Food Engg.*, **75**(3), 308–315 (2006).

[53] Doymaz I., "Thin-layer drying behaviour of mint leaves." *J Food Engg.*, **74**(3), 370–375 (2006).

[54] Kaur P., Kumar P., Arora S., Ghuman B.S., "Quality of dried coriander leaves as affected by pre-treatments and method of drying", *European Food Research and Technology*, **223**(2), 189–194 (2006).

[55] Balasundaram R., Meda V., "Utilization of microwaves in applied natural product chemistry", *CSAE/ASAE* 2003 *Annual Intersectional Meeting, ASAE Paper No. RR*03–0033 (2003).

[56] Kaufmann B., Christen P., "Recent extraction techniques for natural products: Microwave-assisted extraction and pressurized solvent extraction", *Phytochemical Analysis.*, **13**(2), 105–113 (2002).

[57] Ahmed J., Shivhare U.S., Singh G., "Drying characteristics and product quality of coriander leaves", *Food and Bioproducts Processing*, **79**(2), 103–106 (2001).

[58] Pande V.K., Sonune A.V., Philip S.K., "Solar drying of coriander and methi leaves", Solar drying of coriander and Methi leaves. *J Food Sci Technol.*, **37**(6), 592–595 (2000).

[59] Maroto M.C.D., Coello M.S.P., Cabuzedo M.D., "Supercritical carbon dioxide extraction of volatiles from spices: Comparison with simultaneous distillation-extraction" *J Chromatogr A.*, **947**, 23–29 (2002).

[60] Perakis C., Louli V., Mouglas K., "Supercritical fluid extraction of black pepper oil." *J Food Eng.*, **71**, 386–393 (2005).

[61] Araus K, Uquiche E, del Valle J.M., "Matrix effect in supercritical CO_2 extraction of essential oils from plant material", *J Food Eng.*, **92**, 438–447 (2009).

[62] Parthasarathy V.A., Bhageerathy C., Zachariah T.J., Chemistry of Spices *(ed.)*, CABI. ISBN 978-1-84593-420-0 (2008).

[63] Ramasamy R., Prakash M., Keshava K.B., "Aroma characterization of coriander (*Coriandrum sativum* L.) Oil samples", *Euro Fd Research and Technol.*, **225**(3), 367–337 (2007).

[64] Saltveit M.E., "A summary of CA and MA requirements and recommendations for harvested vegetables", *In: 7th Intl. Contr. Atmos. Res. Conf., Vol. 4, Vegetables and Ornamentals.* Univ. Calif., Davis CA, *Postharv Hort Series.*, **18**, 98–117 (1997).

[65] Sukhdev S.H., Suman P.S.K., Gennaro L., Rakesh D.D., (*eds.*), "Extraction Technologies for Medicinal and Aromatic Plants", International Centre for Science and High Technology, Trieste, (2008). https://www.unido.org/fileadmin/user_media/Publications/Pub_free/Extraction_technologies_for_medicinal_and_aromatic_plants.pdf

[66] Srivastava U.S., "Effect of interaction of factors on wilt of coriander caused by *Fusarium oxysporum* Schlecht ex Fr. f. *corianderii* Kulkarni, Nikam & Joshi", *Indian J Agric Sci.*, **42**, 618–620 (1972).

[67] Msaada K., Hosni K., Taarit M.B., Chahed T., Kchouk M.E., Marzouk B., "Changes on essential oil composition of coriander (*Coriandrum sativum* L.) fruits during three stages of maturity", *Food Chem.*, **102**, 1131–1134 (2007).

[68] Manuela Z., Gainer I., Daniela B., "Supercritical CO_2 extraction of essential oil from coriander fruits", *Anul XV (serie nouă)*, **2**, 79–83 (2006).

[69] Anitescu G., Doneanu C., Radulescu V., "Isolation of coriander oil: Comparison between steam distillation and supercritical CO_2 extraction", *Flavour and Fragrance Journal*, **12**(3), 173–176 (1997).

[70] Jasim A., Shivhare U.S., Singh P., "Colour kinetics and rheology of coriander leaf puree and storage characteristics of the paste", *Food Chem.*, **84**, 605–611 (2004).

[71] Nazrul I.B., Jaripa B., Mahbuba S., "Chemical composition of leaf and seed essential oil of *Coriandrum sativum* L. from Bangladesh", *J Pharmacol.*, **4**, 150–153 (2009).
[72] Aharoni N., Reuveni A., Dvir O., "Modified atmospheres in film packages delay senescence and decay of fresh herbs", *Acta Hort.*, **258**, 255–262 (1989).
[73] Aharoni N., Dvir O., Chalupowisz D., Aharon Z., "Coping with postharvest physiology of fresh culinary herbs", *Acta Hort.*, **344**, 69–78 (1993).
[74] Saumendu D.R., Sonika T., Anita N., Minakshi K., Niranjan S., Goutam K.J., "*In vitro* antibiotic activity of volatile oils of *Carum carvi* & *Coriandrum sativum*", *International J of Chemical and Analytical Sci.*, **1**(7), 149–150 (2010).
[75] Loaiza J., Cantwell M., "Postharvest physiology and quality of cilantro (*Coriandrum sativum* L.)", *Hort Science.*, **32**,104–107 (1997).
[76] Cantwell M., "Properties and recommended conditions for storage of fresh fruits and vegetables" *http://postharvest.ucdavis.edu.*, (1997).
[77] Cantwell M.I., Reid M.S., "Postharvest physiology and handling of fresh culinary herbs", *J Hervs Soices Medicinal Plants.*, **1**, 83–127 (1993).

11

Processing of Bitter Gourd (*Momordica charantia*)

SHARANAGOUDA H.[1*] AND ROOPA BAI R.S.[1]

ABSTRACT

Bitter gourd (Momordica charantia) is one of the most popular vegetable in Southeast Asia. Bitter gourd extracts possess antioxidant, antimicrobial, antiviral, antihepatotoxic and antiulcerogenic properties while also having the ability to lower blood sugar. It is mainly used as a hypoglycemic and antidiabetic agent. The bitter principle, for which the fruit is named, is due to the alkaloid momordicin. Bitter gourd fruits are a good source of carbohydrates, proteins, vitamins, and minerals and have the highest nutritive value among cucurbits. The bitter gourd is cooked with other vegetables, stuffed, stir-fried, or added in small quantities to beans and soups to provide a slightly bitter flavour. The value added products such as bitter gourd chips, Bitter gourd pickle, bitter gourd soft drink and health drink are also prepared from bitter gourd. The essential oil extracted from dried bitter gourd seeds consists of twenty-five components, representing 90.9% of the oil.

***Key words*:** Bitter gourd, Dried, Processing, Antioxidant, Antimicrobial, Antiviral, Components

INTRODUCTION

India has been bestowed with wide range of climate and physio-geographical conditions and is most suitable for growing various kinds of horticultural crops such as fruits, vegetables, flowers, nuts, spices and plantation crops.

[1] Department of Processing and Food Engineering, College of Agricultural Engineering, University of Agricultural Sciences, Raichur – 584104, Karnataka, India.
**Corresponding author* E-mail: drsharan.cae@gmail.com, roopa0301@gmail.com

The horticulture production has increased by 30% in the last five years. This has placed India among the foremost countries in horticulture production, just behind China. During 2012-13, its contribution in the world production of fruits and vegetables was 12.6% and 14.0% respectively. Total production of fruits during 2012-13 was 81.2 million tonnes while that of vegetables was 162 million tones[1].

Among the various types of vegetables, bitter gourd is grown extensively throughout India. In terms of area and production, Andhra Pradesh ranks first followed by Odissa, Bihar, Chhattisgarh, Madhya Pradesh, Assam and Mizoram, respectively. The total area under this crop during 2012-13 was 83.0 thousand hectares and the production was about 940.0 thousand metric tones[2].

Bitter gourd (*M. charantia*) is used as a hypoglycemic and antidiabetic agent[3]. The hypoglycemic effect of bitter gourd is equivalent to some of the antidiabetic drugs like tolbutamide. At least three different groups of constituents in bitter gourd have been reported to have hypoglycemic actions which have potential benefits in controlling diabetes mellitus, these include steroidal saponins known as charantin, insulin like peptides and alkaloids[4]. The hypoglycemic activity of the bitter gourd fruit has been shown in both spontaneous and chemically induced diabetes mellitus in experimental animals as well as in human patients[5,6].

The immature fruits called bitter gourd, bitter melon or balsam pear, are harvested at developmental stages up to seed hardening. The bitter principle, for which the fruit is named, is due to the alkaloid *momordicine*, not to *cucurbitacins* as in other members of the *cucurbitaceae*[7]. The bitter gourd is also important for its medicinal properties[8,9].

Fully developed bitter gourd exhibit marked changes during fruit ripening. Increased respiration and ethylene production rates during storage of mature fruits indicate a climacteric behaviour. Mature bitter gourd have high ethylene production rates compared to those of other ripening cucurbit fruits including melons, and ripening can be induced by applying ethylene[10,11] or storing at ambient temperatures[12]. Carotenoids increase greatly during the ripening process, with the fruit changing from green to yellow, and the seed cavity becoming bright red[13]. Coincident with colour changes, the fruit pulp loses bitterness and becomes sweet and the fruit splits or ruptures at the blossom end[8].

Unpeeled immature bitter gourds are sliced for use as a vegetable in various Asian dishes. Good quality bitter gourd should have a fresh appearance and the peel should be of uniform green colour and free from visual defects. The developing fruit should be firm without excessive seed development, and free of defects such as decay and splitting.

The bitter gourd is cooked with other vegetables, stuffed, stir-fried, or added in small quantities to beans and soups to provide a slightly bitter flavour. However, for most food preparation, bitter gourds are blanched, parboiled, or soaked in salt water before cooking to reduce the bitter taste. In addition to frying or cooking (*e.g.,* for curries), the fruits can be dehydrated, pickled or canned. Fruits, flowers, and young shoots are also used as flavouring agents in various dishes. Young *Momordica* shoots and leaves are also cooked and eaten as leafy vegetables, and leaf and fruit extracts are used in the preparation of tea[14, 15].

ORIGIN

Bitter gourd (*Momordica charantia* L.) belongs to the family *Cucurbitaceae*. There are more than 60 species, of which only two are (*charantia* and *cochinchinensis*) important vegetables. It is believed to have originated in the tropics and is widely distributed in China, Malaysia, Indo-Pakistan and tropical Africa. *Momordica charantia* is a cucurbit vine native to Asia and now widely cultivated throughout the world for the immature fruits and sometimes for the tender leafy shoots or the ripe fruits[16].

SCOPE

At present, scientific information reported on chemical and biological properties of bitter gourd remains limited. Although bitter gourd has been widely used as health foods and folk medicine, some of its therapeutic claims still remain unknown. So there is considerable scope for the bitter gourd both as a medicine for its curative properties and as a fruit for its domestic usage.

IMPORTANCE

The fruit of bitter gourd is similar to other cucurbits in terms of nutritional value with the notable exceptions that it is much higher in folate and vitamin C content. The vine tips are an excellent source of vitamin A[17]. The medicinal value of the gourd in the treatment of infectious diseases and diabetes is attracting of scientists worldwide[17]. From the literatures, it is known to contain compounds such as momorcharins, momordenol, momordicilin, momordicins, momordicinin, momordin, momordolol, charantin, charine, cryptoxanthin, cucurbitins, cucurbitacins, cucurbitanes, cycloartenols, diosgenin, elaeostearic acids, erythrodiol, galacturonic acids, gentisic acid, goyaglycosides and goyasaponins[18], caffeic acid and ferulic acid[19], fisetin and isorhamnetin[20]. Its MAP30 protein exhibited anti-HIV and anti-tumor activities[21], momorcharins and momordins showed activities in inactivating ribosomes, momordin 1-c and oleanolic acid glycoside in altering gastrointestinal transit time and blood glucose[18, 22].

PHYSICO-CHEMICAL PROPERTIES OF BITTER GOURD

The physical properties of bitter gourd are given in Table 1. The average fruit length of bitter gourd was 15.74 cm (range 12.50-20.00 cm) with a breadth of 3.20 cm (3.00-3.60 cm) and diameter of 9.23 cm (7.50-10.80 cm). Each fruit on an average weighed 35.60 g (23.60-64.20 g) occupying a volume of 43.00 ml (30.00-80.00 ml) and with a mean bulk density of 0.90 g per ml (0.77-1.07 g/ml). The colour of the fruit was dark green and it contained 20 seeds (14.00-26.00) on an average[23].

Table 1: Physical properties of bitter gourd fruit[23]

Parameters	*Bitter gourd*
Length (cm)	12.50 – 20.00
Breadth (cm)	3.00 – 3.60
Diameter (cm)	7.50 – 10.80
Weight (g)	23.20 – 64.20
Volume (ml)	30.00 – 80.00
Bulk density (g/ml)	0.77 – 1.07
Seed (No.)	14 – 26
Colour	Dark green

Bitter gourd fruits are a good source of carbohydrates, proteins, vitamins, and minerals and have the highest nutritive value among cucurbits[24, 25]. The vitamin C content of Chinese bitter gourd varies significantly (440-780 mg/kg edible portion). Considerable variation in nutrients, including protein, carbohydrates, iron, zinc, calcium, magnesium, phosphorous, and ascorbic acid, has been observed in bitter gourd[26, 27]. Moreover, the crude protein content (11.420.9 g/kg) of bitter gourd fruits is higher than that of tomato and cucumber[28]. The nutritional composition of bitter gourd fruit is given in Table 2.

Table 2: Proximate and nutrient composition of bitter gourd [29]

Proximate value	*Quantity*
Moisture (g/100 g)	83.20
Carbohydrates (g/100 g)	10.60
Proteins (g/g)	2.10
Fibre (g/100 g)	1.70
Calcium (mg/100 g)	23.00
Phosphorus (mg/100 g)	38.00
Potassium (mg/100 g)	171.00
Sodium (mg/100 g)	2.40
Iron (mg/100 g)	2.00
Copper (mg/100 g)	0.19
Manganese (mg/100 g)	0.08
Zinc (mg/100 g)	0.46
B Carotene	126.00
Vitamin C	96.00

Mixture of steroidal saponins known as charantins, insulin-like peptides and alkaloids are the hypoglycemic chemicals of *Momordica charantia* and these chemicals are concentrated in fruits of *Momordica charantia*, therefore fruit of *M. charantia* has shown most effective hypoglycemic property. Major phytochemical in bitter gourd (*Momordica charantia*) fruit and their health benefits is given in Tables 3 and 4.

Momordicine

It is the primary bitter compound of bitter gourd; the structure is given in Fig. 1.

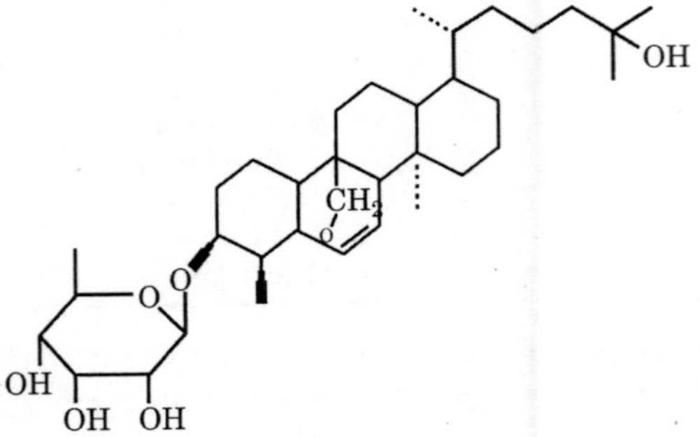

Fig. 1: Basic structure of momordicine, the primary bitter compound of bitter gourd

Steroids, Charantin

Charantin (Fig. 2) is one of the hypoglycemic compounds, which can be isolated from *Momordica charantia* fruit. It is a mixture of two compounds (1:1) sitosteryl glucoside ($C_{35}H_{60}O_6$) and stigmasteryl glucoside ($C_{35}H_{58}O_6$), both of which are steroidal saponins and exhibited hypoglycaemic effects when taken either orally or intravenously.

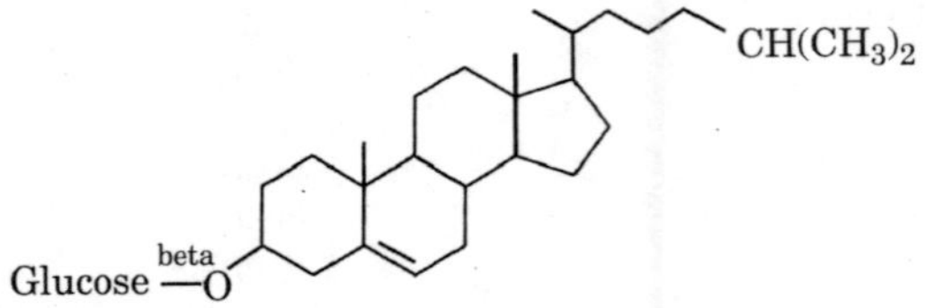

Fig. 2: Structure of charantin

Protein, P-insulin

Protein P-insulin is a polypeptide with molecular weight of about 11,000 Dalton and consists of 166 amino acids. Clinical study revealed that the polypeptide-p-$ZnCl_2$ produced blood sugar lowering effect. Khanna and Mohan[30] reported that besides the fruits, P insulin was also found in seeds and tissue cultures of *Momordica charantia.*

Alkaloids

Seeds of bitter gourd contained pyrimidine nucleoside vicine[31,32]. Vicine induced hypoglycaemia in rats, when administered intraperitoneally.

Table 3: Phytochemicals and constituents of bitter gourd (*Momordica charantia*)

Source	*Phyto chemicals*	*Reference*
Plant body	Momorcharins, momordenol, momordicilin, momordicins, momordicinin, momordin, momordolol, charantin, charine, cryptoxanthin, cucurbitins, cucurbitacins, cucurbitanes, cycloartenols, diosgenin, elaeostearic acids, erythrodiol, galacturonic acids, gentisic acid, goyaglycosides, goyasaponins, multiflorenol,	[33, 34, 35, 36, 37]
Plant body	Glycosides, saponins, alkaloids, fixed oils, triterpenes, proteins and steroids	[38]
Fruit	Momordicine, charantin, polypeptide- p insulin, ascorbigen, Amino acids – aspartic acid, serine, glutamic acid, threonine, glutamic acid, threonine, alanine, g-amino butyric acid and pipecolic acid, luteolin, Fatty acids – Lauric, myristic, palmitic, palmitoleic, stearic, oleic, linoleic, linolenic acid	[27, 39]
Seeds	Urease, Amino acids – valine, threonine methionine, isoleucine, leucine, phenylalanine, glutamic acid	[40]
Plant body	Momorcharins, momordenol, momordicilin, momordicins, momordicinin, momordin, momordolol, charantin, charine, cryptoxanthin, cucurbitins, cucurbitacins, cucurbitanes, cycloartenols, diosgenin, elaeostearic acids, erythrodiol, galacturonic acids, gentisic acid, goyaglycosides, goyasaponins, multiflorenol,	[33, 34, 35, 36, 37]
Plant body	Glycosides, saponins, alkaloids, fixed oils, triterpenes, proteins and steroids	[38]
Fruit	Momordicine, charantin, polypeptide- p insulin, ascorbigen, Amino acids – aspartic acid, serine, glutamic acid, threonine, glutamic acid, threonine, alanine, g-amino butyric acid and pipecolic acid, luteolin, Fatty acids – Lauric, myristic, palmitic, palmitoleic, stearic, oleic, linoleic, linolenic acid	[27, 39]
Seeds	Urease, Amino acids – valine, threonine methionine, isoleucine, leucine, phenylalanine, glutamic acid[40]	

Effect of Peeling and Cooking on Chemical Composition of Bitter Gourd

Nutritional importance of vegetables cannot be neglected in our daily meals. Vegetables are the major source of vitamins and minerals, but vegetable protein is poor in quality. Vitamins and minerals are the chief regulator in

Table 4: Major phytochemicals in bitter gourd (*Momordica charantia*) and their health benefits

Phytochemicals	*Plant parts*	*Usefulness*	*References*
β-momorcharin	Seeds	Glycoprotein that acts as midterm abortifacient	[41]
Vicine	Seeds	Hypoglycaemic glycoalkaloid	[42, 43]
Charatine	Fruits	Non nitrogenous substance having hypoglycaemic principle	[44]
Momordicosides A and B	Seeds	Triterpene glycosides that inhibit tumor growth	[45]
MAP30	Seeds, fruits	Basic protein that inhibits HIV	[46]
Polypeptide-p	Seeds, fruits	Hypoglycaemic peptide, called plant insulin	[47]
Phenols	Seeds	Antioxidants that reduce blood pressure and lower incidence of cancer and cardiovascular diseases	[48]
Carotenoids	Seeds, fruits	Antioxidents that lower the incidence of cancer and cardiovascular diseases	[49]

metabolism in human[50]. Some of the vegetables are used in raw form as salad, but most of them require cooking for the improvement of digestibility and palatability. Some other vegetables require peeling to decrease their useless fibrous[51]. Minerals and other nutrients are affected by both peeling (removal of outer coarse covering) and cooking. Peeling is considered as an inevitable treatment for rendering them more digestible and may result in fairly heavy loss of some nutrient, especially of vitamins. Peeling before boiling, increases the loss of ascorbic acid, folic acid and other vitamins of group B[52]. Methods, temperature and duration of cooking may also effect significantly on the nutritive values of vegetables. Some of the important nutrients such as ascorbic acid and folic acid which are susceptible to oxidation are readily oxidized by brisk cooking. Minerals are also affected by high temperature, in some cases flavour may be lost by brisk cooking. Excessive cooking may cause an adverse affect on the digestibility of the vegetables.

The loss of protein in bitter gourd was estimated 12.85 for peeling, 14.28% for cooking and 22.85% for peeled cooking. The loss of ash for above concentration was found to be 11.11, 36.66 and 60.0% respectively. Percentage losses for the minerals in their peeled, cooked and both peeled and cooked forms were 13.16, 24.37 and 32.16% for calcium, 4.76, 11.56 and 14.96% for magnesium, 9.26, 17.36 and 31.30% for phosphorus and 8.88, 22.22, 27.47% for potassium. The percentage losses of ascorbic acid and folic acid for above mentioned three parameters were 13.26, 53.42, 89.75 and 13.33, 30.52 and 63.15% respectively[53]. The effect of peeling and cooking on compositional analysis of bitter gourd is given in Table 5.

Table 5: Compositional analysis of bitter gourd after peeling and cooking[53]

Parameters	*Raw unpeeled*	*Raw peeled*	*Cooked (unpeeled)*	*Cooked (peeled)*
Protein, g/100 g	0.70	12.85	14.28	22.85
Crude fibre, g/100 g	1.00	20.00	28.00	32.00
Ash, g/100 g	0.90	11.11	36.36	60.00
Calcium, g/100 g	19.00	13.16	24.37	32.16
Magnesium, g/100 g	22.05	4.76	11.56	14.96
Phosphorus, g/100 g	30.25	9.25	17.35	31.30
Potassium, mg/100 g	270.00	8.88	22.22	24.47
Ascorbic acid, mg/100 g	18.25	13.26	53.42	89.75
Folic acid, g/100 g	2.85	13.13	30.52	63.15

Medicinal Properties and Health Benefits of Bitter Gourd

Bitter gourd has been used for centuries in the ancient traditional medicine of India, China, Africa, and Latin America. Bitter gourd extracts possess antioxidant, antimicrobial, antiviral, antihepatotoxic and antiulcerogenic properties while also having the ability to lower blood sugar[18, 38, 54]. These medical activities are attributed to an array of biologically active plant chemicals, including tri-terpenes, pisteins and steroids[18]. Ethno-medical reports of *M. churuntici* indicate that it is used in folkloric medicine for treatment of various ulcers, diabetes, and infections[55, 56, 57]. While the root decoctions have abortifacient properties, leaf and stem decoctions are used in treatment of dysentery, rheumatism and gout[58]. Likewise, the extracted juice from leaf, fruit and even whole plant are routinely used for treatment of wounds, infections, and parasites such as worms, measles, hepatitis, and fever[59]. Medicinal properties of *M. charantia* are summarized in Table 6.

Hypoglycaemic activity

Bitter gourd extracts traditionally used as vegetable insulin possess hypoglycaemic, antioxidative, and antidiabetic agents[60, 61] that are useful in the treatment of diabetes[62]. The hypoglycaemic effects (*i.e.*, blood sugar lowering) of extracts have been well documented in animal[63,64,65,66,67,68,69,70, 71,72,73] and human[5,54,74,] experiments. The beneficial hypoglycaemic properties of fruit pulp, seed, and whole plant extracts have also been documented in rat analyses[70,75], and the medicinal attributes of such extracts have received broad review[58,76,77]. One study cites that there was a significant increase in the number of cells in the pancreas of streptozotocin-induced diabetic rats after 8 weeks of bitter gourd fruit juice treatment[66]. In a parallel *in vivo* clinical human study, oral ingestion of all parts of the bitter gourd plant resulted in low patient toxicity[69]. A mixture of steroidal saponins known as charantin (insulin like peptides), as well as alkaloids, appears to be responsible for the hypoglycaemic actions in bitter gourd

Table 6: Medicinal properties of *M. charantia*

Property	***Active phytochemicals/ extract administered***	***Test system***	***Mechanism of action/ targeted against an organism***	***References***
Anti-viral	MAP 30 (Momordica Anti HIV Protein) in seed and fruit extracts	Viral cell line, H9	Anti -HIV1 (Human Immunodeficiency Virus) activity	[46]
	Anti-HIV proteins MAP30	Human lung fibroblasts	Anti -HSV (Herplex Simplex Virus) activity	[94]
	Alpha and beta momorcharin from seeds, fruits and leaf extracts	Rabbit reticulocyte lysate	Anti- HIV activity	[95]
	Methanol extract	Shrimp	White spot syndrome virus	[96]
Antibacterial	Water, ethanol and methanol extracts from leaves	—	Broad spectrum antimicrobial activity	[97]
	Leaf extract	—	*Eshcherichia coli, Salmonella paratyphi and Shigella dysenterae*	[90]
	Essential oil	—	*Klebsiella pneumoniae, Bacillus megaterium, Bacillus subtilis, Proteus mirabilis, Aspergillus niger, Aspergillus flavus and Escherichia coli*	[91]
Anti-malarial	Aerial part extract	Rat	*Plasmodium vinckei petteri* 279BY (rodent malarial parasite)	[99]
	Leaf extract	Rat	*Plasmodium falciparum*	[100]
Antihelminthic	Lyophilised plant extract -	—	*Caenorhabditis elegans*	[57]
	Plant extract	—	Free living nematodes	[101]
Anti-diabetic	Seed powder	Human	Fall in blood glucose level	[39]
	Charantin administered orally or intravenously	Rabbit	Fall in blood glucose level	[102]
	Chloroform extract Administered intravenously	Alloxan induced rabbit	Fall in blood glucose Level	[103]
	Alcoholic fruit extract	Rat	Fall in plasma glucose level	[104]

extracts. Some studies have shown that at least three components (steroidal saponins, insulin like compounds and alkaloids) were found in bitter gourd plant parts that elicited hypoglycaemic potential and other benefits for sufferers of diabetes mellitus. The hypoglycaemic effect of these chemicals is more pronounced in fruit, where they are present in great abundance. Of the rich mixture of hypoglycaemic compounds in bitter gourd fruit, charantin, vicine, and polypeptide-P are thought to provide the major diabetic medical benefits[78]. Polypeptide-P, previously unidentified insulin like protein similar to bovine insulin, was identified in bitter gourd fruit and seed and in tissue culture[47]. Although the mechanism of action of these hypoglycaemic compounds is still debated, they either regulate insulin release directly or alter glucose metabolism and its insulin like effect.

Anticarcinogenic activity

It has been shown that *M. charantia* fruit juice, peel, pulp, seed and whole fruit extract modulate detoxification pathways in diabetic rats, specifically altering P450 and GSH dependent metabolism[63,79]. Modulation of biotransformation system enzymes may be the cause of anticarcinogenic properties of *M. charantia*. The study by Kumara *et al.*[80] determined the effects of *M. charantia* on the levels of phase I enzymes, which include cytochrome P450 (P450), aniline hydroxylase (ANH) and aminopyrine-N-demethylase (AMD) and to induce the phase II enzymes [glutathione S-transferase (GST)] in rat liver. It was demonstrated that bittergourd fruits contain phases I and II enzyme inducers and compounds capable of repressing some monooxygenases, especially those involved in the metabolic activation of chemical carcinogens. In another study carcinogen induced lipid peroxidation in liver and DNA damage in lymphocytes were reduced following treatment with *M. charantia*. The fruit extract was found to significantly activate the liver enzymes glutathione-S-transferase, glutathione peroxidase and catalase, which showed a depression following exposure to the carcinogen[81]. The results suggest a preventive role of water-soluble constituents of *M. charantia* fruit during carcinogenesis, which is mediated possibly by their modulatory effect on enzymes of the biotransformation and detoxification system of the host.

Antioxidant activity

The antioxidant properties of carotenoids that protect plants during photosynthesis may also protect humans from carcinogens and mitigate free radical effects associated with heart diseases. Natural antioxidants, primarily plant phenolics and polyphenolic compounds in fruits and seeds of bitter gourd are alternatives to synthetic antioxidants for alleviating oxidative deterioration in fruit. For instance, bitter gourd fruit contains as many as 14 carotenoids depending on stage of maturity (5, 6, and 14 in the immature, mature-green, and ripe stage, respectively), where cryptoxanthins becomes the principal chloroplast and chromoplast pigment found in ripe fruit[13]. Other carotenoids, such as 3-carotene, zeaxanthin and lycopene (at ripe stage), and lutein and á-carotene (immature fruit) are also prevalent in the fruits, where they could serve as a model for studying carotenogenesis during ripening[13]. For instance, carotenogenesis in bitter gourd is not affected by temperatures above 30°C[82] in contrast, in tomato, high temperatures inhibit lycopene but not β-carotene synthesis. Likewise, the total carotenoid concentration of bitter gourd seeds can be a 100 fold higher in the ripe than the immature stage, which is exclusively attributable to lycopene (96% of the carotenoids in ripe seeds[49]. Other chemo-preventive antioxidants in plants include vitamin C, vitamin B, phenolic acids, and organosulfur compounds[83]. Bitter gourd is also a rich source of phenolic compounds, where gallic acid, gentisic acid, catechin, chlorogenic acid, and epicatechin are typically abundant. While the concentration of phenolics varies with plant organ (fruit, leaf and root) and cultivar type, the highest content has been found in the Indian white fruited followed by China white-fruited, China green-fruited, and last India green-fruited[48]. These plant phenolic compounds are potentially excellent natural sources of food antioxidants, given their abilities to reduce total cholesterol/triglycerides[76], blood pressure, and the incidence of cancer and cardiovascular diseases.

Anti-fertility effects

Excessive consumption of the fruit and leaves of bitter gourd can reduce sperm production[84]. Bitter gourd ethanol seed extracts have also shown to have potent male anti-fertility effects[76] when administered to dogs[85] and guinea pigs[86].

Antiviral activity

In recent years, a number of chemical components that possess medicinal attributes have been isolated from bitter gourd, such as c-momorcharin, which inactivates ribosome function[87, 88] and stimulates MAP30 (*Momordica* anti-HIV protein) production, which in turn, simultaneously suppresses HIV (human immunodeficiency virus) activity[46]. Interestingly momordicoside A and B present in bitter gourd inhibit tumor growth[45], and several bitter gourd phytochemicals have *in vitro* antiviral activity against viruses including Epstein-Barr, herpes, and HIV viruses[46, 89].

Antimicrobial activity

The leaf extracts of bitter gourd possess antimicrobial activity principally against *Escherichia coli, Staphylococcus, Pseudomonas, Salmonella,*

Streptobacillus, and *Streptococcus*[90]. Moreover, whole plant extracts have shown anti-protozoal activity against *Entamoeba histolytica.* Generally, fresh fruit extracts have similar antibacterial properties, more specifically the fruit extracts of *M. charantia* L. have demonstrated activity against tuberculosis and the stomach ulcer causing bacteria *Helicobacter pylori*[90, 91, 92]. Application of bitter gourd fruit powder to wound sites is similarly effective in stimulating tissue regeneration and wound healing in rats[93].

Processing of Bitter Gourd

Processing may be defined as an activity which is performed to maintain or improve the quality or to change the form or characteristics of an agricultural product. Processing operations of bitter gourd are undertaken to add value after its production. Fruits and vegetables are the reservoir of vital nutrients. Being highly perishable, 20-40% of the total production of fruits and vegetables goes waste from the time of harvesting till they reach the consumers. It is, therefore, necessary to make them available for consumption throughout the year in processed or preserved form and to save the sizeable amount of losses. Fruits and vegetables have great potential for value addition and diversification to give boost to food industry, create employment opportunities and give better returns to the farmers.

Drying and dehydration

Dehydration is supposed to avert spoilage through the removal of water to such a degree that microbial growth is no longer feasible, even at favourable temperature. However at each stage of drying process, there is risk of additional contamination. Dehydrated food stored for fairly long periods after processing will get spoiled by different process *viz*., darkening, insect infestation and microbial deterioration[105]. So it plays an important role in preservation of dry vegetable. Dehydrated bitter gourd rings has the potential to become an important value added product because of relatively inexpensive, easily and quickly cookable and rich in several nutrients, which are essential for human health. The process flow chart for drying of bitter gourd is given in Fig. 3[106].

Pre-treatment for drying

- *Blanching:* Bitter gourds are boiled in water for two minutes to avoid possible enzymatic activities.
- *Brining:* Brine is a mixture of water and salt. Brining is a common technique in cooking of bitter gourd to reduce its bitterness.
- *KMS treatment:* Bitter gourd rings can be treated with 0.1% potassium metabisulphite solution as a chemical additive.

Bitter gourd
↓
Washing
↓
Removal of stalks and tips
↓
Cut into 6 mm thick slices
↓
Blanching in boiling water (8 minutes)
↓
Drying at 60°C in tray dryer or sun drying
↓
Packing
↓
Storage

Fig. 3: Process flow chart for drying of bitter gourd

Drying methods of bitter gourd

Sun drying: Sun drying is the method in which food is directly exposed to sunlight. It is a cheap method of preservation because it uses the natural resource/source of heat *i.e.* sunlight. This method can be used on a commercial scale as well as the village level provided that the climate is hot, relatively dry and free of rainfall during and immediately after the normal harvesting period. The dried product in this method is inferior in quality[107].

Solar cabinet drying: Solar drying can be considered as an elaboration of sun drying and is an efficient system of utilizing solar energy[108, 109, 110, 111]. The tropics and subtropics have abundant solar radiation. Natural convection solar dryers do not require power from the electrical grid or fossil fuels. Hence the obvious option for drying would be the natural convection solar dryers. The cabinet type solar drier suitable for drying fruits and vegetables[112] consists of a solar collector with a transparent cover on the top and a drying unit with an opaque cover on the top. In such a dryer, the product to be dried is contained within a cabinet in a relatively thin bed, which completely spans the cabinet. Air, which is heated in a simple flat plate type solar collector, then flows as a result of the buoyancy forces resulting from the temperature differences up through the product bed thereby producing the drying. Jadhav *et al.* [113] dried the bitter gourd slices in solar cabinet dryer. The optimum process condition was drying of bitter gourd was found as 4.24 min blanching time and 0.49 % KMS which resulted in 20.15 mg/100g chlorophyll retention and 340.55 g hardness. The solar cabinet drying was found to be fastest drying method as compared to open sun and freeze drying methods and showed highest value of effective diffusivity

Tray drying: In tray dryers the bitter gourd is dried on trays which are comparatively thin layers up to few centimetres. The hot air drying characteristics of *Momordica charantia* slices under different temperatures (50, 60, 70 and 80°C) and slice thickness (0.5, 0.75 and 1.00 cm), respectively for colour analysis[114]. The results show that the color of bitter gourd slices changed sharply when temperature was higher than about 70°C. Pre-treated bittergourd slices were dehydrated in a mechanical tray dryer[115]. Bittergourd slices blanched in boiling water were of the best quality followed by 5% salt solution at drying temperature of 60 and 70°C.

Freeze drying: In this method, the products is first freezed on trays in the lower chamber of a freeze drier and the frozen material is dried in the upper chamber under high vacuum. The material dries directly by sublimation of ice without passing through the intermediate liquid stage. The dried product is highly hygroscopic[107].

Post-harvest loss and value addition

The post-harvest loss of bitter gourd is about 25% which is due to ripening and mechanical damage during transport. Reduction of post-harvest loss minimizes the unit production cost and brings down the price gap between grower and consumer. Due to warty nature of the fruit the transport damage is very high. Further polysacks bags being used to pack them cause a severe damage to the fruit. If fruits are carefully transported, post-harvest loss can be minimized to a greater extent. Bitter gourd can be stored at ambient temperature for 4-6 days if they are harvested in a slightly immature stage. However this storage life can further be extended by storing them at 13°C. The maximum storage life at low temperature is limited to 14-16 days.

Value addition of bitter gourd can be done by dehydration. Thin slices can be dehydrated and this technology is adopted in a small scale for domestic purposes. A better quality product can be prepared if driers are used for dehydration. In addition slices of this fruit can be preserved in brine solution. Apart from drying many other value added products such as bitter gourd chips, bitter gourd pickle and bitter gourd soft drinks can be prepared.

Bitter gourd health drink

The fresh and matured bitter gourds were thoroughly washed in water and cut into small pieces. The cut pieces were soaked in water with little salt and half teaspoon turmeric powder for 15 minutes. After that the bitter gourd pieces were removed from the water and grind it in the mixer with enough water. The pulp obtained was strained using strainer. After straining a small amount of lemon juice was added and stirred well. Process flow chart for preparation of bitter gourd health drink is shown in Fig. 4.

Bitter gourd (mature, green)
↓
Washing
↓
Cutting into small pieces
↓
Soaking in water with little salt and turmeric for 15 min.
↓
Removal of the pieces from water
↓
Grinding in a mixer with water
↓
Straining
↓
Addition of lemon juice
↓
Bitter gourd juice

Fig. 4: Process flow chart for preparation of bitter gourd health drink

Bitter gourd chips

The process for manufacturing of bitter gourd chips involves cutting of bitter gourd to 0.25–0.30 cm thick slices, followed by cooking in 0.1% sodium bicarbonate solution at 100°C for 30–40 min. The cooked bitter gourd slices were subjected to osmotic diffusion treatment of 1–2% brine solution at 50–60°C for 2 h. The osmotically diffused bitter gourd slices were dried at 55–60°C for 5–7 hr and fried in refined oil at 160–180°C for 5–7 sec. The fried chips were packed in laminated pouch[116]. Process flow chart for preparation of bitter gourd chips is given in Fig. 5.

Bitter gourd (mature, green)
↓
Washing
↓
Cutting into slices (0.25-0.30 cm)
↓
Cooking in 0.1% Na_2HCO_3 solution (100°C, 30-40 min)
↓
Osmotic diffusion in 1-2 % brine (50-60°C, 2 h)
↓
Drying for 5-7 h (55–60°C)
↓
Frying (160–180°C, 5–7 sec)
↓
Shade drying (for 2 h)
↓
Frying spices in oil
↓
Packing in laminated pouches

Fig. 5: Process flow chart for preparation of bitter gourd chips

Bitter gourd pickle

The green coloured mature bitter gourd is selected for preparation of bitter gourd pickle. The bitter gourds are washed thoroughly with water. The incisions are made longitudinally in the bittergourd. The pulp and seed is removed after incision. Salt is mixed to bitter gourd and kept in sun for 3 h. After that shade drying is done for 2 h and kept aside. The spices are fried in oil and cooled. These spices are mixed with salt and lime juice and filled in bitter gourd and packed in jar. Oil is added to jar. The ingredients required for preparation of 100 kg bitter gourd pickle is given in Table 7. The process flow chart for preparation of bitter gourd pickle is given in Fig. 6[107].

Table 7: Ingredients required for preparation of 100 kg of bitter gourd pickle.

Sl. no	*Ingredient*	*Quantity (kg)*
1	Bitter gourd	100
2	Salt	15
3	Tamarind pulp or amchur	25
4	Red chilli powder	1
5	Cardamom, cumin, aniseed (powdered) each	1
6	Mustard (ground)	10
7	Mustard oil	50

Bitter gourd (mature, green)
↓
Washing
↓
Making incision
↓
Removal of pulp and seed
↓
Mixing a little salt
↓
Keeping in sun (for 3 h)
↓
Draining of water
↓
Shade drying (for 2 h)
↓
Frying spices in oil
↓
Cooling
↓
Mixing with salt and lime juice
↓
Filling into bitter gourds
↓
Packing in jars
↓
Addition of oil after heating and cooling it
↓
Storage

Fig. 6: Process flow chart for preparation of bitter gourd pickle

Bitter gourd soft drinks

Firm bitter gourds having solid, light green colour and without blemishes and any signs of yellow or orange colour on the skin were selected. The firm and fresh bitter gourds were washed thoroughly and cut into 5 mm thick slices. The high sweeteners, citric acid and water were mixed separately and added to a flask containing bitter gourd slices. The top opening of the flasks were tightly closed with cotton bar. Then the samples were heated at optimized time (60 min) and temperature (90°C). After heating, the samples were brought to the room temperature and filtered using cheese cloth. After filtration, the samples were homogenized at high pressure (170 bar) for 3 passes. Potassium meta-bisulfate (KMS) was added to the finished product. Finally the products were bottled, hot sealed and stored[117]. Process flow chart for preparation of bitter gourd soft drink is given in Fig. 7.

Bitter gourd (firm and fresh)
↓
Washing
↓
Cutting into 5mm thickness
↓
Mixing of high sweeteners, citric acid and water with bitter gourd
↓
Keeping in tightly closed flask
↓
Heating the sample (90ºC, 60 min)
↓
Cooling to room temperature
↓
Filtering
↓
Homogenisation (170 bar pressure)
↓
Adding KMS
↓
Filling into bottles
↓
Storage

Fig. 7: Process flow chart for preparation of bitter gourd soft drink

EXTRACTION TECHNIQUES

Medicinal value of bitter gourd has been attributed to its high antioxidant properties due in part to phenols, flavonoids, iso-flavones, terpenes, anthroquinones, and glucosinolates, all of which confer a bitter taste[118]. Removal of active bitter components, through a variety of debittering processes as well as selective breeding can result in loss of possible health benefits. Bitter gourd (*Momordica charantia*) is a very common oriental vegetable in tropical and subtropical areas. It is traditionally regarded as a "cooling" or "fire-reducing" food which are characterized "heating" and "cooling" foods by their effects on PGE_2 production in LPS-stimulated RAW264.7 macrophages, and found that ethyl acetate extract (EAE) of bitter gourd inhibited PGE_2 production[119]. Bitter gourd contain compounds like ethyl acetate extract (EAE), which inhibit lipo-polysaccharide-induced prostaglandin E2 (PGE_2) production in RAW264.7 cells.

Extraction of Phenolic Compounds

1. Subcritical water extraction (SCWE)

SCWE is a new and powerful technique at temperatures between 100 and 374°C and pressure high enough to maintain the liquid state given in Fig. 8. Unique properties of water are namely its disproportionately high boiling point for its mass, a high dielectric constant and high polarity [120]. As the temperature rises, there is a marked and systematic decrease in permittivity, an increase in the diffusion rate and a decrease in the viscosity and surface tension. In consequence, more polar target materials with high solubility in water at ambient conditions are extracted most efficiently at lower temperatures, whereas moderately polar and non-polar targets require a less polar medium induced by elevated temperature. Budrat and Shotipruk[121] found that the extracts obtained by subcritical water extraction methods were significantly higher than solvent extraction and soxhlet extraction.

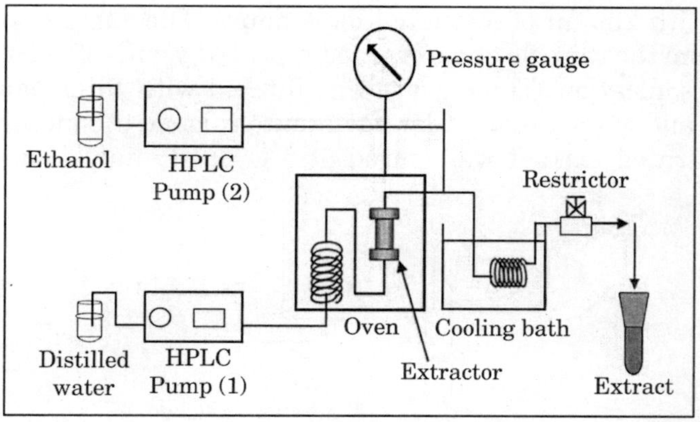

Fig. 8: Diagram of experimental setup subcritical water extraction

2. Solvent extraction

Solvent extraction is one of the most extensively studied and most widely used techniques for the separation and pre-concentration of elements. With proper choice of extracting agents, this technique can achieve group separation or selective separation of trace elements with high efficiencies.

Solvent extraction is a technique in which a solution is brought into contact with a second solvent, essentially immiscible with the first, in order to bring the transfer of one or more solutes into the second solvent. The separations that can be achieved by this method are simple, convenient and rapid to perform; they are clean as much as the small interfacial area certainly precludes any phenomena analogous to the undesirable co-precipitation encountered in precipitation separations.

One gram of the fine ground sample was weighed into a test tube and 10 ml methanol was added and the sample was then extracted for 2 h in an ultrasonic bath (275DAG, Crest, Malaysia) at 65°C. After extraction, the sample was cooled to room temperature and then centrifuged at 1500 rpm for 15 min. The sample residue was extracted repeatedly with 30 ml of methanol. The extract was filtered with filter paper (No. 4, Whatman, England) and evaporated under vacuum to remove the methanol. Then the concentrated extract was stored at 4°C until use[121].

3. *Soxhlet extraction*

Soxhlet extraction is a technique used when the desired compound has a *limited* solubility in a solvent, and the impurity is insoluble in that solvent. It allows for unmonitored and unmanaged operation while efficiently recycling a small amount of solvent to dissolve a larger amount of material.

One gram of the fine ground sample was weighed into a thimble and was extracted with 200 ml of methanol for 4 hours. The sample residue was removed from the thimble and extracted repeatedly with 30 ml of methanol using ultrasonication. The extract was filtered with filter paper (No. 4, Whatman) and evaporated under vacuum to remove the methanol. Then the concentrated extract was stored at 4°C until analysis as shown in Fig. 9[121].

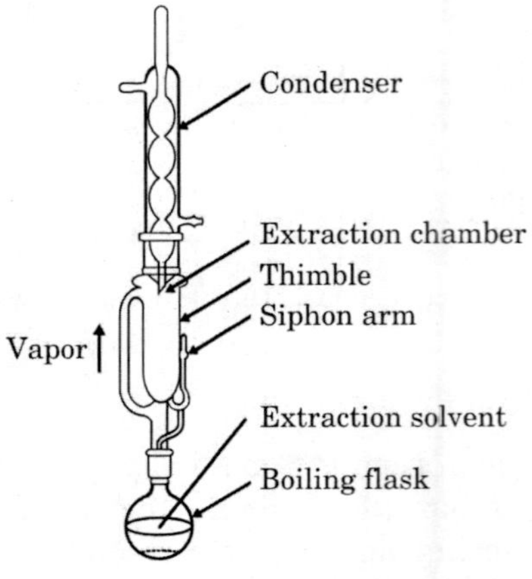

Fig. 9: Diagram of soxhlet apparatus

Extraction of Essential Oil from Bitter Gourd

Essential oil yield of 0.012% can be obtained from the dried seeds of bitter gourd by hydro distillation method using the Clevenger-type apparatus. The essential oil obtained from the seeds of *Momordica charantia* was analysed by GC/MS. The essential oil consists of twenty-five components, representing 90.9% of the oil. The constituents are represented by sesquiterpenes (71.7%), phenylpropanoids (11.0%), and monoterpenes (7.6%), *trans*-nerolidol being the major constituent (61.6%)[122].

STORAGE

The chilling-sensitive nature of bitter melon, with typical injury symptoms of pitting, decay and discoloration occur when the immature fruits are stored below 10°C. These chill-induced changes are associated with increases in respiration and ethylene production to higher than normal rates after transfer to 15°C. Storage at l0-12.5°C maintained acceptable product quality for l0-14 days. Postharvest quality and shelf-life of bitter melon is greatly affected by the developmental stage of the fruit at harvest. Difficulty in accurately determining harvest maturity may lead to undesirable color, firmness, and flavor changes during marketing at temperatures >12.5°C [123].

REFERENCES

[1] Anonymous, Indian Horticulture Database (2013).

[2] Anonymous, Handbook of Horticulture Statistics, Government of India, New Delhi, p. 20 (2014)

[3] Karunanayake, Tennekoon, "Search of novel hypoglycemic agent from medicinal plants", *In*: Sharma A.K. (*ed.*), Diabetes mellitus and its complications, An update, *Macmillan* India Ltd., New Delhi, pp. 192–205 (1993).

[4] Anonymous, Bittermelon. www.canmedbotanics.ml (2000).

[5] Leatherdale B.A., Panesar R.K., Singh C., Atkins T.W., Bailey C.J., Bignell A.H., "Improvement in glucose tolerance due *to Mornordica chorantia* (karats)", *Br Med J. (Clin Res Ed.)*, **282**, 1823–1824 (1981).

[6] Ali L., Khan A.K., Mamun M.I., Mosihuzzaman M., Nahar N., Nur-e- Alam M., Rokeya B., "Studies on hypoglycemic effects of fruit pulp, seed, and whole plant of *Momordica charantia* on normal and diabetic model rats", *Journal of Planta Medica*, **59**, 408–412 (1993).

[7] Walters T.W., Walters D.S.D., "Notes on economic plants. Balsam pear *(Momotdica chamntia,* Cucurbitaceae)" *Econ Bot.*, **42**, 286–288 (1988).

[8] Morton J.F., "The balsam pear - an edible, medicinal and toxic plant", *Econ Bot.*, **21**, 57–68 (1967).

[9] Walters T.W., "Historical overview on domesticated plants in China with special emphasis on the Cucurbitaceae", *Eeon Bot.*, **43**, 297–313 (1989).

[10] Kays S.J., Hayes M.J., "Induction of ripening in the fruits of *Momordica charantia* L by ethylene", *Bop Agric.*, **55**, 167–172 (1978)

[11] Zheng L., "Primary approach on the postharvest physiology of balsam pear", M.S. Thesis, *Beijing Vegetable Research Center,* Beijing, p. 43 (1986).

[12] Mohammed M., Wickham L.D., "Extension of bitter gourd (*Momordica charantia L.*) storage life through the use of reduced temperature and polyethylene wraps", *Journal of Food Quality*, **16**, 371–382 (1993).

[13] Rodriguez D.B., Raymundo L.C., Lee T.C., Simpson K.L., Chichester C.O., "Carotenoid pigment changes in ripening *Momordica charantia* fruits", *Journal of Annalytical Botany*, **40**, 615–624 (1976).

[14] Tindall H.D., "Vegetables in the tropics", *Macmillan,* London (1983).

[15] Reyes M.E.C., Gildemacher B.H., Jansen G.J., *Momordica L.* pp. 206–210. *In*: Siemonsma J.S. and Piluek K. (*eds*.), "Plant resources of South-East Asia: Vegetables", Pudoc Scientific Publishers, Wageningon, the Netherlands (1994).

[16] Yamaguchi M., "World Vegetables", *AVI (Van Nostrand Reinhold Co.),* New York, N.Y. (1983).

[17] Palada M.C., Chang L.C., "Suggested cultural practices for bittergourd. International Cooperators guide", pp. 1–5 (2003).

[18] Grover J.K., Yadav S.P., "Pharmacological actions and potential uses of *Momordica charantia*: A review", *Journal of Ethnopharmacology*, **93**, 123–132 (2004).

[19] Raj S.K., Khan M.S., Singh R., Kumari B., Prakash D., "Occurrence of yellow mosaic geminiviral disease on bitter gourd (*Momordica charantia*) and its impact on phytochemical contents", *International Journal of Food Science and Nutrition*, **56**, 185–192 (2005).

[20] Lako J., Trenerry V.C., Wahlqvist M., Wattanapenpaiboon N., Sotheeswaran S., Premier R., "Phytochemical flavonols, carotenoids and the antioxidant properties of a wide selection of Fijian fruit, vegetables and other readily available foods", *Journal of Food Chemistry*, **101**, 1727–1741 (2007).

[21] Lee-Huang L.S., Huang P.L., Chen H.C., Huang P.L., Bourinbaiar A., Huang H.I., Kung H.F., "Anti-HIV and anti-tumor activities of recombinant MAP30 from bitter melon", *Gene,* **61**, 151–156 (1995).

[22] Chao C.Y., Huang, C.J., "Bitter gourd (*Momordica charantia*) extract activates peroxisome proliferators-activated receptors and upregulates the expression of the acyl-CoA oxidase gene in H4IIEC3 hepatoma cells", *Journal of Biomedical Science*, **10**, 782–791 (2003).

[23] Karthika B., "Physical characters and glycemic response of fresh and dehydrated underutilized cucurbits", *Ph.D Thesis*, UAS Dharwad, (2006).

[24] Miniraj N., Prasanna K.P., Peter K.V., "Bitter gourd *Mornordica* spp. pp. 239–246. *In*: Kalloo G. and Bergh B.O. (*eds*.), "Genetic improvement of vegetable plants", Pergamon Press, Oxford, U. (1993).

[25] Desai U.T., Musmade A.M., "Pumpkins, squashes and gourds", pp. 273–298 (1998).

[26] Kale A.A., Cadakh S.R., Adsule R.N., "Physico-chemical characteristics of improved varieties of bitter gourd (*Momordica chorantio L.*)", *Maharashtra J Hort.,* **5**, 56–59 (1991).

[27] Yuwai K.E., Rao K.S., Kaluwin C., Jones G.P., Rivett D.E., "Chemical composition of *Momordica charantia* L. fruits", *Journal of Agricultural and Food Chemistry,* **39**, 1762–1763 (1991).

[28] Xiang C.P., Wu C.Y., Wang L.P., "Analysis and utilization of nutrient composition in bitter gourd *(Momordica charantia)", J Huazhong Ag Univ.*, **19**, 388–390 (2000).

[29] Gopalan C., Sastri B.V.R., Balalsubramnnian S.C., "Nutritive value of Indian foods, 2nd ed. Hyderabad", *National Institute of Nutrition, ICMR.* Hyderabad (1993).

[30] Khanna P., Mohan S., "Isolation and identification of diosgenin and sterols from fruits and *in vitro* cultures of *Momordica charantia* Linn", *Indian Journal of Experimental Biology,* **11**, 58–60 (1973).

[31] Hardenburg R.E., Watada A.E., Wang C.Y., "The commercial storage of fruits, vegetables, and florist and nursery stocks", *USDA Agric. Handb. No–66. US Dept Agric.*, Washington, DC, p. 130 (1986).

[32] Pantastico E.B., Chattopadhyay T.K., Subramanyam H., "Storage and commercial storage operations. *In*: Pantastico Er.B. (*ed.*), Postharvest Physiology, Handling and Utilization of Tropical and Subtropical Fruits and Vegetables", *AVI, Westport, Conn.*, p. 324 (1975).

[33] Husain J., Tickle I.J., Wood S.P., "Crystal structure of momordin, a type I ribosome inactivating protein from the seeds of *Momordica charantia*", *FEBS Letters,* **342**, 154–158 (1994).

[34] Xie H., Huang S., Deng H., Wu Z., Ji A., "Study on chemical components of *Momordica charantia"*, **21**, 458–459 (1998).

[35] Yuan Y.R., He Y.N., Xiong J.P., Xia Z.X., "Three-dimensional structure of beta-momorcharin at 2.55 A resolution", *Acta Crystallographica Section D Biological Crystallography,* **55**, 1144–1151 (1999).

[36] Parkash A., Ng T.B., Tso W.W., "Purification and characterization of charantin, a napin-like ribosome-inactivating peptide from bittergourd (*Momordica charantia*) seeds", *Journal of Peptide Research,* **59**, 197–202 (2002).

[37] Murakami T., Emoto A., Matsuda H., Yoshikawa M., "Medicinal foodstuffs. Part XXI. Structures of new cucurbitane-type triterpene glycosides, goyaglycosides-a, -b, -c, -d, -e, -f, -g, and -h, and new oleanane-type triterpenesaponins, goyasaponins I, II, and III, from the fresh fruit of Japanese *Momordica charantia*L", *Chemical & Pharmaceutical Bulletin (Tokyo),* **49**, 54–63 (2001).

[38] Raman A., Lau C., "Anti-diabetic properties and phytochemistry of *Momordica charantia* L. (Cucurbitaceae)", *Phytomedicine*, **2**, 349–362 (1996).

[39] Lolitkar M.M., Rao M.R.R., "Pharmacology of a hypoglycaemic principle isolated from the fruits of *Momordica charantia*", Linn. *Indian Journal of Pharmacy,* **28**, 129–133 (1966).

[40] Orlovskaya T.V., Chelombitko V.A., "Amino acid composition of *Momordica charantia* seeds and pericarp", *Chemistry of natural compounds,* p. 43 (2007).

[41] Chan W.Y., Tam P.P.L., Yeung H.W., "The termination of early pregnancy in the mouse by β-momorcharin", *Contraception,* **29**, 91–100 (1954).

[42] Dutta P.K., Chakravarty A.K., Chowdhury U.S., Pakrash S.C., "Vicine, a Favism-inducing toxin from *Momordica charantia Linn* seeds", *Indian J Chem.*, **208**, 669–671 (1981).

[43] Hands C., Singh J., Sharma M.L., Kaul A., Neerja M., Zafar R., "Hypoglycemic principle of *Momordica charantia* Seeds. Indian)", *Nat Prod.,* **6**, 16–19 (1990).

[44] Lotlikar M.M., Rajarama Rao M.R., "Note on a hypoglycemic principle isolated from the fruits of *Mornordica charantia*" *J Univ Bombay.*, **29**, 223–224 (1962).

[45] Okabe H., Miyahara Y., Yamauchi T., Mirahara K., Kawasaki T., "Studies on the constituents of *Momordica charantia* L. 1. Isolation and characterization *of* momordicosides A and B, glycosides *of* a pentahydroxy-cucurbitane triterpene", *Chem Thorne But.,* **28**, 2753–2762 (1980).

[46] Lee-Huang S., Huang P.L., Nara P.L., Chen H.C., Kung H.F., Huang P., Huang H.I., Huang, P.L., "MAP 30: A new inhibitor of HIV-1 infection and replication", *FEBS Letters,* **272,** 12–18 (1990).

[47] Khanna P., Jain S.C., "Hypoglycemic activity of polypeptide.p from a plant source.)", *Nat Products.,* **44**, 648–655 (1981).

[48] Horax R., Hettiarachchy N., Islam S., "Total phenolic contents and phenolic acid constituents in 4 varieties of bitter melons (*Momordic acharantia*) and antioxidant activities of their extracts", *Journal of Food Sciences*, **70**(4), C275–C280 (2005).

[49] Rodriguez D.B., Lee T.C., Chichestet C.O., "Comparative study of the carotenoid composition of the seeds of ripening *Momordica charantia* and tomatoes", *Plant Physiol,* **56,** 626–629 (1975).

[50] Robinson D.S., "Food Biochemistry and Nutritional Value", Longman Scientific and Technical Publisher, New York (1990).

[51] Thane C., Reddy S., "Processing of fruit and vegetables: Effects on carotenoids", *J Nutr and Food Sci.*, **2-3,** 58–65 (1997).

[52] Bennion M., "The Science of Food, 1st Ed.", John Wiley and Sons, New York, p. 210 (1980).

[53] Alvi S., Khan K.M., Sheikh M.A., Shahid M., "Effect of Peeling and Cooking on Nutrients in Vegetables, Pakistan", *Journal of Nutrition,* **2**(3), 189–191 (2003).

[54] Welihinda J., Karunanayake E.M., Sheriff M.H., Jayasinghe K.S., "Effect of *Momordica charantia* on the glucose tolerance in maturity onset diabetes", *J Ethnopharmacol.,* **17**, 277–282 (1986).

[55] Gurbuz I., Akyuz C., Yesilada B., Bilge S., "Anti-ulcerogenic effect *of Momordica charantia* L. fruits on various ulcer models in rats", *J Ethnopharmacol.*, **71**, 77–82 (2000).

[56] Scartezzini P., Speroni B., "Review on some plants of Indian traditional medicine with antioxidant activity", *J Ethnophannacol.,* **71**, 23–43 (2000).

[57] Beloin N., Gbeassorb M., Akpaganab K., Hudsonc J., Soussab K.D., Koumaglob K., Arnasona J.T., "Ethnomedicinal uses of *Momordica charantia* (Cucurbitaceae) in Togo and relation to its phytochemistry and biological activity", *Journal of Ethnopharmacology*, **96**, 49–55 (2005).

[58] Subratty A.H., Fakim A.G., Mabomoodally E., "Bitter melon: An exotic vegetable with medicinal values", *Neu Food Sci.*, **35**, 143–147 (2005).

[59] Behera T.K., Staub J.E., Bohera S., Simon P.W., "Bitter gourd and human health", *Med Aram Plant Sci Biotechnol.*, **1**, 224–226 (2008).

[60] Vikrant V., Grover J.K., Tandon N., Rathi S.S., Gupta N., "lltatment with extracts of *Momordica charantia* and *Eugenic jombolana* prevents hyperglycemia and hyper-insulinemia in fructose fed rats", *J Ethnopharmacol.*, **76**, 139–143 (2001).

[61] Chen Q., Laureen L., Chan L., Li E.T., "Bitter melon *(Memordica charantia)* reduces adiposity, lowers serum insulin and normalizes glucose tolerance in rats fed a high fat diet", *J Nutr.,* **133**, 1086–1093 (2003).

[62] Baynes J.W., "Mechanistic approach to diabetes: Vol. 2," *Ellis Harwood Limited, Chichester,* UK, pp. 203–231 (1995).

[63] Raza H., Ahmed I., Lakhani M.S., Sharma A.K., Pallet D.J., Montague W., "Effect of bitter melon *(Memordica charantia)* fruit juice on the hepatic cvtochrome P450-dependent monooxygenases and g]utathione S-transferases in streptozotocin-induced diabetic rats", *Biochem Pharmacol.,* **52**, 1639–1642 (1996).

[64] Sarkar S., Pranao M., Marita R., "Demonstration of the hypoglycaemic action of *M. charantia* in a validated animal model of diabetes, *Pharmacol Res.,* **33**(1), 1–4 (1996).

[65] Raze H., Ahrned I., John A., Sharma A.K., "Modulation of xenobiotic metabolism and oxidative stress in chronic streptozotocin-induced diabetic rats fed with *Memordica charantia* fruit extract", *J Biochem Mo! Toxicol.*, **14**, 131–139 (2000).

[66] Ahmed I., "Hypo-triglyceridemic and hypo-cholesterolemic effects of anti-diabetic *Memordica charantia* (Karela) fruit extract in streptozotocin-induced diabetic rats" *Journal of Diabetes Research and Clinical Practice*, **51**(3), 155–161 (2001).

[67] Miura T., Itoh C., Iwamoto N., Kato M., Kawai M., Park S.R., Suzuki I., "Hypoglycemic activity of the fruit of the *Memordica charantia* in type 2 diabetic mice", *J Nutr Sci Vitaminol.*, **47**, 340–344 (2001).

[68] Grover J.K., Rathi S.S., Vats V., "Amelioration of experimental diabetic neuropathy and gastropathy in rats following oral administration of plant *(Memordica charantia), Eugenic jombolono, Mucuna pruriens* and *Tiflosporo cordifoha)* extracts", *Indian J Expt Bid.,* **40**, 273–276 (2002).

[69] Rathi S.S., Grover J.K., Vats V., "The effect of *Memordica charantia* and *Mucuna pruriens* in experimentaldiabetes and their effect on key metabolic enzymes involvedin carbohydrate metabolism", *Phytother Res.*, **16,** 236–243 (2002).

[70] Kar A., Choudhary B.K., Bandyopadhyay N.C., "Comparative evaluation of hypoglycaemic activit y of some Indian medicinal plants in alloxan diabetic rats", *J Ethnopharmacol.,* **84**, 105–110 (2003).

[71] Chaturvedi P., George S., Milinganyo M., Tripathi Y.B., "Effect of M. *charantia* on lipid profile and oral glucose tolerance in diabetic rats", *Phytother Res.*, **18**(11), 954–956 (2004).

[72] Sathishsekar U., Subramanian S., "Antioxidant properties *of Memordica charantia* (bitter gourd) seeds on streptozotocin-induced diabetic rats", *Asia Pacific J Clinical Nutt.*, **14**,153–158 (2005).

[73] Shetty K.A., Kumar G.S., Math P.V.S., "Bitter gourd *(Memordica charantia)* modulates activities of intestinal and renal disaccharidases in streptqzotocin-induced diabetic rats", *Molecular Nutt Food Res.*, **49**, 791–795 (2005).

[74] Baldwa V.S., Bhandari G.M., Pangaria A., Goyal R.K., "Clinical trials in patients with diabetes mellitus on an insulin-like compound obtained from plant source", *Upsets J Med Sci.,* **82**, 39–41 (1977).

[75] Jayasooriya A.P., Sakono M., Yukizoki C., Kawano M., Yamamoto K., Fnkuda N., "Effects of *Memordica charantia* powder on serum glucose levels and various lipid parameters in rats fed with cholesterol-free and cholesterol-enriched diets", *J Ethoopharmacol.,* **72**, 331–336 (2000).

[76] Basch B., Gabardi S., Ulhricht C., "Bittermelon *(Memordica charantia):* A review of efficacy and safety", *Am J Health System Pharm.,* **60**, 356–359 (2003).

[77] Krawirikel M.B., Koding G.B., "Bitter gourd *(Memordica charantia*): A dietary approach to hyperglycemia", *Nutr Rev.*, **64**, 331–337 (2006).

[78] Yeh G.Y., Eisenber D.M., Kaptchuk T.J., Phillips R.S., "Systematic reviewof herbs and dietary supplements for glycemic control in diabetes", *Diabetes Care,* **26**, 1277–1294 (2003).

[79] Singh A., Singh S.P., Bamezai R., "Postnatal efficacy of *Memordica charantia* peel, pulp, seed and whole fruit extract in the detoxication pathway of suckling neonates and lactating mice", *Cancer Lett.*, **122**, 121–126 (1998).

[80] Kumara W.R., Ratanavila A., Tepsuwan A., "Effects of neem flowers, Thai and Chinese bitter gourd fruits and sweet basil leaves on hepatic mono oxygenases and glutathione S-transferase activities, and *in vitro* metabolic activation of chemical carcinogens in rats", *Journal of Food Chemical Toxicology,* **36**, 475–484 (1998).

[81] Ganguly C., De S., Das S., "Prevention of carcinogen- induced mouse skin papilloma by whole fruit aqueous extract of *Memordica charantia*", *Eur J Cancer Prev.*, **9**, 283–238 (2009).

[82] Tran T.L.H., Raymundo L.C., "Biosynthesis of carotenoids in bitter melon at high temperature", *Phytochemistry,* **52**, 275–280 (1999).

[83] Simon P.W., "Plant pigments for color and nutrition", *Hort Science.*, **32**, 12–13 (1997).

[84] Prakash A.O., Mathur R., "Screening of Indian plants for antifertility activity", *Indian J Expt Biol.,* **14**, 623–626 (1976).

[85] Dixit V.P., Khanna P., Bhargava S.K., "Effects of *Memordica charantia* fruit extract on the testicular function of dog", *Plante Med.,* **34**, 280–286 (1978).

[86] Udoh P., Ojenikoh C., Udoh V., "Antifertility effect of *Memordica charantia* (bitter gourd) fruit on the gonads on male guinea pigs", *Global J Pure App Sd.,* **7**, 627–631 (2001).

[87] Feng Z., Li W.W., Yeung H.W., Chen S.Z., Wang Y.P., Lin X.Y., Dong Y.C., Wang J.H., "Crystals of α-momorcharin: A new ribosome-inactivating protein", *J Mol Biol.*, **214**, 625–626 (1990).

[88] Leung K.C., Meng Z.Q., Ho W.K.K., "Antigenic determination fragments of α-momorcharin", *Biochem Biophys Arta.,* **1336**, 419–424 (1997).

[89] Nerurkar P.V., Lee Y.K., Linden E.H., Lim S., Pearson L., Frank J., Nerurkar V.R., "Lipid lowering effects of *Memordica charantia* (bitter melon) in HIV-1-protease inhibitor- treated human hepatoma cells, HepG2", *Br J Pharmacol.*, **148**, 1156–1164 (2006).

[90] Omoregbe R.E., Ikuebe O.M., Ihimire I.G., "Antimicrobial activity of some medicinal plants extraction *Escherichia coli, Salmonella paratyphi* and *Shigella dysenteriae*", *African Journal of Medical Science,* **25**, 373–375 (1996).

[91] Hussain H.S.N., Deeni Y.Y., "Plants in Kano ethomedicine: Screening for antimicrobial activity and alkaloids, Int.)", *Pharmacol.*, **29**, 51–56 (1991).
[92] Yesilada E., Gurbuz I., Shibata H., "Screening of Turkish anti-ulcerogenic folk remedies for *anti-Helicobocter pylori* activity", *J Ethnopharmacol.*, **66**, 289–293 (1999).
[93] Parasad G., Tyagi R.K., Bhardwa D.D., Gupta K.K., Chopra A.K., "Screening of antimicrobial potential of certain drug plants for enhancement of their therapeutic spectrum *in vitro*", *Humdard Medicus.*, pp. 77–81 (2002).
[94] Bourinbaiar A.S., Huang S.L., "The activity of plant-derived antiretroviral proteins MAP30 and GAP31 against herpes simplex virus *in vitro*", *Biochemistry and Biophysics Research Communication*, **219**, 923–929 (1996).
[95] Au T.K., Collins R.A., Lam T.L., Ng T.B., Fong W.P., Wan D.C., "The plant ribosome inactivating proteins luffin and saporin are potent inhibitors of HIV-1 integrase", *FEBS Letters.*, **471**, 169–172 (2000).
[96] Balasubramanian G., Sarathi M., Kumar S.R., Hameed A.S.S., "Screening the antiviral activity of Indian medicinal plants against white spot syndrome virus in shrimp", *Journal of Aquaculture.*, **263**, 15–19 (2007).
[97] Khan M.R., Omoloso A.D., "*Memordica charantia* and *Allium sativum*: Broad-spectrum antibacterial activity", *Korean Journal of Pharmacognosy,* **29**, 155–158 (1998).
[98] Ajaji I.A., Jonathan S.G., Adewuyi A., Oderinde R.A., "Antimicrobial screening of the essential oil of some herbal plants from western Nigeria", *World Applied Sciences Journal.*, **3**, 79–81 (2008).
[99] Munoz V., Sauvain M., Bourdy G., Callapa J., Rojas I., Vargas L., Tae A., Deharo E., "The search for natural bioactive compounds through a multidisciplinary approach in Bolivia. Part II. Antimalarial activity of some plants used by Moseteneindians", *Journal of Ethnopharmacology.*, **69**, 139–155 (2000).
[100] Gbeassor M., Kedjagni A.Y., Koumaglo K., De-Souza C., Agbo K., Aklikokou K., Amegbo K.A., "*In vitro* antimalarial activity of six medicinal plants", *Phytotherapy Research.*, **4**, 115–117 (2006).
[101] Das P., Sinhababu S.P., Dam T., "Screening of antihelminthic effects of Indian plant extracts: A preliminary report", *Journal of Alternative and Complementary Medicine,* **12**, 299–301 (2006).
[102] Reddy M.B., Reddy K.R., Reddy M.N., "A survey of plant crude drugs of Anantapur district, Andhra Pradesh, India", *International Journal of Crude Drug Research,* **27**(3), 145–155 (1989).
[103] Ali L., Khan A.K., Mamun M.I., Mosihuzzaman M., Nahar N., Nur-e- Alam M., Rokeya B., "Studies on hypoglycemic effects of fruit pulp, seed, and whole plant of *Memordica charantia* on normal and diabetic model rats", *Planta Medica.*, **59**, 408–412 (1993).
[104] Sharma L.D., Bahga H.S., Srivastava P.S., "*In vitro* anthelmintic screening of indigenous medicinal plants against haemonchuscontortus (rudolphi,1803) cobbold, 1898 of sheep and goats", *Indian Journal of Animal Research,* **5**(1), 33–38 (1971).
[105] Peckham C.G., "Preparation of Food and Food Products 2nd Ed.", *The McMillan Co. Ltd.,* London. p. 420–424 (1969).
[106] Singh I.S., "Post harvest handling and processing of fruits and vegetables", *Westville publishing house,* pp. 116–127, New Delhi (2009).
[107] Srivastava R.P., Kumar S., "Fruits and vegetable preservation principles and practices", *IBDC Publishers,* Lucknow, UP, p. 131 (2012).
[108] Bala B.K., "Drying and Storage of Cereal Grains", *Oxford & IBH Publishing Co. Pvt. Ltd,* India (1997).
[109] Zaman M.A., Bala B.K., "Thin Layer Solar Drying of Rough Rice", *Solar Energy.*, **42**, 167–171 (1989).
[110] Bala B.K., "Solar Drying Systems: Simulation and optimization", *Agrotech Publishing Academy,* India (1998).

[111] Muhlbauer W., "Present status of solar crop drying", *Energy in Agriculture*, **5**, 121–137 (1986).

[112] Sharma V.K., Colangelo A., Spagna G., "Experimental investigation of different solar driers suitable for fruits and vegetable drying", *Renewable Energy*, **6**(4), 413–424 (1995).

[113] Jadhav D.B., Visavale G.L., Sutar P.P., Annapue U.S., Thorat B.N., "Solar cabinet drying of bitter gourd: Optimization of pretreatments and quality evaluation", *International Journal of Food Engineering*, vol. 6 issue 4. pp. 1–20 (2010).

[114] Jie C., Zhou Y., Fang S., Meng Y., Kang X., Xu X., Zuo X., "Mathematical modeling of hot air drying kinetics of *Memordica charantia* slices and its color change", *Advance Journal of Food Science and Technology*, **5**(9), 1214–1219 (2013).

[115] Mudgal V.D., Pande V., "Optimization of pretreatments for dehydration of bittergourd", *Journal of Agricultural Engineering*, **46,** 1 (2009).

[116] Anonymous, Post-harvest management and value-addition DARE/ICAR annual report 2008–2009.

[117] Aziz M.G., Roy J., Sarker M.S., Yusof Y.A., "Isolation and use of bitter gourd polysaccharide in formulating dietetic soft drinks", *African Journal of Agricultural Research*, **6**(23), 5314–5319 (2011).

[118] Drewnowski A., Gomez-Carneros G.C., "Bitter taste, phytonutrients, and the consumer: A review", *American Journal of Clinical Nutrition,* **72**, 1424–1435 (2000).

[119] Huang C.J., Wu M.C., "Differential effects of foods traditionally regarded as 'heating' and 'cooling' on prostaglandin E2 production by a macrophage cell line", *Journal of Biomedical Science,* **9**, 596–606 (2002),

[120] King, J.W., Grabiel, R.D., Isolation of polyphenolic compounds from fruits or vegetables utilizing sub-critical water extraction, US 7208181 B1, US Patent (2007).

[121] Budrat P., Shotipruk A., "Extraction of phenolic compounds from fruits of bitter melon (*Memordica charantia*) with subcritical water extraction and antioxidant activities of these extracts" *Chiang Mai J Sci.,* **35**(1), 123–130 (2008).

[122] Braca A., Siciliano T., Manuela D., Maria P.G., "Chemical composition and antimicrobial activity of *Memordica charantia* seed essential oil", *Fitoterapia.*, **79**, 123–125 (2008).

[123] Zong R.J., Leonard M., Marita C., "Postharvest physiology and quality of bitter melon *(Memordica charantia* L.*)*", *Postharvest Biology and Technology,* **6**, 65–72 (1995).

12

Eggplant Processing and Extraction of Bioactive Components

NEEL R. PARIKH[1], SHIVANAND S. SHIRKOLE[2], PARAG P. SUTAR[3*] AND NAMRATA P. SUTAR[4]

ABSTRACT

Eggplant (Solanum melongena) belongs to the night shaded family Solanaceae and is one of the major world's vegetable. Eggplants are most widely used for the preparation of a wide range of products in different Indian and Western cuisines. The phytochemicals and anthocyanin pigment make eggplant deep in purple, green, white, and yellow color. The eggplant peel is a source of useful products. The peel of purple eggplant fruit contains antioxidants like ascorbic acid and phenolics compounds like nasunin and anthocyanin. Eggplants are rich in dietary fiber, vitamins, and minerals and have many health benefits like very low calories, reduces excess iron from the body, control obesity, reduce glucose level, act against coronary heart disease. Eggplant is used to prepare many products and extract important components that have medicinal and commercial importance. The processing of eggplant is done by different methods like heating, cooking, frying, baking,roasting, canning etc. The seeds of eggplant are used for medicinal purpose while the peel is used to extract the color. The major concern in the processing of eggplant is a rapid enzymatic browning reaction. Though, having greater health benefits of eggplants, it's processing and property of rapid browning reaction has drawn more attention.

[1,4] College of Food Processing Technology and Bio-Energy, Anand Agriculture University, Anand – 388 110, India.

[2,3] Department of Food Process Engineering, National Institute of Technology, Rourkela 769008 – Odisha, India.

**Corresponding author* E-mail: sutarp@nitrkl.ac.in

***Key words*:** Recipes, Nutritional components, Eggplant processing, Eggplant storage.

INTRODUCTION

Eggplant is domesticated from Indian species *Solanum melongena*, a perennial vegetable. The eggplant is a member of night shaded family such as tomatoes, peppers, potatoes and sweet potatoes. It is also known as the aubergine, melongene or guinea squash. In Figure 1, an eggplant tree with fruits is shown. The eggplant fruit is botanically classified as a berry.

Fig. 1: A typical eggplant tree

The eggplant is cultivated as an annual crop in temperate climates. The leading countries in eggplant cultivation are China, Turkey, Egypt, Iran, Japan, Indonesia, India and Italy. The global area under eggplant cultivation has been estimated at 1.85 million ha with total production of eggplant fruit of about 32 million MTs[1]. India accounts for about 8.7 million MTs with an area of 0.53 million hectares under cultivation. Eggplant is also exported in the fresh or frozen form. Spain and the United State of America exports 110090 and 15574 tones of eggplant, respectively are the leading countries in the world[2].

The eggplant grows up to 40 to 150 cm (16 to 57 inch) with large, coarsely lobed leaves that are 10 to 20 cm (4 to 8 inch) long and 5 to 10 cm (2 to 4 inch) broad. It contains numerous small seeds and flesh. The outer skin of eggplant is in different shades like lavender, green, yellow-white and orange. The inside flesh is cream colored with bitter taste due to the presence of seeds. The flesh is edible and contains small amount of nicotinoid alkaloids and has spongy consistency (meaty texture). The eggplant fruit is directly used for cooking purpose due to its several health benefits. Different recipes are made from the fruit and it is also exported in other countries. India ranks 2nd in world production of eggplant and produces 11896000 MT[1].

Many varieties of eggplants are grown throughout the world. These mainly include Chinese, Pea, Indian, Japanese, Thai, Pingtung, Italian, white, Sicilian, African, and Philippine eggplants. Eggplants have good nutritional value along with health benefits, therefore it is popular and used in several recipes. It is rich in dietary fiber, vitamins, minerals and has good amount of antioxidant which is beneficial for human health. Eggplant is used for cooking purpose and extraction of different useful compound like-pigment and phytochemicals. Around the world many popular cuisines like grilled eggplant with meat, eggplant rice, eggplant pickle and chutney, *sambhar, bharta, raita, curry, Italian fried eggplant ball, eggplant salad* and many more cuisines are prepared from eggplants. From eggplant, purple and red pigment are extracted which is used as a dye in textile and paint industries. The extracted pigment is also used in thermal and non-thermal food. As all the fruits require specific storage conditions, eggplant is also requiring specific storage conditions. Eggplant is heat and cold sensitive fruit, the ideal storage temperature is 10°C.

History of Eggplant

The native of eggplant is India as wild eggplants grew in this country, however they were first cultivated in China during 5th century B.C. It has been cultivated in southern and eastern Asia since prehistory, but appears to have become known to the Western world not earlier than circa 1500. Eggplant was introduced to Africa before the Middle Ages and then into Italy, the country with which it has been associated in the 14th century. In early Middle Ages, the Arabs and North Africans brought eggplants into the Mediterranean region from where they spread throughout Europe and the Middle East. In 17th and 18th centuries, the eggplant was known in Britain as the "Jew's apple" because of its great popularity among Sephardi Jews, who may have brought it to England. In Europe, people considered eggplant as a decorative garden plant than food because they thought it was poisonous one. But as new varieties of eggplant developed in the 18th century, eggplant lost its bitter taste and became a popular vegetable in European cuisine. Today it is a popular and important vegetable in the Indian, Italian, Greek, Turkish and French kitchens.

TYPES OF EGGPLANT

There are many different types of eggplants available in various colors like purple to white or green colour. They are described below and shown in Figs. 2 to 12.

Fig. 2: Chinese eggplant

Fig. 3: Pea eggplant

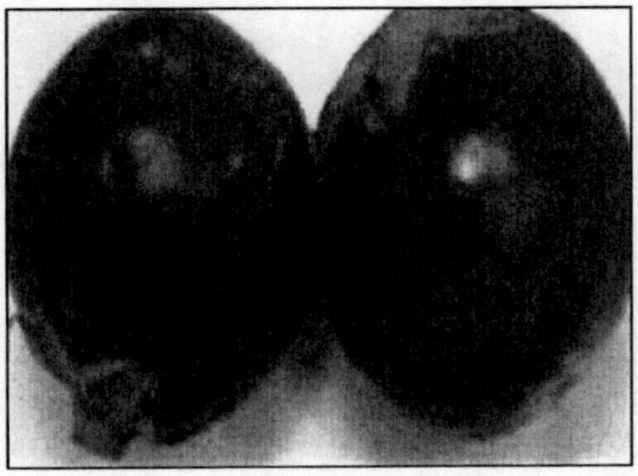

Fig. 4: Indian eggplant

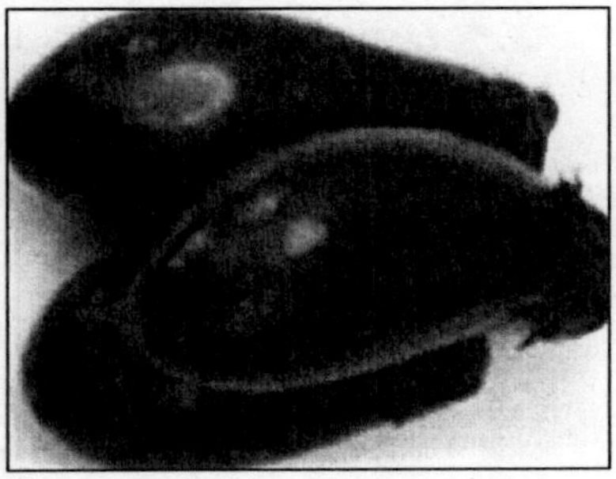

Fig. 5: Japanese eggplant

Fig. 6: Pingtung eggplant

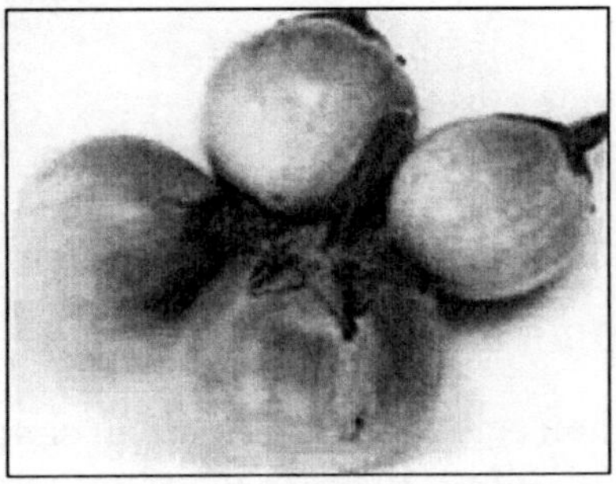

Fig. 7: Thai eggplant

Chinese Eggplant

Chinese eggplants are long but a bit cylindrical. The eggplants are purple in color, sweeter and tender. The skin of the fruitis soft and is best eaten with the peel.

Pea Eggplant

It is Thai eggplant and moderately bitter. They are sold in clusters and look like green peas. It is found in Asian markets.

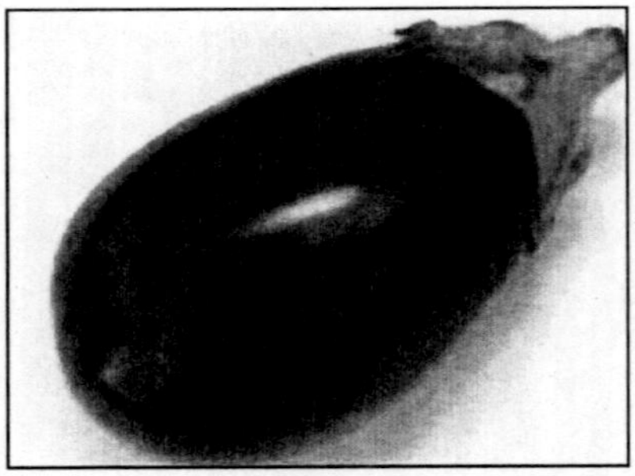

Fig. 8: Italian eggplant

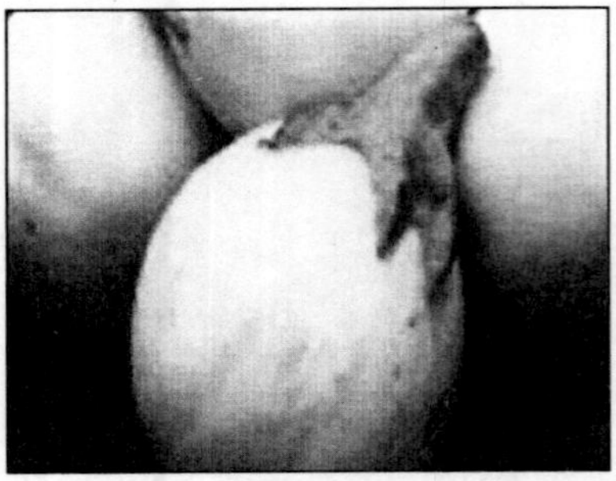

Fig. 9: White eggplant

Fig. 10: Sicilian eggplant

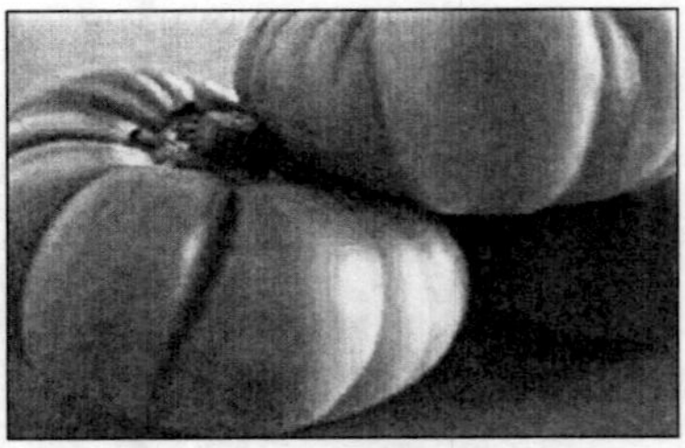

Fig. 11: African eggplant

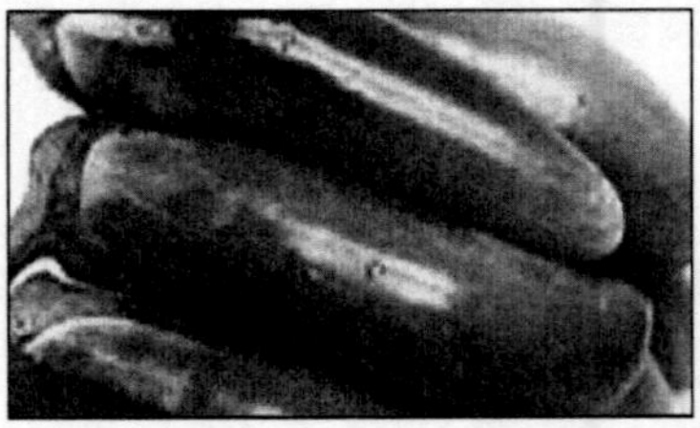

Fig. 12: Philippine eggplant

Indian Eggplant

Indian eggplants are round in shape. They are deep purple to red-purple in color. These types of eggplants are tender with a sweet flavor. They are more fragile than other eggplants, but are perfect for stuffing, baking and used in fish curry.

Japanese Eggplant

Japanese eggplants are similar to the Chinese eggplant. Only it is sweet in taste, has a thin skin and flavor. It is slender and comes in green, pink, white, lavender or purple colour. It can be grilled, steamed, simmered, fried, and pickled. In Japan, tempura eggplant is a popular cuisine which is prepared from the Japanese eggplant.

Pingtung Eggplant

It is very popular in Asia. Pingtung eggplant is long; bright to dark purple, and very tender and sweet in taste. It is named after its native town of Ping Tung in Taiwan from where it originated. It is best prepared when grilled with salt and olive oil.

Thai Eggplant

These are ball shaped small fruits. It is bitter in taste, therefore before utilization for cooking its seeds are removed. It is available in a variety of colors, but normally in green with yellow or white stripes. It is often cubed and used in the curry dishes.

Italian Eggplant

With a deep shade of mauve-purple and some light streaking on the skin, this type of eggplant is smaller and more oval in shape than the regular variety. Italian eggplants have a green calyx.

White Eggplant

The name of white eggplant is due to its white color, and looks like eggs hanging from the plant. It has a tough skin and a delicate flesh. It is smooth and white sometimes it can be round or slightly thinner and longer, like the Italian eggplants. Some popular commercial varieties include albino and white beauty.

Sicilian EggplantThis type of eggplant is small in size with a large base and purple skin streaked with white. It is sometimes called "*Zebra*" or "*Graffiti*" eggplant.

African Eggplant

Many varieties of eggplant are being cultivated in Africa. Some varieties are round in shape and some look like tiny pumpkin. These are available in purple, white and green varieties, but tend to be relatively bitter. Therefore, they are used mainly as an accent ingredient in soups and stews.

Philippine Eggplant

These eggplants are similar in shape to the Japanese and Chinese eggplants. They are distinguished from them by the mix of purple and green color.

COMPOSITION OF EGGPLANT

Proximate and Chemical Composition

Carbohydrates

Carbohydrates not only give energy but also helps in digestion system. It promotes the utilization of fats and reduces wastage of proteins. The carbohydrate present in raw eggplant is 4% and cooked eggplant contains 3%.

Dietary fiber

Dietary fiber is helpful for regulating and facilitating easy bowel movement. It is helpful for the proper functioning of the digestive system. It also acts against coronary heart disease. Eggplant contains 1.3% dietary fiber. It promotes cardiovascular health. And, helps to lower the cholesterol content in the body. The skin of eggplant is rich in fiber, therefore, eggplants are always consumed unpeeled.

Fatty acids

Eggplant contains 0.3% fat which is good for the diabetic people and others suffering from heart and cholesterol problem. Eggplant has low saturated fat, cholesterol and sodium content and contains essential fatty acids like omega-3 and omega-6. These fatty acids help in proper brain functioning and prevent diseases like cardiovascular and cancer.

Vitamins

Vitamins regulates metabolism to convert fat and carbohydrates into energy. It also assists in forming bones and teeth. Eggplant contains vitamin C, A, folic acid, thiamine, riboflavin and other in very small amount. Eggplant is a natural source of vitamin A. It would play an important role for vision and eye health because vitamin A has been recognized as a critical factor in child health and survival[3]. The availability of vitamins in eggplant is given in Table 1.

Table 1: Vitamins per 100g of edible portion of eggplant [4]

Vitamin	*Value per 100 g*
Vitamin C (mg)	12
Thiamine (mg)	0.04
Riboflavin (mg)	0.11
Vitamin A (I.U.)	124.0
Folic acid (mg)	34
β-carotene (mg)	0.74

Minerals

Minerals are the main constituents of bone and teeth. They regulates the metabolism of enzymes, maintain acid-base balance and osmotic pressure of body. Peel of eggplant is rich in minerals. Eggplant contains macro minerals as well as some micro minerals. Macrominerals are- calcium, phosphorus, sulphur, magnesium, sodium, potassium and chlorine. Microminerals are-iron and zinc. Macrominerals help in proper heart functioning with boosting immunity. The availability of minerals in eggplant is given in Table 2.

Table 2: Mineral content per 100g of edible portion[4]

Minerals	*Value per 100 g*
Calcium (mg)	18.0
Magnesium (mg)	15.0
Phosphorus (mg)	47.0
Iron (mg)	0.38
Zinc (mg)	0.32
Sodium (mg)	3.0
Potassium (mg)	2.0
Sulphur (mg)	44.0
Chlorine (mg)	52.0

Arivalagan *et al.*[5] investigated variability in mineral composition of Indian eggplant (*Solanum melongena*) genotypes. The term genotypes represented diversity on the basis of their morphological characteristics, mainly color, shape and size of the fruit. They used twenty-five genotypes of eggplant comprising 20 germplasm accessions and 5 commercial varieties for experiment. Cleaned and surface-dried fruits were sliced longitudinally into small pieces representing the whole sample. The samples were dried at 65 °C in a hot-air oven for 72 h until constant dry weight was achieved and ground into powder with a mechanical grinder for use in estimating mineral content. They observed that, the majority of the germplasm accessions are rich in all macro- and micro-minerals as compared to commercial varieties. Germplasm accessions IC090116, origin from Assam, light purple color and long shape; IC249323 origin from Odisha, green colour and oval shape and IC090146 origin from Assam, green color and oblong shape varieties identified as a potential sources for potassium, iron and zinc.

Eggplants have good nutritional value and health benefits. It is rich in dietary fiber, vitamins and minerals. They contain negligible calories which make eggplants an ultimate component of both, low-fat and weight loss diets. The essential minerals present in eggplants include potassium and magnesium. The overall nutrition composition of eggplant is given in Table 3[4].

Table 3: Composition of eggplant per 100 g edible portion [4]

Nutrient	*Value/100 g (raw)*
Calories, Kcal	24.0
Moisture content (%)	92.7
Carbohydrates (%)	4.0
Protein (%)	1.3
Fat (%)	0.3
Fiber (%)	1.3

Phytochemicals

Phytochemicals are chemical compound that occur naturally in plants, which are responsible for pigment and organoleptic properties. They have potential to affect diseases such as cancer; stroke or metabolic syndrome. Eggplant contains phytochemicals which are flavonoids and chlorogenic acid. The flavonoid nasunin in eggplants has high levels of antioxidant properties, that helps to reduce free radicals and protect the cells in the body. Nasunin is also known for its ability to protect the facts surrounding the cell membrane of the brain.

An estimation of phenolic acids and antioxidant capacity in eggplant pulp extracts was worked out by Luthria[6]. Simple UV spectral scan method was used for this estimation. Four different varieties Orient Express, Orient Charm, Italian Neon, and Calliope Zebra Stripe were selected for this estimation. All eggplant samples were peeled with kitchen peeler (1–2 mm peels) and were cut into small cubes (1–3 cm). The peels and the eggplant pulp samples were stored separately at –60°C under an inert nitrogen atmosphere in zip lock bags. Frozen samples were lyophilized and the freeze-dried samples were ground in a coffee grinder. This dried powdered material was passed through a standard 20 mesh sieve (particle size < 0.825 mm), well mixed and stored under nitrogen at –60°C until extracted and analyzed. These samples were scanned in between 220 and 700 nm using UV spectrophotometer. It was found that, maximum absorption of phenolic acid 5-O-caffeoylquinic acid take place at 330 nm, where the quantity of 5-O-caffeoylquinic acid is higher. He reported that UV spectral scan can be used as a simple high throughput screening low cost method for predicting the antioxidant potential and phenolic acid content of eggplant pulp extracts without peels. Also, it can be used for evaluating the influence of growing conditions, environment, and soil type on the phenolic acid content and antioxidant potential of eggplant pulp extracts.

Antioxidants

Antioxidants play an essential role in the prevention of chronic and degenerative diseases such as cardiovascular diseases and cancer. Eggplant fruit contains ascorbic acid and phenolics, both are powerful antioxidants. Eggplant extracts suppress the development of blood vessels required for tumor growth and metastasis and inhibit inflammation that can lead to atherosclerosis. Extracts from eggplant fruit skin have been demonstrated to possess high capacity in scavenging of superoxide free radicals and inhibition of hydroxyl radical generation by chelating ferrous iron. Nasunin, an anthocyanin isolated from the peel of purple eggplant fruit, is a phenolic compound implicated in both inhibition of hydroxyl radical generation and super oxide scavenging (SOS) activity. Flavonoids isolated from eggplant showed potent antioxidant activity against chromosomal aberrations induced by Doxorubicin. Various parts of the plant are useful in the treatment of inflammatory conditions, cardiac debility, neuralgia, ulcers of nose, cholera, bronchitis and asthma, analgesic and hypolipidemic[3].

Antioxidant activities of four different varieties (long green, purple colored big size, purple colored moderate size and purple colored small size) of eggplants were studied on the basis of scavenging capacity of 2, 2-diphenyl-1-picrylhydrazyl radical (DPPH) by Nisha *et al.* [3]. They found that purple colored small size eggplant showed better antioxidant activity as compared to other varieties of eggplant. Singh *et al.* [7] carried research on polyphenol content and antioxidant activity of hybrid varieties of eggplant cultivated in United State, and Japan namely, Blackbell and Millionaire. Low-density lipoproteins (LDL) antioxidant activity was found to be correlated with the phenolic content of eggplant varieties grown under organic and conventional environments. Blackbell variety had higher or equal phenolic content and antioxidant activity than Millionaire variety. The phenolic and the low-density lipoproteins (LDL)-antioxidant activity of the both varieties showed good correlation with the phenolic content.

Akanitapichat *et al.* [8] studied antioxidant activity of five different varieties of eggplant. They selected uniform purple size, white and green, long green striped green and pale green type eggplants. They reported that purple, white and green and strip green eggplants had higher total phenolic and flavonoids content, while long green eggplants had low total phenolic and flavonoids content.

Concellon *et al.* [9] noticed changes in quality and phenolic antioxidants in dark purple American eggplant as affected by storage at 0°C and 10°C. They harvested eggplant fruit in December. The fruits were washed, air dried and randomly packed in groups of two in plastic (PET) trays and covered with perforated PVC (50 μm thick). Packed fruit were stored at 0°C and 10°C (85–90% RH) for 0, 3, 5, 10 and 14 days. They observed that in the fruit stored at 0°C for 3 days there was increased antioxidant activity, but afterwards the considerable degradation was observed. A steady but constant increase of antioxidant was detected in fruit stored at 10°C.

Nicotine

Nicotine is a potent parasympathomimetic alkaloid found in the nightshade family of plants (*Solanaceae*). It is a colourless to pale yellow oily base with acrid burning taste. Nicotine (3-(1-methyl-2-pyrrolidinyl) pyridine) is a clear, naturally occurring liquid found in several species of eggplants. As body gives response to nicotine in terms of increases in blood pressure, heart rate, and blood flow from the heart, it is important to study the nicotine content of eggplant. Shrivas *et al.* [10] studied nicotine present in white, green and purple eggplant and reported 9.3, 10.3 and 12.5 mg per g, respectively. Relative standard deviation (RSD) in white eggplant, green eggplant and purple eggplant were 3.4, 2.5 and 3.1 percent, respectively. The compound is very hygroscopic and turns brown on exposure to air or light. Nicotine is a clear, naturally occurring liquid found in several species of plant.

Health benefits of eggplant

Nasunin, the potent antioxidant present in eggplant, shields the brain cell membranes from any kind of damage and injury. This is really important as the membrane protects the cells from free radicals, allowing the entry of nutrients and exit of harmful wastes. It receives instructions from messenger molecules telling what activities the different cells should perform. Consumption of eggplant also helps in reducing excess iron from the body. Though, iron is essential for many biological processes in the body, excess accumulation of iron is not beneficial and causes certain harmful effects in the body. Increased accumulation of iron causes rise in the production of free radicals in the body and is associated with high risk of heart disease and cancer. Hence, it is necessary to remove the excess iron to reduce the free radical production. This can be done by the chelating action of nasunin. Chelating iron protects blood cholesterol from peroxidation, reduces the rate of free radical damage in joints and prevents cellular damage that can promote cancer formation.

Eggplant act as antimicrobial, antiviral, antimutagenic and anti LDL, all of which is attributed to the action of a phenolic compound, chlorogenic acid, found in abundance in the vegetable. The acid is one of the most potent free radical scavengers found in the tissues of plants. The potassium content in the vegetable helps to strike a balance in the salt intake and maintain a good hydration level. The potassium level also plays a vital role in regulating the blood pressure of the body. Eggplants are very low in calories and it has high moisture content. Eggplants are used to reduce glucose levels of type II Diabetes. This is attributed to the fact that the vegetable contains low levels of carbohydrates and high fiber levels. Hence, eggplants can be safely and effectively used to control obesity. The plant also has a high source of dietary fiber, which is helpful for the proper functioning of the digestive system, and also acts against coronary heart diseases.

Chemical activity (browning)

Eggplant is widely used for the fresh fruit consumption as well as in food processing industry. But due to lack of the proper pretreatment, freshly cut eggplant undergoes enzymatic browning reaction. Eggplant becomes soft during storage and chilling. The major compound responsible for browning is polyphenol oxidase and softening is owing to pectin methylesterase and polygalacturonase.

The study of browning and softening of eggplant was carried out by Barbagallo *et al.*[11]. To avoid the browning and softening in eggplants, pretreatments like- chemical (calcium citrate and ascorbate), thermal (60°C for 1min) were used by them and various properties like-polyphenol oxidase, pectin methylesterase, polygalacturonase, total phenolics, colour parameters, firmness and total visual quality were monitored during 10 days storage at 4 ± 0.5°C in passive atmosphere (O_2: 20.20%; CO_2: 0.50%). After the minimal processing, they dipped eggplant cubes in Ca salts (citrate or ascorbate). The acceptable result was reported by them with calcium ascorbate solution. They found that the enzymatic activity of eggplant was reduced and overall acceptance was improved. Quality was excellent up to day 4 of storage and acceptable up to day 7.

Mishra *et al*. (2012)[12] studied on browning of fresh cut eggplant. In their experiments, they cut the eggplant fruit using sharp blades into a cube of approximately 3g each. A sharp blade (thickness about 0.04 mm) and conventional vegetable cutting knife (blade thickness about 0.25 mm) were separately used. After cutting, they dipped cubes in water for 10 min, air-dried on a clean surface, packaged in styrofoam trays and wrapped with stretchable film. The sample was stored at ambient (26 ± 2°C), 10 ± 2°C and 4 ± 2°C and examined. They concluded that the sharp blade cut and water-dipped eggplant had markedly less browning, cell death and physical injury to fruit. It resulted into reduced ooze of phenolics compounds and polyphenol oxidase activity and lesser browning compared to eggplant cut with the conventional knife showing significant browning. The shelf life of the sharp blade cut eggplant was extended up to 16 days at 4°C, 12 days at 10°C, and 5 days at 26°C storage temperatures.

POST-HARVEST HANDLING OF EGGPLANT

Eggplant fruit is harvested when the peel of eggplant is deep glossy purple colored and prior to hardening and browning of its seed. Over-mature fruits have a dull color, spongy feel, crinkled skin and wrinkled stem. Traditional teardrop shape fruit varieties are ready for harvest 50 to 70 days after transplantation or one month after fruit set. These varieties can crop for up to five months. Market trends have been towards smaller size fruit 7 to 10 cm diameter and 12 to 15 cm length. Long, slim, cylindrical varieties can be picked when fruits become 1 to 2 cm in diameter and 5 to 10 cm length.

Eggplant fruit is required to be handled very carefully after harvesting to avoid the damage to the fruit itself or to other fruit. After harvesting the fruit is pre-cooled immediately to remove the field heat. To avoid higher ethylene production rate and/or to avoid over ripening, removal of field heat is very essential. Different cooling techniques are used to remove field heat. Pre-cooling is most important method of removing field heat.

Precooling Methods

For pre-cooling of eggplant following methods are most commonly used:

Room cooling

Eggplants are placed in an insulated room equipped with refrigeration units to chill the air. This method can be used with most commodities but is time consuming. Used refrigerated truck bodies makes excellent small storage rooms.

Forced air cooling

Used in conjunction with a cooling room, forced-air cooling is effective for most packaged eggplants, and is 75 to 90% faster than room cooling. Fans should be equipped with a thermostat that automatically shuts them off as soon as the desired product temperature is reached. It is advised not to operate forced-air fans after the fruits have been cooled to its optimum temperature. A low-cost, portable, forced-air pallet cooling system has been developed by people for small quantities of fruits. The optimum temperature of eggplant storage is 8 to 12°C and humidity 85-90% and normal storage life 1 week[13].

PROCESSING OF EGGPLANT

The eggplant processing is done by several methods like- baking, cooking, frying, drying, freezing, fermentation etc. There are several recipes of eggplant throughout the world prepared by above methods or their combination. The brief overview of the processing methods used for recipes is given in Table 4.

Baking

Baking is a food cooking method using prolonged dry heat achieved through convection, rather than by thermal radiation. Baking can be combined with grilling (radiation) to create hybrid method for cooking. The equipments used in baking method are gas oven and baking oven. In baking of eggplant, many cuisines like pizza, sandwich or baked eggplant served with different sauces are prepared. The top green portion of eggplant and peel are removed (not necessary for all food recipes). Eggplant is sliced and covered with flour and bread crumbs and put in microwave oven. Oil drops are put on the slices and baking is done at suitable time-temperature combination. In some countries, gravy of tomato and onion with spices is poured in the baking tray with eggplant slices and grated with cheese. This is served with rice or *chapatti* (unleavened bread)[14, 15, 16].

Flame Cooking

Flame cooking is a food processing method in which food directly comes in contact with fire. This method is easy, but little time consuming for food to cook properly. In this process, oil is applied to the eggplant skin and pricked with fork to pass steam inside the flesh. Eggplant is put directly on medium gas flame to burn outer skin and cook inside flesh. The eggplant is rotated time to time so that it cooks properly from all the sides. After proper cooking, the outer burnt skin of eggplant is removed and the flesh of eggplant is used for making cuisines like-*Bharta* (popular Indian dish by adding tomatoes, onions and spices) or *Raita* (cooked eggplant flesh is mixed with curd and spices)[17].

Frying

Several products are prepared from eggplant using frying. Frying is the cooking of food in oil much more quickly and it gives a characteristic crispiness and texture to the product. There are many different frying techniques-Sautéing, stir frying, pan frying, shallow frying and deep frying. In Sautéing and stir-frying, food is cooked in a thin layer of fat on a hot surface, like frying pan or griddle. Stir frying involves heating quickly at very high temperatures, during which the food is stirred continuously to prevent it from adhering to the cooking surface and burning. Shallow frying is a type of pan frying using only enough fat to immerse approximately one-third to one-half of each piece of the food; fat used in this technique is typically used once. Deep frying is a cooking method in which food is submerged in hot oil. Frying is used for making *Pakoda* (thin slices of eggplant are cut and coated with a batter made from spiced Bengal gram flour and subjected to frying). *Pakoda* are served with tomato sauce and green chutney (coriander leaves and green chilly crush in mixture by adding salt, sugar and lemon juice) [18,19].

Stuffing

In stuffing, an edible substance or mixture is filled in another food item. Many food items like potato, onion, tomato, spices, meat, poultry, cheese, etc. are used as stuffing material for eggplant. In this method, inside flesh or pulp of eggplant is removed to make it hollow and prepared materials are stuffed inside. Eggplant is stuffed with different products like-coconut, spices-onion, tomato, *paneer*, cheese, meat, *arhar* flour and other similar products. It is a conventional recipe in India as well as in western countries. Generally, it is served with curry and/or with *chapatti* (unleavened bread) or rice[16, 20].

Freezing

Freezing is one of the most popular ways of long-term storage of foods. By transforming most of the liquid water existing in food materials into ice, freezing greatly slows down the physical and biochemical changes involved in the deterioration of foods as well as the growth and reproduction of spoilage microorganisms. Generally, freezing preserves the taste, texture and nutritional value of foods better than any other preservation technology. Eggplants, their products, and ready to serve recipes are frozen and can be made available during off season. For freezing, the good quality eggplants are selected and washed properly so that dust and other unwanted ingredients are removed. After washing and peeling, the eggplant are cut in suitable slices and blanched. During blanching, lemon juice is added and the water level is maintained so that eggplant is covered with water. Blanching is a thermal treatment given to eggplant for very short period of time to inactivate enzymes and retain the color. Also, microorganisms are reduced and gases which are present in tissue are also removed. Eggplants are blanched up to 5 minute in boiling water and rapidly cooled in ice water for 5 minutes to avoid over cooking. After cooling, the eggplants can be stored in ziplock bag and vacuum is created in the bag. The packaging bag is kept in freezer. Freezing is done by different techniques for eggplant and its products. These include individual quick freezing and dehydrofreezing.

Fermentaion

Fermentation in food processing, typically is the conversion of carbohydrates to alcohols and carbon dioxide or organic acids using yeasts, bacteria, or a combination thereof, under anaerobic conditions. Fermentation in simple terms is the chemical conversion of sugars into ethanol. The science of fermentation is also known as zymology or zymurgy.

Sesena *et al.*[21] studied contribution of starter culture to the sensory characteristics of fermented Almagro eggplants. Almagro eggplants are pickled products made using a native variety of eggplant grown in Spain, *Solanum melongena* L. var. *esculentum depressum*. This product is highly popular in Spain and in other Mediterranean countries. The eggplants are allowed to undergo spontaneous fermentation, relying on colonization by microbial populations associated with the raw materials, equipment, and local environments. Use of commercial vegetable starters in the manufacture of Almagro eggplants has shown that such starters accelerate the fermentation process and therefore shorten fermentation time. However, an unacceptable and uncharacteristic bitter taste of the product, attributed to the starter employed, was the cause of rejection by most taste panelists. To avoid this problem, it will be necessary to prepare an appropriate starter culture by characterizing the microbial flora present in the spontaneous fermentation of Almagro eggplants. In their work they selected medium size Almagro eggplants. They blanched eggplants and immediately cooled to 20°C. They covered blanched eggplant with 4 L of brine. Lyophilized cultures of *L. brevis* strain, *L. plantarum* strain and *L. fermentum* strain were used to prepare the starter cultures. Three starter culture F1, F2, and F3, were considered. F1 was prepared from strains *L. brevis* and *L. plantarum*; F2 was prepared from strains *L. fermentum* and *L. plantarum*; and F3 was prepared from a combination of all three strains. The starter cultures were added 2 days after the fruits had been placed in the brine. The eggplants were allowed to ferment for 12 days at a controlled temperature of 32 ± 2°C. They reported that, the contributions of the three starter cultures were not the same, although no differences had been observed in the fermentation of the three batches that were made using the starter cultures. The most appropriate starter for the manufacture of Almagro eggplants should contain the obligate heterofermentative species *L. brevis* and the facultative heterofermentative species *L. plantarum*. The heterofermentative species, *L. fermentum*, present in starter culture had found difficulties in growing during fermentation and exerted little influence on the sensory characteristics of the eggplants produced.

Individual quick freezing

Individual Quick Freezing (IQF) is an important technology available in freezing. Quick freezing is only process where virtually all the properties of food stuffs can be preserved. The important feature of this process is ultra-rapid freezing to very low temperatures (–30°C to –40°C) designed to stop the activities of the microorganisms that cause decay and deteriorate food stuffs. It is used to preserve and store raw fruit and vegetables in the same farm-fresh condition for more than a year, with the color, flavor and texture of produce remaining as good as fresh from the farm. In IQF, there is a rapid method of freezing in which each piece of food is frozen individually using technique of fluidization resulting in freezing of fruit and vegetables in 10 to 12 minutes which otherwise takes at least 3 to 4 hours or even more in the blast freezer. This results into better texture and there is no lump or block formation and the product is free flowing. The eggplant is frozen either as pieces or its ready to serve products are frozen. Diced IQF eggplant is stored at –18°C. Eggplant recipes like *Aloo Baingan* (sweet and spicy curry of potatoes and eggplant) and *Baingan Bharta* (flame cooked eggplant, in which outer burned skin is removed and flesh is mixed with onions, tomatoes and spices).

Vacuum dehydrofreezing

Dehydrofreezing is one of the newly developed food freezing techniques in which fresh or minimally processed plant products are dehydrated to an appropriate moisture level prior to freezing. Two major processes, specifically dehydration and freezing, are involved in dehydrofreezing technique; both processes greatly influence the quality of the final product. The dehydration treatment not only reduces the amount of water to be frozen but makes cell structures less susceptible to breakdown by changing cell turgor pressure. As a consequence, the damage to plant cells caused by ice crystal formation and the post-thawing quality degradation such as color change, softening and loss of flavor and nutritional substances may be alleviated. The application of vacuum-dehydration technique can help to overcome the above mentioned drawbacks of the present water removal methods and improve the quality of final product[22].

Some of the experiments on vacuum-dehydrofreezing for the long-term preservation of fresh-cut eggplant were done by Wu *et al.*[22]. They applied three treatments on cut eggplant. In first partial dehydration treatment-samples were vacuum dried to target moisture contents 60 and 80% (w.b.) at 30°C and 2.5 kPa pressures. In freezing-thawing, they applied four freezing-thawing conditions, –40°C air or immersion freezing and 5°C low temperature or 25°C room temperature thawing conditions. The samples at the three moisture levels (60%, 80% and without dehydration) were put into zip top freezer bags before freezing in order to prevent the samples from making direct contact with refrigerating medium. Air freezing was performed at –40°C in a freezer; immersion freezing was achieved by soaking the freezer bags in the refrigerating medium (99.5% ethanol) stored in a sealed beaker, which was placed in the freezer 12 h before the experiment started. The frozen samples were then thawed at 25°C in a constant temperature oven or at 5°C in a domestic refrigerator. After thawing treatment rehydration treatment was applied. The dehydrated samples were rehydrated by soaking in 100 mL distilled water at 25°C. By looking at results it was concluded that, vacuum-dehydration treatment prior to freezing process effectively prevents post-thawing drip loss and inhibit the freezing-thawing deterioration of the eggplant samples to some extent in contrast to conventional freezing. The surface color of dehydro frozen sample will be maintained. The dehydro freezing-thawing-rehydration process caused deterioration in texture of sample, however, the firmness of the dehydrofrozen samples after rehydration was significantly higher than that of the conventionally frozen samples.

Drying

Drying is defined as a process of moisture removal due to simultaneous heat and mass transfer. It is also a classical method of food preservation, which provides longer shelf-life, lighter weight for transportation and smaller space for storage. Egg plant is dried by different method like vacuum, hot air and freeze drying. Vacuum drying, characterized by low processing temperature and low oxygen concentration processing environment, has been widely used to process various food materials, especially those containing heat sensitive substances, to maintain the quality and nutritional value of the products. Wu *et al.* [23] dried fresh eggplant in vacuum chamber at pressures 2.5, 5 and 10 kPa, and drying temperature ranging from 30°C to 50°C. During the experiments, they observed that eggplant undergoes shrinkage. They observed that shrinkage increase as the pressure of drying chamber increases. Hot air drying is one of the most popular methods of drying of agricultural products. Eggplant is dried by hot air at different air temperatures and air velocity and product shape and sizes.

Akpinar and Bicer[24] studied the thin-layer drying behaviour of eggplant slices (6 mm thick layers) in a convective-type cyclone dryer. They conducted experiments at drying air temperatures 55, 65 and 75°C and dry air velocities 1 and 1.5 m/s. In order to estimate and select a suitable form of the drying curve, eight different semi-theoretical and/or empirical models were fitted by them to the experimental data and comparisons made of their coefficients of determination as predicted by non-linear regression analysis. They found that the Page model best described the drying curve of eggplant.

Russo *et al.* (2012) [25] studied on physical properties of hot air dried and rehydrated eggplant. In their work, they cut eggplant in to cylindrical slices with diameter 30 mm and thickness 6 mm. Drying was carried out in conventional dryer at different temperatures 40, 50, 60 and 70°C and air velocity at 2.3 m/s for 270, 225, 165 and 135 min respectively, until the weight was maintained constant. Rehydration was carried out at room temperature for 310 min. After rehydration in water, the samples were spread on absorbent paper for removal of free water on the surface of samples. From the experimental data, they concluded that at 60°C of drying air temperature, the sample was preserved better than other samples. The longer drying time and higher drying temperature results in to the development of wrinkled structure.

Ertekin and Yaldiz[26] conducted some experiments for drying the eggplant using convective method. The drying experiments were carried out in a laboratory dryer with air temperature between 20°C and 120°C and velocity between 0.1 and 3.0 m/s to examine the thin layer drying behavior of eggplants. Their results showed that, increasing the drying air temperature and velocity causes shorter drying times. They also found that in eggplant drying process, pre-treatment must be definitely applied to decrease drying time. Slice thickness do not have any significant effect on the color characteristics. Increasing drying air velocity raises the rehydration ratio, and the highest value was obtained at a drying air temperature of 50°C. In modeling drying data they reported that for eggplant, the model is the best for describing the drying curves of eggplants[27].

The energy consumption by industry is one the important factor to be considered as it decides the cost of final product. Drying is well known process for its high consumption of energy (almost 10% of total energy required for processing). Therefore, attempts are done by researchers which can reduce the drying time of eggplant and save energy. These include use of ultrasound, pretreatments before drying, size reduction etc. In order to properly evaluate and design an ultrasonic application for eggplant, Garcia-Perez *et al.*[28] did experiments on drying of eggplant cylinders (height 20 mm and diameter 24 mm) at 40°C and 1 m/supplying different ultrasonic powers 0, 6, 12, 19, 25, 31, and 37 kW/m^3. They observed that the ultrasonic power had significant ($p<0.05$) influence on both the effective moisture diffusivity and the mass transfer coefficient.

Hybrid Processing (Hurdle Technology)

Many processes are simultaneously or separately used to prepare different products from eggplant. These are flame and steam cooking, roasting and steaming, stuffing and baking, and many more. As a result of these combined methods the taste of product is improve and become delicious. In stuffed eggplant first operation is stuffing and then it may be cooked or baked. Similarly eggplant pickle is preservation technique by adding oil and spices. Pickling is one kind of hurdle technique, a combination of more than two methods of preservation.

Retorting/Canning

Canning is a process where the food is placed in a jar and heated to a temperature that destroy targeted microorganism and to inactivate the enzyme activity. In canning process, the targeted microorganism is *Clostridium botulinum*, produces an extremely potent neurotoxin among the deadliest poisons. Trace amount of this poison cause food-borne illness known as botulism, sufficient to kill people who consume. The spores of *Clostridium botulinum* will not grow if the pH of a food is 4.6 or less. For low-acid foods with a pH value greater than 4.6, these spores must be killed by heating during the canning process. Because these spores are very heat resistant, canned low-acid foods must be pressure-cooked at high temperatures for long periods of time. Eggplant has pH range in between 4.5–5.3, it come in low-acid food category. For eggplant, pressure canning method is used because it is low acid food. For canning of low acid food, temperature of 115.6°C or more are used and process time in range of 20 minutes to several hours. Some of the eggplant recipes, eggplant slices and pickle are canned and exported from India.

CULINARY IMPORTANCE OF EGGPLANT

As eggplant has many benefits, it is widely used in different cuisines worldwide. The brief description of recipes prepared using eggplant is given in Table 4. The ratatouille in the French, or deep fried as in the Italian *Parmigiana di melanzane*, the Turkish *Karniyarik* or Turkish and Greek *Musakka / Moussaka*, and Middle-Eastern and South Asian dishes. In Iranian cuisine, it can be blended with the whey as *Kashk e-bademjan*, tomatoes as *Mirza ghasemior* made into stew as *Khoresh-e-bademjan*. It can be sliced and deep-fried, then served with plain yogurt, (optionally) topped with a tomato and garlic sauce, such as in the Turkish dish *Patlican kizartmasi* (meaning: fried aubergines) or without yogurt as in *Patlican saksuka*. Perhaps the best-known Turkish eggplant dishes are *Imam bayildi* (vegetarian) and *Karniyarik* (with minced meat). Fig. 13 (A to F), some of the differently prepared cuisines are shown. In Table 4, Indian and Western cuisines with its ingredients and method of processing are given.

Fig. 13: (a) Grilled eggplant with meat

Fig. 13: (b) Hot and sour eggplant with garlic

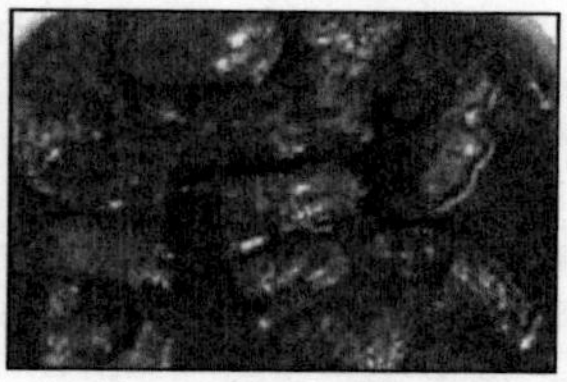

Fig. 13: (c) Eggplant pickle

Fig. 13: (d) Eggplant with rice

Fig. 13: (e) Eggplant slice as a pizza base with pizza topping

Fig. 13: (f) Eggplant chutney

It may also be roasted with its skin until charred, so the pulp can be removed and blended with other ingredients, such as lemon, tahini, and garlic, as in the Middle Eastern *Baba ghanoush* and the similar Greek *Melitzanosalata*. A Spanish dish called *Escalivada* calls for strips of roasted aubergine, sweet pepper, onion and tomato. In Spain, it is typical to find eggplant as *Berenjenas de Almagro*. There, eggplants are also cooked with vinegar, red peppers, paprika and olive oil[15, 29].

As a native plant, it is widely used in Indian cuisine, for example in *Sambhar* (cooked *arhardal* with vegetables, native to South India), *Dalma* (*arhar dal* / pigeon pea preparation with vegetables, native to Orissa),

Chutney (a crush eggplant with spices), *Curry* (vegetables with spices), and Pickles/*Achaar* (eggplant pieces with red chilli, mustard seeds and oil). Owing to its versatile nature and wide use in both every day and festive Indian food, it is often described (under the name eggplant) as the "king of vegetables". Eggplant is stuffed with ground coconut, peanuts, and spices (*masala)*, and then cooked in oil. Indian dish roasted eggplant with spices with or without curd (*Bharta*) is a very popular cuisines and served with rotla (*Chapatti i.e.*, unleavened bread made from bajra), *makke ki roti* (unleavened bread made from maize) or with normal *chapatti* (unleavened bread made from wheat), with or without *lasan ki chutney* (Garlic paste) and buttermilk. Rather than that stuffed eggplant, eggplant rice, *sambhar*, different curries, fried eggplant (eggplant slices coated with Bengal gram flour and fried), *masala* eggplant and many more. A simpler version of the dish, *Begun-pora* (eggplant-charred or burnt), is popular in the east Indian states of Orissa and Bengal, and Bangladesh where the charred pulp is mixed with raw chopped onions, green chilies, salt and mustard oil[16,29,30,31].

Table 4: Eggplant recipes (Indian and western)

Recipe	***Ingredients***	***Method of processing and preservation***
Indian		
Sambhar	Split arhar with vegetables and spices	Steam cooking
Dalma	Split arhar with vegetables	Steam cooking
Chutney	Crushed eggplant with spices	Size reduction and preservation
Curry	Vegetables (potatoes, onions, tomatoes) with spices	Size reduction and preservation
Pickle/Achar	Eggplant with red chili powder, mustard seed and oil	Preservation
Stuffed eggplant	Hollow eggplant filled with different spices, cheese, coconut powder, meat, diced vegetables.	Stuffing
Bharta	Roasted eggplant with onion, tomato and spices	Flame cooking, crushing followed by seasoning with hot oil and spices
Pakora	Thin slices of eggplant covered with Bengal gram flour and spices	Frying
Begun-pora	Eggplant-charred or burnt pulp mixed with chopped onion, green chilies, spices and mustard oil.	Roasting
Brinjal Raita	Roasted eggplant pulp mixed with curd, salt, black paper and red chili powder.	Flame cooking
Western		
Italian Fried Eggplant Balls	Cooked eggplant is pureed with flour; egg and pepperoncini pepper and then dropped by teaspoonfuls into hot oil.	Frying
Eggplant Salad	Baked eggplant, peeled and chopped, then tossed with a balsamic vinaigrette with basil	Baking

Table 4: (*Contd...*)

Table 4: (*Contd...*)

Recipe	*Ingredients*	*Method of processing and preservation*
	and oregano	
Herbed Eggplant Slices	Eggplant slices are twice baked with salt and pepper and herbs sprinkled.	Baking
Chinese Eggplant Fries	Eggplant slices are fried and served with hot and sour sauce	Frying
Flourless Eggplant Pizza	Eggplant slices are made as a pizza base and topped with different topping.	Baking
Baked Eggplant Parmesan	Eggplant slices are coated with bread crumbs and parmesan cheese and baked between layers of tomato sauce and mozzarella cheese.	Baking
Roasted Eggplant and Pasta Caprese	Eggplant, roasted with bell pepper, red onion, tomatoes, and herbs, is seasoned with fresh basil oil, pasta and lots of cheese.	Roasting
Mama's Oh-So-Savory Lamb and Eggplant Casserole	Lamb and eggplant dish is flavored with tomatoes, sweet onion, spices, and fresh lemon juice, and baked	Baking
Garlic Mashed Potatoes with Eggplant	Roasted eggplant is added to garlic mashed potatoes	Roasting
Marinated Grilled Eggplant	Thick eggplant slices are marinated in a basil vinaigrette dressing and grilled to flavorful perfection.	Grilled
Grilled Eggplant and Asparagus Salad	Grilled eggplant and asparagus are tossed with a romaine-endive salad and dressed with simple balsamic vinaigrette.	Grilled
Roasted Eggplant and Garlic Hummus	Roasted eggplant is combined with garbanzo beans, garlic, and olive oil	Roasting
Spicy Eggplant	Eggplant cubes are stir-fried until browned, then simmered in a spicy asian inspired sauce with onion and garlic	Stir-frying
Olive Oil Roasted Eggplant with Lemon	Eggplant slices are roasted in the oven with olive oil and lemon to garnish	Roasting

Component extraction from eggplant

Eggplant is used to extract important components of medicinal and commercial importance. The seed of eggplant is used for medicinal purpose while the peel of eggplant is used to extract the color. Eggplant extracts are used for weight loss and cancer prevention. Some of the important extracts are as follows.

Color

Eggplant peel contains high amount of anthocyanin pigment, which is responsible for purple and red color of fruit. Due to presence of seeds, eggplant is biologically a fruit however it is categorized in vegetables. Eggplant is a good source for extraction of dark purple and purplish red color pigment. For extraction of color most useful method is supercritical carbon-dioxide (SC-CO_2).

Supercritical carbon dioxide is a fluid state of carbon dioxide where it is held at or above its critical temperature and critical pressure. A pure supercritical fluid (SCF) is any compound at a temperature and pressure above the critical values (above critical point). Above the critical temperature of a compound the pure, gaseous component cannot be liquefied regardless of the pressure applied. The critical pressure is the vapor pressure of the gas at the critical temperature. In the supercritical environment only one phase exists. The fluid, as it is termed, is neither a gas nor a liquid and is best described as intermediate to the two extremes. This phase retains solvent power approximating liquids as well as the transport properties common to gases. The basic principle of SCF extraction is that the solubility of a given compound (solute) in a solvent varies with both temperature and pressure. At ambient conditions (25°C and 1 bar) the solubility of a solute in a gas is usually related directly to the vapor pressure of the solute and is generally negligible. Carbon dioxide usually behaves as a gas in air at standard temperature and pressure (STP), or as a solid called dry ice when frozen. If the temperature and pressure are both increased from STP to be at or above the critical point for carbon dioxide, it can adopt properties midway between a gas and a liquid. More specifically, it behaves as a supercritical fluid above its critical temperature (304.25 K) and critical pressure (72.9 atm/7.39 MPa), expanding to fill its container like a gas but with a density like that of a liquid. Supercritical CO_2 is becoming an important commercial and industrial solvent due to its role in chemical extraction in addition to its low toxicity and environmental impact. The relatively low temperature of the process and the stability of CO_2 also allow most compounds to be extracted with little damage or denaturing. In addition, the solubility of many extracted compounds in CO_2 varies with pressure, permitting selective extractions.

One more research on extraction of color from eggplants was carried out by Chatterjee *et al.,*[32]. In his study, they selected local variety '*Muktakeshi*', procured from Jadavpur market of Kolkata, India. Bright purple color eggplants, free from wilts, discolored spots and white fungal patches were chosen for their extraction purpose. They cut the peel (moisture content 87%) in uniform dimensions (10×10 mm). The shredded peels were kept for lyophilization in bench top freeze dryer at –18°C to –20°C prior to extraction. The eggplants peels (10 g) were pre-treated with 10% citric acid solution for 1 min and then loaded to the extraction vessel. Optimization of yield of color from eggplant peels was conducted using a mixed-level factorial design and parameters such as extraction pressure (10, 12, and 15 MPa), temperature (60 and 80°C), time (90 min *i.e.*, 60 min static time + 30 min dynamic time) and 2 L/min flow rate of CO_2. They were found the highest yield 198.4 g/kg of wet weight of peel (17.04 g/kg of dry weight of peel) extracted by SC-CO_2 at 60°C, 10 MPa, 1.5 h extracting time and 2 L/min of CO_2. The eggplant extracted color is used for non-thermal and thermal food jelly and custard, respectively.

A study on recovery of anthocynine pigment from eggplant peel was carried out by Todaro *et al.*[33]. Tartaric and malic acid were used in discontinuous process and they extracted natural red color. Based on this it was concluded that tartaric acid was more efficient than malic acid. The 1.25% tartaric acid at 40°C showed the best extraction yield.

Phytochemicals

Chatterjee *et al*. (2012)[32] analyzed phytochemicals-phenolic, flavonoid and proanthocyanidin compounds extracted from eggplant peel using supercritical carbon dioxide method. They found that at 23°C maximum phenolic content of 7.16 g gallic acid/kg of dry peel weight and maximum proanthocyanidin content of 2.57 g catechin/kg dry peel weight; whereas, maximum flavonoid content of 4.27 g quercetin/kg dry peel weight was obtained in the SC-CO_2 extracted color at 60°C, 15 MPa.

Solid phase

Tokalioglu and Gurbuz[34] did the work on determination of copper and iron in various food samples by the solid phase extraction. Heavy metal composition of foods is of interest because of their essential or toxic nature. Like iron, zinc, copper, chromium, cobalt and manganese are essential, while lead, cadmium, nickel and mercury are toxic at certain levels. Copper and iron play important roles in biological systems. A variety of clinical disorders such as anemia, depression of growth, impaired reproductive performance, heart failure and gastrointestinal disturbances have been associated with a dietary deficiency of copper. Overexposure to copper causes nausea, vomiting, diarrhea, hepatic neurosis, gastrointestinal bleeding, hypertension and dermatitis. Iron is an integral part of many proteins and enzymes and is an essential component of proteins involved in oxygen transport, regulation of cell growth and differentiation. Iron deficiency limits oxygen delivery to cells, leading to fatigue, poor work performance, decreased immunity and death. A recommended dietary allowance (RDA) of 8 and 18 mg of Fe per day has been recommended for the average 19–50 year old males and females, respectively.

In solid phase extraction method developed using N-benzoyl-N-phenylhydroxylamine as a chelating reagent and Amberlite XAD-1180 as an adsorbent was used for the determination of Cu (II) and Fe (III) in various food samples. The eggplants were digested by using nitric acid and hydrogen peroxide. They found that, concentration of Cu (II) and Fe (III) were 1.26 ± 0.25 and 14.5 ± 0.7 μg per gram respectively.

STORAGE OF EGGPLANT

Eggplant fruits are sensitive to heat and cold. The ideal storage condition for eggplant is around 10°C (50°F).The cut eggplant is not stored because its skin is damaged. Uncut and unwashed eggplant is kept in refrigerator for few days in plastic bags. If eggplant is wrapped in plastic film, removal of film is necessary as soon as possible because it inhibits the eggplant from breathing and degrade its freshness. Eggplant needs some specific condition during transportation. Condition for eggplant are desired temperatures 8°C to 12°C (46°F to 54°F) and relative humidity of 90 to 95%. Eggplant's highest freezing point is –0.8°C (30.6°F).

Eggplant is chilling sensitive fruit. Chilling injury has been found eggplant at 10°C and below. Surface scald or bronzing and pitting are symptoms of chilling injury. Chilled eggplants are subject to Alternaria rot when removed from storage or transit. Chilling sensitivity varies by season of harvest, variety and maturity. Over mature fruit harvested in the fall is less sensitive to chilling than mature fruit harvested in the summer. Mature eggplants harvested midsummer are highly susceptible to decay at temperatures higher than 12°C (54°F) and chilling injury below 10°C (50°F). Eggplants shrivel very easily during transport and storage if the relative humidity is not kept at 90%. For eggplant loading and storage fiberboard boxes are used (USDA)[35].

Concellon *et al.*[36] investigated effect of low temperature on eggplant physical and physiological characteristics. They stored fruit at 0°C and 10°C for 15 days. They concluded that eggplants which were stored at 10°C were not damaged by temperature, whereas, fruits stored at 0°C suffered chilling injury. At these temperature flesh tissue revealed ultra-structural damage. Jha *et al.*[37] did the work on surface gloss and weight loss of eggplant during storage. They kept eggplant at 20°C temperature, 80–84% relative humidity for 0–96 h time. They concluded that, during storage weight loss and surface gloss both are reduced and major changes takes place within 48 h.

Fallik *et al.*[38] studied prolonged low-temperature storage of eggplants in polyethylene bags. They stored eggplant fruit in two layers in an unperforated low-density polyethylene (LDPE) bag (30 mm thick, 65 × 70 cm), at 6, 8 and 12°C and relative humidity (RH) of 87–90% for 2, 5, 10, 16 and 20 days. They concluded that, fruit stored at 8°C in PE bag had good storability and shelf life.

Zhang *et al.*[39] carried out research on effects of low temperature soaking on color and texture of green eggplants. They selected green fresh eggplant grown in China. The soaking time and temperature were varied from 10 to 20 min, and from 40 to 70°C. The blanching condition was 3 min at 100°C. They observed pectin esterase and polyphenol oxidase (PPO). They reported that, the optimum temperature of polyethylene was 65°C, but there was no significant difference from 50°C to 65°C, and polyethylene activity could be enhanced by NaCl concentration (0.15–0.25 M). The optimum pH of PPO was at 7.5, but inactivated when the pH was higher than 9.0. The enzymatic browning catalyzed by PPO could occur when soaking temperature was higher than 55°C.

CAUTION

Eggplant contain large amount of oxalates which are harmful to body. It crystallizes which leads to some health problems. This is the reason why people with suspected or untreated kidney or gall bladder problems should avoid the consumption of eggplants. Some allergic problems to eggplant like itchy skin and mouth has been reported. Contacting dermatitis from the plant leaves and hypersensitivity reactions to the flower pollens have also been reported a number of times.

REFERENCES

[1] FAO., Food and agricultural organization. Food and agricultural commodities production. Retrieved from http://faostat.fao.org. (2011).

[2] FAO., Food and agricultural organization. Food and agricultural commodities production. Retrieved from http://faostat.fao.org. (2010).

[3] Nisha P., Nazar P.A., Jayamurthy P., "A comparative study on antioxidant activities of different varieties of *Solanum melongena*", *Food and Chemical Toxicology*, **47**, 2640–2644 (2009).

[4] NIN., Eggplant's Nutritional Values, National Institute of Nutrition, http://www.ninindia.org. (2007).

[5] Arivalagan M., Gangopadhyay K.K., Kumar G., Bhardwaj R., Prasad T.V., Sarkar S.K., Roy A., "Variability in mineral composition of Indian eggplant (*Solanum melongena L.*) genotypes", *Journal of Food Composition and Analysis*, **26**,173–176 (2012).

[6] Luthria D.L., "A simplified UV spectral scan method for the estimation of phenolic acids and antioxidant capacity in eggplant pulp extracts", *Journal of Functional Foods*, **4**, 238–242 (2012).

[7] Singh A.P., Luthria D., Wilson T., Vorsa N., Singh V., Banuelos G.S., Pasakdee S., "Polyphenols content and antioxidant capacity of eggplant pulp", *Food Chemistry*, **114**, 955–961 (2009).

[8] Akanitapichat P., Phraibung K., Nuchklang K., Prompitakkul S., "Antioxidant and hepatoprotective activities of five eggplant varieties", *Food and Chemical Toxicology*, **48**, 3017–3021 (2010).

[9] Concellon A., Zaro M.J., Chaves A.R., Vicente A.R., "Changes in quality and phenolic antioxidants in dark purple American eggplant (*Solanum melongena* L. cv. Lucia) as affected by storage at 0°C and 10°C", *Postharvest Biology and Technology*, **66**, 35–41 (2012).

[10] Shrivas K., Patel D.K., "Liquid-phase microextraction combined with gas chromatography mass spectrometry for rapid determination of nicotine in one-drop of nightshades vegetables and commercial food products", *Food Chemistry*, **122,** 314–318 (2010).

[11] Barbagallo R.N., Chisari M., Caputa G., "Effects of calcium citrate and ascorbate as inhibitors of browning and softening in minimally processed 'Birgah' eggplants", *Postharvest Biology and Technology*, **73**, 107–114 (2012).

[12] Mishra B.B., Gautam S., Sharma A., "Browning of fresh-cut eggplant: Impact of cutting and storage", *Postharvest Biology and Technology*, **67**, 44–51 (2011).

[13] Keith L.S., "Recommended storage temperature and relative humidity compatibility groups", *The OHIO State University*. (2013).

[14] Khanna V., Ramsay G., "Flavors First: An Indian Chef's Culinary Journey", *Lake Isle Press*. (2011).

[15] Brass M., Brass S., "Heirloom cooking with the brass sisters: Recipes you remember and love", *Black Dog & Leventhal Publishers*. (2008).

[16] Roden C., "Arabesque: A Taste of Morocco, Turkey, and Lebanon", *KNOPF,* NY. (2006).

[17] Kapoor S., "Khazana of Indian Vegetarian Recipes", *Popular Prakashan*, India. (2002).

[18] Kapoor S., "Khazana of Indian Recipes", *Popular Prakashan,* India. (1999).

[19] Mehta N., "Indian cuisines", *Nita Mehta publication*. (2005).

[20] Kelley L., "The Silk Road Gourmet: Volume One: Western and Southern Asia", *iUniverse.com*. (2009).

[21] Sesena S., Hurtado I.S., Vinas M.A.G., Palop L., "Contribution of starter culture of the sensory characteristics of fermented Almagro eggplants" *International Journal of Food Microbiology*, **67**, 197–205 (2001).

[22] Wu L., Orikasa T., Tokuyasu K., Shiina T., Tagawa A., "Applicability of vacuum-dehydrofreezing technique for the long-term preservation of fresh-cut eggplant: Effects of process conditions on the quality attributes of the samples", *Journal of Food Engineering*, **91**, 560–565 (2009).

[23] Wu L., Orikasa T., Ogawa Y., Tagawa A., "Vacuum drying characteristics of eggplants", *Journal of Food Engineering*, **83**, 422–429 (2007).

[24] Akpinar E.K., Bicer Y., "Modelling of the drying of eggplants in thinlayers", *International Journal of Food Science & Technology*, **40**(3), 273–281 (2005).

[25] Russo P., Adiletta G., Matteo M.D., "The influence of drying air temperature on the physical properties of dried and rehydrated eggplant", *Food and Bioproducts Processing*, 91(3), 249–256 (2013).

[26] Ertekin C., Yaldiz O., "Drying of eggplant and selection of a suitable thin layer drying model", *Journal of Food Engineering*, **63**, 349–359 (2004).

[27] Midilli, A., Kucuk H., Yapar Z., "A new model for single-layer drying", *Drying Technology*, **20**(7), 1503–1513 (2002).

[28] Perez J.V.G., Ozuna C., Ortuno C., Carcel J.A., Mulet A., "Modeling ultrasonically assisted convective drying of eggplant", *Drying Technology*, **29**(13), 1499–1509 (2011).

[29] Madison D., Smith L., McFarlin P., "Local Flavors: Cooking and Eating from America's Farmers' Markets", *Broadway*. (2002).

[30] Anonymous, The Indian Cookery Book. Retrieved from, http://www.gutenberg.org/dirs/2/8/9/2895/2895-h/2895-h.htm. (2005).

[31] Hoffman N., "Tempting Eggplant Recipes", *Prime Publishing. Retrieved from* http://www.recipelion.com. (2013).

[32] Chatterjee D., Jadhav N.T., Bhattacharjee P., "Solvent and supercritical carbon dioxide extraction of color from eggplants: Characterization and food applications", *LWT - Food Science and Technology*, **51**, 319–324 (2012).

[33] Todaro A., Cimino F., Rapisarda P., Catalano A.E., Barbagallo R.N., Spagna G., "Recovery of anthocyanins from eggplant peel" *Food Chemistry*, **114**, 434–439 (2009).

[34] Tokalýoglu S., Gurbuz F., "Selective determination of copper and iron in various food samples by the solid phase extraction", *Food Chemistry*, **123**, 183–187 (2010).

[35] USDA (2006) Publication AG-414-1, and Agricultural Handbook No. 66. www.ces.ncsu.edu/hil/hil-801.html.

[36] Concellon A., Anon M.C., Chaves A.R., "Effect of low temperature storage on physical and physiological characteristics of eggplant fruit (*Solanum melongena* L.)", *LWT*, **40**, 389–396 (2007).

[37] Jha S.N., Matsuoka T., Miyauchi K., "Surface gloss and weight of eggplant during storage", *Biosystems Engineering*, **81**(4), 407–412 (2002).

[38] Fallik E., Gorodeiski N.T., Grinberg S., Davidson H., "Prolonged low-temperature storage of eggplants in polyethylene bags", *Postharvest Biology and Technology*, **5**, 83–89 (1995).

[39] Zhang M., Chen D., "Effects of low temperature soaking on color and texture of green eggplants", *Journal of Food Engineering*, **74**, 54–59 (2006).

13

Bioactive Compounds From Underground Crops Biosynthesis, Degradation, Extraction and Clinical Importance

ROBIN A. WILSON[1,2*] AND HARSHAD M. MANDGE[1,3]

ABSTRACT

Underground crops constitute a substantial part of the world's food supply, especially in developing countries, where they are the source of most of the daily carbohydrate intake. Underground crops are normally stored and consumed as fresh produce or converted into a more durable product by traditional processes for domestic use. Other uses of underground crops are as animal feed or for industrial processing for products such as starch, distilled spirits, and a range of minor products. Besides a source of carbohydrates, Underground crops are found to be an excellent source of certain bioactive components. These crops have been consumed by man since centuries owing to the health and medicinal benefits conferred by bioactive components present in them. Carrot, beetroot, turmeric and ginger contain betanin, carotenes, curcumin and zingiberene respectively. Although these bioactive components can be availed by fresh consumption of these crops but extraction of bioactive components make them readily available and have found their application in various herbal medicines. This chapter describes the biosynthesis, degradation, extraction and clinical significance of the major bioactive components in four underground crops (beetroot, carrot, turmeric and ginger).

Key words: Beetroot, Carrot, Turmeric, Ginger, Betanin, Carotenes, Curcumin and Zingiberene.

[1] ICAR-Central Institute of Post-Harvest Engineering and Technology, Ludhiana, India.
[2] Victoria University, Melbourne, Australia
[3] Manyawar Shri Kanshiram Ji University of Agriculture and Technology, Banda, India.
**Corresponding author* E-mail: bobzrobin@gmail.com

INTRODUCTION

Underground crops form a major portion of tropical and subtropical crops. Underground crops consist of crops covering several classes of plants. They are grown and consumed in tropical countries as staple foods. Their consumption in developed countries is considerably less. The major underground crops are potato, cassava, sweet potato and yam. Underground crops are mainly categorized into four types *viz.* rhizomes (root stocks), tubers, corms and bulbs. The rhizomes are underground, horizontal stem with reduced non-chlorophyllous leaves. Some rhizomes have thickened portions at their ends, known as tubers and are generally rich in starch. Majority of these crops are rich in starch, however there are number of crops like ginger, turmeric, beet root and carrot etc. that offers valuable bioactive agents other than starch. These bioactive agents are endowed with many medicinal and curative properties. Some of the extracts of these crops have been exploited since ages by mankind. The present chapter deals with biosynthesis degradation, extraction and clinical importance of bioactive compounds obtained from carrot, ginger, turmeric and beetroot.

CARROT

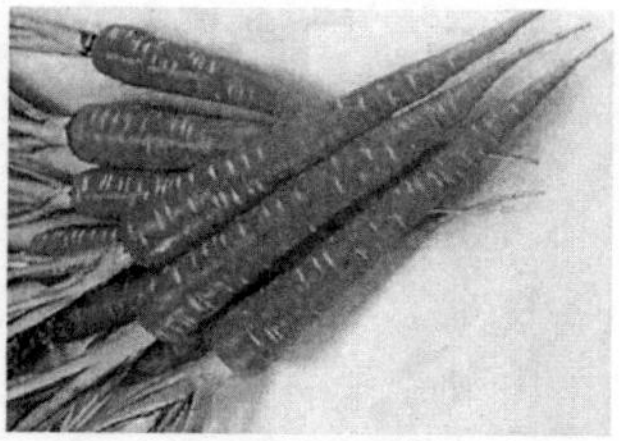

Carrots are richest source of carotenoids among the plant based foods consumed by humans. The characteristic colour of carrots is due to the presence of carotenoid pigments. The major carotenoids in carrot are α and β carotene (Fig. 1), together contributing to about 80-90% of total carotenes, remaining percentage owes to non coloured polygenic pigments[1]. β carotene is the most extensively studied carotene. Carotenoids exist in two isomeric forms *viz. cis* and *trans* carotenes and latter is predominant and nutritionally more potent[2]. Trans form of carotenoids get converted to *cis* form on cooking and processing of carrots[3].

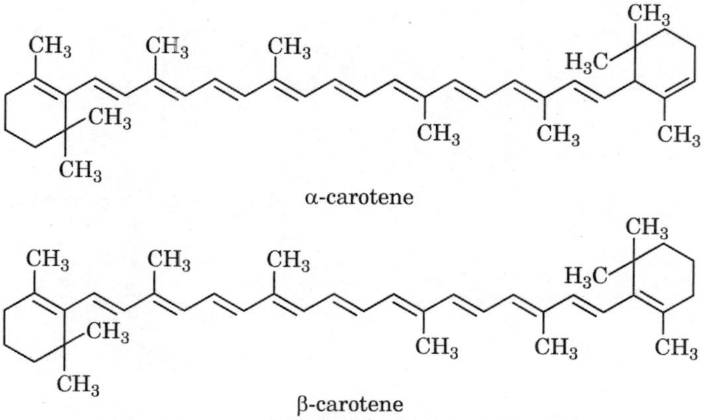

Fig. 1: Structure of α and β carotene of carrot

Biosynthesis of Carotenoid

Carotenoids are synthesised through 2-C-methyl-D-erythritol 4-phosphate (MEP) pathway located in plastids[4, 5, 6]. The pathway begins with reaction between pyruvate and gyceraldehyde-3-phosphate resulting in the formation of 1-deoxy-D-xylulose-5-phosphate. The reaction is catalysed and regulated by 1-deoxy-D-xylulose-5-phosphate synthase (DXS) along with 1-deoxy-D-

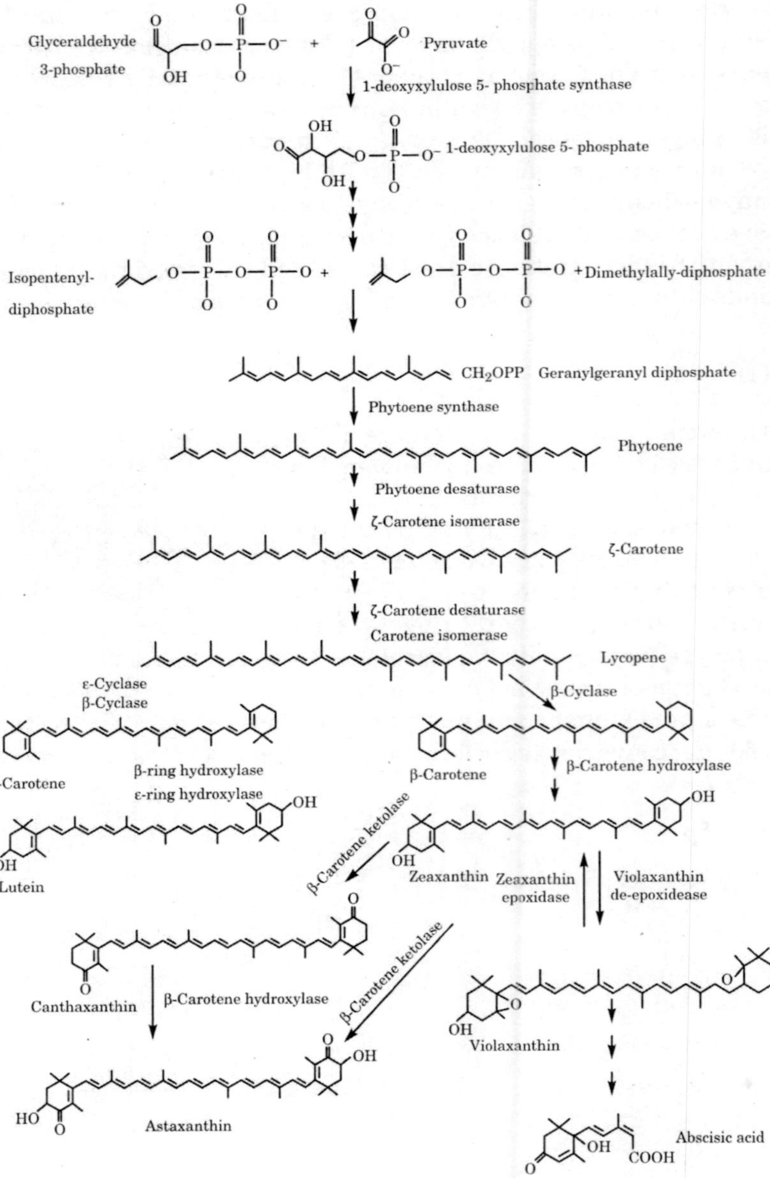

Fig. 2: Biosynthetic pathway of carotenoid[7]

xylulose 5-phosphate reductoisomerase (DXR). The detailed pathway of carotenoid biosynthesis is given in Fig. 2. Condensation of two molecules of Geranyl-geranyl diphosphate (GGPP) by phytoene synthase (PSY) to form phytoene is the first committed step in carotenoid biosynthesis. After phytoene, synthesis is in the following sequence: Phytoene → phytofluene → ζ-carotene → neurosporene → lycopene → γ-carotene and β-carotene. With each step from phytoene to lycopene, a double bond is added to the molecule resulting in lycopene which is a symmetrical molecule with 13 double bonds. Enzymatic cyclization of lycopene lead to the formation of γ-carotene and β-carotene, while addition of oxygen leads to the formation of xanthophylls.

Degradation of Carotenoid

Carotenoids are commonly degraded by isomerization, oxidation and fragmentation pathways. The antioxidant activity of carotenoids is due to conjugated double bond structure. The same structural property makes them unstable. The structure of carotenoids is prone to attack by free radicals like singlet molecular oxygen and other reactive oxygen species[7].

Oxidation is the major cause of carotenoid degradation. Autoxidation of carotenoids is believed to follow free radical route. According to free radical theory the general mechanism of carotenoid oxidation is similar to that of lipid peroxidation[8, 9]. Following mechanism was proposed for oxidation of β-carotene (BC·) by peroxyl radical (ROO·)[10]:

$$\text{BC}\cdot + \text{ROO} \rightarrow \text{ROO-BC}\cdot \quad \text{(a)}$$

$$\text{ROO-BC}\cdot + \text{ROO}\cdot \rightarrow \text{Non radical product} \quad \text{(b)}$$

$$\text{ROO-BC}\cdot + \text{O2} \rightarrow \text{ROO-BC-OO}\cdot \quad \text{(c)}$$

When reaction (a) follows reaction (c), β-carotene is degraded and process is called autoxidation. However when reaction (a) follows reaction (b), free radical is consumed and process is called antioxidant. Channelization of reaction (a) to either (b) or (c) depends on partial pressure of oxygen. High partial pressure of oxygen favours autoxidation whereas low partial pressure of oxygen favours antioxidant reaction.

In photosensitised oxidation, carotenoids are effective quenchers of singlet oxygen. Excited singlet oxygen transfer energy to carotenoid molecule, resulting in triplet state oxygen. Ideally the carotenoid should remain intact, however, a chemical reaction sets in during interaction between carotene and singlet oxygen, destroying carotenoid molecule[11, 12, 13].

Few studies have been carried out to investigate the formation, isolation and structural identification of primary, secondary and tertiary products of carotenoid oxidation. β-carotene in the presence of benzene and tetrachloromethane reacts with oxygen at 30°C in dark resulting in the formation of compounds such as 5,6-epoxy-β-carotene, β apo-13-carotene, Retinal, β-ionone, β-apo-14-carotenal, β-insidenc acetaldehyde, 5,6,5',6',-diepoxy β-carotene, 5,8-epoxy-β-carotene, β-apo-12'-carotenal, β-carotenal, β-cyclocitral, 2,2,6-trimethylcyclohexanone, Dihydroactinidiolid, 5,6-epoxy-β-ionone and 5,8-epoxy-β-carotene[14, 15].

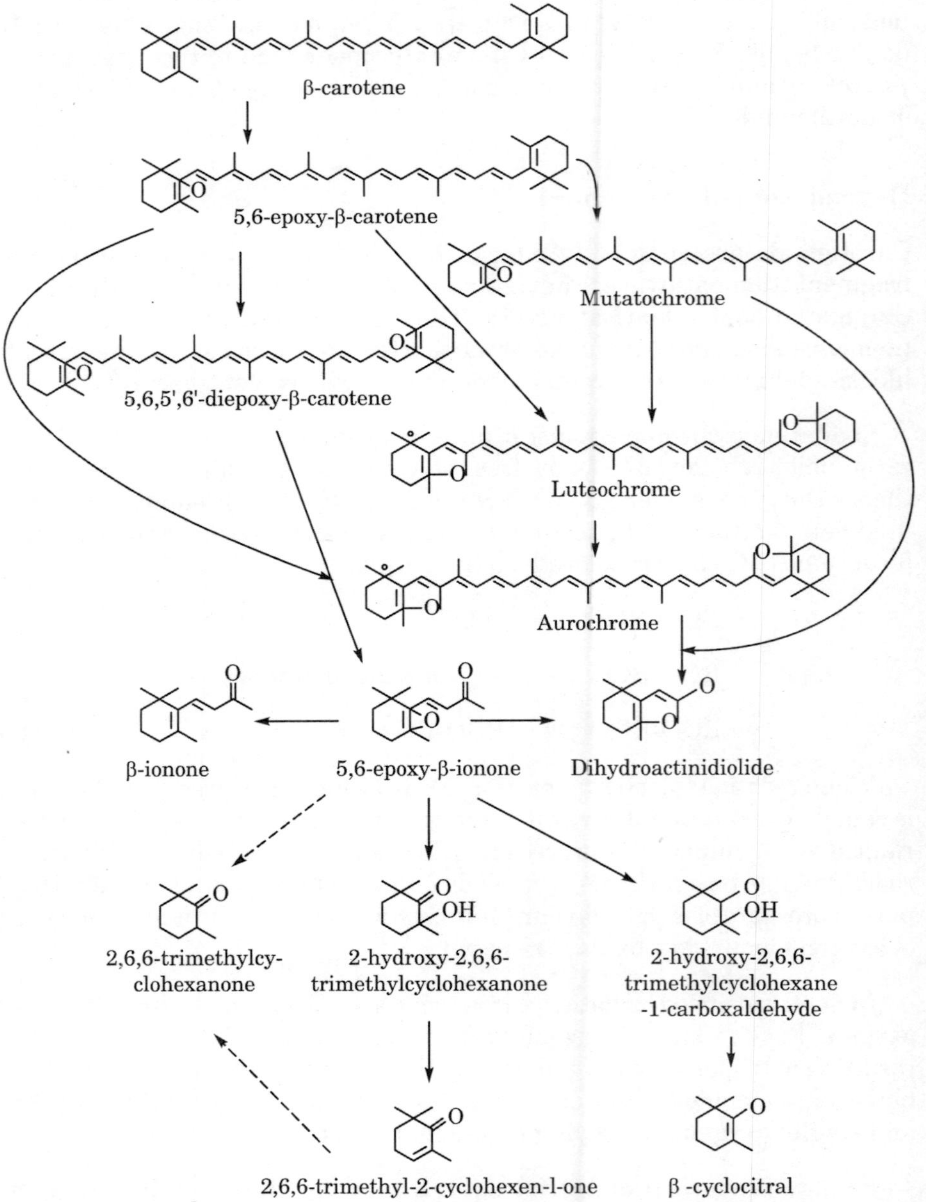

Fig. 3: Formation of volatile and non volatile compounds during heat treatment of β-carotene[14, 15].

Carotenoids are subjected to high temperature conditions during manufacturing of confectionary products. Carotenoids are susceptible to thermal degradation due to the presence of conjugated double bonds. High temperature conditions break double bonded structure resulting in the formation of two kinds of breakdown products *viz.* volatile and non volatile. Non volatile fragments are hard to investigate due to their complex nature. Volatile fragments have been identified using headspace gas chromatography and mass spectrometery[14, 15, 16]. Mechanism of formation of volatile and non volatile compounds during heat treatment of β-carotene have been presented in Fig. 3.

Extraction of Carotenoid

Carotenoids are very complex in structure and are of huge variety. Thus, it is not feasible to extract all the varied carotenoids using a particular extraction method available in the literature. Numbers of techniques have been employed so far to extract carotenoids from food matrix. Liquid-liquid extraction and supercritical fluid extraction are extensively used for extraction of carotenoids.

Liquid-liquid extraction utilizes mixture of two or more organic solvents, mixed in a certain ratio, which enters the food matrix, extracting carotenoids and precipitating other compounds[17]. Carotenoids are hydrophobic in nature and are therefore highly soluble in fats and organic solvents. But the problem does not end here, presence of moisture in the food make the things more difficult. Hence, selection of solvent is very crucial for extraction. Ideally organic solvent used for extraction should be miscible with water which can penetrate the food matrix[18]. Some of water miscible, polar organic solvents are acetone, ethanol and methanol. Mixture of these polar solvents and certain non-polar solvents such as hexane, tetrahydrofuran (THF), diethyl ether and methylene chloride have also been used to extract carotenoids. Selection of organic solvent depends on the food matrix as well as the polarity of the target carotenoids. Below are some examples of solvents used for extraction of carotenoid from different food matrices.

To extract carotenoids from dehydrated plant material, hexane alone or hexane along with acetone in the ratio of 1:1 is suggested[17]. Methanol is also widely used in the extraction of carotenoids either alone or in combination with other non-polar solvents. Methanol and tetrahydrofuran in the ratio of 1:1 (v/v), methanol and chloroform in the ratio of 1:2 (v/v), methanol and dichloromethane in the ratio of 1:2 (v/v) can be used for the extraction of carotenoids from fruits and vegetables[19, 20], fat cured sausages[21], orange juice and tomato products[22] and red pepper and paprika[23], respectively. Mixture of dichloromethane and methanol (2:1, v/v) has been reported to extract lutein, α-carotene, and β–carotene, and lycopene with yield of 99.1%, 99.3%, 98.7% and 101.5% respectively[22]. In some cases, complex mixture of solvents like methanol: ethyl acetate: light petroleum has been reported to extract carotenoids from fruits and vegetables. Likewise hexane: acetone: ethanol mixture has been reported for carotenoid extraction from orange juice[24]. Extraction of carotenoids from tomato juice was carried out using various organic solvent mixtures. Acetone: Hexane (4:6 v/v) and ethanol: hexane (4:3 v/v) showed highest lycopene and β-carotene extraction efficiency as compared to chloroform: methanol (2:1 v/v), dichloromethane: methanol (2:1 v/v) hexane: isopropanol (3:2 v/v) and acetone: petroleum ether (1:1 v/v) mixture[25]. In another comparative study on tomato juice, ethanol: hexane (4:3 v/v) was found to be the best in terms of extraction efficiency in comparison to acetone hexane (3:5 v/v), ethanol : acetone : hexane (2:1:3 v/v/v), ethyl acetate : hexane (1:1 v/v) and ethyl acetate (100%)[26]. Addition of sodium, magnesium, calcium chloride enhances extraction efficiency as these salts neutralize the acid present in the plant tissue[27]. Enzymes like glycosidase, cellulase, hemicellulase and pectinase have been reported to enhance the yield of the carotenoids by hydrolyzing food matrix and facilitating solvent extraction[28,29]. Matrix solid phase dispersion has also been reported to improve extraction efficiency due to less preparation time which reduces oxidation and isomerisation of carotenoids[30].

Super critical fluid extraction (SFE) is a rapid, simple and selective extraction method for extraction of carotenoids[31], vitamin supplements[32], and by products[28]. In addition to this it can be easily automated. Nontoxic CO_2 used in SFE not only facilitate reduced oxidation but also provide higher yield. Due to low critical temperature of CO_2, it is applicable for extraction of heat labile components. However low polarity of CO_2 can be modified by adding certain polarity modifiers such as chloroform, ethanol and methanol which eventually enhance the extraction efficiency[33].

Clinical Importance of Carotenoid

Carrots are one of the best source of carotenoids specially β-carotene. A very essential function of carotenoid in human beings is that of provitamin A (carotenoid to serve as a vitamin A precursor). A number of studies have shown that β-carotene and carotenoids have lipid soluble antioxidant activity[34]. Antioxidant activity of carotenoids is by the virtue of their ability to quench singlet oxygen and entrapment of peroxyl radical. The electron rich conjugated double bond structure facilitates physical quenching of singlet oxygen. Following structural properties impart antioxidant activity to carotenoids[35].

(a) Carotenoid has multiple potential sites for attack of active oxygen.

(b) The energy levels between excited state and ground state of carotenoids are closely spaced facilitating dissipation of excited state energy via small colloidal exchanges with the solvent.

(c) Excited state carotenoids have fewer tendencies to sensitize other molecules.

(d) Resonance state in the excited state carotenoids permits delocalization and stabilization of the excited state.

Antioxidant activity of carotenoids can be enhanced by their association with other antioxidants like vitamin E. Some studies have demonstrated that β-carotene intake promote several biomarkers of immune functions[36, 37, 38]. It has been found that people taking high carotenoid diet are less susceptible to development of lung cancer than the people taking low carotenoid diet[39]. Likewise several studies have reported that carotenoid rich diets are associated with significant reduction in the risk of prostrate cancer particularly more aggressive forms[40].

Carotenoids are insoluble in water and are highly soluble in fats. This inherent property of carotenoids allows them to circulate in the blood in lipoproteins along with cholesterol and other fats. Oxidation of low density lipoproteins (LDL) leads to the development of atherosclerosis. Carotenoids owing to their antioxidant activity prevent LDL oxidation and obviate the development of atherosclerosis[41]. High levels of carotenoids have been found to be associated with significantly lower measure of arterial thickness[42, 43, 44, 45, 46].

Age related degeneration of macula (pigmented area near the centre of the retina) has been a leading cause of blindness. Carotenoids found in retina are leutin and zeaxanthin and their concentration is higher in macula. Carotenoids have long system of alternating single and double bonds which allow them to absorb blue light. Absorption of blue light by carotenoids in macula reduces the amount of blue light reaching the eye and prevents the light induced oxidation damage[47]. Epidemological studies revealed high dietary intake of carotenoids is associated to lower risk of age related macular degeneration[48].

Carotenoids boost immune response by enhancing lymphocyte response to mitogens, increasing natural killer cell activity in aging cells, increasing total white blood cells and CD4/CD8 ratio in HIV infected person[49]. Osteoporosis is known to be associated with oxidative stress which can be counteracted by antioxidant like carotenoid. Several studies have demonstrated the role of carotenoids in improving and maintaining bone health by inhibiting osteoclasts formation and resorption[50,51,52]. Carotenoids have been reported to play role in preventing and curing many human disease such as hypertension, neurodegenerative diseases like Alziemers's disease, Parkinsons disease and vascular dementia[53, 54].

BEET ROOT

Beet root is a rich source of betalains which owes to its deep red colour. On the basis of chemical structure, betalains are categorized into betacynin and betaxanthine (Fig. 4)[55]. Betalains are water soluble nitrogen containing pigments which are synthesised from amino acid tyrosine. The chromophore common to all betalains is betalamic acid. The nature of residue addition to betalamic acid determines the classification of pigments into betacynin and betaxanthine. Betacynin and betaxanthine content ranges between 0.04– 0.21% and 0.02–0.14% respectively[56].

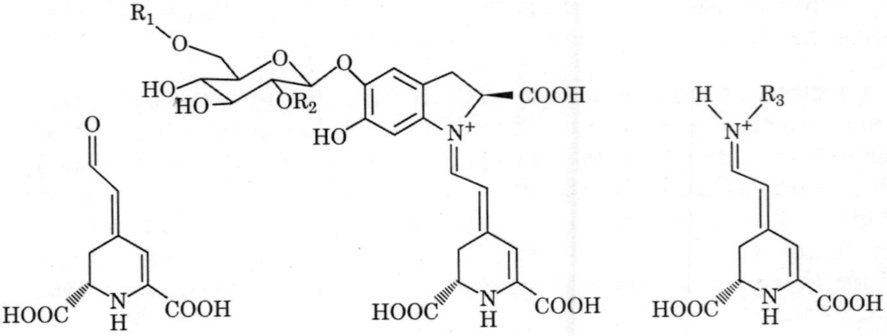

Fig. 4: Chemical structure of (a) betalamic acid, (b) betacynin (R1 = R2 = H), (C) betaxanthine (R3 = amine or amino acid group)[53]

Betanidin is the aglycone of most betacynins. Substitution by glycosylation or acylation of one or both hydroxyl groups at 5 or 6 position in betanidin give rise to various betacynin. Betacynin has a cyclo 3, 4 dihydroxyphelyalanin (cyclo DOPA) residue. The most common betacynin is betanin (Betanidin-5-o-β-glycoside), which is also the major pigment in beetroot[57]. Betacynins absorbs in the range of 535-550 nm. Betaxanthins have different amino acid or amine side chains[58]. Betaxanthins present in the beet root are vulgaxanthin and indicaxanthin. The absorption maxima of betaxanthin is at 480 nm[59].

Biosynthesis of Betalains

Betalains are derived from amino acid tyrosine and the synthesis begins with the hydroxylation of tyrosine to L-DOPA with the help of enzyme tyrosinase. The detailed scheme of betalaine biosynthesis has been shown in Fig. 5[59].

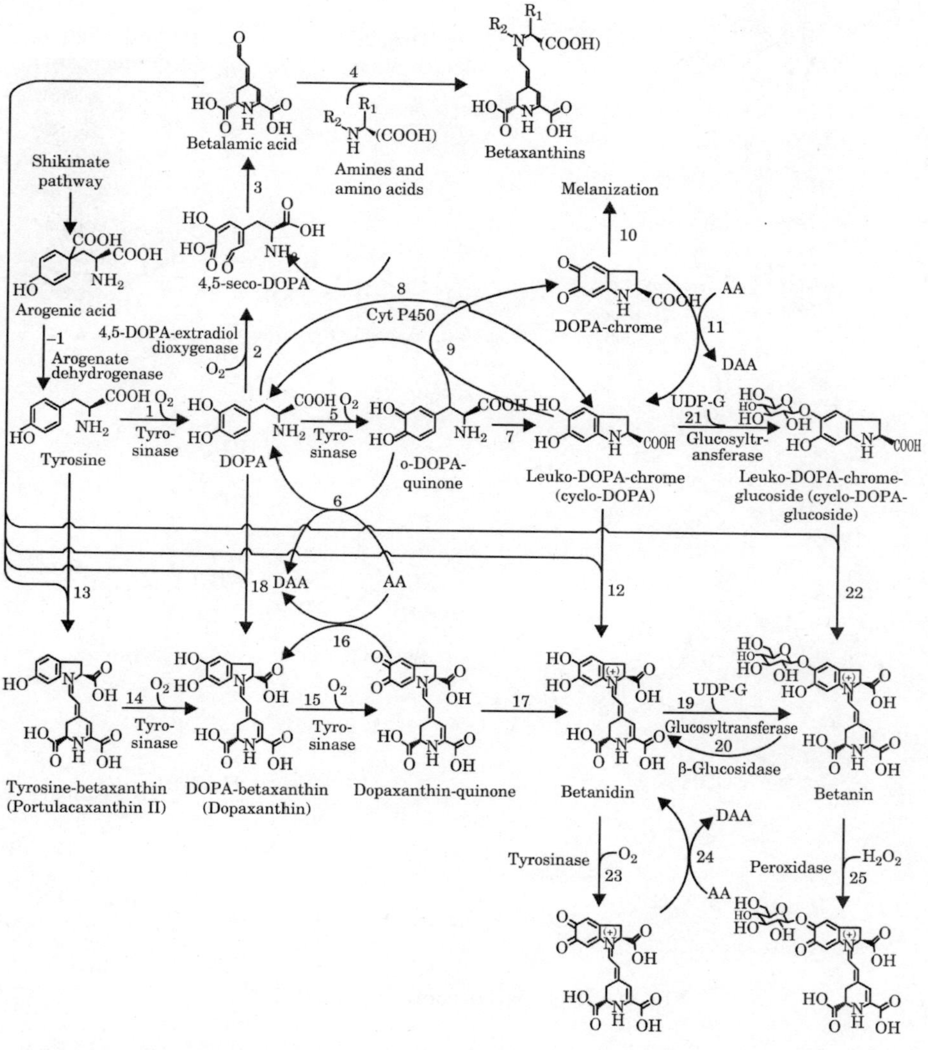

Fig. 5: Biosynthesis of betalains (AA, ascorbic acid; cyt P450, cytochrome P450; DAA, dehydroascorbic acid) [57]

Degradation of Betalains

There are number of factors that can lead to degradation of betalain. Betalains can undergo isomerisation, deglycosylation, hydrolysis, decarboxylation, dehydrogenation, combined decarboxylation and dehydrogenation. Some of the degradation routes have been detailed in Fig. 6[60].

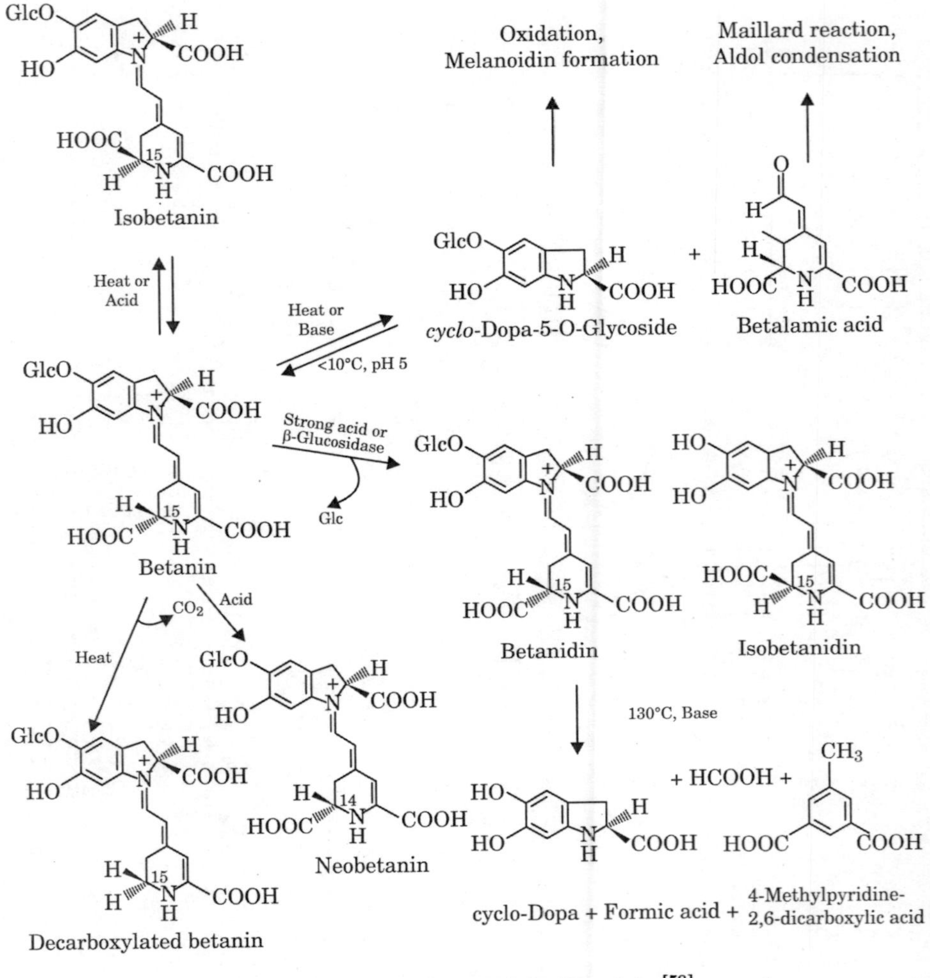

Fig. 6: Degradation routes of betalain [58].

Extraction of Betalains

Due to their solubility in water, betalains are generally homogenised in water and pigments can be extracted. A methanol or ethanol solution (20-50%) is sometimes used for complete extraction [61]. Aqueous extraction leads to better stability of the pigments. However low pH of extraction medium favours betacynin stability and prevent oxidation by polyphenoloxidases[58, 59, 60, 61, 62].

Various methods have been employed to extract betalains *viz.* diffusion-extraction[63], solid–liquid extraction[64], reverse osmosis[65] and ultrafiltration[66, 67]. The above listed methods are more efficient than conventional hydraulic method in terms of betalains recovery. The extraction efficiency can be enhanced by manoeuvring cell membrane permeability. Permeability can be augmented using number of tools like pulsed electric field treatment[68, 69] and gamma irradiation[70]. Pulsed electric field treatment ensures low level of tissue damage and low energy consumption. However gamma irradiation enhanced betanin degradation.

At least 80% of beet juice solids contain fermentable carbohydrates and nitrogen compounds. Fermentation of extract consumes and reduces carbohydrate content enhancing betalain concentration[71]. Enzymatic degradation of hydrocolloids present in the extract enhances the betalain yield[72]. Prior heat treatment of extract also prevents betalain enzymatic degradation[60].

Clinical Importance

Human beings come across innumerable pollutant and anti-nutritional factors in their day to day life. Several such substances cause oxidative stress which is a repercussion of biochemical changes instigated by free radicals. Oxidative stress leads to many dreadful diseases such as cancers, skin ailments and atherosclerosis. It has been revealed by several studies that betalains have antioxidant property[73, 74]. Betalains are potent inhibitors of cancer cell growth and proliferation[75,76]. Antiproliferative effect of betalains has been demonstrated on human myeloid leukaemia cell line (K562) and apotosis induction [76]. Quinone reductase a potent detoxification enzyme associated with cancer chemoprevention is induced by betalains[77].

TURMERIC

Turmeric (*Curcuma longa*) is a perennial, herbaceous, rhizomatous herb of family Zingiberaceae. It is native of southern and south eastern tropical Asia. It is cultivated annually for its rhizomes. Freshly cured rhizomes are either used fresh or boiled for several hours and then dried in sun or in dryer, followed by pulverisation resulting into deep orange-yellow powder[78]. Turmeric has been used for over 5000 years in Indian and Chinese medicines owing to its wound healing, blood clotting, antitumor, anti-inflammatory, anti angiogenesis, antioxidant, immunomodulatory and proapototic properties. Besides its importance as a spice, it is an important component of many cosmetic products and holds an indispensable position in religious ceremonies in Southeast Asia. Because of its brilliant yellow colour, it is also known as "Indian Saffron". The major components of the roots are volatile oil containing cumerone and colouring agents called curcuminoids. Curcumin is the principle curcuminoid present in turmeric (Fig. 7). Desmethoxycurcumin and bis- Desmethoxycurcumin are the other two minor curcuminoids present (Fig. 7)[79, 80, 81]. Volatile oil contains d-α-phellandrene, d-sebinene, cinol, borneol, zingiberene, and sesquiterpenes[82]. There are a variety of sesquiterpenes, like germacrone; termerone; ar-(+)-, α-, and β-termerones; β-bisabolene; α-curcumene; zingiberene; β-sesquiphellanderene; bisacurone; curcumenone; dehydrocurdione; procurcumadiol; bis-acumol; curcumenol; isoprocurcumenol; epiprocurcumenol; procurcumenol; zedoaronediol; and curlone, many of which are specific for a species.

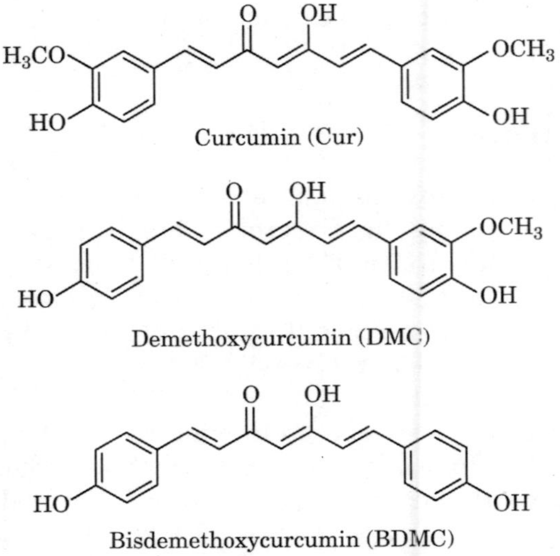

Fig. 7: Structure of curcuminoids present in turmeric [79].

Curcumin [1,7-bis (hydroxyl-3-methoxyphenyl)-1,6-heptadiene-3,5-dione] has gained considerable importance due to its multiple pharmacological activities and therapeutic potential for cancer and Alzheimers disease[83]. Because of its potential to combat number of diseases, myriad researchers have devoted themselves to explore different facets of curcumin. Curcumin has several functional groups. Two α, β-unsaturated carbonyl groups connect aromatic ring systems of phenols. It can exist in several tautomeric forms, including 1,3 diketo form and two equivalent enol forms[84]. Curcumin is soluble in oil and alkali. It is insoluble in hydrocarbon solvents and water at neutral and acidic pH[85].

Biosynthesis of Curcumin

It took more than three decades to prove the biosynthetic mechanism (Fig. 8) of curcumin. Roughley and Whiting[86] suggested two mechanisms for curcumin biosynthesis based on their ^{14}C-labelling experiment. According to first reaction, a chain extension reaction between cinnamic acid and five malonyl CoA, eventually leads to the formation of curcuminoids. Second mechanism suggested condensation of two cinnamic acids to a central carbon provided by malonic acid molecule. Cinnamic acid, utilised by both the above suggested mechanisms is derived from phenylalanine. They were unable to prove the actual mechanism experimentally. Moreover the timing of introduction of functional groups to the aromatic ring, before and after curcuminoid skeleton formation remains unanswered.

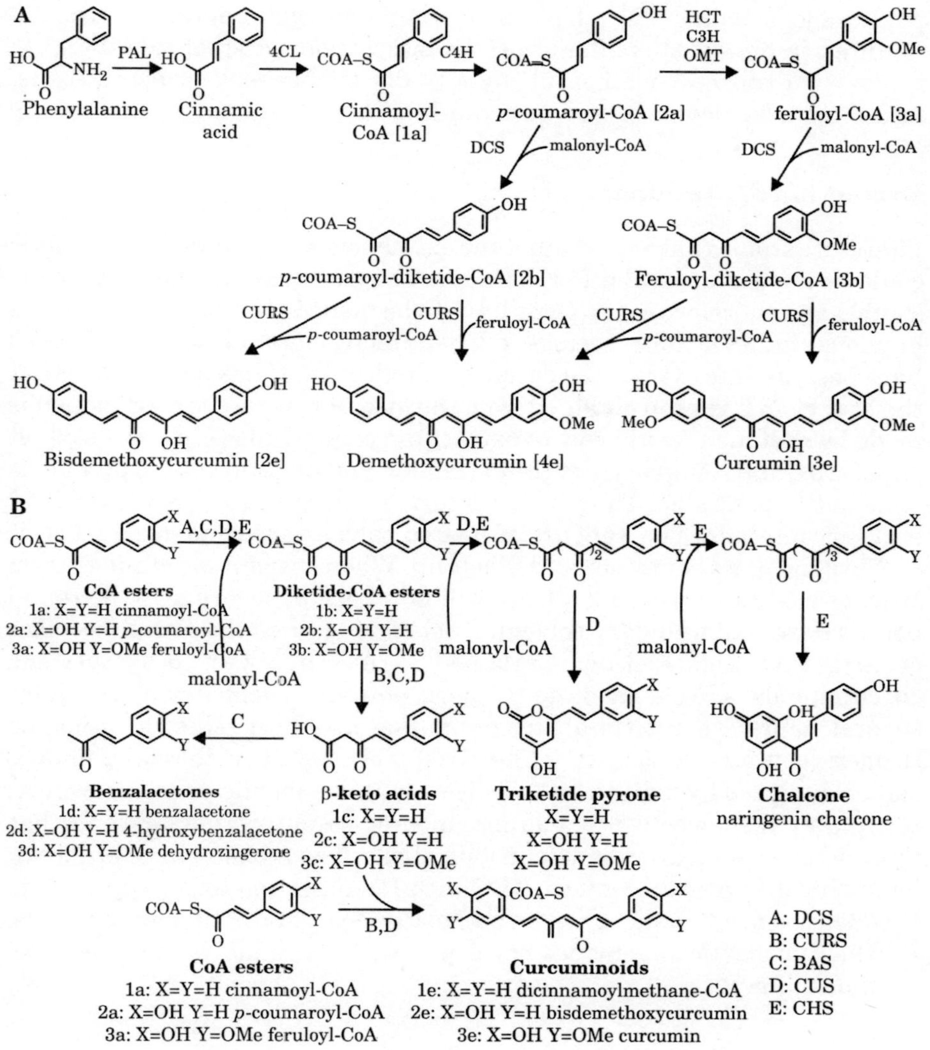

Fig. 8: Biosynthesis of curcuminoids[85]

Katsuyama *et al.*[87] reported the biosynthetic route of curcuminoid with experimental evidence (Fig. 8). The labelling experiment supported the first mechanism in which five malonyl CoA molecules react with cinnamic acid to form curcumin. However, the sequence in which the methoxy and alcohol functional groups are introduced onto the aromatic rings, strongly support the second proposed mechanism. Therefore second pathway was accepted as the actual mechanism for curcumin biosynthesis.

Degradation of Curcumin

Curcumin is quite unstable in aqueous solution and rapidly gets degraded at physiological pH[83, 88, 89]. Up to 90% of curcumin gets degraded within 30 minutes, in phosphate buffer at pH 7.4 into various products identified as trans-6-(4'-hydroxy-3'-methoxyphenyl)-2,4-dioxo-5-hexenal, ferulic aldehyde, ferulic acid, feruloyl methane, and vanillin[90].

Extraction of Curcumin

Literature studies have shown that the extraction of curcumin from turmeric could be done in several different ways. The curcuminoids are not water-soluble and therefore extractions have to be made in nonpolar solvents. In previous studies hexane, acetone, ethylene dichloride and different alcohols have been used to extract the curcumin analogues. It has been shown that the best yield has been obtained from the attempts where the extraction is made by acetone[91]. One way to extract the compounds is to use a soxhlet apparatus, and it implies that the extraction and filtration of the product is done in the same step. The powder of turmeric is put in a paper thimble in a glass container. The vapours of heated solvent are condensed when it reaches the glass container and fills it up. When the volume of the extract reaches a certain level, the filtrate will flow back by leverage to the round bottom flask containing the solvent. By observing when the filtrate becomes colourless, the extraction is stopped[91]. Another way to extract the curcuminoids is to suspend the turmeric powder in acetone and stir it for several hours. To get rid of the remaining unsolved material from the turmeric powder, the solution is filtered off. The quantity of the curcuminoids can be analyzed by using a HPLC with a suitable mobile phase. However, to separate the molecules in a larger amount, column chromatography has to be done with an appropriate mobile phase. The mobile phase is chosen by testing different solvents on TLC with UV-detection to get appropriate R_f (retention factor) values, and an optimized separation of the curcuminoids. Further purification of purified and separated curcuminoids can be done by crystallization[92].

Clinical Importance of Curcumin

Curcumin is bestowed with antioxidant, anti-inflammatory, antiviral and antifungal actions[93]. It inhibits number of agents which are involved in the mechanism of inflammation and is thus a potent anti-inflammatory compound, therefore turmeric is normally recommended in post surgical inflammation and open wound injury[94]. Antioxidant activity of curcumin is comparable to vitamin C and E. Heme oxygenase-1 is an inducible stress protein. An *in vitro* study on endothelial heme oxygenase of aorta incubated for eighteen hours with curcumin demonstrated enhanced cellular resistance to oxidative damage[94].

Numbers of studies[95, 96] have demonstrated hepato-protective effect of curcumin. Hepato-protective effect of curcumin owes to its antioxidant and anti-inflammatory effects. It is effective against number of hepatotoxic agents *viz*. Carbon tetrachloride, galactosamine, paracetamol and *Aspergillus* aflatoxins. Curcumin administered rats with carbon tetrachloride induced liver injury and *Aspergillus parasiticus* infected ducklings demonstrated significantly lower liver injury and reduced (by 90%) production of aflatoxins respectively. Curcumin has been reported to reverse and prevent biliary hyperplasia, fatty liver and cholethiasis[94].

Curcumin and other essential oils present in turmeric have anti-inflammatory activity. Its anti-inflammatory effect is comparable to cortisone and phenylbutazone. In a study on rats with induced arthritis, oral administration of *Curcuma longa* alleviate inflammation[94]. Anti inflammatory activity of curcumin can be attributed to its ability to inhibit synthesis of prostaglandins from arachadonic acid and also inhibit neutrophyl function. Topical application of curcumin also alleviates inflammation and irritation associated with skin ailment and allergies[94]. Recently curcumin has been found to prevent neuroinflammation, a probable cause of Alzheimers disease[97]. Such effect is exhibited by inhibiting transcription factors and coactivators of activator protein[97, 98].

Curcumin has demonstrated anticancer activity over a broad range of cancers viz. colorectal, breast, blood, genitourinary, hematological, thoracic, bone, skin and brain. It has the ability to inhibit carcinogenesis at three stages namely tumor promotion, angiogenesis and tumor growth. Anticarcinogen effect of curcumin owes to its antioxidant activity and its ability to enhance glutathione levels, thereby facilitating hepatic detoxification of mutagens and carcinogens, eventually preventing formation of nitrosamine[94]. Curcumin also induces and promotes apotosis and thus has a potential to act as low cost anticancer, prohealth compound[99].

Curcumin is reported to lower cholesterol and triglyceride levels, decreasing susceptibility of low density lipoproteins to lipid peroxidation and inhibiting platelet aggregation thus protect from cardiovascular diseases. Curcumin lowers cholesterol either by decreased cholesterol uptake in intestine, or increased conversion of cholesterol to bile acids in liver. Prevention of platelet aggregation can be attributed to curcumin's ability to promote prostacyclin synthesis and inhibit thromboxane synthesis[94].

It has been established by several studies that turmeric has considerable antimicrobial activity[94, 100]. Turmeric extract and essential oil has significant effect against bacteria, parasites and pathogenic fungi[94]. Turmeric has several protective effects on gastrointestinal tract. Turmeric inhibits ulcer formation caused by stress, alcohol, indomethacin, pyloric ligation and reserpine[94]. Curcumin also enhance immunity in general. It has been found that number of CD4 T helper and B type immune cells increase in the intestinal lining after ingestion of curcumin[94].

GINGER

Ginger is widely used all around the globe owing to its culinary and curative properties. It is mainly cultivated in tropical countries *viz.* India, Nigeria, Australia, China and Jamaica[101]. Ginger finds extensive use in food due to its spicy, penetrating pungent and slightly biting flavor[102]. The aroma and flavour of ginger is attributed to the chemical composition of its essential oil which consists of sesquiterpene hydrocarbons as well as oxygenated monoterpenes and sesqueterpene compounds. Ginger has been recommended by Ayurvedic and Chinese medicinal system as remedy for nausea, sea or motion sickness, nausea related to pregnancy, vomiting, loss of appetite, stomach cramps, diarrhea, heartburn, colic, flatulence, indigestion, common cold, influenza, cough, arthritis, rheumatic disorders, migraines, headaches, cardiac palpitations, hypertension, and impotence[101]. Perfume and soft drink manufacturing companies also make use of ginger. Gingerole essential oil contains many chemical components out of which 115 have been identified so far. [6]-Gingerole (Fig. 9) has been found to be the principal pungent component of ginger essential oil. In addition to [6]-gingerol, [4]-, [7]-, [8]-, and [10]-gingerol were identified, as well as methyl [4]-gingerol and methyl [8]-gingerol. [4]-, [6]-, [8]-, [10]-, and [12]-shogaol were characterized, as were methyl [4]-, methyl [6]-, and methyl [8]-shogaol. Paradols are 5-deoxygingerols. [6]-Paradol, along with [7]-, [8]-, [9]-, [10]-, [11]-, and [13]- paradols were detected in the fresh ginger, as was methyl [6]- paradol. Other than gingerole, ginger essential oil also contain, zingiberene, ar-curcumene, β-bisabolene, and β-sesquiphellandrene, geranial, neral, and camphene methyl [8]-paradol, methyl [6]-isogingerol, methyl [4]-shogaol, [6]-isoshogaol, two 6-hydroxy-[n]-shogaols (n = 8 and 10), 6-dehydro-[6]-gingerol, three 5-methoxy-[n]-gingerols (n = 4, 8 and 10), 3-acetoxy-[4]-gingerdiol, 5-acetoxy-[6]-gingerdiol (stereoisomer), diacetoxy-[8]-gingerdiol, methyl diacetoxy- [8]-gingerdiol, 6-(4′-hydroxy-3′-methoxyphenyl)-2-nonyl-2-hydroxytetrahydropyran,3-acetoxydihydro-[6]-paradol methyl ether, 1-(4′-hydroxy-3′-methoxyphenyl)-

2-nonadecen-1-one and its methyl ether derivative, 1,7-bis-(4′-hydroxy-3′-methoxyphenyl)-5-methoxyheptan-3-one, 1,7-bis-(4′-hydroxy-3′-methoxyphenyl)-3-hydroxy-5-acetoxyheptane, acetoxy- 3-dihydrodemethoxy-[6]-shogaol, 5-acetoxy-3-deoxy-[6]- gingerol, 1-hydroxy-[6]-paradol, (2E)-geranial acetals of [4]- and [6]-gingerdiols, (2Z)-neral acetal of [6]-gingerdiol, acetaldehyde acetal of [6]-gingerdiol, 1-(4-hydroxy-3- methoxyphenyl)-2,4-dehydro-6-decanone and the cyclic methyl orthoesters of [6]- and [10]-gingerdiols. Gingeroles are responsible for many medicinal properties of ginger. Gineroles are chemically similar to capsaicin, a pungent principal component of chilies[101].

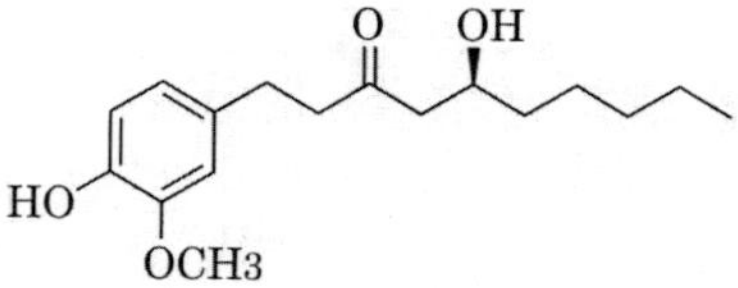

Fig. 9: [6]- Gingerol

Biosynthesis of [6]-Gingerole

Biosynthesis of [6]-ginerole (Fig. 10)[103] begins with phenylanaline. Phenylalanine is converted to feluric acid *via* p-coumaric acid. Feluric acid undergoes condensation with malonyl CoA to form β-Ketoester, which after hydrolysis and condensation with hexanoyl CoA result in 6-dehydro-gingerdione. [6]-Dehydrogingerdione is converted to 6-gingerole through [6]-gingerdione or [6]-dehydrogingerole. [6]-Dehydrogingerdione can be formed through an alternative pathway in which hexanoate and malonyl CoA react to form a compound which on hydrolysis and then condensation with feruloyl CoA yield [6]-dehydrogingerdione[101, 103].

Degradation of Gingerole

Gingeroles are prone to chemical degradation to less pungent products due to poor handling and storage. At temperature above 200°C, gingeroles undergo retro aldol reaction at the β-hydroxy group to yield zingerone and aliphatic aldehyde, such as hexanal. Such reactons lead to quality deterioration owing to the production of off flavours and pungency lossess[104].

Under alkaline conditions at room temperature or under acidic conditions at high temperature, gingeroles undergo dehydration at β-hydroxy keone group to form corresponding less pungent shogoals[105]. Shogoal formation occurs more rapidly during drying of whole ginger than drying of sliced ginger due to longer drying time required in the earlier than later. Shogoals have also been reported to be prone to thermal treatment and acidic conditions and get converted to non pungent polymers. Thus during storage,

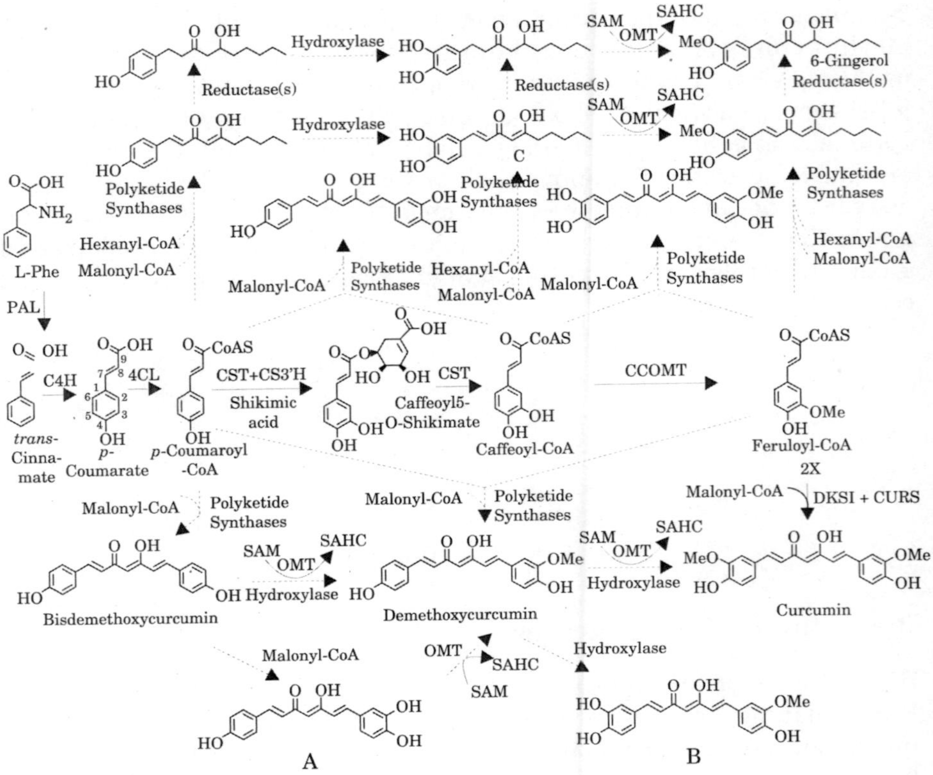

Fig. 10: Biosynthesis of [6]-gingerol[101].

gingeroles get degraded to shogoals which in turn get degraded to non pungent polymers[106].

Extraction of Ginger Essential Oil

There are number of methods that have been used for extraction of ginger bioactive components. Hydro distillation, steam distillation, leaching and pressing are the most commonly used methods. Recently, super critical fluid extraction has been used for extraction of essential oil.

Hydro distillation is usually done in laboratory with steel, copper or glass apparatus connected to a cooling and decanting flask. Oil obtained from water distillation of green ginger was superior to the oil of dry ginger owing to higher content of α-Zingiberene in green oil[107].

Steam distillation is a conventional and commercial method of ginger essential oil extraction. Extraction yield ranges between 0.2% to 3% depending on the source and state of rhizome. Steam distillation yields high level of monoterpenes and low amount of non volatile compounds due to thermal degradation of gingeroles[108].

Oleoresin extracts are obtained using solvent extraction. Gingeroles make up 33% of oleoresins. There are number of solvents that have been reported by different studies *viz.* ethyl ether[109], acetone[110], hexane[111], pentaflouropropane[112], heptaflouropropane[113] and ethyl acetate[114]. Extraction using ethyl acetate is done with dry ginger which is crushed and left with ethyl acetate for 30 minutes. The solution is filtered and evaporated to get oil. Oil is dissolved with 20 ml of ethyl acetate to give stock solution.

Applicability of supercritical fluid extraction for extraction has broadened since 20 years owing to the solvent used which is safe, non combustible, inexpensive, odorless, colourless, tasteless, nontoxic and readily available. Supercritical fluid extraction is over the edge with other techniques because it is time saving, energy saving capability and it gives product of better sensory quality. The yield of ginger essential oil extracted with supercritical fluid extraction depends on the conditions. Recovery of oleoresin is greater with entrainer, high solvent feed ratio and high pressure[108]. The effective component from 435 g ginger powder were extracted by supercritical CO_2 at 80°C and 45 MPa[113]. The components obtained were, 1.1 g ginger wax, 21 g hot gingerin, 9.2 g ginger essential oil and 33.5 g water.

Clinical Importance

Ginger has been well identified as a spice around the globe. Its utilization in Chinese, Ayurvedic and Tibb-Unani herbal medicine has been established since decades[115]. Ginger has been well evaluated in preventing pregnancy induced, postoperative and chemotherapy induced nausea and vomiting. Through the exact mechanism behind its antiemetic effect is unknown. It has been reported that, it apparently inhibits serotonin receptors and exhibit antiemetic effect at gastrointestinal system level and at central nervous system level[116]. Some studies also showed that ginger reduces symptoms of osteoarthritis of knee, alleviate pain and swelling[116]. Few studies suggested that ginger can be used as an antithrombotic and anti-inflammatory agent[117]. Ginger has also been reported to reduce blood pressure[115]. Several studies have reported the anti-inflammatory and analgesic activities of ginger. Its anti-inflammatory action has been credited to its inhibitory action on prostaglandin synthesis, leukotrines synthesis[118] and cytokines synthesis[119]. It has been found that ginger inhibits the induction of several genes involved in inflammatory response[115]. Ginger is quite potent in protecting tissue from chemical and radiation damage. Chemical and radio protective action of ginger can be attributed to its strong antioxidant properties. It is effective against UV induced oxidative damage and many toxic agents like carbon tetrachloride and cisplatin[120, 121]. Ginger extract exhibits strong antibacterial action against *Pseudomonas*, *Aeroginosa*, *Salmonella typhimurium*, *E. coli*, and *Candida albicans*[122]. It has also been reported in several studies that ginger possess antifungal properties. Few reports have also suggested that ginger is also helpful in alleviating the gastrointestinal tract illnesses[123]. Preliminary studies have also suggested that ginger has some antitumor activity thus can be evaluated for its anticancer use. Besides these, it has been found to be effective for the treatment of cataract, rheumatism, nervous diseases, gingivitis, toothache, asthma, stroke, constipation and diabetes[124, 125, 126].

REFERENCES

[1] Baloch A.K., Buckle K.A., Edwards R.A., "Effect of sulphur dioxide and blanching on the stability of carotenoids of dehydrated carrot", *Journal of the Science of Food and Agriculture*, **40**(2), 179–87 (1987).

[2] Rajendran V., Chen B.H., "Isomerization of β-carotene by titanium tetrachloride catalyst", *J Chem Sci.,* **119**(3), 253–8 (2007).

[3] Sant P.,'Ana H.M., Stringheta P.C., BrandÃO SCC, PÁEz HH, QueirÓZ VMVd., "Evaluation of total carotenoids, α- and b-carotene in carrots (*Daucus carota* L.) during home processing", *Fd Sci Tech., (Campinas),* **18**, 39–44 (1998).

[4] Cunningham F.X., Gantt E., "Genes and enzymes of carotenoid biosynthesis in plants", *Annual Review of Plant Physiology and Plant Molecular Biology*, 49, 557–83 (1998).

[5] Farré G., Sanahuja G., Naqvi S., Bai C., Capell T., Zhu C., Christou P., "Travel advice on the road to carotenoids in plants", *Plant Science,* **179**(1–2), 28–48 (2010).

[6] Walter M.H., Strack D., "Carotenoids and their cleavage products: Biosynthesis and functions", *Natural Product Reports,* **28**(4), 663–92 (2011).

[7] Johnson L.E., "Food technology of the antioxidant nutrients", *Critical Reviews in Food Science and Nutrition,* **35**(1–2),149–59 (1995).

[8] Emanuel, Denisov N.M., Maizus E.T., Z.K., "Liquid-phase oxidation of hydrocarbons" *Published by Plenum Press*, (1967).

[9] Karel M., "Lipid oxidation, secondary reactions, and water activity of foods", *Autoxidation in Food and Biological Systems:* Springer, pp. 191–206 (1980).

[10] Burton G.W., Ingold K.U., "Beta-Carotene: An unusual type of lipid antioxidant", *Science (New York, NY),* **224**(4649), 569–73 (1984).

[11] Conn P.F., Schalch W., Truscott T.G., "The singlet oxygen and carotenoid interaction" *Journal of Photochemistry and Photobiology B: Biology,* **11**(1), 41–47 (1991).

[12] Di Mascio P., Kaiser S., Sies H., "Lycopene as the most efficient biological carotenoid singlet oxygen quencher", *Archives of Biochemistry and Biophysics,* **274**(2), 532–538 (1989).

[13] Krinsky N.I., "Antioxidant functions of carotenoids", *Free Radical Biology & Medicine,* **7**(6), 617–635 (1989).

[14] Bonnie T.Y., Choo Y., "Oxidation and thermal degradation of carotenoids", *Journal of Oil Palm Research,* **11**, 62–78 (1999).

[15] Mordi R.C., Walton J.C., Burton G.W., Hughes L., Ingold K.U., Lindsay D.A. "Exploratory study of β-carotene autoxidation", *Tetrahedron Letters,* **32**(33), 4203–4206 (1991).

[16] Kanasawud P., Crouzet J.C. "Mechanism of formation of volatile compounds by thermal degradation of carotenoids in aqueous medium. 1.beta.-Carotene degradation", *J of Agric Food Chem.,* **38**(1), 237–243 (1990).

[17] Rodriguez G.A., "Extraction, isolation, and purification of carotenoids", *Current Protocols in Food Analytical Chem.,* (2001).

[18] O'Neil C.A., Schwartz S.J., "Chromatographic analysis of cis/trans carotenoid isomers", *J of Chromatography.,* **624**(1–2), 235–252 (1992).

[19] Hart D.J., Scott K.J. "Development and evaluation of an HPLC method for the analysis of carotenoids in foods, and the measurement of the carotenoid content of

vegetables and fruits commonly consumed in the UK", *Food Chem.*, **54**(1), 101–111 (1995).

[20] Konings E.J.M., Roomans H.H.S., "Evaluation and validation of an LC method for the analysis of carotenoids in vegetables and fruit", *Food Chem.*, **59**(4), 599–603 (1997).

[21] Oliver J., Palou A., Pons A., "Semi-quantification of carotenoids by high-performance liquid chromatography: Saponification-induced losses in fatty foods", *J of Chromatography* A., **829**(1–2), 393–399 (1998).

[22] De Sio F., Servillo L., Loiuduce R., D.C., "A chromatographic procedure for the determination of carotenoids and chlorophylls in vegetable products", *Acta Alimentaria*, **30**(4), 395–405 (2001).

[23] Weissenberg M., Schaeffler I., Menagem E., Barzilai M., Levy A., "Isocratic non-aqueous reversed-phase high-performance liquid chromatographic separation of capsanthin and capsorubin in red peppers (*Capsicum annuum* L.), paprika and oleoresin", *Journal of Chromatography A.*, **757**(1–2), 89–95 (1997).

[24] JaeHwan L., Steven J.S., "Analysis of Carotenoids and Chlorophylls in Foods. Methods of Analysis of Food Components and Additives, Second Edition", *Chemical & Functional Properties of Food Components: CRC Press*, pp. 231–252 (2011).

[25] Taungbodhitham A.K., Jones G.P., Wahlqvist M.L., Briggs D.R., "Evaluation of extraction method for the analysis of carotenoids in fruits and vegetables", *Fd Chem.*, **63**(4), 577–584 (1998).

[26] Lin C.H., Chen B.H., "Determination of carotenoids in tomato juice by liquid chromatography", *J of Chromatography A.*, **1012**(1), 103–109 (2003).

[27] Su Q., Rowley K.G., Balazs N.D.H., "Carotenoids: Separation methods applicable to biological samples", *Journal of Chromatography B.*, **781**(1–2), 393–418 (2002).

[28] Vargas F.D., Lopez O.P., "Enzymatic treatment to enhance carotenoid content in dehydrated marigold flower meal", *Plant Foods for Human Nutrition* (Dordrecht, Netherlands), **50**(2), 163–169 (1997).

[29] Vargas F.D., López O.P., "Effects of enzymatic treatments on carotenoid extraction from marigold flowers (*Tagetes erecta*)", *Food Chemistry*, **58**(3), 255–258 (1997).

[30] Dachtler M., Glaser T., Kohler K., Albert K., "Combined HPLC-MS and HPLC-NMR on-line coupling for the separation and determination of lutein and zeaxanthin stereoisomers in spinach and in retina", *Analytical Chemistry*, **73**(3), 667–674 (2001).

[31] Barth M.M., Zhou C., Kute K.M., Rosenthal G.A., "Determination of optimum conditions for supercritical fluid extraction of carotenoids from carrot (*Daucus carota* L.) Tissue", *Journal of Agricultural and Food Chemistry*, **43**(11), 2876–2878 (1995).

[32] Burri B.J., Neidlinger T.R., Lo A.O., Kwan C., Wong M.R., "Supercritical fluid extraction and reversed-phase liquid chromatography methods for vitamin A and beta-carotene heterogeneous distribution of vitamin A in the liver", *Journal of Chromatography A.*, **762**(1–2), 201–206 (1997).

[33] Prieto M.S.G., Caja M.M., Herraiz M., Maria G.S., "Supercritical fluid extraction of all-trans-lycopene from tomato", *Journal of Agricultural and Food Chemistry*, **51**(1), 3–7 (2003).

[34] Paiva S.A., Russell R.M., "Beta-carotene and other carotenoids as antioxidants", *Journal of the American College of Nutrition*", **18**(5), 426–433 (1999).

[35] Handelman G.J., "Carotenoids as scavengers of active oxygen species", *In*: Cadenas E. and Packer L. (*eds.*), Handbook of antioxidants, New York: Marcel Dekker, p. 259, (1995).

[36] Hughes D.A., Wright A.J., Finglas P.M., Peerless A.C., Bailey A.L., Astley S.B., Pinder A.C., Southon S., "The effect of beta-carotene supplementation on the immune function of blood monocytes from healthy male nonsmokers", *The Journal of Laboratory and Clinical Medicine*, **129**(3), 309–317 (1997).

[37] Santos M.S., Gaziano J.M., Leka L.S., Beharka A.A., Hennekens C.H., Meydani S.N., "Beta-carotene-induced enhancement of natural killer cell activity in elderly

men: An investigation of the role of cytokines", *The American Journal of Clinical Nutrition,* **68**(1), 164–70 (1998).

[38] Van Poppel G., Spanhaak S., Ockhuizen T., "Effect of beta-carotene on immunological indexes in healthy male smokers", *The American Journal of Clinical Nutrition,* **57**(3), 402–407 (1993).

[39] Michaud D.S., Feskanich D., Rimm E.B., Colditz G.A., Speizer F.E., Willett W.C., Giovannucci E., "Intake of specific carotenoids and risk of lung cancer in 2 prospective US cohorts" *The American Journal of Clinical Nutrition,* **72**(4), 990–997 (2000).

[40] Giovannucci E., "A review of epidemiologic studies of tomatoes, lycopene, and prostate cancer", *Experimental Biology and Medicine (Maywood, NJ)*, **227**(10), 852–859 (2002).

[41] Kritchevsky S.B., "Beta-Carotene, carotenoids and the prevention of coronary heart disease", *The Journal of Nutrition,* **129**(1), 5–8 (1999).

[42] D'Odorico A., Martines D., Kiechl S., Egger G., Oberhollenzer F., Bonvicini P., *et al.* "High plasma levels of alpha- and beta-carotene are associated with a lower risk of atherosclerosis: Results from the Bruneck study", *Atherosclerosis,* **153**(1), 231–239 (2000).

[43] Dwyer J.H., Paul-Labrador M.J., Fan J., Shircore A.M., Merz C.N., Dwyer K.M., "Progression of carotid intima-media thickness and plasma antioxidants: The Los Angeles Atherosclerosis Study" *Arteriosclerosis, Thrombosis, and Vascular Biology,* **24**(2), 313–319 (2004).

[44] Iribarren C., Folsom A.R., Jacobs D.R. Jr., Gross M.D., Belcher J.D., Eckfeldt J.H., "Association of serum vitamin levels, LDL susceptibility to oxidation, and autoantibodies against MDA-LDL with carotid atherosclerosis. A case-control study. The ARIC Study Investigators. Atherosclerosis Risk in Communities. *Arteriosclerosis, Thrombosis, and Vascular Biology,* **17**(6), 1171–1177 (1997).

[45] Mc Quillan B.M., Hung J., Beilby J.P., Nidorf M., Thompson P.L., "Antioxidant vitamins and the risk of carotid atherosclerosis", The Perth Carotid Ultrasound Disease Assessment study (CUDAS). *Journal of the American College of Cardiology,* **38**(7), 1788–1794 (2001).

[46] Rissanen T., Voutilainen S., Nyyssonen K., Salonen R., Salonen J.T., "Low plasma lycopene concentration is associated with increased intima-media thickness of the carotid artery wall", *Arteriosclerosis,Thrombosis, and Vascular Biology,* 20(12), 2677–2681 (2000).

[47] Krinsky N.I., Landrum J.T., Bone R.A., "Biologic mechanisms of the protective role of lutein and zeaxanthin in the eye", *Annual Review of Nutrition,* **23**, 171–201 (2003).

[48] Perlman J.A.M., Millen A.E., Ficek T.L., Hankinson S.E., "The body of evidence to support a protective role for lutein and zeaxanthin in delaying chronic disease Overview", *The J of Nutrition.,* **132**(3), 518s–524s (2002).

[49] Rissanen T.H., Voutilainen S., Nyyssonen K., Salonen R., Kaplan G.A., Salonen J.T., "Serum lycopene concentrations and carotid atherosclerosis: The Kuopio Ischaemic Heart Disease Risk Factor Study", *The American Journal of Clinical Nutrition,* **77**(1), 133–138 (2003).

[50] Ishimi Y., Ohmura M., Wang X., Yamaguchi M., Ikegami S., " Inhibition by carotenoids and retinoic acid of osteoclast-like cell formation induced by bone-resorbing agents *in vitro*", *Journal of Clinical Biochemistry and Nutrition,* **27**(3), 113–122 (1999).

[51] Kim L., Rao A.V., Rao L.G., "Lycopene II—effect on osteoblasts: The carotenoid lycopene stimulates cell proliferation and alkaline phosphatase activity of SaOS-2 cells", *J of Medicinal Food.*, **6**(2), 79–86 (2003).

[52] Rao L.G., Krishnadev N., Banasikowska K., Rao A.V., "Lycopene I–effect on osteoclasts: Lycopene inhibits basal and parathyroid hormone-stimulated osteoclast formation and mineral resorption mediated by reactive oxygen species in rat bone marrow cultures", *J of Medicinal Food.*, **6**(2), 69–78 (2003).

[53] Foy C.J., Passmore A.P., Vahidassr M.D., Young I.S., Lawson J.T., "Plasma chain-breaking antioxidants in Alzheimer's disease, vascular dementia and Parkinson's disease", *QJM : Monthly Journal of the Association of Physicians*, **92**(1), 39–45 (1999).

[54] Paran E., "Reducing hypertension with tomato lycopene", *In*: Rao A.V. (*ed.*), Tomatoes, lycopene and human health, *Scotland: Caledonian Science Press*, pp. 169–182 (2006).

[55] Azeredo H.M.C., "Betalains: Properties, sources, applications, and stability – a review", *International Journal of Food Science & Technology,* **44**(12), 2365–2376 (2009).

[56] Nilsson T., "Studies into the pigments in beetroot (*Beta vulgaris* L. ssp. *vulgaris* var. rubra L.)", *Lantbrukshogskolans Annaler,* **36**, 179–219 (1970).

[57] Stintzing F.C., Carle R., "Functional properties of anthocyanins and betalains in plants, food, and in human nutrition", *Trends in Food Science & Technology*, **15**(1), 19–38 (2004).

[58] Strack D., Vogt T., Schliemann W., "Recent advances in betalain research", *Phytochemistry,* **62**(3), 247–269 (2003).

[59] Herrero F.G., Carmona F.G., "Biosynthesis of betalains: Yellow and violet plant pigments", *Trends in Plant Science*, **18**(6), 334–343 (2013).

[60] Herbach K.M., Stintzing F.C., Carle R., "Impact of thermal treatment on color and pigment pattern of red beet (*Beta vulgaris* L.) Preparations", *J of Food Sci.,* **69**(6), C491–C498 (2004).

[61] Vargas F.D., Jimenez A.R., Lopez O.P., "Natural pigments: Carotenoids, anthocyanins, and betalains–characteristics, biosynthesis, processing, and stability", *Critical Reviews in Food Science and Nutrition,* **40**(3), 173–289 (2000).

[62] Schliemann W., Kobayashi N., Strack D., "The decisive step in betaxanthin biosynthesis is a spontaneous reaction1", *Plant Physiology,* **119**(4), 1217–1232 (1999).

[63] Wiley R.C., Lee Y.N., "Recovery of betalaines from red beets by a diffusion-extraction procedure". *Journal of Food Science*, **43**(4), 1056–1058 (1978).

[64] Lee Y.N., Wiley R.C., "Betalaine yield from a continuous solid-liquid extraction system as influenced by raw product, post-harvest and processing variables", *Journal of Food Science*, **46**(2), 421–424 (1981).

[65] Lee Y.N., Wiley R.C., Sheu M.J., Schlimme D.V., "Purification and concentration of betalaines by ultrafiltration and reverse osmosis", *Journal of Food Science*, **47**(2), 465–471 (1982).

[66] Bayindirli A., Yildiz F., ÖZilgen M., "Modeling of sequential batch ultrafiltration of red beet extract", *Journal of Food Science*, **53**(5), 1418–1421 (1988).

[67] Real S., Cerezal M., "Estudio de los parámetros de operación en la ultrafiltración del jugo de remolacha para colorante", *Aliment Latinoamer*, **206**, 34–40 (1995).

[68] Chalermchat Y., Fincan M., Dejmek P., "Pulsed electric field treatment for solid–liquid extraction of red beetroot pigment: Mathematical modelling of mass transfer", *Journal of Food Engineering*, **64**(2), 229–236 (2004).

[69] Fincan M., de Vito F., Dejmek P., "Pulsed electric field treatment for solid–liquid extraction of red beetroot pigment", *Journal of Food Engineering*, **64**(3), 381–388 (2004).

[70] Nayak C.A., Chethana S., Rastogi N.K., "Raghavarao KSMS. Enhanced mass transfer during solid–liquid extraction of gamma-irradiated red beetroot", *Radiation Physics and Chemistry*, **75**(1), 173–178 (2006).

[71] Drdak M., Altamirano R.C., Rajniakova A., Simko P., Karovicova J., Benkovska D., "Red beet pigment composition. effects of fermentation by different strains of *Saccharomyces cerevisiae*", *J Food Sci.*, **57**(4), 935–936 (1992)

[72] Mobhammer M.R., Stintzing F.C., Carle R., "Colour studies on fruit juice blends from *Opuntia* and *Hylocereus cacti* and betalain-containing model solutions derived therefrom", *Food Res International.*, **38**(8–9), 975–981 (2005).

[73] Escribano J., Pedreño M.A., Carmona F.G., Muñoz R., "Characterization of the antiradical activity of betalains from *Beta vulgaris* L. roots", *Phytochemical Analysis*, **9**(3), 124–127 (1998).

[74] Herrero F.G., Escribano J., Carmona F.G., "Purification and antiradical properties of the structural unit of betalains", *J of Natural Products*, **75**(6), 1030–1036 (2012).

[75] Khan M.I., Sri Harsha P.S.C., Giridhar P., Ravishankar G.A., "Pigment identification, nutritional composition, bioactivity, and *in vitro* cancer cell cytotoxicity of *Rivina humilis* L berries, potential source of betalains ", *LWT - Food Science and Technology,* **47**(2), 315–323 (2012).

[76] Sreekanth D., Arunasree M.K., Roy K.R., Reddy T.C., Reddy G.V., Reddanna P., "Betanin a betacyanin pigment purified from fruits of *Opuntia ficus-indica* induces apoptosis in human chronic myeloid leukemia Cell line-K562", *Phytomedicine : International Journal of Phytotherapy and Phytopharmacology*, **14**(11), 739–746 (2007).

[77] Lee C.H., Wettasinghe M., Bolling B.W., Ji L.L., Parkin K.L.. "Betalains, phase II enzyme-inducing components from red beetroot (*Beta vulgaris* L.) extracts", *Nutrition and Cancer,***53**(1), 91–103 (2005).

[78] Kandiannan K., Thankamani C., Srinivasan V., Rajeev P., Sudhakaran A., "Turmeric (Extension Pamphlet)", *P1-11 Indian Institute of Spices Research Niseema Printers & Publishers*, (2008).

[79] Ruby A.J., Kuttan G., Babu K.D., "Rajasekharan KN, Kuttan R. Anti-tumour and antioxidant activity of natural curcuminoids", *Cancer Letters*, **94**(1), 79–83 (1995).

[80] Selvam R., Subramanian L., Gayathri R., Angayarkanni N., "The anti-oxidant activity of turmeric (*Curcuma longa*)", *Journal of Ethnopharmacology,* **47**(2), 59–67 (1995).

[81] Sandur S.K., Pandey M.K., Sung B., Ahn K.S., Murakami A., Sethi G., Limtrakul P., Badmaev V., Aggarwal B.B., "Curcumin, demethoxycurcumin, bisdemethoxycurcumin, tetrahydrocurcumin and turmerones differentially regulate anti-inflammatory and anti-proliferative responses through a ROS-independent mechanism", *Carcinogenesis,* **28**(8), 1765–1773 (2007).

[82] Ohshiro M., Kuroyanagi M., Ueno A., "Structures of sesquiterpenes from *Curcuma longa*", *Phytochemistry,* **29**(7), 2201–2205 (1990).

[83] Shen L., Ji H.F., "The pharmacology of curcumin: Is it the degradation products?", *Trends in Molecular Medicine,* **18**(3), 138–144 (2012).

[84] Kolev T.M., Velcheva E.A., Stamboliyska B.A., Spiteller M., "DFT and experimental studies of the structure and vibrational spectra of curcumin", *International Journal of Quantum Chemistry,* **102**(6), 1069–1079 (2005).

[85] Kulkarni S., Maske K., Budre M., Mahajan R., "Extraction and purification of curcuminoids from turmeric (*Curcuma longa* L.)", *Int J Pharmacol Pharm Tech.,* **1**, 81–84 (2012).

[86] Roughley P.J., Whiting D.A., "Experiments in the biosynthesis of curcumin", *Journal of the Chemical Society, Perkin Transactions,* **1**, 2379–2388 (1973).

[87] Katsuyama Y., Kita T., Funa N., Horinouchi S., "Curcuminoid biosynthesis by two type III polyketide synthases in the herb *Curcuma longa*", *Journal of Biological Chemistry,* **284**(17), 11160–11170 (2009).

[88] Lin J.K., Pan M.H., Lin-Shiau S.Y., "Recent studies on the biofunctions and biotransformations of curcumin", *BioFactors* (Oxford, England), **13**(1–4), 153–158 (2000).

[89] Tonnesen H.H., Karlsen J., "Studies on curcumin and curcuminoids. VI. Kinetics of curcumin degradation in aqueous solution", *Zeitschrift fur Lebensmittel-Untersuchung und -Forschung*, **180**(5), 402–404 (1985).

[90] Wang Y.J., Pan M.H., Cheng A.L., Lin L.I., Ho Y.S., Hsieh C.Y., Lin J.K., "Stability of curcumin in buffer solutions and characterization of its degradation products", *J of Pharmacy and Biomed Analysis.,* **15**(12), 1867–1876 (1997).

[91] Revathy S., Elumalai S., Benny M., Antony, "Isolation, purification and identification of curcuminoids from turmeric (*Curcuma longa* L.) by column chromatography", *J of Experim Sci.,* **2**(7), 21–25 (2011).

[92] Gupta A.P., Gupta M.M., Kumar S., "Simultaneous determination of curcuminoids in curcuma samples using high performance thin layer chromatography", *Journal of Liquid Chromatography & Related Technologies,* **22**(10), 1561–1569 (1999).

[93] Bharat B.A., Anushree K., Manoj S.A., Shishir S., "Curcumin derived from turmeric (*Curcuma longa*)", Phytopharmaceuticals in Cancer Chemoprevention, *Modern Nutrition Science: CRC Press*, (2004).

[94] Akram M., Uddin A.A.S., Usmanghani K., Hannan A., Mohiuddin E., Asif M., "*Curcuma longa* and curcumin: A review article", *Rom J Biol–Plant Biol.,* **55**(2), 65–70 (2010).

[95] Girish C., Pradhan S.C., "Hepatoprotective activities of picroliv, curcumin, and ellagic acid compared to silymarin on carbon-tetrachloride-induced liver toxicity in mice", *Journal of Pharmacology & Pharmacotherapeutics,* **3**(2), 149–155 (2012).

[96] Cerny D., Lekic N., Vanova K., Muchova L., Horinek A., Kmonickova E., Zídek Z., Kameníková L., Farghali H., "Hepatoprotective effect of curcumin in lipopolysaccharide/-galactosamine model of liver injury in rats: Relationship to HO-1/CO antioxidant system", *Fitoterapia,* **82**(5), 786–791 (2011).

[97] Begum A.N., Jones M.R., Lim G.P., Morihara T., Kim P., Heath D.D., Rock C.L., Pruitt M.A., Yang F., Hudspeth B., Hu S., "Curcumin structure-function, bioavailability, and efficacy in models of neuroinflammation and Alzheimer's disease", *The Journal of Pharmacology and Experimental Therapeutics,* **326** (1), 196–208 (2008).

[98] Weber W.M., Hunsaker L.A.,Gonzales A.M., Heynekamp J.J., Orlando R.A., Deck L.M., Vander Jagt D.L., "TPA-induced up-regulation of activator protein-1 can be inhibited or enhanced by analogs of the natural product curcumin", *Biochemical Pharmacology,* **72**(8), 928–940 (2006).

[99] Pari L., Tewas D., Eckel J., "Role of curcumin in health and disease", *Archives of Physiology and Biochemistry,* **114**(2), 127–149 (2008).

[100] Gul P., Bakht J., "Antimicrobial activity of turmeric extract and its potential use in food industry", *J Food Sci Technol.,* **52**(4), 2272–2279 (2015).

[101] Kubra I.R., Rao L.J., "An impression on current developments in the technology, chemistry, and biological activities of ginger (*Zingiber officinale* Roscoe)", *Critical Reviews in Food Science and Nutrition,* **52**(8), 651–688 (2012).

[102] Nagendra chari K.L., Manasa D., Srinivas P., Sowbhagya H.B., "Enzyme-assisted extraction of bioactive compounds from ginger (*Zingiber officinale* Roscoe)", *Food Chem.,* **139**(1–4), 509–514 (2013).

[103] Koo H.J., McDowell E.T., Ma X., Greer K.A., Kapteyn J., Xie Z., Descour A., Kim H., Yu Y., Kudrna D., Wing R.A., "Ginger and turmeric expressed sequence tags identify signature genes for rhizome identity and development and the biosynthesis of curcuminoids, gingerols and terpenoids", *BMC Plant Biology,* **13**(1), 27 (2013).

[104] Govindarajan V.S., "Ginger—chemistry, technology, and quality evaluation: Part 1", *Critical Reviews in Food Science and Nutrition,* **17**(1), 1–96 (1982).

[105] Surh Y.J., Lee S.S., "Enzymic reduction of [6]-gingerol, a major pungent principle of ginger, in the cell-free preparation of rat liver", *Life Sciences,* **54**(19), PL321–PL326 (1994).

[106] Purseglove J.W., Brown E., Green C., Robbins S., "Spices", Vol. 1: *Longman,* (1981).

[107] Krishnamurthy N., Nambudiri E.S., Mathew A.G., Lewis Y.S., "Essential oil of ginger", *Indian Perfumer,* **14**(1), 1–3 (1970).

[108] Badalyan A.G., Wilkinson G.T., Chun B.S., "Extraction of australian ginger root with carbon dioxide and ethanol entrainer", *The J of Supercritical Fluids,* **13**(1–3), 319–324 (1998).

[109] Mathew A., Krishnamurthy N., Nambudiri E., Lewis Y., "Oil of ginger", *Flavour Industry,* **4**(5), 226–228 (1973).

[110] Jo K.S. "Analysis of gingerol compounds of raw ginger (*Zingiber officinale* Roscoe)

and its paste by high performance liquid chromatography-mass spectrometry (LC/MS)", *Journal-Korean Society of Food Science and Nutrition"*, **29**(5), 747–751 (2000).

[111] Nishimura O., "Enantiomer separation of the characteristic odorants in Japanese fresh rhizomes of *Zingiber officinale* roscoe (ginger) using multidimensional GC system and confirmation of the odour character of each enantiomer by GC–olfactometry", *Flavour and frag J.*, **16**(1), 13–18 (2001).

[112] Hill C.E, Dowdle P.A., Corr S., "Solvent extraction process", PCT Int. Appl. W0 0064,555 (Cl B 01 D1/00), 2 Nov. 2000; G.B. Appl. 1999/9, 136, 22 (1999).

[113] Dowdle P.A., Corr S., Harris H., "Solvent extraction process Pat. WO 2002 036232 A1, Date 20020510, Appl. WO 2001 GB4904, (2002).

[114] Harvey D.J., "Gas chromatographic and mass spectrometric studies of ginger constituents", *Journal of Chromatography A*, **212**(1), 75–84 (1981).

[115] Ali B.H., Blunden G., Tanira M.O., Nemmar A., "Some phytochemical, pharmacological and toxicological properties of ginger (*Zingiber officinale* Roscoe): A review of recent research", *Food and chemical toxicology : An International J Published for the British Industrial Biological Research Associ.*, **46**(2), 409–420 (2008).

[116] White B., "Ginger: An overview", *Am Fam Physician.*, **75**(11), 1689–1691 (2007).

[117] Thomson M., Al-Qattan K.K., Al-Sawan S.M., Alnaqeeb M.A., Khan I., Ali M., "The use of ginger (*Zingiber officinale* Rosc.) as a potential anti-inflammatory and antithrombotic agent", *Prostaglandins, Leukotrienes, and Essential Fatty Acids*, **67**(6), 475–478 (2002).

[118] Kiuchi F., Shibuya M., Sankawa U., "Inhibitors of prostaglandin biosynthesis from ginger", *Chemical & Pharmaceutical Bulletin,* **30** (2), 754–757 (1982).

[119] Grzanna R., Lindmark L., Frondoza C.G., "Ginger—an herbal medicinal product with broad anti-inflammatory actions", *Journal of Medicinal Food,* **8**(2), 125–132 (2005).

[120] Amin A., Hamza A.A., "Effects of Roselle and Ginger on cisplatin-induced reproductive toxicity in rats", *Asian Journal of Andrology*, **8**(5), 607–612 (2006).

[121] Yemitan O.K., Izegbu M.C., "Protective effects of *Zingiber officinale* (Zingiberaceae) against carbon tetrachloride and acetaminophen-induced hepatotoxicity in rats", *Phytotherapy Research : PTR,* **20**(11), 997–1002 (2006).

[122] Jagetia G.C., Baliga M.S., Venkatesh P., Ulloor J.N., "Influence of ginger rhizome (*Zingiber officinale* Rosc) on survival, glutathione and lipid peroxidation in mice after whole-body exposure to gamma radiation", *Radiation Research*, **160** (5), 584–592 (2003).

[123] Afzal M., Al-Hadidi D., Menon M., Pesek J., Dhami M.S., "Ginger: An ethnomedical, chemical and pharmacological review", *Drug Metabolism and Drug Interactions,* **18** (3–4),159–190 (2001).

[124] Awang D.V.C., "Ginger",*Canadian Pharmaceutical Journal,* **125**, 309–311 (1992).

[125] Tapsell L.C., Hemphill I., Cobiac L., Patch C.S., Sullivan D.R., Fenech M., Patch C.S., Roodenrys S., Keogh J.B., Clifton P.M., Williams P.G., Fazio V.A., "Health benefits of herbs and spices: The past, the present, the future", *The Medical Journal of Australia,* **185**(4), S4–24, (2006).

[126] Wang W.H., Wang Z.M., "Zhongguo Zhong yao za zhi = Zhongguo zhongyao zazhi = China [Studies of commonly used traditional medicine-ginger]", *Journal of Chinese materia medica,* **30**(20), 1569–1573 (2005).

14

Recent Advances in Processing of Peas

TANUSHREE MAITY SAXENA[1]* AND ALOK SAXENA[2]

ABSTRACT

Pea (Pisum sativum L.) is widely used in the human diet for a long time being an excellent source of protein, vitamins, minerals and other nutrients, while being low in fat, high in fiber and containing no cholesterol. Pea is an important pod–shaped, cool–season, nutritious leguminous vegetable that is widely cultivated throughout the world. Like any legume crop, pea is an integral component of sustainable agriculture due to its soil enriching and conditioning properties. Peas require careful handling to maintain good quality as they are highly susceptible to changes in the field, after harvest, during processing and storage. Peas are an excellent source of vitamin C, vitamin B1 (thiamine), vitamins B6, B3 (niacin), B2 (riboflavin), pro–vitamin A carotenoids and a range of minerals. Peas are one of the best vegetable sources of protein. The starch and protein contents of the pea seeds range between 30–50% and 20–25%, (db) respectively. Pea starch exhibits high levels of water and oil absorption, good gelling capabilities, and gel clarity should make these ingredients useful in a variety of applications. Pea protein products have been reported to exhibit comparable and complementary functionality to homologous soybean protein products. However, factors such as pH, ionic strength, or the presence of other ingredients will affect the functional properties of pea protein. Not only consumed as fresh, peas are processed to make their availability throughout the year, in the form of dehydrated peas and frozen peas. Peas are dehydrated after blanching to a moisture content of 7–8%. Another process which is used to preserve peas is freezing which preserves peas in their natural form. Peas are frozen by

[1] Fruits and Vegetables Technology Discipline, Defence Food Research Laboratory, Mysore 570011, Karnataka, India

[2] Amity Institute of Food Technology, Amity University Uttar Pradesh, Noida 201313, Uttar Pradesh, India

**Corresponding author* E-mail: tanumft@gmail.com

individual quick freezing (IQF) after blanching. Frozen peas have emerged as one of the leading frozen vegetables in the country. They are highly accepted by the consumers due to their resemblance to fresh ones and convenience to use.

Keywrds: Peas, Pea protein, Processing, Frozen peas, Dehydrated peas, Chlorophyll, Bio-active compounds, Anti-nutritional factors, Quality

---

1. INTRODUCTION

Pea (*Pisum sativum* L.) is one of the world's oldest vegetable crops and belongs to the *Leguminosae* family. It is cultivated for its tender and immature seeds for use as vegetable and mature dry seeds for use as a pulse[1, 2]. Pea was first grown by ancient Romans & Greeks[3]. It originated somewhere in the Ethiopia, Eastern Mediterranean region and south west Asia. Different centers of origin for pea were listed based on genetic diversity, area comprising Central Asia and the Mediterranean[4]. The first cultivation of peas appears to have been in western Asia, from where it spread to Europe, China and India. At present, *Pisum sativum* is found in all temperate countries and in most tropical high–lands.

2. PHYSIOLOGY OF PEA: GROWTH & DEVELOPMENT

Pea cultivars grown in different parts of the world exhibit wide variation in height of stem, branching, pod size, seeds per pod, shelling percentage, smoothness of seeds (smooth/wrinkled) etc. *Pisum sativum* was divided into six sub–species and listed garden pea as *P. sativum* subsp. *hortense* Asch and Graebn and field pea as *P. sativum* subsp. *arvense* Poir[5]. There are two kinds of peas (i) the field pea and (ii) the garden pea. The field peas are small, round and tender, while the garden peas are large and smooth or wrinkled. The field pea is grown as a dry pulse, whereas the garden pea can be harvested immature for culinary purpose in addition to the dry pulse. Two main types of garden peas based on seed characters are smooth seeded peas and wrinkle seeded peas. Among the smooth seeded type, Alaska is the most popular cultivar which was used for the purpose of canning in US during early 1960s. Other cultivars used in canning were Surprise and Resistant Surprise which are sweet type peas. Thomas Laxton was a freezing cultivar with dark green color and tender skin. New cultivars released in 1980s are Minnesota 108, Oregon Sugar pod II, Oregon, M 176, S 423 and B 445–66[3]. Some green peas can be chopped like beans and the entire pod can be used because they lack parchment tissue on the inner side of the pod[6].

The crop is cultivated extensively in countries having a cool climate especially in the Northern Hemisphere. China, India, France, United States of America, Kenya and Egypt are the major producers of pea. China dominated the pea production with a figure of 11500000 MT followed by India in 2012. In India, production of green peas has increased from 2710000 tones in 2000 to 3029400 tones in 2010[7]. According to Indian horticulture database (2012–13) the present area under cultivation of pea crop is 417.87 thousand hectare with production of 3942 thousand tones of peas. The state of Uttar Pradesh is largest grower of green peas followed by Madhya Pradesh, Jharkhand and Himachal Pradesh at second, third and fourth places, respectively. Table 1 shows the area under cultivation of peas and production in different states of India.

Table 1: Statistics on pea production in India

State	*Area "000 Ha*	*Production "000 T*
Uttar Pradesh	171.57	1805.1
Madhya Pradesh	52.03	497.43
Jharkhand	24.13	358.21
Himachal Pradesh	23.15	258.35
Punjab	19.82	202.24
West Bengal	21.80	132.11
Haryana	15.08	105.85
Andhra Pradesh	2.87	100.60
Remaining States	87.42	482.11
Total	417.87	3942

Source: Indian Horticulture Database, 2012–13

Pea is herbaceous annual plant with tap root system. Originally, all the peas are vine plants which creep onto low bushes, rocks, or nearby poles for support as well as protection. Pea is typically a cool season crop and thrives well in cool weather. Green pea cotyledon develops in four phases; (i) cell formation which is characterized by rapid cell division and differentiation; (ii) cotyledon expands by enlargement of parenchyma cells and absorbs sucrose, amino acids, ammonia salts and organic acids; (iii) synthesis of starch, reserve proteins, and lipids and the attainment of maximum fresh weight by the seed; and (iv) maturation and dormancy where starch content and dry weight continue to increase while there is a marked fall in fresh weight, sucrose and respiration as the seed approaches dormancy[15].

3. POST–HARVEST HANDLING & STORAGE OF PEA

Harvesting and handling

As the tender peas with high sugar content fetch premium price in market, care should be taken to harvest pods at correct maturity. The time to harvest is determined by the appearance of the pods. Approximately three weeks following full bloom, peas are ready for harvest. Quality of peas deteriorates with maturity. Hot dry weather during harvest speeds up maturity without corresponding increases in yield. Peas produced on fertile soils maintain good quality for a longer period, thus providing a greater opportunity for orderly harvest of peas. The harvest should be made when the peas are still in prime condition. An inverse relationship was reported between yield and quality after peas reaches a certain maturity[16]. Some growers make two or three pickings of peas for quality of product and high yield, while others make only one by pulling all the pea vines and picking off the pods[9].

Harvesting peas for fresh use

Peas for fresh market are harvested when they are well filled, but still succulent, before they harden and fade in color. The peas should not be hard and starchy. Peas are best picked and shelled just before cooking as the sugar content decreases rapidly after harvest.

Harvesting edible podded peas

These peas are picked when the pods are long and the peas are just developing. Pods, three to five inches long, are produced five to seven days after flowering, and the pea seeds are slim and small. Pods need to be picked every other day to prevent them from developing large seeds and fibrous pods. If the seeds develop, they may be used similar to garden peas[17].

Harvesting peas for processing

Peas for processing are harvested with various types of machines. Transportation to the processing plant and processing should be immediate so this limits distance peas can be grown from the processing plant. Many workers calculated heat units to ascertain maturity and harvesting of peas. Number of degree hours above 4.4°C required to bring a variety to maturity is calculated and it varies from variety to variety. Heat units are calculated as the deviation of the daily mean temperature from the base temperature for growth, in case of peas 40°F[18].

Grading and packing

The over mature pods, flat pods (without seeds) and diseased or infected pods are removed before the peas are packed. Peas can be packed in baskets, hampers and boxes for shipment. Peas for processing are graded into four grades based on the size of shelled peas. The smaller sized pea seed is considered to be best. The size of sieve mesh ranges from 18/64 to 24/64 inches in diameter. However sometimes small peas may be over mature while larger peas may be tender. Tenderometer reading is more acceptable test to judge the quality of peas. Starchiness of seed and organoleptic evaluation of quality of peas are inversely proportional. During maturity, starch content increase and calcium migrates to seed coat and becomes tougher during ripening. Toughness of seeds is determined using tenderometer, especially for processing purposes which assesses pericarp resistance and shear resistance of the developing cotyledons. Peas with low tenderometer reading is offered high price[19]. Brines may be used, in which peas are floated in them, from which their density can be determined[20]. Sucrose–metabolizing enzymes, such as sucrose synthase which catalyses the cleavage of sucrose in the presence of UDP and UDP–glucose and fructose has been strongly correlated with the change in starch concentration in the seed coat during developmental stages of seed filling[21].

Low temperature storage

Shelled peas lose their quality due to increase in starch content, proteins and other insoluble substances. The seed coat also becomes tough with aging. Since the maturation accelerates with heat, proper storage is important immediately after harvest. Peas after harvesting are packed in gunny bags or crates. Fresh unshelled peas can be stored for two weeks if cooled to 0°C immediately after harvest, and held at 0°C with a relative humidity of 90–95%. Peas can be hydrocooled or wetted and vacuum cooled to remove field heat. Packaged ice is effective in keeping peas cold and minimizing wilting[9]. A storage temperature of–1.1°C and 90% relative humidity was recommended for about 10 to 15 days[22, 23].

Controlled atmosphere storage

Refrigeration at 0°C can store peas only up to 7–10 days. Changing the atmospheric composition of the storage environment of peas can extend the storage life of green pea. The storage life of green peas was reported to be extended for about 20 days in controlled atmosphere (CA) storage with 5–10% oxygen and 5–7% carbon dioxide at 0°C[24]. CA storage was found to extend storage–life of green peas for about 9 weeks[25]. Pre–cooling and modified atmosphere packaging (MAP) with polymethyl pentene polymeric films (PMP) of 25 and 35 mm thickness and CA with gas compositions ranging from 5 to 10 kPa O_2 and 5 kPa CO_2 were reported to be the best storage conditions for storage of snow pea pods (*Pisum sativum* L. var. *saccharatum*) at 5 °C. The appearance of pods stored under such conditions was much better than that of air–stored pods (control). Lower O_2 level (2.5 kPa with 5 kPa CO_2) and high CO_2 level (10 kPa with 5 kPa O_2) concentrations were found to have a detrimental effect on quality of stored pods due to development of slight off–flavors[26].

4. COMPOSITION AND NUTRITIONAL VALUE OF PEA

Peas are recognized as a healthy legume with good supplies of high energy and protein for health–conscious diets[27]. Pea is an excellent source of protein, fibers, vitamins, and minerals and is low in fat and contains no cholesterol[28]. However, the amount of these varies with the pea variety and maturity. Pea seeds are rich in protein (23–25%), slowly digestible starch (50%), soluble sugars (5%), fiber, minerals and vitamins. During maturity, the sugar content of pea decreases rapidly and there is a rapid increase in the content of starch and other polysaccharides and insoluble nitrogenous compounds, mainly proteins[29]. Peas are starchy, but high in fiber, protein, vitamins, minerals and lutein. Dry weight is about one–quarter protein and one–quarter sugar. A total starch content of 53.4% was reported in green peas on dry weight basis[30]. The resistant starch (RS), slowly digestible starch (SDS) and rapidly digestible starch (RDS) fractions constitute 32.1, 11.9 and 9.6%, respectively.

Pisum sativum L. is rich in protein (18–30%)[31]. Pea protein contains a balanced profile of amino acids, especially a high content of an essential amino acid, lysine[32]. Yellow field pea seed contains 18–30% proteins[33] and high levels of lysine which can be used to balance the deficiencies of this essential amino acid in cereal–based diets[34]. Amount of sucrose in peas is higher than other sugars. Sweetness of peas is due to fructose & glucose[35]. Compound such as 5– or 6–methylisopropyl–2–methoxypyrazine was reported to give off a dry grass aroma in blanched green peas[36]. The nutritional quality of pea protein has been evaluated in several studies. Pea protein is a good source of essential amino acids, with the exception of a deficiency in the sulfur–containing amino acids, methionine and cysteine. Pea seed peptide fractions have less ability to scavenge free radicals than glutathione, but greater ability to chelate metals and inhibit linoleic acid oxidation[37]. Highest amino acid content in peas was reported when they were at the middle stage of seed development[38]. Pea seed storage proteins are composed of legumin (11S), vicillin (7S) and albumins (2S) and the majority of pea protein isolates contain globular 11S and 7S[39, 40].

Vitamin C content of peas can vary from 20 to 40 mg/100g at harvest, depending upon variety, agronomy, etc.[41]. Ascorbic acid contents of 31, 30, and 26 mg/100 g were reported in three varieties of peas of each season, early season, mid–season and late season, respectively, immediately post–harvest[42]. During storage of peas after harvest vitamin C content continues to degrade. Ascorbic acid content of 39% of the original amount was reported to be lost in 14 days at 4°C and at 20°C, however the amount of total ascorbic acid decreased by 72% of the original vitamin C content after 7 days only[43]. The major constituents of field peas vary with cultivars as well as cultivation year with protein, ash, fat and starch in range of 24.3–31.0, 3.05–4.06, 0.76–3.95 and 33.4–47.5, respectively[44]. Table 2 presents the proximate composition of raw green peas *(Pisum sativum)* in different areas as reported by different authors. About 21.9% protein, 2.34% lipids, 3.0% ash, 10.4% crude fiber, 52.5% carbohydrates, 20.3% insoluble dietary fiber, 1.73 soluble dietary fibers, 2.45% resistant starch and 9.88% moisture was reported in uncooked dry peas[45]. The phyto–chemicals present in raw, canned and frozen pea are presented in Table 3.

Table 2: Proximate composition of green peas

Nutrient/100 g	*United States*[46]	*South Africa*[47]	*Tanzania*[48]
Moisture	78.9	78.2	Not Determined
Protein	5.4	5.5	5.4
Fat	0.4	0.8	0.2
Carbohydrates	12.3	8.3	10.1
Fibre	2.2	6.3	5.5

Table 3: Phyto–chemicals present in raw, canned and frozen pea[49]

Nutrients/100g	*Pea, Raw*	*Pea, Canned*	*Pea, Frozen*
Water, g	79.2	79.6	78.0
Vitamin C, mg	21.9	7.7	20
Thiamin, mg	–	0.110	0.235
Riboflavin, mg	–	0.090	0.160
Total fiber, g	3.7	4.0	4.3
Ca, mg	14.1	16.3	14.6
Mg, mg	42.5	22.4	32.2
K, mg	191.3	163.2	258.2
Na, mg	4.7	215.0	42.7
Fe, mg	2.1	1.2	1.5

The utilization of pea has been limited, due to the presence of certain anti–nutritional factors despite the nutritional potential of peas as a good source of proteins, carbohydrates, vitamins and minerals[50]. Phytic acid, condensed tannins, polyphenols, protease inhibitors (trypsin and chymo–trypsin), a–amylase inhibitors and lectins are the main anti–nutritional factors which reduce the nutritional quality of the protein[51]. The trypsin inhibitor content of peas is similar to that of the field bean (*Vicia faba*), but is only 10% of the level found in soybeans[52–54]. Pea protease inhibitors (PPI), or pea seed trypsin inhibitors (PSTI), are similar in structure to the well–characterized Bowman–Birk inhibitors (BBI) in soybean[55]. Pea cultivars with high trypsin inhibitor activity (TIA) levels in the raw seed may be more suitable for processing because cooking substantially reduces TIA levels, suggesting that adverse effects of consuming TIA would be limited. Conversely, raw peas with low TIA levels would be more suitable for use as animal feed. The concentrations of flavour volatile compounds in peas were reported to be affected by cultivar, crop year and processing conditions[56]. The green hay–like flavour in peas is said to be due to hexanol[57]. Bitter flavours in green peas may be attributed to the presence of the non–volatile saponins[58].

5. PEA STARCH: QUALITY AND EFFECT OF PROCESSING

Pea starch is the second excellent material followed by mung bean starch for processing starch noodles. The general composition of field pea starch of four varieties from Canada is given in Table 4. The morphologies and characteristics of the starches from different phenotypes of pea are different. Pea starch exhibits excellent functional properties[59]. The starch content and amylose content of the peas ranges between 34% and 42.7% of dry matter and between 35% and 38%, respectively. Average diameter of starch particle ranges between 21.4 and 26.1 µm. The typical C–type X–ray diffraction pattern, with relative crystallinity ranges between 36% and 55% and the proportion of B–type crystallites is in range 3.8% –30.4%. The swelling factor of smooth pea starch ranges from 4 to 27 at 50–95°C. But wrinkled pea starch has a lower swelling power[60]. Table 4 shows the composition of field pea starches of four different cultivars grown in Canada, namely, Carneval, Carrera, Grande and Keoma[61]. All the domestic cooking procedures render the legume starches more digestible in comparison to raw seeds. Soaking of dried peas was found to increase the slowly digestible starch fraction, whereas, pressure cooking was found to increase the RDS fraction. However, boiling was found to increase both RDS and SDS fractions[30]. Treatment of pea starch granules with alkali has been reported to decrease the amylose content by about 20–25% after 15 days and 25–30% reduction in peak and final RVA pasting viscosities, while increased the rate of *in–vitro* enzymatic breakdown. However, prolonged alkaline treatment for 30 days did not changed the structural and functional properties of the starch granules[62].

Table 4: Composition of field pea starch (four varieties) from Canada[61]

Parameter	***Value***
Yield (% initial material)	32.7 – 33.7
Moisture, %	9.2 – 13.3
Ash, %	0.03 – 0.14
Nitrogen, %	0.04 – 0.07
Lipid, %	0.05
Apparent amylase content, %	42.9 – 43.7
Total Amylose content, %	48.8 – 49.6
Starch granule size (Round granule) µm	5 – 7

Starch noodles using the isolated legume starches from four varieties of peas and lentils were successfully manufactured at laboratory scale[63]. The optimum processing parameters were less than 50% moisture content of dough, 40 sec cooking and 2 h cooling at 6°C. Pre–treatment of peas by roasting and boiling and processing of seeds into flour could enhance the functional properties and micro–structural characteristics of flour in addition to reducing trypsin inhibitor activity. Boiling pre–treatment could enhance the fat binding capacity, water holding capacity and gelling capacity of pea flour. Such thermal treatments to pulse flours may have very good potential to be used as value–added food ingredients for food applications due to their improved nutritional value and functionality[64]. Nutritional properties of pea can be modified by germination of pea seeds for 3 days, microwave (MW) treatment of germinated seeds and subsequent drying by hot air. The germinated and MW treated peas can be grinded for pea flour[65]. Quality of the extruded products depends upon the conditions of operation such as temperature and the quality of raw material used. Extruded product from pea starch has been found to show highest expansion on extrusion–cooking in twin screw extruder. Extruded product from dehulled and whole field pea flour showed very good puffing properties[66]. Infrared processing for 92 sec is found to significantly improve the starch gelatinization degree (SGD) (from 10.1 to 32.8%) of pea flakes. The SGD could be further improved with steam treatment of peas, prior to IR. However, improvements in SGD are reported to reduce durability of flakes[67]. The physicochemical and functional properties of meal and starch from mechanically and field dried peas were evaluated[68]. Field dried pea meal exhibited greater starch content and lower water binding capacity, swelling power and solubility. However, the mechanically dried pea starches, showed lower retrogradation and higher freeze thaw stabilities than those from field dried peas. They developed noodles from pea starch and showed lower cooking weight and cooking loss than potato starch noodles with insignificant difference in overall acceptability scores. Cooking is also reported to increase the soluble non–starch polysaccharides and digestibility of starch in 2 pea cultivar (Citrina and Warindo) evaluated by the in vitro controlled enzymatic hydrolysis[69]. Various domestic processing and cooking methods including soaking for 6, 12 and 18 hrs, soaking (12 h) followed by dehulling, ordinary and pressure cooking of unsoaked, soaked and soaked–dehulled seeds, and sprouting for varying periods of 12, 24 and 48 h were reported to significantly increase the starch digestibility of peas[70]. Among these, pressure cooking is found to be the most effective method for improving the pea starch digestibility followed by ordinary cooking, sprouting, dehulling and soaking.

6. PEA PROTEIN: QUALITY AND EFFECT OF PROCESSING

Pea protein plays important roles in many foodstuffs, because of its nutritional value, contribution to the texture of food and the surface properties of constitutive protein units[71, 72]. Pea proteins are considered to be an emerging alternative for wheat proteins for the formulation of gluten–free products[73]. The protein content of pea seeds vary in wide ranges due to genetic and environmental factors. Genetically identical pea plants grown in the same year have been reported to contain protein contents in seeds ranging 19.3–25.2 %. Increasing soil phosphorus and decreasing soil potassium increases nitrogen content from 3.5 to 4.9%. A decrease in protein content is also observed in seed with maturation and absence of addition of nitrogen fertilizer[74–76]. The chemical composition of pea samples as a function of content of protein on dry and dehulled basis was also determined[77]. The amino acid content of peas varying in protein content (high and low) is presented in Table 5.

Table 5: Amino acid composition of dry and dehulled peas varying in protein content[77]

Amino Acid(mg of amino acid/g of N for proteins)	*14.5 % protein*	*28.5 % protein*
Aspartic acid	734	736
Threonine	246	215
Serine	256	261
Glutamic acid	994	1062
Proline	238	251
Cystine	64	49
Glycine	299	262
Alanine	299	256
Valine	311	294
Methionine	57	44
Iso–leucine	260	262
Leucine	420	464
Tyrosine	216	197
Phenylalanine	299	287
Histidine	155	152
Lysine	485	442
Arginine	387	559

Pea protein isolates are used in many food applications to improve the texture and the nutritional quality of food products[78]. Owing to their solubility, water and fat binding capacity, gelation, foaming and emulsification capacity, pea protein concentrates are used in meat and sausage substitutes and soups[79, 80]. Pea protein isolates are commonly produced by wet fractionation by dispersing pea flour in water after which the dissolved proteins solution and the suspended starch granules are separated using a hydro–cyclone. The solubilized proteins in solution are then precipitated at their iso–electric point of pH 4.8 and subsequently, the pH is readjusted to 7. A dry protein isolate is obtained with a final drying step (75–90% protein, w/w)[81]. A dry milling and air classification extraction of pea protein isolates was suggested as an alternative to conventional wet extraction with advantages retention of native functionality of proteins and its lower energy and water use[82]. Pea protein isolate produced by iso–electric precipitation and salt extraction has been reported to have low emulsion capacity and stability due to high surface hydrophobicity, low surface charge and low solubility[83]. Pea proteins have emulsifying properties as they are able to decrease the interfacial tension between water and oil and help to stabilize emulsions by forming a rigid membrane at the oil–water interface[84]. Processing such as soaking and cooking has been found to increased protein content, insoluble dietary fiber, total dietary fiber, Ca, Cu, Mn and P in peas whereas ash content, Fe, K, Mg, Zn, sucrose and oligosaccharides were reduced. Cooking is found to be more effective than soaking in reducing oligosaccharides of peas as well as reducing the trypsin inhibitor activity[85].

Soaking increases the protein content by 2.6% to 5.0% while cooking is found to increase the protein by 3.0% to 6.7%. Soaking and cooking both increased the starch and protein content in range of 0.8–2.3% and 2.4–6.7%, respectively, due to the loss of soluble solids during soaking or cooking which increases the concentration of these nutrients in soaked or cooked seeds. Moreover, soaking and cooking showed the reduction in ash content ranging from 1.0% to 7.7% and from 23.2% to 34.4%, respectively[85]. Thermal processing not only increases the nutritional value but also reduces the unacceptable beany flavor, making pulses more palatable. It improves the functionality of pulse such as solubility, foaming, gelling, water binding and fat binding properties which can affect processing applications, food quality and acceptance[86]. Protein isolates from pea were separated into water–soluble (WS), salt–soluble (SS), alkaline–soluble (AS) and ethanol–soluble (ES) fractions and evaluated the functionality of these fractions[87]. They found that AS fraction was the most abundant about 87% of the proteins and the emulsions formed with ES fraction were the most stable. The protein solubility and foaming capacity at all the pH values was found to be high in WS fraction. Pea proteins have also been used to make low cost protein based films which showed similar physicochemical properties to those found for soy protein, whey protein or zein[88, 89]. Pea protein films containing lysozyme at different concentrations has been found show antimicrobial activity and is effective against *Listeria monocytogenes*[90].

7. ADVANCES IN PEAS PROCESSING

The three main uses of pea cultivars have been distinguished into field pea grown for their dry seeds, garden pea grown for their immature green seeds and sugar pea grown for the immature pods. Garden pea (green) is used in various culinary preparations. Applications of green peas are in soup mixes, purees and processed products. Tender pea seeds are also used in soups. Green peas are eaten cooked as a vegetable, and are marketed fresh, canned, or frozen while ripe dried peas are used whole, split, or made into flour. In some parts of the world, dried peas are consumed as dhal/dal, roasted, parched or boiled. Some cultivars are grown for their tender green pods, which are eaten cooked or raw. Field peas are primarily dried for storage and require long periods of simmering to cook them thoroughly[91]. The pods of peas should be of uniform maturity and should shell out easily. Peas were also among one of the first vegetables to be canned commercially. Peas have also been found to be an important vegetable crop for the canning industry[92]. Beginning in the 1920"s, peas, along with other vegetables began to be frozen as a means of preservation, but did not become popular until the 1950"s with the advent of the refrigerator freezer. Peas for canning have to be rapidly selected by means of tenderometer. The optimum canning stage of peas is reported to be at a tenderometer reading of 120 and at an alcohol–insoluble solid content of 13.4 %[93]. Canned, frozen and dehydrated peas are very common for use during off–season.

Apart from whole consumption of pea seed, starch and proteins extracted from peas also find application in food processing. Pea starch has been used to prepare noodles by high temperature extrusion process. The noodles thus prepared exhibited significantly lower cooking loss, higher cooked weight and superior quality than the noodles prepared from lentil starch or *mung* bean. Textural analysis showed higher firmness values, but lower surface stickiness in cooked pea starch noodles than cooked lentil starch noodles[94]. Similarly durum wheat semolina noodles fortified with pea proteins have also been developed[95]. Pasta was fortified with split pea flour (35%, db) by an adaptation of the pasta making process at a pilot–scale to improve the nutritional quality of pasta by enrichment in proteins and fibers, vitamins, minerals and amino acids[96]. Pea protein isolates are valuable for the production of palatable meat analogues opening up a wide range of products for different consumer requirements. Pea protein isolates can be successfully texturized at 55% moisture content to highly fibrous extrudates for the development of whole–muscle meat alternatives by controlling the cooking temperature and provide more benefits over soy[97]. Macaroni has also been developed from flour from green and yellow mature pea seeds after cooking the dough for 15 and 30 min with added emulsifiers. Addition of emulsifiers was found to improve the dough quality and changed the solubility of protein fractions by altering the soluble fraction molecular weight distribution and promoting penetration of low molecular weight fractions into the protein network[98].

Radiation processing at final stages could be used to make safe products for direct consumption. Irradiation (1 and 2 kGy doses) of minimally processed pea sprouts do not affect the quality parameters such as vitamin C content, total carotenoids content, color and texture over a storage period of 12 days at 4 and 8°C[99]. They reported that irradiation had no significant effect on the ratings of sensory attributes in garden pea sprouts and did not alter the overall acceptability of the irradiated sprouts up to 12 days. Slight decrease was found in taste scores of peas irradiated with higher dosage after 11 days of irradiation[100]. The sweetness and softness of peas increased with increase in dose of irradiation after the storage period. The control samples and peas irradiated at 1×10^5 rad were found to spoil after 11 days. However, those irradiated at 3 and 5×10^5 rad were green and not moldy. In a study, electron spin resonance technique has been found to be useful in detection of irradiated pea even at the end of a storage period of 75 days due to high radiation yield at 10 kGy and stability of the induced free radical at room temperature[101].

8. EFFECT OF PROCESSING ON BIOACTIVE COMPOUNDS OF PEAS

Peas are not only a good source of macro–nutrients but also have high contents of micro–nutrients such as ascorbic acid, β–carotene, thiamine and riboflavin[102]. They are also a good source of a wide variety of phenolic compounds such as kaempferol, quercetin, ferulic acid, and vanillic acid etc[103]. Phenolic compounds possess antioxidant activity due to their ability to act as hydrogen donors to break the oxidative free radical chain reactions. Antioxidants protect against reactive oxygen species in the human body and have a number of health effects, such as reducing the incidence of cancer and cardiovascular diseases[104]. Pea superoxide dismutase can act as a carrier for certain pea phenolic compounds and facilitate a protein–bound thermostable antioxidant[105]. The quercetin content in peas has been reported to be 0.15 mg/100 g[106]. The water–soluble fraction of blanched and frozen peas has a total anti–oxidant capacity of 0.61μmol/g and the water–insoluble fraction has a value of 0.23 μmol/g. Ascorbic acid accounts for a large part of the water–soluble antioxidant capacity[102]. Anti–oxidant activity of seed coat is more than the cotyledons of the pea due to the presence of phenolic compounds with flavonoid structures. Seed coats of dark peas contain flavonol's and flavone glycosides and hydroxybenzoic and hydroxycinnamic compounds[107]. Even though the seed coat represents only 10–11% of the total weight of the seed in grain legumes, the phenolic compounds of this part contributes more to the antioxidant capacity of the entire seed[108].

Ascorbic acid is a pH and light sensitive compound which is oxidizes to de–hydro ascorbic acid and irreversibly hydrolyses into diketogluconic acid, which has no vitamin activity[109]. It is also affected by naturally occurring enzyme ascorbic acid oxidase. The average Ascorbic acid and de–hydro ascorbic acid contents of unblanched fresh peas has been reported to be 26.1 mg/100 g and 2.2 mg/100 g, respectively. Thermal treatment such as blanching has been found to decrease the ascorbic acid content of green peas. Blanching at 70C for 4 min. decreased the ascorbic and de–hydro ascorbic acid contents of peas to 25.9 mg/100 g and 2 mg/100 g. However, blanching at 80 °C for 2 min. decreased the ascorbic and de–hydro ascorbic acid contents of peas to 25.7 mg/100 g and 1.7 mg/100 g, respectively. Frozen storage of green peas also affects the ascorbic acid content which could be minimized by a prior blanching step. The ascorbic acid loss in green peas after storage of 12 months has been reported to be 90.6%, 44.9% and 35.3% in un–blanched, 70°C, 4 min. blanched and 80°C, 2 min. blanched samples, respectively[110]. Ascorbic acid content of cooked peas has shown a rapid decrease of 39% from the original amount (88 mg/100 g dry matter) within 14 days at 4°C. At higher storage temperature (20°C) the amount of total ascorbic acid decreased by 72% of the original vitamin C content, after 7 days only[111]. However, less than 10% of the original ascorbic acid was reported to decrease in blanched peas after 12 months of deep–frozen storage[112].

Processing adversely affects the anti–oxidant activity of peas especially dry peas if soaked, boiled and steamed. These processing steps causes significant decrease in total phenolic content (TPC) and free radical scavenging activity in dry green and yellow peas as compared to original unprocessed peas. However, pressure boiling and pressure steaming could cause significant increase in oxygen radical absorbing capacity as compared to atmospheric processes[113]. Though steamed peas showed decrease in anti–oxidant activity but still contained substantial amounts of antioxidants compared to boiled peas. Therefore, steaming has been recommended as an ideal preparatory step for dry peas in domestic and industrial processes. It not only preserves the antioxidant components, but also decrease the cooking time.

9. EFFECT OF PROCESSING ON CHLOROPHYLLS OF PEAS

The characteristic green color of peas is due to the presence of chlorophylls, which are highly susceptible to many chemical or enzymatic degradation reactions such as pheophytinization, epimerization, pyrolysis, hydroxylation, oxidation or photo–oxidation[114]. Chlorophyll a and chlorophyll b are the major chlorophylls in plants, which occur in the approximate ratio of 3:1[115]. Chlorophylls are susceptible to degradation during processing due to the actions of enzymes, weak acids, oxygen, light and heat which lead to the formation of a large number of degradation products resulting in discolouration of green colour of vegetable to olive green. The general phenomenon associated with the discoloration is transformation of chlorophylls to their corresponding pheophytins by displacement of the magnesium ion in the porphyrin ring with hydrogen ions[116]. The change from bright green to dull olive–green or olive–yellow colour of a vegetable is ultimately perceived by the consumer as a loss of quality. Chlorophyll a has been reported to be thermally less stable than chlorophyll b[117, 118]. Lower tissue pH values and higher process temperatures increase the formation of pheophytins in processed vegetables[119]. Extensive heating may also form a derivative known as pyropheophytins by the loss of the carbomethoxy group from pheophytins[120] Enzyme chlorophyllase break the phytol chain of chlorophyll to form chlorophyllide which undergo loss of the magnesium and form pheophorbide in the presence of acid. In another mechanism, chlorophyllase hydrolyzes the phytol chain of pheophytins giving rise to pheophorbides[121, 122]. The optimum temperature for chlorophyllase activity in vegetables ranges between 60 to 82.2°C. Chlorophyllase looses its activity when heated to 100 °C[123, 124].

Chlorophyll content of green peas has also been reported to be affected by frozen storage. Chlorophyll–a and chlorophyll–b contents of fresh unblanched peas were found to be 164.7 and 75.8 mg/kg, which decreased to 94.15 and 59.55 mg/kg, respectively, after 12 months of frozen storage. Blanching prior to frozen storage adversely affected the stability of Chl–a and Chl–b in peas. Chl–a and Chl–b were converted into their corresponding epimers at ratios of 2.17–2.80% and 5.84–6.72% following blanching. In this study, 8.35–8.62 mg/kg of pheophytin–a was found in blanched peas only after 12 months of storage[110]. High pressure processing has been recommended over thermal processing not only to sterilize the green vegetables but also to avoid the chlorophyll degradation[125].

10. EFFECT OF PROCESSING ON ANTI–NUTRITIONAL FACTORS AND OTHER COMPONENTS OF PEAS

The volatile compounds in pea were found to reduce significantly on cooking pea seeds in boiled water. Higher aldehydes were observed in milled–dehulled seeds, followed by milled–whole seeds. Cooking whole and dehulled seeds increased the total relative peak area of alcohols and ketones[56]. Only 0.15 mg quercetin per 100 g in blanched peas was reported which was found to degrade as effect of different types of processing such as boiling, frying and microwave heating[126]. Another flavonoid, kaempferol was not at all detected in the peas.

The importance of extrusion conditions (148°C, 25% moisture and 100 rpm) was reported for the improvement of nutritional quality of pea seeds and minimization of tannin, trypsin, chymotrypsin, α–amylase inhibitors and haemagglutinating activity, without modifying protein content[51]. Moist heat treatment was reported to reduce protease inhibitor activity in peas by up to 86%[127]. Heat–labile anti–nutritional factors, such as trypsin inhibitors, are less important in human diets as cooking and processing are normally carried out before consumption[128]. High hydrostatic pressure has also found application in reducing the anti–nutritional factors in dried peas. Oligosaccharides and phytic acid were significantly reduced through pressurization by up to 68 and 36% in peas, respectively. Trypsin inhibitor activity decreased by up to 100 % and protein digestibility increased by up to 4.3% in peas when treated at 600 MPa and 60 °C[129]. Precipitation and ultra–filtration processes were investigated for their potential to produce protein products from aqueous wrinkled pea extracts[130]. The products obtained by ultra–filtration process were superior to those obtained by precipitation, in terms of their solubility, their fat binding capacity, their emulsifying activity and their foam–forming capacity.

Six varieties of pea were reported with trypsin inhibitor activity (TIA) in range from 1.30 to 2.01 mg g^{-1} sample[85]. They found that soaking peas in distilled water increases the TIA (3.2–19.3%) due to high protein concentration in the soaked seeds than in the raw seeds. However, cooking of peas improved the nutritional quality and resulted in a significant reduction (62.3–77.8%) of TIA suggesting sufficient cooking of peas is necessary to not only soften the peas but also to inactivate or reduce TIA. Cooking also caused a significant reduction (5.3–10.8%) of phytic acid reduction in oligosaccharide levels with a mean decrease of 29.5% for raffinose, 46.5% for stachyose and 28.9% for verbascose. Germination of pea seeds for 3 days is found to decrease the alpha–galactoside content and improve the nitrogen absorption[131]. Germination of peas for 2, 4 or 6 days with or without light is also reported to reduce the factors causing flatulence and improve the utilization of available proteins and carbohydrates[132].

11. DEHYDRATION OF PEAS

Peas are mostly consumed in dry mature state as the dried seeds could be easily store and hence offer more practical usage as a food particularly in poor economies[133]. They are primarily used for human consumption and livestock feed. Dry pea crop consists mainly of green and yellow peas. The global production of dry peas was estimated to be 9830016 tonnes in 2012. Among the world's pulses, field peas rank second to dry beans in production and consumption. They are typically consumed after cooking, both in the form of whole seeds and decorticated splits in various types of foods. For industrial purposes they are usually dried with hot air in static bed or thin layer[134]. For drying of peas fresh and green pea pods are thoroughly washed in water. Then pea seeds are separated and cleaned with the help of pea podder. Then they are pricked as pricking facilitates quick and uniform drying of peas. Then they are blanched and sulphited to retain colour, taste and texture in the final product. Blanched peas are then dried in a drier wherein moisture is reduced to 7–8%. Drying time is around 3 hours. Finally dried peas are graded and packed. The initial seed moisture content of peas must be reduced to about 8% before storage. This moisture content reduces the deterioration rate of peas during storage and reduces attack by moulds and insects. The seed should be stored in a dry and cool place, free of pests and protected from absorbing moisture from the surroundings. The literature contains references about the convective drying of green peas in fluidized bed as well as under atmospheric freeze drying conditions[135]. Amylose content in dry pea cultivar grown in US is in the range of 32.2–41.1%. Granules morphology is elliptical or spherical with smooth surfaces. Starch samples displayed "C patterns"[136].

Drying of peas in solar–assisted spouted bed was found to be highly efficient than open sun drying. The drying rate in solar–assisted spouted bed has been reported to be about 3.5 times of drying rate for open sun drying. Change in air temperature was between 20 °C and 27.4 °C during open sun drying while temperature of air at the inlet of solar–assisted spouted bed dryer varied between 35.3°C and 65.5°C. Solar–assisted spouted bed drying was found to produce better product with less shrinkage and bulk density and high rehydration capacities[137]. Hydration properties are very important in the processing of pea. Varieties of pea suitable for processing must be able to absorb water consistently to an acceptable level (>110% on a dry weight basis). The initial moisture on a dry weight basis may be an important consideration in terms of hydration capacity. It is expected that higher initial moisture content in pea is desired for thermal processing since this would lead to improved hydration in processed products, but at the same time, moisture must be low enough to prevent mold growth[138].

Multi–stage heat pump fluidized bed atmospheric freeze drying (HP FB AFD) and microwave vacuum drying (MVD) were applied to evaluate the quality of dried peas. MVD was found to increase the drying rate (0.59 l/min) compared to the hot air convective drying (0.20 l/min) with minimal deformation in structure of peas. Both HP FB AFD and MVD generated high product quality due to low material temperature during AFD and low pressure during MVD[139]. A thin–layer drying of three varieties of green peas was reported at different temperature (55–75°C) and air velocity of 100 m/min[140]. The green peas dried at 60°C were found to be of best quality on the basis of sensory evaluation and rehydration ratio. The Thomson model was found to represent thin–layer drying kinetics within 99.9% accuracy. In this study, green pea drying is investigated experimentally in a laboratory–scale spouted bed dryer. Another mathematical model was developed to investigate the operating parameters such as inlet air temperature, particle size, and flow rate of the drying air on the performance of the laboratory–scale spouted bed dryer[141]. An initial osmotic dehydration step was found to shorten the subsequent air drying time by 55–70%, depending on the osmotic solution and process conditions. The dehydrated product thus obtained had a more attractive color and high organoleptic acceptability[142]. Moisture–dependent physical properties of pea seed namely, bulk density decreased from 712.1 to 647.5 kg/m^3 and the true density from 1160.5 to 1085.0 kg/m^3 respectively, with an increase in moisture content from 10.1% to 35.1% db. In the same moisture range studies on rewetted pea seed showed that the thousand seed mass increased from 177.7 to 214.1 g, the projected area from 30.84 to 44.08 mm^2, the sphericity from 0.836 to 0.851, the porosity from 38.64% to 40.32% and the terminal velocity from 9.0 to 9.4 m/s[143].

12. FREEZING OF PEAS

Freezing is considered to be the simplest and natural way to preserve vegetables for retaining quality during long–term storage[144]. Frozen foods are highly acceptable due to several advantages inclusive of retention of better sensory attributes such as flavor with minimal alterations in the composition of heat sensitive vital nutrients[145]. Green peas were the first vegetable to be available in a frozen form in the 1950s and the first to have its consumption, in a "fresh" form, during off season[146]. Frozen peas rank third behind processed potato products and corn in their importance amongst frozen vegetables[147]. Consumers prefer green peas in the frozen state as they are far more convenient in terms of both speed and ease of preparation than unprocessed green peas[148]. Freezing allows the retention of fresh qualities of fruits and vegetables for long periods, extending their availability well beyond the normal season[149]. However, enzymatic action deteriorates the texture, color, flavor and nutritional quality of raw vegetables even at subzero temperatures[150]. Therefore, blanching is a critical step prior to freezing to inactivate enzyme[151].

Green peas for commercial freezing should be processed within hours of harvesting[152]. Tenderometer reading is used to assess the maturity of peas to determine time of harvest. Tenderometer readings between TR 115 and TR 120 are considered to be optimum for freezing[153]. Upon reaching the factory the peas are inspected for adherence to the set quality specifications and the extraneous vegetable matter is removed[154]. After washing of peas, blanching and subsequent cooling is carried out. Blanching is a thermal process in which plant tissue is exposed to heat, usually steam or hot water for prescribed time at a specified temperature to inactivate enzymes and to reduce microbial load. The blanched material should be rapidly cooled down to minimize the degradation of heat–labile nutrients. Steam blanching takes more time than the water method, but helps retain water–soluble vitamins and minerals[155]. Blanching of green peas has been reported to be 1.5 min at 100°C[156]. Blanching as a pre–freezing stem serves to minimize the quality losses during frozen storage[157]. One minute blanching was reported for immature sweet peas for retention of nutrients[158]. However, the longer the blanching time, the retention of ascorbic acid was lower in the final product. Green peas are quick frozen and the most common freezers are fluidized or fixed bed freezers. Freezing rates in a fluidized bed are so rapid that green peas are brought down to an internal temperature of –18°C in 4 min. This may be due to the fluidization which helps in exposing the whole pea surface to the freezing medium. A general process flow diagram to develop frozen pea is presented in Fig. 1.

Packaging of frozen peas is done to provide effective functional barrier to contamination. The packaging material should have sufficient impact and compressive strength to withstand the mechanical stress which is most likely to occur during transportation and handling. A good frozen vegetable package will prevent oxidation of product; prevent air and light from entering the package; should prevent foreign odor and flavor absorption and should not be affected by physical damage caused by compression during storage and transportation. Cartons coated with wax or polyethylene, plastic films such as deep freeze grades of polyethylene or thermoformed high impact polystyrene are considered to be appropriate materials for frozen vegetable packaging[159].

Quality consideration for freezing of green peas

The quality of the freshly harvested pea decides the quality of final frozen peas. Some of the critical parameters to judge the quality of peas to be processed are color, appearance, size and texture.

Chosen cultivar for freezing
↓
Maturity assessment
↓
Harvesting and transportation to the processing unit
↓
Inspection and removal of extraneous vegetable matter
↓
De–stoning and washing by floatation
↓
Blanching
↓
Inspection of blanching efficiency
↓
Freezing
↓
Quality grading and packaging
↓
Storage and transportation
↓
Marketing

Fig. 1: General flow chart for the production of frozen peas[149]

Appearance and color

Visual appearance is considered a main source of information for quality grading. Seed color tones are important for pea seed quality. Single individual pea plant or pod produces seeds differing in their appearance. The darker green colored seeds of dry green pea had better seedling vigor and germination ability[160]. This could be useful information to the grower of seeds for frozen pea on selection of suitable cultivar. Sequential development and spatial heterogeneity of the pod position can lead to significant differences in shape, size, weight and color of the seeds in grain legumes[161].

Size

Uniform size of pea seeds is required to produce a high quality product. Size along with specific gravity, visual color and seed skin character has been used as a criterion for sorting frozen green grass pea into maturity groups[162]. Green pea quality has been assessed by the sizes of seeds by passing through sieves with different mesh sizes. Four mesh size levels were tested to sort pea seeds within the range 6.0 mm to >10.2 mm[35]. Three mesh sizes were evaluated to sort the pea seeds; 6.00 to 8.75 mm for small seeds; 8.75 to 10.20 mm for medium seeds and >10.20 mm for large seeds[163]. Size has been linked to pea maturity with smaller sized seeds being harvested earlier than larger seeds. The smaller sized less mature seeds are expected to be tenderer than their larger more mature counterparts[164].

Changes in quality of frozen peas during frozen storage

Freezing of fruits and vegetables can have deleterious effects on texture and may cause excessive soluble solids and water loss during thawing. Water expands upon freezing and the ice crystals cause the cell walls to rupture[145]. Vegetable texture is mainly dependent on the turgidity which is the ability to retain water inside cells. Storage of most frozen foods is usually carried out at –18°C and even at this temperature a fraction of the food constituent water remains liquid and free for freeze damage reactions[165]. Frozen peas can be stored at –18°C for over a year without any significant loss in quality[147]. Frozen peas should be stored at temperatures lower than the glass transition temperature (*Tg*). The glass transition temperature of frozen green peas was found to be –20 °C for old peas and –26 °C for younger peas[166]. The difference in *Tg* was attributed to a higher *Tg* of old peas than the younger peas due to less juice quantity in them.

Inefficient blanching, freezing and improper post processing distribution, storage and home–handling are main causes of quality deterioration of frozen material[167]. Around 5–30% of total ascorbic acid content loss was reported during deep–frozen storage, dependent on the cultivar[168]. The amount of folic acid in peas was also reported to be unchanged during freezer storage of 18 months at –20°C. A blanching time of 2.0 min at 80°C was suggested to inactivate 90% of initial POD activity to retain ascorbic acid and chlorophyll pigments during 12 months storage at 18 °C[169]. The quality of deep–frozen peas did not change significantly during a storage period over 12 months[43]. Losses of around 22–55% were reported after 9 months of storage at –18°C[170]. A loss of approximately 30% on process was reported after one day of freezing in frozen peas[42]. The Vining pea varieties such as Avola showed the greatest processing losses while Bikini showed the least. The ascorbic acid content decrease through the 12 month period was less than 10%.

Color loss during frozen storage of green plant tissue is attributed to the fading of the vivid green color of chlorophyll to an olive brown, characteristic of pheophytin which ultimately is perceived by the consumer as a loss of quality[171]. It was suggested that a chlorophyll–pheophytin ratio may be used as a suitable quality index for frozen peas[172]. Storage of both blanched, unfrozen and blanched, cooked frozen green peas for three months at –18 °C were reported not to change the sensory characteristics of the peas.

13. CONCLUSIONS

Peas are extensively grown in temperate zones being a cool–season crop, but it is restricted to cooler altitudes in the tropics and winter season in the subtropics. Apart from being an important source of food and feed, pea plays a role in soil fertility restoration as a suitable rotation crop that fixes atmospheric nitrogen. Peas can be used fresh or in dried forms. They are very good candidates for freezing which has offered the consumers fresh peas during off–seasons. Share of frozen pea is highest in processing and marketing of frozen foods. Pea protein concentrates and isolates can be used in a number of food applications such as addition to meat, fish, processed foods, soups and sauces to bind fat and/or water, and protein enrichment of baked foods, cereals and snacks. They could replace casein and whey proteins in sports nutrition, weight management products and functional foods. Starch extracted from pea is also more economical than mung starch for the preparation of instant noodles.

REFERENCES

[1] Ambrose M.J., "From near east centre of origin the prized pea migrates throughout world", *Diversity*, **11**, 118–119 (1995).

[2] Zohary D., Hopf M., *"Domestication of Plants in the Old World"*, Oxford University Press: Oxford, UK, (2000).

[3] Salunkhe D.K., Desai B.B., *"Peas and beans in Postharvest biotechnology of vegetables*, CRC Press, *Florida.*, **1**, 125 (2000).

[4] Vavilov N.I., "Studies on the origin of cultivated plants, Trudy Byuro Prikl", *Bot.*, **16**, 139–248 (1926).

[5] Hendrick U.P., Hall F.H., Hawthorn, L.R., Berger A., *Vegetables of New York*, **1**, New York, N.Y. State Exp. Stn. Rep. (1928).

[6] Manay S.N., Shadaksharaswamy M., *"Foods: Facts and Principles"*. **2** New Age International (P) Ltd. 294 (2005)

[7] FAO (2008). www.fao.org

[8] Gane A.J., *"Vining Peas in England"*, Published by the Processors and Growers Research Organization & Grower Books. London, U.K. **53** (1972).

[9] Thompson H.C., Kelly W.C., *"Beans and peas"*, **5**, Tata McGraw–Hill Pub. Comp. Ltd. N.Y. 431 (1985)

[10] Boswell V.R., "Factors influencing yield and quality of peas – biophysical and bio–chemical studies". *Md Bull.*, **306** (1929).

[11] Khunt K.A., Desai D.B., "Economic feasibility and marketing of perennial vegetables in South Gujarat". *Finan Agric.*, **28,** 9–14. (1996)

[12] Thakur, D.S., Thakur D.R., Sharma K.D., "Economics of off–season vegetable production and marketing in hills", *Ind J Agric Marketing.*, **8,** 72–82 (1994).

[13] Chaubey C.N., "Note on effect of delayed sowing on yield of pea (*Pisum sativum* L.)", *Ind J Agric Res.*, **11**(2), 119–121 (1977).

[14] Smykal P., Aubert G., Burstin J., Coyne C.J., Ellis N.T.H., Flavell A.J., Ford R., Hybl M., Macas J., Neumann P., McPhee K.E., Redden R.J., Rubiales D., Weller J.L., Warkentin T.D., "Pea (*Pisum sativum* L.) in the genomic era", Agronomy, **2**, 74–115 (2012).

[15] Basterrechea M., Hicks J.R., "Effect of maturity on carbohydrate changes in sugar snap pea pods during storage", *Scientia Horti.*, **48,** 1–8 (1991).

[16] Pollard L.H., Wilcox E.B., Peterson H.B., *"Maturity studies with canning peas"*, Utah Exp. Stn. Bull, **328** (1947).

[17] Sandstead R., Becker R., Ackerman R., *"Production of Processing Peas"*, Cornell University, Ithaca, N.Y. **8** (1977).

[18] Peirce L.C., *"Legumes* in *Vegetables characteristics, production and marketing"*. John Wiley & sons. 349 (1987)

[19] Lee F.A., *"Objective methods for determining the maturity of peas, with special reference to the frozen product"*, N.Y. Exp. Stn. Tech. Bull, 256 (1941).

[20] Sivasankar B., "*Food Processing and Preservation*", PHI Learning Pvt. Ltd., pp.175–177 (2002).

[21] Dejardin A., Rochat C., Maugenest S, Boutin J.P., "Purification, characterization and physiological role of sucrose synthase in the pea seed coat (*Pisum sativum L.*)", *Planta*, **201**(2), 128–137 (1997).

[22] Ryall L., Lipton W.J., "Handling, transportation and storage of fruits and vegetables", Vol 1, *Vegetables and melons*, AVI Publishing, Westport, Conn (1972).

[23] Pantastico B., Chattopadhyay T.K., Subramanayam H., "Storage and commercial storage operations", *Postharv physio handl utilize trop sub–trop fruits veg*, B. Pantastico, Ed. AVI Publishing, Westport, Conn. 314 (1975).

[24] Tomkins R.G., "Peas kept for 20 days in gas store", *Grower*, **48**, 226 (1957).

[25] Wager H.C., "Physiological studies of the storage of green peas", *J Sci Food Agric.*, **15**, 245 (1964).

[26] Juan A.T., Pariasca A., Miyazaki T., H. Hisaka, H. Nakagawa and T. Sato. "Effect of modified atmosphere packaging (MAP) and controlled atmosphere (CA) storage on the quality of snow pea pods (*Pisum sativum* L. var. saccharatum)". *Postharvest Biolo Technol.*, **21**, 213–223 (2000).

[27] Davies S.J., Gouveia A., "Enhancing the nutritional value of pea seed meals (*Pisum sativum*) by thermal treatment or specific isogenic selection with comparison to soybean meal for African catfish, Clarias gariepinus", *Aquaculture,* **283**, 116–122. (2008).

[28] Sharareh, H., "Effects of various processing conditions on nutritional qualities of legumes", *The Emerging Bean Magazine,* **27**, (2006).

[29] Worthington J.M., Burns E.E., "Post harvest changes in southern peas", *Am Soc Hortic Sci.*, 65th Mktg. Prog. **100** (1968).

[30] Eyaru R., Shrestha A.K., Arcot, J., "Effect of various processing techniques on digestibility of starch in red kidney bean (*Phaseolus vulgaris*) and two varieties of peas (*Pisum sativum*)", *Food Res Int.*, **42**, 956–962 (2009).

[31] Gueguen J., Barbot J., "Quantitative and qualitative variability of pea (*Pisum sativum L.*) protein composition", *J Sci Food Agric.,* **42**, 209–224 (1988).

[32] Nunes M.C., Ravmundo A., Sousa I., "Rheological behaviour and microstructure of pea protein/ kappa–carrageenan/ starch gels with different setting conditions", *Food hydro.*, **20**(1), 106–113 (2006).

[33] Kaur M., Sandhu K.S., Singh N., "Comparative study of the functional, thermal and pasting properties of flours from different field pea (*Pisum sativum* L.) and pigeon pea (*Cajanus cajan* L.) cultivars", *Food Chem.*, **104**, 259–267 (2007).

[34] Tian S., Kyle W.S.A., Small D.M., "Pilot scale isolation of proteins from field peas (*Pisum Sativum*) for use as food ingredients", *Int J Food Sci Technol.*, **34**, 33–39 (1999).

[35] Bech A.C., Juhl H.L., Hansen M., Martens M., Andersen L., "Quality of peas modelled by a structural equation system", *Food Quality Preference,* **11**, 275–281 (2000).

[36] Jakobsen H.B., Hansen M., Christensen M.R., Brockhoff P.B., Olsen C.E., "Aroma volatiles of blanched green peas (*Pisum sativum* L.)", *J Agric Food Chem.*, **46** (9), 3727–3734 (1998).

[37] Pownall T.L., Udenigwe C.C., Aluko R.E., "Amino acid composition and antioxidant properties of pea seed (*Pisum sativum* L.) enzymatic protein hydrolysate fractions", *J Agric Food Chem.,* **58**(8), 4712–4718 (2010).

[38] Shukla J.P., Shukla H.S., Pandey N.C., "Studies on the changing pattern of amino acids in developing pea seeds of *Pisum sativum* L.", *Prog Hortic.*, **11**, 71 (1979).

[39] Gatehouse J.A., Lycett W., Croy R.R.D., Boulter D., "The post–translational proteolysis of the subunits of vicilin from pea (*Pisum sativum* L)", *Biochem J.*, **207**, 629–632 (1982).

[40] O"Kane F.E., Vereijken J.M., Gruppen H., Van Boekel M.A.J.S., "Gelation behaviour of protein isolates extracted from 5 cultivars of *Pisum sativum* L", *J Food Sci*, **70**, 132–137 (2005).

[41] Lee C.Y., Downing D.L., Tredale H.D., Chapman J.A., "The variations of ascorbic acid content in vegetable processing", *Food Chem*, **1**, 15–22 (1976).

[42] Favell D.J., "A comparison of the vitamin C content of fresh and frozen vegetables", *Food Chem*, **62**(1), 54–64 (1998).

[43] Berger M., Kuchler T., Maaben A., Busch–Stockfisch M., Steinhart H., "Correlations of ingredients with sensory attributes in green beans and peas under different storage conditions", *Food Chem.,* **103,** 875–884 (2007).

[44] Nikolopoulou D., Grigorakis K., Stasini M., Alexis M.N., Iliadis K., "Differences in chemical composition of field pea (*Pisum sativum*) cultivars: Effects of cultivation area and year", *Food Chem.*, **103**, 847–852 (2007).

[45] Costa G.E.A., Monici K.S.Q., Reis S.M.P.A., Oliveira A.C., "Chemical composition, dietary fibre and resistant starch contents of raw and cooked pea, common bean, chickpea and lentil legumes", *Food Chem.,* **94**, 327–330 (2006).

[46] Chin H.B., Dudek J.A., "Composition and nutritive value of raw and processed vegetables", *Commercial Vegetable Processing,* pp. 647–678 (1988).

[47] Langenhoven M.L., Kruger M., Gouws E., Faber M., *"MRC Food Composition Tables"* 4th Edition, National Program: Interventional Medical Research Council, Parow, South Africa, (1998).

[48] Lukmanji Z., Hertzmark E., Mlingi N., Assey V., Ndossi G., Fawzi W., *"Tanzania Food Composition Tables"* Dar es Salaam, Tanzania (2008).

[49] Makhlouf J., Zee J., Tremblay N., Belanger A., Michaud M.H., Gosselin A., "Some nutritional characteristics of beans, sweet corn and peas (raw, canned and frozen) produced in the province of Quebec", *Food Res Int.*, **28**(3), 253–259 (1995).

[50] Gatel F., Grosjean F., "Composition and nutritive value of peas for pigs: A review of European results", *Live Stock Production Science*, **26**, 155–175 (1990).

[51] Alonso R., Orue E., Marzo F., "Effects of extrusion and conventional processing methods on protein and anti–nutritional factor contents in pea seeds", *Food Chem.*, **63**(4), 505–512 (1998).

[52] An J., Bacon J.R., Fenwick G.R., "Standardization of the extraction procedure for the analysis of trypsin inhibitors in pea seeds", *In*: vander Poel A. F. B., Huisman J., and Saini H.S. (*eds.*), Recent advances of research in anti–nutritional factors in legume seeds, Wageningen Pers. pp. 31–35 (1993).

[53] Hove E.L., King S., "Trypsin inhibitor contents of lupin seeds and other grain legumes", *New Zealand J Agric Res.*, **22**(1), 41–42 (1979).

[54] Valdebouze P., Bergeron E., Gaborit T., Delort–Laval J., "Content and distribution of trypsin inhibitors and hemagglutinins in some legume seeds", *Canadian J Plant Sci.*, **60**(2), 695–701 (1980).

[55] Morrison S.C., Savage G.P., Morton J.D., Russell A.C., "Identification and stability of trypsin inhibitor isoforms in pea (*Pisum sativum* L.) cultivars grown in New Zealand", *Food Chem.,* **100**, 1–7 (2007).

[56] Azarnia S., Boye J.I., Warkentin T., Malcolmson L., Sabik H., Bellido A.S., "Volatile flavour profile changes in selected field pea cultivars as affected by crop year and processing", *Food Chem.*, **124**, 326–335 (2011).

[57] Murray K.E., Whitfield F.B., "The occurrence of 3–alkyl–2–methoxypyrazines in raw vegetables", *J Sci Food Agric.,* **26**, 973–986 (1975).

[58] Wienberg L., Martens M., Brockhoff P.B., "Sensory quality criteria for cold versus warm green peas studied by multivariate data analysis", *J Food Quality.*, **23**, 565–581 (2000).

[59] Ratnayake W.S., Hoover R., "Pea starch: Composition, structure and properties: A review", *Starch*, **54**(6), 217–234 (2002).

[60] Wang N., Toews R., "Certain physicochemical and functional properties of fibre fractions from pulses", *Food Res Int.*, **44**, 2515–2523 (2011).

[61] Ratnayake W.S., Hoover R., Shahidi F., Perera C., Jane J., "Composition, molecular structure, and physicochemical properties of starches from four field pea (*Pisum sativum L.*) cultivars", *Food Chem.*, **74**, 189–202 (2001).

[62] Wang S., Copeland L., "Effect of alkali treatment on structure and function of pea starch granules", *Food Chem.*, **135**, 1635–1642 (2012).

[63] Rask H., "Evaluation of Canadian pea/lentil starch extraction and noodle preparation", *Pulse Crops Canada Association Saskatoon,* pp. 34–37 (2004).

[64] Ma Z., Boye J.I., Simpson B.K., Prasher S.O., Monpetit D., Malcolmson L., "Thermal processing effects on the functional properties and microstructure of lentil, chickpea, and pea flours", *Food Res Int.*, **44**, 2534–2544 (2011).

[65] Kadlec P., Rubecova A., Hinkova A., Kaasova J., Bubnik Z., Pour V., "Processing of yellow pea by germination, microwave treatment and drying", *Inn Food Sci Emerg Technol.*, **2,** 133–137 (2001).

[66] Gujska E., Czarnecki Z., Khan K., "High temperature extrusion of pinto bean (Phaseolus vulgaris) and field pea (*Pisum sativum*) flours", *Polish J Food Nutr Sci.*, **5**(2), 51–60 (1996).

[67] Yang S.C., Zandstra T., Poel A.F.B., "Starch gelatinization and physical quality of pea flakes in canine dinners as affected by soaking, steam treatment and infrared radiation", *J Anim Physiol Anim Nutri.*, **92**(3), 310–315 (2008).

[68] Singh B., Nagi H.P.S., Sekhon K.S., Singh N., "Studies on the functional characteristics of flour/starch from wrinkled peas (*Pisum sativum*)", *Int J Food Properties.*, **8**(1), 35–48 (2005).

[69] Periago M.J., Englyst H.N., Hudson G.J., "The influence of thermal processing on the non–starch polysaccharide (NSP) content and in vitro digestibility of starch in peas (*Pisum sativum* L.)", *LWT– Food Sci Technol.*, **29**(1/2), 33–40 (1996).

[70] Bishnoi S., Khetarpaul N., "Effect of domestic processing and cooking methods on in–vitro starch digestibility of different pea cultivars (*Pisum sativum*)", *Food Chem.*, **47**(2), 177–182 (1993).

[71] Sun X.D., Arntfield S.D., "Gelation properties of salt–extracted pea protein induced by heat treatment", *Food Res Int.*, **43**, 509–515 (2010).

[72] Gharsallaoui A., Yamauchi K., Chambin O., Cases E., Saurel R., "Effect of high methoxyl pectin on pea protein in aqueous solution and at oil/water interface", *Carb Poly.*, **80**, 817–827 (2010).

[73] Mariotti M., Lucisano M., Pagani M.A., Ng P.K.W., "The role of corn starch, amaranth flour, pea isolate, and psyllium flour on the rheological properties and the ultrastructure of gluten–free doughs", *Food Res Int.*, **42**(8), 963–975 (2009).

[74] Gottschalk, W., "Seed Protein Production of Pisum Mutants", *Seed Proteins: Biochemistry, Genetics, Nutritive Value*, **2,** 377. (2012).

[75] Eppendorfer W.H., Bille S.W., "Amino acid composition as a function of total–N in pea seeds grown on two soils with P and K additions", *Plant and Soil*, **41**(1), 33–39. (1974).

[76] Holl F.B., Vose J.R., "Carbohydrate and protein accumulation in the developing field pea seed", *Canadian Journal of Plant Science*, **60**(4), 1109–1114 (1980).

[77] Reichert R.D., MacKenzie S.L., "Composition of peas (*Pisum sativum*) varying widely in protein content", *J. Agric. Food Chem.,* **30**, 312–317 (1982).

[78] Shand P.J., Ya H., Pietrasik Z., Wanasundara P.K.J.P.D., "Physicochemical and textural properties of heat–induced pea protein isolate gels", *Food Chem.*, **102**, 1119–1130 (2007).

[79] Gonzalez–Perez S., Arellano J.B., *"Handbook of Hydrocolloids", Cambridge: Woodhead Publishing Limited* (2009).

[80] Swanson B.G., "Pea and lentil protein extraction and functionality", *J Am Oil Chem Society.*, **67**(5), 276–280 (1990).

[81] Boye J., Zare F., Pletch A., "Pulse proteins: Processing, characterization, functional properties and applications in food and feed", *Food Res Int.,* **43**, 414–431 (2010).

[82] Pascalle J.M., Pelgrom A., Vissers A.M., Boom R.M., Schutyser M.A.I., "Dry fractionation for production of functional pea protein concentrates", *Food Res Int.*, **53**, 232–239 (2013).

[83] Karaca A.S., Low N., Nickerson M., "Emulsifying properties of chickpea, faba bean, lentil and pea proteins produced by isoelectric precipitation and salt extraction", *Food Res Int.*, **44**, 2742–2750 (2011).

[84] Ducel V., Richard J., Popineau Y., Boury F., "Adsorption kinetics and rheological interfacial properties of plant proteins at the oil–water interface", *Biomacromolecules*, **5**(6), 2088–2093 (2004).

[85] Wang N., Hatcher D.W., Gawalko E.J., "Effect of variety and processing on nutrients and certain anti–nutrients in field peas (*Pisum sativum*)", *Food Chem,* **111**, 132–138 (2008).

[86] Mahajan A., Dua S., "Salts and pH induced changes in functional properties of Amaranth (Amaranthus tricolor L.) seed meal", *Cereal Chem.,* **79**(6), 834–837 (2002).

[87] Adebiyi A.P., Aluko R.E., "Functional properties of protein fractions obtained from commercial yellow field pea (*Pisum sativum* L.) seed protein isolate", *Food Chem.*, **128**, 902–908 (2011).

[88] Choi W.S., Han J.H., "Film–forming mechanism and heat denaturation effects on the physical and chemical properties of pea–protein–isolate edible films", *J Food Sci.*, **67**(4), 1399–1406 (2002).

[89] Choi W.S., Han J.H., "Physical and mechanical properties of pea–protein–based edible films", *J Food Sci.*, **66**(2), 319–322 (2001).

[90] Fabra M.J., Gonzalez L.S., Chiralt A., "Lysozyme release from isolate pea protein and starch based films and their antimicrobial properties", *LWT– Food Sci Technol.*, **55**, 22–26 (2014).

[91] Messiaen C.M., Seif A.A., Jarso M., Eneni G., "*Plant Resources of Tropical Africa 2, Vegetables",* Wageningen, Netherlands: Prota Foundation, Backhuys Publishers. (2004).

[92] Jokanovic M.R., Jovicevic D., Tepic A.N., Vujicic B.L., "Suitability of some green pea (Pisum sativum L.) varieties for processing" , *Periodica Technol., APTEFF*, **37**, 13–20 (2006).

[93] Rodriquez R., Raina B.L., Pantastico B., Bhatti M.B., "*Postharvest physiology, handling and utilization of tropical and sub–tropical fruits and vegetables"*, Pantastico Er. B. (*ed.*), AVI publishing, Westernport Conn., 467 (1975).

[94] Wang N., Warkentin T.D., Vandenberg B., Bing D.J., "Physicochemical properties of starches from various pea and lentil varieties, and characteristics of their noodles prepared by high temperature extrusion", *Food Res Int.*, **55**, 119–127 (2014).

[95] Mercier S., Villeneuve S., Mondor M., Marchais L.P.D., "Evolution of porosity, shrinkage and density of pasta fortified with pea protein concentrate during drying", *LWT – Food Sci Technol.,* **44**, 883–890 (2011).

[96] Petitot M., Boyer L., Minier C., Micard V., "Fortification of pasta with split pea and faba bean flours: Pasta processing and quality evaluation", *Food Res Int.*, **43**, 634–641 (2010).

[97] Osen R., Toelstede S., Wild F., Eisner P., Weisz U.S., "High moisture extrusion cooking of pea protein isolates: Raw material characteristics, extruder responses, and texture properties", *J Food Eng.*, **127**, 67–74 (2014).

[98] Frias J., Kovacs E., Sotomayor C., Hedley C., Vidal–Valverde C., "Processing peas for producing macaroni", *LUF–Food Res Technol.,* **204**(1), 66–71 (1997).

[99] Hajare S.N., Saroj S.D., Dhokane V.S., Shashidhar R., Jayant R., "Effect of radiation processing on nutritional and sensory quality of minimally processed green gram and garden pea sprouts", *Rad Phy Chem.*, **76**, 1642–1649 (2007).

[100] Salunkhe D.K., "Gamma radiation effects on fruits and vegetables", *Econ Bot.,* **15**(1), 28 (1961).

[101] Polat M., Korkmaz M., "The effect of temperature and storage time on the resonance signals of irradiated pea (*Pisum sativum* L.)", *Food Res Int.*, **36**, 857–862 (1982).

[102] Nilsson J., Stegmark R., Akesson B., "Effect of processing on major flavonoids in processed onions, green beans, and peas", *Food Chem.,* **86**, 501–507 (2004).

[103] Mendez J., Lojo M.I., "Phenolic and indole constituents of edible peas", *J Food Sci.*, **36**, 871–872 (1971).

[104] Diplock A.T., Charleux J.L., Crozier W.G., Kok F.J., Evans C.R., Roberfroid M., Stahl W., Ribes J.V., "Functional food science and defence against reactive oxidative species", *British J Nutri.*, **80**, S77–S112 (1998).

[105] Nice D.J., Robinson D.S., Holden M.A., "Characterization of a heat–stable antioxidant co–purified with the superoxide dismutase activity from dried peas", *Food Chem.*, **52**, 393–397 (1995).

[106] Ewald C., Fjelkner–Modig S., Johansson K., Sjoholm I., Akesson B., "Effect of processing on major flavonoids in processed onions, green beans, and peas", *Food Chem.*, **64,** 231–235 (1999).

[107] Troszynska A., Ciska E., "Phenolic compounds of seed coats of white and coloured varieties of pea (*Pisum sativum* L.) and their total antioxidant activity", *Czech J Food Sci.*, **20**, 15–22 (2002).

[108] Duenas M., Hernandez T., Estrella I., "Assessment of in vitro antioxidant capacity of the seed coat and the cotyledon of legumes in relation to their phenolic contents", *Food Chem*, **98**, 95–103 (2006).

[109] Gokmen V., Kahraman N., Demir N., Acar J., "Enzymatically validated liquid chromatographic method for the determination of ascorbic and dehydroascorbic acids in fruit and vegetables", *J Chrom.*, **881**, 309–316 (2000)

[110] Gokmen V., Bahceci K.S., Serpen A., Acar J., "Study of lipoxygenase and peroxidase as blanching indicator enzymes in peas: Change of enzyme activity, ascorbic acid and chlorophylls during frozen storage", *LWT*, **38**, 903–908 (2005).

[111] Berger M., Kuchler T., Maaben A., Busch–Stockfisch M., Steinhart H., "Correlations of ingredients with sensory attributes in green beans and peas under different storage conditions", *Food Chem.*, **103**, 875–884 (2007).

[112] Favell D.J., "A comparison of the vitamin C content of fresh and frozen vegetables", *Food Chem.*, **62**, 59–64 (1998).

[113] Xu B., Chang S.K., "Effect of soaking, boiling, and steaming on total phenolic contentand antioxidant activities of cool season food legumes", *Food Chem.*, **110**(1), 1–13 (2008)

[114] Mangos T.J., Berger R.G., "Determination of major chlorophyll degradation products", *ZLF– Int J Food Res Technol.*, **204**, 345–350 (1997)

[115] Elbe J.H.V., Schwartz S.J., "Colorants", *In*: Fennema O.R. (*ed.*), New York: Marcel Dekker, *Food Chem.*, pp. 651–722 (1996).

[116] Minguez–Mosquera M.I., Garrido–Fernandez J., Gandul–Rojas B., "Pigment changes in olives during fermentation and brine storage", *J Agric Food Chem.*, **37,** 8–11 (1989).

[117] Canjura F.L., Schwartz S.J., Nunes R.V., "Degradation kinetics of chlorophylls and chlorophyllides", *J Food Sci.*, **56**, 1639–1643 (1991).

[118] Lajollo F.M., Tannenbaum S.R., Labuza T.P., "Reaction at limited water concentration. 2. Chlorophyll degradation", *J Food Sci.*, **36**, 850–853 (1971).

[119] LaBorde L.F., Von Elbe J.H., "Zinc complex formation in heated vegetable purees", *J Agric Food Chem.*, **38**(2), 484–487 (1990).

[120] Schwartz S.J., Elbe J.H.V., "Kinetics of chlorophyll degradation to pyropheophytin in vegetables", *J Food Sci.*, **48**, 1303–1306 (1983).

[121] Heaton J.W., Marangoni A.G., "Chlorophyll degradation in processed foods and senescent plant tissues", *Trends Food Sci Technol.*, **7**, 8–15 (1996).

[122] White R.C., Jones I.D., Gibbs E., "Determination of Chlorophylls, Chlorophyllides, Pheophytins, and Pheophorbides in Plant Materiala", *J Food Sci.*, **28**, 431–436 (1963).

[123] Clydesdale F.M., Francis F.J., "Chlorophyll changes in thermally processed spinach as influenced by enzyme conversion and pH adjustment", *Food Technol.*, **22**, 135 (1968)

[124] Buckle K.A., Edwards R.A., "Chlorophyll, colour and pH changes in HTST processed green pea puree", *J Food Technol.*, **5**, 173–178 (1970).
[125] Sanchez C., Baranda A.B., Maranon I.M., "The effect of high pressure and high temperature processing on carotenoids and chlorophylls content in some vegetables", *Food Chem.*, **163**, 37–45 (2014).
[126] Ewalda C., Fjelkner–Modig S., Johansson K., Sjoholm I., Akesson B., "Effect of processing on major flavonoids in processed onions, green beans, and peas", *Food Chem.*, **64**, 231–235 (1999).
[127] Sayeed S., Njaa L.R., "Effect of a Bangladeshi home–cooking procedure on the amino acid content, trypsin inhibitor activity and in vitro digestibility of some legume seeds", *Plant Foods Hum Nutri.*, **35**, 379–388 (1985).
[128] Savage G.P., Deo S., "The nutritional value of peas (Pisum sativum): A literature review", *Nutri Abst Rev.*, **59**(2), 65–88 (1989).
[129] Martin G.L., Weiglhofer K., Phuong T.P.T., Berghofer E., "High hydrostatic pressure influences antinutritional factors and in vitro protein digestibility of split peas and whole white beans", *LWT – Food Sci Technol.,* **51**, 331–336 (2013).
[130] Fuhrmeister H., Meuser F., "Impact of processing on functional properties of protein products from wrinkled peas", *J Food Eng.*, **56**, 119–129 (2003).
[131] Urbano G., Lopez–Jurado M., Frejnagel S., Gomez–Villalva E., Porres J.M., Frias J., Idal–Valverde C., Aranda P., "Nutritional assessment of raw and germinated pea (Pisum sativum L.) protein and carbohydrate by in vitro and in vivo techniques", *Nutrition.*, **21**(2), 230–239 (2005).
[132] Urbano G., Aranda P., Vilchez A., Aranda C., Cabrera L., Porres J.M., Lopez–Jurado M., "Effects of germination on the composition and nutritive value of proteins in Pisum sativum, L", *Food Chem.*, **93**(4), 671–679 (2005).
[133] Alonso R., Grant G., Dewey P., Marzo F., "Nutritional assessment invitro and invivo of raw and extruded peas (*Pisum sativum* L.)", *J Agric Food Chem.*, **48**, 2286–2290 (2000).
[134] Jayaraman K.S., Gupta D.K.D., "Dehydration of fruits and vegetables–recent developments in principles and techniques", *Drying Technol.*, **10**, 1–50 (1992).
[135] Garcia Pascual P., Alves Filho O., Eikevik T.M., Strommen I., "Freeze drying of peas at atmospheric pressure", *Aliment Equipos Tecnologia*, **23**(192), 74–78 (2004).
[136] Simsek S., Tulbek M.C., Yao Y., Schatz B., "Starch characteristics of dry peas (Pisum sativum L.) grown in the USA", *Food Chem.,* **115**, 832–838 (2009).
[137] Sahin S., Sumnu G., Tunaboyu F., "Usage of solar–assisted spouted bed drier in drying of pea", *Food Bioprod Process*, **91**(3), 271–278 (2013).
[138] An D., Arntfield S.D., Beta T., Cenkowski S., "Hydration properties of different varieties of Canadian field peas (Pisum sativum) from different locations", *Food Res Int.*, **43**, 520–525 (2010).
[139] Zielinska M., Zapotoczny P., Alves–Filho O., Eikevik T.M., Blaszczak W., "A multi–stage combined heat pump and microwave vacuum drying of green peas", *J Food Eng.*, **115**, 347–356 (2013).
[140] Pardeshi I.L., Arorab S., Borkera P.A., "thin–layer drying of green peas and selection of a suitable thin–layer drying model", *Drying Technol: An Intl J.*, **27**(2), 288–295 (2009).
[141] Ghalavand Y., Hatamipour M.S., Rahimi A., "Experimental and parametric study on drying of green peas in spouted bed", *Int J Food Sci Technol.*, **45**(12), 2546–2552 (2010).
[142] Kaymak F., Ertekin A., Cakaloz T., "Osmotic dehydration of peas II. Influence of osmosis on drying behavior and product quality", *J Food Process Preserv.*, **20**(2), 105–119 (1996).
[143] Yalcin I., Ozarslan C., Akbas T., "Physical properties of pea (*Pisum sativum*) seed", *J Food Eng.*, **79**, 731–735 (2007).

[144] Cano M.P., "Vegetables", *In*: *Freezing effects on food quality*, Published by Marcel Dekker, New York (1996).
[145] Maity T., Chauhan O.P., Shah A., Raju P.S., Bawa A.S., "Quality characteristics and glass transition temperature of hydrocolloid pre–treated frozen pre–cut carrot", *Int J Food Prop.*, **14**, 17–28 (2011).
[146] Green K., Foster C., "Give peas a chance: Transformations in food consumption and production systems", *Technol Forecast Social Change.*, **72**, 663–679 (2005).
[147] Deshpande S.S., Adsule R.N., "Garden Pea", *In*: handbook in vegetable science and technology: Production, composition, processing and storage. Marcel and Dekker Inc. New York, pp. 433–456 (1998).
[148] Mcewan J.A., Clayton D., *"Consumer perception of quality attributes of frozen peas: A qualitative approach"*, United Kingdom: Campden and Chorleywood Food Research Association, R & D Report, **71** (1998).
[149] Arthey D., "Freezing of vegetables and fruits", *Frozen Food. Technology*. Blackie Academic and Professional, an imprint of Chapman and Hall, Glasgow, UK, pp. 237–269 (1993).
[150] Williams D.C., Lim M.H., Chen A.O., Pangborn R.M., Whitaker J.R., "Blanching of vegetables for freezing – which indicator enzyme to choose", *Food Technol.*, **40**, 130–140 (1986).
[151] Barrett D.M., Garcia E.L., Russell G.F., Ramirez E., Shirazi A., "Blanch time and cultivar effects on quality of frozen and stored corn and broccoli", *J Food Sci.*, **65**(3), 534–540 (2000).
[152] Siddiq M., Pascall M.A., "Peas, sweet corn and green beans", *Handbook of Vegetables and Vegetable Processing*. Blackwell Publishing Limited, Iowa, USA, pp. 605–623 (2010).
[153] Rutledge P., "Preparation procedures", *Vegetable Processing*. Blackie and Son Limited, Glasgow, UK, pp. 42–68 (1991).
[154] Barbosa Canovas G.V., Altunakar B., Mejia–Lorio D.J., *"Freezing of Fruits and Vegetables: An Agribusiness Alternative for Rural and Semi–Rural Areas"*. Chapter 1. Food and Agriculture Organization of the United Nations, Rome, pp. 1–36 (2005).
[155] Luh B.S., Lorenzo M.C., "Freezing of Vegetables", In *Commercial Vegetable Processing*. 2nd edition, Van Nostrand Reinhold, New York, pp. 343–384 (1988).
[156] Barrett D.M., Theerakulkait C., "Quality indicators in blanched, frozen and stored vegetables", *Food Technol.*, **62**, 64–65 (1995).
[157] Heberlein D.G., Ptak L.R., Medoff S., Clifcorn L.E., "Quality and nutritive value of peas as affected by blanching", *Food Technol.*, **4**, 109–113 (1950).
[158] Paine F.A., Paine H.Y., *"A Handbook of Food Packaging"*, 2nd ed, Chapman and Hall, London, UK, pp. 248–264 (1992).
[159] Atak M., Kaya M.D., Kaya G., Kaya M., Khawar K.M., "Dark green colored seeds increase the seed vigor and germination ability in dry green pea (*Pisum sativum* L.)", *Pak J Botany.*, **40**(6), 2345–2354 (2008).
[160] Fenner M., "Environmental influences on seed size and composition", *Horticultural Reviews*, **13**, 183–213 (2010).
[161] Kmiecik W., Korus A., Lisiewska Z., "Evaluation of physico–chemical and sensory quality of frozen green grass pea (*Lathyrus sativus* L.)", *Int J Food Sci Technol.*, **39**, 149–155 (2004).
[162] Edelenbos M., Thybo A., Erichsen L., Wienberg L., Andersen L., "Relevant measurements of green pea texture", *J Food Qual.*, **24**, 91–110 (2001).
[163] Sayre B.C., Tapley W.T., Barton D.W., *"Variety comparison of peas used for canning and freezing"*, New York Agriculture Experiment Station, Cornell University, Geneva, New York, p. 758 (1953).
[164] Torregiani D., Lucas T., Raoh–Wack A.I., "The pre–treatment of fruits and vegetables", *Managing Frozen Foods*. CRC Press, Cambridge, UK, pp. 7–80 (2000).

[165] Lim M., Wu H., Breckell M., Birch J., "Influence of the glass transition and storage temperature of frozen peas on the loss of quality attributes", *Int J Food Sci Technol.*, **41**, 507–512 (2006).

[166] Giannakourou M.C., Taoukis P.S., "Stability of dehydrofrozen green peas pretreated with non conventional osmotic agents", *J Food Sci.*, **68**, 2002–2010 (2003).

[167] Puupponen–Pimia R., Hakkinen S.T., Aarni M., Suortti T., Lampi A.M., Eurola M., Pironen V., Nuutila A.M., Oksman–Caldentey K.M., "Blanching and long–term freezing affect various bioactive compounds of vegetables in different ways", *J Sci Food Agric.*, **83**, 1389–1402 (2003).

[168] Gokmen V., Baceci K.S., Serpen A., Acar J., "Study of lipoxygenase and peroxidase as blanching indicator enzymes in peas: Change of enzymes activity, ascorbic acid and chlorophylls during frozen storage", *LWT– Food Sci Technol.*, **38**, 903–908 (2005).

[169] Bognar A.A., "Vitaminabbau bei der Be– und Verarbeitung von Gemu se", *Lebensmitteltechnik,* **10**, 541–554 (1987).

[170] Martins R.C., Silva C.L.M., "Modeling color and chlorophyll losses of frozen green beans (*Phaseolus vulgaris*, L.)", *Int J Refrig.*, **25**, 966–974 (2002).

[171] Lynch L.J., Mitchell R.S., Casmir D.J., "The chemistry and technology of the preservation of green peas", *Advances in Food Science*, **9**, 61–65 (1959).

[172] Jokanovic M.R., Jovicevic D., Tepic A.N., Vujicic B.L., "Suitability of some green pea (*Pisum sativum* L.) varieties for processing", *Acta Period Technol.,* **37**, 13–20 (2006).

15

Bioactive Compounds in *Broccoli*: Extraction and Processing

Pankaj B. Pathare[1]*, Debabandya Mohapatra[2]

ABSTRACT

Healthier diets, safety and convenience are targets correlated by current consumers to high nutritional and sensory quality plant products. Bioactive compounds are extra nutritional constituents that typically occur in small quantities in foods. Brassica species are very rich in health–promoting phytochemical constituents, including phenolic compounds, vitamin C and glucosinolates. Broccoli (Brassica oleracea var. italica) offers many heath promoting properties owing to its content of antioxidant and anti–carcinogenic compounds. These bioactive compounds are extracted mainly by solvent extraction process, with water, methanol, ethanol, ethylene chloride, hexane, acetone etc. Solvent less extraction processes like super critical fluid extraction, microwave and ultrasound aided extraction methods are gaining worldwide acceptance and can be suitably modified to extract bioactive compounds from broccoli florets, stems, stalks, leaves, seeds and sprouts. The concentration and bioavailability of polyphenols, glucosinolates, sulforaphane and selenium depend on plant biochemistry, cultivation strategy and type of processing. It is well known that phenolic compounds play an important role in human health owing to their effects as antioxidants and anti–carcinogens. Seasonal nature of vegetable production and sometimes the impossibility of consuming fresh vegetables have prompted the use of processed vegetables. The processing of fruit and vegetables, however, almost always has an unfavourable effect on their nutritional values. In this chapter, the main bioactive compounds of broccoli are reviewed. Additionally, the way these properties are affected by processing is discussed. The information

[1] Postharvest Technology Research Laboratory, South African Research Chair in Postharvest Technology, Faculty of AgriSciences, Stellenbosch University, South Africa
[2] Agricultural Produce Processing Division, Central Institute of Agricultural Engineering (CIAE), Bhopal, India.
**Corresponding author* E-mail: pbpathare@gmail.com

provided will be beneficial for further commercialization and exploration of these novel technologies.

Keywords: Bioactive Compounds, Broccoli, Extraction, Processing, Storage

I. INTRODUCTION

In the last few years, great attention has been paid to the bioactive compounds due to their ability to promote benefits for human health, such as the reduction in the incidence of some degenerative diseases like cancer and diabetes[1,2], reduction in risk factors of cardiovascular diseases[3, 4], antioxidant, anti–mutagenic, anti–allergenic, anti–inflammatory, and anti–microbial effects[5, 6, 7]. Due to these countless beneficial characteristics for human health, researches have been intensified aiming to find fruits, vegetables, plants, agricultural and agro–industrial residues as sources of bioactive phenolic compounds. Broccoli (*Brassica oleracea* var. *italica)* has acquired a considerable relevance as a health–promoting food. The sprouting broccoli is thought to originally come from the eastern Mediterranean area and to be introduced to Europe, especially to Italy, in the medieval times[8]. The healthy food attributes assigned to broccoli are a consequence of its high contents of bioactive phytochemicals such as nitrogen–sulfur compounds (glucosinolates and isothiocyanates), phenolic compounds (chlorogenic and sinapic acid derivates and flavonoids) and nutrients (vitamin C, E, A, K, and so on, and essential minerals: N, K, Ca, Fe, and so on), the consumption of which is beneficial for the prevention of chronic disorders, such as carcinogenic and cardiovascular pathologies[9,10,11]. Broccoli is also rich in polyphenols, a large group of phytochemicals that are often considered the most abundant antioxidants in the diet[12]. Polyphenols cause interference with oxidation of lipids and other molecules by the rapid donation of hydrogen atoms to free radicals[13].

Broccolis are good source of flavonoids like quercetin (30–40 mg/kg fresh weight), kaempferol (40–72 mg/kg fresh weight), and flavones like Luteolin (8 mg/g fresh weight)[14]. From a dietary point of view glucosinates are restricted to plants of the Brassica species. Several of these species are widely consumed by humans as cooked or salad vegetables such as cabbage, brussels sprouts, cauliflower, broccoli, turnip, radish and watercress or condiments including horseradish and white mustard[15]. The predominant glucosinates in broccoli is glucoraphanin, constituting more 50% of the total glucosinolates (GSL). When crucifers are chopped or crushed, a hydrolyzing enzyme, myrosinase, comes into contact with the glucosinolates, releasing glucose and leaving an unstable thiono compound that can rearrange to form the active isothiocyanate or less active nitriles or thiocyanates[10]. Sulforaphane, an isothiocyanate derived from glucoraphanin and present at high levels in broccoli have anti–cancerous properties[15]. Although the antioxidant potential of broccoli is known, as yet little practical effort to utilize their byproducts for phenolic recovery has been reported. This might be caused by three limiting factors often overlooked in scientific studies: the effectiveness of recovery and extraction, the marketability of resulting extracts and the practical suitability for the food, cosmetic or pharmaceutical products[16].

According to the American Food Standards Agency and the USDA, broccoli is a good candidate as source of essential minerals for human consumption[17]. Bioactive compounds are secondary metabolites present ubiquitously in the plant kingdom and are considered as non–nutritional but vital ingredients for the maintenance of human health[18]. There is strong epidemiological evidence that a high consumption of brassica vegetables, including broccoli, is associated with a decreased risk for lung, stomach, colon and rectal cancers, most likely due to their glucosinolate content[19].

Broccoli is mostly consumed as a processed food, and then its functional properties, as well as its physical characteristics, can be affected to different extents[20]. The glucosinolate content in vegetables depends on the cooking method, degree of raw material disintegration, and on the raw material itself[21,22]. Glucosinolates are a group of secondary plant metabolites that are found in the Brassicaceae family. Upon hydrolysis by the endogenous enzyme myrosinase, a large number of compounds can be formed of which some are potentially anti–carcinogenic, while others are largely inactive. Furthermore, some bioactive compounds are unstable. Therefore, it is not only important to determine the type and amount of glucosinolates present or hydrolysed in a given plant, but it is also relevant to investigate the type of hydrolysis products[23]. Glucosinolates are primarily lost from broccoli tissue through leaching into the cooking water but the rate and extent of loss depends on the amount of water, the type of cooking used and the cooking time[24]. The present chapter is aimed to provide a comprehensive review of different techniques for extraction of bioactive compounds from broccoli. The effect of different processing methods on bioactive compounds is also discussed in detail.

II. EXTRACTION METHODS FOR BROCCOLI BIOACTIVES

Extraction methods have pivotal role to play in the isolation, identification and use of phenolic compounds and this area is now garnering enough interest among the researchers. There is no single standard method of extraction that can be applied to all the plant materials for the extraction of bioactive compounds. The phenolic compounds have been extracted by grinding, drying or lyophilising fruits, vegetables and herbs or only by soaking fresh plants with subsequent solvent extraction. These methodologies imply the co–extraction of non–phenolic substances, such as sugars, organic acids and proteins, requiring subsequent purification processes. Conventional extraction as heating, boiling, or refluxing can be used to extract natural phenolic compounds; however, the disadvantages are the loss of polyphenols due to ionisation, hydrolysis and oxidation during extraction, as well as the long extraction time[25]. Several methods of pertaining to extraction of bioactive compounds have been developed to enhance the efficiency of extraction such as solid state fermentation process, pressurised solvents, supercritical fluid extraction, sub critical fluid extraction, pressurised liquid extraction, enzyme, pulsed electric fields (PEF), ultrasound waves and microwave assisted extraction[26,27,28,29,30,31,32,33]. These techniques reduce the total extraction time and in some cases solvent less extraction is possible. Among all these methods sonication is cheap and effective method for extraction of bioactive compounds[26]. Uses of ionized water, hot water, hydrotropic solution instead of inorganic solvents are also being practiced for extraction process, which reduces the solvent use and cost[34]. The chemical is analysed by HPLC and Micellar Electrokinetic Capillary Chromatography (MECC), however the extract has to be clean so as not to choke the system and to prevent system failure. The usual practice of extraction includes high speed blending in boiling in water or by solvents with solvents. Therefore, in these following sections, different protocols developed for the extraction of bioactive compounds from different parts of broccoli are discussed in brief.

Solid–liquid Extraction

Solid–liquid extraction, or leaching can be defined as a mass transport phenomenon in which solids contained in a solid matrix migrate into a solvent brought into contact with the matrix. Mass transport phenomena can be enhanced by changes in concentration gradients, diffusion coefficients or boundary. It is a unit operation extensively used to recover many important food components. The efficiency of this extraction method depends on the temperature of the solvent, liquid–solid ratio, particle size, flow rate and time of contact between solvent and solute[25].

Common solvents

Ethyl acetate, hot water, methanol, ethanol, diethyl ether, petroleum ether, chloroform, sodium hydroxide, n–Hexane, methyl chloride, isooctane, acetone, dichloromethane are some of the solvents used by various researchers for extraction of bioactive compounds[25,30]. Among all these solvents methanol and ethanol are predominantly used. Methanol has better extraction efficiency however, due to its toxic nature, its use is restricted and for most industrial use ethanol is preferred[25].

Various researchers have used different solvents to extract the bioactive compounds from different parts of broccoli *i.e.,* seed, sprouts, inflorescence, stalk, stem, leaves etc. These methods are described in the following paragraphs, chronologically. The different content of bioactive compounds extracted by different method is enlisted in Table 1.

Bertelli *et al.*[35] used solvent extraction method, followed by purification of broccoli florets, leaves and stalk extracts using solid phase extraction (SPE) and reverse phase HPLC analysis. To obtain information about the usefulness of acid hydrolysis of glucosinolates, the florets were left to autolyze at room temperature or were treated with concentrated hydrochloric acid, then analysed. Freeze dried powders were extracted with methylene chloride at room temperature, the first time with 30 ml g^{-1} for 7 h, the second time with 10 ml g^{-1} for 4 h, whereas the stalks and leaves were treated at only pH with HCl and the organic extracts were pooled and evaporated in a rotary evaporator (35°C) followed by repeated washing with water. The aqueous layers were filtered, pooled and saturated with sodium chloride and then re–extracted with methylene chloride. The methylene chloride layers were dried over anhydrous sodium sulfate and stored at –20°C until processed. The organic broccoli extracts were purified by SPE silica gel (SiOH) 3 ml disposable columns; the extraction cartridge was inserted on SPE vacuum column. Sulforaphane was extracted by passing 3 ml of organic extract through the cartridge, washing the cartridge with 3ml of ethylacetate, which was then discarded, and eluting the sulforaphane with 3ml of methanol. Acidified extraction method yielded better result (Table 1).

Methanol is one of the widely used solvents for the extraction of bioactive compounds from the plant tissues. Several researchers have used pure methanol and aqueous methanol for extraction of bioactive compounds from freeze dried powders of broccoli[36, 37] with methanol (70%), whereas, Heimler *et al.*[38] had extracted phenolic compounds from broccoli by using ethanol. Frozen tissues ground in a mortar with a pestle under liquid nitrogen and 1.5 g of tissue was extracted in 20 mL of 70% ethanol (pH 3.2, by formic acid) overnight. Peschel *et al.*[16], compared the extraction ability of different solvents like water, methanol, ethanol, acetone and hexane. The broccoli remains (stems and stalks), were dried and crushed to fine powder and were subjected to extraction by various water, ethanol, methanol, hexane and acetone in a 10:1 solvent: raw material ratio at 25C, for 4 h, except in case of hexane, which was for 6 h in soxhlet apparatus. The extraction yield of harvest remains of Broccoli in water, methanol, ethanol, acetone and hexane were 21.8, 6.5, 2.1, 1.2, and 2.1%, respectively. This shows that water extraction method is superior to other chemical solvents in terms of total yield.

Trenerry *et al.*[39] when used an aqueous extraction method by loading a 5 ml aliquot of an aqueous sample extract onto an activated C18 SPE cartridge, the glucoraphanin passed straight through. It was observed that, non–polar compounds which could potentially interfere with the separation were trapped on the column. They have modified the method by washing

Table 1: Different solvents and methods for the extraction on the bioactive components from different parts of broccoli

Plant parts	***Solvent/ Method***	***Bioactive compound***	***Quantity***	***Reference(s)***
Broccoli	Water and methylene chloride	Sulforaphane	12.69–17.99 g/g	Bertelli *et al.*[35]
inflorescence	Water + HCl (methylene chloride)		31.82–48.28 g/g	
	70% methanol	Total glucosinolates	3–28.3 g/g DW	Vallejo *et al.*[36]
		Total phenolics	60 mg/kg DW	
	70% ethanol	Total phenolic content	12.85 ± 0.195 mg	Heimler *et al.*[38]
		Flavonoids tannin content	GAE/g DW	
			6.71 ± 0.467 mg catechin/g DW	
			0.41 ± 0.033 mg catechin/g DW	
	Aqueous extraction with additional wash with deionized water	Glucoraphanin	71 mg/100 g	Trenerry *et al.*[39]
	70% methnol	Total polyphenols	289 ± 6.12 mg/100 g FW	Sikora *et al.*[67]
		Total carotenoids	0.26 ± 0.046 mg/100 g FW	
	Acetone:water (80:20)	Quercetin		
		Kaempferol total phenolics	0.03–10.85 mg/ 100 g FW	Koh *et al.*[41]
			0.24–13.20 mg/ 100 g FW	
			48.15–157.77 mg GAE/ 100 g FW	
	80% acetone in 0.2% formic acid	Total phenolics	102.1 ± 5.9 mg GAE/100 g FW	C?íz *et al.*[42]
	Boiling water	Total	1.85–36.68 g/g FW	Wang *et al.*[40]
		Glucosinolates	6.33 g/g FW	
		Aliphatic GS	6.94 g/g FW	
		indole GS	0.08 mol/g	
		aromatic GS		
	Acetone/water/acetic acid (70:29.5:0.5)	Total phenolic content	99 mg GAE/ 100 g FW	Isabelle *et al.*[43]
		Neoxanthin	2.38 g/g FW	
		Violaxanthin	3.34 g/g FW	

Table 1: (*Contd...*)

Table 1: (*Contd...*)

Plant parts	*Solvent/ Method*	*Bioactive compound*	*Quantity*	*Reference(s)*
		Lutein	4.42 g/g FW	
		Zeaxanthin	0.53 g/g FW	
		beta-Carotene	5.12 g/g FW	
		δTocopherol	0.04 g/g FW	
		γTocopherol	2.96 g/g FW	
		αTocopherol	13.11 g/g FW	
		δTocotrienol	0.02 g/g FW	
		γTocotrienol	0.03 g/g FW	
	Tetrahydrofuran + methanol-acetic acid-water mixture (50:3.7:46.3)	FRAP (Lipophilic) FRAP	11.30 ± 0.21 mol Fe(II)/g FW	Deng *et al.*[46]
		(Hydrophilic)	2.15 ± 0.28 mol Fe(II)/g FW	
		FRAP total values	13.45 ± 0.49 mol Fe(II)/g FW	
		TEAC (Lipophilic)	13.10 ± 0.15 mol Trolox/g FW	
		TEAC (Hydrophilic)	0.87 ± 0.02 mol Trolox/g FW	
		TEAC (total values)	13.97 ± 0.17 mol Trolox/g FW	
		TPC (Lipophilic) TPC	8.60 ± 0.15 mg GAE/g FW	
		(Hydrophilic) total	1.24 ± 0.02 mg GAE/g FW	
		phenolics contents	9.84 ± 0.17 mg GAE/g FW	
	Methanol:water (70:30) with or without 1.49 g/ L trifluoroacetic acid	Total glucosinolate (GL) content	0.75 0.05 mol/ g FW (organic) 0.35 0.2 mol/ g (conventional)	Miranda Rossetto *et al.*[48]
Leaves	Methanol: water (70:30) with or without 1.49 g/ L trifluoroacetic acid	Total glucosinolate	1.0 ± 0.4 mol/g (organic) 2.0 ± 0.1 ± 0.05 mol/g (conventional)	Miranda Rossetto *et al.*[48]
	Water+HCl (methylene chloride)	Sulfloraphane	110.03 g/g	Bertelli *et al.*[35]
	SC-CO_2 with 35% methanol as modifier, 100-250 bar	Free Amino acids	719–60,491 mg/kg lyophilized sample	Arnáiz *et al.*[51]

(*Contd...*)

Table 1: (*Contd...*)

Plant parts	*Solvent/ Method*	*Bioactive compound*	*Quantity*	*Reference(s)*
Stalks/stems	Water ethanol methanol hexane acetone	total phenolics contents	29.87 ± 1.58 mg/ g GAE, (water), 25.58 ± 2.51mg/g GAE (methanol), 28.31 ± 1.69 mg/ g GAE (ethanol), 36.18 ± 1.89mg/ g GAE (acetone), 33.45 ± 2.32 mg/ g GAE (hexane)	Peschel *et al.* [16]
	Water+HCl (methylene chloride)	Sulfloraphane	17.45 g/g	Bertelli *et al.* [35]
	Methanol	Antioxidant power (FRAP) DTPH FCR	761 ± 9 mg TE/ 100 g DW 512 ± 24 mg TE/ 100 g DW 494 ± 6 mg GAE/100 g DW	Wijngaard *et al.* [37]
	methanol:water (70:30) with or without 1.49 g/ L trifluoroacetic acid	Total glucosinolates	1.5 0.4 mol /g(organic)	Miranda Rossetto *et al.* [48]
Broccoli seeds	Aqueous extraction+ wash with deionized water	Glucoraphanin	2.1 g/100 g	Trenerry *et al.* [39]
	HPLC	Total glucosinolates	88- 113 mol/g DM	Martinez Villaluenga *et al.* [52]
Sprouts	HPLC	Total glucosinolates	79 mol/g DM	Martinez-Villaluenga *et al.* [52]
	Phosphate-citrate buffer and extraction with methyl chloride, pH=4,	Isothiocyanates	705.34 ± 18.11 mg/100 g FW	Guo *et al.* [47]

the C18 column with deionised water to recover glucoraphanin. This procedure, even though it separated glucoraphanin from potential interfering compounds, did not allow for the inclusion of a concentration step. The steps followed were as follows: the extract from broccoli was achieved by boiling 10 g materials in water for 5 minutes, followed by blending and sonication for 5 minutes each. The filtrate was loaded on to C18 Sep–pak cartridge for the extraction of glucosinolate fraction. As per the modified method: the cartridges were washed with 5 ml of deionised water, the C18 Sep–pak cartridge discarded and the protonated amino cartridge washed with 5 ml of methanol. The glucosinolate fraction was removed from the protonated amino cartridge with 8 ml of freshly prepared 2% v/v solution of concentrated ammonium hydroxide in methanol. The solvent was evaporated to dryness under a stream of nitrogen at room temperature. The dry residue was made up with 2 ml deionised water. The following observations were made: Strong anion exchange (SAX) SPE was not suitable to isolate and concentrate glucoraphanin using the C18 extraction technique without washing with deionized water as glucoraphanin was too tightly bound to the packing material. Glucoraphanin was quantitatively recovered when the experiment was repeated with a protonated 500 mg amino propyl cation exchange cartridge. The residue, after Weak anion exchange (WAX) treatment, was dissolved in deionised water (2 ml for broccoli floret extract and 5 ml for seeds) and assayed by MECC. Both electropherograms were cleaner than those recorded for the original extracts. C18 and WAX cleanup provided solutions that were cleaner than the original extracts. The aliquots of broccoli extracts were passed through activated C18 cartridges attached to the top of protonated WAX cartridges. The cartridges were washed with water, the C18 cartridges removed and the WAX cartridges washed with methanol followed by 2% v/v of concentrated ammonium hydroxide in methanol. The residues that remained after solvent removal were dissolved in deionised water and assayed by MECC, gave better result. Wang *et al.*[40] also extracted the glucosinolates from broccoli of different cultivars with boiling water followed by 10 min centrifugation at 4000 G. The extract was purified by loading on to a polypropylene column filled with diethylaminoethyl DEAE–Sephadex A–25), which had been activated with 0.5 M pyridine acetate. It was observed that the pure lines have more glucosinolate content than the commercial varieties. The phenolic compounds from frozen broccoli tissue by acetone/water solvent (80/20 v/v) was extracted[41]. For extraction of bioactive compounds from broccoli inflorescence researcher used 80% acetone in 0.2 % formic acid[42]. Isabelle *et al.*[43] modified the method by using a solvent mixture (acetone/water/acetic acid: 70: 29.5: 0.5) followed by sonication for 15 minutes. Michiels *et al.*[44] studied the effect of different extraction conditions like solvent composition, temperature, extraction time (duration), solvent–to–solid ratio and storage conditions on the antioxidant capacity of broccoli. Various extraction parameters were tested: solvent mixture, solvent volume (2–100 mL per g fresh weight), extraction temperature (4, 25, 50, 70°C), extraction time (20, 40, 60 min), and number of extractions (1–3). The mean extraction efficiency for phenolic compounds was highest when the volume used was equal to (+141%) or higher than (+132%) 40 mL. The highest antioxidant capacities were recorded with 40 mL solvent. The two solvents acetone/water/acetic acid (70/28/2, v/v/v) and methanol/acetic acid (99.5/0.5, v/v) extracted the same amount of phenolics from broccoli. As temperature increased extraction yield also increased.

Survay *et al.*[45] extracted glucosinolate Cinnamoyl derivative [6'–O–trans–(43– hydroxy cinnamoyl)–4–(methylsulphinyl) butyl glucosinolate] from fresh broccoli florets. The fresh *B. italica* florets were extracted with triple deionized boiling water for 24 h followed by 30 min sonication at ambient temperature. After 24 h the filtrate was collected and vacuum evaporated. This concentrated aqueous extract was applied onto a glass column filled with Diaion PK228 that was previously activated with methanol followed by 1% acetic acid. It was subsequently washed with methanol (1000 mL), water (1000 mL), and finally with 2% ammonium hydroxide in methanol. The target compound was purified through HPLC by using Atlantis dC18 column with water and acetonitrile as mobile phases. Deng *et al.*[46] studied the antioxidant capacities and total phenolic contents of lipophilic and hydrophilic extracts of 56 commonly consumed vegetables including broccoli. The extraction process involved extraction from the fresh samples washed with deionized water. Hydrophilic and lipophilic components of ground vegetables were extracted with 5 mL of tetrahydrofuran in a shaking water bath (100 rpm, 37 °C) for 30 min. Then the mixture was centrifuged at 4,200g for 30 min, and the supernatant was collected. The extraction was repeated once with 5mL of tetrahydrofuran under the same condition, and the two supernatants were combined into lipophilic fraction. Subsequently, the residue was extracted twice with methanol–acetic acid–water mixture (50 : 3.7 : 46.3, v/v/v) in a shaking water bath (100 rpm, 37°C) for 30 min and the two supernatants were combined into hydrophilic fractions.

Guo *et al.*[47] optimized the solvent extraction technique for the isocyanates from broccoli sprouts with methyl chloride and found that at pH 4 and addition of ETDA and ascorbic acid improves the extraction of isocyanates from glucosinolates. Miranda Rossetto *et al.*[48] extracted phenolic compounds from different parts of organically cultivated and conventionally cultivated broccoli. Organic and conventionally grown broccoli plants were harvested at the same physiological phase of maturation. Plants were morphologically separated into inflorescence (I), leaves (L) and stalks (S). The samples were washed and dried and frozen by dipping in liquid nitrogen at –20°C. For extraction, the samples were homogenized by using mortar and pestle in 5ml methanol: water (70: 30) solution with or without the presence of 1.49 g/L trifluoroacetic acid (TFA). The extraction was carried out at 70°C for 30 minutes. After cooling, the extracts were centrifuged at 8000 g for 20 min. The collected supernatants were filtered and transferred to glass flasks at 40°C until solvent was completely evaporated (approximately 72 h). The dry glucosinolate–containing precipitate was reconstituted with 1 mL of 0.2

mol/ L HEPES–KOH (pH 7.0). Total glucosinolate concentrations were determined enzymatically in a thioglucosidase solution containing myrosinase. Glucosinolates tend to accumulate in higher amounts in vegetables that were cultivated with organic procedures. The authors attributed the high glucosinolate concentration to the extraction medium, which contained TFA. It was also found that the discarded part of the plants have higher amount of glucosinolates.

Super Critical Fluid Extraction

Due to the increasing concern about the quality and safety of food products and the stricter regulations about the residual level of solvents, the use of supercritical fluid technology as a primary alternative for the extraction, fractionation, and isolation of active ingredients has increased. As a solvent–free process, supercritical fluid extraction (SFE) with carbon dioxide is a common choice for the functional foods and nutraceutical sectors, as its physicochemical properties (higher diffusivity, lower viscosity, and lower surface tension than conventional solvents) facilitate mass transfer and allow an environmentally friendly operation[49].

A simple SFE consists basically of two major steps: extraction of the soluble substances from the solid substratum by the SCF solvent and separation of these compounds from the supercritical solvent after the expansion. In the first step, the solvent is fed into (solvent pump and heat exchange) the extractor and uniformly distributed throughout the fixed bed formed by the solid substratum. During the extraction, the solvent flows through the fixed bed and dissolves the soluble compounds. The solute–solvent mixture is separated in the flash tanks by rapidly reducing the pressure, increasing the temperature, or both. The solvent is cooled and recompressed and returns to the extractor. Different structures of solid substratum have been used, from leaves to bark or wood, where the samples were prepared by drying, lyophilizing and milling etc, which increases the surface area to facilitate the extraction rate by reducing the mass transfer resistance. The extraction phenomenon can be described as follows: (1) the solid substratum absorbs the supercritical solvent, promoting the dilatation of the cellular structures. This step facilitates the solvent flow by decreasing the mass transfer resistance; (2) concurrently, the soluble compounds are dissolved by the solvent; (3) the dissolved compounds are transferred by diffusion to the surface of the solid; and (4) the compounds are transported by the solvent and then removed from the extractor[50].

However, one of the CO_2 drawbacks is its non–polar nature that does not favor the extraction of polar substances such as amino acids. Therefore, Arnáiz *et al.*[51] had used methanol as organic modifier to improve the extraction of amino acids from broccoli leaves. The lyophilized leaves of broccoli were loaded on a SPE unit, having two preparative pumps; one to provide methanol (35%) and the other one to supply the CO_2, while a pressure and temperature between 100–250 bar, and 50–80°C were maintained. It was observed that at low pressure even though the temperature increased the yield decreased which was attributed to the low solubility at lesser pressure. A pressure of 250 bar at 70°C and extraction time of 30 minutes, with an organic modifier (methanol, 35%) was chosen to be the optimum condition. SC–CO_2 gave comparable result as in case of methanol: water, but the extraction yield was highest in case of water (double than methanol: water and SC–CO_2, extraction) as it is a better solvent for polar substances.

Hplc Method for Separation

HPLC methods have been in use for separation of bioactive compounds. Martinez–Villaluenga *et al.*[52] isolated the bioactive compounds from broccoli seed and sprouts by HPLC. Extracts were prepared from 200 mg of freeze–dried samples with 70% boiling methanol. Desulfo–GLS were separated in the HPLC system in a LiChrospher 100 RP–18 (254 mm, 5lm) column (Merck, Darmstadt, Germany) using an auto injector and a 1.2 ml/min flow rate by eluting with a gradient of double distilled water (A) and 20% acetonitrile (B).

High Speed Counter Current Chromatography (HSCCC) Method For Purification And Separation

Liang *et al.*[53] developed a high–speed counter current chromatography (HSCCC) method. Without any initial cleanup steps, sulforaphane was successfully purified from the ethyl acetate extract of the broccoli seed meal by HSCCC. The separation was performed with two–phase solvent systems: n–hexane/ethyl acetate/methanol/water (1:5:1:5, v/v/v/v). From 850 mg of the ethyl acetate extract, 186 mg of sulforaphane was isolated with the solvent system. The purified compound was over 97% purity as determined by HPLC analysis. This method eliminated multiple steps during the extraction, separation, and purification processes, which often generate large amounts of organic solvent wastage.

III. EFFECT OF PROCESSING ON BIOACTIVE COMPOUNDS

Vegetable industrial processing such as blanching, canning, sterilization and freezing, as well as domestic cooking, is expected to affect the content, composition, antioxidant activity and bioavailability of antioxidants. In addition, operations such as cutting and slicing may induce a rapid enzymatic depletion of several naturally occurring antioxidants as a result of cellular disruption which allows contacts of substrates and enzymes[54]. Broccoli is mainly consumed cooked, and domestic handling and processing, including cooking processes, have been shown to considerably affect its health–

promoting compounds[24, 55, 56, 57]. In general, vegetables are prepared at home on the basis of convenience and taste preference rather than retention of nutrient and health–promoting compounds[58]. Cooking in water decreased significantly the content of polyphenols, glucosinolates, and vitamin C compounds in broccoli[59]. Temperature, humidity, or light may induce some reactions during storage, sometimes leading to a loss of quality[60]. Blanching, cooking and freezing affect the content of glucosinolates, sulforaphane, and polyphenols in broccoli[61]. During the cooking process only 14–28% of the individual glucosides were retained in the cooked tissue, the remainder being largely leached into the cooking water with only a small loss attributed to the formation of the respective aglycones[62]. Blanching and cooking affect the content of glucosinolates in broccoli[63], since these compounds can be hydrolyzed by the action of myrosinase present in the plant tissue[64]. Besides, chopping and crushing facilitates the release of myrosinase, thus stimulating the formation of active isothiocyanates, such as the anticarcinogenic sulforaphane[10].

Effect of Cooking Method

Cooking as a domestic processing method has a great impact on food nutrients[65]. It is known that cooking induces significant changes in chemical composition, affecting the bioavailability and content of chemopreventive compounds in vegetables. Cooking methods were shown to affect the contents of nutrient and health–promoting compounds such as vitamin C, carotenoids, polyphenols, and glucosinolates in broccoli[56, 63, 66, 67]. The post–harvest process that has arguably the most effect on glucosinolate and other phytochemical content is cooking[24]. Many nutrients, vitamins and phytochemicals are known to significantly decline during cooking. Ascorbic acid, carotenoids and phenolics in broccoli all decreased significantly with time during boiling or microwave cooking[68]. The effect of different cooking methods on phenolic compounds in broccoli is shown in Table 2.

Glucosinolate content in broccoli heads also declined significantly during cooking. Sones *et al.* [69] found that the total glucosinolate content of broccoli boiled for 10 min was approximately 40% less than for fresh broccoli. Glucosinolates were primarily lost from broccoli tissue through leaching into the cooking water, but the rate of loss was dependent on the type of cooking used[21, 56, 64, 70]. Vallejo *et al.* [56] used four domestic cooking processes, high pressure boiling, steam cooking, microwaving and low pressure boiling (conventional) to evaluate the glucosinolates in the edible portions of fresh harvested broccoli florets before and after cooking and in the cooking water. Microwave cooking showed the high loss of total glucosinolates (74%) in comparison with the rest of treatments. The loss of total glucosinolate from the fresh raw broccoli was significantly high during high pressure (33%) and conventional boiling (55%) due to leaching into the cooking water. On the other hand, steaming had minimal effects on glucosinolates. They

Table 2: Effect of cooking on bioactive compounds in broccoli

Cooking process	*Plant parts*	*Time (min)*	*Determined value*	*%Recovery*	*Reference*
Conventional	Floret	15	Total flavonol glycosides	82	Price *et al.*[62]
			Quercetin 3-O-sophoroside	72	
			Kaempferol-3-O-sophoroside	86	
			Isoquercitrin	85	
			Kaempferol-3-O-glucoside	76	
			Kaempferol-diglucoside	81	
Conventional	Edible part	5	Flavonoids	80.1	Vallejo *et al.*[57]
			Caffeoyl-quinic derivatives	60.1	
			Sinapic and feruoyl derivatives	50.7	
Microwave	Edible part	5	Flavonoids	97.2	Vallejo *et al.*[57]
			Caffeoyl-quinic derivatives	87.4	
			Sinapic and feruoyl derivatives	76.2	
Steaming	Edible part	3.5	Flavonoids	11.1	Vallejo *et al.*[57]
			Caffeoyl-quinic derivatives	8.1	
			Sinapic and feruoyl derivatives	0.3	
High pressure	Edible part	3	Flavonoids	53.4	Vallejo *et al.*[57]
			Caffeoyl-quinic derivatives	49.3	
			Sinapic and feruoyl derivatives	41.9	
Boiling		10	Total phenolic content	Decrease 41.3	Gawlik-Dziki[73]
Boiled	Organic/edible	6 min	Soluble polyphenols	24	Faller & Fialho[12]
Microwave		4 min	Soluble polyphenols	28.9	
Steaming		8 min	Soluble polyphenols	37	
Boiled	Conventional /edible	6	Soluble polyphenols	26.1	
Microwave		4	Soluble polyphenols	+9.1	

Table 2: (*Contd...*)

Table 2: (*Contd...*)

Cooking process	*Plant parts*	*Time (min)*	*Determined value*	*%Recovery*	*Reference*
Steaming		8	Soluble polyphenols	4.4	
Steaming		5	Hydrophilic extract	+6.4	Roy *et al.*[76]
		10		+18.2	
		5	Lipophilic extract	8.96	
		10		20.32%	
Conventional	Stem	1	Total phenolics	22.25	Zhang & Hamauzu[68]
		5	Total phenolics	42.2	
	Floret	1	Total phenolics	47.5	
		5		71.9	
Microwave	Stem	1		17.8	Zhang & Hamauzu[68]
		5		44.4	
	Floret	1		46.3	
		5		71.6	
Boiling	Edible part	5	Total phenolic	6	Turkmen *et al.*[107]
Steaming		7.5		+18	
Microwave		1.5		+25	
Steamed	Edible part	10	Total polyphenols	+37.1	Gliszczynska-Swiglo *et al.*[59]
			Flavonoids	+34.4	
Water-cooked	Edible part	5	Total polyphenols	12.5	
			Flavonoids	72.4	

conclude that a large quantity of glucosinolates will be consumed in steamed broccoli when compared to the other cooking processes.

Hot water immersion of broccoli for a brief period of time usually between 1 and 10 min, at temperature ranging from 55°C to 100 °C and using drinking water, affects the overall quality during storage[71]. However, this effect is minimized by using the lowest temperature as possible. Sanjuan *et al.*[72] applied different blanching treatments to broccoli stems before drying, and found that firmness and chlorophyll content was impaired. Hydrothermal treatment of broccoli florets negatively affected its antioxidant properties, mainly when the treatment was applied to fresh vegetables[73].

Pellegrini *et al.*[74] studied the effect of cooking methods namely boiling, microwave treatment, basket and oven steaming on fresh and frozen broccoli. In the case of fresh broccoli, steaming methods did not affect the content of carotenoids; whereas during microwave heating a significant decrease of both lutein and β–carotene (35 and 60%, respectively) was observed. Boiling also resulted in a significant decrease of lutein (33%). In the case of frozen broccoli, all carotenoids were negatively affected by all of the cooking methods applied. This trend in frozen broccoli could be related to the blanching step leading to a softening of the vegetable matrix; further cooking leads to a loss of these compounds. Yuan *et al.*[75] investigated the effects of five domestic cooking methods, including steaming, microwaving, boiling, stir–frying, and stir–frying followed by boiling (stir–frying/boiling), on the nutrients and health–promoting compounds. Total aliphatic and indole glucosinolates were significantly modified by all cooking treatments but not by steaming. In general, steaming led to the lowest loss of total glucosinolates, while stir–frying and stir–frying/boiling presented the highest loss.

Gliszczyńska–Świglo *et al.*[59] reported the effect of water– and steam–cooking on the content of vitamin C, polyphenols, carotenoids, tocopherols and glucosinolates, as well as on the antioxidant activity of broccoli. The results indicated that steam–cooking of broccoli results in an increase in polyphenols, as well as the main glucosinolates and their total content as compared with fresh broccoli, whereas cooking in water has the opposite effect. Both water– and steam–cooking of broccoli results in an increase in β–carotene, lutein, and α– and γ–tocopherols as compared with fresh broccoli. Similar effects of steaming and water–cooking of broccoli on their antioxidant activity were observed.

Roy *et al.*[76] evaluated the level of total phenolics and total flavonoid content in fresh and steam processed broccoli samples. Total phenolic content and total flavonoid content increased in steamed–processed broccoli. Zhang and Hamauzu[68] compared the phenolics content and the antioxidant activity of conventional and microwave cooked broccoli. No significant differences were detected in both properties due to the cooking process;

however, a significant decrease in polyphenols was observed in cooked compared to fresh broccoli.

Faller and Fialho reported that cooking lead to an increase in the antioxidant capability of broccoli, and that soluble polyphenols were more negatively affected than hydrolysable polyphenols by boiling and microwaving. However, steaming resulted in a significant increase in the recovery of hydrolysable polyphenols. Besides, the directly proportional relationship between polyphenols content and antioxidant activity was not affected by different cooking methods, such as boiling, steaming and microwaving. Steam blanching of broccoli may result in an increased antioxidant activity, thus improving the health–promoting properties of this vegetable.

A significant decrease in total glucosinolates (GSL) content was detected in cooked broccoli[24] mostly attributed to its loss from vegetal tissue through leaching into the cooking water, and being dependent on the cooking conditions[64]. Despite some thermal degradation of glucosinolates it is unlikely that thermal degradation is the main cause of GSL loss during cooking for periods shorter than 10 min[77].

On the other hand, Gliszczyńska–Świglo *et al.*[59] showed that steam cooking of broccoli increased the total content of glucosinolates comparing to fresh broccoli. This situation may be attributed to an increase in the extractability of antioxidant and anticarcinogenic compounds triggered by processing rather than a real increase in their total content. Nevertheless it could be speculated that this improved extractability increases the bioavailability of these compounds *in vivo*. As a consequence, steam processed broccoli should have enhanced antioxidant (through phenols and flavonoids) and anticarcinogenic (through glucosinolates) properties.

Porter[13] investigated the effects of boiling and microwaving on the antioxidant properties of green broccoli and purple–sprouting broccoli. Results showed antioxidant components of cooked broccoli to be quite different from uncooked broccoli. The antioxidant content of broccoli was retained and/or enhanced more after microwaving than after boiling. Cooking in water caused a leaching effect of antioxidants, and this increased with cooking time. Purple–sprouting broccoli was found to contain higher contents of antioxidant compounds compared with green broccoli, but tended to show higher sensitivity to cooking treatments. Cooking methods should be carefully considered in current dietary recommendations.

Lisiewska *et al.*[78] reported that cooked *Brassicae* leaves contained 78% of the total amino acid content found in fresh leaves, while the frozen products contained between 76% and 78%. The effect of technological processing of broccoli on the content of non proteinogenic amino acids, including SMSeC, has not been studied so far. However, in view of the results reported by Lisiewska *et al.*[78], it could be expected that the content of these compounds is affected in the same way. Then, in order to maximize the health–promoting properties of broccoli, the effect of processing on SMSeC content should be considered as a key factor.

Effect of Storage Conditions

In recent years, demand on broccoli has increased greatly, as freshly consumed or in salads ready to eat. However, broccoli is a perishable commodity with a reduced shelf life after harvest. The quality loss is mainly due to surface dehydration and loss of green colour with floret yellowing, although differences among cultivars in terms of shelf life have been reported[79]. Apart from processing, storage conditions have significant effect on phytochemicals such as ascorbic acid and phenols[80].

Freezing is one of the most popular and effective methods for food preservation, allowing to keep the nutritional and the sensorial properties of the raw materials for extensive periods of time. However, before freezing, foodstuffs are usually subjected to pre–treatments such as blanching, which affect the food properties in a major or minor way. PuupponenPimiä *et al.*[81] studied the effects of blanching/freezing and long–term freezer storage of broccoli on total phenolics. Total phenolics and flavonol contents of broccoli decreased considerably during long–term storage. Patras *et al.*[80] investigated the effect of blast freezing and blanching in combination followed by chilling, on, phenols of broccoli. Phenols are unaffected by blanched frozen and unblanched frozen treatments. However, Gêbczyñski and Lisiewska[61] studied the process of freezing (–20, –30 °C) in broccoli florets and found significant decrease in the level of polyphenol content.

Gębczyński and Lisiewska[61] investigated the effect of freezing of blanched or cooked broccoli on the content of some antioxidant compounds. They found that cooking before freezing negatively affected the polyphenols content and the antioxidant activity of broccoli, compared to the blanched frozen vegetable. Besides, the authors reported a steady decrease in these properties along storage during 12 months at –20°C or –30°C. Furthermore, the frozen material contained up to 66% less polyphenols than the raw vegetable, and it showed a decrease of 33% in the antioxidant activity.

Ninfali and Bacchiocca[82] compared the antioxidant capability and total polyphenols content of fresh and frozen broccoli. They found a 58.4% decrease in polyphenols content and a 12.4% decrease in antioxidant activity. Cieœlik *et al.*[63] studied the effect of freezing on the glucosinolates content in broccoli. They found that freezing had a minor effect on GSL content, as compared to blanching and boiling. Consequently, the loss of bioactive properties in frozen broccoli seems to be due to the pre–treatments more than to the freezing process itself. Freezing affects the retention of some minerals of broccoli. Kmiecik *et al* [83] found that freezing after blanching produced a significant decrease in the content of phosphorus (30% from the content in the fresh vegetable), potassium (67%) and magnesium (51%), while the content of calcium remained unaffected. Freezing after cooking resulted in much less marked losses of minerals.

Refrigerated storage of brassica has produced more variable results. Glucoraphanin in broccoli heads stored at 4° C in open boxes or packaged in perforated polymeric bags is unaltered for ≤1 week[84], while a reduction of 31 and 16% in glucoraphanin and total glucosinolate concentrations respectively has been observed after refrigerating broccoli for 5 days[85]. Force *et al.* [86] found no significant changes in individual glucosinolate concentrations in 7 day–old seed–sprouts of broccoli during storage at 4°C for 3 weeks.

Serrano *et al.* [87] packaged broccoli heads were using 3 types of polypropylene films: macro–perforated (Ma–P), micro–perforated (Mi–P) and non–perforated (No–P), and then stored at 1 °C for 28 days to study the effects of modified atmosphere packaging (MAP) by comparison with non–wrapped heads. Results revealed that deterioration, a rapid decrease in total antioxidant activity (TAA), ascorbic acid and total phenolic compound concentration was observed. Conversely, in those heads packaged under MAP, especially for Mi–P and No–P, total phenolic compounds remained almost unchanged during the whole period.

Zapata *et al.* [88] evaluated the effect of heat treatment on senescence and antioxidant compounds evolution during storage at 20°C was evaluated in organic and conventionally grown broccoli. Phenolic concentration and antioxidant activity (in both hydrophilic and lipophilic fractions) also decreased during storage, although these decreases were higher in conventional than in organic broccolis, and no differences were found attributable to heat treatment. Heat treatment was effective in delaying broccoli senescence, manifested by chlorophyll retention. In addition, organic broccoli maintained higher concentrations of bioactive compounds (ascorbic acid and phenolics) and antioxidant potential during storage than conventional broccoli, with higher potential health beneficial effects.

Rangkadilok *et al.* [84] investigated the effect of post–harvest and packaging conditions on glucoraphanin concentrations in freshly–harvested broccoli heads. Glucoraphanin concentration is reduced by 56% after storing broccoli for 3 day in an open environment or for 7 day in polymeric bags, as a result of hydrolysis of glucoraphanin by plant myrosinase to a greater extent in the open environment. Similarly, Rodrigues & Rosa[85] have reported a decline of 82 and 79% in glucoraphanin and total glucosinolate concentrations respectively when freshly–harvested broccoli inflorescences are stored at room temperature for 5 days. Vallejo *et al.* [89] have shown a lower reduction of aliphatic glucosinolates (reduced by 40%) compared with indole glucosinolates (reduced by 78%) after broccoli inflorescences, wrapped in polymeric film, are stored for a week at 1°C.

Effect of Drying/Heating

Drying is an ancient process used to preserve foods and its main objective is the removal of water to the level at which microbial spoilage and deterioration reactions are minimized. During hot–air drying, vegetables undergo physical, structural, chemical and nutritional changes that can affect quality attributes. Hot–air drying implies a thermal treatment and thermal degradation of polyphenols is expected but the decomposition of polyphenols was proven to depend on the food matrix and the processing conditions. Mrkìc *et al.*[90] investigated the effect of temperature (50–100°C) and air flow rate (1.20–2.25m/s) in a pilot tray dryer on the bioactive compounds content and antioxidant activity of broccoli. They found that temperature had a negative effect on total polyphenols content, and that air flow rate had a positive effect on the antioxidant activity of dehydrated broccoli. High–temperature short–time drying processes kept the highest antioxidant activity of broccoli. Drying of intact broccoli at 50–65°C maintained glucosinolates and myrosinase activity and it is only when the product was re–hydrated that glucosinolates were hydrolysed[77].

Matusheski *et al.*[91] examined the effects of heating broccoli florets and sprouts on sulforaphane and sulforaphane nitrile formation, to determine if broccoli contains epithiospecifier protein (ESP) activity, then to correlate heat–dependent changes in ESP activity, sulforaphane content and bioactivity. Heating fresh broccoli florets or broccoli sprouts to 60°C prior to homogenization simultaneously increased sulforaphane formation and decreased sulforaphane nitrile formation. A significant loss of ESP activity paralleled the decrease in sulforaphane nitrile formation. Heating to 70°C and above decreased the formation of both products in broccoli florets, but not in broccoli sprouts.

Mahn *et al.*[92] optimized the convective drying of broccoli. Maximum antioxidant activity was obtained at 60°C, air flow rate of 4m/s and 1.5cm particle diameter, resulting in a 70% reduction in free radical scavenging ability and a 29% increase in total reductive capability, compared to the fresh vegetable. Reyes *et al.*[20] investigated the effect of air temperature and flow rate, plate rotational speed, and initial broccoli particle size in a fluidized pulsed bed dryer on moisture content, diffusivity, antioxidant capacity, polyphenols content, and selenium concentration of conventional and selenium–enriched broccoli. The main loss of antioxidant capacity and total polyphenols content for conventional and selenium–enriched broccoli was due to blanching, and after drying these properties were only slightly affected. Pareto analysis shows air flow rate and the interaction between particle size and air temperature had a significantly negative effect on the total polyphenols content of broccoli, whereas the interaction between air flow rate and both particle size and air temperature had significantly positive effects on this response. The interaction between temperature and air flow rate had a significantly positive effect on the total polyphenols content of selenium–enriched broccoli.

Mahn *et al.*[93] optimized the convective dehydration of broccoli to maximize the content of SMSeC, total polyphenols and the free radical scavenging capacity. The optimization of a process to obtain dehydrated broccoli florets with high SMSeC content was attained. The process consisted of two steps: a short hydrothermal pre–treatment, followed by freeze–drying. Both steps were optimized to yield the maximum anti–radical power, total polyphenols content and SMSeC concentration in the processed vegetable. The optimal blanching conditions were immersion in water at 72 °C during 1 min, and the optimal freeze–drying conditions were slow freezing rate (in a household freezer at "20°C), atmospheric pressure (96.13 kPa), and use of IR. Under these process conditions it was possible to keep 25% ARP, 18% polyphenols content and 174% SMSeC compared to the fresh vegetable. The apparent increase in SMSeC concentration after processing can be attributed to the higher extractability of the processed broccoli. We believe that the results presented here will help in establishing the most adequate processing conditions that allow keeping the functional properties of broccoli as a processed food.

Sosiñska and Obiedziñski [94] investigated the influence of thermal treatment, pickling and fermentation on the content of glucobrassicin and its selected breakdown products in broccoli. The highest content of glucobrassicin and the lowest content of its breakdown products were found in vegetables which underwent thermal treatment such as boiling and steam cooking. The lowest content of glucobrassicin and the highest content of its breakdown products were present in raw, pickled and fermented vegetables in which the myrosinase remained active. Ascorbigen was the major glucobrassicin breakdown product and its content was negatively correlated with glucobrassicin content in broccoli. After heating of vegetables in 60°C for 20 min (causing thermal deactivation of epithiospecific protein) no presence of indole–3–acetonitrile was found.

Other Preservation Methods

Seasonal nature of vegetable production and sometimes the impossibility of consuming fresh vegetables have prompted the use of processed vegetables. The processing of fruit and vegetables, however, almost always has an unfavourable effect on their nutritional values. Chipurura *et al.*[95] analysed the effect of a combination of UV–C and heat treatment on quality and senescence of fresh–cut broccoli florets stored at 0°C. Treated samples showed higher phenolic content and antioxidant capacity. The results suggest that a combined treatment with heat and UV–C may reduce senescence, tissue damage and helps to maintain a better quality of the product during storage at 0°C.

Gomes *et al.*[96] studied the electron–beam irradiation of fresh broccoli florets, and found that this method produced a significant decrease in the content of some antioxidant compounds during storage. One–hundred broccoli heads were irradiated at 1, 2, and 3 kGy with a 10 MeV linear accelerator at 22°C.

Xu *et al.*[97] investigated the effect of 1–methylcyclopropene (1–MCP) combined with 6–benzylaminopurine (6–BA) treatment on antioxidant enzymes and bioactive compounds in broccoli florets. Superoxide dismutase, ascobate peroxidase and catalase activities increased while the activity of peroxidase decreased during storage in treated samples in comparison with the controls. The combined treatment enhanced the biosynthesis of glucosinolate and the formation of the anticarcinogen sulforaphane, which improved the health benefit of broccoli. Lemoine *et al.*[98] studied the effect of UV–C (8 kJ m^{-2}) treatment on refrigerated storage of broccoli, and reported that treated broccoli exhibited a higher phenolic and ascorbic acid contents as well as higher antioxidant activity than controls.

Ghawi *et al.*[99] investigated the effect of adding mustard seeds to processed broccoli with a view to intensify the formation of sulforaphane. Thermal degradation of broccoli glucoraphanin was investigated in addition to the effects of thermal processing on the formation of sulforaphane and sulforaphane nitrile. Limited thermal degradation of glucoraphanin (less than 12%) was observed when broccoli was placed in vacuum sealed bag (sous vide) and cooked in a water bath at 100 °C for 8 and 12 min. Boiling broccoli in water prevented the formation of any significant levels of sulforaphane due to inactivated myrosinase. Addition of powdered mustard seeds to the heat processed broccoli significantly increased the formation of sulforaphane.

Van Eylen *et al.*[23] studied the effect of combined pressure–temperature treatments (100–500 MPa, 20–40°C) on the glucosinolate conversion and the kind of hydrolysis products in broccoli, both during treatment and after autolysis. The results, showed that high pressure can induce glucosinolate hydrolysis during treatment, promote the formation of isothiocyanates after treatment and that relatively more indole oligomers are formed during treatment than during autolysis. These results indicate that pressure treatment limits the loss of glucosinolates and its health beneficial products. Sulforaphane, a naturally occurring cancer chemopreventive, is the hydrolysis product of glucoraphanin, the main glucosinolate in broccoli. The hydrolysis requires myrosinase isoenzyme to be present in sufficient activity; however, processing leads to its denaturation and hence reduced hydrolysis[99].

Conaway *et al.*[100] reported that the bioavailability of isothiocyanates from raw broccoli was three times greater than that from steamed broccoli where myrosinase was inactivated. It is interesting to note that the same processing conditions do not affect glucosinolate content substantially if leaching into processing water is avoided. Total phenolic compounds, glucosinolates and sulforaphane in broccoli were increased after treating with ethanol[101].

IV. MODELLING APPROACHES FOR PREDICTING CHANGES IN BIOACTIVES

Empirical models are among the most common types of models that may be used to predict changes of bioactive compounds of during drying. These models simply express relationships between the content of a particular compound and variables or parameters involved in the drying process[102]. Knowledge regarding the kinetics of physicochemical properties is essential to predict quality losses during the blanching process. Kinetic data will facilitate the inclusion of these quality aspects into the design of optimal processing conditions, which will be important for the development of new food products. A number of kinetic models such as zero order, first order and fractional conversion (FC) first order have been used for the phytochemical contents, of fruits and vegetables[103].

According to Sarvan *et al.*[104], the concentration of glucosinolates compounds is strongly affected by processing of the vegetables. Various mechanisms are responsible for these changes: Lysis of plant cells and compartments, diffusion and leaching of glucosinolates and myrosinase (a plant enzyme able to hydrolyze glucosinolates) into the cooking water, enzymatic hydrolysis of glucosinolates, inactivation of myrosinase and thermal degradation of glucosinolates. The study presents a dynamic mathematical model that includes these mechanisms with their estimated parameters and the effect of temperature on them. Simulations made by the model for several process conditions show losses of glucosinolates as a consequence of domestic boiling of 69%, microwaving of 8%, industrial blanching of 37% and by industrial sterilization of 82%. The model can assist in adapting the processing conditions like the time–temperature profile and the vegetable–water ratio to optimize industrial and domestic processing of *Brassica* vegetables in terms of health benefits. In addition this model can help to add the effect of vegetable preparation practices on the quantitative glucosinolate intake in epidemiological studies.

Goncalves *et al.*[105] studied the effect of water blanching treatment on total phenolic content at five temperature. Blanching caused a significant decrease of broccoli total phenolic content, probably due to thermal degradation and leaching phenomena to water. Total phenolic values decreased with treatment intensity. An Arrhenius first–order kinetics model was used to model the thermal degradation of total phenolic content. Activation energy value (75.4 kJ mol^{-1}) is in the range normally expected for nutrients nonenzymatic degradation reactions. Van Eylen *et al.*[106] studied the stability of endogenous myrosinase in broccoli juice during combined temperature"pressure treatments. The thermal inactivation of the myrosinase could be described by the consecutive step model, while during combined isobaric"isothermal treatment, broccoli myrosinase inactivation followed first–order kinetics. Kinetic information can be used for process evaluation and optimisation to increase the health effect of broccoli.

V. SUMMARY

It could be summarized that different solvents like methyl chloride, pure methanol and aqueous methanol, ethanol and aqueous ethanol, water, hexane, tetrahydrofuran, acetone: water and water are able to extract bioactive compounds from different parts of broccoli. Extraction by organic solvents improved in acidic medium. It is apparent that use of different solvents, maintaining solvent to material ratio, temperature, and different pre–treatments like sonication has improved the solvent extraction yield as well as clarity of the extract from broccoli. SC–CO_2 can also be a good alternative; however it works in presence of organic modifier like methanol for extraction of polar substances. Water has performed well in terms of yield of bioactive compounds. It was also observed that the wastes like leaves, stem and stalks of the broccoli have more bioactive compounds than the inflorescence and the glucosinolate content vary with genotypes, and season. Other novel systems like pulse electric field, microwave and enzyme assisted extraction system has worked for other plant materials and such methods can be adopted to extract the bioactive compounds from broccoli, which will have industrial application. HPLC and High Speed Counter current Chromatography methods can be adopted to extract, separate and purify the bioactive compounds from broccoli.

Considering the increasing demands of consumers for healthy products, and of producers for emerging and effective technologies to prevent deterioration of fresh and minimally processed fruits and vegetables, it is important to evaluate the effect of new treatments on the phytochemical status of the treated produce, because phytochemical content and bioactive compound must be contemplated as one more major quality attribute. Brassica vegetables are consumed all over the year as the ingredients of different salads or after cooking of raw and frozen vegetables. Besides consumer preferences, the selected cooking method is an important factor affecting not only the food chemical composition, but also the intake of bioactive compounds under normal dietary conditions. Limited information on kinetic evaluation of bioactive compounds upon different processing and storage in terms of empirical models and structured models is available.

VI. REFERENCES

[1] Conforti F., Menichini F., Formisano C., Rigano D., Senatore F., Arnold N.A., "Comparative chemical composition, free radical–scavenging and cytotoxic properties of essential oils of six Stachys species from different regions of the Mediterranean Area", *Food Chemistry,* **116**(4), 898–905 (2009).

[2] Kim G.N., Shin J.G., Jang H.D., "Antioxidant and antidiabetic activity of Dangyuja (Citrus grandis Osbeck) extract treated with Aspergillus saitoi", *Food Chemistry,* **117**(1), 35–41 (2009)

[3] Jiménez J.P., Serrano J., Tabernero M., Arranz S., Rubio M.E.D., Diz L.G., *et al.* "Effects of grape antioxidant dietary fiber in cardiovascular disease risk factors", *Nutrition,* **24**(7), 646–53 (2008).

[4] Etherton P.M.K., Hecker K.D., Bonanome A., Coval S.M., Binkoski A.E., Hilpert K.F., *et al.* "Bioactive compounds in foods: Their role in the prevention of cardiovascular disease and cancer", *The American Journal of Medicine,* **113**(9),71–88 (2002).

[5] Balasundram N., Sundram K., Samman S., "Phenolic compounds in plants and agri–industrial by–products: Antioxidant activity, occurrence, and potential uses", *Food Chemistry.,* **99**(1), 191–203 (2006)

[6] Ham S.S., Kim S.H., Moon S.Y., Chung M.J., Cui C.B., Han E.K., Chung C.K., Choe M., "Antimutagenic effects of subfractions of Chaga mushroom (Inonotus obliquus) extract", *Mutation Research / Genetic Toxicology and Environmental Mutagenesis,* **672**(1), 55–59 (2009).

[7] Parvathy K., Negi P., Srinivas P., "Antioxidant, antimutagenic and antibacterial activities of curcumin–â–diglucoside", *Food Chemistry,* **115**(1), 265–271 (2009).

[8] Gray A., "Taxonomy and evolution of broccoli (Brassica oleracea var. italica)", *Economic Botany,* **36**(4), 397–410 (1982).

[9] Premier R., "Phytochemical composition: A paradigm shift for food-health considerations", *Asia Pacific Journal of Clinical Nutrition,* **11**, S197–S201 (2002).

[10] Jeffery E., Brown A., Kurilich A., Keck A., Matusheski N., Klein B., *et al.* "Variation in content of bioactive components in broccoli", *Journal of Food Composition and Analysis,* **16**(3), 323–330 (2003).

[11] Moreno D.A., Carvajal M., Berenguer C.L., Viguera C.G., "Chemical and biological characterisation of nutraceutical compounds of broccoli", *Journal of Pharmaceutical and Biomedical Analysis,* **41**(5), 1508–1522 (2006).

[12] Faller A., Fialho E., "The antioxidant capacity and polyphenol content of organic and conventional retail vegetables after domestic cooking", *Food Research International,* **42**(1), 210–215 (2009).

[13] Porter Y., "Antioxidant properties of green broccoli and purple–sprouting broccoli under different cooking conditions", *Bioscience Horizons,* **5**, http://biohorizons.oxfordjournals.org/content/5/hzs004.full (2012).

[14] Williams D.J., Edwards D., Hamernig I., Jian L., James A.P., Johnson S.K., Tapsell L.C., "Vegetables containing phytochemicals with potential anti–obesity properties: A review" *Food Research International,* **52**(1), 323–333 (2013).

[15] Padilla G., Cartea M.E., Velasco P., de Haro A., Ordás A., Variation of glucosinolates in vegetable crops of Brassica rapa", *Phytochemistry,* **68**(4),536–545 (2007).

[16] Peschel W., Sánchez–Rabaneda F., Diekmann W., Plescher A., Gartzía I., Jiménez D., Lamuela–Raventos R., Buxaderas S., Codina C., "An industrial approach in the search of natural antioxidants from vegetable and fruit wastes", *Food Chemistry,* **97**(1), 137–150 (2006).

[17] White P.J., Broadley M., "Historical variation in the mineral composition of edible horticultural products", *Journal of Horticultural Science & Biotechnology,* **80**(6), 660–667 (2005).

[18] Patil B.S., Jayaprakasha G., Murthy K.C., Vikram A., "Bioactive compounds: Historical perspectives, opportunities, and challenges", *Journal of Agricultural and Food Chemistry,* **57**(18), 8142–8160 (2009).

[19] Poppel G.V., Verhoeven D.T., Verhagen H., Goldbohm R.A., "Brassica vegetables and cancer prevention– Epidemiology and mechanisms", *Advances in Experimental Medicine and Biology,* **472**, 159 –168 (1999).

[20] Reyes A., Mahn A., Guzmán C., Antoniz D., "Analysis of the Drying of Broccoli Florets in a Fluidized Pulsed Bed", *Drying Technology,* **30**(11–12), 1368–1376 (2012).

[21] Rosa E.A., Heaney R.K., "The effect of cooking and processing on the glucosinolate content: Studies on four varieties of Portuguese cabbage and hybrid white cabbage", *Journal of the Science of Food and Agriculture*, **62**(3), 259–65 (1993).

[22] Srisangnam C., Salunkhe D.K., Reddy N.R., Dull G.G., "Quality of cabbage Part II: Physical, chemical, and biochemical modification in processing treatments to improve flavor of blanched cabbage (Brassica Oleracea L.)", *Journal of Food Quality*, **3**(4), 233–250 (1980).

[23] Van Eylen D., Bellostas N., Strobel B.W., Oey I., Hendrickx M., Van Loey A., Sørensen H., Sørensen J.C., "Influence of pressure/temperature treatments on glucosinolate conversion in broccoli (Brassica oleraceae L. cv Italica) heads", *Food Chemistry*, **112**(3), 646–653 (2009).

[24] Jones R.B., Faragher J.D., Winkler S., " A review of the influence of postharvest treatments on quality and glucosinolate content in broccoli (Brassica oleracea var. italica) heads ", *Postharvest Biology and Technology,* **41**(1), 1–8 (2006).

[25] Ignat I., Volf I., Popa V.I., "A critical review of methods for characterisation of polyphenolic compounds in fruits and vegetables", *Food Chemistry*, **126**(4), 1821–1835 (2011).

[26] Vinatoru M., "An overview of the ultrasonically assisted extraction of bioactive principles from herbs", *Ultrasonics sonochemistry*, **8**(3), 303–313 (2001).

[27] Rodríguez–Meizoso I., Jaime L., Santoyo S., Señoráns F.J., Cifuentes A., Ibáñez E., "Subcritical water extraction and characterization of bioactive compounds from Haematococcus pluvialis microalga", *Journal of Pharmaceutical and Biomedical Analysis,* **51**(2), 456–463(2010).

[28] Martins S., Mussatto S.I., Avila G.M., Montañez–Saenz J.M., Aguilar C.N., Teixeira J.A., "Bioactive phenolic compounds: Production and extraction by solid–state fermentation: A review," *Biotechnology Advances*, **29**(3), 365–373 (2011).

[29] Wijngaard H., Hossain M.B., Rai D.K., Brunton N., "Techniques to extract bioactive compounds from food by–products of plant origin", *Food Research International,* **46**(2), 505–513 (2012).

[30] Tanongkankit Y., Sablani S.S., Chiewchan N., Devahastin S., "Microwave–assisted extraction of sulforaphane from white cabbages: Effects of extraction condition, solvent and sample pretreatment", *Journal of Food Engineering*, **117**(1), 151–157 (2013).

[31] Puri M., Sharma D., Barrow C.J., "Enzyme–assisted extraction of bioactives from plants", *Trends in Biotechnology,* **30**(1), 37–44 (2012).

[32] Chari K.L.N., Manasa D., Srinivas P., Sowbhagya H.B., "Enzyme–assisted extraction of bioactive compounds from ginger (Zingiber officinale Roscoe)", *Food Chemistry*, **139**(1–4), 509–514 (2013).

[33] Azmir J., Zaidul I.S., Rahman M.M., Sharif K.M., Mohamed A., Sahena F., Jahurul M.H., Ghafoor K., Norulaini N.A., Omar A.K., "Techniques for extraction of bioactive compounds from plant materials: a review", *Journal of Food Engineering*, **117**(4), 426–436 (2013).

[34] Dandekar D.V., Jayaprakasha G., Patil B.S., " Hydrotropic extraction of bioactive limonin from sour orange (Citrus aurantium L.) seeds", *Food Chemistry,* **109**(3), 515–520 (2008).

[35] Bertelli D., Plessi M., Braghiroli D., Monzani A., "Separation by solid phase extraction and quantification by reverse phase HPLC of sulforaphane in broccoli", *Food Chemistry*, **63**(3), 417–421 (1998).

[36] Vallejo F., Tomás-Barberán F.A., GarcíaViguera C., "Potential bioactive compounds in health promotion from broccoli cultivars grown in Spain", *Journal of the Science of Food and Agriculture,* **82**(11), 1293–1297 (2002).

[37] Wijngaard H.H., Rößle C., Brunton N., "A survey of Irish fruit and vegetable waste and by–products as a source of polyphenolic antioxidants", *Food Chemistry*, **116**(1), 202–207 (2009).

[38] Heimler D., Vignolini P., Dini M.G., Vincieri F.F., Romani A., " Antiradical activity and polyphenol composition of local Brassicaceae edible varieties ", *Food Chemistry*, **99**(3), 464–469 (2006).

[39] Trenerry V.C., Caridi D., Elkins A., Donkor O., Jones R., "The determination of glucoraphanin in broccoli seeds and florets by solid phase extraction and micellar electrokinetic capillary chromatography", *Food Chemistry*, **98**(1), 179–187 (2006).

[40] Wang J., Gu H., Yu H., Zhao Z., Sheng X., Zhang X., " Genotypic variation of glucosinolates in broccoli (Brassica oleracea var. italica) florets from China", *Food Chemistry,* **133**(3), 735–741 (2012).

[41] Koh E., Wimalasiri K., Chassy A., Mitchell A., "Content of ascorbic acid, quercetin, kaempferol and total phenolics in commercial broccoli" *Journal of Food Composition and Analysis,* **22**(7), 637–643 (2009).

[42] Èíž M., Èížová H., Denev P., Kratchanova M., Slavov A., Lojek A., "Different methods for control and comparison of the antioxidant properties of vegetables", *Food Control,* **21**(4), 518–523 (2010).

[43] Isabelle M., Lee B.L., Lim M.T., Koh W. P., Huang D., Ong C.N., "Antioxidant activity and profiles of common vegetables in Singapore", *Food Chemistry*, **120**(4), 993–1003 (2010).

[44] Michiels J.A., Kevers C., Pincemail J., Defraigne J.O., Dommes J., "Extraction conditions can greatly influence antioxidant capacity assays in plant food matrices", *Food Chemistry*, **130**(4),986–93 (2012).

[45] Survay N.S., Kumar B., Upadhyaya C.P., KoE Y., Lee C.H., Choi J.N., Yoon D.Y., Jung Y.S., Park S.W., "Characterization of a cinnamoyl derivative from broccoli (Brassica oleraceaL. var. italica) florets", *Fitoterapia,* **81**(8), 1062–1066 (2010).

[46] Deng G.F., Lin X., Xu X.R., Gao L.L., Xie J.F., Li H.B., "Antioxidant capacities and total phenolic contents of 56 vegetables", *Journal of Functional Foods*, **5**(1), 260–266 (2013).

[47] Guo Q., Guo L., Wang Z., Zhuang Y., Gu Z., "Response surface optimization and identification of isothiocyanates produced from broccoli sprouts", *Food Chemistry*, **141**(3), 1580–1586 (2013).

[48] Rossetto M.R.M., Shiga T.M., Vianello F., Lima G.P.P., "Analysis of total glucosinolates and chromatographically purified benzylglucosinolate in organic and conventional vegetables" *LWT – Food Science and Technology,* **50**(1), 247–252 (2013).

[49] Reinoso B.D., Moure A., Domínguez H., Parajó J.C., "Supercritical CO_2 extraction and purification of compounds with antioxidant activity" *Journal of Agricultural and Food Chemistry",* **54**(7), 2441–2469 (2006).

[50] Pereira C.G., Meireles M.A.A., "Supercritical fluid extraction of bioactive compounds: Fundamentals, applications and economic perspectives", *Food and Bioprocess Technology,* **3**(3), 340–372 (2010).

[51] Arnáiz E., Bernal J., Martín M., Nozal M., Bernal J., Toribio L. "Supercritical fluid extraction of free amino acids from broccoli leaves", *Journal of Chromatography A.,* **1250**, 49–53 (2012).

[52] Martinez–Villaluenga C., Peñas E., Ciska E., Piskula M.K., Kozlowska H., Vidal–Valverde C., Frias J., "Time dependence of bioactive compounds and antioxidant capacity during germination of different cultivars of broccoli and radish seeds", *Food Chemistry*, **120**(3), 710–716 (2010).

[53] Liang H., Li C., Yuan Q., Vriesekoop F., "Application of high–speed countercurrent chromatography for the isolation of sulforaphane from broccoli seed meal", *Journal of Agricultural and Food Chemistry*, **56**(17), 7746–7749 (2008).

[54] Podsêdek A., "Natural antioxidants and antioxidant capacity of Brassica vegetables: A review", *LWT–Food Science and Technology,* **40**(1), 1–11 (2007).

[55] Nunn M.D., Giraud D.W., Parkhurst A.M., Hamouz F.L., Driskell J.A., "Effects of cooking methods on sensory qualities and carotenoid retention in selected vegetables1", *Journal of Food Quality*, **29**(5), 445–457 (2006).

[56] Vallejo F., Barberán F.T., Viguera C.G., Glucosinolates and vitamin C content in edible parts of broccoli florets after domestic cooking. *European Food Research and Technology*, **215**(4), 310–316 (2002).

[57] Vallejo F., Barberán F.A.T., Viguera C.G., "Phenolic compound contents in edible parts of broccoli inflorescences after domestic cooking", *Journal of the Science of Food and Agriculture,* **83**(14), 1511–1516 (2003).

[58] Masrizal M.A., Giraud D.W., Driskell J.A., "Retention of vitamin C, iron, and beta–carotene in vegetables prepared using different cooking methods", *Journal of Food Quality,* **20**(5), 403–418 (1997).

[59] Œwig³o A.G., Ciska E., Lemañska K.P., Chmielewski J., Borkowski T., Tyrakowska B., "Changes in the content of health–promoting compounds and antioxidant activity of broccoli after domestic processing", *Food Additives and Contaminants*, **23**(11), 1088–1098 (2006).

[60] Sanjuan N., Bon J., Clemente G., Mulet A., "Changes in the quality of dehydrated broccoli florets during storage", *Journal of Food Engineering,* **62**(1), 15–21 (2004).

[61] Gêbczyñski P., Lisiewska Z., "Comparison of the level of selected antioxidative compounds in frozen broccoli produced using traditional and modified methods", *Innovative Food Science & Emerging Technologies*, **7**(3), 239–245 (2006).

[62] Price K.R., Casuscelli F., Colquhoun I.J., Rhodes M.J., "Composition and content of flavonol glycosides in broccoli florets (Brassica olearacea) and their fate during cooking", *Journal of the Science of Food and Agriculture*, **77**, 468–472 (1998).

[63] Cieœlik E., Leszczyñska T., Florkiewicz A.F., Sikora E., Pisulewski P.M., "Effects of some technological processes on glucosinolate contents in cruciferous vegetables", *Food Chemistry,* **105**(3), 976–981 (2007).

[64] McNaughton S., Marks G., "Development of a food composition database for the estimation of dietary intakes of glucosinolates, the biologically active constituents of cruciferous vegetables" *British Journal of Nutrition,* **90**(3), 687–698 (2003).

[65] Berenguer C.L., Carvajal M., Moreno D.A., Viguera C.G., "Effects of microwave cooking conditions on bioactive compounds present in broccoli inflorescences", *Journal of Agricultural and Food Chemistry,* **55**(24), 10001–10007 (2007).

[66] Lin C.H., Chang C.Y., "Textural change and antioxidant properties of broccoli under different cooking treatments", *Food Chemistry*, **90**(1), 9–15 (2005).

[67] Sikora E., Cieœlik E., Leszczyñska T., Florkiewicz A.F., Pisulewski P.M., "The antioxidant activity of selected cruciferous vegetables subjected to aquathermal processing", *Food Chemistry*, **107**(1), 55–59 (2008).

[68] Zhang D., Hamauzu Y., "Phenolics, ascorbic acid, carotenoids and antioxidant activity of broccoli and their changes during conventional and microwave cooking", *Food Chemistry,* **88**(4), 503–509 (2004).

[69] Sones K., Heaney R.K., Fenwick G.R., "An estimate of the mean daily intake of glucosinolates from cruciferous vegetables in the UK", *Journal of the Science of Food and Agriculture*, **35**(6), 712–720 (1984).

[70] Howard L.A., Jeffery E.H., Wallig M.A., Klein B.P., "Retention of phytochemicals in fresh and processed broccoli", *Journal of Food Science*, **62**(6), 1098–1104 (2006).

[71] Dong H., Jiang Y., Wang Y., Liu R., Guan H., "Effects of hot water immersion on storage quality of fresh broccoli heads", *Food Technology and Biotechnology,* **42**(2), 135–139 (2004).

[72] Sanjuan N., Benedito J., Clemente G., Mulet A., "The influence of blanching pretreatments on the quality of dehydrated broccoli stems/Influencia del tipo de escaldado en la calidad de tallos de bróculi deshidratados", *Food Science and Technology International,* **6**(3), 227–234 (2000).

[73] Gawlik–Dziki U., "Effect of hydrothermal treatment on the antioxidant properties of broccoli (Brassica oleracea var. botrytis italica) florets", *Food Chemistry,* **109**(2), 393–401 (2008).

[74] Pellegrini N., Chiavaro E., Gardana C., Mazzeo T., Contino D., Gallo M., Riso P., Fogliano V., Porrini M., "Effect of Different Cooking Methods on Color, Phytochemical Concentration, and Antioxidant Capacity of Raw and Frozen Brassica Vegetables", *Journal of Agricultural and Food Chemistry,* **58**(7), 4310–4321 (2010).

[75] Yuan G.f., Sun B., Yuan J., Wang Q.M., "Effects of different cooking methods on health–promoting compounds of broccoli", *Journal of Zhejiang University–Science B,* **10**(8), 580–588 (2009).

[76] Roy M.K., Juneja L.R., Isobe S., Tsushida T., " Steam processed broccoli (Brassica oleracea) has higher antioxidant activity in chemical and cellular assay systems", *Food Chemistry*, **114**(1), 263–269 (2009).

[77] Rosa E., Heaney R., Fenwick G., Portas C., "Glucosinolates in crop plants", *Horticultural Reviews*, **19**, 199–215 (1997).

[78] Lisiewska Z., Kmiecik W., Korus A., "The amino acid composition of kale (Brassica oleracea L. var. acephala), fresh and after culinary and technological processing", *Food Chemistry*, **108**(2), 642–648 (2008).

[79] Toivonen P.M., Sweeney M., "Differences in chlorophyll loss at 13°C for two broccoli (Brassica oleracea L.) cultivars associated with antioxidant enzyme activities", *Journal of Agricultural and Food Chemistry*, 46(1), 20–24 (1998).

[80] Patras A., Tiwari B.K., Brunton N.P., "Influence of blanching and low temperature preservation strategies on antioxidant activity and phytochemical content of carrots, green beans and broccoli", *LWT – Food Science and Technology,* **44**(1), 299–306 (2011).

[81] PuupponenPimiä R., Häkkinen S.T., Aarni M., Suortti T., Lampi A.M., Eurola M., Piironen V., Nuutila A.M., OksmanCaldentey K.M., "Blanching and long term freezing affect various bioactive compounds of vegetables in different ways", *Journal of the Science of Food and Agriculture,* **83**(14),1389–1402 (2003).

[82] Ninfali P., Bacchiocca M., "Polyphenols and antioxidant capacity of vegetables under fresh and frozen conditions", *Journal of Agricultural and Food Chemistry*, **51**(8), 2222–2226 (2003).

[83] Kmiecik W., Lisiewska Z., Korus A., "Retention of mineral constituents in frozen brassicas depending on the method of preliminary processing of the raw material and preparation of frozen products for consumption", *European Food Research and Technology,* **224**(5), 573–579 (2007).

[84] Rangkadilok N., Tomkins B., Nicolas M.E., Robert R., Bennett R.N., Eagling D.R., Taylor P.W., "The effect of post–harvest and packaging treatments on glucoraphanin concentration in broccoli (Brassica oleracea var. italica)", *Journal of Agricultural and Food Chemistry*, **50**(25),7386–7391 (2002).

[85] Rodrigues A.S., Rosa E.A.S., "Effect of post-harvest treatments on the level of glucosinolates in broccoli", *Journal of the Science of Food and Agriculture,* **79**(7), 1028–1032 (1999).

[86] Force L.E., O'Hare T.J., Wong L.S., Irving D.E., "Impact of cold storage on glucosinolate levels in seed–sprouts of broccoli, rocket, white radish and kohl–rabi", *Postharvest Biol Technol.*, **44**(2), 175–178 (2007).

[87] Serrano M., Romero D.M., Guillen F., Castillo S., Valero D., "Maintenance of broccoli quality and functional properties during cold storage as affected by modified atmosphere packaging", *Postharvest Biology and Technology,* **39**(1), 61–68 (2006).

[88] Zapata P.J., Tucker G.A., Valero D., Serrano M., "Quality parameters and antioxidant properties in organic and conventionally grown broccoli after pre–storage hot water treatment", *Journal of the Science of Food and Agriculture*, **93**(5), 1140–1146 (2013).

[89] Vallejo F., Barberán F.T., Viguera C.G., "Health–promoting compounds in broccoli as influenced by refrigerated transport and retail sale period", *Journal of Agricultural and Food Chemistry*", **51**(10), 3029–3034 (2003).

[90] Mrkìc V., Cocci E., Rosa M.D., Sacchetti G., "Effect of drying conditions on bioactive compounds and antioxidant activity of broccoli (Brassica oleracea L.)", *Journal of the Science of Food and Agriculture*, **86**(10), 1559–1566 (2006).

[91] Matusheski N.V., Juvik J.A., Jeffery E.H., "Heating decreases epithiospecifier protein activity and increases sulforaphane formation in broccoli", *Phytochemistry,* **65**(9), 1273–1281 (2004).

[92] Mahn A.V., Antoine P., Reyes A., "Optimization of drying kinetics and quality parameters of broccoli florets", *International Journal of Food Engineering,* **7**(2) DOI: 10.2202/1556–3758.2181 (2011).

[93] Mahn A., Zamorano M., Barrientos H., Reyes A., "Optimization of a process to obtain selenium–enriched freeze–dried broccoli with high antioxidant properties", *LWT–Food Science and Technology,* **47**(2), 267–273 (2012).

[94] Sosiñska E., Obiedziñski M.W., "Effect of processing on the content of glucobrassicin and its degradation products in broccoli and cauliflower", *Food Control*, **22**(8), 1348–1356 (2011).

[95] Chipurura B., Muchuweti M., Manditseraa F., "Effects of thermal treatment on the phenolic content and antioxidant activity of some vegetables", *Asian J Clin Nutr.,* **2**, 93–100 (2010).

[96] Gomes C., Da Silva P., Chimbombi E., Kim J., Perez E.C., Moreira R., " Electron–beam irradiation of fresh broccoli heads (Brassica oleracea L. italica)", *LWT–Food Science and Technology*. **41**(10), 1828–1833 (2008).

[97] Xu F., Chen X., Yang Z., Jin P., Wang K., Shang H., Wang X., Zheng Y., "Maintaining quality and bioactive compounds of broccoli by combined treatment with 1methylcyclopropene and 6benzylaminopurine" *Journal of the Science of Food and Agriculture*, **93**(5), 1156–1161 (2012).

[98] Lemoine M.L., Civello P.M., Martinez G.A., Chaves A.R., "Influence of postharvest UVC treatment on refrigerated storage of minimally processed broccoli (Brassica oleracea var. Italica)", *Journal of the Science of Food and Agriculture*, **87**(6), 1132–1139 (2007).

[99] Ghawi S.K., Methven L., Niranjan K. " The potential to intensify sulforaphane formation in cooked broccoli (Brassica oleracea var. italica) using mustard seeds (Sinapis alba)", *Food Chemistry*, **138** (2–3),1734–1741 (2013).

[100] Conaway C.C., Getahun S.M., Liebes L.L., Pusateri D.J., Topham D.K., Botero–Omary M., Chung F.L., "Disposition of glucosinolates and sulforaphane in humans after ingestion of steamed and fresh broccoli", *Nutrition and Cancer*, 38(2), 168–178 (2000).

[101] Xu F., Chen X., Jin P., Wang X., Wang J., Zheng Y., "Effect of ethanol treatment on quality and antioxidant activity in postharvest broccoli florets", *European Food Research and Technology,* **235**(5), 793–800 (2012).

[102] Devahastin S., Niamnuy C., "Invited review: Modelling quality changes of fruits and vegetables during drying: a review", *International Journal of Food Science & Technology,* **45**(9),1755–67 (2010).

[103] Gonçalves E., Pinheiro J., Abreu M., Brandão T., Silva C.L., " Carrot (Daucus carota L.) peroxidase inactivation, phenolic content and physical changes kinetics due to blanching", *Journal of Food Engineering*, **97**(4),574–581 (2010).

[104] Sarvan I., Verkerk R., Dekker M., " Modelling the fate of glucosinolates during thermal processing of brassica vegetables", *LWT–Food Science and Technology,* **49**(2), 178–83 (2012).

[105] Goncalves E.M., Pinheiro J., Alegria C., Abreu M., BrandaÞo T.R., Silva C.L., "Degradation kinetics of peroxidase enzyme, phenolic content, and physical and sensorial characteristics in broccoli (Brassica oleracea L. ssp. Italica) during blanching", *Journal of Agricultural and Food Chemistry*, **57**(12), 5370–5375 (2009).

[106] Eylen D.V., Oey I., Hendrickx M., Van Loey A., "Kinetics of the stability of broccoli (Brassica oleracea Cv. Italica) myrosinase and isothiocyanates in broccoli juice during pressure/temperature treatments", *Journal of Agricultural and Food Chemistry,* **55**(6), 2163–2170 (2007).

[107] Turkmen N., Sari F., Velioglu Y.S., "The effect of cooking methods on total phenolics and antioxidant activity of selected green vegetables", *Food chemistry*, **93**(4), 713–718 (2005).

16

Bamboo Shoots: Processing, Value Addition, Bioactive Compounds, Food Safety and Market Potential

P. SURESH KUMAR[1]*, YOGESHWAR SINGH[2], G.C. WAKCHAURE[2] A. SANGEETHA[3], K. USHA KUMARI[4] AND V.K. CHOUDHARY[5]

ABSTRACT

Bamboo is a fast growing, wide spread, renewable, versatile, low–or–no cost, environment enhancing resource with potential to improve livelihood security, in both rural and urban areas. Bamboo is an important resource in the Indian socio–economic–cultural–ecological–climate functional context with about 1500 recorded uses. India is one of the richest countries, next to China in bamboo resources with about 136 species out of the total 1250 species under 75 genera found in the world. The shoots (new bamboo culms that come out of the ground) are used in numerous Asian dishes and broths in raw, canned, boiled, marinated, fermented, frozen, liquid and medicinal forms and nowadays also available in supermarkets. In India, particularly in the North Eastern Himalaya, they form a part of the traditional diet. The market for bamboo shoots is growing steadily and more and more people are developing a taste for them. Bamboo shoots are now an important food crop in the international market as bamboo shoot contains more moisture (89.3%), being low in fat (0.41 g /100g), high in dietary fibre (3.90%) and rich in mineral content (1.03%), like an ideal vegetable have been used traditionally by tribal's for decades, world over. For enhancing its business potential, research on various aspects of bamboo shoot as food is being carried out in Japan, Taiwan, Thailand, and Asian countries and several products are

[1] ICAR–National Research Centre for Banana, Tiruchirappalli, Tamil Nadu, India.
[2] ICAR–National Institute of Abiotic Stress Management, Baramati, Maharashtra, India.
[3] Tamil Nadu Agricultural University, HC&RI, Periyakulam, Tamil Nadu, India.
[4] YSR Horticulture University, Tadepalligudem, Andhra Pradesh, India.
[5] ICAR–National Institute of Biotic Stress Management, Raipur, Chhattisgarh, India.
**Corresponding author* E-mail: psureshars@gmail.com; psureshars@rediffmail.com

available in the market. The consumption pattern of bamboo shoots in most of the countries is traditional, non–standardized, seasonal and region specific with little value addition. In view of the seasonal availability of bamboo shoots, the post–harvest preservation system for handling cyanogenic toxicity in raw shoot while keeping nutrients intact and enhancement of shelf life of the value added products assume great significance for the business potential of this natural product. The current research on bamboo shoot is very fascinating leading to the emergence of new avenues of bamboo utilization including food and pharmaceutical potential. The integration of traditional processes after scientific validation would go a long way in developing a suitable improved system for storage and preservation of this perishable commodity for rural entrepreneurship.

***Key words*:** Bamboo, Processing, Traditional foods, Hydrogen cyanide, Nutraceuticals

---

I. INTRODUCTION

Increasing health awareness among the human population is creating a genuine need for adopting a nutritionally complete diet, with increasing personalized value of convenience, cost and taste[1, 2]. This trend, conspicuous in affluent countries, is spreading to the developing countries as well. Various terms such as "designer foods"[3], "functional foods"[4] and "nutraceuticals" have been coined to designate foods for health promotion and disease prevention. Broadly, all refer to food or food ingredients that provide medicinal or health benefits including prevention of diseases. Bamboo shoots, with their high nutritive value and bioactive compounds, hold great promise for utilization as a health food. Unlike most other agricultural crops, bamboos develop naturally with very little artificial selection, are fairly resistant to diseases, insects, climatic injuries, free from residual toxicity as they grow well without the use of fertilizers and pesticides, and are protected from the surrounding pollutants by several layers of tightly clasped sheaths[5].

Bamboo is a fast growing, wide–spread, renewable, versatile, low–or–no cost, environment enhancing resource with potential to improve livelihood security in the years to come, in both rural and urban areas [6]. Bamboo is a member of flowering perennial evergreen plants in the grass family Poaceae, subfamily Bambusoideae, tribe with a unique rhizome–dependent system. Bamboos are of notable economic and cultural significance in South Asia, Southeast Asia and East Asia[7]. The leaves have been used as fodder for livestock by the Japanese for hundreds of years. It is also a very important food for the giant pandas in China, because they survive only on bamboos. Many bamboo species are popularly used as ornamental plants to beautify homes and gardens. In addition, due to their characteristic growth habits, particularly its interwoven system of rhizomes and roots that performs the function of cohesion. Bamboos have enormous potential for alleviating many environmental conditions such as soil erosion control, water conservation, land rehabilitation, and carbon sequestration[8]. They are intermingled with the tradition and culture of rural and tribal populations and are an integral part of their cultural, social, and economic conditions[9,10] from times immemorial due to which they have been variously called as "The Cradle to Coffin Plant" "The Poor Man's Timber", "Friend of the People", "Green Gasoline", "The Plant with Thousand Faces" and "The Green Gold." Because of its multifarious utility, both in the traditional way for the rural people as well as in modern society, bamboo is becoming a very important plant worldwide. There are more than 1500 different documented traditional uses of bamboo[11,12,13].

Energy, climate change, food security, and health care are the major challenges humanity is facing throughout the world. Bamboo grass, which can grow successfully on wasteland with minimal agronomical inputs, would help in meeting these challenges to a significant extent. For various reasons, this grass has fascinated people; it has been called a "miracle plant" or an "ordinary plant with extraordinary qualities." From an "orphan crop," bamboo is now emerging as one of the most important Non–Timber Forest Products (NTFPs) in the world. The natural distribution of bamboo encompasses mainly the tropical, subtropical, and mild temperature zones of the world, with the tropical belt having the maximum number (320 species) of bamboo species. More than 1,250 species belonging to 75 genera are being reported worldwide, to which India has contributed more than 125 species belonging to 23 genera[14, 15]. Fifty–eight species of bamboo belonging to 10 genera are distributed in the northeastern states of India alone and the tribal communities of the region use young succulent shoots of various bamboo species as vegetable, pickles, salad[16].

India is the third–largest country in the world, that is, next to China (300 species) and Japan (237 species) in terms of diversity of bamboo species. Other than these countries, 90 bamboo species in Burma, 55 in Philippines, 50 in Thailand, 44 in Malaysia, 33 in Bangladesh, 31 in Indonesia, 30 in Nepal, 30 in Sri Lanka[14] are found. With unmatched abilities, they are used from fiber to food to floor, from musical instruments to housing and in making various articles of everyday use. India is also one of the leading bamboo producing countries of the world, second to China which produces 32.3 million tons/year[5], covering an area of around 10.03 million ha, which accounts for 12.8% of the total forest area in the country. The shoots (new bamboo culms that come out of the ground) are used in numerous Asian dishes and broths, and are available in supermarkets in various sliced forms, both fresh and canned version. Bamboo shoots are now an important food crop in the international market. Bamboo shoot being low in fat, high in dietary fibre and rich in mineral content, like an ideal vegetable has been used traditionally by tribals for decades, world over[6]. The empty hollow part in the stalks of larger bamboo is often used to cook food in many Asian cultures.

There are a number of bamboo species available in India and many of them are used for edible purposes. *Dendrocalamus longispathus, D. brandisii, Bambusa balcooa, B. polymorpha, B. pallida* and *Melocanna baccifera* are used for edible purpose in North Eastern part of India[7]. Other than these, *Arundinaria aristata, A. hirsuta, B. arundinacea, B. glaucescens, B. longispiculata, B. vulgaris, Cephalostachyum capitatum, C. fuchsianum, D. hookeri and Oxytenanthera albociliata* are the edible species found in southern India[17]. *D. strictus* and *B. bambos* are commonly occurring species of central India. The other species are *B. nutans, B. tulda, D. giganteus* and *D. hamiltonii.* However, *Dendrocalamus asper* is an important edible species of Thailand[18, 19, 20] has been introduced in India for shoot production. In central India, bamboos are not commercially cultivated for its edible shoot production. However, in some places it is also harvested from cultivated sources (plantations and home gardens). Different types of preparations like bamboo shoot curry, chutney, bamboo candy, pickle, fried shoots, rice, *keema*, manchurian (oriental cuisine), soup, canned bamboo juice and bamboo beer are made from bamboo shoots[21, 22, 23, 24]. They are also used as biofertilizer, bioinsecticide and as medicine for stomach disorders[19]. Juice of fermented shoots or bamboo vinegar, stored for about 50–60 days is used for flavouring vegetables[20].

Although the bamboo shoot has been used in tribal diet since from ancient period, yet the further potential of the bamboo shoot as food remains unexplored. Few studies revealed its richness in nutritive components like protein and fiber content and thus the bamboo shoot could be helpful in mitigating the problem of malnutrition[9, 23]. This chapter therefore tries to look into the insight of bamboo shoot as a food resource with a holistic perspective. Exotic food products and traditional dishes of bamboo shoots being eaten at different locations have been compiled and discussed. The antinutritional constituents present in different bamboo shoot species studied by various investigators have also been discussed. The nutritional values of different species of bamboo shoots and the effect of processing methods on its nutritive quality have also been looked into. The feasibility of integrating raw/processed bamboo shoot in the modern diet and lifestyle for enhancing food–nutritional security is also explored. Future R & D areas in relation to the above aspects for enhancing the business potential of bamboo shoot and quality control have been identified.

II. BAMBOO SHOOT: AN ALTERNATE SOURCE OF NUTRITION AROUND THE WORLD

Bamboo shoot is a young, tender stalk emerging from nodes of the (pseudo–) rhizome of the bamboo plants. The edible part consists of meristematic tissue with regions of rapid cell division and differentiation, which is enveloped in protective, non–edible leaf sheaths[25]. They are frequently used in Asian cuisine. The shooting period of a bamboo varies from species to species. In general, the temperate–climate–bamboos are runners, which shoot in the spring, while the tropical and sub–tropical species are clumpers, which shoot in the late summer and autumn (fall). Bamboo shoots are tender, soft, crispy, generally ivory yellow in color. The sheaths covering the shoots are black, brown, yellow or purple. The white meat that is revealed, once the culm sheath is peeled off, turns yellowish when it is cooked. The edible tender shoots look like coiled springs and have an astringent flavor.

Commercially canned bamboo shoot is common, but fresh, locally grown shoot has far better flavour and texture and may increase the share of fresh shoots in future business. The presence of good amount of vitamins, carbohydrates, proteins, and minerals in bamboo shoot and their easy availability to common man may help in solving the nutritional deficiency of the rural population[26]. In the northeast states of India, and in some other parts of the country, bamboo shoot is used as a traditional cuisine–fresh, dried, and pickled. This indicates the vast potential of bamboo shoot as a food resource.

Depending upon species, bamboo shoots are usually 20–30 cm long and taper to a point. A bamboo shoot at the time of harvest normally weighs more than 1 kg. However, their size and weight depend considerably upon the location, depth, pH and nutrition of the soil, irrigation and drainage conditions, climate, rainfall, temperature and soil type and fertility. Cold tolerance is a limiting factor in the growth of certain bamboo species[27, 28]. To get tender white colour portion of bamboo shoot, hard outer layer is peeled off. The top most part of the shoot is conical in shape. Its tender crunchy texture and delicate flavor is priced by connoisseurs. But most canned bamboo shoots lack flavor and texture but are convenient to use.

Edible bamboo shoots are of two types, namely, winter and spring shoots. As compared to the winter shoots, the spring shoots are larger and tougher. They come in both fresh and canned forms although the canned bamboo shoots are easier to find as compared to the fresh ones. The bamboo shoot in its fermented state forms an important ingredient in cuisines across the Himalayas. In Assam, India, for example, it is called *khorisa*. In Nepal, a delicacy popular across ethnic boundaries consists of bamboo shoots fermented with turmeric and oil, and cooked with potatoes into a dish that usually accompanies rice (*alu tama* in Nepali). In Indonesia, they are sliced thin and then boiled with *santan* (thick coconut milk) and spices to make a dish called *gulai rebung*[29]. Other recipes using bamboo shoots are *sayur lodeh* (mixed vegetables in coconut milk) and *lun pia* (fried wrapped bamboo shoots with vegetables). The shoots of some species contain toxins that need to be leached or boiled out before they can be eaten safely. The pith of the young shoots is used in making pickled bamboo which is used as a condiment. The tender bamboo shoots are grated into juliennes and fermented to prepare *kardi* in Sambalpur, Orissa, India. This fermented bamboo shoot is used in various culinary preparations and is notably added to a sour vegetable soup called the *amil*. It is also made into pancakes using rice flour as a binding agent. In Vietnamese cuisine, shredded bamboo shoots are used alone or with other vegetable in many stir–fried vegetable dishes. It may also be used as the sole vegetable ingredient in pork chop soup. In Philippine cuisine, they are called "*labong*" [30, 31]. Two most popular dish for this is the "*ginataang labong*" (*labong* with coconut milk and chilies) and "*dinengdeng na labong*" (*labong* in fish *bagoong* with string beans, *saluyot*, and *tinapa*).

Nutrient Composition & Health Benefits

Bamboo shoots are valuable in pharmaceutical and food processing industries and can be processed into beverages, medicines, additives or health foods. The nutritional value of edible shoots of different bamboo species has been worked out by several workers[26, 32, 33, 35, 36, 38, 39, 40]. Table 1 represents the chemical composition of commonly edible bamboo shoots. Water, protein, carbohydrate, mineral and hydrocyanic acid contents of bamboo shoots vary from 85.98– 92.37%, 1.98–3.29%, 3.89–9.94%, 0.89–1.14% and 0.032–0.13%, respectively.

Bamboo shoots are low in calories, high in dietary fiber, and rich in various nutrients. The main nutrients in bamboo shoots are protein, carbohydrates, amino acids, minerals, fat, sugar, fiber, and inorganic salts. They are rich in protein, containing between 1.49 and 4.04 (average 2.65 g) per 100 g of fresh bamboo shoots. Fresh shoots are a good source of thiamine, niacin, vitamin A, vitamin B6, and vitamin E[33,41]. Fat content is

Table 1: Proximate analysis on selected species of bamboo shoots [16]

Species	*Moisture, %*	*Minerals (as total ash), %*	*Phosphorus, mg/100g*	*Calcium, mg/100 g*	*Iron, mg/100 g*	*HCN, %*
Bambusa balcooa	91.48	0.99	30.99	24.01	1.04	0.071
Bambusa polymorpha	91.65	0.91	15.06	180.69	1.44	0.032
Melocanna bambusoides	91.22	0.98	14.28	47.58	0.89	0.056
Dendrocalamus strictus	85.98	1.14	58.13	139.5	2.74	0.13
Dendrocalamus hamiltonii	92.37	1.01	27.76	44.16	1.65	0.07
Dendrocalamus giganteus	91.19	0.89	12.57	26.93	1.09	0.044
Bambusa pallida	92.29	1.12	32.27	21.17	1.16	0.106

comparatively low (0.26% to 0.94%) and the shoots contain important essential fatty acids. The total sugar content (2.5% on average) is lower than most of other vegetables. Bamboo shoots are low in cholesterol and saturated fats contents (total fats 0.5%), are high in carbohydrate (5.70%), protein (3.9%), minerals (1.1%) and moisture (88.8%) [45, 46]. It is also a good source of Vitamin *E* (*α*–Tocopherol), Vitamin *C*, *B6*, thiamin, riboflavin, niacin and dietary fibers like hemicelluloses, cellulose, pectin, lignin.

Protein content is highest in *Dendrocalamus hamiltonii* (Nees and Arn. ex Munro) (3.72 g/100 g fresh weight). Carbohydrate content ranged from 4.32 to 6.92 g/100 g fresh weight with *Bambusa tulda* (Roxb.) having the highest. Fiber content was maximum in *B. kingiana* (Gamble) (4.49 g/100 g fresh weight). Shoots of *B. bambos* ([Linn.] Voss) and *B. tulda* (Roxb.) have sweet flavor, are tender, and tasty and are the most popular, being liked by all consumers. Of the 14, five (5) species, *B. tulda*, *B. bambos*, *D. asper* ([Schultes f.] Becker ex Heyne), *D. giganteus* (Munro), and *D. hamiltonii* are nutritionally better than the rest and have high potential for the necessary massive commercial production of shoots[6, 35, 40]. Freshly emerged shoots are nutritionally richer than the fermented and canned shoots. Bamboo shoots, though having lower vitamin E content, have a higher amount of vitamin C in comparison to the other green vegetables.

The shoots have a good profile of minerals, consisting mainly of potassium (K), calcium (Ca), manganese, zinc, chromium, copper, iron (Fe), plus lower amounts of phosphorus (P), and selenium[33,40]. Bamboo shoots have a comparatively higher K content than most of the vegetables except spinach, *Spinacea oleracea* (Linnaeus) (558 mg/100 g) and potatoes, *Solanum tuberosum* (Linnaeus) (421 mg/100 g). The sodium (Na) contents in cucumber, *Cucumis sativa*(Linnaeus), *S. tuberosum*, brinjal, *S. melongena* (Linnaeus), and ladies finger, *Abelmoschus esculantus* [(Linnaeus) *Moench*.] are lower than that of bamboo shoots. Magnesium (Mg) is a life–supporting element and has an indispensable role in body metabolism. Its content was highest in the shoots of *D. giganteus* (10.09 mg/100 g) that is close to that of *S. melongena* (10.0 mg/100 g fresh weight). Children and pregnant and nursing women require high amounts of iron[44]. A higher amount of selenium, commonly known as "Miracle life element" is also present in some species of bamboo shoots.

Bamboo shoots are soft and crispy and develop an acrid flavor, if not harvested as soon as they come out of the ground. They contain a potentially toxic glycoside of α –hydroxynitrile, called taxiphyllin, which is turned on by the hydrolytic enzyme: α–glycosidase, upon disruption of the plant cell[47]. Taxiphyllin further breaks down to form cyanohydrins and sugar, which rapidly decomposes into hydrocyanic acid and an aldehyde or a ketone. In *D. giganteus*, it varies upto 894 mg/kg[43], in *M. bambusoides*, 0.14 mg/g, in *B. pallida*, 0.04 mg/g, respectively. The new shoots are almost free from acridity and are brilliant for human consumption. Homogentisic acid is, however, also responsible for the pungent taste of the shoots[48]. They contain 17 amino acids, 8 of which are essential for the human body[42, 43]. Tyrosine amounts to 57% to 67% of the total amino acid content[35]. But the taste also depends on the total sugar content, total amino acid content like aspartic acid (Asp), glutamic acid (Glu), glycine (Gly) and tannin contents; while the amino acids increase the deliciousness of bamboo shoots, tannins decrease the same by increasing the offensive taste[49].

Bamboo Shoot as Medicine

Bamboo shoots have been regarded as a traditional Chinese medicinal component for more than 2000 years, and according to archaic Chinese medicinal books, such as "Ben Chao Qui Zheng" "Ben Jing Feng Yuan "Yao Pin Hua Yi and "Jing Yue" were proclaimed to be beneficial to human health, by promoting motion and peristalsis of the intestine, helping digestion, preventing and curing cardiovascular diseases (CVDs) and cancers.

Shoots of *B. arundinacea*/*B. bambos* contain choline, betain, nuclease, urease, cyanogens, glucosides and are used in the treatment of diarrhoea, thread worm and cough; shoots and dried pith of *D. strictus* contain silicious matter and have tonic and stringent action. The juice of pressed bamboo shoots possesses protease activity that helps in digestion of proteins. Boiled bamboo shoots are used as appetizers and the decoction of shoots are used for cleaning wounds and maggot infected sores, ulcers etc; mixed with palm–jaggery, it is known to induce parturition and abortion[50]. In Java, sap from inside the shoots of *B. vulgaris* is used for curing jaundice. Bamboo is filled with antimicrobial qualities and its shoots are used in preparation of steroidal drugs[51].

In South Asian countries, bamboos have been utilized for traditional medicine treatments to relieve hypertension, sweating, and paralysis. However, little scientific evidence has supported such claims until now. Modern research has revealed that bamboo shoots have a number of health benefits, from cancer prevention and weight loss to lowering cholesterol level, improving appetite and digestion. It is also low in sugar and therefore can be used by persons on sugar–restricted diets. The shoots also contain anti carcinogenic agents, and making them a regular part of a diet effectively reduces the free radicals that can produce harmful carcinogens. Hong *et al.*,[52] studied the effects of pyrolysates–derived from 3 bamboo species, *Phyllostachys bambusoides* (Siebold et Zucc.), *P- nigra* ([Lodd. ex Lindl.] Munro), and *P. pubescens* (Mazel ex J. Houz.), and indicated that pyrolysates may have antiapoptotic effects and can be useful as a supplement for ischemic injury treatment. Several antimicrobials and antioxidants have been isolated by supercritical CO_2 and subsequent hydrothermal treatment of the residues from moso bamboo, including an ethoxyquin, a sesquiterpene, and a cyclohexanone derivative. Bamboo shoots are also reported to have anticancer, antibacterial, and antiviral activity due to the presence of lignans that are an important component of fiber[33, 53]. Because of its high content of K, bamboo helps to maintain normal blood pressure and is labeled as a heart–protective vegetable. Its relatively high content of up to 4% cellulose increases the peristaltic movement of the intestines and helps digestion. It also prevents constipation and decreases body fat.

With different flavones and glycosides, bamboo shoots have excellent anti–microbial qualities and can be extracted to make capsules and tablets. In the traditional system of Indian medicine, the silicious concretions found in the shoots are called '*banslochan*'; and in the Indo–Persian and Tibettan system of medicine, it is called '*tabashir*' or '*tawashir*'; commonly in English, it is called 'bamboo manna.' Earlier obtained from *M. bambusoides*, it is known for its unique healing properties, but is very hard to get. Shoots contain flavonoids, phenols and **phenolic acids** which possess antioxidant activity[54,55]. They also contain lethal concentration of the anti–nutrient (cyanogen)[56]. Anti carcinogenic constituents present in shoots makes them a regular part of diet. It is believed that bamboo extract provide anti–inflammatory effects[57, 58]. Bamboo–derived pyrolysates possess antimicrobial and antifungal activities[53] which protect neurons from **oxidative stress**[59]. Bamboo shoots are also a good source of phytosterols that are the precursors of many pharmaceutically active steroids found in plants[51, 60] and act as nutraceuticals[61]. Bamboo shoot juice possesses protease activity which helps in digestion of proteins[62,63]. In Java, sap from inside the shoots of *B. vulgaris* is used for curing jaundice. In most Indian literature, *Bambusa arundinacea* is described as the source of bamboo manna[63]. Phytosterols have cholesterol–lowering activity[64]. Bamboo shoots are a good source of edible fiber (6 to 8 g/100 g fresh weight), which helps in lowering the blood cholesterol. Fat content is extremely low in bamboo shoots (2.46 g/100 g) that are, therefore, very suitable for people interested in losing weight. The high cellulosic content of bamboo shoots stimulates appetite. Being crisp, crunchy, and tender with a sweet flavor, shoots have a unique and delicious taste that function as an appetizer.

Decoction of the shoots are used for cleaning wounds and maggot infected sores, ulcers etc. Bamboo shoots, mixed with palm–jaggery, are known to induce parturition or abortion[33, 63]. Bamboo salt tablets used in Korea are known to help treat certain internal maladies. These are prepared by sealing salt in bamboo shoots using uncontaminated yellow clay and then baking them eight times in a specially designed furnace. This allows the salt to absorb highly therapeutic trace elements including Cu, Zn and Fe and also eliminates any impurities and heavy metals from the salt that can cause any damage to the body. It serves as a natural detoxifying agent with a strong anti–microbial property and also provides energy and nutrients to the body[65]. Korea markets a delicious bamboo sea salt, which is used as an alternative to table salt in cooking[66].

Bioactive Compounds and Extraction

Nowadays, with the prospects of feeding a fast growing population in the 21st century, agriculture by–products can be valorised as a source of valuable nutraceuticals and they are identified as the ultimate substrates for the development of new products with market value[2,3]. Bioactive compounds are normally accumulated in all parts of plants, but their concentration varies according to the part of plants. Increasing research has focused on natural active compounds extracted from medicinal plants. Plasma pharmaco chemistry techniques are proven to be effective tools for bioactive compound screening in medical plants. If the antibiotic resistance could not be overcome, there will be no cure for tomorrow. Using natural antimicrobial agent had some benefits. Association of antibiotics and plant extracts showed synergistic antibacterial activity against antibiotic–resistant bacteria.

Mass transfer principles state that the concentration gradient between solid and solvent is the driving force for mass transfer[28]. An increase in solid–to–solvent ratio results in an increase in concentration gradient, which results in an increase in the diffusion rate thus permitting greater extraction of solids by the solvent[17, 29]. High leaching–out rates of bioactive compounds when the extraction solvent was increased. Ethanol concentration had significant effects on the yields of TPC, TFC, TF, DPPH radical scavenging capacity, flavanolgallates (EGCG and ECG) and iriflophenone 3–C–β–glucoside. A prolonged extraction time may cause increased exposure to light and oxygen, which will eventually lead to the oxidation of phenolics unless reducing agents were added to the solvent system[27, 31, 32]. Even though 60% (v/v) ethanol resulted in the highest yields of antioxidant compounds, includingTPC, TFC, TF, and flavanol gallates as well as the highest iriflophenone 3–C–β–glucoside at 80% (v/v) ethanol, the DPPH radical scavenging activity was observed to have insignificant differences between40%, 60% and 80% (v/v) ethanol. Therefore, from an economic perspective, a 40% (v/v) ethanol concentration was eventually chosen for the evaluation of solid–to–solvent ratios.

The antibacterial activity of the bamboo leaves has been studied previously. Fresh leaves of bamboo were more effective in inhibiting *S. aureus* than penicillin. Bamboo leaves are rich in flavonoids and the main functional components of extract from bamboo leaf are flavone *C*–glycosides[16, 17]. According to the literature flavone *C*–glycosides, including isoorientin, orientin, vitexin and isovitexin, were poorly absorbed in the gastrointestinal tract of rat. Flavone *C*–glycosides largely reach the colon where they can be degraded into various aromatic acid metabolites by the microflora. The flavone *C*–glycosides cannot be absorbed in the rats' blood after oral administration, whereas *p*–coumaric acid can be absorbed in blood[18]. The ability of *p*–coumaric acid to protect rat's heart against doxorubicin–induced oxidative stress has been proved[19]. The result of microdilution assay showed that both ethanolic and methanol–ethanolic extracts of *D. asper* leaves could inhibit the growth of all tested *E. coli.*

There were 20 organic compounds in methanol–ethanolic extract of *D. asper* leaves which had been identified using py–GC/MS, and 15 of them could be identified with similarity index ≥90%. They could be grouped as fatty acids and the esters (54.72%), long chain alcohols (30.11%), derivatives of phenol (2.52%), aliphatic hydrocarbon (1.04%), and cyclic hydrocarbon (0.83%). Py–GC/MS predicted that there were 32 organic compounds in ethanolic extract of *D. asper* leaves, and 21 of them could be identified with similarity index e"90%. The inhibition activity of caprylic acid toward *E. coli* and total anaerobic bacteria in bovine rumen fluid depended on pH. Caprylic compound, such as caprylic glycol has been known about its antimicrobial activity against *S. aureus, E. coli, P. aeruginosa,* and *C. albicans.*

Bambusa textilis McClure is a traditional Chinese medicinal plant used to treat chronic fever and infectious diseases. Three potential bioactive compounds, including (*E*)–*p*coumaric acid, (*Z*)–*p*–coumaric acid, and apigenin–8–*C*–β–D–(2"–*O*–α–L–rhamnosyl)–glucopyranoside, were detected in both the leaves of *B. textilis* McClure and rabbit plasma. Of the three compounds, apigenin–8–*C*–β–D–(2"–*O*–α–L–rhamnosyl) glucopyranoside was identified based on its UV, MS, and NMR spectra.

III. BAMBOO AS FOOD: MARKET POTENTIAL

The freshly harvested shoot is cream yellow in color, has a strong smell and is sweet in taste. However, all species of bamboo shoots available worldwide are not edible. The edible bamboo shoots have a matchless taste and flavor. In many parts of the world, bamboo shoots form a part of the conventional cuisine and are consumed in various forms *viz.*, raw, dried, canned, boiled, fermented or medicinal[30, 67, 68, 69, 70, 71]. Bamboo grows naturally however, it is also cultivated in homesteads and farms, and is one of the underutilized natural resources in the international scenario. The total potential of bamboo worldwide is estimated at $10 billion. In international markets, China earns 6,500 million Indian rupees every year from export of edible bamboo shoots, with import of USA at around 44,000 tonnes accounting for 14.5% of the total world import[72]. Every year USA imports 30,000 tonnes of canned bamboo shoots from Taiwan, Thailand, India and China for domestic consumption as food items[67]. In Singapore, mostly consumed are the canned shoots; however, frozen cooked shoots are also used[73]. *Dendrocalamus asper*, and *Dendrocalamus lactiferous* and *Bambusa oldhami* are the most important edible species in Thailand and Taiwan[74], respectively. The import of Australia is estimated about 8,000 tonnes per annum[75]. Taiwan consumes about 80,000 tonnes of bamboo shoots annually constituting a value of 2,500 million Indian rupees, covering

30,000 ha of land of bamboo shoots under cultivation, producing total 380,000 tonnes of bamboo shoots per year[76]. In Japan, the present annual consumption of bamboo shoots is 3 kg per person, compared to 1.2 kg per person in 1950s[77]. Estimated bamboo shoot annual import and/or export for major countries are shown in Table 2. In view of growing scientific interest in bamboo shoots for food, pharmaceuticals, etc., it can create opportunities for rural entrepreneurship. Bamboo shoots are exported to United States in canned form (*e.g.*, in the early 1990s, 30,000 t/year supply of bamboo shoot). Exports from Taiwan are worth Indian Rupee 250 crore annually, and those from Thailand Indian Rupee 150 crore, with much of this going to meet Japanese demand[78]. Bamboo shoot and its products have been used as food in China, Japan, Thailand, Australia, and New Zealand. Some of the food products from bamboo shoot commercially available in these countries are bamboo shoot pickle, bamboo shoot sweet soup, bamboo shoot masala curry (kadi), bamboo shoot soup, bamboo shoot russian salad, bamboo shoot pulav, bamboo shoot palak, bamboo shoot manchurian, bamboo shoot *keema*, bamboo shoot *kadhi*, bamboo shoot *halwa*, bamboo shoot fry, bamboo shoot *chutney*, bamboo shoot *bhajji*, canned bamboo shoots, bamboo shoot candies, bamboo shoot sweet pickles, fermented bamboo shoots, bamboo shoot powder, bamboo shoot juice, steamed bamboo shoot rolls, bamboo beer, etc.

Table 2: Estimated bamboo shoot annual import/export for major countries[77]

Country	***Import (tons)***		***Export (tons)***	
	Canned	***Fresh/frozen***	***Canned***	***Fresh***
Australia	12000	–	–	–
Japan	130000	4000	–	–
USA	44000	–	–	–
Taiwan	5000	–	38500	1500
China	–	–	143000	7000
Thailand	–	–	68000	–
Others	99000	6000	40500	1500
Total	290000	10000	290000	10000

At present, over two million tonnes of edible bamboo shoots are consumed in the world every year[77, 79]. Statistics show that about 26.2, 435 and 426.8 tonnes of bamboo shoots are harvested annually in the north eastern states of India like Sikkim, Meghalaya and Mizoram, respectively, where about 20–30 million tonnes of bamboo shoots are utilized for production of canned bamboo shoots annually[5,6]. India's size of domestic bamboo economy currently is estimated at 2,000 million Indian rupees. The market potential of bamboo in India is estimated to be 450 million Indian rupees at present, which is likely to increase to 26,000 million Indian rupees by 2015, thus enabling five million families of artisans and farmers to cross over the poverty line[80].

The bamboo shoots are consumed in the form of fermented–slice, crushed–fermented moist, crushed–fermented dry, fermented whole shoot, roasted whole shoot and boiled whole shoot in different states of the NE region[6]. The consumption of fermented, roasted and boiled shoots was estimated to be the highest in Arunachal Pradesh (481 tonnes/year) and the lowest in Nagaland (19.5 tonnes/year). Cost–return analysis for sales of these bamboo products revealed a net income of 23 million rupees per annum (US$ 502950) from the entire region with the highest (17.5 million rupees/year or US$ 38 270) in Arunachal Pradesh and the lowest in Sikkim (0.47 million rupees/year or US$ 10 280). Considering the bamboo shoot quality and safety aspect together, suitable processing and preservation methods integrating indigenous knowledge system and modern scientific inputs, needs to be developed. It is hoped that, it would add a new dimension to business activity, especially from a natural autarkic economy to a market–orientated industry in the developing countries.

IV. BAMBOO SHOOT PRESERVATION

Since fresh bamboo shoot is an active organic mass performing metabolic activities, the external factors like temperature, humidity, microorganisms, and storage conditions exert influence on its shelf life. Water is necessary to keep bamboo shoot physiologically active and fresh. If it looses water, the freshness is also lost and the quality deteriorates[44]. Water losses resulting in enhanced enzyme activity and saccharine content make bamboo shoot prone to hydrolysis, thus becoming susceptible to infection and rotting. A low temperature storage condition is therefore required to inhibit water transpiration and microbial activity. Thus, the water content, temperature, humidity, and enzymatic activity are all interrelated factors responsible for shoot respiration and have to be taken into account for developing a shoot storage system in the food industry.

Good techniques for bamboo shoot processing and preservation are the selection and grading of bamboo shoots; 10–20% salt solution; cooking thoroughly; rationally preparing preservative solution; aseptic operation; storage at low temperature; and keeping away from light. It is interesting to note that refining and washing don't have any significant effects on the loss of protein, cellulose, and phosphorus, while a slight effect on fat, the total sugar, and a significant effect on amino acid was found[81]. Also, salt boiling, adding the appropriate amount of citric acid, blocking the air invasive, and low–light storage of shoots can extend the shelf life of bamboo shoot[82].

Low temperature inhibits growth and high temperature destroys the spoilage agents and inactivates enzymes. Hua[83] emphasized the requirement of low temperature and proper packaging to reduce transpiration of bamboo shoot. Liu[65] attributed the loss of quality of bamboo shoot due to the high respiration rate (>4.08 mmol CO_2 $kg^{-1}h^{-1}$) at elevated storage temperature

(20°C). After harvest of bamboo shoot, the fiber content increased quickly from the cut end towards the tip. The fiber content could be lowered by keeping it under high humidity or low temperature conditions or both. A rapid increase of phenylalanine ammonialyase (PAL) and Peroxide (PO) activity (both are responsible for enzymatic browning) was also found in the harvested bamboo shoot, while no significant change in polyphenol oxidase (PPO) was observed. The investigation of physiological and biochemical changes in apical, middle, and basal sections of bamboo shoots revealed that wounding from minimal processing induced a high respiration rate, increased PAL, PPO activity (except in base), and contents of Malondialdehyde (MDA) and phenolic compounds, but decreased total sugar, reducing sugar, lignin, and cellulose contents[84]. During the later stage of storage, the respiration rate decreased gradually, while the total sugar and reducing sugar accumulated. PPO activity and phenolic content declined. The MDA content increased sharply after processing and remained at high level during storage[54, 85].

Tamang and Tamang[71] attempted to explore the traditional knowledge of biopreservation of bamboo shoots in Northeast India. Lactic acid fermentation is the main mechanism involved in the biopreservation process by lowering the pH and increasing the acidity of products. Once the hard outer layers and lower parts are trimmed away, the shoot can be slit vertically in halves. Then, starting from the base, they are cut into ½ cm thick slices, keeping the top 6–8 cm in one piece. The slices and tips are then immersed in a big bowl of water and left to soak for 24–36 hours, changing the water a couple of times. This process removes the slight bitterness of bamboo shoots.

Vacuum cooling combined with hydrocooling and vacuum drying processes could increase the cooling speed, and prolong the preservation period of bamboo shoots in low–temperature storage. Compared to hydrocooling, other systems like forced–air cooling, and hydrocooling combined with forced–air cooling, the vacuum cooling combined with hydrocooling and vacuum drying process for bamboo shoots have the advantages of a lower number of bacteria, higher stability, longer preservation period, and better appearance. Kong *et al.*[86] studied peeled bamboo shoots (*Phyllostachys praecox*) and Law[87] investigated the browning and lignification of peeled bamboo shoots (*Phyllostachys praecox* f. *prevelnalis*) at 10°C . The malondialdehyde (MDA) content of bamboo shoots under MAP (0.04 mm thick LDPE bag, 2% O_2, 5% CO_2, and 93% N_2) with low O_2 level was lower than that in the control during early storage, and also POD and PAL activities were significantly inhibited. MAP treatment could also significantly inhibit the lignification of the bamboo shoots by preventing the formation of cellulose and lignin contents. It has been reported[88] that if the contents (dry basis) of cellulose and the lignification of bamboo shoots are more than 46.64% and 15.63%, respectively, it preserves its edible characteristics. MAP treatment prevented the bamboo shoots from being damaged or senescence by the inhibition of the activity of POD, PAL and the accumulation of MDA. Though MAP treatment increased the activity of PPO, the low O_2 level in the LDPE packaging inhibited the enzymatic reaction effectively. Kleinhenz and Midmore[89] found that the visual quality of shoots stored at 8°C declined after 6 days of storage, while in those stored at 1°C such a change in quality was not notable even after one month of storage. The optimum packaging was recommended at 1°C with semi–permeable materials (micro–perforated LPDE bag or LPDE film) in which the shelf life could be extended beyond 28 days.

Quality and Food Safety

The bamboo shoots of some species, owing to the presence of cyanides, develop an acrid flavor[18, 51, 90]. The new shoots are free from acrid taste and are brilliant for human consumption. HCN is a crash product of cyanogenic glycosides which breakdown upon disruption of the plant cell. The quantity of cyanides in bamboo shoots, however, varies depending upon the species, for instance, the amount of cyanides is 894 mg/kg in *Dendrocalamus giganteus*[43]. The shoots of the giant bamboo (*Cathariostachys madagascariensis*) contain cyanide. Despite this, the golden bamboo lemur ingests many times the quantity of toxin that would kill a human. It has been reported that up to 0.16% of total cyanide is contained in the tip of the shoot, reducing to 0.01% in the base[91]. Table 1 shows the hydrogen cyanide (HCN) content (mg/g of bamboo shoot) of some of the commonly edible bamboo shoot species. Cyanogenic glycosides can produce both acute and chronic toxicity, but degrades readily by boiling in water. Nearly 70% of HCN is removed by boiling bamboo shoots for 20 min at 98°C and about 96% is removed by boiling at this temperature for longer interval[43].

Cyanogenic glycosides are nitrogeneous phytoanticipins and are used by various plants as effective defensive mechanism against predators[92]. A mechanisms responsible for the formation of HCN has been formulated by Miller and Conn[93], and it has been found that in most of the species it is the degradation of the cyanogenic glycosides that produces HCN[94]; and the enzyme responsible for this are found out to be α–cyanoalanine synthase which is found in a number of plant species apart from Rhodanese (thiosulphate–cyanide sulphur-transferase) EC Bamboo shoots contain 0.3 to 0.8% HCN[26, 95, 96].

Out of which, up to 0.16% of the total cyanide is contained in the tip, reducing to 0.01% in the base[91], with highest in leaves of young plants, but dropping rapidly after pollination. However, subsequent processing helps in fighting the cyanide concentration, though incomplete cooking result in glycoside hydrolysis and higher release of HCN, but the total amount of

HCN in the shoots can be eliminated/detoxified by boiling/cooking for two hours.

Cyanogenic glycosides were assessed by various authors[83,88,97,98, 99, 100, 101]. Subsequent detoxification and potential toxicity of cyanoglycosides resulting in acute cyanide poisoning in human, bird, fish, wildlife and livestock has been documented. The intermediate degradation of cyanogenic glycosides and their products – the cyanohydrins – are also reviewed[102, 103, 104, 105]. The potential toxic compound in bamboo shoot is cyanogenic glycosides, that is, Taxiphyllin, which are amino acid–derived plant constituents and can be defined chemically as glycosides of α–hydroxynitriles and belong to the secondary metabolites of plants. They are composed of a α–hydroxynitrile type aglycone and a sugar moiety (mostly D–glucose)[106]. Cyanohydric acid is released from the cyanogenic glycosides when fresh plant material is macerated as in chewing, which allows the enzymes and the cyanogenic glycosides to come together, releasing cyanohydric acid (HCN) and a ketone or aldehyde[100]. The second step can also be catalyzed by the hydroxynitrilelyase, which is widespread in cyanogenic plants. In the intact plant, the enzyme and the cyanogenic glycosides remain separated, but if the plant tissue is damaged, both are put in contact and HCN is released[92]. Hydrogen cyanide after oral administration is readily absorbed (it is also readily absorbed after inhalation and through the skin and eyes) and is rapidly distributed in the blood. So, approximately 50–60 mg of free cyanide from bamboo shoot constitutes a lethal dose for an adult man[107].

The cyanide content is reported to decrease substantially following harvesting. Bamboo shoots contain up to 0.16% total cyanide in the tip reducing to 0.01% in the base[91]. The toxic content varies with different parts, geographical and climatic conditions, soil conditions, and also with the method chosen for analysis. Recently, Satya *et al.*[45] reported the variation of toxic content due to aging in fresh shoots of four species of edible bamboos (*Bambusa balcooa, B. bambos, B. tulda, and Dendrocalamus giganteus*) at three different stages of growth. Also, homogentisic acid is responsible for the disagreeable pungent taste of shoot[48]. Satya *et al.*[45] have reported cyanide content in 4 Indian bamboo species, namely *D. strictus, B. tulda, B. vulgaris, and B. balcoa*. Raw bamboo shoots of *B. vulgaris and D. strictus* contained higher cynogenic glucoside content of 200 and 386 mg/kg fresh weight.

Methods of removing HCN from bamboo shoot include processing (boiling, canning, soaking, and fermentation) before consumption. Boiling bamboo shoot in an open vessel for 3–4 hours can reduce the toxicity by 97%[43]. Cyanide content has been reported to reduce with prolonged fermentation by lowering the pH through microbial activity. Surprisingly, literature is silent on the effect of preservation and storage on the toxic content in bamboo shoot. Recently, Satya *et al.*[45] reported that substantial reduction up to 82.7% in toxicity of *B. balcoa* and *B. tulda* stored at "18°C for 2 months.

The significant reduction in the toxic content may be due to degradation of the toxic compound taxiphyllin into HCN. Cyanide content, naturally, is reported to decrease substantially following harvesting[39]. Different indigenous methods of reducing acidity/bitterness from fresh bamboo shoots has been reported and some of them include chopping of tender shoots into small pieces, partial drying of fresh shoots, boiling in water/salt water and draining or keeping shoots in hot water for 10–15 min or in water for a week at ambient temperature, etc. *Adi* women of Arunachal Pradesh used banana leaves for semi–fermentation of shoots and pressed under stones near water stream for 3 – 4 months to reduce bitterness[81, 108]. Similarly, Bhatt *et al.*[6] reported unique traditional processing of bamboo shoot fermentation to reduce the cyanide percentage. Ferreira[43] reported the optimum cooking conditions that resulted in 97% reduction of HCN were 98–102°C for 148–180 min. Subsequently, Tripathi[26] mentioned that removal of HCN can be done by steaming bamboo shoot. Bhargava *et al.*[48] reported removal of this during cooking shoots by changing water several times or by pre–soaking for a long time by subsequent changing 2% salt solution. Wongsakpairod[110] reported superheated steam drying under low temperature removes HCN from bamboo shoot as Taxiphyllin decomposes at around 116°C.

V. BAMBOO PROCESSING: TRADITIONAL AND MODERN METHODS

Processing and value addition refers to any activity that enhances the value of product in the market thereby increasing its utility and profit. Value addition in bamboo shoots can be done by making different edible products; this will lead to cultivation of bamboo shoots by the farmers and help in their income generation. There are numerous ways in which bamboo shoots are processed and prepared. The major processing methods are shown in Fig. 1. During processing, the accompanying physical and chemical changes alter the nutritional integrity of the shoots. Changes in nutrient components in shoots after boiling, fermentation, canning, and during aging have been studied[38, 39, 40]. Protein and sugar contents in the shoots decrease after boiling. The ash content also decreases on boiling, fermentation, and canning. The reduction varies from 20% after boiling, 15% after canning, and 12% after fermentation. The carbohydrate content ranges from 4.09 to 6.91 g/100 g fresh weight, and though there was an increase in the content after boiling, there was a substantial increase up to 72% after fermentation and canning[39]. The fiber content does not seem to change after boiling[38] but it increases significantly after fermentation and canning. Changes in the nutrient components in fresh, fermented, and canned shoots of a commercially important bamboo, *D. giganteus*, were determined[39]. The study revealed that the freshly harvested shoots are richer in nutrient components compared to canned and fermented shoots. There was an overall

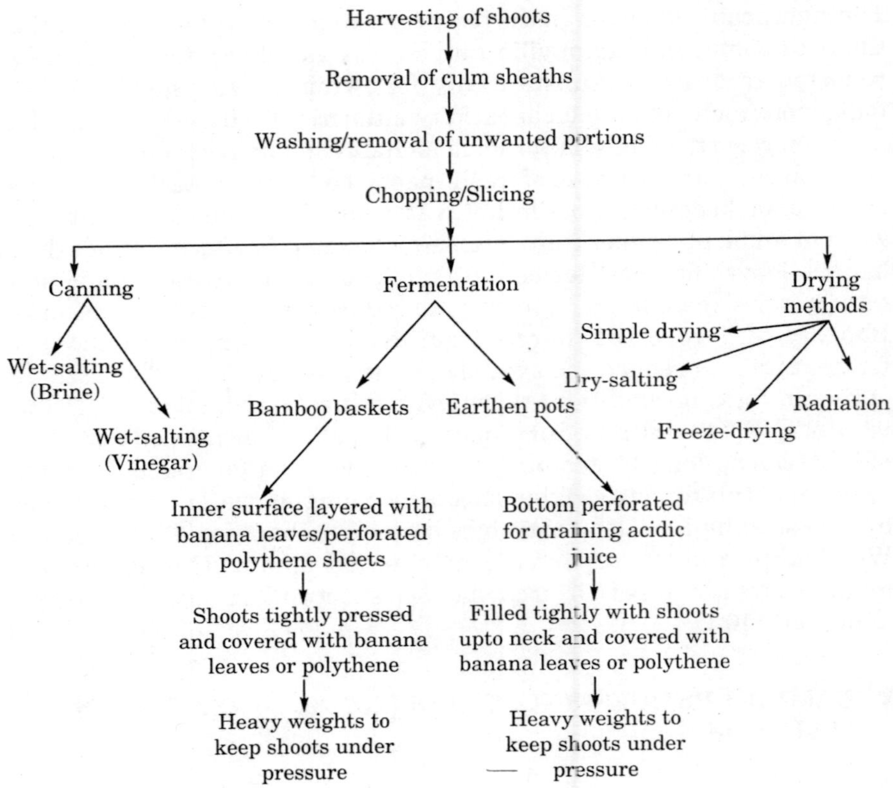

Fig. 1: Outline of bamboo shoot processing

decrease of the nutrient components, except the dietary fiber content, during fermentation and canning. Fresh shoots have higher quantities of macronutrients such as amino acids, proteins, carbohydrates, fat, and fiber than the fermented and canned shoots, except for vitamins C and E and minerals such as Ca, Fe, K, and P. Vitamin C and E content were highest in raw shoots (3.28 mg and 0.69 mg/100 g fresh weight) followed by canned (1.8 mg and 0.3 mg/100 g fresh weight) and fermented shoots (1.09 mg and 0.21 mg/100 g fresh weight). The moisture content was highest in the canned shoots. Thus, freshly harvested shoots, being nutritionally richer than the fermented shoots are recommended for consumption as food. Changes in nutrient components were also observed during the aging and harvesting time of emerging juvenile shoots[40]. A study conducted on 5 commercially important bamboos, *B. bambos*, *B. tulda*, *D. asper*, *D. giganteus*, and *D. hamiltonii*, showed that whereas nutrient components of the shoots were depleted with aging, the dietary fiber, and moisture contents increased. Vitamin and mineral contents also decreased in the older shoots. This indicates that freshly emerging shoots are nutritionally superior to the older emerged shoots[103].

Fresh Bamboo Shoots as Foods

People consume fresh bamboo shoots in various forms. Bamboo shoot can be eaten fresh after boiling. In Indonesia, bamboo shoots are eaten with thick coconut milk and spices, which are called *gulei rebung*; sometimes also mixed with other vegetables, called *sayur lade*[6]. Sweet pickles, chutney and candies are prepared from the pith of the bamboo shoots in the region. The sap of young stalks tapped during the rainy season is simply made into a soft drink in China[77]. In Manipur, the fresh bamboo shoots are taken with dry fish[71]. The edible bamboo species in Western Ghats of India are extensively used as snacks, fried food stuffs, and curries[6]. Tama, a non–fermentated bamboo shoot curry is very familiar among the people of Sikkim[71]. In Singapore, people consume bamboo shoots in form of canned or frozen[73].

Traditional Bamboo Based Food Products

In north east part of India, bamboo is part and parcel of peoples day to day life activities especially in mid and lower altitude regions of Arunachal Pradesh (Fig. 2). *Byapu* is the most famous bamboo product of *apatani* tribe. *Papu Sududanii* is one of the delicious products of *apatani* tribes in which egg is poured into the bamboo shoots[110]. The egg is well fried with typical bamboo flavour. *Paroyo Sududanii* (from chicken) is also prepared in the same way[111]. The sap of young stalks tapped during the rainy season may be fermented to make *ulanzi* (a sweet wine) or simply made into a soft drink. Bamboo leaves are also used as wrappers for steamed dumplings which usually contain glutinous rice and other ingredients. Pickled bamboo shoots (Nepali: *tama*) are cooked with black–eyed beans as a delicacy food in Nepal. Fresh bamboo shoots are sliced and pickled with mustard seeds and turmeric and kept in glass jar in direct sunlight for the best taste. It is used alongside many dried beans in cooking during winter months. Baby shoots (Nepali: *tusa*) of a very different variety of bamboo (Nepali: Nigalo) native to Nepal is cooked as a curry in Hilly regions. Similarly, steamed tea is sometimes rammed into bamboo hollows to produce compressed forms of *Pu–erh* tea.

Fig. 2: Fresh bamboo shoots for sale in market

Mesu is a traditional fermented bamboo shoot pickle with sour acidic taste. Mainly prepared in Sikkim and Darjeeling hills in India. Locally grown young edible shoots *Choya bans* (*D.hamiltonii*), *D. sikkimensis and Bambusa tulda* are defoliated, chopped finely and pressed tightly in a green hollow bamboo stem[16,71,112]. *L. plantorum, L. brevis, L. curvatus, Leuconostoc citreum, Pedicoccus pentosaceus* are believed to responsible for fermentation.

Tender bamboo shoots are collected and cleaned, outer sheaths are removed and from inner white portion Soibum, a fermented product is produced. It is prepared either ***Inside a pit:*** In this method, a pit is dug and a basket made by using bamboo of the shape of the pit is placed into it. Wild *colocasia* leaves is put in and around the pit in a thick layer of about 2–3 inch, then the bamboo shoots are made into longitudinal shreds and kept in air tight condition. The fermented bamboo shoot is ready for sale or for making curry within 3 to 5 days. The water collected from this can also be preserved and is used again in new fermentation of bamboo shoots. Bamboo varieties such as *Sanaibi, Nath, Unal, Longa wa, Meiribob, Ooii* etc are used for this purpose. Care should be taken to avoid varieties such as *Utang* and *Knok* as they are not edible. *Sanaibi* is the best bamboo shoot variety for preparation of fermented bamboo shoots, which is followed by *nath* variety. But *Unal* variety gives better texture and appearance[16]. ***In an earthen pot*:** This method of the fermentation of bamboo shoot is the same as said above, the only difference is that instead of fermenting the bamboo shoots inside a pit, an earthen pot is used where a hole is made at the bottom for the drainage of excess water during fermentation. During fermentation in an earthen pot some people prefer to add *Heibung* for enhancing the fermentation and also for a sourer taste. ***In open condition:*** wild *colocasia* leaves are used in thick layers. The sliced bamboo shoots are placed on top of it and are again covered with these leaves. Fermentation is allowed to take place in this way. ***Dipping in water:*** For this method a special bamboo variety is used *i.e. nath* variety. Locally it is called *nath ki soibum.* This method is commonly used by the people of Bishempur district of Manipur. As the *nath* bamboo is very small and long, it is sliced longitudinally as whole for about 2–3 cm and it is placed in a container (plastic buckets are generally employed these days) where water is poured just to dipped the contains. It is covered and kept till sale.

The *soibum* in general (fermented or dried) are consumed by all the people of Manipur devoid of any caste or tribe. It is eaten as raw with fermented fish or boiled and cooked form with any meat or fish or vegetables. It forms an important food part in all the festivals observed by the *Manipuries*[16]. Nowadays, in some small scale industries the bamboo shoot are blanched after slicing into small pieces in hot water to reduce its enzymatic activity and is treated with KMS (1%) for 10 min. After which it is sun dried and packed in an air tight container and is send for sale. *Soidon, Ekung, Eup, Hirring, Lung–Sieg* are some other bio–preserved bamboo shoots of North eastern region of India.

The tender bamboo shoots (*Dendrocalamus hamiltonii*) are collected and made in to strips after removing the outer green colour sheaths to prepare bamboo *tenga*. The strips are boiled for twenty minutes; excess water drained and kept in air tight container for fermentation. The fermentation process takes about a week's time.

Dried Bamboo Shoots

The process for drying is tedious as it requires fine chopping of bamboo shoots. However, due to its high cost in market it is prepared by people and send to market for sale (Fig. 3). Like bamboo *tenga*, finely chopped bamboo shoots are boiled, fermented and then sun dried. This material is also added in to several traditional meat and fish products prepared by different tribal communities. The prepared dried bamboo shoot is called as *Ikung* by *Adi* and *Galo* tribes. *Apatanis* call it as *hikku*. Locally grown tender bamboo shoots (*Dendrocalamus hamiltonii* or *Bambusa balcooa*) are chopped into small pieces after removing the leaf sheath. The bamboo baskets are laid in the previously dug pits and lined with leaves and then sealed. Stones are kept in the top to drain excess water from the bamboo shoots and fermented for 3 months and stored in air tight container for year. *L. plantarum, L. brevis, L. casei and Tetragenococcus halophilus* are some of the organisms involved in fermentation. The prepared product has 52.1% carbohydrate and energy value of 363.0 kcal/100 g. This can be stored up to 3 years without reduction in quality of the produce. The dried *ikung* is grinded into fine powder called *Ipe* and used as seasoning material for many preparation. *Eup*, a dried product of *nishi*, is prepared in the same way like *ekung*. This is the dried product in which colour changes from whitish to chocolate brown. It is consumed as a curry along with meat, fish or vegetables. *Hirring* is also a fermented bamboo shoot prepared by the *Apatani* of Arunachal Pradesh. *Nishi* tribes call it as *huch* or *hitak*. Only tip of the tender bamboo are either cut longitudinally into

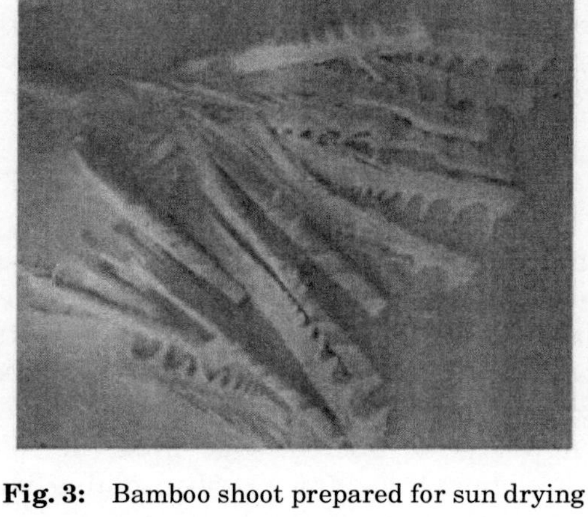

Fig. 3: Bamboo shoot prepared for sun drying

2–3 pieces or whole shoots are flattened by crushing and are kept in bamboo baskets lined with leaves. Rest os the process is followed as like *ikung*. Literature revealed *L.plantarum* and *Lactococcus lactis* as the microbes responsible for fermentation. The prepared product has 0.81% acidity, 2.7% fat (DWB), 49.3% carbohydrate and food value of 353.5 kcal/100 g[79].

Industrial drying is the oldest and simplest food processing technology in the food industry. Extensive research work has been carried out on drying technology. However, very little work is reported on drying of bamboo shoots. Nevertheless, it has been reported that there is a 95.1% decrease in the moisture content of dried bamboo shoots[100]. Water content of fresh bamboo shoots is 92.6/100 g of fresh weight compared to 4.6/100 g in dried bamboo shoots. Bhattacharya[84] reported a multi–stage vacuum pressure reserving technique combined with hydrocooling could reduce the temperature of bamboo shoots efficiently. The author suggested that the process has the advantage of minimizing the energy consumed by the vacuum system. A comparative study of various methods for drying of bamboo shoot along with their advantages and disadvantages has been systematically reported by Majumdar[113] and Satya *et al.*[100]. Bal *et al.*[81] studied the effect of microwave power on drying kinetics of bamboo shoots. The authors suggested that increasing microwave output power, the effective moisture diffusivity values increased from 4.153×10^{-10} to 22.835×10^{-10} m^2s^{-1}. It may be noted that in selecting or designing an appropriate drying technology, it is important to explore energy, environmental and cost issues.

Different technologies may be appropriate at different geographical locations depending on local socio–cultural conditions. Osmotic dehydration of bamboo has also been tried[114]. The sugar concentration could be optimized with the range of 55–58°Bx, the temperature with 55°C, slice thickness equal to 8 mm and Fruit to Syrup ratio with 1:4 can be chosen to obtain higher values for different process variables for osmotic dehydration of bamboo slices (Fig. 4). Quality attributes of end product using multi–stage drying technologies, such as solar–assisted heat pump drying, solar drying with thermal energy storage, microwave assisted vacuum drying, refractive drying and supper heated steam drying may be explored for drying bamboo shoots[114].

Bamboo Fibre in Food Industry

Application of bamboo fibre in various industries is presented in Table 3. Many companies use bamboo fiber in their products. *Vitacel*, a bamboo fiber product is used in fruit juices, beverages, spices, tablets, bakery goods, pasta, fragile products, batters, shredded cheese, sauces, mustard, and ketchup. CFF GmbH (Cellulose Fiber Fabric), based in Gehren, Germany, has developed a new type of fiber derived from the fiber–rich parts of bamboos. It consists of more than 90% water–insoluble fiber that cannot be

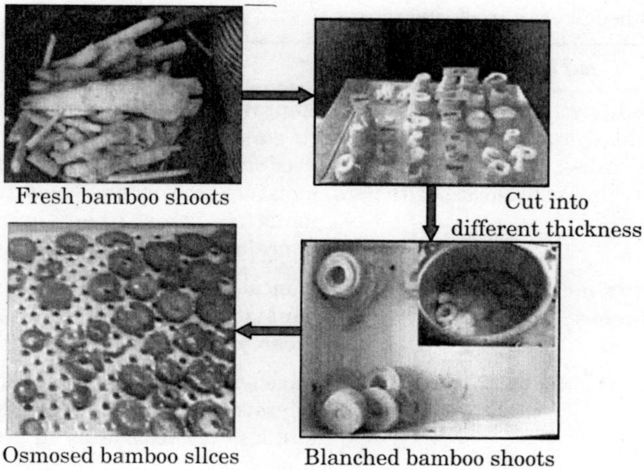

Fig. 4: Steps in osmotic dehydration of bamboo slices [112]

broken down by the human digestive system[98]; it bulks in the stomach, removes undesired metabolic by–products, and shortens intestinal transit time. InterFiber (Zhejiang, China), a Chinese company, is a producer and supplier of high quality, insoluble dietary fiber derived from natural plant sources such as wheat, bamboo, and cellulose[100].

Fermented Shoots

Sheaths from the freshly harvested shoots are peeled off, cleaned in water, and cut into small slices. The shoots are subjected to heavy weights like stones or wooden logs to keep them under pressure. If using pots, sun–dried pots are filled tightly with shoots up to the neck and covered; the bottom is perforated for draining the acidic fermented juice produced during fermentation. The shoots are kept in this condition for 6 to 12 months. Major species used for fermentation include *B. tulda*, *B. vulgaris*, *B. balcooa*, *B. pallida*, *D. hamiltonii*, *D. hookeri*, *D. flagellifer*, *D. giganteus*, and *M. baccifera*[6, 115, 116]. No preservative is used and after complete fermentation, the shoots can be stored 1 to 2 years before marketing, thus making them available throughout the seasons. The microorganisms involved in fermentation of bamboo shoots are *Bacillus subtilis, B. licheniformis, B. coagulans*, and *Micrococcus luteus*[51]. It has been shown that the fermented succulent shoots of *B. tulda* and *D. giganteus* are an enriched source of phytosterols and can, therefore, be used as starting material in the production of steroidal drugs[51, 60]. Sitosterols, the most widely distributed phytosterol, are abundant in fermented shoots and have come into prominence because of their easy microbiological conversion into androstadienedione that is an intermediate product in the synthesis of estrone[51]. Succulent bamboo shoots of various species are harvested and

Table 3: Bamboo fiber in the food industry[73, 83, 100].

Food items	*Benefit*
1. Bakery products, fragile dry bakery products such as pretzels, ice cream cones and cookies, cakes, wafers, baked, and fried tortilla chips, noncaloric health bars	Improves dough yield and consistency due to water binding capacity, decrease of product breakage or crumbling; controls moisture loss in high- and intermediate-moisture foods, fiber enrichment with noncaloric fiber in health nutrition bars.
2. Dairy products, milk, yoghurt, ice cream, shredded cheeses	Noncaloric fiber enrichment, viscosity and consistency improvement, stabilizer, creamy mouthfeel.
3. Meat and aquatic products	Excellent water retention capacity, texture improvement and binding, longer freshness and less fat absorption in product during processing.
4. Health beverages	Noncaloric fiber enrichment, viscosity and consistency improvement, stabilizer creamy mouth feel.
5. Miscellaneous sauces, dressings, juices, ketchup, mustard, low–calorie dressings, pasta	Noncaloric fiber enrichment, viscosity and consistency improvement, stabilizer creamy mouth feel.

fermented by the Tibeto–Mongoloids of Asia[117]. "*Soyadawadawa*," a moist fermented product with the strong odor of ammonia, is commonly consumed in Nigeria[118]. "*Lungsiej*" prepared from *Dendrocalamus hamiltonii* species of bamboo, is a traditional fermented product used as a condiment in the preparation of many Khasi dishes in Meghalaya. Fermentation of sliced bamboo shoots is carried out in two different ways: inside the bamboo cylinder or inside the glass bottle[119].

Bamboo cylinders are made by cutting the bamboo nodes in such a way that one side is closed while the other side remains open. The sliced bamboo shoots are filled inside the bamboo cylinder to its full capacity and the open side is closed with the help of leaves and sealed by tying up the rim with thread or grass. The bamboo cylinders are then immersed in the nearby stream upside down for a period of about 1–2 months for fermentation. Fermentation is also done inside the glass bottle. In this process, the sliced bamboo shoots are filled inside the glass bottle and then water is added till they are submerged. Then the bottle is closed tightly with the cap and kept in the corner of a room or above the cooking place of the house for fermentation. The keeping quality of fermented bamboo shoots inside the bottle is better than the fermentation inside the bamboo cylinders. Shoots in the bottle can be kept up to 1 year where–as in the case of bamboo cylinders, the product has to be consumed within 1 or 2 months. Fermentation of the bamboo shoots in glass bottles is more popular among the people of urban areas, whereas fermentation of bamboo shoots inside the bamboo cylinder is preferred by the farmers and rural people. The dry matter contents were 8.1 g/100 g in *lungsiej*[119]. The protein contents of the fermented foods were marginally higher than their unfermented counterparts. The reason for this could be due to the presence of microorganisms which grow during the process of fermentation. In fermented bamboo, the contents of protein were significantly higher (8.5 g/100 g) in comparison to 3.9 g/100 g in unfermented bamboo shoots. The fat content was found to be 0.6 g/100 g in *lungsiej* as compared to 0.5 g/100 g in fresh bamboo shoots. *Lungsiej* contained iron levels of nearly 1 mg/100 g on fresh weight basis. High magnesium levels were also found in *lungsiej* (294 mg/100 g).

Bamboo Canning

Even though fresh shoots are far superior in taste and texture, bamboo shoots are commonly available in the canned form. The Japanese were the first to develop and use modern technology for the processing of fresh shoots. The shoots are cut in various ways—into halves, thin sliced, stripped, diced, or canned as whole after processing (Fig. 5). The processing technique to produce canned bamboo shoot involves crucial steps that must be followed in order to obtain good quality products. This involves boiling the sliced bamboo shoots in water for about 4 h or for 40 to 60 min at 120°C, cooling, and then storage in brine solution containing usually around 5% to 10% NaCl or more, depending on the species used, and 1% citric acid. There are more than 80 companies manufacturing 49 products of canned shoots under various brand names in the world. China, Taiwan, and Thailand are the leading countries supplying canned bamboo products to the world market. The canned shoots find markets in Singapore, Malaysia, Vietnam, United States, Canada, Europe, Australia, and New Zealand[90]. In India, there are a few companies such as Tai Industries Limited, Kolkata, that imports and markets canned bamboo shoots from Bhutan and Thailand. Canned shoots find their usage mainly in 5–star hotels and upscale specialty restaurants serving Chinese, Thai, Japanese, Korean, and other continental and Asian dishes. Though fresh shoots are cheaper, the efforts involved in processing fresh shoots have made canned products a convenient choice in some hotels and restaurants. Thus, convenience to use and availability are the 2 important considerations favoring canned products. The drawbacks of canned shoots are poor shape, loss of aroma and crunchiness, and unknown source of shoots thereby questionable in quality and prompt use within 3 to 4 days after opening the container. In some countries, good species are cultivated to produce shoots of low fiber, and good texture, color, and taste. However, except for the dietary fiber, there is an overall decrease of the nutrient components, especially vitamins and minerals, during canning[39].

Style of cut	***Appearance***	***Description of shoot after cut***
Whole		Fleshy portion occupies whole length of shoot at over
Half	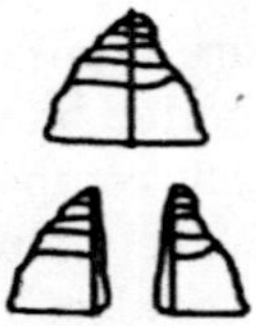	Whole cut lengthwise into two halves
Top or tail whole		Long shoots almost without fleshy portion, 10.15 cm in length
Tip		Shoot tail part without fleshy portion, 4.10 cm in length
Lump	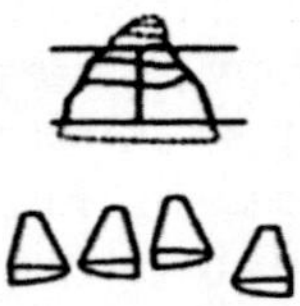	Shoot without tip and cut lengthwise into quarters
Sticed		Flesh portion cut from long part of whole and cut into about 40 × 10 × 3 mm for length, width, and thickness, respectively.
Strip		Fleshy portion cut from long part of whole and cut into about 60.80 × 2.4 × 2.4 mm for length, width, and thickness,
Diced		Fleshy portion cut from long part of whole and cut into about 10 × 10 × 10 cm

Fig. 5: Ways of cutting bamboo shoots for canning* [120]

Bamboo Shoot Based Powder

Bamboo shoot, from its constituents, indicates that it contains considerable amount of carbohydrate and dietary fibers. During boiling, the polysaccharides get hydrolyzed into simple sugars and gives sweet taste to the shoots[38]. The key advantages of using dried bamboo shoot powder is its low moisture content which may allow its direct use into various dry food items, and preparing chutney and beverages. Other advantages of using dried bamboo shoot based powder may include free–flowing, ease of handling and weighing, reduced storage space, ease of cleaning and sanitary aspects. In Japan, bamboo shoot–based powder is used as an essence in cookies and various other food items. Japanese use bamboo powder in standard bread flour and also recommends a 3–8% addition of the powder to any food products[86]. In China, bamboo juice produced by pressure–cooking, is used to make beverages and specific liquors, apart from medicines[121]. With a characteristic bamboo aroma and beer flavor, bamboo juice beers show a good number of health benefits by lowering blood lipids and fighting heart ailments[33, 100].

VI. FUTURE PROSPECTS

Scientific advances, awareness of personal health deterioration, increasing healthcare costs, busy lifestyles, insufficient exercise, and technical advances in the food industry have stimulated the advances in functional foods[107, 122, 123]. Today, foods are not only intended to satisfy hunger and provide necessary nutrients for humans but also to prevent nutrition–related diseases and improve physical and mental well–being of consumers[124, 125, 126]. Bamboo shoots are finding a new place in the spectrum of plant foods used to enhance the quality of life. Functional foods containing phytosterols are now available in the form of margarines, spreads, and salad dressings. Present sources of phytosterols are limited being derived from vegetable oil and tall oil, a by–product of paper production from wood, and it is estimated that 2500 tons of these oils need to be refined to produce 1 ton of phytosterols[127]. Bamboo, being a fast growing plant, growing 1 m/day, can be an important source of phytosterols. For the future, the addition of plant sterols to foods could be an important public health policy if new technology and economies of scale can lower the cost and enable a greater demand[87].

Bamboo is well placed to address the food security through bamboo–based agro–forestry systems by maintaining the fertility of adjoining agricultural lands, and as a direct food source like edible bamboo shoots. Bamboo shoots hold the prospect of value added economic activities at industrial and society levels through cultivation, processing, packaging and commercialization. However, the preparation of various bamboo shoot–based food products is traditional, local, unorganized and based on the taste of the local people. There is no standardized process technology for preservation of the raw bamboo shoots into various food items in an organized manner. This calls for the development of appropriate technologies for preservation of bamboo shoots in various forms. In India, there is yet neither major approach for promotion of shoots nor is there well thought–out market or supply chain for the raw or processed bamboo shoots.

In contrast to the Asian cuisine where bamboo shoots are a routine component of everyday meals, in the western countries, bamboo shoots are available only in canned form, imported from other countries. But, since the local produce is not sufficient to meet the market demand, bamboo shoots in Western markets and Australia are usually imported in processed form. The bamboo shoot–based industry is expanding rapidly in many Asian countries, mainly in China, Japan, Korea, the Philippines, and Thailand, where bamboo farming for edible shoot production is well–established and generating good income for the growers. Canned bamboo shoots occupy an important place in Thailand's global trade providing export earnings of about US$ 30 million annually. In India, despite the rich bamboo genetic resources, edible bamboo species in the country are presently not raised for shoot production, being confined to only certain pockets of the country. There is a perceptible, untapped demand for bamboo shoots in countries such as Japan, Denmark, Australia, the Philippines, Malaysia, and Singapore. Japan imports 0.13 million tons of canned shoots and 3400 tons of fresh shoots annually, whereas Australia imports about 6 to 12000 tons of canned shoots annually[78]. India has a substantial price advantage in edible bamboo exports. The local market price of fresh bamboo shoots during the peak season is generally U.S. dollar 0.23 to 0.27 per kg. In contrast, bamboo prices in Japan range from U.S. dollar 1.5 to 3.0. During the off–season, prices shoot up to U.S. dollar 4 to 6 per kg. This price factor is exploited by Chinese bamboo growers who manipulate their shoot production season by laying steam pipes under the soil surface of their bamboo fields. Export markets have their own preferences as far as the shape, size, and weight of bamboo shoots are concerned. India with its vast bamboo resources has considerable scope for cultivation of edible bamboo species in several areas of the country. If bamboos are popularized as food and commercial production is taken to a large scale, it will contribute enormously to rural economies and boost export earnings. Focus should be directed for following aspects for sustainable development of a bamboo shoot–based food industry.

- Screening and selection of suitable bamboo species for production and value addition of various products
- Scientific cultivation of bamboo
- Bio-prospecting and standardization of extraction procedures for bioactive compounds
- Standardization of appropriate processing and value addition technologies
- Pre and postharvest management of bamboo shoots
- Bio-prospecting and tapping nutraceutical potential of bamboo shoots
- Minimal processing and selection of suitable packages and packaging methods for fresh and processed produces
- Standardization of fermentation technology of edible bamboo shoots
- Standardization of equipment design and Product manufacturing
- Identification of niche markets
- Policy initiatives for bamboo growers and processors

VII. CONCLUSIONS

Bamboo shoot is one of the common food items in many countries and its popularity is growing day–by–day, as main or supplementary food stuff. Bamboo shoot has tremendous export potential which remains unexplored so far. It is interesting to note that in spite of the fact that bamboo shoot has been an integral part of the diet of the tribal community, scientific validation of traditional processing methods in terms of food quality and safety has not been attempted yet. The integration of traditional processes after scientific validation would go a long way in developing a suitable improved system for storage and preservation of this perishable commodity for rural entrepreneurship. Bamboo shoots are getting popular as a nutritional supplement among various communities. Hence, development of different products from bamboo shoots will be useful in optimum utilization of bamboo shoots which will lead to sustainable management of bamboo resources. Also, processing techniques to take care of the food safety aspect would enhance the export potential of this wonderful product. In this context, efficacy of discarded toxic extract as bio–pesticide needs to be explored. The use of hybrid technologies, such as solar–assisted heat pump dryer, solar dryer with thermal energy storage, microwave assisted drying may be more cost effective to get the desired quality products. It should however, be noted that, in selecting an appropriate process technology for the bamboo shoots, it is important to examine energy, environment as well as cost issues. To promote and to revitalize the bamboo industry co–ordination and promotion of tourism, medicinal uses, health care, biotechnology, green energy and high quality agriculture is of paramount importance.

VIII. REFERENCES

[1] German J.B., Yeretzian *C.,* Watzke *H.J.*, "Personalizing food for health and preference", *Food Tech.*, **58***(*12*),* 26–31 (2004).

[2] Wahlgvist M.L., "Requirement for healthy nutrition: Integrating food sustainability, food variety and health", *J Food Sci.*, **68**, 16–*1*8 (2004).

[3] Caragay A.B., "Cancer–preventive foods and ingredients", *Food Technol.,* **46***(*4*),* 65–*6*8 (1992*).*

[4] Thomas P.R., Earl *R.,* "Enhancing the food supply. *Opportunities in the nutrition and food sciences", Washington D.C., Natl. Academy Press.* pp. 98–142 (1994).

[5] Bhatt B.P., Singh L.B., Singh K., Sachan M.S., *Bamboo Sci Culture.*, **17**(1), 24–29 (2003).

[6] Bhatt B.P., Singh L.B., Sachan M.S., Singh K., "Some commercial edible bamboo species of North East India: Production, indigenous uses, cost–benefit and management strategies", *J Bamboo Rattan.*, **4**(1), 13–31 (2005).

[7] Satya S., Singhal P., Bal L.M., Sudhakar P., "Bamboo shoot: A potential source of food security", *Mediterranean J Nut Metabolism.*, **5**(1), 1–10 (2012).

[8] Benzhi Z., Maoyi *F.,* Jinzhong *X.,* Xiaosheng *Y.,* Zhengcai *L.*, "Advances in Plant Phiosology Vol.8", *J For Res.,* **16***(*2*),* 143–*14*7 (2005).

[9] Akao Y., Seki N., Nakagawa Y., Yi H., Matsumoto K., Ito Y., Ito K., Funaoka M., Maruyama W., Naoi M., Nozawa Y., "A highly bioactive lignophenol derivative from bamboo lignin exhibit a potent activity to suppress apoptosis induced by oxidative stress in human neuroblastoma SH–SY5Ycells", *Bioorganic Med Chem.,* **12**(18), 4791–4801 (2004).

[10] Tewari D.N., "Bamboo as Poverty alleviator", *Ind For.,* **114**, 610–*6*12 (1988).

[11] INBAR, "The plant with a thousand faces"*, Intl Network Bamboo Rattan.*, **5**, 13 (1997).

[12] Shrestha K., "Distribution and status of bamboos in Nepal"*, In:* Rao A.N. and Rao V.R. *(eds.), Proceedings of a training course cum workshop,* 1998 May 10–17; Kumming

and Xishuangbanna, Yunnan, China and IPGRI–APO, Serdang, Malaysia: Bamboo–conservation, *Diversity, Ecogeography, Resource, Utilization and Technology*, (1999). http://www2.bioversityinternational.org/publications/ Web_version/572/ ch29.htm#istribution %20 and %20 status %20 of %20 bamboos %20 in %20 Nepal %20 Keshab %20 Shrestha.

[13] Xuhe C., "Promotion of bamboo for poverty alleviation and economic development", *J Bamboo Rattan.*, **2**(4), 345–350 (2003).

[14] Sharma Y.M.L., "Bamboos in Asian Pacific region", *In*: Bamboo Research in Asia, Lessard G. and Chouniard A. (*eds.*), World Publications, Singapore, pp. 99–120 (1980).

[15] Biswas S., Rao V.R., Rao A.N., "Diversity and genetic resource of Indian bamboos and the strategies for their conservation", *In:* Bamboo and Genetic Resources and Use, *Eds.*, Proceedings of the first INBAR Biodiversity, Genetic resources, and Conservation Working Group, Singapore, pp. 29–34 (1994).

[16] Kumar P.S., Bhgawati R., Devi P., "Osmotic dehydration of Bamboo shoots: A new minimal processing technology and way to income generation for farmers of Arunachal Pradesh", *ICFOST 2009, Souvenir:* XX Indian convention of food scientists and technologists 21–23 Dec., Bangalore, p. 141 (2009).

[17] Shanmughavel P., "Cultivation potential of culinary bamboos in Southern India", *Nat Prod Radiance.*, **3**, 237–239 (2004).

[18] Fu M.Y., Ma N.X., Qiu F.G., "Bamboo production and scientific research in Thailand. (Chinese)", *J. Bamboo Res.*, **6**, 54–61 (1987).

[19] ERG, Report on process, market and business opportunity report on edible bamboo shoot prepared by Engineering Resources Group, Bangalore in association with CPF, FRESH and Delphi. Engineering Resource Group, pp. 4–7 (2003).

[20] Sharma T.P., Borthakur S.K., "The use of bamboo shoots as traditional foods of Northeast India and the potential", *Indian J Trad Knowl.*, **7**, 594–597 (2008).

[21] Wang D., Shen S.J., "Bamboos of China", *Timber Press, Portland, Oregon* (1987).

[22] Wang H.X., Ng *T.B.*, "Dendrocin, a distinctive anti–fungal protein from bamboo shoots", *Biochem Biophys Res Commun.*, **307**, 750–755 (2003).

[23] Lee A.W.C., Xuesong B., Audiman P.B., "Flexural properties of bamboo– reinforced southern pine OSB beams", *For Prod J.*, **47**(6), 74–78 (1997).

[24] Lee A.W.C., Xuesong B., Perry N.P., "Selected physical and mechanical properties of giant timber grown in. Sauth Carolina", *For Prod J.*, **44**(9), 40–46 (1994).

[25] Kleinhenz V., Gosbee *M.*, Elsmore *S.*, Lyall *T.W.*, Blackburn *K.*, Harrower *K.*, Midmore *D.J.*, *"Storage methods for extending shelf life of fresh, edible, bamboo shoots [Bambusa Oldhami (Munro)]", Postharvest Bio Tech.*, **19**, 253–264 (2000).

[26] Tripathi Y.C., "Food and nutrition potential of bamboo", *MFP News,* **8**(1), 10–11, (1998).

[27] Kigomo B., "Guidelines for growing bamboo", KFRI *Guideline Series: No 4, Kenya Forestry Research Institute, Nairobi, Kenya*, (2007).

[28] Dollo M., Samal P.K, Sundriyal R.C., Kumar K., "Environmentally sustainable traditional natural resource management and conservation in Ziro Valley, Arunachal Himalaya, India", *J Am Sci.*, **5**(5), 41–52 (2009).

[29] Tamang J.P., Tamang N., Thapa S., Dewan S., Tamang B., Yonzan H., Rai A.K., Chattri R., Chakrabarty J., Kharel N., "Microorganism and ethnic value of ethnic fermented foods and alcoholic beverages from North East, India", *Indian J Tradit Knowle.*, **11**(1), 7–25 (2012).

[30] Tamang J.P., Sarkar P.K., Heseltine J., "Traditional fermented foods and beverages of Darjeeling and Sikkim", *J Sci Food Agric.*, **44**, 375–379 (1988).

[31] FAO. Global forest resources assessment 2005: India country report on bamboo resources, Working Paper No. 118, Forestry Department, Food and Agriculture Organization, Rome, Italy (2006).

[32] Giri S.S., Janmejoy L., "Nutrient composition of three edible bamboo species of Manipur", *Front Biol.*, **4**, 53–56 (1992).

[33] Shi Q.T., Yang K.S., "Study on relationship between nutrients in bamboo shoots and human health", *Proceedings* of the International Symposium on Industrial Use of Bamboo, December 7–11, 1992, *International Tropical Timber Organization and Chinese Academy, Beijing, China,* pp. 338–346 (1992).

[34] Chen C.J., Qiu *E.F.*, Huang *R.Z.*, Fan *H.H.*, Jiang *J.X.*, "Shoot nutrient content of *Phyllostachys pubescens* of different provenances", *J Bamboo Res.,* **18**, 6–11 (1999).

[35] Sharma M.L., Nirmala C., David R.E., "Variations in nutrient and nutritional components of juvenile bamboo shoots," *Pun Univ Res J.,* **54**, 101–104 (2004).

[36] Xu S., Cao *W.*, Song *Y.*, Fang *L.*, "Analysis and evaluation of protein and amino acid nutritional components of different species of bamboo shoots", *Food Sci.*, **26**, 222–227 (2005).

[37] Kumbhare V., Bhargava A., "Effect of processing on nutritional value of central. Indian bamboo shoots", *J Food Sci Tech.,* **44**(1), 29–31 (2007).

[38] Nirmala C., David *E.*, Sharma *M.L.*, "Changes in nutrient components during ageing of emerging juvenile bamboo shoots", *Int J Food Sci Nut.*, **58**, 345–352 (2007).

[39] Nirmala C., Sharma *M.L*, David *E.*, "A Comparative study of nutrient components of freshly emerged, fermented and canned Bamboo shoots of *Dendrocalamus giganteus* Munro", *J Am Bamboo Soc.*, **2**, 33–39 (2008).

[40] Xia N.H., "Analysis of nutritive constituents of bamboo shoots in Guangdong", *Acta Botanica Austro Sinica.*, **4**, 199–206 (1989).

[41] Qiu F.G., "The recent development of bamboo foods", Proceedings of the International Symposium on Industrial Use of Bamboo, December 7–11, 1992, *International Timber Organization and Chinese Academy of Forestry, Beijing, China*, pp. 333–337 (1992).

[42] Ferreira V.L.P., Azzini *A.*, de Figueriredo *I.B.*, Salgado *A.L.B.*, Barbieri *M.K.*, "Evaluation of bamboo shoots for human consumption", *Coletanea do Instituto de Tecnologia de Alimento, Brazil*, **16**, 23–36 (1995).

[43] Bal L.M., Singhal P., Satya S., Naik S. N., Kar A., "Bamboo shoot preservation for enhancing its business potential and local economy: A review", *Critical Rev Food Sci Nut.*, **52**(9), 804–814 (2012). http://dx.doi.org/10.1080/10408398.2010.511321

[44] Park E.J., Jhon *D.Y.*, *"Effects of bamboo shoot consumption on lipid profiles and bowel function in health young women", Nutrition,* **25**,723–728, (2009).

[45] Satya S., Singhal P., Bal L.M., Sudhakar P., "Bamboo shoot: A potential source of food security", *Mediterranean J Nut Metabol.*, **5**, 1–10 (2012).

[46] Nahrstedt A.F., "Cyanogenesis and food plants", *In*: van Beek T.A. and Breteler H. (*eds.*), *Proceedings of the International Symposium on Phytochemistry and Agriculture,* 22–24, *Wageningen. Oxford, Oxford University Press*, pp. 107–129 (1993).

[47] Bhargava A., Kumbhare V., Srivastava A., Sahai A., "Bamboo parts and seeds for additional source of nutrition", *J Food Sci Technol.*, **33**(2), 145–146 (1996).

[48] Xia B., "Studies on nutrient and chemical components of shoots of *Arundinaria oleosa*", *Master's thesis, Nanjing Forestry University, China,* (2006).

[49] RFRI (2008). Bamboo as food and medicine, Report of Rain Forest Research Institute (RFRI), Jorhat, India. www.icfre.gov.in/new/rfri/Bamboo %20food %20medicine 221206.pdf

[50] Sarangthem K., Singh *T.N.*, "Microbial bioconversion of metabolites from fermented succulent bamboo shoots into phytosterols", *Curr Sci.,* **84**, 1544–1547 (2003).

[51] Kumar P.S., Bhagawati R., Ngachan S.V., "Technologies developed for scientific cultivation of horticultural crops in Arunachal Pradesh", *ICAR RC NEH region, AP Centre, Basar, Arunachal Pradesh*, p. 33 (2010).

[52] Hong E.J., Jung *E.M.*, Lee *G.S.*, Kim *J.Y.*, Na *K.J.*, Park M.J., Kang H.Y., Choi K.C., Seong Y.H., Choi I.G., "Protective effects of the pyrolyzates derived from bamboo against neuronal damage and haematoaggregation", *J Pharmacol.*, (2010). *doi:* 10.1016/jep.2010.01.045

[53] Akao Y.N., Seki Y., Nakagawa H., Yi K., Matusumoto Y., Ito Y., Ito K., Funaoka M., Maruyama W., Naoi M., Nozawa Y., "A highly bioactive lignophenol derivative from bamboo lignin exhibit a potent activity to suppress apoptosis induced by oxidative stress in human neuroblastoma SH–SY5Y cells", *Bioorganic Med Chem.*, **12** (18), 4791–4801 (2004).

[54] Pandey A.K., Ojha V., Yadav S., Sahu S.K., "Phytochemical evaluation and radical scavenging activity of *Bauhinia variegata*, *Saraca asoka* and *Terminalia arjuna* Barks", Res *J Phytochem.*, **2**, 89–97 (2011).

[55] EFSA, Opinion of the scientific panel on food additives flavourings, processing aids and materials in contact with food (AFC) on hydrocyanic acids in flavourings and other food ingredients with flavouring properties. *EFSA J.*, **105**, 1–2 (2004).

[56] Srivastava R.C., "Bamboo, new raw material for phytosterols", *Curr Sci.*, **59**, 1333–1334 (1990).

[57] Quitain A., Katoh *S.*, Moriyoshi *T.*, "Recovery of Oil Components of Okara by Ethanol–Modified Supercritical Carbon Dioxide Extraction", *Indus Engin Chem Res.*, **43**, 1056–1060 (2004).

[58] Gupta V.K., Kumria R., Garg M., Gupta M., "Recent updates on free radicals scavenging flavonoids: An overview", *Asian J Plant Sci.*, **9**, 108–117 (2010).

[59] Choudhury D., Sahu J.K., Sharma G.D., "Biochemistry of bitterness in bamboo shoots", *Assam Univ J Sci Technol Phys Sci Technol.*, **6**(2), 105–111 (2010).

[60] Miettinen T.A., Gylling H., "Non–Nutritive Bioactive Constituents of Plants: Phytosterols", *Int. J. Vitam. Nut. Res.*, **73**, 127–134 (2003).

[61] Puri H. S., "Rasayana: Ayurvedic herbs for longevity and rejuvenation", CRC press pp. 71–73 (2003).

[62] Brufau G., Canela *M.A.*, Rafecas *M.*, *"Phytosterols: physiologic and metabolic aspects related to cholesterol–lowering properties", Nutr Res.*, **28**(4), 217–225 (2008).

[63] Cost B., "Asian ingredients: A guide to the foodstuffs of China, Japan, Korea, Thailand and Vietnam", *Harper Collins Publishers, USA,* (1988).

[64] Daphne L., "Bamboo shoots: Delicious to eat, easy to sell", *Washington Tilth. Autumn*, pp. 7–9, (1996).

[65] Liu M.S., "Bamboo shoot", *In*: *Encyclopedia of Food Science and Technology, Vol. 3, Hui, Y.H., (ed.)*, Wiley, New York, pp. 177–180 (1992).

[66] Caitlin B., Miles C., "Investigating bamboo as an alternative crop in the Maritime Pacific Northwest", *Pacific Northwest Sustain Agric.*, **12**(2), 4–6 (2000).

[67] Bal L.M., Sahu J.K., Prusty S.R., "Opportunity of bamboo shoots for nutritional security and socio–economical prosperity of north eastern region of India", *Proceedings of the conference on Agricultural Engineering inputs for the development of the NR region. Assam University, Silchar* on 3rd December, 2008, pp. 108–114 (2008).

[68] Pandey S.K., Pandey S., "Bamboo shoots for the 21st century", *Int For Rev.*, **10**(2), 134–146 (2008).

[69] Lobovikov M., "Bamboo and rattan products and trade", *J Bamboo Rattan.*, **2**(4), 397–406 (2003).

[70] Pan C., "Market opportunities for fresh and processed Asian vegetables" *Canberra, RIRDC Research Paper* No. 95/14, p. 117 (1995).

[71] *Tamang B., Tamang J.P., "Traditional knowledge of biopreservation of perishable vegetable and bamboo shoots in Northeast India as food resources", Indian J Tradit Knowle.*, **8**(1), 89–95 (*2009*).

[72] Cahill A., "Field day to explore edible bamboo shoot market", *News Release, Dept of Primary Industries Queensland*, (1999).

[73] Yang Q., Duan Z., Wang Z., He K., Sun Q., Peng Z., "Bamboo resources, utilization and *ex–situ* conservation in Xishuangbanna, South–eastern China", *J Forest Resour.*, **19**(1), 79–83 (2008).

[74] Vaiphei S.L., Bamboo's economic value to the northeast. ManipurOnline, (2005). www.manipuronline.com/Economy/January2006/bamboo18_1.htm.

[75] Farooquee N.A., Dollo M., Kala C.P., "Traditional Wisdom of Apatani Community in the management and sharing of natural resources in North Eastern India", *In*: Misra K.K. (*ed.*), Traditional knowledge in contemporary societies: Challenges and opportunities. *Pratibha Prakashan,* Delhi, pp. 110–126 (2007).

[76] Tai K.Y., "The management and utilization of shoot–producing bamboos in Taiwan *Q.J. Chin For.*, 18(2), 1–46 (1985).

[77] Scurlock J.M.O., Dayton *D.C.,* Hames *B.*, "Bamboo: An overlooked biomass resource", *Biomass Bioenergy*, **19**, 229–244 (2000).

[78] Bal L.M., Kar A., Satya S., Naik S.N., "Drying kinetics and effective moisture diffusivity of bamboo shoot slices undergoing microwave drying", *Int J Food Sci Technol.,* **45**(11), 2321–2328 (2010).

[79] Hu C.Y., Zhang Y., Kitts D.D., "Evaluation of antioxidant and prooxidant activities of bamboo *Phyllostachys nigra* var. Henonis leaf extract *in vitro*", *J Agric Food Chem.,* **48**, 3170–3176 (2000).

[80] Fujimura M., Ideguchi M., Minami Y., Watanabi K., Tadera K., "Amino acid sequence and anti microbial activities of chit–in–binding peptides, Pp–AMP1 and Pp–AMP2, from Japanese bamboo shoots", *Biosci Biotechnol Biochem.*, **69**, 642–645 (2005).

[81] Zhi Min L., *Journal of Zhejiang Forestry Science and Technology*, **23**(6), 15–21 (2003).

[82] Lu B., Wu *X.,* Shi *J.,* Dong *Y.,* Zhang Y., "Toxicology and safety of antioxidant of bamboo leaves. Part 2: Developmental toxicity test in rats with antioxidant of bamboo leaves", *Food Chem Toxicol.,* **44***(*10*),* 1739–*1*743 (2006).

[83] Hua X., "Bamboo shoot cultivation and management in Japan", *Symposium on Bamboo Professional Commission of Zhaihiang Forestry Society,* **16**, (1987).

[84] Bhattacharya S., "Tropical Bamboo: Molecular profiling and genetic diversity study", *Lambert Academic Publishing*, ISBN 978–3–8383–7422–2, (2010).

[85] Wang J.W. "Study on ageing physiology of postharvest of bamboo shoots", *Forest Res.*, **15**(6), 687–692 (2002).

[86] Kong F.C., Lu S.M., Wang Q., "Study on storage of peeled bamboo shoots (*Phyllostachy*) with modied atmosphere packaging", *Storage and Process*, **6**, 24–25 (2003).

[87] Law M., "Plant sterol and stanol margarines and health", *Br Med J.,* **320***,* 861–*8*64 (2000).

[88] Kleinhenz V., Midmore D.J., "Improved management practices for culinary bamboo shoots: Local and export markets", *A Report for the Rural Industries Research and Development Corporation, Publication* No., 02/035. Project No. UCQ–9A (2002).

[89] Midmore D., "Culinary Bamboo Shoots", *In:* The new rural industries, Hyde K.W. (*ed.*), Rural Industries Research and Development Corp, Canberra, pp. 188–196 (1998).

[90] Haque M.R., Bradbury J.H., "Total cyanide determination of plants and foods using the picrate and acid hydrolysis methods", *Food Chem.*, **77**, 107–114 (2002).

[91] Francisco I.A., Pinotti P.H.M., "Cyanogenic glycosides in plants", *Brazilian Archives of Biol. Technol.*, **43**(5), 116–120 (2000).

[92] Miller J.M., Conn E.E., "Metabolism of hydrogen cyanide by higher plants", *Plant Physiol.*, **65**, 1199–1202 (1980).

[93] Blumenthal S.G., Hendrickson H., Abrol Y.P., Conn E.E., "Cyanide metabolism in higher plants III. The biosynthesis of beta–cyanolanine", *J Biol Chem.*, **243**, 5302–5307 (1968).

[94] Poulton J.E., Keeler R.F., Tu A. T., "Plant and fungal toxins", *In: Handbook of Natural Toxins,* Vol. 1, p. 117. *eds.* Marcel Dekker, New York, (1983).

[95] Anonymous. Cyanogenic glycosides in cassava and bamboo shoots, a human health risk assessment. *Technical report series no. 28. Food Standards Australia New Zealand* (2004).

[96] JECFA (1993). Cyanogenic glycosides. *In*: Toxicological evaluation of certain food additives and naturally occurring toxicants. Geneva, world health organization, 39th meeting of the joint FAO/ WHO expert committee on food additives (WHO Food Additives Series 30). http://www.inchem.org/documents/jecfa/jecmono/v30je18.htm.

[97] Speijers G., "Cyanogenic glycosides", *Food Additive Series No. 30. Geneva: JECFA,* (1993).

[98] ATSDR Toxicological profile for cyanide. www.atsdt.cde.gov/toxprofiles/ tp8.html (2006).

[99] Satya S., Bal L.M., Singhal P., Naik S.N., "Bamboo shoot processing: Food quality and safety aspect: A review", *Trends in Food Science & Technology,* **21**(4), 181–189 (2010).

[100] NMBA (2009). Bamboo shoot composition. National mission on bamboo application. http://www.bambootech.org/subsubTOP.asp?subsubid=89&subid= 29&sname = USAGE.

[101] WHO Toxicological evaluation of certain food additives and naturally occurring toxicants. WHO food additive series 30. World health organization, Geneva. (1993).

[102] Majak W., "Metabolism and absorption of toxic glycosides by ruminants", *J Range Manage.,* **45**, 67–71 (1992).

[103] Brimer L., Rosling H.A., "Microdiffusion method with solid state detection for determination of cyanogic glycosides from cassava in human urine", *Food Chem Toxi.*, **31**, 599–603 (1993).

[104] Hernández T., Lundquist P., Oliveira L., Christiá R.P., Rodriguez E., Rosling H.A., "Fate in humans of dietary intake of cyanogenic glycosides from roots of sweet cassava consumed in Cuba", *Nat Toxins.,* **3**, 114–117 (1995).

[105] Vetter J., "Poisonous plants", *Gyogyszereszet*, **43**, 80–96 (1999).

[106] Hasler C.M., "Functional foods: The Western perspective", *Nutr Rev.,* **54**, S6–S10 (1996).

[107] Sundriyal M., Sundriyal R.C., "Wild edible plants of the Sikkim Himalaya: Marketing, value addition and implications for management", *Economic Bot.*, **58**(2), 300–315 (2004).

[108] Singh R.K., Singh A., Sureja A.K., "Traditional foods of monpa tribe of west Kameng, Arunachal Pradesh", *Indian J Tradit Knowle.*, **6**(1), 25–36 (2007).

[109] Wongsakpairod T., "Bamboo shoot drying using superheated steam", *M Eng Thesis, King Mongkut's University of Technology*, Thonburi, Bangkok, Thailand (2000).

[110] Devi P., Kumar P.S., "Traditional, ethnic and fermented foods of different tribes of Manipur", *Indian J Tradit Knowle.*, **11**(1), 70–77 (2012).

[111] Kumar P.S., Bhgawati R., Kanwat M., Alone R.A., Bhuyan M., "Bamboo processing: A new way of income generation for livelihood of Arunachal Pradesh", *National seminar on sustainable hill agriculture. Abstracts of papers.* Manipur, Imphal Feb 28–Mar 2, p. 40 (2009).

[112] Kumar P.S., Kanwat M., Choudhary V.K., "Mathematical modeling and thin–layer drying kinetics of bamboo slices on convective tray drying at varying temperature", *J Food Process Preserv.,* **37**(5), 914–923 (2012). doi:10.1111/j.1745–4549.2012.00725.x

[113] Majumdar A.S., "Some innovative drying technologies for dehydration of foods", *Int J Postharvest Technol Innov.*, **1**, 76–92 (2006).

[114] Giri S.S., Janmejoy *L.*, "Effect of bamboo shoot fermentation and aging on nutritional and sensory qualities of soibum", *J Food Sci Tech.,* **37***(4),* 423–*4*26 (2000).

[115] Jeyaram K., Singh *A.T.,* Romi *W.,* Devi *R.W.,* Singh *M.W.,* Dayanidhi *H.,* Singh *R.R.N.,* Tamang *J.P., Traditional fermented foods of Manipur. Indian J Tradit Knowle.,* **8**, 115–*1*21 (2009).

[116] Lu B., Wu *X.,* Tie *X.,* Zhang *Y.,* Zhang *Y.*, "Toxicology and safety of anti–oxidant of bamboo leaves. Part 1: *Acute and subchronic toxicity studies on anti–oxidant of bamboo leaves*", *Food Chem Toxicol.*, **43***(5),* 783–792 (2005).

[117] Rubatzky V.E., Yamaguchi M., "World vegetables: Principles, production and nutritive values", *New York: Chapman & Hall*, pp. 658–660 (1997).

[118] Murugkar D.A., Subbulakshmi G., "Preparation techniques and nutritive value of fermented foods from the khasi tribes of meghalaya", *Ecology of Food and Nutrition*, **45**(1), 27–38 (2006). http://dx.doi.org/10.1080/03670240500408336

[119] Menrad K., "Market and marketing of functional food in Europe", *J Food Eng.*, **56**, 181–*188* (2003).

[120] Qing Y., Zhu B.D., Zheng L.W., Kai H.H., Qi X.S., Zhen H.P., "Bamboo resources, utilization and *ex–situ* conservation in Xishuangbanna, south–eastern China", *J Forest Res.*, **19**(1), 79–83 (2008).

[121] Siro I., Kapolna *E.*, Kapolna *B.*, Lugasi *A.*, "Functional food. Product development, marketing and consumer acceptance – A review", *Appetite*, **51**, 456–*4*67 (2008).

[122] Granato D., Branco *G.F.*, Nazzaro *F.*, Cruz *A.G.*, Faria *J.A.F.*, "Functional foods and nondairy probiotic food development: Trends, concepts, and products", *Com Rev Food Sci Food Safety.*, **9**, 292–302 (2010).

[123] Roberfroid M.B.,"Concepts and strategy of functional food science: The European perspective", *Amer J Clinical Nut.*, **71**, S1660–*S1664 (*2000).

[124] Kim N.K., Cho S.H., Lee S.D., Ryu J.S., Shim K.H., "Functional properties and antimicrobial activity of bamboo (*Phyllostachys* sp.) extracts", *Korean J Food Preser.*, **8**, 475–80 *(*2001).

[125] Kumar P.S., Bhagawati R., Choudhary V.K., Devi P., Ronya T., "Effect of boron and molybdenum on growth, yield and quality of cauliflower in mid altitude condition of Arunachal Pradesh, *Ind food packer.*, **64**(3), 52–58 (2010).

[126] Ko W.C., Hui Y.H., GhazalaS., Graham D.M., Murrell K.D., Nip W.K., "Canned Chinese bamboo shoots, water chestnuts, mushrooms, and imitation vegetarian products", *In: editors* Chapter 5, *Handbook of vegetable preservation and processing. New York: Marcel Dekker, CRC Press. p.* 95 *(*2003).

[127] Jones D.A., "Why are so many food plants cyanogenic", *Phytochemistry*, **47**, 155–162 (1998).

17

Processing of Curry Leaf and Extraction of Various Compounds

RAVINDRA NAIK[1]* AND DAWN C.P. AMBROSE[1]

ABSTRACT

Curry (Murraya koenigii.) is one of most consumed leaves in various diets of Indian cuisines and other Asian countries. It is rich in many bioactive compounds like polyphenols, alkaloids and flavonoids. Curry leaves, a part of spicing up dishes is not only used for garnishing, but also for therapeutic use. They are rich in medicinal, nutraceutical properties and have even cosmetic uses. It shows bioactive functions like antioxidant, anticancer, antimicrobial, antidiabetic and hepatoprotective properties. In addition to the basic bioactive compounds, are also rich in essential oil compound namely coumarine, bicyclomahanimbicine, mahanimbicine, phebalosin are also present in it. In this chapter we are exploring the post harvest processing methods of curry leaves and its medicinal uses.

Key words: *Murraya koenigii*, Bioactive compounds, Curry leaf, Extraction of Compounds

I. INTRODUCTION

Curry leaf (*Murraya koenigii* Spreng) or curry patte (Hindi) is a perennial leaf vegetable. It belongs to the family *Rutaceae*. It is an aromatic more or less deciduous shrub or small tree, found almost throughout India and Andaman Islands up to an altitude of 1500 metres. The leaves of the plant are used extensively for seasoning and flavoring dishes. It is widely used in Kerala and in South Indian states as an important ingredient in Indian

[1] ICAR Central institute of Agricultural Engineering, Regional centre Coimbatore - 641007, Tamil Nadu, India.

**Corresponding author* E-mail: naikravindra@gmail.com

cuisine. It is also used as a medicine in local ayurvedic systems in West Bengal, Madhya Pradesh and in the South and South Western states, namely, Maharashtra, Tamil Nadu, Kerala and Andhra Pradesh. The leaves of the plant are extensively used as flavoring agents in curries like 'dhal', 'South Indian Sambar', 'rasam' and chutneys. Ground curry leaf with spices could be used as an excellent preserve. Apart from cooking, the curry leaf has a number of medicinal uses. Fresh leaves on steam distillation under pressure yield a volatile oil. The leaves contain a residual Glucoside named as Koenigin. They strengthen the functions of stomach and promote its action.

II. ORIGIN

Curry leaf trees are naturalized in forests and waste land throughout the Indian subcontinent except in the higher parts of the Himalayas. From the Ravi River in Pakistan its distribution extends eastwards towards Assam in India and Chittagong in Bangladesh, and southwards to Tamil Nadu in India. The plants were spread to Malaysia, South Africa and Reunion Island with South Asian immigrants [1]. Various names of curry leaves in given in Table 1.

HISTORY

The curry leaf tree is native to India, Sri Lanka, Bangladesh and the Andaman Islands. Later spread by Indian migrants, they now grow in other areas of the world where Indian immigrants settled. Widely cultivated, the leaves are particularly associated with South Indian cuisines. The use of curry leaves as a flavoring for vegetables is described in early Tamil literature dating back to the 1st to 4th centuries AD. Its use is also mentioned a few centuries later in Kannada literature. Curry leaves are still closely associated with South India where the word 'curry' originates from the Tamil 'kari' for spiced sauces. An alternative name for curry leaf throughout India is kari–pattha[1].

CULTIVATION AND COLLECTION

Flowering starts from the middle of April and ends in the middle of May. The peak flowering season under the Sanwara (H.P.) conditions was observed to be the last week of April. The fruiting season was observed to continue from the middle of July to the end of August. The peak fruiting season, however, was found to continue from the last week of July to the 1st week of August. Curry leaf is native to India. Pinnate leaves are used in many South Indian curries. Fertilize with palm or citrus fertilizer to promote leaf production. The Fig. 1 shows healthy curry leaf field[1].

Fig. 1: Curry leaf (*Murraya koenigii*) field

III. MORPHOLOGICAL CHARACTERS

The curry leaf tree is a small spreading shrub, about 2.5 metres high. It has dark green to brownish coloured main stem, with numerous dots on it. Its bark can be peeled off longitudinally, exposing the white wood underneath. The girth of the main stem is 16 cm. Leaves are exstipulate, bipinnately compound which is 30 cm long, each bearing 24 leaflets. It has reticulate venation and the leaflets are lanceolate which are 4.9 cm long and 1.8 cm broad, having 0.5 cm–long petiole. Flowers are bisexual, white, funnel–shaped, sweetly scented, stalked, complete, ebracteate, regular, actinomorphic, pentamerous, and hypogynous. The average diameter of a fully opened flower is 1.12 cm. Inflorescence, a terminal cyme, each bearing 60 to 90 flowers. Fruits are round to oblong, 1.4 to 1.6 cm long, 1 to 1.2 cm in diameter; weight, 880 mg; volume, 895 microlitres. Fully ripe fruits are black with a very shining surface; the pulp in wistaria blue; the number of fruits per cluster varying from 32 to 80. Seed, one in each fruit, 11 mm long, 8 mm in diameter, colour spinach green; weight, 445 mg; volume, 460 microlitres[1]. The different parts of curry leaf are mentioned in Fig. 2.

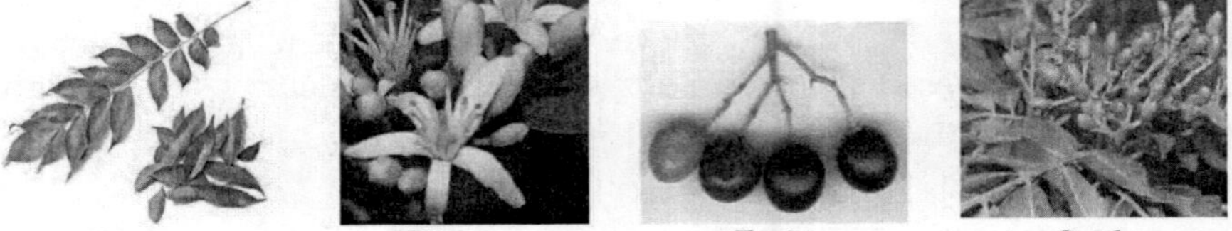

Leaves *Flowers* *Fruits* *Seeds*

Fig. 2: Various parts of curry leaf

IV. PRODUCTION

The curry leaf bushes are usually cultivated in one or two numbers in the home–nutrition or kitchen garden. But in south India especially in Tamil Nadu and Karnataka this is grown as a cash crop. In Tamil Nadu, it is cultivated in Coimbatore, Salem and Trichy Districts. Cultivating even in an area of half an acre fetches very sizeable income to the farmer. It is commonly found in forests, often as a gregarious growth along the foot of the Himalayas up to Sikkim and Assam. It is also found aromatic leaves and for ornamental value throughout India. It is native to Sri Lanka but also found in Burma, Indo–China, South China and Hawaii. It is used as a medicine in local ayurvedic systems in West Bengal, Madhya Pradesh and in the South and South Western states, namely, Maharashtra, Tamil Nadu, Kerala and Andhra Pradesh. Curry leaf is exported as curry leaf and as curry leaf oil from India[2].

Table 1: Various names of curry leaves[3]

	Indian Names
Assamese	Narsinghs, Bisharhari
Hindi	Kathnim, Mitha neem, Curry or kurry patta, Gandhela, Bareanga
Bengali	Barsanga, Kariphulli
Gujarati	Goranimb, Kadhilimbdo
Kannada	Karibevu
Malayalam	Karriveppilei
Marathi	Karhinimb, Poospala, Gandla, Jhirang
Oriya	Barsan, Basango, Bhuraunga
Punjabi	Curry patta
Sanskrit	Krishna nimbi
Tamil	Karivempu, Karuveppilei
Telugu	Karepaku
Name in international languages	
French	Feuilles de cari
German	Curryblatter
English	Curry leaves
Chinese	Ga lei yihp
Dutch	Kerriebladeren
Spanish	Hoja

V. CHEMICAL COMPOSITION

Fresh curry leaves on steam distillation yield 2.5% volatile oil which is used as a fixative in soap industry[4]. The antioxidant and anti carcinogenic effect of curry leaves have been studied and it has been reported that the curry leaves have a high potential as a reducer of the toxicity of carcinogen[5]. The leaves contain the following free amino acids: asparagine, glycine, serine, aspartic acid, glutamic acid, theonine, alanine, proline, tyrosine, tryptophan, amino butyric acid, phenylalanine, leucine, isoleucine, and traces of ornithine, lysine, arginine and histidine. The leaves also contain a crystalline glucocide, koenigin and a resin. Table 2 shows the nutritional information of curry leaves.

Table 2: Nutritional composition of curry leaves[1]

Element	*Quantity*
Moisture	66.30 %
Protein	6.10 %
Fat (ether extract)	1.0 %
Carbohydrate	18.70 %
Fibre	6.40 %
Mineral matter	4.20 %
Calcium	810 mg/100 g of edible portion
Phosphorous	600 mg/100 g of edible portion
Iron	3.1 mg/100 g of edible portion
Carotene (as Vitamin A)	12600 IU/100 g
Nicotinic acid	2.3 mg/100 g
Vitamin C	4 mg/100 g
Thiamine and riboflavin	Absent

The fruit is edible. It yields 0.76% of a yellow volatile oil with neroli-like odour and pepper-like taste, accompanied by an agreeable sensation of coolness on the tongue. The fruit is reported to contain koenigin. Yellow clear and transparent oil is procured from the seeds which are known as limbolee oil. The constituents of the oil are α-phellandrene (4.6%), *d*-sabinene (9.2%), *d*-α-pinene (5.5%), dipentene (6.8%), α-terpineol (3.2%), caryophyllene (26.3%), isosafrol (4.4%), cadinene (18.2%), cadinol (12.8%), lauric acid (2.7%) and palmitic acid 3.4%[6]. The chemical composition of green (fresh) curry leaves at 3 stages of maturity namely tender, medium and mature was determined on moisture free basis. The composition of leaves at these three stages is given in Table 3[7].

It is seen that with advancing maturity, there was a gradual decrease in volatile oil and oleoresin (acetone extracted) while there was a progressive increase in all other items of analysis. Vacuum self drying of curry leaves gave a better product of greenish color than dried by any other method. The free amino acids present in the leaves are asparagine, serine, aspartic acid, glutamic acid, threonine, alanine, proline, tyrosine, tryptophan, alpha amino butyric acid, phenylalanine, leucine, and isoleucine, traces of lysine, arginine and histidine[8].

VI. COMPONENTS RESPONSIBLE FOR FLAVOUR

The major constituents responsible for the aroma and flavor have been reported as pinene, sabinene, caryophyllene, cadinol and cadinene. Essential oils from *M. koenigii* serves as an important part in soap making ingredients, lotions, massage oils, diffusers, potpourri, scent, air fresheners, body fragrance, perfume oils, aromatherapy products, bath oils, towel scenting, spa's, incense, facial steams, hair treatments, and more[1].

Table 3: Chemical composition at various maturity stages of curry leaves

Element	*Quantity (%)*		
	Tender	*Medium*	*Mature*
Protein	5.44	6.44	7.19
Fat	3.30	4.74	6.15
Sugar	14.9	17.9	18.9
Starch	11.4	14.2	14.6
Crude fibre	5.80	6.2	6.20
Volatile oil	0.82	0.55	0.48
Acetone extractive (Oleoresin)	1.60	1.40	1.30
Total ash (mineral matter)	12.54	12.7	13.1
Acid insoluble ash	1.2	1.3	1.35

Volatile Oil

Fresh leaves on steam distillation under pressure (6.33 kg/cm^2) yield 2.6% of a volatile oil which may be used as a fixative for a heavy type of soap perfume distillation. At ordinary pressure it gives very poor yields of oil, while distillation with superheated steam at 220°C, it yields dark coloured foul smelling oil. Rectified curry leaf oil is deep yellow in color with strong spicy odour and pungent clove like taste. Essential oil obtained by simple water distillation has the characteristics: Yield: 1%; color: light yellow; smell: aromatic with spicy note; specific gravity at (25°C): 0.8589; optical rotation: –36°14; acid value: 1.17; saponification value: 8.18; value after acetylation: 52.81%. Figs. 3 and 4 show the volatile compounds and the Carbazole structures, respectively[9].

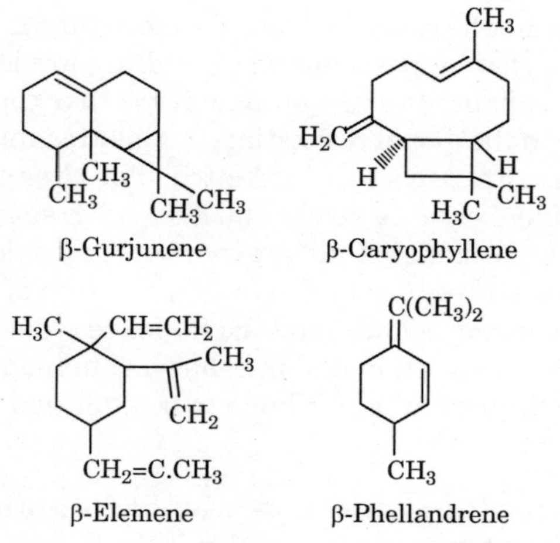

Fig. 3: Volatile compounds from *Murraya koenigii*

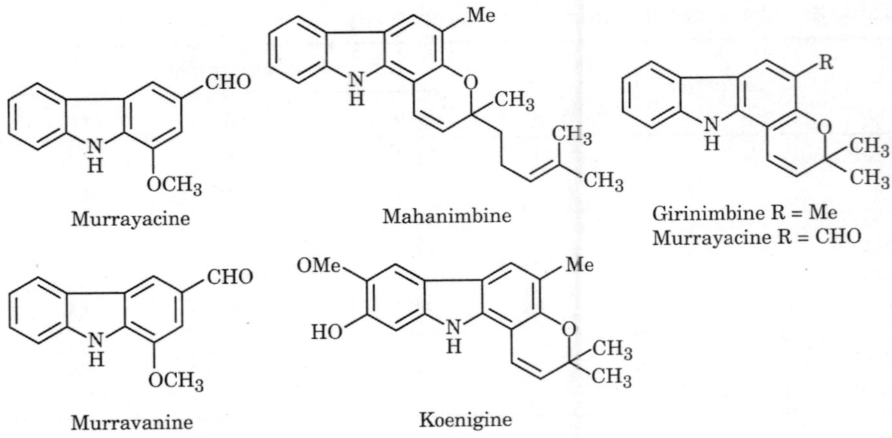

Fig. 4: Carbazole alkaloids from *Murraya koenigii*

The major chemical constituents are l–caryo–phyllene; 6–7%; l–cadinelle: 5.2%; and unidentified residue: 3.2%. Thus, further research is needed to have a clear picture of aromatic principles of curry leaves. Using modern techniques, scientists from the Central Food Technological Research Institute, Mysore, Karnataka, India have reported the presence of beta-pinene in curry leaf oil[10].

PROCESSING OF CURRY LEAVES

Need for Processing of Curry Leaves

Though the leaves are available throughout the year, they are highly perishable and cannot be retained fresh for more than a day. There is need to spread the research on post–harvest handling in view of its enhanced global demand. To enable cheaper shipment to other countries, which will play a pivotal role in the global marketing, simpler techniques for extending the shelf life during transport is warranted. The changing demographics nationwide that have created ready market and greater demand for this spice, there is a potential demand for export of leafy spices from India, if a suitable method is evolved for drying with retention of flavor[2]. It's huge demand in India and abroad has made curry leaves commodity of immense trade value. Proper cultivation and post harvest management practices of this commercially important plant have to be analyzed to fetch maximum returns[1].

A group of agricultural scientists who had scrutinized a village in Periyanaickenpalayam, Coimbatore of Tamil Nadu, India has reported that curry leaf is a crop of commercial importance and it could be profitably exported if farm technology was upgraded[11].

Technologies Developed for Curry Leaf Processing by CIAE–IEP Centre

Curry leaves stripping

The existing method of stripping the curry leaf from the harvested leaflets is done manually with bare hands by labor in an unhygienic way and there is every possible risk of contamination during handling which could carry over to the processed product in the food chain. Further, since the product is being used in the export market both as fresh product and in the powder/ masala industries, there is also possibility of contamination being carried over. The curry leaf stripping operation should be made easier and hygienic in operation. Curry leaf stripper make the operation easier and to reduce the human drudgery. A power operated curry leaf stripper has been developed for stripping the curry leaves. The capacity of the machine is about 40–50 kg/h with a stripping efficiency of about 90–95%. The machine consists of power source, pulling mechanism, stripping mechanism and collecting tray as shown in Fig. 5[12].

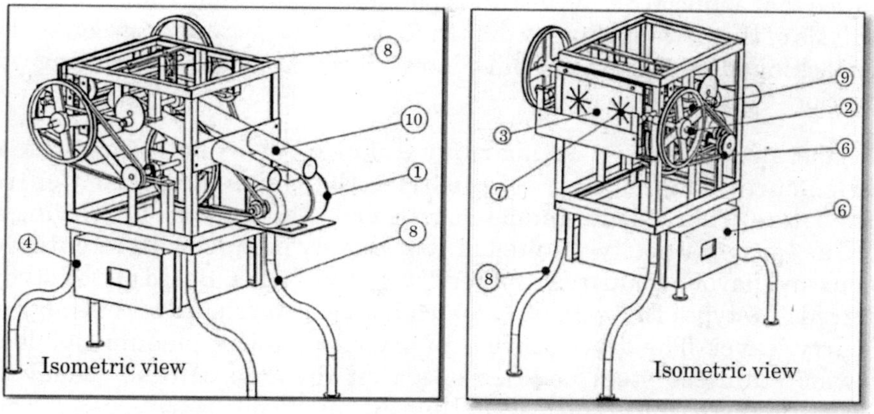

1 = Power source–Motor; 2 = Pulling mechanism; 3 = Stripping mechanism; 4 = Collecting tray; 5 = Striped rollers; 6 = Power transmission system; 7 = Nylon plates with holes; 8 = Stand; 9 = Reciprocating rocker arm; 10 = Stem outlet pipe

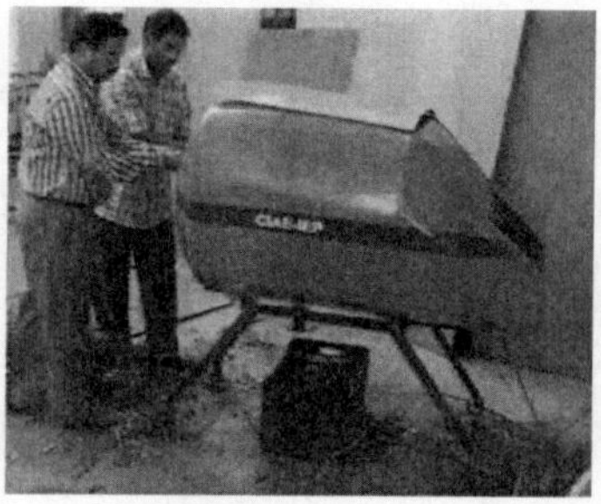

Fig. 5: Isometric view of power operated curry leaf striper

Drying of curry leaves

Curry leaves may be air dried or oven dried, producing leaves that have a longer shelf life. Some recipes require the baking or toasting of fresh curry leaves before the leaves are added as a flavoring. Dried leaves are also available commercially[1].

There is not much loss of volatile oil during drying either in sun/shade or in cross flow dryer. The leaves retain their flavor after scientifically dried and hence these are marketed both in fresh and dried forms. Drying at 50°C is recommended as the best technology for conversion of fresh leaves into dry powder. Higher temperatures during drying deteriorated powder quality. The dried curry leaf is found to have use in the masala industries and also is being exported. The dehydration of curry leaf in different types of dryers *viz.*, cross–flow, through flow and vacuum dryers were studied. It was found from the studies that vacuum shelf–dried product gave a better green color compared to other dryers[13].

The flavor characteristics of curry leaves dried under various conditions showed that artificial drying in a cross flow drier is superior. The organoleptic analysis of the powder showed that though the characteristic green colour was lacking, product resulted out of cross flow drying and oven drying was superior[14].

Laboratory studies of drying curry leaf at ICAR – Central Institute of Agricultural Engineering – Industrial Extension Regional Centre, Coimbatore, Tamil Nadu, India has revealed that fluidized bed drying at 45°C at 4m/s air velocity resulted in good quality product with retention of colour and flavour compared to other drying conditions. Based on the study, a forced flow type of dryer of 50 kg/batch has been developed for bulk drying of curry leaves. The dryer consists of drying chamber, plenum chamber, blower, stainless steel paddles operated through a hand wheel for intermittent stirring of the leaves while drying[15].

The dehydration of curry leaves in cabinet drier at 50°C for 3 hours and 30 minutes was found to be more suitable than drying in shade, sun and solar drier for the preparation of curry leaves powder. The optimum pretreatment to retain the green colour in the processed curry leaves powder was standardized. Curry leaves dipped in the cold solution containing magnesium oxide, potassium Meta bisulphite and sodium bicarbonate was found to be the best pretreatment. Table 4 gives the comparative nutritional profile between fresh and the dehydrated curry leaves. The therapeutic values (TP and TAA) of the dried samples were significantly less than that of fresh leaves. This may be due to the long drying time or higher temperature drying, which removes the volatile compounds[16].

Table 4: Comparative nutrient content of fresh and dehydrated curry leaves[2].

Nutrients	*Value of fresh curry leaves (100g)*	*Value of dehydrated curry leaves (100g)*
Protein, g	6.0	12.00
Fat, g	1.0	5.40
Carbohydrate, g	18.7	64.31
Calcium, mg	830	2040
Iron, mg	0.93	12.00
Beta carotene, μg	7560	5292

Packaging

One of the most common post harvest problems of green vegetables and fresh herbs is associated with their rapid senescence, a process which is thought to be regulated by ethylene, the senescing harmone.

The effect of storage temperature on the nutritional value of curry leaf revealed that there was a lesser loss of β-carotene and apparently no loss of α-tocopherol and chlorophyll concentration in curry leaf frozen at –15°C. However, the loss of lutein in the frozen samples was higher perhaps due to destruction of lutein during freezing process. The loss of β-Carotene, α-tocopherol and chlorophyll when air dried or oven dried may be attributed to loss of these compounds due to oxidation[17].

Storage of fresh curry leaves

Marketing of perishable commodities often requires some storage to balance day to day fluctuations between the product harvested and sale for long time storage to extend marketing beyond the end of harvest season. Among horticultural products, leafy vegetables are one that leads to spoilage of product rapidly. There is change in developmental, structural, physiological and biochemical processes leading to the senescence of leaves. The most obvious symptoms of senescence in the leaves are loss of fresh weight, drying, shriveling, flavor changes and losses in chlorophyll, ascorbic acid and protein content[18].

The shelf life of fresh curry leaf could be extended upto four days under ambient condition when prepackaged in 20 micron PP bag of 5mm dia. vent covering 0.1% vent area. The same could be sold at the retail outlets. Group of self help groups were trained and encouraged to prepack curry leaf for retail outlets. For export of fresh curry leaf, the refrigeration was found to extend the shelf life upto 16 days in 75 micron PE bag of 5mm dia. covering a vent area of 0.1%[15].

Essential oil extraction from curry leaves

An essential oil is a liquid that is generally steam or hydro–distilled from flowers, leaves, bark and roots of plants and trees and are the compounds responsible for the aroma and flavor associated with herbs, spices, and perfumes. Essential oils from *M. koenigii* serves as an important part in soap making ingredients, lotions, massage oils, diffusers, potpourri, scent, air fresheners, body fragrance, perfume oils, aromatherapy products, bath oils, towel scenting, spa's, incense, facial steams, hair treatments, and more. This extraction method is a combination of solvent extraction and ultrasonic extraction method[1]. Steam distillation is used in the extraction of essential oil from the plant material. It is a special type of distillation or a separation process for temperature sensitive materials like oils, resins, hydrocarbons, etc. which are insoluble in water and may decompose at their boiling point. Various methods have being employed these days for extracting essential oils from different spices and herbs. It has been reported that, hydro distillation and solvent extraction are the common extraction methods being used to extract *M. koenigii* leaves[19].

Hydro distillation

Hydro distillation is one of the oldest methods of extraction used. The spice is fully immersed in hot water. The result is a soup, which carries aromatic molecules of the plant. The method is not much in use these days, because of the risk of overheating the plant and subsequent loss of the oil. The fundamental nature of steam distillation is that it enables a compound or mixture of compounds to be distilled at a temperature substantially below that of the boiling point(s) of the individual constituent(s). Essential oil contains components with boiling points up to 200°C or higher temperatures. In the presence of steam or boiling water, however, these substances are volatilized at a temperature close to 100°C, at atmospheric pressure[20].

Solvent extraction

Many herbs and spices cannot be extracted from distillation method, so solvent extraction is the safest method for extracting high quality oil. In this process, the spices or herbs plants are immersed in the solvent and the 'separation' is performed chemically. These include pigments, volatile molecules and non–aromatic waxes. The herbs and spices are then subjected to low pressure distillation and the volatile oil is then separately collected. It should be noted that, even with the most advanced techniques, absolutes extracted in this manner do contain traces of solvent. In solvent extraction, solvent selection is important. A solvent is a liquid or gas that dissolves a solid, liquid, or gaseous solute, resulting in a solution. Solvents usually have a low boiling point and evaporate easily or can be removed by distillation, leaving the dissolved substance behind[20].

Ultrasound–assisted extraction (UAE)

Ultrasonic radiation, which has frequencies higher than 20 kHz, facilitates the extraction of organic and inorganic compounds from solid matrices using liquid solvents. Sonication is the production of sound waves that create cavitation bubbles near the sample tissue, which break down to disrupt cell walls, thereby releasing cell contents[21]. An appropriate solvent is mixed with a sample and sonicated under controlled temperature for a specified time. Extract recovery is influenced not only by sonication time, temperature and solvent selection, but also by wave frequency and ultrasonic wave distribution. Ultrasound has been used in both static and dynamic modes to extract phenolics from plant materials. A static system is a closed–vessel extraction for which no continuous transfer of solvent occurs.

In dynamic extraction, fresh solvent is supplied continuously, which allows efficient adsorption of analytes and their effective transfer from the extraction vessel. Continuous transfer of extracted analytes prevents degradation of any thermo–labile compounds by the heat associated with sonication[22].

Curry leaf puree

Curry leaves were separated from stems, washed in running tap water and blanched for 2 min in hot water at 90°C. The blanching time was based on preliminary trials that resulted in complete inactivation of peroxidase enzyme in curry leaves. Curry leaves were cooled immediately in chilled water, drained and comminuted in a laboratory size grinder fitted with a 14 mesh (1.19 mm aperture size) screen to obtain a puree of uniform size. The puree was stored at 0°C, immediately, until further use[23].

Value added products from curry leaf powder

For the preparation of value added products, the curry leaves were dried in a hot–air oven at a temperature of 40°C for 4 h. The dried curry leaves were powdered in grinder and the dry powder was stored in an airtight container and placed inside a refrigerator at a temperature of 8°C until further use. Dried curry leaves were incorporated in various commonly prepared dishes at different levels by weight. The selected products were chapatti (unleavened Indian bread), seasoned potatoes and a spice mixture. The foods and spice mixture were analyzed by 53 panel members for color, appearance, texture, odor and taste. The sensory responses for chapatti indicate that the incorporation of Curry Leaf Powder (CLP) reduces the sensory scores for the assessed attributes. Scores were lower as the level of CLP incorporation increased. It was finalised that the 10% incorporation in chapatti and spice mix, 5% in potatoes gave a good result[24].

Value added products *viz.*, cookies, bread, *paruppupodi*, and *adaimix* were formulated by incorporating curry leaves powder at different combinations. Results showed that curry leaves powder incorporated at a level of 4.0, 4.0, 25 and 15 per cent respectively in cookies, bread, *paruppupodi* and *adaimix* was found to be highly acceptable by the judges and the score card ranged between 8.0 and 9.0. For 100 kg of fresh curry leaves yielded 19.0 kg of powder. The cost of the prepared curry leaves powder was Rs. 21 per 100 gram[16].

Traditional uses

The bark and the roots are used as a stimulant by the physicians. They are also used externally to cure eruptions and the bites of poisonous animals. The green leaves are stated to be eaten raw for curing dysentery, and the infusion of the washed leaves stops vomiting. Curry leaves are also used in calcium deficiency. It has Vitamin A, Vitamin B, Vitamin C, Vitamin B2, Calcium and iron in plenty. Its nutritional value benefits both the young and the old alike. Women who suffer from calcium deficiency, osteoporosis etc can find an ideal natural calcium supplement in curry leaves. Fresh juice of curry leaves, with lime juice and sugar, is an effective medicine in the treatment of morning sickness, nausea and vomiting due to indigestion. One or two teaspoons of juice of these leaves mixed with a teaspoon of lime juice may be taken in these conditions. The curry leaves, ground to a fine paste and mixed with buttermilk, can also be taken on an empty stomach with beneficial results in case of stomach upsets. Also used as laxative. Boils and similar eruptions appear on skin during summer. Most of the boils tend to subside over time, but some may persist and remain painful. Curry leaves come handy in treating such conditions. A paste made of curry leaves is applied on these persistent boils for quick relief. Along with mint leaves and coriander leaves, curry leaves can be used in treating excessive burn conditions. Curry leaves can be used with effective result to treat burn, bruises and skin eruption. Cataract development can be prevented by using fresh juice of curry leaves. Kidney pain can be cured by using juice of root of *Murraya koenigii*. It can be used in preventing premature greying of hair[1].

PHARMACOLOGICAL AND MEDICINAL USES

Curry leaves are rich in many minerals and trace minerals such as Iron, zinc and copper. Therefore, researchers recommended that people with diabetes may benefit from the addition of curry leaves in the diet. Minerals found in curry leaf extract are important for maintaining normoglycemia, or the normal glucose content of the blood. This is done by the activation of pancreatic beta cells, which are responsible for the creation of insulin. While the nutrients in curry account for only about 1 to 2 percent of the required daily intake for these elements, they are bioavailable, or readily usable by the body. Therefore, the researchers suggested that curry leaves may be useful for the management of diabetes[1].

A scrutiny of literature reveals some notable pharmacological activities of the plant such as activity on heart, anti diabetic and cholesterol reducing property, antimicrobial activity, antiulcer activity, antioxidative property, cytotoxic activity, anti diarrhea activity, phagocytic activity[25]. It also revealed hepato–protective activity against ethanol–induced hepatotoxicity. Chronic ethanol consumption diminishes the cellular antioxidant levels through free radical induced injury causing hepatitis and cirrhosis with mortality in severe case[26].

Antioxidant Properties

Carbazole alkaloids isolated form curry leaf namely, euchrestine B, bismurraya–foline E, mahanine, mahanimbicine, mahanimbine etc. exhibited antioxidative properties[27, 28].

Antimicrobial Properties

Curry leaf has been shown to possess more antibacterial activity and less antifungal activity, due to compounds such as isomahanine and murrayanol[29]. Alkaloids like 1–formyl–3–mehtoxy–6–methyl carbazole and 6, 7–dimethoxy–1–hydroxy–3–methyl carbazole from the leaves was found to be active against Gram–positive and Gram–negative bacteria and fungi[30]. Literature survey revealed that methanolic extract of 21 plant species were screened for *in vitro* anti bacterial activity against multi resistant bacteria isolates including Gram positive and Gram negative strains. Study showed that *M. koenigii* shown maximum antibacterial activity[31].

Antidiabetic Properties

It is reported that curry leaves slow down the starch to glucose breakdown in diabetic patients. It has been recommended to promote them as preferable food adjuvants for diabetic patients[32]. The methanol extract of curry leaf was more effective in lowering the blood glucose levels compared with the aqueous extract[33]. There are a large number of reports on the use of curry leaf for diabetes.

Pesticidal Properties

The essential oil of curry leaf is composed of many chemicals with a multispectral activity. The monoterpene hydrocarbons and sesquiterpene hydrocarbons play a major role in this activity. Its leaves also play a major role in antifungal activity against pathogenic fungi. The essential oils had shown fungitoxity of more than 50% at 2000 ppm concentration. It has great potential for use in mosquito control programmes [34].

Role in Preventing Alzheimer's Disease

Alzheimer's disease (AD) is characterized by signs of significant oxidative stress and associated loss of cholinergic cells. In studies conducted by Mani *et al.*[35] the total alkaloidal extracts of *Murrya koenigii* leaves in doses of 20 and 40 mg/kg p.o., improved the values of protective antioxidants like glutathione peroxidase, reduced glutathione, glutathione reductase, superoxide dismutase and catalase in brain homogenate. Besides reducing brain cholinesterase activity, *Murrya koenigii* extracts significantly improved cognitive functions as evidenced by the significant increase in the memory scores[35].

Cytotoxic Activity

Carbazole alkaloids isolated from the stems of *M. koenigii* have effects on the growth of the human leukemia cell line HL–60. Also the carbazole alkaloids, mahanine, Pyrafoline–D and murrafoline–I showed significant cytotoxicity against HL–60 cells and induced the loss of mitochondrial membrane potential[36].

Vasodilating Activity

Crude aqueous leaf extract of *M. koenigii* was prepared which showed a dose dependent negative chrono tropic effect on cardiovascular system of heart preparations which might be due to its direct actions on the heart and blood vessels. The aqueous leaf extract possesses vasodilatory effect which is independent of muscarinic, histaminergic and β-adrenergic receptor as it increased the number of drops/minute in hind limb perfusion experiment and also does not possess β-adrenergic receptor antagonistic activity[37].

Hypocholesterolemic Activity

Hypocholesterolemic activity was checked, which was done by using crude ethanol extract of plant leaves of *M. koenigii*. The experiment was confirmed by observing a decrease in cholesterol level in dose dependent manner. The dose of 500 mg/kg was found more efficient than the 300 mg/kg and was comparable with the standard cholesterol reducing agent, Simvastatin. Carbazole alkaloids a major photochemical constituent of plant found to have various biological activities antioxidant, antidiabetic, antimicrobial, lipid lowering etc[38].

Antiulcer Activity

Crude aqueous extract of leaves showed antiulcer activity which was evaluated by using models of acute gastric lesions induced by ethanol induced, aspirin induced, cold restrain stress and pylorus ligation. Doses of 200 mg/kg and 400 mg/kg of aqueous extract showed efficient reduction in lesion index, total affected area and percentage of lesion in comparison with control group in the ethanol induced, aspirin induced, cold restrain stress induced ulcer and pylorus ligation models. These observations provide a confirmation about aqueous extract of leaves of *M. koenigii* can act as good antiulcer drug[39].

Anti Diarrheal Activity

The bioassay guided fractionation of the *n*-hexane extract of the seeds of *M. koenigii* resulted in the isolation of three pure compounds of bioactive carbazole alkaloids, kurryam, koenimbine and koenine. Of the three compounds kurryam and koenimbine exhibited significant inhibitory activity against castor oil-induced diarrhea. The compounds also produced a significant reduction in gastro-intestinal motility in the charcoal meal test[40].

Radioprotective and Chemoprotective Activity

The study was aimed to investigate the radio protective and chemo protective effect of *M. koenigii* methanolic extract after irradiation. Gama Radiation of 4Gy produced a significant increase in the percent aberrant metaphases and different types of aberrations compared to same treated control. Radiation increased all types of aberrations like fragment chromatid and chromosome breaks, rings and dicentrics. Quantitatively, methanolic extract pretreated group showed similar aberrations only in one group. But treatment with methanolic extract of *M. koenigii* before radiation significantly reduced percentage aberrant metaphases and different types of aberrations. Thus, the study on leaf extract clearly revealed that a single dose of 100 mg/kg methanolic extract of *M. koenigii* can significantly decrease the chromosomal damage caused by irradiation and cyclophosphamide and also showed bone marrow protection[41].

Antiamnesic Activity

Transfer latency (TL) was measured using elevated plus model. Standard cholinergic agent, Piracetam 400 mg/kg was comparable to the petroleum ether extract of plant leaves 300 and 500 mg/kg in improving the learning and memory of aged mice and was reversed from the effect of scopolamine. A 15 days pre–treatment with petroleum ether extract of leaves 300 and 500 mg/kg reversed the effect of sodium nitrite which was comparable with standard Piracetam 400 mg/kg. Thus *M. koenigii* leaves extract could be said to improve the learning capabilities of the aged mice in hypoxic condition as indicated by a better performance of animals in the learning task. The results of cholinesterase assay showed that 15 days treatments with petroleum ether 300 and 500 mg/kg extract of *M. koenigii* leaves remarkable reduce the brain cholinesterase activity compared with those of their control groups. The standard, Doneprezil 0.5mg/kg, reduced more cholinesterase activity[38].

Antihelminthic Activity

The antihelminthic activity was performed on adult Indian earth worm 'Pheretima posithuma'. The organism was chosen due to its resemblance with the intestine round worm parasites of human beings. Petroleum ether and alcoholic extracts were selected for the activity and Piperazine citrate as standard. Ether and methanolic extracts were taken at 25 mg/ml, 50 mg/ml and 100 mg/ml concentrations. Worms were kept under observation to notice the time taken to paralysis and death in individual worms. Paralysis was said to occur when the worms do not revive even in normal saline. Death was concluded when the worms lost their motility followed by the fading of body color. Among them alcoholic extract of plant leaves at a dose of 100 mg/ml had significant antihelminthic activity where as petroleum ether showed moderate activity[42].

Memory Enhancing

It was observed that ethanolic extract of leaves lowered serum cholesterol inhibited brain acetylcholinesterase enzyme and thereby elevated the acetylcholine concentration in brain homogenate and ultimately improved memory. Extract was used in two different concentrations, *viz.* 300 mg/ml and 400 mg/ml. Thus, a combination of anticholinesterase and cholesterol lowering effect exhibited by leaves extract may be the factors responsible for this memory improving effect observed in the study[38].

Wound Healing Effect

Male albino rates were used to check the wound healing activity by screening with ethanolic extract of leaves of *M. koenigii*. In the excision, wound healing model reveals that three groups which were taken for wound healing activity showed a decrease in wound area from day to day. Incision model showed a significant increase in tensile strength of the 12–day old wound due to treatment with *M. koenigii*. Thus, the leaves of *Murraya koenigii* were proved to possess significant wound healing capacity[43].

Skin pigmenting

The formulation of cream of essential oil of leaf of *M. koenigii* was found to have sun protection factor. It was postulated that cream parameters complied as per official acceptance criteria's but the SPF. Sun pigmenting factor for curry leaf oil cream formulation showed minimum sun protection activity for sunlight and erythema. The cream was found useful in maintaining the natural skin pigmentation or it can be used as additives in other formulations to enhance the activity[26].

OTHER USES

The major constituents of the oil are ciscaryophyllene (11.74%), dispentene (11.3%), α-eudesmol (9.61%) isocaryoyphyllene (8.41%) and β-elemene (7.09%). The composition of the oil suggests that it many find application as a fixative in soap and detergent perfumes[30].

Plant Growth Inhibitors

The growth–inhibitory properties are comparable to those of other coumarins, such as psoralen and xanthotoxin[44].

Synthesis of Silver Nano Particles

Biological synthesis of silver nanoparticles using *Murraya koenigii* leaf extract was investigated by Laura Christensen *et al*[45]. Effect of broth concentration in reducution mechanism has been reported upon. In their study the rapid reduction of silver (Ag^+) ions was monitored using UV–visible spectrophotometry and it shows the formation of silver nanoparticles within 15 minutes. Transmission electron microscopy (TEM) and atomic force microscopy (AFM) analysis showed that the synthesized silver nanoparticles are varied from 10–25 nm and have the spherical shape. Further the XRD analysis confirms the nanocrystalline phase of silver with FCC crystal structure. From this study, it was found that the increasing curry leaves broth concentration increases the rate of reduction and decreases the particle size.

VII. CONCLUSION

M. koenigii (Linn), commonly known as the curry leaf plant, is highly valued for its characteristic aroma and medicinal properties. Its leaves are used extensively for culinary purposes, especially in curries and chutney. Curry leaf plant is a very easy source of spice to enhance the palatability of the food and is also a good source of major nutrients and some valuable volatile oils. The various notable pharmacological activities of the plant such as activity on heart, anti diabetic and cholesterol reducing property, antimicrobial activity, antiulcer activity, antioxidative property, cytotoxic activity, anti diarrhea activity, phagocytic activity. The chemical composition of the fresh leaves of *Murraya koenigii* consists of volatile oil. The major components in curry leaf are α-pinene, β-caryophyllene, (E)-β-ocimene, linalool and β-phellandrene. *M. koenigii* is a rich source of carbazole alkaloids. Its leaves, roots and bark are a tonic, stomachic and carminative. It is shown to possess a hypo cholesterol effect and many other health benefits. The plant can be cultivated in each home garden as well as commercially like any other crop 10-12 years after planting, we can go on getting good income from the land. Curry leaf trees can also be planted as an inter crop among other seasonal or perennial crops. Being a perennial crop the yearly expenditures is low. Various products has been developed from curry leaves powders like cookies, bread, dhal mixes, puree, etc. Dehydrated curry leaves also available in the markets. The dehydrated curry leaves can be successfully incorporated in various products which are beneficial to health. The incorporation of levels of curry leaves can increase the value of iron, calcium and β-carotene in all the products. Along with the medicinal properties it has a great scope for value addition in various forms.

VIII. REFERENCES

[1] Singh S.K., More, Mohan S.M., "Curry leaves (*Murraya koenigii* Linn. Sprengal) – a miracle plant", *Ind J Sci.*, 4(1), 46–52 (2014).

[2] Khatoon J.A., Verma N., Chacko., Sheikh S., "Utilization of dehydrated curry leaves in different food products", 2(4), 508–511 (2011).

[3] SPICE Board of India. www.indianspices.com

[4] Kumar K.J.L., Dassharma K., Mohandas A., "Curry leaf –an inevitable spice of Indian cuisine", Spice India, pp. 8–9 (2003).

[5] Khanum K.R., Kumar A., Sudarshana K.K.R., Viswanathan K.R., "Anticarcinogenic effects of curry leaves in dimethylhydrazine –treated rats", *Plant Foods Hum Nutr.*, 55, 347–355 (2000).

[6] Walde G.S., Joythirmay T., Rao P.G.P., Shivaswamy R., Srinivas P., "Flavour volatiles of leaves, fruits and seed cotyledons of *Murraya koenigii* L", *Flav Frag J.,* 20, 169–72 (2005).

[7] Bisht A., Negi J., "Chemical composition of leaf essential oil of *Murraya konegii* (Linn.) Spreng", *Guru Drone J Pharm Res.*, 2(3), 1–6 (2014).

[8] Mishra G., Ajay, Rahul, Sumit, Paras, "Comprehensive review: *Murraya koenigii* Linn", *Asian J Pharma Life Sci.*, 1(4), 2231 – 4423 (2011).

[9] Rain V.K., Lal R.K., Tripathi S., Khan M., Syamasundar K.V., Srivastava S.K., "Essential oil composition of genetically diverse stocks of *Murraya koenigii* from India", *Flavour Fragr J.,* 17, 404 (2002).

[10] Curry Leaf – Agri Academy, www.agriacademy.info/curry–leaf.html

[11] The Hindu, "A look at the science of farming", 29th Nov (2004).

[12] Naik R., Annamalai S.J.K., Dawn A.C.P., "Development of power operated curry

leaf stripper", *J Agri mechan Asia, Africa and Latin America*, 44(1), 69–72 (2013).

[13] Prakash V., Natarajan C.P., "Studies on curry leaf (*Murraya koenigii* L.)", *J Food Sci & Technol.*, 11, 284–286 (1974).

[14] Omanakutty A.M., Rajaraman K., Sankarikutty B., Sumathikutty M.A., "Processing of curry leaves", *Ind Fd Pack.*, 38(4), 32–36 (1984).

[15] Dawn A.P., Naik R., "Investigation on extending the shelf life of fresh curry leaf (curry patha)", *Fd pack com.*, 1(9), 16–18 (2007).

[16] Vennila P., "Processing of curry leaves powder", agritech.tnau.ac.in/postharvest/pdf/ curry_leaves_powder.pdf (2014).

[17] Palaniswamy U.R., Stuart J.D., Caporuscio C.A., "Effect of storage temperature on the nutritional value of curry leaf", *Trends in New Crops and New Uses*, pp. 567–569 (2002).

[18] Halevy H., Mayak S., "Senescence and post harvest physiology of cut flowers", Part II, *Hort Rev.*, pp. 59–143 (1981).

[19] Palaniswamy U.R., "Asian horticultural crops and human dietetics", *Hort Tech.*, 11, 504–509 (2001).

[20] Iskandar Z., Dikui M.B., "Extraction of essential oil from *Murraya koenigii* leaves using ultrasonic–assisted solvent extraction method" Faculty of chemical engineering and natural resources. University Malaysia Pahang (2009).

[21] Blasco M., Domeno C., Lopez P., Nerin C., "Behaviour of different lichen species as biomonitors of air pollution by PAHs in natural ecosystems", *J Environ.*, 13, 2588–2596 (2011).

[22] Khoddami A., Wilkes M.A., Roberts T.H., "Techniques for analysis of plant phenolic compounds", *Molecules.*, 18, 2328–2375 (2013).

[23] Sharon M.E.M., Abirami C.V.K., Alagusundaram K., "Modelling thermal degradation kinetics of colour in curry leaf paste", *Asian J Sci Techn.*, 5(3), 272–275 (2014).

[24] Shanthala M., Prakash J., "Acceptability of curry leaf (*Murraya koenigii*) incorporated products and attitude toward consumption", Department of Studies in Food Science and Nutrition, University of Mysore, Manasagangotri (2005).

[25] Syam S., Abdul A.B., Sukari M.A., Mohan S., Abdelwahab S.I., Wah T.S., "The growth suppressing effects of girinimbine on HepG2 involve induction of apoptosis and cell cycle arrest", *Molecules.*, 16(8), 7155–70 (2011).

[26] Patil R.A., Padmaja M.L., Pramod B.D., Hiray A.Y., "Antinociceptive activity of acute and chronic administration of *Murraya koenigii* L. leaves in experimental animal models", *Ind J Pharma.*, 44(1), 15–19 (2012).

[27] Nakatani N., "Phenolic antioxidants from herbs and spices", *Biofactors,* 13(1–4), 141–146 (2000).

[28] Tachibana Y., Kikuzaki H., Lajis N., Nakatani N., "Antioxidative activity of carbazoles from *Murraya koenigii* Leaves", *J Agri Fd Chem.*, 49(11), 5589–5594 (2001).

[29] Aqil F., Ahmed I., "Broad–spectrum antibacterial and antifungal properties of certain traditionally used Indian medicinal plants", *World J micro biotech.,* 19(6), 653–657 (2003).

[30] Saini C.S., Reddy G.B.S., "*Murraya koenigii*", *IOSR J Pharma Biol Sci.*, 7(6), 15–18 (2013).

[31] Panghal M., Kaushal V., Jaya P., "*In vitro* antimicrobial activity of ten medicinal plants against clinical isolates of oral cancer cases", *Annals of Clinical Micro Antimicrobials.*, 10, 21 (2011).

[32] Grover J.K., Yadav S., Vats V., "Medicinal plants of India with anti–diabetic potential", *J Ethano pharma.,* 81, 81–100 (2002).

[33] Vinuthan M.K., Girish V., Kumara M., Narayanaswamya, Veena T., "Lipid lowering effect of aqueous leaves extract of *Murraya koenigii* (curry leaf) on alloxan induced male diabetic rats", *Pharmacognosy Magazine*, 3(10), 112–115 (2007).

[34] Ray P., Srivastava S., "Curry Leaf (*Murraya koenigii*) the aromatic biopesticide", *J Interacademicia.*, 10, 231–235 (2006).

[35] Mani V., Ramasamy K., Ahmad A., Parle M., Shah S.A., Majeed A.B., "Protective effects of total alkaloidal extract from *Murraya koenigii* leaves on experimentally induced dementia", *Food Chem Toxicol.*, 50, 1036–1044 (2012).

[36] Ito C., Itoigawa M., Nakao K., Murata T., Tsuboi M., Kaneda N., Furukawa H., "Induction of apoptosis by carbazole alkaloids isolated from *Murraya koenigii*", *Phytomedicine*, 13, 359–365 (2006).

[37] Kadam S.H., Dombe S., Naikwadi P., Patil M., "Cardiovascular effects of aqueous extract of *Murraya koenigii* on isolated perfused frog heart preparation", *J Pharma Res.*, 42, 462–463 (2011).

[38] Tembhurne S.V., Sakarkar D.M., "Beneficial effects of ethanolic extract of *Murraya koenigii* Linn. Leaves in cognitive deficit aged mice involving possible anticholinesterase and cholesterol lowering mechanism", *Int J pharmtech Res.,* 21, 181–188 (2010).

[39] Sharma P., Gali V., Bhandari A., Singh S., Ghule S., Agrawal S.,"Antiulcer activity of leaves extract of *Murraya koenigii* in experimentally induced ulcer in rats", *Pharmacology online,* 2, 818–824 (2011).

[40] Mandal S., Nayak A., Kar M., Banerjee S.K., Das A., Upadhyay S.N., Singh R.K., Banerji A., Banerji J., "Antidiarrhoeal activity of carbazole alkaloids from *Murraya koenigii* Spreng rutaceae seeds", *Fitoterapia.*, 81, 72–74 (2010).

[41] Goswami R.B., Khare P., Singh S., Goswami N., Thomas P., Devi P.U., "Studies on antigenotoxic of *Murraya koenigii", Inte J Pharma Recent Res.*, 22, 65– 68 (2010).

[42] Khuntia T.K., Panda D.S., "Evaluation of antibacterial, antifungal and anthelmintic activity of *Murraya koenigii* Spreng", *An Intern J Pharma Sci.*, 22,105–110 (2011).

[43] Anand T., Kalaiselvan A., Gokulakrishnan K., "Wound healing activity of *Murraya koenigii* in male albino rats", *Intern J Curr Res.*, 32, 425–427 (2011).

[44] Ghasemzadeh A., Jaafar H.Z.E., Rahmat A., Devarajan T., "Evaluation of bioactive compounds, pharmaceutical quality and anticancer activity of curry leaf (*Murraya koenigii* L.)", *Evidence–Based Complementary and Alternative Medicine*, 8 (2014). Article ID 873803, http://dx.doi.org/10.1155/2014/873803

[45] Christensen L., Vivekanandhan S., Misra M., Mohanty A.K., "Biosynthesis of silver nanoparticles using *Murraya koenigii* (curry leaf): An investigation on the effect of broth concentration in reduction mechanism and particle size", *Advanced Materials Letters*, 2(6), 429–434 (2011).

—

18

Recent Advances in Mushrooms Processing

G. C. Wakchaure[1]*, Kamlesh K. Meena[1] and P. Suresh Kumar[1]

ABSTRACT

Mushrooms being delicacy edible fleshy fungi, a rich source of good quality vegetative proteins, essential amino acids, minerals and vitamins with low calories neutraceutical food produced form agro–forest, industrial and household organic / inorganic wastes. Out of 20 genera of mushrooms, very few viz., button (Agaricus bisporus), oyster (Pleurotus spp.), milky (Calocybe indica), shiitake (Lentinula edodes) and paddy straw mushroom (Volvariella volvacea) are cultivated across the world for commercial mushroom farming. As the mushrooms are highly perishable (> 90% moisture content) causes it an extremely short life. Hence, the research should be taken to solve their processing problems to get stable products is an important. Also the study on appropriate postharvest management practices to extent the shelf life of each type of mushroom should be considered for priority in the production and marketing chain. There are many short term and long term post–harvest and preservation approaches like suitable harvest index, washing and grading, post–harvest handling, pre–cooling, packaging and packaging materials, modified atmosphere packaging (MAP), controlled atmosphere packaging (CAP), steeping and canning preservation, drying / dehydration technologies are well explained in this chapter. All these approaches help to maintain mushroom quality and extend the shelf life through development of various medicinal and value added products. Also to boost India's the mushroom production and marketing of preserved mushroom in international market; strength, weaknesses, opportunities and threats (SWOT) analysis of mushroom industries, social awareness and financial help from the various private / government agencies is need of hour.

[1] National Institute of Abiotic Stress Management (NIASM), Baramati, Pune–413115, Maharashtra, India.

**Corresponding author* E-mail: goraksha.wakchaure@gmail.com

***Key words*:** Processing techniques, Quality, Mushroom, White button, Oyster mushroom, Milky, Paddy straw, Drying, Products, Medicinal, Industries

---

I. INTRODUCTION

Mushrooms as an important horticultural cash crop and its cultivation are one of the largest money turning enterprises in the world. Mushrooms being nutraceutical food are potent and vital as it produced from recyclable agro–wastes/agro–byproducts and the requirement of land is not pre–requisite in its cultivation, as it is grown indoors in protected houses with environment control with intensive space utilization. Indoor commercial cultivation of mushrooms utilizes the vertical space and is regarded as the highest protein producer per unit area and time i.e. almost 100 times more than the conventional agriculture and animal husbandry. This hi–tech horticulture venture has a promising scope to meet the food shortages without undue pressure on land[1]. Mushrooms, basically a form of fleshy edible fungi, are a rich source of good quality proteins, having most of the essential amino acids, minerals and vitamins with low calories.

Mushrooms have been shown to produce several biologically active compounds that are usually associated with cell wall, and these have been suggested to contribute to enhancement of immunity and tumor–retarding effects. Among the local communities, mushrooms may represent potential sources of antibacterial drugs, since in the early days, screening for antibiotics started with mushrooms and proved to be successful. Sixty antimicrobial compounds have been isolated from mushrooms; however, only the compounds from microscopic fungi have been present in the market as antibiotics until now.

Almost 20 genera of mushrooms are being cultivated worldwide, but only few types, *viz.,* white button mushroom (*Agaricus bisporus*), oyster mushroom (*Pleurotus* spp.), shiitake mushroom (*Lentinula edodes*), milky mushroom (*Calocybe indica*) and paddy straw mushroom (*Volvariella volvacea*) are grown commercially across the world. Also as per FAO database mushrooms have been classified as *FAO Stat* code 0449 and have been defined as those inter alia: *Boletus edulis; Agaricus campestris, Morchella spp. and Tuber magatum* for marketing purpose. The current global mushroom production is about 3.5 million tons and China, USA, Netherlands, Poland, Spain, France, Italy, Ireland, Canada and UK are the leading mushroom producers. Out of the total (1.25 lakhs tons) India's mushroom production, white button mushroom contributes majorly about 85%, oyster mushroom 10% and remaining 5% share coming from the other mushrooms. Punjab, Uttarakhand, Haryana and Himachal Pradesh are the leading mushroom producers. The mushrooms cultivation in general and button mushroom in particulars is highly scientific and engineering activities requires highly technical knowledge and environmental facilities right from the spawn, compost and casing soil preparation, mushroom fruit body formation to post–harvest processing of mushrooms. It becomes a money–spinning job among most unemployed youths, housewives and small and marginal farmers. The conversion of spent mushroom substrate, a byproduct of mushroom industries into the organic fertilizers, briquettes for fuels and many other useful products is also one of the important agro–processing operations.

In India, mushroom production is done round the year in an environmentally controlled units and seasonal growers grows the mushroom in bamboo and straw huts during the winters season and the supply at the local market exceeds causing less profit due to fall in price and spoilage. The increased in mushroom production demands proper post–harvest infrastructure to increase shelf life and marketability. Mushrooms are highly perishable vegetable commodity and get spoiled after harvesting of matured fruit bodies due to veil opening, browning, wilting, liquefaction, microbial activities, loss of texture, aroma, flavor, etc, making it unsalable[2]. Most of the mushrooms, being high in moisture content 87–93% and delicate in texture cannot be stored for more than 24 hours at the ambient conditions prevailing in the tropics[3]. The shelf life of freshly harvested mushrooms varies from few hours to 4 days depending upon the types of mushroom e.g. shelf life of freshly harvested paddy straw, button, oyster and milky mushrooms are noted as 4–6 h, 24 h, 1–2 days and 2–4 days, respectively at ambient temperature of 25–37°C. Enzymatic browning is believed to be the primary cause of spoilage in fresh mushroom. Later, it was suggested that mushroom spoilage might be caused by the action of bacteria on the mushroom tissue and browning of mushrooms was due to a combination of auto–enzymatic and microbial action on the tissue[4,5,6]. To overcome these problems, especially during peak season, suitable post–harvest management/practices are to be followed to increase the shelf life and marketability of mushrooms. Since sound post–harvest practices have been developed to extend the shelf life of the fresh mushrooms and preserve mushroom for long terms.

II. PROCESSING TECHNIQUES TO MAINTAIN MUSHROOMS QUALITY

As far as mushroom processing technologies are concerned, it can be classified as short term (< 6 months) and long term (6–12 months) techniques depending on the storage period of mushrooms (Fig. 1). These processing techniques could be applied for all types mushrooms with modification in processing operations. The sun drying of mushrooms is one of the simplest and oldest methods followed by the growers from the time immemorial. Due to the difficulties in drying of some of the mushrooms, new preservation technologies like canning, steeping, pickling, individual quick freezing, modified atmospheric packaging, mechanical and chemical drying (freeze drying, fluidized bed drying, batch type cabinet drying and osmotic drying), irradiation treatment, medicinal value added products of mushrooms have been developed to improve the shelf life and consumption of mushrooms[7,8].

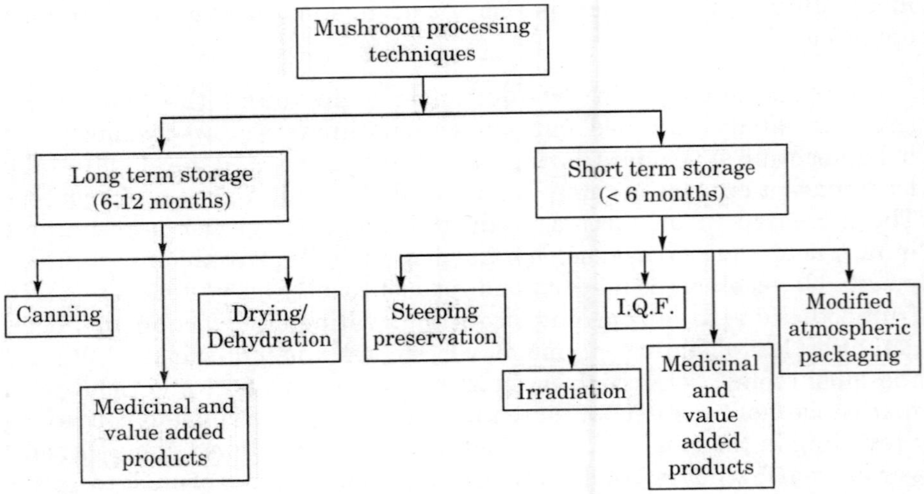

Fig. 1: Mushroom processing technologies

In recent years, rising demand of quality fresh vegetables including mushrooms has been reflected in the prices and marketability of the produce. Indian mushrooms market is largely the contribution of small and marginal farmers with limited resources who are dependent on local market for the sale of their produce. The rate of respiration of the harvested mushrooms varies largely with temperature in comparison to the other horticultural crops and this result in a shorter post–harvest life[9]. Many short–term and long–term storage measures are followed to retard the deterioration in quality at the level of mushroom grower till it reaches the consumer table. These post–harvest measures are almost same for all types of mushrooms (e.g. button, oyster, paddy straw and milky mushroom) with minute variation in processing operation depending upon types of mushrooms. By following proper post–harvest measures, the shelf life and quality of harvested mushrooms can be extended. The general post–harvest practices followed for mushrooms processing is given in Fig. 2.

Mushrooms are rich in protein, dietary fiber, vitamins and minerals. Chemical compositions of different mushrooms are given in Table 1[92]. In addition to these, mushroom is an excellent source of vitamins especially C and B (Folic acid, Thiamine, Riboflavin and Niacin). Minerals viz., potassium,

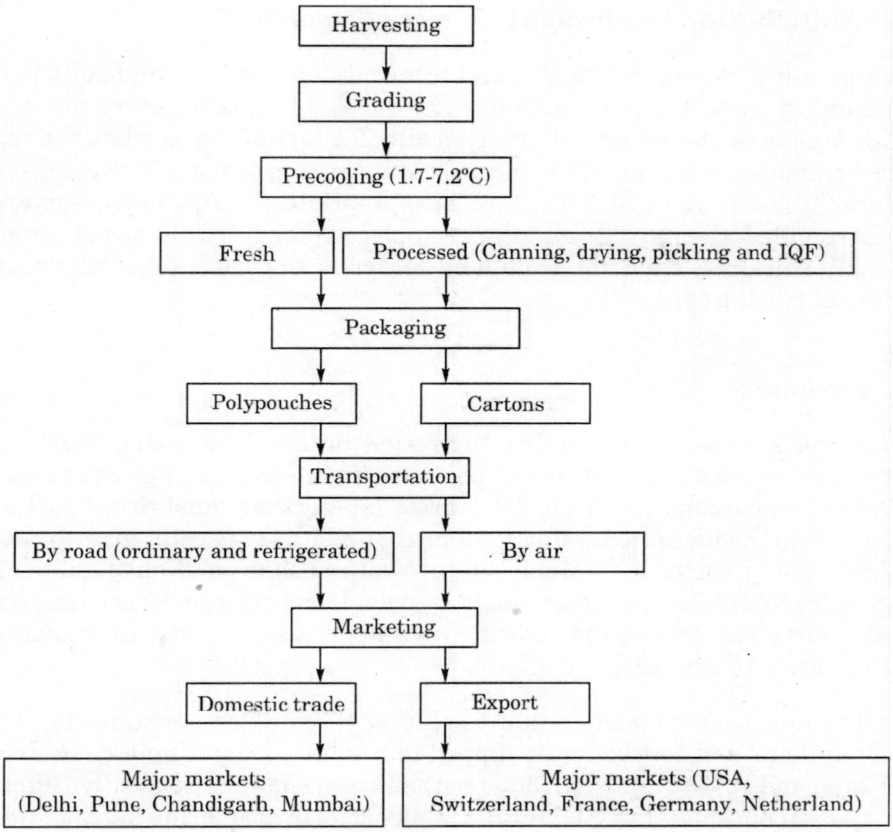

Fig. 2: Mushroom post–harvest cultivation practices

sodium and phosphorous are higher in fruit bodies of the mushroom. It also contains other essential minerals (Cu, Zn, Mg) in traces but deficient in iron and calcium.

Table 1: Chemical compositions of different mushrooms (dry weight basis g/100 g)

Mushroom	***Carbohydrate***	***Fibre***	***Protein***	***Fat***	***Ash***	***Energy cal***
Agaricus bisporus	46.17	20.90	33.48	3.10	5.70	499
Pleurotus sajor–caju	63.40	48.60	19.23	2.70	6.32	412
Lentinula edodes	47.60	28.80	32.93	3.73	5.20	387
Pleurotus ostreatus	57.60	8.70	30.40	2.20	9.80	265
Volvariella volvacea	54.80	5.50	37.50	2.60	1.10	305
Calocybe indica	64.26	3.40	17.69	4.10	7.43	391
Flammulina velutipes	73.10	3.70	17.60	1.90	7.40	378
Auricularia auricula	82.80	19.80	4.20	8.30	4.70	351

A. White Button Mushroom (*Agaricus Bisporus*)

White button dominates Indian and International markets and sufficient amount of work has been done to reduce losses in quality of mushroom. This mushroom is generally harvested after 21 days of casing when the cap size diameter is about 30–45 mm[10]. Button mushrooms harvested by maturity and not by size of the caps. At appropriate maturity stage caps are well rounded, the partial veil intact completely, stipe should have a small length: thickness ratio and should be sufficient to permit some trimming without cutting caps.

1. *Washing*

Washing is optional for freshly harvested button mushroom. Washing becomes necessary to remove soil particles if non–peat casing soil is used. Washed mushrooms generally deteriorate rapidly than mushrooms packed dry, due to the increased water content that results in greater growth rate of spoilage by bacteria[11]. Mushroom growers wash in solution of reducing agents to retard the browning caused by polyphenol oxidase. Hence, various anti–microbial as well as reductant compounds are used in washing mushrooms to extend the shelf life.

The mushroom dipped in dilute solution of hydrogen peroxide (1:3) for half an hour and consequently dipped in a solution containing citric acid (0.25%) and sulphur dioxide (500 ppm) mixture showed the positive effect on mushroom shelf life[12] Oxine, a stabilized form of chlorine dioxide, was very effective in controlling bacterial growth and colour deterioration when used at a level of 50 ppm or higher with a two minute or longer wash period at 12°C[13]. The use of sodium hypochlorite (100 ppm) and calcium chloride (0.55%) with oxine (100 ppm) resulted in increased antibacterial effectiveness. Use of calcium chloride and oxine also resulted in lower cap opening and firmer mushrooms during storage. Washing fresh mushrooms in water containing sodium sulfite solutions results in lower bacterial numbers and an improved initial appearance, but more rapid bacterial growth and browning occurred during subsequent storage compared to unwashed controls[14]. Mushrooms washed in hard water (150 ppm calcium carbonate) reduced bacterial growth and there was less colour deterioration during storage. Washing mushrooms in a solution consisting of oxine (50 ppm), sodium erythorbate (0.1%), and calcium chloride (0.5%) resulted in significantly lower bacterial populations and less colour deterioration during the storage later. An Indian mushroom grower commonly utilizes 0.05% potassium metabisulphite for washing of mushrooms, improves the initial whiteness, which lasted longer during the storage as shown in Fig. 3[15]. Even though many farmers are adapting this approach of washing, but selling clean unwashed properly packed mushroom may be better option as available in International market, as many people prefer mushroom not just because of health benefit, but also considered it a more chemical free food.

Fig. 3: Washing with 0.05% sodium metabisulphite

2. *Grading*

Mushroom grading is very important for marketing point of view. In general, the grading means segregation of mushrooms into various grade standards as per market demand (Fig. 4). The various grades standards including USDA grade standards[16] are used in across the world market based on cap and step size, color, disease spots and free from damage caused by insects, bruising, discoloration, feathering or other means. Usually based on the cap diameter mushroom is graded into grade A (large 45 mm), B (medium 32–45 mm) and C (small 19–32 mm). In India for dried edible mushrooms, the Directorate of Marketing & Inspection (D.M.I.) had formulated the various grade standards as Mushrooms Grading & Marking Rules, 1972 under the Agricultural Produce (Grading & Marking) Act, 1937.

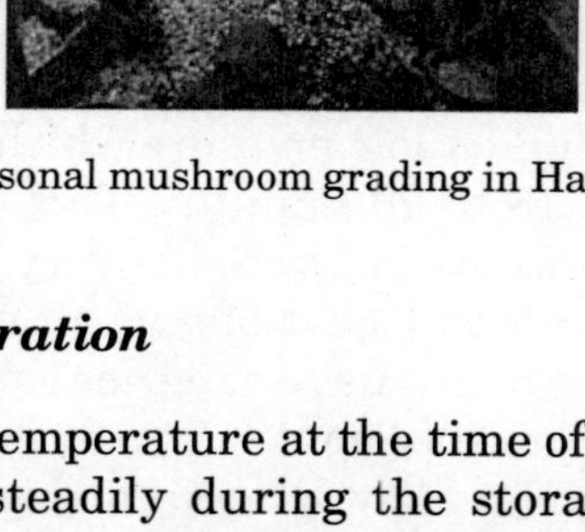

Fig. 4: Seasonal mushroom grading in Haryana, India

3. *Pre-cooling/refrigeration*

The button mushroom temperature at the time of picking varies between 16–18°C, and it rises steadily during the storage due to atmospheric temperature and respiration. This heat generated causes deterioration in quality mushroom and spoilage. Hence, the heat should be removed immediately after the harvest by cooling mushroom as quickly as possible at temperature 4–5°C. The selection and choice of the cooling system depends upon the quantity to be handled, which may be a refrigerator for a small grower to a cold room with all facilities for a commercial grower. Using evaporative cooling, hydro–cooling, forced–chill air, ice bank or vacuum cooling systems, mushrooms can be cooled up to 2–4°C[17]. Air spray moist chillers can also cool the mushrooms rapidly. The temperature can be lowered by 16–18°C in an hour without any moisture loss.

Temperature of the mushrooms increases through respiration after picking at ambient temperature and the respiratory rate increases with the increase in the storage temperature. It has been estimated that mushrooms at a temperature of 10°C have 3.5 times higher respiratory capacity than those at temperature of 0°C, which necessitates immediate shifting of mushrooms to the refrigerated zone[18].

The pre–cooled mushrooms can be either used for fresh consumption or for further processing wherein the produce is kept in a food grade plastic bag and stored in cooling unit. The size and shape of the packs also play an important role in cooling room. Packs with more than 10 kg mushrooms or with 15 cm thick layers of mushroom causes spoilage problem. Vertical flowing of air is more suitable for cooling. The mushrooms should not be stored in the same cooler along with fruits as the gases produced by fruits e.g. ethylene cause discoloration of mushrooms. This forced–chill air–cooling system is time consuming and vacuum cooling is becoming more popular. To ensure high quality mushrooms in the market place with enhanced shelf–life, these must be cooled as quickly as possible after picking and kept cool throughout the cold chain[19].

Mushrooms can be held for 7–9 days upon rapid cooling and storage at 0–1 °C with 95% RH as optimum storage conditions. Storage at 2°C shortens storage–life to 3–5 days by accelerating surface browning, stipe elongation, and veil opening[20]. High RH is essential to prevent desiccation and loss of glossiness. Moisture loss is correlated with stipe blackening and veil opening.

3.1. *Vacuum cooling*

In vacuum cooling, water existing in cell walls and inter hyphal spaces of mushroom is evaporated under low pressure which lowers the temperature from ambient to 2°C in 15 to 20 minutes. This is a uniform and faster process, where mushrooms are subjected to very low pressure and water evaporates giving off the latent heat of vaporization, thus cooling itself. The vacuum–cooled mushrooms have superior colour (enzymic browning) than conventional–cooled mushrooms [21]. The freshly harvested mushrooms (whole/sliced) keep their quality longer throughout their shelf–life if immediately packed, vacuum–cooled and then transported at a low temperature. The major disadvantages of this system are the high capital cost; and an inevitable loss of fresh weight during the process of cooling. Filling and emptying the cooling chamber introduces another operation and expenses into the marketing chain.

3.2. *Ice–bank cooling*

With a view to reduce the weight loss during the conventional vacuum cooling, quick ice bank cooling of mushrooms is now in vogue in some Europeans countries wherein a stack of mushrooms is passed through forced draft of chilled but humidified air from the ice–bank[22].

4. *Mushroom packaging*

Packing is essential to protect, improves aesthetic look and help in advertisement of produce during marketing. The various packaging materials and packaging technologies *viz.*, modified atmosphere packaging, modified humidity packaging and controlled atmospheric storages are used for mushrooms[23,24].

4.1. *Modified atmosphere packaging (MAP)*

MAP helps in extending the shelf life and maintenance of quality of perishable produce *viz.*, mushroom by creation of appropriate gaseous atmosphere around the surroundings of produce packaged in plastic films[25,26]. The modified atmosphere is created in a sealed package of a fresh product by respiratory gas exchange, namely oxygen (O_2) intake and carbon dioxide (CO_2) evolution. When the rate of gas permeation through the packaging material equals respiratory gas exchange, equilibrium concentrations of O_2 and CO_2 are consequently established[27]. The equilibrium depends on temperature, respiration rate of specific product, product weight, O_2 and CO_2 permeabilities of the packaging material, free volume in the package and film area[28]. In this technique, the natural process of respiration of the produce in conjunction with the restricted gas exchange through a polymeric films *viz.*, low density polyethylene (LDPE), normal and oriented polypropylene (PP) is used to control the in–pack oxygen and carbon dioxide[29, 30].

Based on mode of modified atmosphere creation, mushroom MAP can be classified into two methods: active and passive packaging. In passive MAP, the product is just sealed in a polymeric package. Due to the respiration of the fresh product and permeation of gases into the package, the atmosphere is modified. In active MAP, air is flushed into the package initially, so that the steady state atmosphere is reached quickly after packaging. In passive MAP, it takes a long time to reach the steady state conditions within the package[31].

MAP of mushrooms has been shown successfully to delay senescence and maintain quality after harvest[32,33,34,35,36,37]. Shelf life of mushrooms can be increased by over wrapping them with PVC films[38]. Thickness of 100 gauge polythene bags with 0.5% venting area are recommended for packing mushroom in case of refrigerated storage[39]. For transporting mushroom to long distance, polystyrene or pulp–board punnets should be used instead of using polythene bags. The punnets are over–wrapped with differentially permeable poly vinyl chloride (PVC) or poly acetate films (Fig. 5). They create modified atmosphere in punnets producing an atmosphere of about 10% CO_2 and 2% O_2. The optimum atmosphere for storage of mushrooms is found to be 2.5–5% CO_2 and 5–10% O_2[15].

Fig. 5: MAP of mushroom in polythene bags and pulp board punnet over wrapped with PVC film

In MAP, the packaging material plays a vital role in modifying the inside atmosphere around the mushroom and the mushroom quality as well. Among the various packaging materials *viz.*, polyvinylidene chloride coated, oriented nylon, anti–fogging, wrap or vacuum packing film, the antifogging film maintained the quality of mushrooms for 24 days in modified atmosphere packing[40]. The best packaging material polyethylene extends the shelf life of fresh mushrooms to 15 days under MAP conditions[41].

In MAP, the shelf life of the mushroom can further be extended by supplementing some preservatives in addition to modifying the atmosphere inside the package. The various supplementary packaging materials viz., activated carbon, sorbitol, chitosan, potassium permanganate can be used in MAP for maintaining the quality of oyster/button mushrooms at ambient temperature in MAP[42]. In a study conducted in Finland about the washing and use of a humidity absorber (Silica gel) in the packages during the modified atmosphere and it was found that washing in chlorinated water and incorporation of dehumidifiers decreased the microbial contamination and greater shelf–life in *Agaricus bisporus*. The commercially available food grade moisture absorbers such as montmorillonite clay and silica gel used to extent the shelf–life of mushrooms in packs[43]. Sorbital maintained the best colour in mushroom when it was packed and stored along with fresh mushroom trays over wrapped with PVC films[44].

4.2. *Modified humidity packaging (MHP)*

In conventional packing, most polymeric films used have comparatively lower water vapour transmission rates relative to transpiration rates of highly perishable fresh produce like mushroom. This resulted in creation of nearly saturated conditions inside the packages. The high in package relative humidity (IPRH) can causes condensation of water vapour with in a package and it allow microbial growth. This may leads either increase or decrease the spoilage depending on the product, their transpiration coefficients and water potentials[45]. There are two possible approaches to obtain the desired IPRH *i.e.* perforation of the package which precludes the possibility of achieving modified atmosphere conditions withi2150*9 5n the package and use of in–package water absorbing compounds like calcium chloride which can maintain the required RH[46]. MAP in combination with MHP further improved the shelf–life of fresh mushrooms. An IRPH of 87–90% is desirable for best colour in mushrooms during storage[47].

4.3. *Controlled atmospheric storage (CAS)*

In CAS packaging, the O_2 and CO_2 concentrations are altered inside the package and because of that respiration rate gets altered. CAS packaging helps to reduce browning of fresh mushroom caused due to enzymatic action and extend the shelf life of mushroom. Effect of CA on shelf of mushroom showed that mushrooms stored in a 5% CO_2 atmosphere for 7 days did not break their veil but their texture was very soft and spongy. Texture losses decreased when CO_2 concentrations increased[35]. Mushrooms derive a moderate benefit from CA storage under 3 to 21% O_2 + 5 to 15% CO_2[48]. CA of 3% O_2 and 5–10% CO_2 extends storage–life to 12–15 days at 0°C[49] and reduces cap opening, browning and stipe elongation. Mushroom quality was maintained in 8% O_2 and 10% CO_2[50]. Storage under low O_2 and high CO_2 inhibits cap opening and internal browning, but causes yellowing of the cap surface. Levels of $O_2 < 1\%$ can favor growth of *Clostridium botulinum*, the development of off–odors and off–flavors, as well as cap opening and stipe elongation. For this reason, CA is not commonly used[51]. Even mushroom grown in controlled atmosphere (CA) with use NaCl in irrigation water improves mushroom quality, yield and colour of the canned mushroom product. Also mushrooms exposed to lower airflow rates in the growing room remain whiter than those exposed to higher airflow rates[52,53]. Addition of proprietary form of stabilized chlorine dioxide (oxine, 150–200 ppm) to pre–harvest watering applications on mushrooms had lower incidence of blotch and retained whiter colour during storage. The commercially available food grade moisture absorbers viz., montmorillonite clay and silica gel can be used to extent the shelf life of mushrooms in packaging[43]. Mushrooms treated with honey (0.5–1%) for 18 hours, air dried to remove surface moisture, packed in 100 gauge polythene pouches with 0.5% venting area, increased the shelf life by more than a week over control at 3–5°C and 2–3 days at room temperature[39].

Presently controlled atmosphere packaging (CAP) and modified atmosphere packaging (MAP) are catching up fast for all types of fruits and vegetables, in which the concentrations of CO_2 and O_2 are attained in the environment where the product package alters the respiration rate of the product. These packaging techniques have to be effectively used for packing mushrooms to have improved shelf life. Even where simple polythene bags are used, it is important to make desired number of holes for proper humidity control.

5. Mushroom packing, packaging materials and transportation

Usually good packaging sells a product; a mediocre package can interfere with sale of an otherwise excellent product. Improved and attractive packaging is an important but partially neglected area in mushrooms– it is still unprinted plain polypouches whereas attractive and labelled over–wrapped trays are in vogue in the developed countries. For local markets because of short transportation, mushroom generally packed in polythene bags of 200–400 g sizes. For long distance transport, pulp board punnets wrapped with PVC films should be used instead of polythene bags. This over–wrappings of punnets help in creating modified atmosphere in punnets with 10% CO_2 and 2% O_2 and mushrooms maintain their fresh look for 3 days at 18°C[54]. Sometimes, pre–cooled mushrooms are packed in insulated containers having ice in it, so that mushrooms remain fresh, healthy during long transport. Also sometimes the effect of pre-cooling is lost partially in post pre–cooling period. Hence, for long distant transport, speedy refrigerated transport system is essential even if costlier; in which CFB cartons contained with polythene pouches of mushrooms may be transported by refrigerated trucks. For short distance markets, pre–cooled fresh mushrooms are dispatched in wooden boxes with sufficient crushed ice in polypacks. For local retailer's markets in India, mushrooms are packed in polypouches and dispatched by cycle, cycle–rickshaws, auto–rickshaws etc[10].

The standard types of packaging boxes with their sizes and capacities[1,18]. used for mushrooms during transportation are shown in Table 2.

Table 2: Types of packaging boxes for mushroom

Material	*Size (mm)*	*Capacity (kg)*
Plastic punnets	130 × 130 × 72	0.40
Cardboard chips	305 × 125 × 118	1.82
Plastic tray	330 × 280 × 145	2.30
Expanded polystyrene	400 × 333 × 167	4.56

6. Individual quick freezing (I.Q.F.)

Individual quick freezing (I.Q.F.) is popular processing method followed in large mushroom units[55]. In this process, harvested mushrooms are washed at processing units after receipt from farms, and then inspected, sliced and graded according to quality. After water cooling of blanched mushrooms, it is subjected to tunnel for freezing stage. During this stage, these are cooled in a system having temperature of –40°C and the core areas of mushroom pieces acquire a temperature around –18°C. Consequently, packed in multi–layer polybags and stored in a cold storage having temperature of –20°C to –25°C[10].

7. *Steeping preservation*

This is one of the simple and more economical method[56] in which the mushrooms can be preserved for short period up to 6 months by steeping them in solution of salt or acids (Fig. 6). The common practice is that cleaned mushrooms are washed in water or chemical added water. Their after blanching of mushroom in brine solution for 5 minutes is generally done before filling them into plastic cans. Brine solution (10–20% salt) is then added into the cans or containers. Steeping of water blanched mushrooms in 1% potassium meta bisulphite (KMS) along with 2% citric acid (overnight), before drying which helps to improves the color, texture and reconstitution properties[57]. For steeping purpose as a complete mushroom, sliced into two halves, cap or stipe can be used as either separate or mixed form. Solution consisting of 2% sodium chloride, 2% citric acid, 2% sodium bicarbonate and 0.15% KMS is used for steeping preservation of blanched mushrooms for 8–10 days at 21–28°C[58]. Chemical solution of 2% salt, 2% sugar, 0.3% citric acid, 0.1% KMS and 1% ascorbic acid is also used for steeping preservation of mushrooms[59]. Steeping preservation of mushrooms helps to reduce spoilage and extend shelf life of mushrooms[60,61].

Fig. 6: Steeping of fresh button mushroom

8. *Radiation preservation*

Radiation preservation is the process of exposing mushroom to ionizing radiation in order to destroy microorganisms or insects that could be present in the mushrooms and some time to improve the functional properties of mushrooms or to eliminate toxins, with the least compromise on sensory and nutritive quality. The irradiation of mushrooms can be a safe and cost effective method to enhance shelf–life as well as to ensure hygienic and sensory quality[62]. Contamination and shelf life of mushroom can be reduced by using lower doses of γ–radiation. Irradiation given to mushroom immediately after harvesting has found optimum benefits out of it. Irradiation can potentially delay the maturation i.e. development of cap, stalk, gill and spore and also reduces the loss of water, colour, flavour texture and delays the quality losses. Cobalt 60 is used as a common source of γ rays. A dose of 10 KGy (Kilo Gray) will completely destroy microorganisms. The shelf life of button mushroom (*Agaricus bisporus*) can be enhanced upto a period of 10 days, which can be achieved by application of γ ray close to 2 KGy at storage temperature of 10°C. Irradiation helps to reduce the incident of fungal/bacterial infection and retards to the breakdown of mannitol trehalose. Also the application of irradiation in mushrooms significantly losses the flavor of mushroom. The cap opening is also delayed by irradiation. Amino acid in fresh mushroom is preserved by γ irradiation, irradiation at low levels proved better than irradiation levels of 1 and 2 KGy.

9. Canning

Canning is long term preservation technique in which mushrooms can be stored for up to a year and most of the national/international trade in mushrooms is done in this form. It is an established process of preserving mushroom pieces in brine, butter, oil, vinegar etc in a canning unit at correct pressure and time[63]. Canning involves six–seven unit operations namely cleaning, blanching, filling, sterilization, cooling, labelling and packaging[10]. For these operations, mushrooms are graded, cleaned, blanched (pre–cooked), filled in brine solution of cans and ultimately sterilized with heat and cooled through water spray and labelled for storage. In details, to produce good quality canned mushrooms, these should be processed as soon as possible after the harvest. If delay is inevitable; mushrooms should be stored at 4–5°C till processed. The mushrooms with a stipe length of one cm are preferred and based on the demand mushrooms are canned whole, sliced and stems and pieces[64]. Well graded fresh mushrooms white in color, without dark marks on either caps or stems are preferred for canning. Mushrooms are washed 3–4 times in cold running water to remove adhering substances. Use of iron free water with 0.1% citric acid prevents discoloration. Thereafter blanching is normally done to inhibit polyphenol oxidase enzymes activity and to inactivate microorganisms. It also removes the gases from the mushroom tissue and reduces bacterial counts. The mushrooms are blanched in stainless steel kettles filled with a boiling solution of 0.1% citric acid and 1% common salt. The blanching time ranges from 5–6 minutes at 95–100°C[54,65]. The mushrooms after blanching are filled in sterilized tin cans (A–2½ and A–1 tall can sizes containing approximately 440 and 220 g drained mushroom weight respectively)[54]. Brine solution (2% salt with 0.1% citric acid or 100 ppm ascorbic acid) is added to the mushroom–filled cans at 5.5: 4.4 ratios after bringing its temperature to 90°C. After filling, the cans are exhausted by passing them in exhaust box for 10–15 minutes, so that temperature in the centre of cans is reaches up to 85°C. Then the cans are sealed hermetically with double seamer and kept in upside down position. After exhausting of cans, sterilization of cans is needed. Sterilization is the process of heating the cans up to 118°C to prevent the spoilage by microorganisms during storage[54]. The cans cooled immediately after sterilization process to stop the over–cooking and to prevent stack burning. Cooling can be done by placing the cans in a cold–water tank. Thereafter the clean and dry cans are labeled manually or mechanically and packed in strong wooden crates or corrugated cardboard cartons. The cans are stored in cool and dry place before dispatching to market. In a hot country like India, where the ambient temperature is high during the several months in a year, basement stores are useful, especially during the summer months. The effect of canning and freezing together showed the least negative impact on antioxidant capacity, however canning alone have significant loss of 23–71% antioxidant capacity[66]. General flowchart of canning preservation and canned button mushroom are shown in the Figs. 7 & 8, respectively.

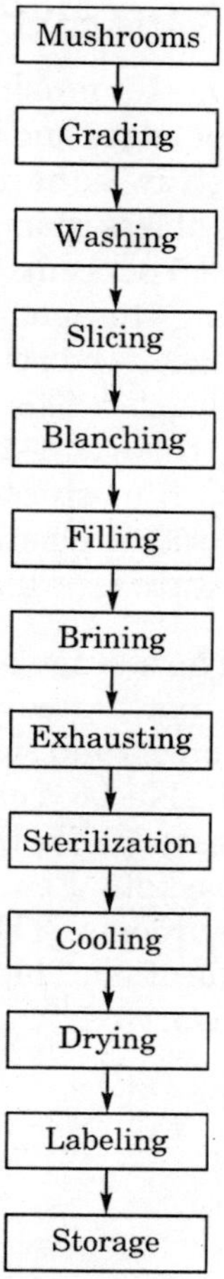

Fig. 7: Flow chart (canning)

Fig. 8: Canned button mushroom

The all above mentioned post–harvest unit operations *viz.*, washing, grading, pre–cooling, packaging and transportation; steeping, irradiation and canning used for freshly harvested button mushroom, which can be applicable for other freshly harvested mushrooms (oyster, milky and paddy straw etc.) with slight modification.

B. Oyster Mushroom *(Pleurotus Spp.)*

The oyster mushroom is harvested when the fruiting body becomes curled under edges and well–formed gills as shown in Fig. 9[10]. During harvesting the straw adhered to mushroom is removed and are packed in polythene bags of less than 100 gauge thickness with perforations having vent area of about 5 per cent. Though these perforations cause slight reduction in weight during storage, it helps in maintaining the freshness and firmness of the produce. The rough handling should be avoided which causes mushroom spoilage[15]. The best method for fresh storage of oyster mushroom is to keep them at 8–10°C in packed container wrapped in plastic film. It's called 'PVC film storage'. Wrappping mushrooms with such microporous or perforated plastic film can improve their storage life, as this reduces the moisture loss and preserves the quality of mushrooms[67].

The storage of oyster mushroom at very low temperatures especially in non–perforated polypacks results in condensation of water with increased sliminess and softening of the texture. Cooling with positive ventilation is desirable *i.e.* cold air should be directed through the packed produce is necessary. For transporting '*dhingri*', the fruit bodies are stacked in trays or baskets. Few polypouches containing crushed ice are kept along with mushrooms. The tray is then covered with thin polythene sheet with perforation. The pre–packed polythene packs with perforations may also be transported[18].

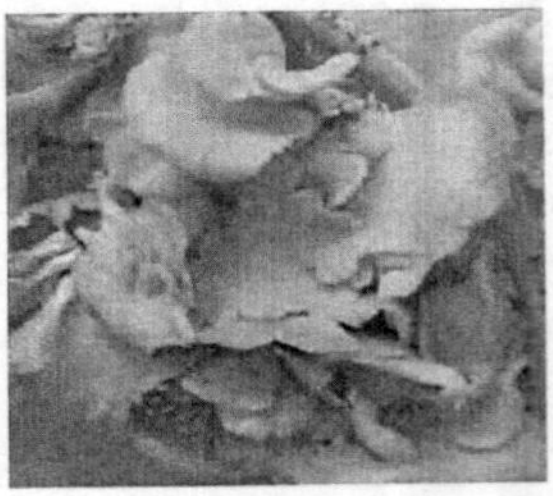

Fig. 9: Matured pink oyster mushroom

C. Milky Mushroom *(Calocybe indica)*

Milky mushroom is the new introduction from India to world and its production catching up fast in different parts of the country especially in Tamil Nadu[68] during the summer months and the mushroom has revolutionized the so called off–season mushroom growing. Fresh mushroom market is largely catered by seasonal growers who do not have cold–chain storage and transport facilities and sell the produce in highly localized market. This mushroom has very good shelf life of 3–4 days without loss of color and appearance. Washing, packaging, pre–cooling and refrigeration, transports and storage of fresh milky mushroom if needed, are almost same as the button mushroom.

D. Paddy Straw Mushroom *(Volvariella volvacea)*

Paddy straw mushroom is very good in taste and in India cultivated in humid and hot regions of Odisha under natural conditions (Fig. 10). This mushroom is packed in polythene bags. As low temperature 2–4°C storage causes frost injury and deterioration in quality, the best way of storage is at 10–15°C in polythene bags with perforations[69]. Mushrooms packed in bamboo baskets with an aeration channel at the center and dry ice wrapped in paper placed above the mushrooms, is in practice for transportation in Taiwan. Packing in wooden cases for transport by rail or boat is practiced in China[54]. In general the shelf life of this mushroom is very less 4–6 h and since mushrooms are sold on the day of harvest.

Fig. 10: Matured paddy straw mushroom (Curtesy photo by: DMR Solan)

III. DRYING OF MUSHROOMS

Drying is one of the oldest techniques known to the mankind for preservation of food commodities/vegetables for long duration. Drying reduces the bulk volume, thus facilitating easy handling, transportation and storage[70]. It is the process of removal of moisture from the product to such a low level that microbial and biochemical activities are checked due to reduced water activity which makes the products suitable for safe storage and protection against the attack by microorganisms during the storage. Mushroom contains about 90% moisture at the time of harvesting and dried to a moisture level down below 10–12 per cent[71]. At a drying temperature of 55–60°C, the insects and microbes on the mushrooms will be killed in few hours, which gives us the dehydrated final product of lower moisture content with longer shelf-life. The temperature, moisture of the mushroom and humidity of the air affect the color of the dried product[72]. Mushroom dried at higher temperature looses texture, flavor, colour along with reduced rehydrability. Although wide temperature range of 37.8–70°C with finishing temperature upto 82.2°C was reported in literature for mushroom drying. For commercial use mushrooms are dried at temperature less than 65.5°C and moisture level of 5% after being blanched for 2–5 minutes[73]. These dehydrated mushrooms are used as an important ingredient in several food formulations including soup powder, pasta salads, snack seasonings, stuffing, casseroles, meat and rice dishes[74].

Recently with advances in drying technologies various drying methods such as sun/solar drying, cabinet drying, dehumidified air–cabinet drying, osmo–air drying, freeze–drying, fluidized/ sprouted bed drying and microwave drying are efficiently used for almost all types of mushrooms[7,75,76].

1. Sun Drying

Although sun drying is traditional slow process but more economical used for long terms storage for most of the mushrooms *e.g.*, oyster, shiitake, paddy straw, *Auricularia* etc except the button mushroom[19]. In case of button mushroom it is the blackening and irreversible change of texture, which often discourages the use of this otherwise simple technique of preservation. However, sliced button mushroom can be sun dried with controlled condition[77] for powder preparation (Fig. 11). In fact, sun drying is more famous for oyster mushrooms (Fig. 12) as of huge demand of dried oyster mushroom in international trade. At optimal temperature 28–32°C, relative humidity 45–52% gives the best quality dried mushroom products in 18–20 h[18,78].

Fig. 11: Sun drying of sliced button mushroom

Fig. 12: Sun drying of oyster mushroom

2. Solar Drying

Effective utilization of solar energy for drying of mushrooms is one of challenging task especially design of suitable solar dryer. Sun is one of the cheapest sources of solar energy during clean climatic conditions. The factors affecting the utilization of solar energy are sunlight, temperature and relative humidity. This is controlled method commonly used for oyster mushroom (Fig. 13) and gives superior quality and colour of dried mushroom products in a relatively shorter period (10–12 h) as of sun dried mushroom products. The designed and developed a solar dryer for oyster mushroom with a capacity of 2.5 kg / day shows that the oyster mushrooms treated with 0.5% KMS for 15 min and 0.5% sodium benzoate for 15 min had the same amount of nutrients (protein, sugars and amino acids) as fresh mushrooms. The highest rehydration ratio (5.25) and best organoleptic quality was obtained for the sample treated with 0.5% KMS[79]. Solar dryer is effective for dehydration of edible mushrooms with suitable qualitative and provides necessary proteins and carbohydrates contents for human nutrition[80].

Fig. 13: Solar dried oyster mushroom

3. Cabinet Drying

This is a batch type drying operation especially applicable for oyster, milky and paddy straw mushrooms. For drying of mushrooms, freshly harvested whole/ sliced mushrooms are spread uniformly in a stainless trays and thereafter kept these trays one above the other inside the cabinet dryer (Fig. 14). After closing the insulated cabinet dryer, hot air is circulated either horizontal/vertical direction through the filled mushroom trays and finally air goes outside through side/top opening provided[81]. Mushrooms can be dried at 60°C for 8 hours to get good quality dried product. But non uniformity in drying of mushroom is a measure drawback of this drying method. The cabinet tray dried oyster mushroom samples at 60°C and 1% KMS pretreatment were found to be best quality[82].

Fig. 14: Cabinet dryer

4. Vacuum Freeze Drying

Vacuum freeze drying (V.F.D.) is a further development in mushroom processing technology. In this process the original shape, quality, colour size, texture, freshness properties of thermal sensitive produce are retained and the shrinkage, which is a problem with other drying methods, is negligible. This processing technique involves the cooling of mushroom much below the freezing point *i.e.*, –40°C where moisture present in mushroom is converted to tiny ice molecules which further directly sublime into vapour when subjected to vacuum with a slight rise in temperature resulting a dried end product[10]. Mushrooms washed with 0.05% KMS followed by freeze drying at –60°C for 16 h gave the best product but costly operation. For freeze drying of sliced mushrooms, slices are immersed in a solution of 0.05% KMS and 2% salt for about 30 minute. The pretreated mushrooms are then blanched in boiling water for two min followed by cooling. The product is frozen at –22°C for one min. The frozen mushrooms are dried to moisture content of 3% in a freeze dryer and packed under vacuum[83]. The quality of vacuum freeze dried button and oyster mushrooms are shown in Figs. 15 and 16, respectively.

Fig. 15: Freeze dried button mushroom

Fig. 16: Freeze dried oyster mushroom

5. Fluidized Bed Drying

The dehydrated product obtained by fluidized bed drying rank next to the freeze–dried product in quality. This method of drying is used for almost all types of mushrooms either whole or sliced forms. In this method, mushrooms are being dried under fluidized condition. The whole/sliced mushrooms used for drying is fluidized by hot air of 45–65°C temperature with high velocity of 2–2.5 m/s to causes suspension[7, 84]. In this drier, higher rate of moisture migration takes place. Since every surface of product is exposed to drying air, uniform drying of products takes place. Sliced button and oyster mushrooms dried at 50°C for 7 h in the fluidized bed drier yielded good quality dried product[70]. The quality of fluidized bed dried button and oyster mushrooms are shown in Fig. 17 and 18, respectively.

Fig. 17: Fluidized bed dried button mushroom

Fig. 18: Fluidized bed dried oyster mushroom

6. Other Drying Methods

The other drying methods *viz.*, dehumidified, osmotic and microwave drying are also applicable for drying of mushrooms[82,85]. In dehumidified drying, the ambient air is passed through a desiccated medium *e.g.* silica gel or hot plate which absorbs moisture from the moving air. As a result, the relative humidity of air is reduced and at the same time there is an increase in its temperature. When such air comes in contact with fresh mushrooms, transfer of moisture from mushroom to drying air takes place. For energy efficient operation same air is re–circulated again in dehumidified dryer[86]. In osmotic drying whole/sliced mushrooms are blanched initially with sucrose and 2–5% salt concentrations and then dried at suitable temperature[87]. Similarly, few mushroom species are dried in microwave oven.

IV. MUSHROOM VALUE ADDED PRODUCTS

Worldwide, focus of mushroom industry is predominantly on trade of the fresh produce rather than the real value–addition. Domestically, almost entire mushrooms are consumed in fresh form while most of the export is in the preserved form (canned, steeped and dried). Current era is characterized by greater awareness about mushrooms and many other vegetables as a nutraceutical product with the demand for the readymade or ready–to–make food products. Mushrooms being delicacy and high moisture cannot be stored for more than 24 h at the ambient conditions of the subtropics/tropics which leads microbial spoilage of the product making it unsalable. Effective processing techniques will not only diminish the post–harvest losses but also result in greater remuneration to the growers as well as processors. Value can be added to the mushrooms at various levels, right from grading to the readymade snacks or the main–course item. Real value–added products in the Indian market are the mushrooms pickle and mushroom soup powder. Technologies for production of some other products like mushroom based biscuits, nuggets, preserve, noodles, *papad*, candies, cookies, ketchup, *murabba*, chips, chutney and readymade mushroom curry in retort pouches have been developed but are yet to be popularized [88, 89]. However, recently many private firms are coming forward for development of various value added products from mushrooms because of its increasing awareness. Attractive packaging of the value–added products is yet another area which may be called the secondary value–addition. While small growers may add value by grading and packaging, industry may go for the processed products for better returns as well as improvement in the demand, which shall have cascading positive effect on the production. Various mushroom value added products developed from mushrooms and their commercial packages are shown in Fig. 19[18].

Button mushroom soup powder

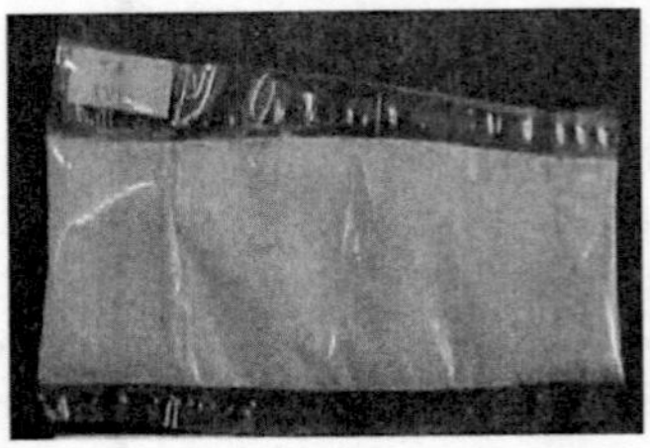

Oyster mushroom soup powder

Oyster mushroom pickle

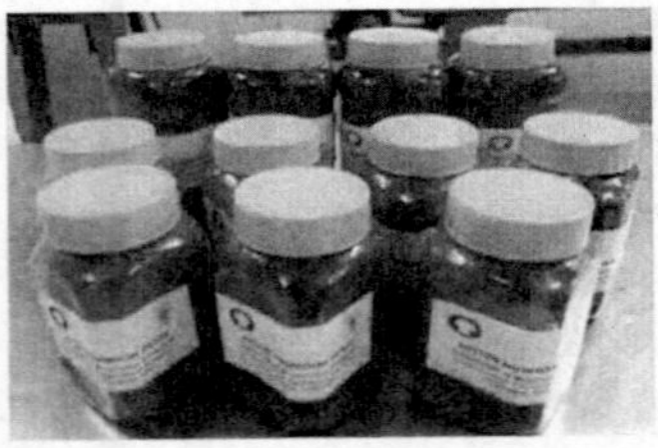

Button mushroom pickle

Button mushroom biscuits

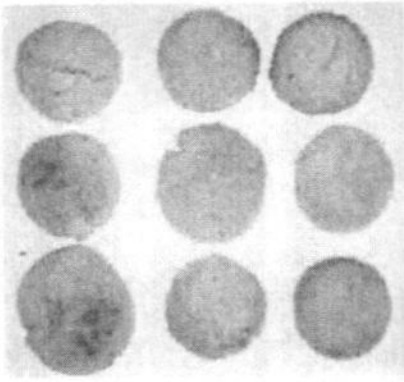

Oyster mushroom biscuits

Mushroom Nuggets

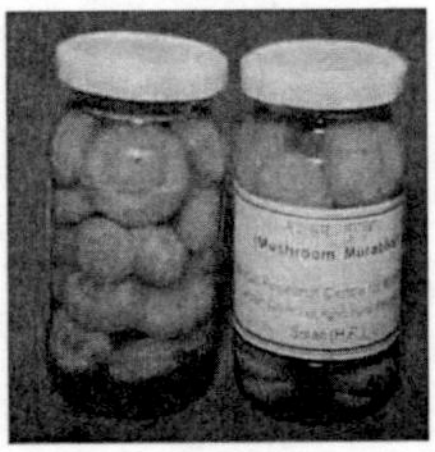

Mushroom *Murabba*

Mushroom Ketchup

Mushroom curry

Mushroom *Papad* Mushroom soup powder

Fig. 19: Various mushroom value added products (Courtesy Photos by: DMR, Solan)

V. MUSHROOMS MEDICINAL PRODUCTS

Mushrooms are considered to be a complete, health food and suitable for all age groups, child to aged people. The nutritional value of mushrooms is affected by numerous factors such as species, stage of development and environmental conditions. Mushrooms are an excellent source of proteins, dietary fiber, vitamins especially C, B (folic acid, thiamine, riboflavin and niacin), antioxidants, minerals *viz.*, potassium, sodium and phosphorus[90,91]. It also contains other essential minerals (Cu, Zn, Mg etc.) in traces but deficient in iron and calcium. Most of the edible fungi have been commonly known for their immense health benefits and extensively used in folk medicine. Specific biochemical compounds in mushrooms are responsible for improving human health in many ways. These bioactive compounds include polysaccharides, tri–terpenoids, low molecular weight proteins, glycoproteins and immunomodulating compounds. Hence mushrooms have been shown to promote immune function; boost health; lower the risk of cancer; inhibit tumor growth; help balancing blood sugar; ward off viruses, bacteria, and fungi; reduce inflammation; and support the body's detoxification mechanisms[92]. Increasing recognition of mushrooms in complementing conventional medicines is also well known for fighting many diseases. Medicinal functions of few important are given in the Table 3[92].

Table 3: Medicinal mushrooms and their medicinal functions

Mushrooms	***Compounds***	***Medicinal properties***
Ganoderma lucidum	Ganoderic acid	Augments immune system Liver protection
	Beta–glucan	Antibiotic properties Inhibits cholesterol synthesis
Lentinula edodes	Eritadenine Lentinan	Lower cholesterol Anti–cancer agent
A. bisporus	Lectins	Enhance insulin secretion
P. sajor–caju	Lovastatin	Lower cholesterol
G. frondosa	Polysaccharide Lectins	Increases insulin secretion Decrease blood glucose
Auricularia auricula	Acidic polysaccharides	Decrease blood glucose
Flammulina velutipes	Ergothioneine Proflamin	Antioxidant Anti cancer activity
Trametes versicolor	Polysaccharide–K (Kresin)	Decrease immune system depression
Cordyceps sinensis	Cordycepin	Cure lung infections Hypoglycemic activity Cellular health properties Anti–depressant activity

Some extracted medicinal products viz., Reishi Gano, Corilus, real natures/life revival juices derived from *Ganoderma Lucidum* lowers blood glucose, acts as antioxidants, immunity booster, cell regenerators and reduces gynecological problems in pregnant women's (Fig. 20). Similarly Cordyceps Cs a medicinal products derived from *Cordyceps sinensis* is used as energy and endurance[93].

Fig. 20: *Ganoderma Lucidum* medicinal products

Mushrooms are not only source of nutritious food but also act as source of bioactive metabolites and prolific resource for drugs. Knowledge advancement in biochemistry, biotechnology and molecular biology boosts application of mushrooms in medical sciences. From a holistic consideration, the edible mushrooms and its by–products may offer highly palatable, nutritious and healthy food besides its pharmacological benefits[92].

VI. FUTURES OF MUSHROOM INDUSTRIES

China, USA, Netherlands, Poland, Spain, France, Italy, Ireland, Canada and UK are the major producers and out of it China itself contributing about 46% of the world mushroom production. USA and many European countries majorly produces button mushroom. However, in Asian countries the scenario is different and other species are also cultivated for commercial production. The major importing countries are UK, Germany, France, Netherlands, Belgium, Russian Federation & Japan. In broad world mushroom producing countries are classified in into three major hubs i.e. Asian and other East Asian countries as one hub produces and consuming almost all kind of mushrooms exports to the other two European and American button mushroom dominating hubs[94]. Earlier, majorly mushrooms were exported either in preserved/processed form but of late the scenario is changing. Fresh mushrooms are preferred over preserved ones in EU and American countries. Presently, India produces about 1.5 lakh metric tons mushrooms, which is about 3% of the world mushroom production. Because of high tech machinery with environmental control equipments imported and indigenous mushroom mechanization in 1990, India's mushrooms production were drastically increased and started export of canned mushrooms into international market of Europe and USA. Despite of huge biomass production and country's subtropical climatic conditions more favorable for cultivation of all kind of mushrooms, still India's mushrooms share in domestic and international market is almost negligible. Also for Indian mushrooms, China is the main competitor in international market. However, rising demand of fresh mushroom and improved prices in domestic as well as international market, India has a good chance to stands in mushroom business if we view the following strength, weaknesses, opportunities and threats (SWOT) analysis of Indian mushroom industry[10].

Strength

- Climatic diversification– suitable for all kind of mushroom cultivation
- Abundant raw material and man power coupled with government support
- Strategic location

Weaknesses

- Higher cost of finances, packaging materials, energy and transportation
- Poor quality of raw materials and higher cost machineries

Opportunities

- Decline production of other countries and breaking down of international trade barrier
- Increasing awareness and liking for mushrooms in domestic market
- Well adaptation of modern technologies and PHT management

Threats

- Competition from China and other countries
- Demand shifted towards the fresh consumption

VII. CONCLUSIONS

Serious efforts in mushroom research for commercial cultivation especially for development of suitable low cost compost, disease resistant high yielding spawns, new varieties and educating the growers are in recent origin. However, still it felt to develop suitable mushroom harvesting techniques. For efficient production and post–harvest management/practices of mushrooms, growers at seasonal/small/household levels are needed to be trained. Special subsidy on mushroom transportation cost and provision for refrigerated vans in hilly areas helps to minimize losses during transport of highly perishable fresh mushrooms. In order to protect small farmers from the exploitation by middlemen, an efficient centralized marketing system with proper processing units may be needed set up surrounding the potential/ growing areas. Also the mushroom enterprise is a capital–intensive venture need financial assistance is available from NABARD, State Govt financial institutions, public sector banks with lower interest in India. Besides this, different incentives are also available from Department of Food Processing Industries, NHB, and State Govt. agencies engaged in agricultural development. In general, small seasonal growers are unable to avail of these facilities and therefore they may unite to form cooperatives or SHGs for better marketing and solving their problems. Overall, the India has to establish "A quality produce of India" for mushrooms in foreign market to exploit present declining trend of mushroom production in many other developing/ developed countries.

Acknowledgement

The authors would like to thanks all the information sources in the public domain used in the compilation of this book chapter.

REFERENCES

[1] Singh M., Vjay B., Kamal S., Wakchaure G.C., "Mushrooms Cultivation, Marketing and Consumption, *Indian Council of Agricultural Research,* Chambaghat, Solan, India, p. 246 (2011).

[2] Azad K.C., Srivastava M.P., Singh R.C., Sharma P.C., "Commercial preservation of mushrooms –1. A technical profile of canning and its economics", *Indian J of Mush.,* **XII–XIII**, p. 21 (1987).

[3] Arumugnathan T., Hemkar A.K., Rai R.D., "Studies on drying characteristics and effect of pretreatments on the quality of sun dried oyster mushrooms pleurotus florida", *Mushroom Res.*, **13**(1), 3 (2004).

[4] Burton K.S., Quality investigations into mushroom browning. *Mush J.*, **158,** 68–70 (1986).

[5] Choi S.W., Sapers G.M., "Effects of washing on polyphenols and polyphenol oxidase in commercial mushrooms(agaricus bisporus)", *J Fd Sci.,* **42**, 2286–2290 (1994).

[6] Estanga E.G., "Visible and hyperspectral Imaging Systems for the detection and discrimination of mechanical and microbiological damage of mushrooms", Dublin Inst. Technol., Dublin (2010).

[7] Kulshreshtha M., Deepti A., Vipul, "Effect of drying conditions on Mushroom quality", *J Eng Sci Tech.,* **4(1)**, 90 (2009).

[8] Tulek Y., "Drying kinetics of oyster mushroom (*Pleurotus ostreatus*) in a convective hot air dryer", *J Agri Sci Tech.,* **13**, 655–664 (2011).

[9] Iqbal T., Rodrigues F.A., Mahajan P.V., Kerry J.P., "Effect of time, temperature, and slicing on respiration rate of mushrooms", *J Food Sci.*, **74**(6), E298–303 (2009).

[10] Anonymous, Directorate of Marketing and Inspection, Nagpur (2003).

[11] Byrnes V.C., O'Beirne D., "Effects of washing treatment on microbial and sensory qualities of modified atmosphere (MA) packaged fresh sliced mushroom (Agaricus bisporus)", Postharvest *Biol Tech.,* **48**, 283–294 (2008).

[12] Pruthi J.S., Manan J.K., Raina B.L., Teotia M.S., "Improvement in whiteness and extension of shelf life of fresh and processed mushrooms (*Agaricus bisporus and*

volrariella volvacea)", *Ind Fd Packer.*, **38**(2), 55 (1984).

[13] Beelman R.B., "Factors influencing postharvest quality and shelf–life of fresh mushrooms", *Mush News.,* **35(7)**, 12–18 (1987).

[14] Guthrie B.D., Bellman R.B., "Control of bacterial deterioration in fresh washed mushrooms", *Mush Sci.*, **12**(2), 689–699 (1989).

[15] Saxena S., Rai R.D., "Storage of button mushrooms (*Agaricus bisporus*)– the effect of temperature, perforation of packs and pretreatment with potassium metabisulphite", *Mush J Tropics.*, **8**, 15–22 (1988).

[16] United States standards for grades of mushrooms for processing, USDA, (2003).

[17] Adamicki F, Department of vegetable storage, research institute of vegetable crops Skiemiewice, Poland, http://www.ba.ars.usda.gov.,(2012).

[18] Rai R.D., Arumuganathan T., "Postharvest technology of mushroom", Director, NRC forMushroom (ICAR), Chambaghat, solan, *Technical Bulletin*, Solan, Himachal Pradesh, p. 84 (2008).

[19] Arumuganathan T., Rai R.D., Upadhyay R.C., Sharma S.R., "Machinery equipments and instruments in mushroom production and processing. *In*: *Emerging Areas of Mushroom Research and Production",* NRCM, Solan, pp. 282–296 (2003).

[20] Umiecka L., "Effect the material quality, treatment, package methods, and storage conditions on export quality of several mushroom races", *Biul Warzywniczy.*, **29**, 271–292 (1988).

[21] Burton K.S., Frost C.E., and Atkey P.T., "Effect of vacuum cooling on mushroom browning", *Int J Food Sci Technol.*, **22**, 599–60 (1987).

[22] Tasnadi G., Racz J., Royse, "New developments of cultivation technologies and new products in a Central Europeon Mushroom Company, Hungary. *In*: Mushroom Biology and Mushroom Products, University Park, pp. 223– 232 (1996).

[23] *Ares G., Lareo C., Lema P.,* "Modified atmosphere packaging for postharvest storage of mushrooms. A review", *Fresh produce.,* **1**(1), 32–40 *(2007).*

[24] Briones G.L., Varoquaux P., Chambroy Y., Bouquant J., Bureau G., Pascat B., "Storage of common mushrooms under controlled atmosphere", *Int J Fd Sci Tech.,* **27**(5), 493–505 (1992).

[25] Moleyar V., Narasimhan P., "Modified atmosphere packaging of vegetables: An appraisal", *J. Food Sci Tech.,* **31**, 267–278 (1994).

[26] Philips C.A., "Review, modified atmosphere packaging and its effect on the microbiological quality and safety of produce", *Int J Fd Sci Tech.,* **31**, 463–479 (1996).

[27] Fishman S., Rodov V., Peretz J., Ben–Yehoshua S., "Modified atmosphere packaging of fruits and vegetables", *J Fd Sci.,* **60**, 1078–1083 (1995).

[28] Zakia B., Rajarathnam S., Shashirekha M.N., Ghosh P.K., "Advances in Mushroom Biology and Production", Mushroom Society of India, NRCM, Solan, p. 321 (1997).

[29] Greengrass J., "Films para envasado en atmósfera modicada. *In*: Parry R. (*ed.*), Envasado de los alimentos en atmósfera modicada. A.Madrid Vicente Ediciones", A Madrid Vicente Edicions, Madrid, Espana, pp. 79–118 (1995).

[30] Schlimme D.V., Rooney M.L., "Envasado de frutas y hortalizas mínima– Procesadas y Refrigeradas, Editorial Acribia S.A", Zaragoza, pp. 131–178 (1997).

[31] Farber J.N., Harris L.J., Parish M.E., Beuchat L.R., Suslow T.V., Gorney J.R., Garett E.H., Bustta F.F., "Microbiological safety of controlled and modified atmosphere packaging of fresh and fresh cut produce", *Comprehensive Review in Food Sci & Food Safety.*, **2**, 142–160 (2003).

[32] Tano K., Arul J., Doyon G., Castaigne F., "Atmospheric composition and quality of fresh mushrooms in modified atmosphere packages as affected by storage temperature", *J Food Sci.*, **64**(6), 1073–1077 (1999).

[33] Roy S., Anantheswaran R.C., Beelman R.B., "Modified atmospheres and modified humidity packaging of fresh mushrooms", *J Food Sci.,* **61**(2), 391–97 (1996).

[34] Saray T., Balla C., Horti K., Sass P., "The importance of packaging and modified atmosphere in maintaining the quality of cultivated mushrooms (*Agaricus bisporus* L.) stored in chill chain.", *Acta Hort.*, **368**, 322–326 (1994).

[35] Briones G.L., Varoquaaux P., Bureau G., Pascat B., "Modified atmosphere packaging of common mushroom", *Int J Fd Sci Tech.,* **28**(1), 57–68 (1993).

[36] Burton K.S., Maher M.J., "Modified atmosphere packaging of mushrooms– review and recent developments", *Mush Sci.,* **13**(2), 683–688 (1991).

[37] Henze J., "Storage and transport of *Pleurotus* mushrooms in atmospheres with high concentrations", *Acta Hort.*, **258**, 579–584 (1989).

[38] Gormley T.R., Mac Canna C., "Packaging and shelf life of mushrooms", *Irish J Agric Res.*, **6(2)**, 255 (1967).

[39] Chopra S.K., Naquash G.S., Chadha T.R., "National Symposium on Production and Conversation Forestry", Solan (1985).

[40] Chi J.H., Taemoon H., Yang H.K., Ha T.M., Kim Y.H., "Effects of packing materials for the keeping freshness of *Pleurotus ostreatus* and *Flammulina velutipes*", *RDA J Indus crop Sci.,* **40**(2), 58–64, (1998).

[41] Chi J.H., Ha T.M., Ho K.Y., Cheul J.Y., Kim Y.H., Ju Y.C.,"Effects of storage temperature and packing method for keeping freshness of fresh mushrooms", *RDA J Agri Sci Farm Mgt, Agri Engin Sericulture Farm Products Utilization.,* **38**(1), 915–921 (1996).

[42] Cho S.S., Ha T., "The effect of supplementary package materials for keeping freshness of fresh mushroom at ambient temperature", *RDA J Indus Crop Sci.,* **40**, 52–57 (1998).

[43] Anantheswaran S., Sunkara R., "Use of commercial moisture adsorbers to increase the shelf life of fresh mushrooms", *Mushroom News*, **44**, 24–27 (1996).

[44] Roy S., Anantheswaran R.C., Beelman R.B., "Sorbitol increases shelf life of fresh mushrooms stored in conventional packages", *J Fd Sci.,* **60**(6), 1254–1259 (1995).

[45] Cook R.J., Papendick R.I., "Role of water potential in microbial (bacteria and fungi) growth and development of plant disease, with special reference to postharvest pathology", *Hort Sci.*, **13**(5), 559 (1978).

[46] Eaves C.A., "A modified atmosphere system for the package of stored fruit", *J Hort Sci.,* **35**(2), 110, (1960).

[47] Roy S., Anantheswaran R.C., Beelman R.B., "Modified atmosphere and modified humidity packaging of fresh mushrooms", *J Fd Sci.,* **61**(2), 391–397, (1996).

[48] Saltveit M.E., "A summary of CA and MA requirements and recommendations for harvested vegetables. *In*: Seventh International Controlled Atmosphere Research Conference. Vegetables and Ornamentals", University of California, Davis, *Postharv Hort Series.,* **18**(4), 98–117 (1997).

[49] Suslow T., Cantwell M., "Mushrooms Fresh Produce Facts", http://www.postharvest.ucdavis.edu., (1998).

[50] Zheng Y.H., Xi Y.F., "Preliminary study on colour fixation and controlled atmosphere storage of fresh mushrooms", *J Zbej Agr Univ.,* **20**, 165–168 (1994).

[51] Anonymous, "A Guide for Shipment of Perishable Cargo in Refrigerated Containers", Controlled Atmosphere Handbook, *Carier Corp.,* Syracuse, NY, (1995).

[52] Ahlawat O.P., Rai R.D., Ahlawat K., Verma R.N., "Control of browning in white button mushroom (*Agaricus bisporus)", Ind Fd Packer.,* **54**(1)**,** 60–64 (2000).

[53] Simons, Suzanne S., Beelman R.B., "Influence of calcium chloride added to irrigation water on crop yield and quality attributes of fresh mushrooms", In Sci. & cultivation of edible fungi, Elliott (*ed.*), *Rotterdam Balkena*, **14**, 741–748 (1995).

[54] Saxena S., Rai R.D., "Postharvest Technology of Mushrooms", *Technical Bulletin No.2*, NRCM, Solan (1990).

[55] Hui Y.H., Cornillon P., Lagaretta I.G., Lim M.H., Murrel K.D., Nip W.K., "Handbook of frozen foods", Marcel, Dekker, Inc., New York, (2004).

[56] Suman B.C., Sharma V. P., "Mushroom Cultivation India", *Daya Publishing House*, New Delhi, (2007)

[57] Pruthi J.S., Gopalakrishnan M., Bhat A.V., "Studies on the dehydration of tropical paddystraw mushroom (*Volvariella volvacea*)", *Ind Fd Packer.,* **32**(2), 7–16 (1978).

[58] Kapoor J.N., "Mushroom cultivation", ICAR, New Delhi, **15** (1989).

[59] Singh S., Kumar C.G., Singh S., "Production, processing and consumption patterns of mushrooms", *Indian Food Industry*, **14**(6), 38–47 (1995).

[60] Bano Z., Singh N.S., "Steeping preservation of an edible mushroom (Agaricus bisporus)", *J. Food Sci. & Technol.*, **9**(1), 13–15 (1972).

[61] Sethi V., Bahl N., Bhagwan J., "Steeping preservation of mushrooms", Symposium of impact of pollution in and from food industry and its management, *C.F.T.R.I., Mysore* (1989).

[62] Akram K., Kwon J.H., "Food irradiation for mushrooms: A review", *J of the Korean Soc. for Applied Biological Chemistry.*, **53**(3), 257–265 (2010).

[63] William S., Food Safety, University of Minnesota Extension educator (2010).

[64] Beelman R.B., Edwards C.E., "Variability in the protein content and the cultivated mushrooms (*Agaricus bisporus*)", *Mushroom News,* **37**(7), 17 (1989).

[65] Tanga A.Q., "Weight loss in mushroom during canning", *Indian Mush Sci.,* **1**, 225–232 (1974).

[66] Murcia M.A., Martinez–tome M., Jimnez A.M., Vera A.M., Honrubia M., Parras P., "Antioxidant activity of edible fungi (truffles and mushrooms): Losses during industrial processing", *Food Proc.,* **10**, 1614–1622 (2002).

[67] Kim B.S., "Oyster Mushroom Cultivation, part–II, Mushroom growers handbook", http:// www.alohamedicinals.com (2013).

[68] *The Mushroom growers News Letter, The mushroom company, USA. http://* www.mushroomcompany.com *(2013).*

[69] Ahlawat O.P., Kumar P., Rai R.D., Tewari R.P., "Variations in biochemical properties of different strainsof *Volvariella volvacea* (BullFr.) Sing under different conditions", *Ind J Micobiol.,* **46**(1), 31–37 (2006).

[70] Wakchaure G.C., Manikandan K., Mani I., Shirur M., "Kinetics of Thin Layer Drying of Button Mushroom", *J Agri Eng.,* **47**(4), 41 (2010).

[71] Arumugnathan T., Hemkar A.K., Rai R.D. "Studies on drying characteristics and effect of pretreatments on the quality of sun–dried oyster mushroom", *Pleurotus florida , Mush. Res.,* **13**(1)*,* 35–38 (2004).

[72] Yapar S., Helvaci S.S., Peker S., "Drying behavior of mushroom slices", *Drying Technol.,* **8**(1), 77 (1990).

[73] Luh B.S., Woodroof J.G., "Commercial Vegetable processing". *The* AVI Publishing Company Inc., West Port, Connecticut, pp. 395–396 (1975).

[74] Tuley L.,"Swell time for dehydrated vegetables", *Int Food Ingredients.,* **4**, 23–27 (1996).

[75] Drying methods: Methods and Tips, http:// www.mushroom–appreciation.com, (2013).

[76] Kuo M., Alternative Drying Methods, http:// www.mushroomexpert.com, (2013).

[77] Linsao K., Ishan, Maluniu, Thomscher, "How to Dry Mushrooms", http:// www.wikihow.com, (2013).

[78] Rama V., John P.J., "Effects of Metods of Drying and Pretreatments on Quality of Dehydrated Mushroom", *Indian Food Packer*, **54**(5), 59–64, (2000).

[79] Nehru C., Kumar V.J.F., Maheshwari C.U., Gothandapani L., "Solar drying characteristics of oyster mushroom", *Mush Res.,* **4**(1), 27–30 (1995).

[80] Madlopa A., Mponda T.J., "Dehydrating edible mushrooms using a locally made solar dryer", *Malawi J Agri Sci.*, **2**(1), 49–55 (2003).

[81] Mudahar G.S., Bains G.S., "Pre–treatment effect on quality of dehydrated Agaricus bisporus mushrooms", *Indian Fd Packer.*, **28**(5), 19–27 (1982).

[82] Naik S., Ramachandra M., Rajashekharappa K.S., Tulasidas T.N., Murali K., Mallesha B.C., "Drying of oyster mushroom *Plurotus florida* in different dryers", *J Dairying Fd Home Sci.*, **25**(2), 79–86 (2006).

[83] Kannaiyan S., Ramaswamy K., "A handbook of edible mushrooms", *Todays & tomorrow's printers and publishers*, pp. 44–50 (1980).

[84] Singh S.K., Narain M., Kumbhar B.K., "Drying kinetics of button mushroom in fluidized bed". *J Agric Eng.,* **47**(1), 69–75 (2007).

[85] Sahni C.K., Khurdiya D.S., Dalal M.A., Maini S.B., "Microwave processing of foods–potential and prospects",. *Indian Food Packer*, **5**, 32–41 (1997).

[86] Sahay K.M., Singh K.K., "Unit operations of Agricultural Processing", Vikas publishing House Pvt. Ltd., New Delhi *Agricultural processing*, pp. 386 (1994).

[87] Rezagah M.E., Kashaninejad M., Mirzaei H., Khomeiri M., "Osmotic dehydration of button mushroom Fickian diffusion in slab configuration", *Latin American Applied Res.*, **40**, 23 (2010).

[88] Wakchaure G.C., Shirur M., Manikandan K., Rana L., "Developmnt and evaluation of oyster mushroom value added products", *Mush Res.,* **19**(1), 40–44 (2010).

[89] Singh R.P., Dang R.L., Bhatia A.K., Gupta A.K., "Water binding additives and canned mushroom yield", *Indian Food Packer*., **36**(1), 32 (1982).

[90] Thekkuttuparambil A.A., Janardhanan K.K., "Indian medicinal mushrooms as a source of antioxidant and antitumor agents", *J. Clin Biochem Nutr.,* **40**(3), 157 (2007).

[91] Chang S.T., Miles G.M., "Mushrooms: Cultivation, Nutritional Value, Medicinal Effect, and Environmental Impact", *CRC Press*, (2004).

[92] Singh M., Vijay B., Shwet Kamal., Wakchaure, G.C., "Nutritional and medicinal values of mushroom", Mushroom cultivation, marketing and consumption, *Directorate of Mushroom Res.,* Solan, Himachal Pradesh, India, pp.11–14 (2011).

[93] Cordyceps Essence, http:// www.lohongka.com, (2013).

[94] Wakchaure G.C., Singh M., Bhuvnesh Vijay, Shwet Kamal., "Production and Marketing of Mushrooms: Global and National Scenario", Mushrooms–cultivation, marketing and consumption, Publisher: *Directorate of Mushroom Research,* Solan, Himachal Pradesh, India, pp.15–22 (2011).

19

Green Chillies: Importance and Processing

I.L. PARDESHI[1]* AND P.U. JADHAV[1]

ABSTRACT

Green chilies are major source of Vitamin C and Vitamin A besides important minerals like Calcium, Phosphorus, Sulphur and Iron. India grows green chillies over 5700 ha producing 59100 tons, annually. Chilli crop normally yields 2 to 2.5 tonnes dry chilli and 7.5 to 10 tonnes green chilli from one hectare in India. Maximum chillies are produced in Andhra Pradesh followed by Karnataka, Orissa, West Bengal, Maharashtra etc. Chilli peppers are ready for harvest 3–6 weeks after flowering, depending on the fruit maturity desired. Green chillies are harvested earlier than orange or red chillies. Harvesting is done by hand or with the aid of a small knife. Some cultivars are so pungent that the persons doing harvesting needs to wear gloves to avoid blistering of skin on their hands. Chilli peppers are cleaned and sometimes graded before sale as a fresh spice. The chillies are graded apparently by characteristics like colour, size, moisture and stalk of the pods besides seed and fruit (pod) ratio, seed size and hardness, thickness of the skin of the pod and pungency. Green chilli contains nearly 85% moisture content. Green chillies can be dried upto moisture of about 8 to 10%. In India, sun–drying is common, which often takes place on mats in fields and by roadsides. The improved drying practices improve the quality of the green chillies. While drying, Chilli loses most of its vitamin C, it increases the vitamin A content by 100 times. Vitamin A is a powerful anti–oxidant and anti–inflammatory agent. Dried chillies when stored are, often attacked by drugstore beetles like Stegobium paniceum (Linn) and the cigarette beetle like Lasioderma serricorne. The probable infestation due to pest during storage can be

[1] Agricultural Engineering, College of Agriculture, Dr. Panjabrao Deshmukh Krishi Vidyapeeth, Akola, Maharashtra – 444 104, India.
**Corresponding author* E-mail: ilpardeshi@gmail.com

eliminated by drying under sun or otherwise by fumigation. These can be stored for months before selling, or can be further processed into sauces or as ground, canned, pickled or bottled chillies.

Key words: Green and red chilli, Chemical, Constituents, Medicinal, Dye, Grading, Packaging, Storage, Diseases, Processing, Shelf life, Nutritional quality

1. INTRODUCTION

Chilli (*Capsicum annuum* L.) is one of the most important spices cum vegetable crop grown in India with great export potential. It is grown almost throughout the country. There are more than 400 different varieties of chillies found all over the world. It is also called as hot pepper, cayenne pepper, sweet pepper, bell pepper, chili, chilies, aji, pimiento, paprika, and capsicum. Chili peppers occur in a number of varieties that vary greatly from mild to very hot, which is determined by capsaicin content. It is a warm–season crop and a member of the *Solanaceae* family. Different varieties are grown for vegetables, spices, condiments, sauces and pickles.

Chilli occupies an important place in Indian diet. It is an indispensable item in the kitchen, as it is consumed daily as a condiment in one form or the other. Green chilies are major source of Vitamin C and Vitamin A besides important minerals like calcium, phosphorus, sulphur and iron.

World demand and supply of chilli peppers have been steadily increasing. Between 2000 and 2007, fresh chilli and peppers increased on average 6.6 per cent per year in the world and 6.3 per cent per year in Europe. Bulk of chillies is consumed in India. Only 1 to 2% is exported to various countries worth 5 to 10 crore rupees.

Chilli is also known as hot peppers and Capsicums also referred as bell peppers or sweet peppers. The native home of chilli is considered to be New Mexico with secondary origin of Guatemala. They are known from pre–historic remains of Peru. They were widely cultivated in central and South America. Till the discovery of American Chilli, the crop was unknown to Europe. Chilli is carried to the old world by the early explorers being introduced into Spain in 1493 and later spread rapidly to Europe. The Portuguese brought capsicum from Brazil to India during the year 1584. Chilli is a fruit of the plants '*Capsicum annuum*' and '*C. frutescens*' that come from the genus 'Capsicum,' belonging to the family of '*Solanaceae*,' which also includes tomato and potato. These fruits are small in size and known for their sharp acidic flavour and colour. Some chili peppers are dried and sold individually or tied together in ornamental arrangements. Peppers are eaten raw in salads and salsa, processed by canning, freezing, pickling, and dehydrated and powdered to produce paprika and chilli powder.

Unlike the U.S., most of the European paprika is mildly pungent, while most of the Indian varieties such as *Naga Jolokia, Ellachipur Sanman, Guntur Sanman, Hindpur, Jwala, Kanthari, Nalcheti* are highly pungent. The medium pungent *sannam* and the mildly pungent *mundu* chilli varieties of India are internationally recognized as the finest in quality. Chilli hotness is measured in Scoville Heat Units (SHU) which is originally a subjective measure but today, chilli hotness is more frequently determined by HPLC (high performance liquid chromatography), and these results can be correlated with the traditional Scoville ratings: the conversion generally accepted is that 15 Scoville units is equal to 1 ppm capsaicin plus capsaicinoids. It is found that the *Naga Jolokia* variety of Indian chilli contains maximum capsaicin and dihydrocapsaicin contributing to a pungency of 8,55,000 SHU, which seems to be the hottest chilli known so far. Mostly, red and dried chilli powders are used for culinary purpose whereas green chilli in processed form is getting attention nowadays for the cause that green chilli has more Vitamin C and antioxidant properties and finds a different place for preparation of Indian recipes.

Chili peppers are harvested manually. Chilli peppers are ready for harvest 3–6 weeks after flowering, depending on the fruit maturity desired. Green chillies are harvested earlier than orange or red chillies. Harvesting is done by hand or with the aid of a small knife. Some cultivars are so pungent that the persons doing harvesting needs to wear gloves to avoid blistering of skin on their hands. Chilli peppers are cleaned and sometimes graded before sale as a fresh spice. The chillies are graded apparently by characteristics like colour, size, moisture and stalk of the pods besides seed and fruit (pod) ratio, seed size and hardness, thickness of the skin of the pod and pungency.

2. WORLD-WIDE PRODUCTION OF CHILLIES

Indian Scenario

India is the world leader in chilli production followed by China and Pakistan (FAO). India grows green chillies over 5,700 ha producing 59,100 tons, annually. Chilli crop normally yields 2 to 2.5 tonnes dry chili and 7.5 to 10 tonnes green chilli from one hectare in India. Although chilli is cultivated in all Indian states, maximum chilli production is by Andhra Pradesh followed by Karnataka, Odisha, West Bengal and Maharashtra.

World Scenario

Asia stands at the top in producing 65.8% of world green chillies and pepper followed by Europe, which stands 2nd contributing 12.1% and Africa 3rd with 9.5% of world production. European production is mainly of mild pungent chillies. Chillies produced in Asia are mainly of hot types, where as African countries produce both hot and mild types (paprika) and European production is predominantly of mild type. India, China, Spain, Thailand, Mexico, USA, South Korea, Turkey, Ethiopia, Uganda, Kenya and Tanzania form the major producers of chilli.

3. IMPORTANCE

3.1 Comparative Account of Green and Red Chilli

Green chilli is beneficial in providing protection against development of cancers and ulcers. Red chilli, on the other hand enhances the risk of developing stomach ulcers and cancers of stomach, gall bladder, esophagus and liver. The risk of gastric cancer may be linked to the high consumption of red chilli. Furthermore, Red chillies cause acidity, constipation as well as piles after a prolonged use. The Scoville heat units (SHU) for red chilli (4949.08) are also greater than Green Chilli (2216.58). Therefore, pungency of Red chilli is higher as comparison to its Green counterpart. Green colour implies the presence of beta carotene, flavonoids, anti–oxidant and endorphins.

3.2 Chemical Constituents

Green Chilli comprises of chemical constituents such as capsaicinoids, flavonoids, vitamins, minerals, carotenoids, steroids, steroidal glycosides, polyphenols and macronutrients. Capsaicinoids Capsaicin (oleoresin), dihydrocapsaicin, homocapsaicin I, homocapsaicin II, norcapsaicin, nor-norcapsaicin, homodihydrocapsaicin I, homodihydrocapsaicin II, nordihydrocapsaicin and nornordihydrocapsaicin are different capsaicinoids present of Green Chilli[1]. The active ingredient in a green chilli is called capsaicin, (8–methyul–N–vanillyl–6–nonenamide)[1], a phenolic compound related to vanillin, which gives the strong pungent taste to the capsicum. Of the capsaicinoid fraction, capsaicin (48.6%) is quantitatively followed by 6, 7–dihydrocapsaicin (36%), nordihydrocapsaicin (7.4%), homodihydrocapsaicin (2%), and homocapsaicin (2%). Capsaicinoids and capsaicin are collectively found in amounts of 0.1% to 1%, with quantities varying according to soil and climate. Total flavonoids present in Green Chilli are 83.5 mg/kg of dry weight. They include myrecetin, luteolin, both flavones and flavonone glycoside conjugates[2].

Vitamins

Green chilli is rich in vitamins A and C. Vitamin–C content of a Green chilli is 6 times more than that of an orange fruit. Green chilli also contains vitamins B2 (Riboflavin), vitamin B6 (pyridoxine), niacin, folate and vitamin E. Recently, Russian scientists have identified vitamin–P in green chilli[3].

Minerals

Green chilli is a good source of potassium, manganese, iron and magnesium.

Essential amino acids

Tryptophan, lysine and phenylalanine are also found in green chilli. Green chilli is rich in carotenoids. Some of the phytonutrients present in green chili are alpha–carotene, beta–carotene, lutein and zeaxanthin.

Steroids and Steroidal Glycosides

Green chilli has lanosterol and lanostenol as steroids. Capsicosides A, D and proto–degalactotigonin are steroidal glycosides present in green chilli Polyphenols Polyphenols obtained from green chilli (*Capsicum frutescens*) showed anti–oxidant and vasodilator effects[4].

Macronutrients

A 1/2 cup (30–40 g) of raw green chili peppers provides 1.5 g of protein, 0.15 g of fat, 7.1 g of carbohydrates and 1.1 g of fiber[3].

Capsaicin

Pungency in chilli is due to the alkaloid "capsaicin" contained in the pericarp and placenta of fruits, it produces mild to intense spice when eaten. Capsaicin is a potent inhibitor of neuropeptide associated with inflammatory processes. The hotter the chili pepper, the more capsaicin it contains. The hottest varieties include *Naga Jolokia, habanero* and *Scotch bonnet* peppers. *Jalapeños* are next in their heat and capsaicin content, followed by the milder varieties, including *Spanish pimentos*, and *Anaheim* and *Hungarian* cherry peppers. Capsaicin is being studied as an effective treatment for sensory nerve fiber disorders, including pain associated with arthritis, psoriasis, and diabetic neuropathy. When animals injected with a substance that causes inflammatory arthritis were fed a diet containing capsaicin, they had delayed onset of arthritis, and also significantly reduced paw inflammation. Capsaicin is the active ingredient present in green chilli. It helps in dissolving blood clot and is used in the preparation of ointments and tinctures for their astringent, anti–puritic, counter–irritant and analgesic properties.

3.3 Medicinal Uses

The green chillies have different medicinal uses as natural pain killer, immunity booster, cardiovascular benefits, anti–inflammatory agent, gastro protective agent, anti–rhinitic agent, anti–obesity agent, enhancement of insulin sensitivity, anti–cancer agent, anti–epileptic agent, anti–psoriatic agent, etc.[3].

Benefits in cardiovascular diseases

Green chilli reduces blood cholesterol, triglyceride levels, and platelet aggregation. Thus it lowers the risk of heart attack, stroke and pulmonary embolism.

Enhancement of insulin sensitivity

Diabetes mellitus (DM) is a major chronic disease caused by an improper balance of glucose homeostasis[5]. Diabetes has emerged as a major healthcare problem in India. According to Diabetes Atlas published by the International Diabetes Federation (IDF), there were an estimated 40 million persons with diabetes in India in 2007 and this number is predicted to rise to almost 70 million people by 2025. The countries with the largest number of diabetic people will be India, China and USA by 2030. It is estimated that every fifth person with diabetes will be an Indian[6]. Around 3.2 million deaths every year are attributable to complications of diabetes; there are six deaths every minute[7]. It is now believed that inhibition of α-amylase and β-glucosidases enzymes involved in the digestion and absorption of carbohydrates can significantly decrease the postprandial increase of blood glucose level after a mixed carbohydrate diet, and therefore can be an important strategy in the management of type 2 diabetes[8]. A main drawback of using currently used α-glucosidase and β-amylase inhibitors, such as acarbose, are side effects such as abdominal distention, flatulence, meteorism and possibly diarrhea[9]. Dietary management of DM and hypertension by consuming plant food and their ingredients could be a more effective strategy because of likelihood of high compliance and absence of side effects. Plant based foods, such as pepper, have nutritional benefits with low calories. If person's diet contains chilli, the amount of insulin required to lower the blood glucose is diminished. When consumed on regular basis, it may even drop the requirement of insulin level to the lowest[3].

Carotenoids

Carotenoids play an important role in fruit colouring and act as antioxidants, reacting with free radicals, mainly peroxide radicals and singlet molecular oxygen[10]. Lycopene is a powerful natural antioxidant that acts as the most efficient singlet oxygen quencher *in vitro* among common carotenoids[11] and as a determinant factor in reducing the mortality from several cancers[12]. For another major carotenoid in pepper, β-carotene, there is much *in vitro* evidence of its interaction with free radicals, acting as a chain–breaking anti–oxidant and as a scavenger and quencher of singlet oxygen[13].

L–Asparaginase

L–Asparaginase (L–asparagine amidohydrolase) acquired some clinical importance in 1961, when the antitumour effect of guinea pig serum, originally discovered by Kidd[14], and was traced by Broome[15] to the presence of this enzyme. The enzyme from *Escherichia coli* was clinically used to treat patients suffering from asparagines–dependent leukemias and lymphomas. A number of undesirable side effects, ascribed to the presence of contaminating bacterial endo–toxins in the enzyme preparations were observed[16]. Hence, efforts attempted to prepare this enzyme from an entirely different source plant material and study its biochemical and biological properties. Very little work has been carried out on L–asparaginase from higher plants. A survey of several plants revealed that green chillies (*Capsicum annuum* L.) contained appreciable amounts of this enzyme[17].

3.4 Natural Dye

The synthetic dyes which are of wide commercial importance cause severe atmospheric and environmental pollution. The investigation was carried out

Table 1: Nutritional value of Chilli[3]

(g/ 100 g dry matter)

Parameters	***Chillies dry (Red)***	***Chillies (Green)***
Moisture content	11.11	599.3
Protein	16.66	20.27
Fat	6.88	4.19
Minerals	6.77	6.99
Fibre	33.55	47.55
Carbohydrates	35.11	20.97
Calcium	0.177	0.209
Phosphorus	0.411	0.559
Iron	0.0025	0.0307
Vitamins		
Carotene	0.000383	0.00122
Thiamine	0.00103	0.00132
Riboflavin	0.000477	0.002727
Niacin	0.0105	0.00629
Vitamin C	0.0555	0.776
Minerals & Trace Elements		
Sodium	0.0155	–
Potassium	0.588	–
Phytin Phosphorus	0.0788	0.0489
Magnesium	–	1.90
Copper	–	0.00979
Manganese	–	0.00965
Molybdenum	–	0.00049
Zinc	–	0.01245
Chromium	–	0.00028
Oxalic acid	–	0.4685
Energy (K cal per 100 g dry matter)	273–203	–

to extract natural dye from green chilli (*Capsicum annum*)[17]. The main coloring component in chilli is oleoresin. The dye was extracted using solvent extraction method. The mordants are used in fixing the dye to the fabric as well as improve the fastness properties of the dyed cloth. Different shades of yellow dye were obtained from the dye extracted from chilli skin. These different shades are obtained from a single dye using the two mordants in different ratios. Good light fastness, good rub fastness and moderate wash fastness was observed in fabrics dyed with the dye extracted from Green chilli.

3.5 Composition of Chilli

The various important composition of chilli are indicated in Table 1.

4. POST HARVEST MANAGEMENT

4.1 Harvest Maturity

Chilli is generally harvested when ripe, but it also can be harvested at a green, immature stage[18] (Fig. 1). Maturity of green chili can be based on size, firmness, and color. For fruit to be harvested ripe, at least 50% color should have developed (Aguilar, undated). The large sweet peppers usually are picked while they are still green colour but full grown when sold in the market.

Fig. 1: Hand picking Chilli

Chili is non–climacteric as it does not ripen after harvest. Chili harvested unripe has no capacity to complete ripening unless postharvest conditions are favorable. In a study by Krajayklang *et al.*[19], it was found that *Capsicum annuum* L. paprika and cayenne chilli fruit harvested green or deep green failed to attain full red colour under room conditions, while fruit harvested at or after the colour break stage (when the color begins to change) developed full red color within 7–9 days. In a study on bell pepper harvested mature–green, the fruit was able to develop deep red colour when stored under evaporative cooling conditions[20].

4.2 Time of Harvesting

Harvesting can be done weekly, preferably during the early part of the day. Harvesting during or just after rain is not recommended, as wet conditions favour microbial growth and enhance fruit breakdown. If chilli must be harvested when the sun is up, the fruit should placed under shade to dissipate heat before it is packed.

4.3 Harvesting Method

Chilli is picked by hand (Fig. 2). The fruit is harvested by removing it from the branch and ensuring that the stem remains intact and attached to the fruit. Only fruit with the required color and size should be harvested. Overripe soft fruit is removed from the plant and graded out to be used for other purposes, if possible. During harvesting, pickers should wear gloves to protect their hands because the oils (capsaicin) in the fruit can cause severe burns; pickers should take care not to touch their face or eyes. Harvested fruits can be placed directly into plastic field crates or into smaller plastic buckets, which are then transferred to crates at the side of the field. Cotton waist bags also can be used to collect the peppers and carry them to field crates.

Fig. 2: Green chilli harvester

Mechanical harvesting is not advisable for chilli destined for the fresh market because it may injure the fruit. However, mechanically harvested fruit can be used for processing. The harvester designed by New Mexico State University (NMSU) Manufacturing Technology and Engineering Center (M–TEC) for green chilli harvesting gives 80 per cent undamaged chilli.

***Field handling*:** The harvested fruit should be kept in shaded conditions, and protected from sun, wind and rain. Plastic crates are recommended when transporting from the field to the packing house to minimize damage. Sacks or mesh bags do not provide sufficient protection for the fruit during transport.

4.4 Grading

There are no recognized grade standards for chilli. Good quality fruit should be of uniform shape, size, and color typical of the variety. Fruit with defects such as cracks, decay, mechanical damage, and sunburn should be sorted out and rejected. Undersized, shriveled, dull–looking, pitted, or softening fruit also should be rejected.

For different purposes, the varieties of chillies are chosen by the end user. End users are mainly of two types *viz.* domestic retail users and industrial wholesale users. Industrial users who usually prepare chilli powder and puree give preference for colour and pungency, fleshly skin and less seeds. Whereas, the domestic users prefer all varieties for varied uses. There are several local and conventional grades followed by the farmers, village merchants and itinerant merchants. The visual assessment of grades by looking at the lots/heaps and by picking hand full of pods and analyzing them to enable the traders to adequate and assess the prices both in open and closed auction.

4.5 Packaging

Packaging is an important function for every produce and holds importance for marketing of the produce. It is a practice to protect the produce from any damage during storage, transportation and other marketing aspects. It is required at every stage of marketing from the producer to the consumer. In recent years, packaging plays an important role in marketing of produce. Good packaging of chilli not only facilitates convenience in transportation and storage but also attracts consumer to pay more. The packaging reduces the marketing cost and protects the quality.

The good packaging material must possess the following qualities:

- It must protect quality and quantity.
- It must prevent spoilage during transit and storage.
- It must tell information about quality, variety, date of packing, weight and price etc.
- It must be convenient in handling operations.
- It must be cheap, clean and attractive.
- It must be biodegradable.
- It must free from adverse chemicals.
- It should be useful after the first use.

4.6 Storage

Freshly harvested peppers must be stored between 7 to 10°C and 95% relative humidity. The typical storage life of peppers under these conditions is 3–5 weeks. Peppers are sensitive to chilling injury when exposed to temperatures below 7°C. Symptoms of chilling injury include pitting and water–soaked tissue. Physiological disorders in peppers include blossom–end rot and pepper speck. Blossom–end rot is characterized by dark, sunken lesions at the blossom end of the fruit. Pepper speck occurs as spot–like lesions that penetrate the fruit walls. Fruit showing either of these disorders should not be stored and should be discarded.

In a study on the chili cultivar '*Jinda*,' it was found that high–density polyethylene bags were more effective in prolonging shelf life than polypropylene and polyvinyl chloride films[21]. The fruit was still of acceptable appearance at the end of the 28–day storage period at 10°C. Evaporative cooling storage is a simple method with potential for chilli, particularly for red color development. Its effectiveness in extending shelf life and promoting red peel colour development of mature–green harvested fruit has been demonstrated in bell peppers[20].

4.7 Storage Diseases

The major post–harvest diseases of peppers are Alternaria and Botrytis rots. Phytophthora rots can also occur if the pepper fruit are exposed to prolonged periods of heavy rainfall. Fruit infected by Phytophthora have water–soaked lesions.

Fungal rots during storage can be controlled by:

1. Promptly cooling of harvested fruit
2. Avoiding bruising and injury
3. Post–harvest hot water dips at 52–55°C for 2 minutes
4. Post–harvest application of registered antimicrobials (*e.g.*, O-phenylphenol at 98 g/L).

5. PROCESSING

Chilli is generally found to be used in three forms, namely, fresh green chillies, red ground and raw red. Usually, red chillies are dried in the open sun without any pre–treatment. Notably, there are very limited research reviews on the drying, processing and preservation of green chillies. The local variety green chillies are found to be wasted at the farm level due to the lack of proper processing and preservation technology. The occasional sluts of green chillies cause economical losses to the growers whereas the scarcity periods make the green chillies too costly, that invites the necessity of introducing the appropriate processing of green chillies in production catchment itself. Thus, the processing and preservation of green chillies are very important in order to minimize the field losses as well as enhance the value of green chillies. Processing of green chilies has vast potential as it would avoid long drying operation and reduces losses and also can provide the chilli powder with better nutritional quality and controlled pungency.

5.1 Green Chilli Powder

The green chilli is highly perishable but at the same time available throughout the year. The increasing demand of processed ready to eat and ready to cook products has resulted in growing industry of Indian spices. Dehydrated and chemically pretreated green chilli powder has extended shelf life and can be easily incorporated in various homemade recipes like red chilli powder.

Sample preparation: Grading is done manually to remove the undersized, black and spoiled chillies to get the good quality green chilli powder from the fresh green chillies. Before pretreatment, the waste material *i.e.*, stalk portion is removed manually. The chillies were blanched in hot water at 90°C for 3 minutes. The chillies were given horizontal cuts of 1cm in order to improve the drying rate after blanching. The green chillies were pretreated with the solution containing: 1% ascorbic acid, 0.3% sodium metabisulphite and 1% calcium chloride after blanching by giving 10 minutes soaking in above solutions. The ratio of the green chilli: pretreatment solution is 250 g: 1 Litre.

Drying of green chilli: Pre treated chillies are then surface dried and transferred to the drying in the hot air at 60°C.

Powdering of dehydrated chilli: After complete drying, the dehydrated chilli samples obtained were taken for grinding in food processor mixer grinder to make powder of dried green chillies.

As drying air temperatures using a lab–scaled dryer increased, decrease in ascorbic acid was noted. Ascorbic acid was degraded by higher temperatures and the degradation product (L–dehydroascorbic acid, DHAA) could participate in Strecker degradation with amino acid, producing a browning pigment. Not only does the high temperature of drying air affect the loss of ascorbic acid, but a long period of drying time can also introduce a significant loss of ascorbic acid. While drying, Chilli loses most of its vitamin C, it increases the vitamin A content by 100 times[3]. Vitamin A is a powerful anti-oxidant and anti-inflammatory agent.

5.2 Shelf Life and Nutritional Quality of Green Chilli (*Capsicum annuum* L.) Powder

The powder prepared from the chilli pods without pedicle, sliced into two parts along the length and treated with 0.01% KMS was found to be better than that of the other samples in terms of the shelf life and sensory quality with acceptable nutritional values when it was packed in HDPE. Processing of green chillies in the form of powder and preserving it in low–cost poly packages were identified to be potential techniques to minimize field wastage of green chillies, and hence, chilli growers could benefit from these techniques. This powder can be used with many food adjuncts in place of fresh green chillies and red chilli powder. Further research is recommended for standardizing the efficient and sustainable technique for processing and preserving green chilli powder using other mechanical drying methods such as vacuum, oven drying, freeze drying, fluidized bed drying, heat pump drying and others, both in domestic and industrial scales.

A lab scale hot air oven using a temperature of 70°C provided darker brown color and less green color as shown in spectrophotometric absorbance at 540 nm. Not only has the drying temperature affected the nutrient content but also the drying time. Using Sodium Metasulphite (NaMS) at drying air temperature of 60°C provided more bright green colour as sulfite inhibits the non–enzymatic browning reaction. In addition, soaking in NaMS combined with $CaCl_2$ produced highest colour value. Reducing sugar could not be preserved by soaking in chemical solutions. An increase in ascorbic acid content was found when treated with ascorbic acid solutions for both the temperatures of drying[22].

5.3 Chilli Pickles

Pickles are very popular across the country and they are regularly consumed by almost all households. Restaurants, *dhabas*, caterers etc. are bulk consumers. There are many varieties of pickles with certain regional specialities as well. They are generally spicy but some are sweet also and used as stable enrichers.

The process of food preservation in common salt or in vinegar or in both is called pickling. Pickling with the help of vinegar and oils has been in practice from time immemorial in this country. In modern days, pickles of different fruits and vegetables are prepared with the mixture of salt, oil, vinegar and spices etc. In pickles, oil acts as preservative[23]. Green chilli is used as spice in cooking food, raw in salad and pickle[24].

Green chillies are washed in water and then dried under the sunlight. After cutting the top and bottom portion, they are cured in brine solution for 2½ to 3 days. After that, oil, turmeric powder and other suitable ingredients (depending upon local palate) are added and mixed thoroughly and then pickle is packed in bottles and plastic pouches.

5.4 Puree

Chillies were washed in running tap water, destalked manually and blanched for 3 min in hot water at 100°C. Preliminary trials demonstrated that blanching at 100°C for 3 min resulted in complete inactivation of chlorophyllase and peroxidase enzymes in green chillies. Chillies were cooled immediately in chilled water and drained. The blanched chillies were then comminuted in a grinder through 14 mesh (1.19 mm) screen to obtain a puree of uniform size. The puree was stored immediately at 0°C till further use.

5.5 Chilli Shrimp Paste

Chilli shrimp paste (CSP) is a regional favourite condiment, usually consumed uncooked as a side dish with meal or raw vegetables. It is known as 'sambal belacan' in Malaysia. This paste is prepared by crushing the fresh chilies, toasted fermented shrimp paste 'belacan', calamansi juice, sugar, and salt[25]. These ingredients contain natural microflora such as mold, yeasts and specific bacterial genera such as *Bacillus, Clostridium, Enterobacter, Escherichia, Micrococcus, Proteus, Streptococcus, Acetobacter, Pediococcus, Lactobacillus, Sarcina, Staphylococcus, Brevibacterium–like, Flavobacterium*–like, and *Corynebacterium*–like bacteria[26,27]. Toasting of fermented shrimp paste before mixing with other ingredients can reduce its microbial load. Since the ingredients used contain natural microflora and that CSP is usually consumed uncooked, this paste can be easily spoiled over time. This is supported by the fact that shelf life of CSP is not more than three days when kept in a refrigerator[28].

5.6 Pepper Sauce

For preparing Pepper sauce, the hot peppers are washed and drained followed by cutting off stems. The peppers may be sliced in two to three pieces. The sliced material is packed in hot jars, leaving some space at the top. The boiling hot cider vinegar is added into the jars leaving space at the top. Jars are sealed by lid and heated in boiling water for 10 min at sea level or for 15 min at altitudes of 300 to 1800 m or for 20 min at altitudes higher than 1800 m.

5.7 Pepper Jelly

Hot peppers, cored and chopped, medium green or red bell peppers, cored and chopped, white vinegar (5%), sugar, liquid pectin, green food coloring

1. Put half the peppers and half the vinegar into a blender, cover and process until peppers are liquefied. Repeat with remaining peppers and vinegar.
2. Combine the peppers and vinegar mixture with the sugar in a saucepan and boil slowly for 10 min. Remove from heat.
3. Add liquid pectin and boil hard for 1 min.

4. Skim foam off the top of the jelly and add a few drops of green food colouring. Pour jelly immediately into canning jars, leaving 6 mm of headspace.
5. Wipe jar rims with a clean, damp cloth and secure lids and ring bands.
6. Process the jars of jelly in a boiling water bath.

5.8 Canned Chilli

Selecting chillies: Select chillies that are mature, heavy for their size, smooth and symmetrical, bright green in colour, fresh, and crisp. Avoid misshapen pods, shriveled skin, mold, soft spots, and bruises.

Blistering: The tough outer skin must be removed from chillies. Blistering the skin by one of the following methods makes removal easy. Wash and drain out the water and make a small slit in the side to allow steam to escape. Very hot heat source is utilized for this purpose. Turn chillies frequently to prevent scorching and ensure even blistering. Remove from heat and spread chillies on a flat surface in a single layer to cool before peeling. For easier peeling, place in a pan and cover with a damp towel for a few minutes. The following are three heat–source methods for blistering chillies:

- *Oven or broiler method*: Place chillies in a hot oven or broiler 204 to 232°C for 6–8 minutes until skin blisters so that it can be pulled away from the flesh.
- *Range top method*: Place chillies on a hot electric or gas burner after covering burner with a layer of heavy wire mesh.
- *Outdoor grill method*: Place chillies on a charcoal grill about 13–15 mm above glowing coals.

Salt: Use salt as per taste. Salt may be omitted, as it is used only for flavor. For best results, do not use salt substitutes for canning because heat causes some substitutes to become bitter or develop a metallic taste. Instead, add salt substitutes just before serving.

5.9 Green Chilli Wine

Green chilli wine uses a mild green chilli as the key ingredient to create a unique New Mexican flair. It is semi–sweet and very smooth. It has sharp clarity with a faint aroma of green chilli. It's crisp and refreshing with just a touch of sweet.

5.10 Frozen Peppers

Peppers lose their crispness when frozen and thawed. Frozen peppers are useful in cooked dishes where texture is not important but the pepper's flavour is desirable. Unlike most vegetables, they do not require blanching before freezing. Peppers can be sliced or diced, quick frozen on a cookie sheet and packed into moisture and vapor–proof containers. Freeze at or below –18°C. Frozen peppers will maintain their quality for 9 months.

5.11 Green Chilli Cashew

Chilli cashews are widely served as snacks with tea or coffee across the world. These cashews are acknowledged for their features like crispiness, purity and long shelf life. Cashews are blended with the flavours of green chilli and lemon. Chilli adds spicy taste whereas lime offers sour taste to the nuts that can be eaten as healthy snack anytime.

6. CONTRAINDICATIONS AND PRECAUTIONS

Chilli is contraindicated in patients, who are hypersensitive to capsicum or its products. It is contraindicated in pregnant ladies, as it stimulates the uterus. One should be careful, while handling chillies, since it can irritate or may even burn eyes. If consumed in large amounts, it could prove toxic, and may burn the mouth. But at lower doses, it seems to possess anti–cancer properties. Casein (a milk product) can help to suppress the Green Chilli fire[3].

7. CONCLUSION

Green chilli makes an integral component of Indian diet. Salads, pickles, sauces, chutneys, curries and spicy snacks are almost incomplete without this green small sized vegetable. Green chillies cleanse the skin and fair the complexion by stimulating perspiration. Green chilli or *Capsicum frutescens* is a natural, spicy and hot food item belonging to potato family. Its active ingredient Capsaicin, responsible for its intense hot taste is mainly located in seeds or in the white pith, where seeds are attached. A small chilli fruit provides lots of dietary fibers, vitamins–A, C, B_2, B_6, E, along with several minerals and macronutrients. It possesses numerous medicinal properties such as anti–cancer, anti–inflammatory, anti–oxidant, anti–hemorrhoidal anti–obesity, gastro–protective, anti–pyretic and analgesic. It provides relief against various ailments like rhinitis, sinusitis, migraine, diabetes and arthritis. Thus, inclusion of Green Chilli preferentially in our daily diet offers plenty of health benefits.

8. REFERENCES

[1] Cordell G.A., Araujo O.E., "Capsaicin: Identification, nomenclature and pharmacotherapy", *Ann Pharmacother*, **27**(3), 330–336 (1993).

[2] Miean K.H., Flavonoid M.S., "Myricetin, quercetin, kaempferol, luteolin and apigenin, content of edible tropical plants", *J Agric Food Che.*, **49**(6), 3106–3112 (2001).

[3] Tandel B.M., Shah K.A., Nayaka P., Timbadia C.K., FLD Impact analysis on scientific cultivation of chilli, *Gujarat Journal of Extension Education,* **22**, 68–69 (2011).

[4] Abeywardena M., Runnie I., izar M.N., Momamed S., Head R., "Polyphenol–enriched extract of oil palm fronds (Elaeis guineensis) promotes vascular relaxation *via* endothelium dependent mechanisms", *Asia Pacific J Clin Nutr.*, **11**, S467–S472 (2002).

[5] Rosetti L., De fronzo R., Gherzi R., Stein P., Andraghetti G., Falzetti G., Schulman G., Kelin R.E., Cordera R., "Effect of metformin treatment on insulin action in diabetic rats: *In vivo* and *in vitro* co–relations" , *Metablosim*, **39**, 425–35 (1990).

[6] Gupta R., "Diabetes in India: Current Status", Online available at www.expresshealthcare.in (2008).

[7] WHO (2004) Diabetes action now programme – A Report, Available online at http://www.who.int/diabetes/actionnow/en/DANbooklet.pdf, pp. 4

[8] Puls W., Keup U., Krause H.P., Thomas G., Hoffmeister F., "Glucosidase inhibition. A new approach to the treatment of diabetes, obesity and hyperliprproteinemia", *Naturwissenschaften,* **64**, 536–537 (1977).

[9] Puls W., Keup U., "Inhibition of sucrase by TRIS in rat and man, demonstrated by oral loading tests with sucrose", *Metabolism.*, **24**(1), 93–98 (1975).

[10] Namiki M., "Antioxidants/antimutagens in food", CRC *Critical Reviews in Food Science and Nutrition,* **29**, 273–300 (1990).

[11] Masceo P.D., Kaiser S., Sies H., "Lycopene as the most efficient biological carotenoid singlet oxygen quencher", *Archives of Biochemistry and Biophysics,* **274**(2), 532–538 (1989).

[12] Zhang Y., John M., Wallace, David S., Battisti, "ENSO–like Interdecadal Variability: 1900–93", *J Climate.*, **10**, 1004–1020 (1997).

[13] Conn P.F., Lambert C., Land E.J., Schalch W., Truscott T.G., "Free Radical Research Commun. 16, 401Cordell GA, Araujo OE. Capsaicin: Identification, nomenclature and pharmacotherapy", *Ann Pharmacother.*, **27**(3), 330–336 (1992).

[14] Kidd J.G., "Regression of transplanted lymphomas induced *in vivo* by means of normal guinea pig serum. I. Course of transplanted cancers of various kinds in mice and rats given guinea pig serum, horse serum, or rabbit serum", *J Exp Med.,* **98,** 565–583 (1953).

[15] Broome J.D., "Evidence that the L–Asparaginase activity of guinea pig serum is responsible for its antilymphoma effects", *Nature.,* **191**, 1114–1115 (1961).

[16] Burchenal J.H., Clarkson B.D., Dowling M.D., Gee T., Haghbin M., Tan C.T.C. "Experimental and clinical studies of L–asparaginase in combination therapy", *In*: La L–Asparaginase (Bernard J., Boiron M., Jacquillet C., Weil M. and Levy D. *eds.*), *Editions De Centre National de la Recherche Scientifique,* Paris, pp. 243–248 (1970).

[17] Kulkarni S.S., Bodake U.M., Pathade G.R., "Extraction of natural dye from chili (*Capsicum annum*) for textile coloration", *Universal Journal of Environmental Research and Technology,* **1**, 58–63 (2011).

[18] Berke T., Black L.L., Talekar N.S., Wang J.F., Gniffke P., Morris R., "Suggested cultural practices for chili pepper", *AVRDC Pub*. 03–575, p. 8 (2004).

[19] Krajayklang M., Klieber A., Dry P.R., "Colour at harvest and post–harvest behaviour influence paprika and chili spice quality", *Postharv Biol Technol.,* **20**(3), 269–278 (2000).

[20] Acedo Jr A.L., "Storage life of vegetables in simple evaporative coolers", *Trop Sci.,* **37**, 169–175 (1997).

[21] Suwannasopon P., "Modified atmosphere packaging of chili", *Kasetsart University, Bangkok. BS Thesis.* (2004).

[22] Sarker M.S.H., Hasan S.M.K., Aziz M.G., Islam M.T., Azam S.M.R., Roy S., Ibrahim M.N., "The effect of processing treatments on the shelf life and nutritional quality of

green chilli (*Capsicum annuum* L.) Powder", *Pertanika Journal of Tropical Agriculture Science*, **35**(4), 843–852 (2012).

[23] Shahjahan M., Miaruddin M., Rahman M.M., "Pickling and fermentation of fruits and vegetables' *In*: Proceedings of international workshop on agro processing techniques of fruits and vegetables for income generation. Bangladesh Agricultural Research Institute, Joydebpur, Gazipur. (2005).

[24] Pruthi J.S., "Spices and Condiments", *National Book Trust,* New Delhi, India. p. 269 (1976).

[25] Sobhi B., Adzahan N.M., Karim M.S.A., Karim R., "Physicochemical and sensory properties of a traditional chilli shrimp paste", *Journal of Food, Agriculture and Environment.,* **8**(1), 38–40 (2010).

[26] Steinkraus K.H., "Hand book of indigenous fermented foods" New York: Marcel Dekker, Inc (1996).

[27] Saraya S., Temsiririrkkul R., Manamuti C., Wongkrajang Y., Cheewansirisuk C., "Sappan wood extract used as preservative in chili paste", *Mahidol Uni J Pharma Sci.*, **36**(1–4), 38–40 (2009).

[28] Passmore J., "The letts companion to Asian food and cooking", Singapore: Kyodo Printing Co Ltd Press (1991).

20

Developments in Processing, Quality Evaluation Techniques and by–Product Utilization of Onion and Onion Products

DATTATREYA M. KADAM[1]* AND ARUN KUMAR T V[1]

ABSTRACT

Onions (Allium cepa L.) is very popular crop worldwide and an important vegetable and spice along with immense medicinal and therapeutic property. Onions form natural part of the daily diet for most of the population all over the world. Because of its chemical composition, apart from taste and flavor, onions are also associated with imparting a number of health benefits to its users. Onion, as an essential ingredient for the cuisine, is an everyday kitchen need. Since it is a seasonal crop, storage of onion bulbs becomes mandatory to meet the year round demand from the consumers. Field–dried onions can be stored up to two months under ambient conditions in well ventilated trays on pallets or in a field windbreak. Onions are stored in the field, as a cheap and effective method, in many countries with dry climate. Onions are susceptible to a number of physiological disorders during storage such as, watery scales, translucent scales and freezing injury. Symptoms of watery scales include a thick leathery skin with watery glassy scales below; freezing injury resembles watery scales and is characterized by soft water soaked scales. Sprouting and root formation accelerate deterioration besides dictating marketability of onion bulbs. Processed onion products have considerable advantages to food industry, including reduced bulk resulting in lower transportation and distribution costs, stable price and availability, convenient dispersion in food products compared to chopped or sliced fresh form etc. Currently onion products are marketed in a variety of forms. They include dehydrated onion pieces, dried flakes, onion powder, onion paste,

[1] ICAR–Central Institute of Post–Harvest Engineering and Technology (ICAR–CIPHET), P.O. PAU, Ludhiana–141004, Punjab (India)

**Corresponding author* E-mail: kadam1k@yahoo.com

onion flavoring, encapsulated flavors, oleoresins and essential oils, onion salt, pickled onions, canned, frozen, minimally processed and packaged onion.

Key words: Harvesting, Storage, Sprouting, Process, Products, Dehydrated onion, Flavour oil, Solar drying, Minimal processing, Waste

---

INTRODUCTION

Onion (*Allium cepa* L.), belongs to the family *Liliaceae*, is one of the oldest vegetables in continuous cultivation dating back to at least 4700 BC or more. Onion was probably first domesticated in the mountainous regions of Turkmenistan and north Iran. Therefore, south–west Asia is regarded as being the primary centre of domestication and variability. Other regions of great diversity like the Mediterranean are considered secondary centres[1]. *Allium cepa* is cultivated mainly as a biennial, but some types are treated as perennials. It is propagated by seeds, bulbs or sets (small bulbs). Very wide variation in shape, size and colour is observed in this species. Onion is a biennial crop but cultivated as annual and harvested at various stages of maturity either to provide green leafy vegetables or the bulbs. Due to its significant medicinal property, nutritional worth and energy value, onions impart numerous health benefits to the consumers. The onion has many culinary uses. It can also be processed into dry products such as rings, flakes, and powder for the food processing industry.

World Scenario

As an important vegetable and spice along with immense medicinal and therapeutic property, onion is a very popular crop worldwide. Tropical countries accounts for nearly 50% of global onion production. Although some tropical countries are net importers, export potential of onions is developing in several tropical regions as the bulbs can be transported for considerable distances without deteriorating[2]. World dry onion production stood around 85 million tons during the year 2013, with the crop being grown across 4.4 million hectares in over 134 countries. China, India and US alone accounts for more than half the onions produced in the world. Production statistics along with the area and percent share in world production of top onion producing countries are indexed in Table 1.

Chemical Composition

Onions form natural part of the daily diet for most of the population all over the world. Because of its chemical composition, apart from taste and flavor, onions are also associated with imparting a number of health benefits to its

Table 1: Global production statistics of onion during 2013

Country	*Area (million Ha)*	*Production (million tons)*	*% share in the global production*
People's Republic of China	1.02	22.34	26.04
India	1.21	19.29	22.48
United States	0.06	3.16	3.68
Turkey	0.06	1.90	2.20
Pakistan	0.12	1.66	1.90
World	4.44	85.79	–

Source: FAO–STAT

users. Onion nutritional composition is very complex. Table 2 shows nutrient composition of the onion.

Table 2: Nutrient composition of onion[3, 4]

Proximate composition	*Content (per 100 g of edible portion)*
Moisture	88.6 – 92.8%
Protein	0.9 – 1.8 g
Carbohydrates	5.2 – 9.0 g
Crude fiber	0.9 – 1.2 g
Fat	Trace – 0.2 g
Ash	0.6 g
Elements	***mg per 100 g fresh weight***
Calcium	190 – 540
Phosphorus	200 – 430
Potassium	80 – 110
Magnesium	81 – 150
Iron	1.8 – 2.6
Zinc	1.5 – 2.8
Copper	0.05 – 0.64
Manganese	0.5 – 1.0
Sulfur	50 – 51
Sodium	31 – 50
Vitamins	***mg per 100 g fresh weight***
Thiamin (Vitamin B_1)	0.3
Riboflavin (Vitamin B_2)	0.02 – 0.05
Niacin	0.2
Ascorbic acid (Vitamin C)	8.0 – 10.0
Vitamin D	0.3

Besides, onion contains compounds characteristic of its flavour such as thiol, dimethyldisulfide, diallylsulfide, disulfides, and thiosulfate, as well as small amounts of citric acid, o–hydrocinnamaldehyde, caffeic acid, asafetidic acid, protocatechuric acid, polysaccharide A, polysaccharide B, quercetin 3, 4′–diglycoside, thymine, and several amino acids. Medical reports have shown that onion can protect against cancer, fight fungi and bacteria, promote cardiovascular health, reduce high blood pressure and insulin resistance, aid in weight loss, possess antioxidant activity, fight chronic bronchitis, infections, fever etc[5].

It has been shown that onion is one of the major sources of dietary flavonoids. Onion flavonol content and antioxidant activity have been related to human health promoting effects. Quercetin and quercetin glucosides (quercetin–4′–glucoside and quercetin–3, 4′–diglucoside) have been reported the main onion flavonols[6]. Onion is rich in other bioactive compounds such as fruto–oligosaccharides and sulfur compounds[7]. The nutritive value, pungency and the quantity of lachrymatory factor of onion bulbs vary in different genotypes. The white onion varieties with high dry matter content and pungency are considered suitable for dehydration purposes while the varieties with high contents of non–reducing sugars and low protein contents are best for storage purposes.

Flavour and Pungency of Onion

Widespread use of onion lies in the flavour that it imparts to various other dishes. Distinctive flavour and pungency of onion is due to sulphur containing compounds available in the scales of bulbs. The sulphur is in the form of various non–protein amino acids which include the precursors of volatile flavor compounds. These precursors are odourless, non–volatile amino acids of general name S–alk(en)yl cysteine sulphoxides. Intact onion cells have no odour, but when cells are disrupted the enzyme alliinase is released. This enzyme hydrolyses the S–alk(en)yl cysteine sulphoxides to produce pyruvate, ammonia and many volatile sulphur compounds associated with flavour and odour[8]. Boiled onion does not produce flavour as the enzyme is destroyed before having access to the flavour precursors. The tear–producing character of onion on cutting is known as the lacrimatory factor. Lacrimator is formed enzymatically during the hydrolysis of S–propenyl cysteine sulphoxide[4].

Classification of Onion

Onions are usually classified by their day length requirement for bulbing. Within the group, there are early and late maturing cultivars. Onions can also be grouped into mild or pungent:

1. ***Short–day type*** (European onion): bulbing in response to 10 –11 h day length, mild, soft–fleshed bulbs for fresh use.
2. ***Intermediate–day type***: bulbing in response to 12 – 13 h day length, pungent, soft fleshed for fresh use.
3. ***Long–day type*** (Spanish onion): bulbing in response to 14 or more hours of day length, pungent, hard, good for storage.

Bulbs vary in skin colour, shape, and size. There are more yellow or brown onion cultivars than red or white cultivars.

HARVESTING

Time/Stage of Harvest

For any agricultural commodity, stage of harvest is a prime factor that substantial influences post-harvest life and onion has no exception. Harvest maturity stage in onion significantly affects initial bulb weight, respiration and incidence of sprouting, decay and cumulative weight loss [9, 10]. Highest yields are achieved when plants remain intact until the leaves are completely dry. However, for long storage life, it is now common in developed countries to harvest the crop when 50–90% of the tops have fallen [11]. Bulbs harvested at 90% fallen tops and field–cured had the worst skin retention. Recommendations were to harvest at 70–90% tops down and allow the foliage to dry in the field before topping. Late lifting (at 90% or more tops down) of European storage–type onions results in increased sprouting and rooting, storage rots and weight loss and higher incidence of watery scale. Bulbs harvested too soon (*i.e.* when still developing) have low levels of growth inhibitors, high moisture content in the foliage leaves and bulb necks (providing an environment which favours pathogen infestation) and thicker necks, are not yet fully dormant and thus are simply unsuitable for storage [12].

It is important to consider the type of onions and their subsequent use before finalizing the harvesting stage. The timing of harvesting should be decided according to these considerations: relatively early harvesting favours better skin retention while later harvesting maximizes yields. In Europe and USA, the consensus is that the optimum harvest time for storage onions is at 80–90% tops down [11].

Harvesting

Large volumes of onions are harvested by hand in many parts of the world, including sweet onions in the southern USA and elsewhere if soils are difficult to manage. Manual harvesting is time consuming and uneconomical, hence this process has gradually been mechanized. Mechanical harvesters for onions were first adapted from potato harvesters but, in many cases, they were not successful because of soil characteristics or bulb damage. Any mechanical onion–harvester must keep onion damage to a minimum. This is important if the crop is for fresh consumption, and even more if it is intended for storage[11]. Mechanized lifting is the common practice in temperate climates. Undercutting onions (*i.e.,* running a blade below bulb level to separate the bulbs from most of the roots) is commonly done prior to lifting. Initially, under cutters with a flat blade were used; later, oscillating blades and rotating bar under cutters were introduced [13]. The principles of operation of a mechanical harvester for sweet onions was developed that included grasping the onion plants by their tops, lifting them from the ground, shaking soil from the roots, severing the leaves above each bulb and then delivering bulbs to a container and tops to the ground[14].

When onions are to be stored in bins or in bulk, tops are usually removed, in order to improve air flow through the bulbs and reduce trash in the store. Topping can be carried out before, during or after harvest and before store loading. In temperate countries, it is often done before harvest, provided that a heated, forced–air ventilated store can be used for immediate curing. In the tropics, where controlled drying facilities are rare, it is safer to top after most of the leaves have dried [15]. In India, sun–curing of bulbs with tops or storing bulbs with dried foliage minimized storage losses in Kharif (summer wet season) onions compared with other treatments [16]. Topping before undercutting can reduce storability if the bulbs are not dried immediately [17].

Curing

Curing is an essential post-harvest unit operation to extend the shelf life of roots and tubers, when the produce is intended for long term storage. In onion and garlic, curing is the drying of external scales together with colour development and neck closure. During curing, the thin outer layers of the bulb are dried to form one or more complete dry skins, which act as a barrier to water loss, microbial infection and also decreases the incidence of neck rot during storage [18]. Heat and good ventilation, preferably with low humidity, are the essentials for curing of onion bulbs. Initial curing (surface and neck drying) may take between a few hours and several days under forced, heated air ventilation, depending on the temperature and RH of the ventilating air and the stage of maturation of the bulbs. Onions are considered cured when neck is tight and the outer scales are dried until they rustle. This condition is reached when onions have lost 3 to 5% of their weight. After curing, the outermost layer, which may be contaminated with soil, usually falls away exposing the dry under–layer, which should have an attractive appearance.

Wind–rowing is the conventional way of field–curing onions. In this method, uprooted onion bulbs are laid on their sides, with the leaves of one row covering the bulbs of the next row in order to promote thorough drying of the tops while protecting the bulbs from undue sunburn, and are left in the field for 1 or 2 weeks. This method is generally followed in dry and warm conditions. Under wet conditions, it may be necessary to dry onions on racks or trays under cover because drying takes longer and the bulbs risk becoming watermarked and are prone to rot in subsequent storage. In hot climates, the bulbs are sometimes covered with straw or leaves while curing. In temperate regions, good practice is to cure the onions in storage facilities by forced circulation of hot and humid air. Initially by passing air at 30–32°C through bulk stores or bins. After 3–5 days the heat is gradually lowered to the safer level of 27°C and 70–75% RH for the completion of drying over about 20 days this continues gradually until the outer skins reach the 'rustling dry' stage on top of the stack [18, 19].

Uncontrolled curing, either field or artificial, is detrimental to the quality and storability of the bulbs. Late harvesting (at 100% tops down), followed by long drying at 30°C results in translucent scales, high internal CO_2 levels and low O_2 levels in European onions [20]. On the other hand, late harvesting and prolonged field–curing increases the incidence of leathery scales, a severe quality defect.

STORAGE

Onion, as an essential ingredient for the cuisine, is an everyday kitchen need. Since it is a seasonal crop, storage of onion bulbs becomes mandatory to meet the year round demand from the consumers. Field–dried onions can be stored up to two months under ambient conditions in well ventilated trays on pallets or in a field windbreak. Onions are stored in the field, as a cheap and effective method, in many countries with dry climate. *Clamp* is one of the traditional on–farm storage methods of onion in which cured bulbs are piled on the ground in an elongated conical heap of 1 – 2.5 m wide (height depends on the angle of repose). Clamps are often covered with a tarpaulin, straw, earth or plastic materials to protect bulbs from condensation and occasional rains. *Cellar* is another kind of traditional onion storage structure which is essentially an underground or partly underground room often beneath a house. Cellars have good insulation that can facilitate cooling in warm ambient conditions and protection from excessively low temperatures in cold climates. Several types of traditional onion storage structures have been developed by the farmers in tropical countries[18]. In tropics onions can be successfully stored for short duration at 25–30°C and 75–85% humidity levels. However for long duration, this is uneconomical as the weight loss due to desiccation and sprouting makes bulbs unmarketable [21, 22]. Warade *et al.* [23] has described traditional Indian onion storage structures made of locally available materials. But these structures lack ventilation facilities due to which considerable storage losses were encountered. Air ventilation improves the removal of excessive humidity and heat. It helps to keep the outside layers of the onions dry and the bulbs dormant. Naturally/forced–air ventilated structures can be successfully used for long term storage of onions. Such stores must be designed and built specifically for each intended location [23].

Cold storage can be used for long–term storage of onion. Since onion is a non perishable commodity, it can be stored for about 3–7 months (12–28 weeks) at 5–9°C. For large scale commercial storage 0°C with 70–75% relative humidity is recommended. Maintaining temperature and relative humidity is essential to check significant storage losses. High humidity may lead to the development of roots and also promote rotting on the other hand elevated temperatures aggravates sprouting and encourage pathological disorders such as *Botrytis* rots [2]. Advance techniques like modified/controlled atmospheric storage can also be used for storage of onion which reduces usage of chemicals for sprout inhibition by manipulating gas composition [24]. The levels of CO_2 and O_2 combinations required vary with the temperature of the store (Table 3). Generally, high CO_2 (0–5%) and low O_2 (1–3%) coupled with low temperature storage has been shown to reduce sprouting and root growth [2, 25]. Abscisic acid (ABA) is a naturally occurring phytohormone found in all onion tissues and varies according to plant age and tissue type. The concentration of ABA declines exponentially during CA storage and hence its concentration at harvest (measured on a fresh weight basis) is an indicator of storage life under CA conditions [26]. The negative effects of CA storage includes reduced pungency and flavor due to reduction in flavour precursors (Pyruvic acid and total S–alkyl–L–cy– steine sulfoxide content) and CO_2 injury or internal spoilage of onion, a physiological disorder induced by high CO_2 levels coupled with low temperature[24, 27]. Further, the response of onion bulb under CA conditions is greatly influenced by the genetic makeup and hence differs with cultivars. Therefore, experimental trials should be conducted to determine the optimum levels of gas compositions for safe storage of local varieties.

Table 3: Recommended CA composition for storage of onion bulbs [28]

CO_2(%)	*O_2 (%)*	*Temperature (°C)*
0	1–2	–
0	1–2	0–5
0–5	1–2	0–5
0–5	0–1	0–5
5	3	1
5	3	1
5	3	–
5	5	4–5
10	3	4–5

Physiological Disorders and Diseases During Storage

Onions are susceptible to a number of physiological disorders during storage such as, watery scales, translucent scales and freezing injury. Symptoms of watery scales include a thick leathery skin with watery glassy scales below; freezing injury resembles watery scales and is characterized by soft water soaked scales. Translucent scales are characterized by a water–soaked translucent appearance. These physiological disorders often become entry points for fungal and bacterial rots. For control of these problems, effective curing and prompt storage are critical [29].

Black mould, caused by *Aspergillus niger*, commonly occurs on onions stored at temperatures above about 25°C, with an optimum at about 30°C, especially at > 80% RH[30]. This is characterized by black discolouration at the necks or sometimes on the outer scales of onions. Bruised onions are more susceptible to this fungus [29, 31]. Pre–harvest treatment of bulbs with sulphur dioxide at 1% (v/v) for 72 h or heat treatment at 50°C for 3 h can reduce losses due to *A. niger* in store [32], sulphur dust treatment is an effective long–term storage control method [33]. Neck rot (*Botrytis allii*), the causal organism of neck rot, grows optimally at 21°C and is therefore a problem in temperate climates, such as northern Europe and Canada. Symptoms occur after 8 to 10 weeks in store, with a softening and rotting of neck tissues. Numerous small (1–5 mm) black sclerotia develop beneath the outer dry skins [31]. Bruising of onion bulbs during harvesting, storing under humid conditions, and exposing the inner tissues due to breakage of outer scales increases the incidence of neck rot[34]. Post-harvest treatment by heated–air drying at 30–32°C during the early stage of storage discourages the pathogen development [11]. Stored onions are also susceptible to *blue mould*, caused by *Penicillium*. These moulds induce watery soft rot of onion tissues and/or blue–green discolouration at the neck or other tissues. Minimizing mechanical damage and proper curing often reduces the incidence of this fungus [34].

Bacterial soft rots caused by *Pseudomonas* and *Erwinia* often occur during storage of onions. Onions infected by bacterial soft rots often appear healthy on the outside but when cut open some of the inner scales are brown, water–soaked and have a cooked appearance. A characteristic foul smell often occurs and the centre core of the onion often slips out when pressure is applied at the base of the onion. A bacterial rot caused by *Pseudomonas* infects outer scales and are characterized by yellow slime which produce a sour odour. Bacterial soft root can be controlled by: (1) pre–harvest application of a registered fungicide such as Rovral; (2) harvesting at proper maturity; (3) minimizing bruising of bulbs; (4) discarding defective onions; (5) prompt and effective curing, and; (6) storing as quickly as possible [29].

Storage Losses

Storage losses in onion is a biggest concern for farmers and market dealers. Storage losses in the range of 50–90% were reported depending upon genotype and storage conditions. Sprouting (20–40%), physiological loss in weight (PLW) *i.e.*, moisture loss and shrinkage (30–40%), rotting (20–30%) comprise major storage losses in onion. Good harvesting practices, proper curing and storage will minimize the PLW. Generally high temperature coupled with high relative humidity promotes rotting and is observed to be severe during June–July months in tropical countries. Application of post–harvest fungicidal sprays reported to reduce the rotting.

Sprouting losses are usually recorded at the end of storage period or when exposed to high temperature of humid air. Noticeable sprouting losses are observed because of storage of poor quality bulbs having less rest and dormant period and also having thick neck. Comparatively, more sprouting losses are recorded in dark red and white onion cultivars than the light red onion cultivars

Sprouting

Sprouting and root formation accelerate deterioration besides dictating marketability of onion bulbs. After development, onion bulbs enter into a rest/dormant period. This is characterized by reduced physiological activity with non response to environmental conditions. In other words, they do not sprout even when they are placed under ideal conditions of temperature and humidity. During rest, endogenous sprout inhibitors like abscisic acid predominate over promoters like gibberellins, auxins and others [35]. This balance changes with the length of storage to get into a II–dormant period where bulbs sprout or form roots if placed under favorable environmental conditions. There are no clear–cut boundaries between these stages. Instead, there is a slow transition from one to the other as the balance between promoters and inhibitors change. The sprouting occurs at minimal abscisic acid concentration which decreases during post–harvest storage [26].

Inhibition of Sprouting

Storage life of onion bulbs is limited by the rate of sprout elongation in the bulb. Sprouting of both potatoes and onions is a problem in temperate countries, where they are stored for up to eight months. Many varieties, with a long dormancy period, have been developed but its adoptability for a particular geographical location is a concern. Refrigeration and CA conditions can give substantial extension of storage owing to its inhibitory effect on sprouting. The technology is, however, quite expensive to install and needs well trained technical staff to be operated effectively, hence seldom used. Researchers across the globe have also found many ways to suppress sprouting during storage.

Maleic hydrazide (MH), a synthetic sprout suppressant, is used to extend the storage life of onion [27]. Use of MH, a minimum concentration of 20 ppm [37], for extending storage life of onion bulbs by inhibiting sprouting was demonstrated in early 1950s [38]. It hinders the sprout development by inhibiting mitosis in the meristematic region [39] and also reduces water loss by reducing respiration rate [40]. This chemical is generally applied as pre–harvest spray 2–3 weeks before harvest when enough green foliage is present to absorb the chemical, because the effectiveness depends on its translocation into the inner meristeam [24]. The efficiency of MH declines when the onion is stored at relatively high temperatures [41]. Maleic hydrazide does not have universal clearance because of which various researchers have tried different chemicals and natural substances to inhibit sprouting during storage.

Ethephon (2–chloroethyl phosphonic acid) is another chemical that is proved to reduce sprout incidence when applied as pre–harvest spray [42, 43]. Post–curing application of ethylene[44] or holding bulbs in ethylene environment reduces sprout growth compared to those held in air [45]. Treating cured bulbs for 24 h with 1–methylcyclopropene (1–MCP), ethylene–binding inhibitor, suppress sprout growth in cold stored onion bulbs up to 25 weeks [26, 44]. The fumigating onion bulbs with sulphur dust before store suppress sprouting and root development for about 16 weeks under naturally ventilated condition[46]. Many other chemicals, such as boron, lime, carbendazim, iprodione and streptocycline, have been tried to improve storage life on onion. But the fungicide application is not in wide use because of consumer unattractiveness [18, 48, 49].

Sprouting can be inhibited by exposing onion bulbs to ionizing radiations (gamma radiations). The irradiation dosage is measured in terms of kilograys (kGy). World Health Organization (WHO) has approved gamma radiations (0.15 kGy) to reduce sprout incidence during storage. Irradiation dosage in the range of 20–150 Gy is sufficient to prevent sprout growth without significantly effecting taste and nutritional composition. However, higher levels can reduce the flavoring compounds [50, 37]. Thomas *et al.* [51] found that the effect of irradiation on sprout inhibition was influenced by the physiological state of the bulbs at the time of irradiation, radiation dose and the storage temperature after irradiation. Irradiation is more effective when the bulb is fully dormant; hence treatment is preferred immediately after curing [51]. If the internal sprout development is suppressed by other means such as low temperature storage, then the irradiation can be postponed by some extent [37].No toxic effects of this treatment have been reported and also many countries have authorized the sale of irradiated onions.

PROCESSED PRODUCTS FROM ONION

Onion (*Allium cepa*), commercially is by far the most important semi–perishable crop/spice food product having short storage life, as compared with the other spice bulb crops [52]. Onion is generally consumed as fresh vegetable, nevertheless undersize/sorted onions can be processed into high value products which can provide opportunities for emerging agribusinesses. Onion is used as both vegetable and spice or condiment, when dehydrated. Even in ancient times, it was used as seasonings, as foods and also medicine. Onions stored for long periods are subject to storage rot, sprouting, rooting, and loss of water making dehydration necessary for long–term storage [53]. Convenience, freshness and ready to use nature of primary processed vegetables and bulbs have led to their increasing demand in recent years. Recently, the market demand for primary/minimally processed fruits and vegetables have undergone an important rise because of busy lifestyles, increasing purchasing power and increasingly health–conscious consumers[54]. Moreover, these primary processed vegetables and bulbs occupy less storage space, are easy to handle and transport and ready to use. Many foods with small physical, chemical and even microbiological defects may be perfectly safe to eat but, because of regulations, consumer appeal and long–term health benefits, food quality needs to be acceptable to avoid dramatic economic losses. Storage losses of about 25–30% have been reported in onion and hence value addition of raw onion in the form of processed products is an alternate. Therefore, primary processing by cleaning, pre–treatment with permitted chemicals such as potassium meta–bisulphite (KMS)/sodium chloride followed by drying and packaging are used as a means to improve shelf life of dehydrated onion slices [52]. Further, processed products have considerable advantages to food industry, including reduced bulk resulting in lower transportation and distribution costs, stable price and availability, convenient dispersion in food products compared to chopped or sliced fresh form etc. Currently onion products are marketed in a variety of forms. They include dehydrated onion pieces, dried flakes, onion powder, onion paste, onion flavoring, encapsulated flavors, oleoresins and essential oils, onion salt, pickled onions, canned, frozen, minimally processed and packaged onion.

Dehydrated Onion

Free water is removed from the vegetables during the drying process so that microorganisms do not survive and reproduce. Simultaneously, the solids such as sugar and organic acids are concentrated thereby exerting osmotic pressure to further inhibit the microorganisms. Onion is dehydrated in the form of flakes, rings, kibbles and powder. Dehydrated onion products are generally used as seasoning in production of catsup, chilli sauce, meat casseroles, sauces, soup, mayonnaise etc. especially at food service outlets because of its convenience in storage, preparation and use. Important attributes for dehydration of onion are: high total soluble solids (>20%); high degree of pungency; high insoluble solids; low reducing to non–reducing sugars ratio to avoid caramelization and good storage quality. In general, white onion varieties are preferred for dehydration purpose. White Creole, Southport White, Dehydrator No. 8, Dehydrator No. 14, VH–12, etc are some of the important cultivars for dehydration. The coloured varieties like dark red, red and yellow are also used for dehydration purposes but the quality of dehydrated produce is inferior to white varieties.

Dehydration of onion

- Raw material is first sorted for infected, diseased and bruised.
- Topping and rooting (remove the bulb neck and bottom end of the onions) and peeling (removal of skin)
- Slicing as per requirement
- Dip the slices in chlorinated water (25 ml 3.25% "javel" in 5 L of water for 2 to 3 minutes). Rinse slices under water, drain excess water
- Soak the slices in a solution of 0.1% sodium metabisulphite and 1% citric acid for 5 minutes (1.0 litre of solution is required for 1.0 kg peeled onion) and rinse in clean water
- Drying at 60-65°C to a final content of about 6.0–8.0%
- Sort and grade dehydrated onion according to size and colour
- Pack in hermetic jars or laminated low density polyethylene (LDPE) bags

On a domestic scale, onion dehydration is performed by sun drying. Various types of solar driers have been designed for dehydration of onion [55]. However, controlled drying under optimum temperature for a specific time gives good quality product. Cabinet drying at 55–60°C for 10–15 hours or 60–65°C for 8–10 hours[52] gives a better quality dehydrated product than sun drying and drying in solar huts [4]. Commercial dehydration is achieved by forced hot air with the total process divided into three stages: drying at 75, 65 and 55–60°C [56]. Oven dried onion slices retains better mineral contents than sun and microwave assisted dried slices [57]. Ventilated oven drying of onion slices with varying temperature from 30–60°C reduced drying time from 7 h to 3 h representing a substantial energy saving but affected the quality of the product to some extent [58].

Kadam *et al.* [52] reported that, 0.25% and 0.50% KMS pre–treated onion slices dried in tray and greenhouse, respectively, were found to be best after a storage period of 6 months. The greenhouse–dried samples, pre–treated with chemicals were subjected to longer drying period of about 16–18 sunshine hours as compared with the 8–9 h in tray dryer. Greenhouse dried onion slices were at par with the tray–dried samples with respect to microbial load and it is beneficial to use solar energy to produce dehydrated onion slices then mechanical dryer using electrical energy. Therefore, the dehydrated onion slices generated in this study may be deemed safe as far as national and international standards of microbial safety are concerned [52].

Color is the most important quality parameter of dehydrated onion. This deciding parameter is influence by the many reactions and the most important being non–enzymatic browning during drying process. Many researchers have studied the kinetics of non–enzymatic browning in onion slices during drying process. This reaction follows zero order kinetics and rate of reaction varies as quadratic of water activity, maximum browning observed in water activity range between 0.6–0.7. The reaction was temperature dependent, following the Arrhenius relation[59, 60].

Onion Pickle

Onion pickles are prepared out of two types, namely (i) brown or dark red onion 28–45 mm diameter and white or silver skin (pearl or cocktail) onions between 10 and 28 mm in diameter [56]. Silver skin white onions give pickles with a better appearance. Peeled onions are first allowed to ferment in brine solution during which sugars from the bulbs are converted to lactic acid and a small amount of acetic acid and alcohol.

Onions for pickling are first peeled and then allowed to ferment in 10% brine for 24–96 h. During fermentation, sugars in the bulbs are converted to lactic acid plus small amounts of acetic acid and alcohol. The fermentation is controlled by adding small quantities of lactic acid. A decrease in pH to less than 4.6 in the bulbs within a few days prevents the growth of botulism bacteria. The pickled bulbs are bottled in vinegar, possibly darkened with caramel and pasteurized at 80°C[37].

Frozen Onion Rings

Pre–processing steps such as removing bulb neck, bottom end and skin followed by chlorinated water treatment is same as that of dehydrated onions. After rinsing, following are the steps to prepare frozen onions using a blast freezer.

- Place the onion rings in a Blast chiller set at 1 and cool for 15 minutes and then transfer the chilled rings in the Blast freezer set at (–50°C) and leave for 15 minutes or until slices harden,
- Sort and grade rings according to diameter size and
- Seal in freezer bags and store at (–18°C).

Shelf life of the frozen onion rings is reported to be one year. Frozen onion rings are used without thawing. For preparing frozen onion cubes, after chlorine dip onion are cut into quarters and blended to paste. This paste is then poured in to ice cube trays and frozen in a blast freezer.

Onion Chutney

Well cured onion is used owing to its strong pungency and flavour, tamarind (12% w/w), dry chillies and salt (2% w/w).

Preparation

- Select good quality onions sets and remove bulb neck and bottom–end
- Blanch for 1–2 minutes in boiling water and cool in cold water for 5 minutes
- Peel onions and wash thoroughly under running water
- Prepare tamarind pulp, soak tamarind in hot water (volume is equal to weight of tamarind) for easy removal of seeds and collect the filtrate.
- Add the salt, dry chillies and tamarind filtrate to the onion. Blend the mix into a paste using a pre sterilized blender.
- In a pan, warm 10% v/w oil, add in the chutney. Cook under low heat for 15 minutes, stirring the mixture continuously
- Hot fill the chutney in pre sterilized jars leaving a headspace of 5–10 mm with thin layer of oil on the top
- Seal jars. Place the jars in water at 70°C and pasteurize at 85°C for 10 minutes or in boiling water for 10 mins. Allow to cool, label and store away from direct light.

Yield: ≈ 65%; Shelf life: 15 months

OTHER PRODUCTS FROM ONION

Concentrated flavour oil: Concentrated oils can be used to impart the flavour of onions to processed foods. Onion oil is produced by boiling a mixture of the minced bulbs after allowing it to stand for several hours and extracting the oils from the condensed steam, a process known as steam distillation. A kilogram of fresh onion is reported to yield onion oil in the range of 0.02–0.3 g depending upon the extraction process and cultivar. One gram of oil is equivalent to 4.4 kg fresh onions or 500 g onion powder[4]. Onion oil contains numerous volatile components, mostly alkyl and allyl di– and tri–sulfides.

Gas chromatography and gas chromatography–mass spectrometry analysis indicated that it contains about 10–21% of dipropyl trisulfide, varying with type and cultivation area. Various kinds of sulfides occupied 53–62.1% of the oils, chief constituents being di–n–propyl disulfide and dimethyl–n–propyl disulfide[61,62].

Onion oil has found its application in non–alcoholic beverages, ice creams, confectionery, baked goods, condiments, meats and pickles [56]. Furthermore, the oil–soluble aromatics are extracted into food–grade solvents, and this can then be evaporated to concentrate the flavour substances and recover the solvent. Because these aromatic oils are so concentrated, they are sometimes dispersed into a solid phase using salt, dextrose or flours as a carrier. Alternatively, they can be micro–encapsulated using edible polymers, *e.g.*, gum arabic. These forms of dilution and dispersal simplify the mixing and incorporation of the flavour oils into manufactured foods [37]. Essential oil form onion is reported as a natural antimicrobial and antioxidant agents for food systems[63,64,65]. Onion oil exhibits moderate antioxidant activities (0.51–0.67 mg/mL as IC_{50} value) and antimicrobial activity against referenced strains. The reducing power of the oil depends on the dosage and is inferior to butylated hydroxytoluene, which is known to be a strong reducing agent[63]. Romeilah *et al.* assessed and reported essential oil obtained from onion bulbs as a potential natural antiviral (against herpes simplex virus1) and antioxidant substance [66]. Another attractive reported application of essential onion oil is as a potential corrosion inhibitor of tin plates which attracts its applicability in canning industry [67].

***Vinegar*:** Vinegar can be produced from juice of red onion *cv.* Kurenai by batch culture using yeast and *Acetobacter aceti* [68]. The total amino acid, potassium content and organic acid contents of onion vinegar are much higher, while sodium was lower than in other conventional vinegars. An effective two–step fermentation system combining a repeated batch process using flocculating yeast with a charcoal pellet bioreactor has been developed for onion vinegar production [69]. The commercial feasibility of this new type of product needs to be assessed. In onion producing countries like India, China and Pakistan lump sum quantity of red onions go to waste during the lean season and these can best be utilized for value addition.

***Onion juice*:** Onion juice with low flavour component is another processed product. Ground pulp of onion is flash heated (140–160ºC) and then cooled at 40ºC. The product is evaporated to 72–75% solids to facilitate preservation. Juice yield of 66, 53, 72 and 82% were reported by various extraction methods pressing, rotary crushing, centrifuge and enzyme assisted, respectively. Effect of heating temperature and time on juice yield was also studied [70]. Onion juice yield varied from 73–83% when extracted at 80 and 100°C for 10, 20 and 30 min. During processing of onion juice, aromatic components may be removed so that the product has a low flavour profile[71]. Quercetin–related substances, reported functional components of onion, are rich in onion juice (2.26 mg) than the residue (1.57 mg) obtained from fresh onions. Further the functional components of onion juice were found moderately thermo–stable, only 33% reduction was observed upon heating at 80°C for 10 min. Quercetin is stable at 200°C for 60 min but highly sensitive to light [72].

***Onion salt*:** Onion salt is prepared by mixing 19–20% onion powder with 78% free flowing pulverized refined table salt and 1–2% anti–caking agent which prevents water absorption, and caking, etc [73].

Solar drying and dehydration

Drying involves the removal of moisture contained in fruits and vegetables in order to preserve. Although preservation for enhanced shelf life is the primary reason for drying, it also lowers the product mass and volume. The reduction in mass and volume improves the efficiency of packaging, storing and transportation. Traditionally, fruits and vegetables are dried in open sunlight, which is weather dependent and also prone to microbial and other contamination. To get best quality, dried product hot air industrial dryers are quite prevalent but require huge amount of energy. Steep rise in energy cost on one hand and strict legislation of environmental pollution on the other hand have created greater demand for energy efficient drying processes in the food industry [74]. Avoiding wastage of costlier energy can benefit the food industry and it has been reported that improving energy efficiency by 1% could result to 10% increase in profits [75]. Use of nonconventional source of energy (renewable) for drying of vegetables and application of pretreatments of preservatives for improvement in quality retention of dried product has been a major research goal for better efficiency of drying and quality of dried product [76]. Solar energy, being abundantly available in tropical and subtropical regions of the world, being environmentally benign and which could not be monopolized can effectively is utilized for drying of agri–produce, if harvested properly [77].

Sun drying results in the product rich in colour and translucent in appearance, but the process has several disadvantages such as time consuming, weather dependent, labour intensive etc [78]. The components of onion which are not affected by artificial drying include ash, fat, protein and fibre where as thermo–sensible components such as sugars, acidity and vitamin C are considerably influenced by the method of drying [58]. Innovative techniques that increase the drying rate and enhance product quality considerably gained the attention of researchers and industries to explore its feasibility for drying of onion.

Conventional convective hot air drying takes more time causing many undesirable changes in the material [79]. Infrared radiation (IR) drying is a kind of drying where the material is dried by the absorption of IR energy unlike convective drying where heat is transferred by an external medium such as heat [80]. Hot air heating in combination of electromagnetic radiation is more effective than radiation or hot air alone because of its synergistic effect. Pathare and Sharma [81] also studied the effect of IR energy in combination with hot air by varying infrared intensity (26.5–44.2 kW/m^2), air temperature (35–45°C) and air velocity (1–1.5 m/sec) for drying of onion slices (6 mm). The average effective moisture diffusivity and activation energy raged between 0.2514×10^{-10} to 0.3233×10^{-10} m^2/sec and 5.06 to 10.63 kJ/mol. Higher drying rates was observed for higher radiation intensity and inlet air temperature. The effective moisture diffusivity during infrared–

convective drying was influenced by the moisture content and increased with decreasing moisture content. However, coupling IR with hot air will reduce the drying time but heat damage to the sensible components is unavoidable in this process [81]. This problem can be addressed by the vacuum drying technology in which product is exposed to elevated temperature under negative pressure. Under vacuum, increased vapour pressure gradient between the inside and outside of the sample enhances the mass transfer, at considerably low temperatures thereby preserving essential thermo–liable components [82, 83]. Mitra *et al.* [84] assess the applicability of vacuum drying in drying and preservation of onion with better quality. In this study, onion slices of 1, 3 and 5 mm thick were dried in a laboratory scale vacuum dryer operating at 710 mm Hg and 50–70°C temperature. Temperature and slice thickness was found to have positive and negative significant influence on drying rate. Effective moisture diffusivity which was in the range of 1.32 $\times 10^{-10}$ –1.09 $\times 10^{-10}$ increased with increase in temperature and thickness[84].

Microwave drying (MD) is one among the innovative methods gaining popularity because of its inherent advantages over conventional heating such as reducing the drying time of biological material without quality loss, cost effectiveness and feasibility. Drying characteristics and the quality degradation in terms of colour and total phenolics content of onion slices during sun, oven and microwave drying was studied by Arslan and Ozcan[57]. MD took shorter time to dry the product than sun and oven drying methods and exhibited effective diffusivity of fifty folds as that of sun and oven drying (50°C). Further, MD with low output power (210 W) found to retain phenolics of the product better than sun and oven drying and also retained optimum colour values. High power out (700W) results in discolouration of the product.

Minimal processing

Minimal processing of raw fruits and vegetables has two purposes: (a) keeping the produce fresh, without losing its nutritional quality and (b) ensuring a product shelf–life sufficient to make distribution feasible within a region of consumption[85]. As minimal processing, fresh peeled onion bulbs can be shredded and packed is packed in PP–O, 40 mm, PA/PE, 70–100 mm or any comparable packaging material[86]. Brief steps for the minimal processing of fresh onion in the form of shreds are given below:

Decontamination is a challenge in the minimal processing of any fresh cut fruits and vegetables. Retaining flavonoids is a biggest issue in fresh cut onions. The impact of common disinfection treatments (sodium hypochlorite, or amukine, hydrogen peroxide, sodium dichloroisocyanate and UV–C irradiation) on the flavonols and anthocyanins content of onions, as well as the role of structural features on their degradation revealed that the onion flavonoids decreased by chemically disinfectant procedures of fresh–cut onion,

Processing guidelines for cut onion[86]

Processing temperature	0–5°C
Raw material	Suitable variety or raw material lot should be selected using a rapid storage test on prepared produce at room temperature
Pretreatment	Stems, damaged and contaminated parts, as well as spoiled onions must be removed.
Peeling	Peeling with knife or with pressurized air (dry onions).
Washing	Mild washing immediately after peeling. The temperature of washing water should be 0–5°C. Microbiological quality of washing water must be excellent.It is preferable to use active chlorine 0.01% in washing water.
Cutting	The cutting should be done with sharp knives immediately after washing. The shelf–life of cut onion is shorter, the smaller the pieces.
Washing and centrifugation	No washing or centrifugation for cut onion
Packaging	Immediately after cutting. Proper packaging gas is normal air or gas mixture 5% O_2 + 5–20% CO_2 + 75–90% N_2, and the headspace volume of a package 2 l per kg onion. Suitable oxygen permeability of packaging materials is between 1200 (*e.g.*, oriented polypropylene) and 2900 (50 mm LD polyethylene) cm^3/m^2 per 24 h, 101.3 kPa, 23°C, RH 0%. If cutting grade is small (*i.e.*, big cuts), quite impermeable materials can also be used, *e.g.*, 80 mm nylon–polyethylene, the permeability of which is 70 cm^3/m^2 per 24 h, 101.3 kPa, 23°C, RH 0%
Storage	0–5°C, preferably in the dark.
Other remark	Good Manufacturing Practices must be followed (hygiene, low temperatures and disinfection)
Shelf–life	Cut onion has very poor stability, the shelf–life is only 3 days at 5°C

major culprit being solubility in immersion water. Further it was noticed that the dry decontaminantion treatments like UV–C irradiation, maintain and improves the initial flavonoid content [87].

Extraction of flavonoids

Flavonoids play a beneficial role in the maintenance of normal physiological functions in the human body [88]. Further, flavonoids reported to regulate the expression of genes involved in the inflammatory response, the metabolism of carcinogens and the antioxidant defence [89]. Anthocyanins (cyanidin and peonidin glycosides) and flavonols (quercetin, isorhamnetin, kaempferol and their glycosides) are two major flavonoids found in onion[90,91], of which the most abundant is quercetin and quercetin glucosides (quercetin–4′–glucoside and quercetin–3,4′–diglucoside) accounting for 85% of total flavonoid content[6, 92].

Quercetin is usually found in higher concentration in the outer dry skin, where it exists mostly in the aglycone form and imparts a yellow or brown colour [93]. Quercetin is typically obtained by acid hydrolysis of quercetin glycosides extracted from onion tissues by solvent extraction assisted by heating, boiling, pressing, blending, maceration and mechanical fragmentation of plant material [94, 95]. Aggressive mechanical damage during the conventional process results in degradation of chemically sensitive phenols. Also, conventional methods results in degradation of some chemically sensitive phenols due to intensive mechanical disruption in addition to longer extraction time. The extensive use of organic solvents in the conventional process promotes the release of oxidative enzymes which accounts for further degradation[96]. More flavonoid compounds (anthocyanins), 21–36% depending on the species, can be extracted if degradation is prevented during the extraction process, but a longer extraction times would be required to reach the peak extraction level [97].

Many efficient and automated methods have been developed to cope with the industrial demand for natural antioxidants. Novel techniques of extraction includes, ultrasound assisted extraction [98], microwave assisted extraction [96, 99], supercritical fluid extraction [100] and accelerated solvent extraction [101]. The improved methods are developed mainly to increase the yield of extraction and the process efficiency by reducing the process time and solvent and energy consumption [102]. Major advantages of advanced extraction methods are: (i) can be carried out in absence of light and oxygen there by preserving the sensitive components (ii) reduced organic solvent consumption (iii) as these methods employ elevated temperatures and/or pressures, extraction time is drastically reduced. Several researchers have demonstrated applicability of these improved methods for extraction of antioxidant compounds from onion. The evaluated five different extraction methods *viz.* water bath, pressurized liquid extraction, ultrasonication, ultrasonic liquid processor and microwave extraction for extraction of flavonoids in bulb onions using 60% aqueous methanol as solvent. The extraction efficiencies were lowest for the conventional water bath extraction compared to other methods [103].

Pressurized fluid extraction (PFE) is one of the novel methods and it allows fast extraction, small solvent consumption, and automated extraction procedures. It is known by several names as pressurized liquid extraction, pressurized solvent extraction and accelerated solvent extraction. The efficiency and reproducibility of this method is alike microwave assisted and ultrasonic liquid processor method, the cleanest extracts obtained during the process and possible of carrying the extraction in inert atmosphere renders this process more advantageous [100]. Pressurized hot water extraction is a sister concept of PFE, where the solvent is mainly water, and is a safer method to extract anthocyanins from red onion, provided a careful control over degradation reaction during extraction [97].

Microwave hydro–diffusion and gravity (MHG) method is a novel and green extraction technique that utilizes hydro–diffusion phenomenon generated by microwaves to extract essential components from bio–materials [104]. This method combines the effect of microwaves and the earth's gravitational force at atmospheric pressure to achieve the desired effect without the use of solvent. Microwaves, by virtue of their cell disruption ability, enhance the diffusion of secondary metabolites by softening the tissue and increasing cell permeability [105, 106]. MHG, at a power level of 500 W, offers advantage over conventional solid–liquid extraction in terms of shorter extraction time (23 min as against 45 min with 80% methanol) and quality onion juice with comparable phenolic content (58.29 mg GAE/g DW as against 64.81 mg). With MHG it is possible to extract 330.46 mg of flavonol content from 1 kg of fresh peeled onion (yellow variety) along with crude onion juice which is similar to fresh unprocessed product [107]. Further, extracts of red, yellow, white and grelot onion varieties obtained under MHG exhibited the highest antioxidant activities in all the tests against conventional solvent extraction [96].

Extraction of oil

Partial water solubility of onion constituents demands a special recovery technique for better yield and quality of oil. Industrial techniques for extracting the aromatic flavor constituents from unheated homogenates of bulbs have been developed.

Supercritical fluid extraction (SFE) is a unit operation that exploits the dissolving power of fluids at temperatures and pressures above the critical point [108] and is an alternative to steam distillation and solvent extraction to obtain the oil and flavour compounds from onion [109]. Supercritical carbon–dioxide extraction can be successfully applied for producing oleoresin from dried onion. The extraction parameters such as pressure, temperature and time affect the yield and quality of oleoresin. From quality point of view, oleoresin with stronger onion characteristics, 300 or 450 bar pressure and the 65°C temperature are the most suitable conditions for extraction of oleoresin from dried onion [110]. Dron *et al.* compared different methods of extraction (liquid CO_2 extraction, SFE with CO_2, SFE with CO_2 and entrainers, and steam distillation–solvent extraction with dichloromethane) of onion flavour oil. SFE of onion juice was performed in a 2 l volume extractor using ethyl alcohol as entrainer reported to enhance the onion oil yield [111]. Gas chromatography and gas chromatography–mass spectrometry analysis of the extracts indicated that the CO_2 extracts had fresh onion–like flavour. Furthermore, pretreatment like concentration of onion juice by reverse osmosis (from 10 to 18°Brix) improves the efficiency of SFE without altering the characteristic fresh aroma of the product [112]. However, the complicated equipment, critical operation conditions and high operation cost limits the industrial application of supercritical CO_2 extraction in the production of onion volatile oils [113].

Comparison of the yield and the concentration of sulphur of onion extracts obtained by different methods [109]

Method	*Yield (w/w%)*	*Sulphur content (g/kg extract)*
Steam distillation	0.032 ± 0.004	393.0 ± 20.0
Soxhlet extraction Hexane	0.85 ± 0.07	37.8 ± 2.0
Ethyl alcohol	67.20 ± 3.50	7.9 ± 0.5
Supercritical CO_2 extraction Overall	0.66 ± 0.02	46.7 ± 3.0
First separator	0.43 ± 0.06	21.8 ± 2.2
Second separator	0.23 ± 0.06	87.7 ± 7.8

Rectification is an efficient technique for gas–liquid separation and is being widely applied in the fields of chemical, petrochemical engineering, etc. Rectification extraction is best suitable for enriching volatile oils from plants or herbs in which volatile oils are water–soluble and with low–content [114]. Multi–rectification, a process equivalent to multi–stage distillation, to extract volatile oils from onions of different varieties and growing areas of China was applied by Zhang *et al.* [113]. Golden yellow, clear and sticky onion volatile oil with the yield of 0.033 mL/kg extracted by coupling the lab and the pilot equipment based on multi–rectification process affirm the suitability of the this novel method for production of high quality onion oil [113].

UTILIZATION OF ONION WASTE

Transforming fresh onion into various products such as dehydrated onion, frozen rings/flakes, onion pickle, onion chutney, onion salt, onion juice etc. generates processing waste. The onion processing waste consists of brown skins and outer fleshy scales which is not suitable as animal feed due to its strong taste and it also carries the risk of spreading crop disease [37]. The waste may be transformable into a valuable source of pectic substances and dietary fibre–rich food–grade products using appropriate technologies like extraction and cooling [115]. Onion peel (skin) is reported to be a good and easily accessible source of nutraceutical compounds. The extracts/fractions of red onion peel have an antioxidant and anti–mutagenic properties [116]. The amount of quercentin related substances are richer in inedible portion (317.99 mg/100 g) than the edible portion (4.1 mg/100 g) [70]. The quercentin–rich onion peel extract have beneficial antiobesity effects [117].

Martino *et al.* [118] extracted of–free quercetin from red and yellow onion skins by ethanol–modified supercritical carbon dioxide extraction process. About 0.024 and 0.020 g/kg of quercetin was reported in red and yellow onion skins, respectively, at 5700 psi, 40°C, an average of 7.6% (molar concentration) of ethanol, and with an extraction period of 2.5 h [118]. Though ethanol is a GRAS (Generally–Recognized–as–Safe) solvent, its utilization in this application is restricted by the long extraction time and the some legal statutes that exist in many countries.

Ko *et al.* [119] investigated the application of subcritical water extraction (SWE) for quercetin from onion skin and evaluated the effect of key operation conditions by varying the temperature (100–190°C), extraction time (5–30 min), and mixture ratio of onion skin and diatomaceous earth (0.5:3.5–2:2) under high pressure (90–131 bar). SWE is advantages over other novel extraction techniques because of its suitability for nonpolar components, environmental safety, efficiency and lower operation cost. Quercetin yield of about 16.29 ± 0.75 mg/g onion skin was reported when extracted at 165°C, mixture ratio of 1.5:2.5 for onion skin and diatomaceous earth and extraction time of 15 min. Yield of quercetin by SWE is reported to be 8, 6 and 4 times higher than those of ethanol, methanol, and water–at–boiling–point extraction methods, respectively [119].

The highest yield of several extraction processes for extraction of quercetin from onion skin and also effect of independent variables such as concentration of ethanol, provided power or the temperature, and reaction time was evaluated [102]. Conventional solvent extraction (CSE), microwave assisted extraction (MAE), and ultrasound assisted extraction (UAE) were studied and optimized using response surface methodology. From 1 g sample of dried onion skin, the highest yield of each method could be achieved at 16.5 min of process time under 59.2°C for CSE with 59.3% ethanol, 117 s for MAE with 69.7% ethanol, and 21.7 min for UAE using power of 606.4 W with 43.8% ethanol. Among three methods, MAE was found to be the most productive method with yield of 20.3 and 30.8% higher than UAE and CSE, respectively.

The basic factors those are likely to govern the extractability of polyphenols from outer dry and semi–dry layers and the apical trimmings of brown–skin onion bulbs were examined [120]. In this study, reusable and non–toxic solutions composed of water/ethanol/citric acid were employed as extracting media and response surface methodology was used to optimize pH, extraction time and ethanol concentration to establish a sustainable process. Highest total phenolics yield of 9342 ± 1435 mg gallic acid equivalents per 100 g dry weight was obtained under optimal conditions of 60% ethanol, pH 2 and 4.2 h of extraction time [120].

Sprouting is one of the major storage concern which renders onion unmarketable. The chemical composition and biological nature of sprouted onion with respect to the antioxidant and anti–inflammatory activities were investigated by Takahashi and Shibamoto[121]. Onion sprouts were transformed into powder form by freeze drying followed by extraction of volatile components by conventional steam distillation method. Antioxidant and anti–inflammatory activities of onion sprouts were comparable with those of onions including flavor and taste. The chemical composition of onion sprouts were also identical to that of onion asserting that the sprouted onion can be consumed in the same way as onion.

Preparation of onion peel extract: Onion peel is washed for about three times in tap water and extracted for three hours at 50°C using 60% aqueous ethanol solution followed by filtering by a filter press. The filtrate is concentrated using a vacuum concentrator and then processed to powder by means of a freeze dryer. The quality of onion peel extract is usually expressed in terms of milligram quercetin, along with mixed flavonoids per gram of extract.

CONCLUSION

Onion (*Allium cepa* L.), one of the oldest vegetables in continuous cultivation, is probably first domesticated in the mountainous regions of Turkmenistan and north Iran. It is an important vegetable and spice crop with immense medicinal and therapeutic property. Due to its significant medicinal property, nutritional worth and energy value, onions impart numerous health benefits to the consumers. Onions can also be processed into dry products such as rings, flakes, and powder for the food processing industry. Onion is a biennial crop but cultivated as annual and harvested at various stages of maturity either to provide green leafy vegetables or the bulbs. Being a seasonal crop, storage becomes mandatory to meet the year round demand from the consumers. Storage losses, accounting up to 90%, is a biggest concern for farmers and market dealers. Curing is the most important post harvest unit operation that determines storage and/or shelf life on onion. Advanced techniques like irradiation are standardized for inhibiting sprouting, a major storage concern, in onion. Other methods like modified/controlled atmospheric storage were also assessed and being used for onion storage reducing the usage of chemicals. Though onion is generally consumed as fresh vegetable it can be processed into high value products providing opportunities for emerging agribusinesses. Processed products have considerable advantages to food industry, including reduced bulk resulting in lower transportation and distribution costs, stable price and availability, etc. Currently onion products in the market include dehydrated onion pieces, dried flakes, onion powder, onion paste, onion flavoring, encapsulated flavors, oleoresins and essential oils, onion salt, pickled onions, canned, frozen, minimally processed and packaged onion. Further, products such as concentrated flavor oil, vinegar, onion juice, salt etc expands product diversity. Flavonoids play a beneficial role in the maintenance of normal physiological functions in the human body. Anthocyanins and flavonols are two major flavonoids found in onion. Many efficient and automated methods have been developed to cope with the industrial demand for natural antioxidants. Novel techniques of extraction like ultrasound assisted–, microwave assisted extraction, supercritical fluid extraction and accelerated solvent extraction accelerated and widened the scope onion processing in recent times. Processing waste or byproducts of onion processing is reported to be a good and easily accessible source of nutraceutical compounds. The extracts/fractions of red onion peel have an antioxidant and anti–mutagenic properties. Overall, onion could be seen as diversified crop from both fresh consumption and processing sense. In spite of substantial studies regarding storage of onion during off season, safe and economical storage in subtropical countries remains unsolved. Existing techniques/technologies may be further refined to extend the storability of fresh dry onions. Also processing methods and protocols may be improvised and updated with latest energy efficient and economical processes.

REFERENCES

[1] James L.B., "Onions and other vegetable Alliums, 2nd edition", CABI, UK (2008).

[2] Linus U.O., "Onions: Post–harvest operation", Mejia D. (*ed.*), FAO– *Post–harvest Compendium.www.fao.org / fileadmin / user_upload / inpho / docs / Post_Harvest_ Compendium_–_*Onion.pdf (2003).

[3] Hui Y.H., Ghazala S., Graham D.M., Murrell K.D., Nip W.K., "Handbook of vegetable preservation and processing", Chapter 1 and 22. Marcel Dekker, Inc., New York, U.S.A. (2004).

[4] Lawande K.E., Peter, K.V., "Onion", Handbook of Herbs and Spices, Chapter 21, Woodhead Publishing Limited, New York (2000).

[5] Bhattacharjee S., Abida S., Muhammad H.S., Muhammed A.I., Asaduzzaman M.M.A., "Analysis of the proximate composition and energy values of two varieties of onion (*Allium cepa* L.) bulbs of different origin: A comparative study", *Int J Nutr Fd Sci.*, **2**(5), 246–253 (2013).

[6] Eduvigis R., Concepcion S.M., Rosana L., Begona de A., Cano M.P., "Onion high–pressure processing: Flavonol content and antioxidant activity", *LWT – Fd Sci Tech.*, **42**, 835–841 (2009).

[7] Roldan, E., Sanchez–Moreno C., Ancos de B., Cano M.P., "Characterization of onion (*Allium cepa* L.) by–products as food ingredients with antioxidant and ant browning properties", *Food Chem.*, **108**(3), 907–916 (2008).

[8] Lancaster J.E., Boland M.J., "Flavour biochemistry– Onions and Allied Crops", Chapter 3 Vol. III, *CRC Press*, B, oca Raton, Florida (1990).

[9] Sargent S.A., Zoellner J.J., Stoffella P.J., Maynard D.N., "Harvest maturity affects storage quality of fresh, short–day onions", Proceedings Annual Meeting. *Florida State Horti Soc.*, **104**, 64–68 (1991).

[10] Sargent S.A., Stoffella P.J., Maynard D.N., "Harvest date affects yield and postharvest quality of non dried short day onion", *Hort Sci.,* **36**, 112–115 (2001).

[11] Rabinowitch H.D., Currah L., "Allium crop science: Recent advances", Chapter 9 and 10. CABI publishing, New York (2002).

[12] Isenberg F.M.R., Ludford P.M., Thomas T.H., Weichmann J "Hormonal alterations during the postharvest period", *In: Post–Harvest Physiology of Vegetables*, Marcel Dekker, New York, and Basle, Switzerland, pp. 45–94 (1987).

[13] Maw B.W., Smittle D.A., "Undercutting onions", *Hort Sci.,* **21**, 432–434 (1986).

[14] Maw B.W., Smittle D.A., Mullinix B.G., Cundiff J.S., "Design and evaluation of principles for mechanically harvesting sweet onions", *Trans ASAE.*, **41**, 517–524 (1998).

[15] Brice J., Currah L., Malins A., Bancroft R., "Onion storage in the tropics", NRI publications, university of Greenwich, UK, p. 120 (1997).
[16] Pandey U.B., Singh L., Singh S.P., Mishra P.K., Singh L., "Studies on the effect of curing on storage life of Kharif onion" (*Allium cepa* L.), *Newsletter Associated Agric Dev Foundation.*, **12**(3), 14–16 (1992).
[17] Fustos Z., Gilinger M. P., Ombodi A., "Effects of postharvest handling and cultivars on keeping quality of onions (*Allium cepa* L.) in storage", *Acta Horticulturae.,* **368**, 212–219 (1994).
[18] Gubb I.R., MacTavish H.S., "Onion Pre– and Postharvest Considerations in Allium Crop Science: Recent Advances (*Ed.* Rabinowitch and Currah)", *CABI publishing,* New York, pp. 234–265 (2002).
[19] Gubb I.R., MacTavish H.S., "Onion Pre– and Postharvest Considerations in Allium Crop Science: Recent Advances (*eds.*), Rabinowitch and Currah", CABI publishing, New York, pp. 234–265 (2002).
[20] Solberg S., Dragland S., "Effects of harvesting and drying methods on internal atmosphere, outer scale appearance, and storage of bulb onions (*Allium cepa* L.)", *J Veg Crop Prod.*, **4**(2), 23–25 (1998).
[21] Stow J.P., "Effects of humidity on losses of bulb onions (*Allium cepa*) stored at high temperature", *Exp Agri.*, **11**, 81–87 (1975).
[22] Thompson A.K., Booth R.H., Proctor F.J., "Onion storage in the tropics", *Tropical Sci.*, **14**(1), 19–34 (1972).
[23] Warade S.D., Desale S.B., Shinde K.G., "Effects of different storage recommendations on storability of onion bulbs", *J Maharashtra Agri Univ.*, **22**, 283–285 (1997).
[24] Priya E.P.B., SinjaV.R., Alice R.P.J.S., Shanmugasundaram S., Alagusundaram K., "Storage of onions–A review", *Agricultural Reviews.*, **35**(4), 239–249 (2014).
[25] Hardenburg R.E., Watada A.E., Wang C.Y., "The commercial storage of fruits, vegetables and florist and nursery stocks", USDA, ARS, *Agriculture Handbook,* p: 66 (1990).
[26] Chope G.A., Terry L.A., White P.J., "Effect of controlled atmosphere storage on abscisic acid concentration and other biochemical attributes of onion bulbs", *Postharvest biology Tech.*, **39**(3), 233–242 (2006).
[27] Uddin M.M., MacTavish H.S., "Controlled atmosphere and regular storage–induced changes in S–alk (en) yl–L–cysteine sulfoxides and alliinase activity in onion bulbs (*Allium cepa* L. cv. Hysam)", *Postharvest Biology and Tech.,* **28**(2), 239–245 (2003).
[28] Thompson A.K., "Postharvest technology of fruits and vegetables", London: *Blackwell Science Ltd.*, (1996).
[29] Shika A., Doug W., "Onions: Post–harvest handling and storage", University of Saskatchewan, http://www.usask.ca/agriculture/plantsci/vegetable/resources/factsheet/ postharvonions.pdf (2001).
[30] Hayden N.J., Maude R.B., Proctor F.J., "Strategies for the control of black mould (*Aspergillus niger*) on stored tropical onions", *Acta Horticulturae.*, **358**, 271–274 (1994).
[31] Hayden N.J., Maude R.B., "The use of integrated pre– and post–harvest strategies for the control of fungal pathogens of stored temperate onions", *Acta Horticulturae.,* **433**, 475–479 (1997).
[32] Thamizharasi V., Narasimham P., "Growth of *Aspergillus niger* on onion bulbs and its control by heat and sulphur dioxide treatments", *Tropical Science.*, **33**, 45–55 (1992).
[33] Padule, D.N., Lohate, S.R., Kotecha, P.M., "Control of spoilage of onion bulbs by postharvest fungicidal treatments during storage", *Onion Newsletter for the Tropics.*, **7**, 44–48 (1996).
[34] Maude R.B., "Onions in Post–Harvest Pathology of Fruits and Vegetables (*ed.*) Dennis C.", Academic Press, New York, pp. 73–101 (1983).

[35] Anonymous., "Detailed feasibility report–fruit and vegetable cleaning, grading and packaging unit", *Ministry of Economic Affairs, Royal Government of Bhutan, (www.moea.gov.bt/documents/files/pub9tz7451nu.pdf)* (2015).

[36] Chope G.A., Terry L.A., White P.J., "The effect of 1–methylcyclopropene (1–MCP) on the physical and biochemical characteristics of onion cv. SS1 bulbs during storage", *Postharvest Biology and Technology,* **44**(2), 131–140 (2007).

[37] Brewster J.L., "Onions and other vegetable Alliums (2nd Ed)", CAB International, Wallingford, UK (2008).

[38] Isenberg F.M., "The use of maleic hydrazide on onions", *Proceedings Am Soc Horti Sci.*, **66**, 331–333 (1956).

[39] Isenberg F.M., "The use of maleic hydrazide on onions", *Proceedings Am Soc Horti Sci.*, **66**, 331–333 (1956).

[40] Salama A., Hicks J., "Respiration and fresh weight of onion bulbs as affected by storage temperature, humidity and maleic hydrazide, *Tropical Sci.,* p. 238 (1987).

[41] Thomas T.H., "The role of growth substance in the regulation of onion bulb dormancy", *J Expl Botany.*, **20**, 124–137 (1969).

[42] Pedeliski A.V., "Extension of onion blub dormancy by plant growth regulators and its importance in water permeability determinations", *Hortic Sci.*, **8**, 416 (1973).

[43] Adamicki F., "Effects of pre–harvest treatments and storage conditions on quality and shelf–life of onions", *Acta Hortic.*, **688**, 26–33 (2004).

[44] Cools K., Chope G.A., Hammond J.P., Thompson A.J., Terry L.A., "Ethylene and 1–methylcyclopropene differentially regulate gene expression during onion sprout suppression", *Plant Physiology.*, **156**(3), 1639–1652 (2011).

[45] Bufler G., "Exogenous ethylene inhibits sprout growth in onion bulbs", *Ann Bot* (Lond), **103**, 23–28 (2009).

[46] Abbey I., "Effect of poultry manure and post production application of fungicide on the shelf life of onion cv. Bawku Red", *Crop Res.*, **20**(1), 87–92 (2000).

[47] Alphonse M., "Response of stored onions to different boron treatments", *Alexandria J Agri Res.*, **42**, 171–183 (1997).

[48] Srivastava P.K., Gupta R.P., Tiwari B.K., Sharma R.C., "Effect of systemic fungicides/ bactericides on the control of diseases of onions in storage", *News Letter, Nat Horti Res Dev Found.*, **17**(3), 4–6 (1997).

[49] Srivastava P.K., Tiwari B.K., "Effect of pre–harvest fungicidal spray on the control of storage diseases of onion", News Letter, *Nat Horti Res Dev Found.*, **17**(2), 4–6 (1997).

[50] Benkeblia N., Varoquaux P., Gouble B., Attou S.G., "Respiratory parameters of onion bubs (*Allium cepa* L.) during storage: Effects of ionizing radiation and temperature", *J Sci Fd Agric.*, **80**, 1772–1778 (2000).

[51] Thomas P., Srirangarajan A., Limaye S., "Studies on sprout inhibition of onions by gamma irradiation—I. Influence of time interval between harvest and irradiation, radiation dose and environmental conditions on sprouting", *Radiation Botany,* **15**(3), 215–222 (1975).

[52] Kadam D.M., Nangare D.D., Oberoi H.S., "Influence of pre–treatments on microbial load of stored dehydrated onion slices", *Int J Fd Sci Tech.*, **44**(10), 1902–1908 (2009). doi:10.1111/j.1365–2621.2009. 01980.x

[53] Akbari S.H., Patel, N.C., "Studies on drying characteristics of onion", *NHRDF News Letter*, **23**, 7–12 (2003).

[54] Baldwin E.A., Carriedo M.O.N., Baker R.A., "Use of edible coatings to preserve quality of lightly (and slightly) processed products", *Critical Reviews in Food Science and Nutrition.,* **35**, 509–524 (1995).

[55] Pawar V., Singh N., Dev D., Kulkarni D., Ingale U., "Solar drying of white onion flakes. *Ind Food Packer.*, pp. 15–24 (1988).

[56] Fenwick G.R., Hanley A.B., "Chemical composition", Chapter 2 of *Onions and Allied Crops, Vol. III.* Brewster J.L. and Robinowitch H.D. (*Eds.*), CRC Press, Boca Raton, Florida (1990).

[57] Arslan D., Ozcan M.M., "Study the effect of sun, oven and microwave drying on quality of onion slices", *LWT–Food Science and Technology*, **43**(7), 1121–1127 (2010).

[58] Mota C., Luciano C., Dias A., Barroca M.J., Guiné R., "Convective drying of onion: Kinetics and nutritional evaluation", *Food and Bioproducts Processing*, **88**(2), 115–123 (2010).

[59] Rapusas R., Driscoll R., "Kinetics of non–enzymatic browning in onion slices during isothermal heating", *Journal of Food Engineering.*, **24**(3), 417–429 (1995).

[60] Kaymak E.F., Gedik A., "Kinetic modeling of quality deterioration in onions during drying and storage", *J Food Eng.*, **68**, 443–453 (2005).

[61] Shaath N.A., Flores F.B., "Egyptian onion oil", *Developments in Food Science,* **40**, 443–453 (1998).

[62] Rogers R.D., Zhang J., Marinsky J.A., Marcus Y., "*Ion Exchange and Solvent Extraction*", Dekker, New York, **13**, 141 (1997).

[63] Ye C.L., Dai D.H., Hu W.L., "Antimicrobial and antioxidant activities of the essential oil from onion (*Allium cepa* L.)", *Food Control,* **30**(1), 48–53 (2013).

[64] Santas J., Almajano M.P., Carbó R., "Antimicrobial and antioxidant activity of crude onion (*Allium cepa*, L.) extracts", *Int J Food Sci Tech.,* **45**(2), 403–409 (2010).

[65] Benkeblia N., "Antimicrobial activity of essential oil extracts of various onions (*Allium cepa*) and garlic (*Allium sativum*)", *LWT–Food Science and Technology,* **37**(2), 263–268 (2004).

[66] Romeilah R.M., Fayed S.A., Mahmoud G.I., "Chemical compositions, antiviral and antioxidant activities of seven essential oils", *J Appl Sci Res.,* **6**(1), 50–62 (2010).

[67] Grassino A.N., Grabariæ Z., Pezzani A., Fasanaro G., Voi A.L., "Influence of essential onion oil on tin and chromium dissolution from tinplate", *Food Chem Toxicology.,* **47**(7), 1556–1561 (2009).

[68] Horiuchi J., Kanno T., Kobayashi M., "New vinegar production from onions", *J Biosci Bioengin.*, **88**(1), 107–109 (1999).

[69] Horiuchi J., Kanno T., Kobayashi M., "Effective onion vinegar production by a two–step fermentation system", (2000) http://www.areu.mu/fruit/processing/onion%20process

[70] Kang S.K., Kim Y.D., Hyun K.H., Kim Y.W., Song B.H., Shin S.C., Park Y.K., "Development of separating techniques on quercetin–related substances in onion (*Allium cepa* L.) 1. Contents and stability of quercetin–related substances in onion", *J Korean Soc Food Sci Nutr.,* **27**, 682–686 (1998).

[71] Arrell K.T., "Spices, Condiments and Seasonings", AVI Publishing, Westport, CT (1985).

[72] Kang S.K., Kim Y.D., Hyun K.H., Kim Y.W., Seo J.S., Park Y.K., "Development of separating techniques on quercetin–related substances in onion (*Allium cepa* L.) 2. optimal extracting condition of quercetin–related substances in onion", *J Korean Soc Food Sci Nutr.,* **27**, 687–692 (1998).

[73] Pruthi J.S., "Spices and Condiments Onion", National Book Trust, India, pp. 173–175 (1987).

[74] Kadam D.M., Nangare D.D., Singh R., Kumar S., "Low cost greenhouse technology for drying onion (*Allium cepa* L) Slices", *J Food Process Engin.*, **34**(1), 67–82 (2011) doi: 10.1111/j.1745–4530.2008.00337.x

[75] Beedie M., "Energy savings–a question of quality", *S. Afr. J. Fd Sci Tech,* **48**(30), 14–16 (1995).

[76] Cohen J.S., Yang T.C.S., "Progress in food dehydration", *Trends Food Sci. Tech.*, **6**, 20–25 (1995).

[77] Imre L.L., Mujumdar A.S., "*Solar drying in handbook of industrial drying*", Marcel Dekker, New York, NY, p. 357 (1987).

[78] Maskan A., Kaya S., Maskan M., "Hot air and sun drying of grape leather (pestil)", *J Food Engin.*, **54**, 81–88 (2002).

[79] Lewicki P.P., "Effect of pre drying, drying and rehydration on plant tissue properties. A review", *Int J Food Prop.*, **1**, 1–22 (1998).

[80] Bal S., Wratten F.T., Chesnen J.L., Faulkner M.D., "An analytical and experimental study of radiant heating of rice grain", *Trans ASAE.*, **13**(5), 644–652 (1970).

[81] Pathare P.B., Sharma G., "Effective moisture diffusivity of onion slices undergoing infrared convective drying", *Biosystems Engin.*, **93**(3), 285–291 (2006).

[82] Jaya S., Das H., "A vacuum drying model for mango pulp", *Drying Tech.,* **21**(7), 1215–1234 (2003).

[83] Pere C., Rodier E., "Microwave vacuum drying of porous media: Experimental study and qualitative considerations of internal transfers", *Chem Engin Processing: Process Intensification.,* **41**(5), 427–436 (2002).

[84] Mitra J., Shrivastava S., Rao P.S., "Vacuum dehydration kinetics of onion slices", *Food Bioproducts process.,* **89**(1), 1–9 (2011).

[85] Huxsoll C.C., Bolin H.R., "Processing and distribution alternatives for minimally processed fruits and vegetables", *Food Tech.*, **43**, 124–128 (1989).

[86] Wim J., "Fruit and vegetable processing: Improving quality", Chapter 3, 14. *Woodhead Publishing Limited, Cambridge* (2002).

[87] Rodrìgues A.S., Pérez–Gregorio R.M., García–Falcón M.S., Símal G.J., "Effect of curing and cooking on flavonols and anthocyanins in traditional varieties of onion bulbs", *Food Res Int.,* **42**(9), 1331–1336 (2009).

[88] Li Y.L., Gan G.P., Zhang H.Z., Wu H.Z., Li C.L., Huang Y.P., Liu Y.W., Liu J.W., "A flavonoid glycoside isolated from *Smilax china* L. rhizome *in vitro* anticancer effects on human cancer cell lines", *J Ethnopharmacol.*, **113**, 115–124 (2007).

[89] Havsteen B.H.,"The biochemistry and medical significance of the flavonoids", *Pharma Therap.*, **96**(2–3), 67–202 (2002).

[90] Donner H., Gao L., Mazza G., "Separation and characterization of simple and malonylated anthocyanins in red onions, *Allium cepa* L", *Food Res Int.*, **30**(8), 637–643 (1997).

[91] Griffiths G., Trueman L., Crowther T., Thomas B., Smith B., "Onions–a global benefit to health", *Phytotherapy Res.*, **16**, 603–615 (2002).

[92] Bonaccorsi P., Caristi C., Gargiulli C., Leuzzi U., "Flavonol glucosides in *Allium species*: A comparative study by means of HPLC–DAD–ESI–MS–MS", *Food Chem.*, **107**(4), 1668–1673 (2008).

[93] Herrmann, K., "Flavonols and flavones in food plants: A review", *J Food Tech.*, **11**, 433–448 (1976).

[94] Gorinstein S., Leontowicz H., Leontowicz M., Namiesnik J., Najman K., Drzewiecki J., Cvikrová M., Martincová O., Katrich E., Trakhtenberg S., "Comparison of the main bioactive compounds and antioxidant activities in garlic and white and red onions after treatment protocols", *J Agri Food Chem.*, **56**, 4418–4426 (2008).

[95] Patil B.S., Pike L.M., Yoo K.S., "Variation in quercetin content in different colored onions (*Allium cepa* L.)", *J Am Soc Horti Sci.*, **120**, 909–913 (1995).

[96] Chemat F., Zill–e–Huma., Khan M.K., "Applications of ultrasound in food technology: Processing, preservation and extraction", *Ultrasonics Sonochemistry*, **18**, 813–835. *Acta Scientiarum Polonorum,* **6**(3): 89–99 (2011).

[97] Petersson E.V., Liu J., Sjöberg P.J., Danielsson R., Turner C., "Pressurized hot water extraction of anthocyanins from red onion: A study on extraction and degradation rates", *Analytica Chimica Acta.*, **663**(1), 27–32 (2010).

[98] Chemat S., Lagha A., Ait A.H., Bartels P.V., Chemat F., "Comparison of conventional and ultrasound assisted extraction of carvone and limonene from caraway seeds", *Flav Frag J.,* **19**(3), 188–195 (2004).

[99] Brachet A., Christen P., Veuthey J.L., "Focused microwave assisted extraction of cocaine and benzoylecgonine from coca leaves", *Phytochem Anal.*, **13**(3),162–169 (2002).

[100] Calvey E.M., Matusik J.E., White K.D., Betz J.M., Block E., Littlejoh M.H., Naganathan S., Putman D., "Off–line supercritical fluid extraction of thiosulfinates from garlic and onion", *J Agri Food Chem.*, **42**(6), 1335–1341 (1994).

[101] Brachet A., Rudaz S., Mateus L., Christen P., Veuthey J.L., "Optimization of accelerated solvent extraction of cocaine and benzoylecgonine from coca leaves, *J Sep Sci.*, **24**(10–11), 865–873 (2001).

[102] Jin E.Y., Lim S., Kim O., Park S., Jang Y.S., Chung M.S., Park H., Shim K.S., Choi Y.J., "Optimization of various extraction methods for quercetin from onion skin using response surface methodology", *Food Sci Biotech.,* **20**(6), 1727–1733 (2011).

[103] Soltoft M., Christensen J.H., Nielsen J., Knuthsen P., "Pressurised liquid extraction of flavonoids in onions. Method development and validation", *Talanta,* 80(1), 269–278 (2009).

[104] Bousbia N., Vian M.A., Ferhat M.A., Meklati B.Y., Chemat F., "A new process for extraction of essential oil from Citrus peels: Microwave hydrodiffusion and gravity", *J Food Engin.,* **90**(3), 409–413 (2009).

[105] Chemat F., Lucchesi M., "*Microwave–assisted extraction of essential oils*", *In*: Microwaves in organic synthesis – 2nd Edition (*ed.*), Loupy A. Wiley–vch Verlag GmbH and Co., pp. 959–985 (2006).

[106] Chemat F., Vian M.A., Visinoni F., "Microwave hydro diffusion for isolation of natural products", *Euro Patent EP.,* **1**(955), 749 (2008).

[107] Chemat F., Abert–Vian M., Zill–e–Huma Y.J., "Microwave assisted separations: Green chemistry in action", *Green chemistry research trends,* Nova Science Publishers, New York, pp. 33–62 (2009).

[108] Saengcharoenrat C., Guyer D.E., "Effects of supercritical carbon dioxide conditions on onion oil desorption", *J Food Engin.*, **63**(1), 33–37 (2004).

[109] Simandi B., Sass–Kiss S.A., Czukor B., Deak A., Prechl A., Csordas A., Sawinsky J., "Pilot–scale extraction and fractional separation of onion oleoresin using supercritical carbon dioxide", *J Food Engin.*, **46**(3), 183–188 (2000).

[110] Sass K., Simandi, BGao., Y., Boross F., Vamos–Falusi Z., "Study on the pilot–scale extraction of onion oleoresin using supercritical CO_2", *J Sci Food Agri.*, 76, 320–326 (1998).

[111] Dron A., Guyer D.E., Gage D.A., Lira C.T., "Yield and quality of onion flavor oil obtained by supercritical fluid extraction and other methods", *J Agri Food Chem.*, **42**(6), 1335–1341 (1997).

[112] Nuss J.S., Guyer D.E., Gage D.A., "Concentration of onion juice volatiles by reverse osmosis and its effects on supercritical CO_2 extraction", *J Food Process Engin.,* **20**(2), 125–139 (1997).

[113] Zhang L.G., Xue Z.B., Ni L.J., Ma D.S., "Extraction and quality analysis of volatile oils from onions by coupling pilot and laboratory equipment based on multi–rectification", *Sep Purifi Tech.,* **137**, 36–42 (2014).

[114] Zhang L.G., Zhang C., Ni L.J., Yang Y.J., Wang C.M., "Rectification extraction of Chinese herbs' volatile oils and comparison with conventional steam distillation", *Sep Purifi Tech.,* **77**(2), 261–268 (2011).

[115] Ng A., Lecain S., Parker M.L., Smith A.C., Waldron K.W., "Modification of cell–wall polymers of onion waste. III. Effect of extrusion–cooking on cell–wall material of outer fleshy tissues", *Carbo Polymers.*, **39**, 341–349 (1999).

[116] Singh B.N., Singh B., Singh R., Prakash D., Singh D., Sarma B., Upadhyay G., Singh H., "Polyphenolics from various extracts/fractions of red onion (*Allium cepa*) peel with potent antioxidant and antimutagenic activities", *Food Chem Toxico.*, **47**(6), 1161–1167 (2009).

[117] Moon J., Hyun J.D., Kim O.Y., Shin M.J., "Antiobesity effects of quercetin–rich onion peel extract on the differentiation of 3T3–L1 preadipocytes and the adipogenesis in high fat–fed rats", *Food Chem Toxico.,* **58**, 347–354 (2013).

[118] Martino K.G., Guyer D., "Supercritical fluid extraction of quercetin from onion skins", *J Food Process Engin.*, **27**(1), 17–28 (2004).
[119] Ko M.J., Cheigh C.I., Cho S.W., Chung M.S., "Subcritical water extraction of flavonol quercetin from onion skin", *J Food Engin.,* **102**(4), 327–333 (2011).
[120] Kiassos E., Mylonaki S., Makris D.P., Kefalas P., "Implementation of response surface methodology to optimise extraction of onion (*Allium cepa*) solid waste phenolics", *Innov Food Sci Emerg Tech.*, **10**(2), 246–252 (2009).
[121] Takahashi M., Shibamoto T., "Chemical compositions and antioxidant/anti–inflammatory activities of steam distillate from freeze–dried onion (*Allium cepa* L.) sprout", *J Agri Food Chem.,* **56**(22), 10462–10467 (2008).

—

21

Carrot Processing and By–Product Utilization

SURYA TUSHIR[1]* AND MANJU BALA[1]

ABSTRACT

Carrot is one of the ancient root vegetable cultivated during cool season worldwide. The edible part is an enlarged taproot. These are generally consumed either fresh, as a salad, or cooked. Carrots are goldmine of nutrients and contain Vitamin A, B and C particularly rich in carotene (pro–vitamin A). Large quantities of carrots are also processed, either alone or in mixtures with other vegetables, by canning, freezing or dehydration. These could commercially be converted into nutritionally rich processed products like juice, dried powder, candy, pickle and gazrailla etc. Carrot is a significant source of phytonutrinents including phenolics, polyacetylenes and carotenoids. Carrot pomace is a by–product obtained during carrot juice processing known to contain more than 50% of β–carotene which could be utilized for the supplementation of products like cake, bread and preparation of several types of functional products. The present chapter highlights the nutritional composition, active components, carrot harvesting, processing, products development and by–products utilization of carrot along with their potential application.

Key words: Antioxidants, By-product, β-carotene, Carrot, Dietary fiber, Carrot pomace, Functional products, Pro Vitamin A, Carrot juice, Vitamins, Carotenoids, Active components, Phenolics, Drying, Processing.

[1] ICAR–Central Institute of Post–Harvest Engineering and Technology (ICAR–CIPHET) Ludhiana, Punjab, India.

**Corresponding author* E-mail: surya.kadian@gmail.com

A. INTRODUCTION

Carrot is a herbaceous root vegetable, *Daucus carota subsp. sativus*, in the parsley family (Apiaceae or Umbelliferae), which also includes the similar parsnip. The term carrot applies to the long, edible, usually tapering taproot. These taproots are commonly orange in color, but also found in variety of colors depending on the cultivar, including white, pink, yellow, or purple. They have a crisp texture when fresh. These colors still exist, with orange–red color being the most popular today. Carrots can be eaten raw, whole, chopped, grated, or added to salads for color or texture. They are also often chopped and boiled, fried or steamed, and cooked in soups and stews, as well as fine baby foods and selected pet foods. It is a rich source of β–carotene and contains other vitamins, like thiamine, riboflavin, vitamin B–complex and minerals. Carrot is believed to have originated in Afghanistan which remains the centre of diversity of *D.carota*. Cultivation of carrots for medicinal purposes began 2000 to 3000 years ago. They were known to the Greeks and the Romans and their early uses were mainly medicinal. Carrots were introduced to China by the thirteenth century and cultivation spread from the Middle East to Italy, Spain and throughout Europe by the fourteenth century. Eventually, white and orange types were selected. Orange types, first grown in the Netherlands during the seventeenth century, were brought to North America by early settlers. The Orange types were popular with Native Americans and production currently exists worldwide. The greatest development and improvement of the original wild carrot that had thin, long roots took place in France. Carrots are now a popular vegetable grown all over the world. It is among the top ten most economically important vegetable crops in the world in terms of both area of production and market value. One of the reasons for the widespread production is that carrots are the major single source of Vitamin A in the diets of many cultures.

Carrots can grow best when planted in cooler climates. It is difficult to establish carrots in summer because heat, rain and early blights are major causes of crop failure and quality reduction. Globally, China was by far the largest producer and accounted for 45.8% of the global output[1] followed by Russia, United States of America, Uzbekistan, Ukraine, Poland and the United Kingdom. In India, area under carrot production is 64.27 hectares with a production of 1144.54MT in 2012–2013 [2] with Uttar Pradesh, Assam, Karnataka, Andra Pradesh, Punjab and Haryana being the major carrot producing states. Carrots are produced for a variety of uses. They are also a good source of other vitamins, minerals and fiber. In recent years, the consumption of carrot and its products have increased steadily due to their recognition as an important source of natural antioxidants. Carrot has many medicinal properties. It increases quantity of urine and helps in elimination of uric acid. It has cooling effect and is beneficial for people suffering from gall stones, constipation and heat troubles. Falcarinol, a seventeen–carbon diyne fatty alcohol was isolated from carrot and red ginseng (*Panax ginseng*). It was shown to have potent anticancer properties on primary mammary epithelial cells [3].

B. ABOUT THE PLANT

Classification

Domain: *Eukaryota*
Kingdom: *Plantae*
Phylum: *Magnoliophyta*
Class: *Magnoliopsida*
Order: *Apiales*
Family: *Apiaceae*
Genus: *Daucus*
Species: *Daucus carota*

Carrot **(2n – 2 × 18)** also known as *Daucus carota* is a member of the *Apiaceae* family and thus related to parsley, dill and celery. Carrot is an annual or biennial herb with an erect or branched stem (30–120 cm high) arising from a thick fleshy root. Edible portion is the fleshy tap root composed of an outer cortex (phloem) and inner core (xylem). They have a crisp texture when fresh.

Nutritional Composition

Consumers identify Carrots as one of the top ten foods that has "health benefits beyond basic nutrition" [4]. The moisture content of carrot varies from 86 to 89%[5]. Carrots are a good source of carbohydrates and minerals like Ca, P, K, N, Fe and Mg. Carrots are rich in carotenoids which are used to make vitamin A during digestion, and vitamins B, C, D, and E. They're also high in folic acid, and fiber. Table 1 shows the nutritional composition of Carrot.

C. CARROT HARVESTING AND POST-HARVEST HANDLING

Roots grown on ridges are usually harvested after loosening soil with a spade and by pulling out roots by grasping top. In flat ground, top is removed close to ground and roots are dug out with a spade. A light irrigation is usually given before harvesting for easy uprooting. Yield varies with season, climate, varieties etc. Fresh Carrot roots can be stored for 3–4 days under ordinary conditions and for six months at 0°C and 93–98% RH.

Table 1: Nutritional composition of carrot

Sl. no.	*Name of Parameters, Unit*	*Amount*
1.	Moisture, %	88.8
2.	Protein, %	0.7
3.	Fat, %	0.5
4.	Carbohydrate, %	6.0
5.	Total Sugar, %	5.6
6.	Crude fibre, %	2.4
7.	Ca, mg/100g	34.0
8.	Fe, mg/100g	0.4
9.	P, mg/100g	25.0
10.	Na, mg/100g	40.0
11.	K, mg/100g	240
12.	Mg, mg/100g	9.0
13.	Cu, mg/100g	0.02
14.	Zn, mg/100g	0.20
15.	Carotenes, mg/100g	5.33
16.	Thiamine, mg/100g	0.04
17.	Riboflavin, m/100g	0.02
18.	Niacin, mg/100g	0.20
19.	Vitamin C, mg/100g	4.0
20.	Energy value, kj/100g	126

Washing

Washing is an operation that generally is the point of departure of any fruit and vegetable production process. In a small–scale operation, this activity is normally carried out in basins with recirculating water, or simply with still water that is continuously replaced. After harvesting, the carrots are placed in a shade at the field in crates before washing. Carrots should be washed carefully by taking at most care not to cause any damage to the root surface. The operation consists of eliminating the dirt sticking to the material before it enters the processing line, thus avoiding complications deriving from the possible contamination of the raw material. The washing must be performed using clean water, which should be as pure as possible, and if necessary should be made potable by adding 10 ml of 10% sodium hypochlorite solution for every 100 litre of water. It is advisable to use implements that allow for an adequate cleaning of the material, so that no traces of dirt are left in the subsequent phases.

Sorting

Once the raw material is clean, it must then undergo the selection phase. At this stage, the material that will really be used in the process will be separated from material presenting some sort of defect, which will become second–choice and will be used for a different purpose, or will simply be eliminated. In the case of a semi–mechanized small–scale plant, the selection is carried out on a table suitable for this process or on a conveyor belt. Such a process will entail the removal of all of the carrots that do not have uniform characteristics compared to the rest of the lot, in terms of ripeness, color, shape and size, or which present mechanical or microbiological damage. Uniformity is a significant quality factor, since it is of utmost importance for the material to be even and uniform. The function of the sorting process is precisely that of securing such homogeneity. A high throughput variable aperture grader with parallel rollers was developed [6] that was the simplest sizing machine developed for agricultural produce. These had been popular for some time for crops such as carrots and parsnips, which were of cylindrical or tapered profile and would tend to be damaged and poorly sized through a square–mesh screen.

Grading

Grading also improves packaging, handling and other post–harvest operations. Grading is basically separating the material in different homogenous groups according to its specific characteristics like size, shape, color and on quality basis. It saves time and energy in different processing operations and reduces the handling losses during the transportation. Carrots are graded according to their size, diameter or length. The carrot should ideally, be of uniform and similar shape for the variety, well formed, smooth, firm but of tender texture. It should also be free from harvest damage, decay, disease or insects attack.

Packaging

Carrots are packed in baskets or gunny bags before sending to market. These are sometimes marketed along with their tops to indicate freshness.

Storage

Temperature is the most important factor affecting the storage of carrots. Carrots can be stored for 3–4 days under room temperature. However in cold storage, topped carrots can be stored at temperatures of 0 – 4.4°C with 93–98% relative humidity for 6 months. Untopped carrots can be stored only for 10–14 days at 0°C.

Marketing

Carrots together with their leaves are still mostly sold in bunches of five to ten. When sent to distant markets, carrots are sold without their leaves and in small bags (Fig. 1). For processing, the leaves are cut off in the field and the carrots are placed in bags or plastic containers and sent to the factory.

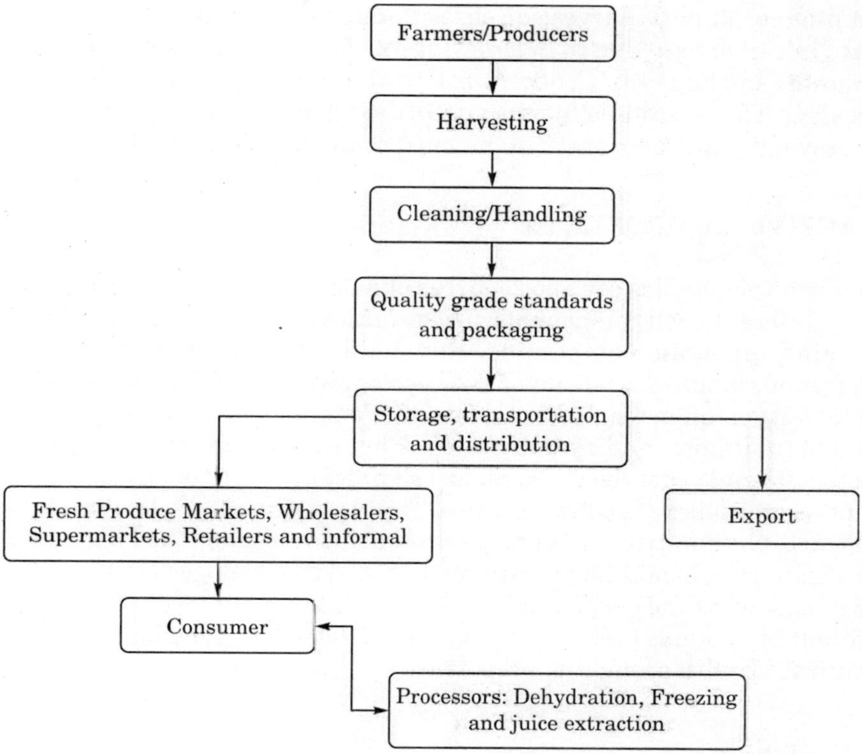

Fig. 1: Schematic view of market value chain for carrots.

The producers of carrots (farmers); pack house owners (cleans, grade and quality control); cold storage and transport facilities (store and transport carrots on behalf of farmers); traders in carrots (market and sell carrots); processors (add value to carrots and process carrots to other usable forms); and end users (consumers).

Factors Causing Storage Loss

Carrots can be stored for several months in the refrigerator or over winter in a moist, cool place. Carrot has low metabolic activity at low temperatures, due to low respiration rate [7]. Processes such as transpiration, respiration, activation of growth and attacks by pathogens lead not only to quantitative losses but quality losses, which can destroy the marketability of the product or remain invisible. A common trend in sugar composition is for the hexose content to increase and the sucrose content to decrease, especially during the first months of storage [8]. Carotene content is little affected during storage [9]. However, carrot is sensitive to wilting if not protected from water loss. In commercial refrigerated stores, storage diseases, mainly caused by pathogenic fungi, pose the greatest risk. Carrots are susceptible to a number of post–harvest diseases and disorders. Major post–harvest losses in carrots are due to *Sclerotinia* rot, *Botrytis* rot, Bacterial soft rot (*Erwinia*) and Sour rot (*Geotrichum*). Of these, Sclerotinia rot is the most prevalent. The Sclerotinia fungus infects the carrot roots in the field through the crown [10] and they are also the most common storage pathogens.

C. ACTIVE COMPONENTS

Fruits and vegetables are good sources of natural antioxidants containing many different such components. These antioxidants include carotenoids, vitamins, phenolic compounds, flavonoids, dietary glutathione and endogenous metabolites which have been shown to function as singlet and triplet oxygen quenchers, free radical scavengers, peroxide decomposers, enzyme inhibitors, and synergists [11]. Phenolic compounds are found in most fruits and vegetables. Carrot is a significant source of phytonutrients including phenolics [12], polyacetylenes [13] and carotenoids [14]. *In vitro* studies indicated phytonutrients such as carotenoids and phenolics may play a significant role, in addition to vitamin in protecting biological systems from the effects of oxidative stress [15]. Due to appreciable level of variety of different compounds present, carrots are considered as functional food with significant health promoting properties [16].

(i) Phenolics

Phenolics are ubiquitous plant components that are primarily derived from phenylalanine *via* phenylproponoid metabolism [17]. Phenolics in carrots are present throughout the roots but are highly concentrated in periderm tissue [18]. Two major classes of phenolics are hydroxycinnamic acids and para–hydroxybenzoic acids [19]. Among them, chlorogenic acid is a major hydroxycinnamic acid, representing from 42.2% to 61.8% of total phenolic compounds detected in different carrot tissues [20]. The presence of phenolic compounds in carrots contributes to their sensory qualities, like colour [21], bitterness [22], or aroma [23]. The quality of vegetables mostly depends on the components accumulated in fresh matrix, peel and cortex [24]. The response of phenolic compounds could be used as a good indicator to evaluate the vegetables quality during processing and storage. A number of carrot species with appreciable levels of antioxidant activity against the DPPH radical were identified as potential sources of free radical scavenging compounds[25].

(ii) Carotenoids

Carotenoids are naturally occurring colored compounds that are abundant as pigments in plants. Around 500 to 600 specific carotenoids have been identified, of which only about twenty–four commonly occur in human foodstuff. In general carotenoids in food are classified into carotenes and zanthophylls, which give attractive red or yellow colour and contribute to food quality. Structurally, the carotenoids may be acyclic or contain a ring of 5 or 6 carbons at one or both ends of the molecule [26]. The principal carotenoids of foods are β–carotene, β–cryptoxanthin, lycopene, lutein and violaxanthin. The carotenoids of fruit, vegetables and animal products are usually fat–soluble and are associated with lipid fractions. Carotenoids are important micronutrinents for human health [27]. Absorption efficiency of carotenoids is known to be affected by the presence or absence of other component in the diet such as dietary fat and proteins [28]. Many factors influence the absorption and utilization of pro–vitamin A such as the amount, type , physical form of the carotenoids in the diet, intake of fat, vitamin E, fiber, protein and zinc status, existence of certain diseases and parasite infestation [29].

The importance of carotenoids in foods goes beyond their role as natural pigments Fig. 2 showing various health promoting functions attributed to carotenoids. Biological functions and actions have been increasingly attributed to these compounds. β–carotene has an important nutritional role as the principal precursor of vitamin A, which is involved in vision, cell differentiation, synthesis of glycoprotein, mucus secretions from epithelial tissues, reproduction, overall growth and development of bones. On a worldwide basis, about 60% of dietary vitamin A is estimated to come from pro–vitamin A [30]. Carotenoids are also present intracellular and may be involved in the regulation of gene expression or effect cell functions like inhibition of monocyte adhesion and platelet activation [31]. The main physiological function of carotenoids is as precursor of vitamin A [32].These biological effects are independent of the pro–vitamin A activity and have been attributed to the antioxidant property of carotenoids, through deactivation of free radicals and singlet oxygen quenching [33].

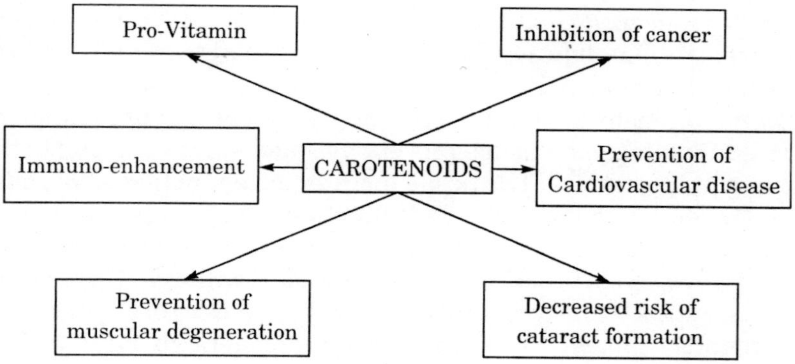

Fig. 2: Health promoting functions attributed to carotenoids

(a) β-carotene

β–carotene is a red–orange pigment found in algae, plants and fruits, especially dunaliella, carrots and colorful vegetables. The name beta–carotene comes from the Greek beta and Latin carota (carrot). It is the yellow/orange pigment that gives vegetables and fruits their rich colors. Beta–carotene's chemical formula $C_{40}H_{56}$ was discovered in 1907 (Fig. 3). Besides carrot β–carotene has been found in most vegetables and fruits such as kale, pumpkin, spinach, papaya, apricots, and peaches. Absorption of β–carotene is enhanced if eaten with fats, as carotenes are fat soluble. It is a group of natural chemicals also referred to as carotenoids. Carrot rich in β–carotene, ascorbic acid and tocopherol and is classified as vitaminized food [34]. It is an organic compound and chemically classified as a hydrocarbon and specifically as a terpenoid (isoprenoid), reflecting its derivation from isoprene units.

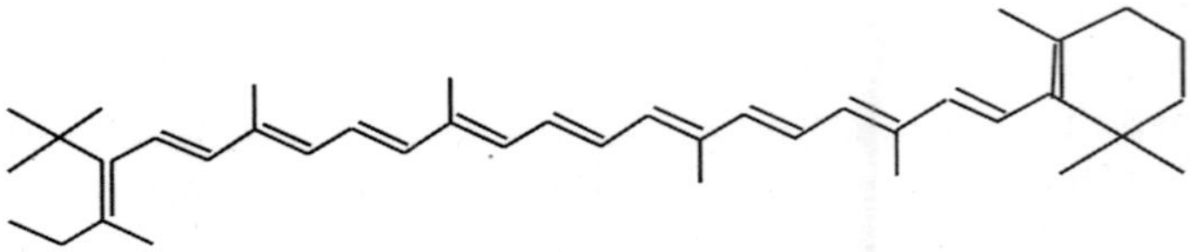

Fig. 3: Structure of β–carotene

The effect of preservation and retention on β–carotene content of some indigenous vegetables and fruits in India [35], Pakistan [36], Thailand [37], Bangladesh [38] and Malaysia [39] have been evaluated. Blanching followed by freezing is an effective preservation technique for the retention of β–carotene in fruits and vegetables. Frozen food products maintain most of the physical, chemical and sensory properties of β–carotene. In past decade, carotenoids such as β–carotene have attracted considerable attention because of their possible protective effects against some types of cancers [40]. The nutritional properties of β–carotene have been studied extensively, and its role as an antioxidant has been widely reported [41]. The presence of high concentration of antioxidant especially β–carotene may account for the biological and medicinal properties of carrots. In biological systems, β–carotene functions as free radical trapping agent and oxygen quencher and has anti–mutagenic, chemo–preventive, photo–protective and immune–enhancing properties [42]. The tissue maintenance function of vitamin A is apparently related to its antioxidant function, most or all of which can be taken over by β–carotene.β–carotene is the principal precursor of vitamin A, which is involved in vision, cell differentiation, synthesis of glycoprotein's, mucus secretion from the epithelial cells, and overall growth and development of bones [43]. Diets that are deficient in vitamin A have precipitated the death of children from measles, diarrhea, and other diseases because of impaired immunity [44]. However, humans are incapable of synthesizing this health related component and therefore the uptake of β–carotene through the diet is necessary [45].

It has been reported several times that β–carotene exists in crystalline form in the carrot chromoplast [46] and that these crystals can be associated with proteins and/or residual membranes [47]. This intracellular and membrane bound localization implies that β–carotene needs to be released during food processing and during the digestion in order to be absorbable in the gut. However, the presence of cell walls may affect the release of β–carotene [48]. Pigments of spray–dried carrot pulp waste rich in carotenoids proved to be prone to degradation during storage, depending on storage time and temperature. It was suggested that the stability of carotenoid powder can be greatly enhanced by employing appropriate packaging methods and storage conditions [49]. Freeze–dried powder showed higher pigment stability during storage than the spray–dried product [50].

Compared to vitamin–supplemented drinks composed of carrot juice as a natural source of β–carotene, drinks containing synthetic β–carotene formulations suffered from higher isomerisation rates which have been associated with nutritional consequences such as reduced pro–vitamin A activity [51]. Apart from color, cloud stability of functional drinks represents an important quality criterion which primarily determinates customer's decision to buy such products.

(iii) Dietary Fibers

It is an indigestible complex carbohydrate found in structural components of plants. They cannot be absorbed by the body and therefore have no calorific value. However, the health benefits of eating fiber rich diet are immense including prevention of constipation, regulation of blood sugar, protection against heart diseases, reducing high levels of and prevention of certain forms of cancers. The composition of dietary fiber constituents in the fresh carrot on the dry weight basis as pectin (7.41%), hemicelluloses (9.14%), cellulose (80.94%) and lignin (2.48%)[52]. Dietary fibers are not only desirable for their nutritional properties but also for their functional technological properties and because of these they could be used as food ingredients [53].

CARROT PROCESSING

Carrot processing includes canning, freezing and dehydration. Carrot can be eaten raw, whole, chopped, grated or added to salads for color or texture. They are also often chopped and boiled, fried or steamed, and cooked in soups and stews, as well as fine baby foods and select pet foods. Grated carrots are used in carrot cakes, as well as carrot pudding. The green leaves are edible as a vegetable, but are rarely eaten by humans. Baby carrots have been a popular ready to eat snack food available in supermarkets.

Carrot juice is also widely marketed, especially as a health drink, either stand–alone or blended with other fruits and vegetables. Carrot value chain is shown below in Fig. 4.

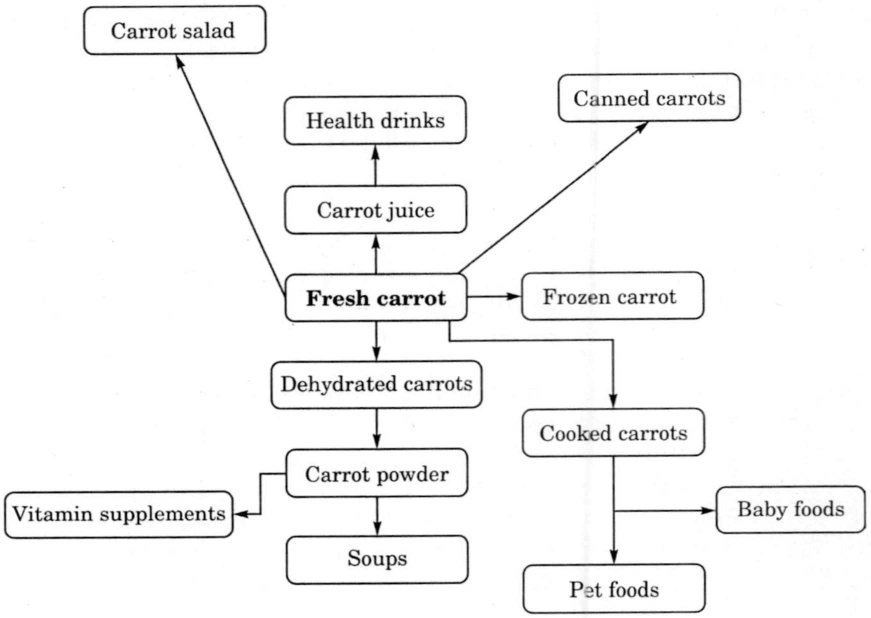

Fig. 4: Carrot value chain tree explaining its uses

Drying of Carrot

Drying of vegetables has been practiced as a means of improving storability for many centuries. The dried products can be stored for months or even years without appreciable loss of nutrients as compared to fresh vegetables which can be kept for few days only under ambient conditions. Drying of vegetables also reduces the bulk weight thus facilitating ease of transportation and reduction in storage space requirement.

A number of researches have conducted for drying of fruits and vegetables, so that the moisture content can be reduced and other value added product can be prepared further. Studies on the trans beta–carotene content in fresh carrots, blanched fresh carrots and carrots dehydrated by different processes was reported [54].The air and freeze drying of carrot was studied in which the beta–carotene retention, color, texture and flavor were reported as better in case of freeze–drying in comparison to air dried sample [55]. The optimal conditions for drying of carrot slices using a combination of an initial period of freeze drying, a short microwave treatment and a final period of vacuum or air drying was determined by [56]. Different blanching (steam, water and microwave) and drying (cabinet and fluidized–bed–drying) methods on the stability and composition of total carotenoids in carrots [57] in which total carotenoid losses were higher in unblanched carrots and in fluidized bed–dried samples compared to blanched and cabinet–dried samples. Development of non–enzymatic browning during storage was also influenced by the blanching treatments. It was concluded that steam–blanching prior to drying of carrots minimizes loss of carotenoids compared to microwave–blanching and water–blanching. The drying properties of carrots and their suitability for producing various value added products was studied [58]. Chemical properties of carrots indicated their suitability for drying and the feasibility of using carrot shreds for further processing. Drying of carrot was reported by different methods *viz.* vacuum–microwave method [59] amd in fluidizer [60]. The drying kinetics of carrot cubes was studied in which convective air drying characteristics of carrot cubes were evaluated in a cabinet dryer [61]. A process for the dehydration of the carrots was described in which the carrots were dried to about 10% moisture and transferred to portable finishing bins to complete dehydration at 44.4°C [62]. The methods of preparation and improvement in color, taste and flavor of dehydrated carrots have been reported by other workers [63, 64, 65, 66]. Freeze drying provides dried product with porous structure and little or no shrinkage, better taste retention, and on rehydration the food resembling the original [67]. The flavor of freeze dried carrot is better than air dehydrated products [68] however main disadvantage of freeze drying is its high cost [69]. The effect of different drying technologies such as hot air, vacuum and combination drying, it was suggested that the combination drying technique could keep the carotenoids of carrot well with in short drying zone [70]. β carotene degradation in carrot is comparatively less in vacuum drying and low super heated steam drying as compared to conventional air drying [71]. The nutritional quality of carrot powder and grits (Fig. 5) based food supplements have been reported to have good source of crude protein, crude fiber, iron, calcium, β–carotene and dietary fiber [72]. Blanching of carrot before dehydration results in higher retention of β–carotene [73].

Selection of carrots
↓
Washing & peeling
↓
Cutting into 0.3 cm cubes
↓
Water blanching (containing 0.2% potassium metabisulphate, 2% starch & 1% salt) at 92°C for 3 min
↓
Dehydration (45–50°C for 10–11 h)
↓
Grinding of cubes
↓
Sieving of powder
↓
Carrot Powder

Fig. 5: Process protocol for carrot powder preparation [72]

The evaluation of β–carotene content of dehydrated carrots (Table 2) exhibited that the shreds lost the most β–carotene followed by powder and chops during 3 months storage [74].

Table 2: β–carotene content on dehydration of carrot [74]

Products	*β-carotene, (mg/100 g)*	*Loss of carotene (%)*
Fresh carrot	39.6 ± 0.81	–
Dehydrated carrot chops	24.7 ± 0.73	37.0
Dehydrated carrot shreds	22.5 ± 0.68	43.0
Carrot powder	23.9 ± 0.24	40.0

(F) BY–PRODUCT UTILIZATION

The exploitation of by–products of fruit and vegetable processing as a source of functional compounds and their application in food is a promising field which requires interdisciplinary research of food technologists, food chemists, nutritionists and toxicologists. Carrot pomace is a by–product obtained during processing of carrot juice which is discarded contributing to environmental pollution [75]. The juice yield in carrots is only 60–70%, and even up to 80% of carotene may be lost with left over carrot pomace. So far the left over pomace, received after juice extraction of carrots, does not find proper utilization. The nutritional characteristics of carrot pomace have been reported by several researchers. It has good residual amount of all the vitamins, minerals and dietary fiber [76], uronic acids and neutral sugars [77], besides containing high amounts of carotenoids and phenolic compounds that could contribute to its application as functional ingredient, improving the quality of foods and helping to reduce the environmental problem. The carrot pomace contains high level of moisture, which needs to be reduced to safe storage level. A number of efforts have been done for utilization of carrot pomace for foods such as pickle, cake, bread [78], Drying or dehydration is the useful means to increase the shelf life of perishable food for further use [79, 80] reported the utilization of a carotene–rich functional food ingredient recovered through mechanical and enzymatic breakdown of the tissue of carrot pomace. According to reports carrot pomace is a source of functional ingredients [81]. Studies carried out on the feasibility of using carrot peels, by–products from food industry, as a starting raw material to produce antioxidant dietary fiber powder [82]. But the Consumer acceptability of such products needs to be accepted especially the sensory parameters [83]. Composition of Carrot Pomace on dry weight basis [84] is given in Table 3.

Table 3: Composition of carrot pomace on dry weight basis [84]

Composition	*% (dry weight basis)*
Protein	4–5
Reducing sugars	8–9
Minerals	5–6
Dietary fibre	37–48

Hot air oven drying characteristics of carrot pomace was reported in which optimal drying was observed at 65° C on the basis of β–carotene and ascorbic acid retention [85]. According to suggestions the carrot pomace containing relatively more total soluble solids, total and reducing sugars, non–reducing sugar, acidity and ascorbic acid can be utilized for the preparation of good quality toffees [86] (Fig. 6). Storage studies on dried carotenoid powder extracted from carrot waste revealed that isomerization of carotenoids occurs readily under high storage temperature (45°C) or extended exposure to light [87].Dehydration of carrots during the growing season is one of the important alternatives of preservation for further development of value added products throughout the year. Processing of carrot into products such as pickle, candy, halwa, juice etc can make its utilization throughout the year. Carrot pomace containing more than 50% of carotenoids and fibers could be utilized for developing value added products. Thus nutritious products at reasonable prices can be developed by utilization of carrot pomace.

Carrot pomace
↓
Addition of sugar and fat
↓
Heating upto 80°Brix
↓
Addition of citric acid and common salt
↓
Spreading on aluminum trays
↓
Cooling to room temperature
↓
Cutting into small pieces
↓
Toffees

Fig. 6: Flow sheet for the preparation of carrot pomace toffess[86]

(G) PRODUCT DEVELOPMENT

Carrots are processed into variety of products like juice, beverages, candy and halwa etc [88]. Canning is one of the methods of preserving food in which the food contents are processed and sealed in an airtight container. Canning provides a shelf life typically ranging from one to five years, although under specific circumstances it can be much longer proper canning of fresh vegetables is an excellent way to preserve vegetables for later use. Firmness is an important attribute for canning. Usually tender and small carrots are used for canning.

Blanching removes trapped air and metabolic gases within vegetable cells and replaces them with water, forming a semi–continuous water phase that favors a more uniform crystal growth during freezing. Although, in this case, enzyme inactivation also takes place, it is not relevant because any remaining activity is destroyed on retorting. Blanching facilitates peeling and dicing, and is also accompanied by microbial load reduction. Blanching of diced carrot at 71°C for 3–6 min results in better quality of canned product than that of done at 87.5°C for short time [89]. Improvement in color and quality of canned carrots by heat treatment and by the use of chemicals has been reported [90]. Carrots are canned in various forms like diced, halved, quartered or whole. Thermal processing increased the amount of carotenoids in products [91]. Cooking, however, partially dissolves cellulose–thickened cell walls, freeing up nutrients by breaking the cell membranes. According to reports the total carotenoids increases after steam blanching [92].

Carrot Juice

Carrot juice has particularly high content of β–carotene [93] a source of vitamin A, but it is also high in B–complex vitamins like folate and many minerals including calcium, copper, magnesium, potassium, phosphorus, and iron. Carrot juice and blends are among the most popular non–alcoholic beverages and steady increase in carrot juice consumption has been reported from various countries [94]. Carrot juice is also reported to have its use with other fruit juices in blended form [95].Extraction, canning and storage of carrot juice have been described by various workers [96]. Due to low yields associated with carrot juice production, industry is using new technologies for extracting juice such as depolymerizing enzymes, mash heating and decanter technology. Yield and quality of carrot juice extracted by pressing is varied with the pre–treatment conditions such as pH, temperature and time [97]. The effect of pre–treatment conditions on the physico–chemical parameters of carrot juice and the effects of different blanching solutions and blanching times (1–5 min) on the quality of carrot juice have been studied [98].

Pickle

Pickling is the process of preserving, even expanding the lifespan of food by either anaerobic fermentation in brine or immersion in vinegar. Natural fermentation at room temperature, by lactic acid bacteria, produces the required acidity. Carrots are pickled by lactic acid fermentation. Carrots could be preserved in good condition for 6 months at room temperature even in non air tight containers using acidified brine with potassium meta bisulphate and the product can be used for manufacturing good quality pickles [99].

Preserve

Carrot candy can be prepared by covering small whole carrots or slices of carrots with sugar or heavy sugar syrup so that total soluble solids content increases to 70–75°C [100]. Carrots have been processed to obtain intermediate moisture foods containing about 55% moisture [101]. At low temperatures (1–3°C), the ready to serve product remained acceptable for 6 months in a glass container with β–carotene retention of 40%.

Carrot Kheer /Halwa/Gajrailla

It is one of the popular sweet dishes of India. Carrot halwa is prepared by cooking shredded carrots with moderately frying in hydrogenated oil or milk powder [102] as shown in Fig. 7.

HEALTH BENEFITS

Carrots are perhaps best known for their rich supply of the antioxidant nutrient that was actually named for them: beta–carotene. However, these delicious root vegetables are the source not only of beta–carotene, but also of a wide variety of antioxidants and other health–supporting nutrients.

Carrot
↓
Washing
↓
Trimming and peeling
↓
Slicing (2.5 cm length and 4 mm thick)
↓
Blanching in boiling water containing 0.12% potassium metabisulphate
↓
Soaking in 10% sugar solution for 2 h
↓
Draining of slices
↓
Drying in cabinet drier (60-70°C)
↓
Addition of ingredients
Sugar, skim milk powder, cashew nut, cardamom & corn flour
↓
Carrot kheer mix
↓
Packing and storage

Fig. 7: Flow sheet for preparation of carrot kheer mix [102]

Antioxidant Benefits

All varieties of carrots contain valuable amounts of antioxidant nutrients. Included here are traditional antioxidants like vitamin C, as well as phytonutrient antioxidants like beta–carotene. The list of carrot phytonutrient antioxidants is by no means limited to beta–carotene, however. This list includes:

- Carotenoids
  - alpha–carotene
  - beta–carotene
  - lutein
- Hydroxycinnamic acids
  - caffeic acid
  - coumaric acid
  - ferulic acid
- Anthocyanindins
  - cyanidins
  - malvidins

Different varieties of carrots contain differing amounts of these antioxidant phytonutrients. Red and purple carrots, for example, are best known for the rich anthocyanin content. Oranges are particularly outstanding in terms of beta–carotene, which accounts for 65% of their total carotenoid content. In yellow carrots, 50% of the total carotenoids come from lutein.

The synergistic effect of carrot antioxidants is a great example of a whole food and its uniqueness as a source of nourishment. Yet in addition to the diverse mixture of carrot antioxidants, there is yet another category of carrot phytonutrient that is believed to help explain carrot protection against cardiovascular disease. That category is polyacetylenes. Polyacetylenes are unique phytonutrients made from metabolism of particular fatty acids (often involving crepenynic acid, stearolic acid and tariric acid). They are particularly common in the *Apiaceae / Umbelliferae* family of plants (which includes carrot). The two best–researched polyacetylenes in carrot are falcarinol and falcarindiol. So in addition to the unique mix of antioxidants in carrot, polyacetylenes may play a key role in the cardiovascular protection provided by this amazing food. Carrots are simply too rich in digestive tract–supporting fiber, antioxidant nutrients, and unique phytonutrients like falcarinol to be neutral when it comes to support of the lower digestive tract and colon cancer protection.

(H) OPPORTUNITIES AND CHALLENGES

(i) Opportunities

Carrots have found increasing favors among consumers. Several health benefits have promoted marketing of this vegetable as a convenient and good tasting snack food or juice product. Carrots are popular as snacks, side dishes, salads ingredients, juice mixtures and dessert mixtures (*e.g.*, carrot cake and carrot pudding). The rise of the fresh cut industry has meant some of the misshapen and otherwise imperfect carrots have an alternative profitable outlet. Carrot that would not have made the grade in a standard pack of fresh carrots do not have to be sent to freezers or canners to be cut, diced or juiced. Today, the cutting and peeling process for various fresh–cut carrot products allows a majority of the raw carrots destined for the fresh market to become fresh market products. As a way of minimizing post harvest losses carrots are also utilized as animal fodder.

(ii) Challenges

The major concern is the production of a disease–free high quality clean carrot root. Other essential part of carrot production and marketing is the need to define the quality standards necessary for consumer acceptance. Product physical defects can be readily measured, but the flavor and sweetness characteristics must be defined and the parameters affecting these important factors understood. Numerous root diseases affect carrot but proper cultural practices can keep them under control. Carrot growers also face strong market competition from the national and international producers. Undoubtedly, functional foods represent an important, innovative and rapidly growing part of the overall food market. However, their design, *i.e.*, their complex matrix and their composition of bioactive principles, requires careful assessment of potential risks which might arise from isolated compounds recovered from by–products. Since functional foods are on the boundary between foods and drugs, their regulation still proves difficult. In any case, consumer protection must have priority over economic interests, and health claims need to be substantiated by standardized, scientifically sound and reliable studies.

REFERENCES

[1] FAOSTAT database. Food and Agriculture Organization of the United Nations. 16 January 2013. Retrieved 2013–03–21.

[2] Indian Horticulture Database 2013.

[3] Nowick E.A., "*Daucus carota*", *University of Nebraska–Lincoln* (2007).

[4] Scribner C., "Encyclopedia of Food and Culture", New York, p. 6 (2003).

[5] Anonymous, "The wealth of India: Raw materials, Vol 3", *Council of Scientific and Industrial Research,* New Delhi, pp. 20–21 (1952).

[6] Hutchison P.S., McRae D.C., "A variable grader for fruit and vegetables", Dept. Note S1N/293, *Scot Inst Agric Engng.,* Penicuik, (unpubl.) (1980).

[7] Stoll A., Schieber A., Carle R., Pfannhauser W., Fenwick G.R., Khokhar S., "Biologically active phytochemicals in food", Cambridge, *The Royal Society of Chemistry,* pp. 525–527 (2001).

[8] Phan C., Hsu H., "Physical and chemical changes occurring in the carrot root during growth", *Canadian J of Plant Sci.,* **53**, 629–634 (1973).

[9] Fritz D., Habben J., "Determination of ripeness of carrot (*Daucus carota* L.)", *Acta Horticulturae.*, **52**, 231–238 (1975).

[10] Lewis B.G., Davies W.P., Garrod B., "Wound–healing in carrot roots in relation to infection by *Mycocentrospora acerina*", *Annals of Applied Biology*, **99**, 35–42 (1981).

[11] Larson R.A., "The antioxidants of higher plants", *Phyto chem.*, **4**, 969–78 (1988).
[12] Babic I., Amiot M.J., Ngugen–The C., Aubert S., "Changes in phenolic content in fresh, ready–to–use and shredded carrots during storage", *J Fd Sci.*, **58**, 351–356 (1993).
[13] Hansen S.L., Purup S., Christensen L.P., "Bioactivity of falcarinol and the influence of processing and storage on its content in carrots (*Daucus carota L*)", *J Sci Fd Agric.*, **83**,1010–1017 (2003).
[14] Block G., "Nutrient source of pro–vitamin A carotenoids in American diet", *Am J Epidemiol.*, **139**, 290–293 (1994).
[15] Kalt W., "Effects of production and processing factor on major fruit and vegetable antioxidants", *J Food Sci.*, **70**, 11–19 (2005).
[16] Hager T.J., Howard L.R., "Processing effects on carrot phytonutrients", *Hortic Sci.*, **41**, 74–79 (2006).
[17] Dixon R.A., Paiva N.L., "Stress induced phenolpropanoid metabolism", *Plant Cell.*, **7**, 1085–1097 (1995).
[18] Mercier J.J., Arul J., Julien C., "Effect of food preparation on the isocoumarin 6–methoxymellein content of UV–treated carrots", *Food Res Int.*, **27**, 401–404 (1994).
[19] Babic I., Amiot M.J., Ngugen T.C., Aubert S., "Changes in phenolic content in fresh, ready–to–use and shredded carrots during storage", *J Food Sci.*, **58**, 351–356 (1993).
[20] Wang H., Hu X., Chem F., Wu J., Zhang Z., Liao X., Wang Z., "Kinetics analysis of non enzymatic browning in carrot juice concentrate during storame", *Eur Food Res Techn.*, **223**, 282–289 (2006).
[21] Zhang D., Hamauzee Y., "Phenolic compounds and their antioxidant properties in different tissues of carrots (*Daucus carota* L.)", *Food Agric Environ.*, **2**, 95–101 (2004).
[22] Kreutzmann S., Christensen L.P., Edelenbos M., "Investigation of bitterness in carrots (*Daucus carota* L.) based on quantitative chemical and sensory analyses", *LWT– Food Science and Technology.*, **41**(2), 193–205 (2008).
[23] Naczk M., Shahidi F., "Phenolic compounds in plant foods: Chemistry and health benefits", *Nutraceutical Food.*, **8**(2), 200–18 (2003).
[24] Goncalves E.M., Pinheiro J., Abreu M., Brandao T.R.S., Silva C.L.M., "Carrot (*Daucus carota L.*) peroxidase inactivation, phenolic content and physical changes kinetics due to blanching", *J Food Engin.*, **97**, 574–81 (2010).
[25] Pimia R.P., Hakkinen S.T., Aarni M., Suortti T., Lampi A.M., Eurola M., Piironen V., Nuutila A.M., Caldentey K.M.O., "Blanching and long–term freezing affect various bioactive compounds of vegetables in different ways", *J Sci Food Agric.*, **83**, 1389–1402 (2003).
[26] Carle R., Schiber A., "Recovery and characterization of functional compounds from by–products of fruit and vegetable processing—effect of processing on the nutritional quality of food", *Karlsruhe.*, **75**, 21–23 (2001).
[27] Castermiller J.J.M., West C.E., "Bioavailability and bioconversion of carotenoids", *Ann Rev Nutr.*, **18**, 19–38 (1998).
[28] Bendich A., "Recent advances in clinical research involving carotenoids", *Pure Appl Chem.,* **66**, 1017–1024 (1994).
[29] Rodriguez–Amaya, D.B., "Carotenoids and food preparation: The retention of pro–vitamin A carotenoids in prepared, processed and stored foods", INC/OMNI project, John Snow, pp. 1–99 (1997).
[30] Simpson K.L., Katayama T., Chichester C.O., "Carotenoids in fish feeds", *In:* Bauernfeind J.C. (*ed.*), Carotenoids as colorants and Vitamin A precursors, Academic Press, New York, pp. 463–538 (1981).
[31] Devraj S., Li D., Jialal I., "The effects of α–tocopherol supplementation on monocyte cell function: Decreased lipid oxidation, interleukin 1b secretion and monocyte adhesion to endothelium", *J Clin Invest.*, **98**, 756–763 (1996).

[32] Block G., "Nutrient source of pro–vitamin A carotenoids in American diet", *Am J Epidemiol.,* **139**, 290–293 (1994).
[33] Krinsky N.I., Mathews M.M., Taylor R.F., "*Carotenoids in medicine, In: Carotenoids: Chemistry and Biology*", New York, Plenum Press, pp. 279–291 (1990).
[34] Hashimoto T., Nagayama T., "Chemical composition of ready–to–eat fresh carrot", *J Food Hyg Soc Japan.*, **39**, 324–328 (2004).
[35] Dikshit S.N., Udipi S.A., Rao A., Manohar V., "Separation of carotenoids and estimation of carotene content of selected Indian Food and food preparations by HPLC", *J Food Sci Tech.,* **25**, 39–41 (1988).
[36] Nagra S.A, Kahn S., "Vitamin A (Beta– carotene) losses in Pakistani cooking", *J Sci Food Agric.*, **46**, 249–251 (1988).
[37] Speek A.J., Saichua S.S., Schreurs W.H.P., "Total carotenoids and Beta– carotene contents of Thai vegetables and effects of processing", *Food Chem.*, **27**, 245–251 (1988).
[38] Rahman M.M., Wahed M.A., Ali M.A., "Beta–Carotene losses during different methods of cooking green leafy vegetables in Bangladesh", *J Food Composition Anal.*, **3**, 47–53 (1990).
[39] Tee E.S., Lim C.L., "The carotenoid composition and content of Malaysian vegetables and fruits by the AOAC and HPLC methods", *Food Chem.*, **3**, 309–339 (1991).
[40] Bast A., Van den Berg H., Van der Plas R.M., Haenen G.R.M. "β–Carotene as antioxidant", *Eur J Clin Nutr.*, **50**, 554–556 (1996).
[41] Deshpande S.S., Deshpande U.S., Salunkhe D.K., Madhavi D.L., "*Nutritional and health aspects of food antioxidants, Food antioxidants—technological, toxicological and health perspectives*", Marcel Dekker, New York, pp. 361–382 (1995).
[42] Wolf G., Kritchevsky D., Slater R.B.A., "Vitamin A", *In: Human Nutrition – A Comprehensive treatise. Nutrition and the adult: micro–nutrients*", New York, Plenum Press, **3**B, 97–203 (1980).
[43] Sommer A., Hussain G., Tarwotjo I., "Increased mortality in children with mild Vitamin A deficiency", *Lancet.*, **1**, 585–585 (1983).
[44] Amaya D.B.R., Kimura M., "Harvest Plus handbook for carotenoid analysis", Washington, International Food Policy Research Institute (2004).
[45] Jaramillo F.M.E., González–Cruz L., Cornejo–Mazón M., Dorantes–Alvarez L., Gutiérrez–López G.F., Hernández–Sánchez H., "Effect of thermal treatment on the antioxidant activity and content of carotenoids and phenolic compounds of cactus pear cladodes (*Opuntia ficus–indica*)", *Food Sci Tech International.*, **9**(4), 271–278 (2003).
[46] Bast A., Van den Berg H., Van der Plas R.M., Haenen G.R.M., "β–Carotene as antioxidant", *Eur J Clin Nutr.,* **50**, 554–556 (1996)
[47] Aguilera J.M., Why food microstructure, *J Food Engin.*, **67**(1–2), 3–11, (2005).
[48] Chen B.H., Tang Y.C., "Processing and stability of carotenoid powder from carrot pulp waste", *J Agric Food Chem.*, **46**, 2312–2318 (1998).
[49] Freedman D.S., Patel D.A., Srinivasan S.R., Chen W., Tang R., Bond M.G., Berenson G.S., "The Contribution of Childhood Obesity to Adult Carotid Intima–Media Thickness: The Bogalusa Heart Study," *Int J Obesity.*, **32**(5), 749–756 (2008).
[50] Bendich A., Langseth L., "Safety of vitamin A", *Am J Clin Nut.*, **49**(2), 358–371 (1989).
[51] Nawirska A., Kwasniewska M., "Dietary fiber fractions from fruit and vegetable processing waste", *Food Chem.*, **91**, 221–225 (2005).
[52] Thebaudin J.Y., Lefebvre A.C., Harrington M., Bourgeois C.M., "Dietary fibers: Nutritional and technology interest", *Trends Food Sci Technol.*, **8**, 41–48 (1997).
[53] Pazarincevic T.J., Baras J., "Effects of drying processes on trans beta–carotene content of dehydrated carrots", *Hrana–i–Ishrana.*, **11**(7–8), 325–330 (1970).
[54] Mellor J.D., Bell G.A., "Freeze drying the basic process", *Encyclopaedia Food Sci Technol Nutr.,* 2nd Edn., Academic, St Louis, pp. 2697–2701 (1993).

[55] Litvin, S., Mannheim, C.H., Miltz, J., "Dehydration of carrots by a combination of freeze drying, microwave heating and air or vacuum drying", *J Food Engin.*, **36**(1), 103–111 (1998).

[56] Sharma G.K., Semwal A.D., Arya S.S., "Effect of processing treatments on the carotenoids composition of dehydrated carrots", *J Food Sci Tech.*, **37**(2), 196–200 (2000).

[57] Machewad G.M., Kulkarni D.N., Pawar, V.D., Surve V.D., "Studies on dehydration of carrot (*Daucus carota* L.)", *J Food Sci Tech.*, **40**(4), 406–408 (2003).

[58] Stepiri B., "Effect of vacuum–microwave drying on selected mechanical and rheological properties of carrot", *Biosystem Engin.*, **99**, 234–238 (2008).

[59] Maguer M.L., Mazza G., "An engineering analysis of the drying of carrots in a vibro fluidizer", *In: 'Engineering and food, Vol. I. Engineering sciences in the food industry'* Conference–Proceedings at Dept. of Food Sci., Univ. of Alberta, Edmonton, Alberta T6G 2P5, Canada, pp. 235–243 (1986).

[60] Doymaz I., "Drying kinetics of white mulberry", *J Food Engin.,* **61**, 341–346 (2004).

[61] Cruess W.V., "*Commercial fruit and vegetable products*", McGraw Hill, New York, 4th Ed. (1958).

[62] Feinberg B., Schwimmer S.R., Reeve R., Juilly M., Arsdal Van W.B., Copley M.J., Morgan A.J., "Vegetables" *In: Food dehydration*", AVI, Westport (1964).

[63] Santo M.S., Leka L., Fotouhi N., Meydani M., Hennekens G.H., Meydani S.N., Wu D., Gaziano J.M., "Natural killer cell activity in elderly men is enhanced by β–carotene supplementation", *Am J Clin Nutr.,* **64**, 772–777 (1996).

[64] Luh B.S., Woodroof J.C., "Commercial vegetables processing", AVI, Westport, 388 (1982).

[65] Mudahar G.S., Buhr R.J., Jen J.J., "Infiltrated biopolymers effect on quality of dehydrated carrots", *J Food Sci.,* **57**, 526–529 (1992).

[66] Mellor J.D., Bell G.A., "Freeze drying the basic process", *Encyclopedia Food Sci Tech Nutr.,* 2nd Edn. Academic, St Louis, pp. 2697–2701 (1993).

[67] Kalra C.L., Kulkarni S.G., Berry S.K., "The carrot—a most popular root vegetable", *Ind Food Pack.*, **41**(6), 46–73 (1987).

[68] Krokida M.K., Philippopoulos C., "Volatility of apples during air and freeze drying", *J Food Eng.*, **73**, 135–141 (2006).

[69] Zhang–Xue T., Zhao–Yong R., Mingon Y., "Effects of drying technologies on the content of moisture and carotenoids in carrot pomace", *Sci Agric Sin.*, **40**, 995–100 (2007).

[70] Suvarnakuta P., Devahastin M.S., Arun S., "Drying kinetics and β–carotene degradation in carrot undergoing different drying processes", *J Food Sci.,* **70**, 520–526 (2005).

[71] Singh P., Kulshrestha K., "Nutritional quality of food supplements based on carrot powder and grits", *J Food Sci Tech.*, **45**, 99–101 (2008).

[72] Negi P.S., Roy S.K., "The effect of blanching on quality attributes of dehydrated carrots during long term storage", *Eur Food Res Technol.*, **212**, 445–448 (2001).

[73] Suman M., Kumari K., "A study on sensory evaluation, beta–carotene retention and shelf–life of dehydrated carrot products", *J Food Sci Technol.*, **39**, 677–681 (2002).

[74] Alklint C., "Carrot juice processing—Effects on various quality aspects", *Bulletin,* Lund Institute of Tech, Lund University, Sweden (2003)

[75] Nocolle C., Cardinault N., Aprikian O., Busserolles J., Grolier P., Rock E., Demigne C., Mazur A., Scalbert A., Amouroux P., Remesy C., "Effect of carrot intake on cholesterol metabolism and antioxidant status in cholesterol fed rats", *Eur J Nutr.,* **42**, 254–26 (2003).

[76] Stoll T., Schweiggert U., Schieber A., Carle R., "Application of hydrolyzed carrot pomace as a functional food ingredient to beverages", *J Food Agric Environ.,* **1**, 88 (2003).

[77] Filipini M., "Preparation, application and evaluation of functional food additives from organic residues using carrot pomace and wheat bread as the model system", *Bulletin*, Institute for Food Tech., University of Bonn, Bonn (2001).

[78] Roberts J.S., Kidd D.R., Zakour O.P., "Drying kinetics of grapes seeds", *J Food Engin.*, **89**, 480–465 (2008).

[79] Stoll T., Schweiggert U., Schieber A., Carle R., "Application of hydrolyzed carrot pomace as a functional food ingredient to beverages", *J Fd Agric Environ.*, **1**, 88–92 (2003).

[80] Schweiggert, U., "Carrot pomace as a source of functional ingredients", *Flussiges Obst.*, **71**(3), 136–140 (2004).

[81] Chantaro P., Devahastin S., Chiewchan N., "Production of antioxidant high dietary fiber powder from carrot peels", *LWT–Food Sci Tech.*, **41**, 1987–1994 (2008).

[82] Stoll T., Schweiggert U., Schieber A., Carle R., "Application of hydrolyzed carrot pomace as a functional food ingredient to beverages", *J Food Agric Environ.*, **1**, 88–92 (2003) .

[83] Bao B., Chang K.C., "Carrot pulp chemical composition, colour, and water–holding capacity as affected by blanching", *J Food Sci.,* **59**(6), 1159–1161 (1994).

[84] Upadhyay A., Sharma H.K., Sarkar B.C., "Characterization of dehydration kinetics of carrot pomace", *Agri Engin Int: The CIGR Ejournal.*, FP 07 35. Vol. X (2008).

[85] Jagtap S.A., Adsule R.N., Chavan U.D., "Utilization of carrot pomace for preparation of toffees", *Beverage Food World.*, **27**, 30–32 (2000).

[86] Chen B.H., Tang Y.C., "Processing and stability of carotenoid powder from carrot pulp waste", *J Agri Food Chem.*, **46,** 2312–2318 (1998).

[87] Kalra C.L., Kulkarni S.G., Berry S.K., "The carrot—a most popular root vegetable", *Ind Food Pack.*, **41**(6), 46–73 (1987).

[88] Dan A., Jain N.L., "New blanching medium for dehydration of carrot", *Ind Food Pack.*, **25**(4), 10–13 (1971).

[89] Edwards C.G., Lee C.Y., "Measurements of pro–vitamin A carotenoids in fresh and canned carrots and green peas", *J Food Sci.*, **51**, 534–535 (1986).

[90] De Sa M.C., Amaya D.B.R., "Optimization of HPLC quantification of carotenoids in cooked green vegetables—comparison of analytical and calculated data", *J Food Compos Anal.,* **17**, 37–51 (2004).

[91] Pimia R.P., Hakkinen S.T., Aarni M., Suortti T., Lampi A.M., Eurola M., Piironen V., Nuutila A.M., Caldentey K.M.O., "Blanching and long–term freezing affect various bioactive compounds of vegetables in different ways", *J Sci Food Agric.*, **83**,1389–1402 (2003).

[92] Chen B.H., Tang Y.C., "Processing and stability of carotenoid powder from carrot pulp waste", *J Agric Food Chem.*, **46**, 2312–2318 (1998).

[93] Schieber A., Stintzing F.C., Carle R., "By–products of plant food processing as a source of functional compounds—recent developments", *Trends Food Sci Technol.*, **12**, 401–405 (2001).

[94] Stoll T., Schieber A., Carle R., Pfannhauser G.R., Fenwick W., Khokhar S., "Carrot pomace—an underestimated by–product ? Biologically–active phytochemicals in food: Analysis, metabolism, bioavailability and function, Cambridge, UK: *Royal Society of Chem.*, pp. 525–527 (2001).

[95] Ohsawa K., Chinen C., Takanami S., Kuribayashi T., Kurokouchi K., "Studies on effective utilisation of carrot pomace and effective utilization to cake, dressing and pickles", *Res Report Nagano State Laboratory Food Technol.*, **22**, 24–28 (1995).

[96] Sharma H.K., Kaur J., Sarkar B.C., Singh C., Singh B., "Effect of pre–treatment conditions on physico–chemical parameters of carrot juice", *Int J Food Sci Technol.*, **44**, 1–9 (2009).

[97] Bin–Lim S., Kyung–Jwa M., "Effect of blanching conditions on the quality of carrot juice", *J the Korean Society Food Sci Nutrition.*, **25**, 680–686 (1996).

[98] Pruthi J.S., Saxena A.K., Mann J.K., "Studies on the determination of optimum conditions of preservation of fresh vegetables in acidified sulphited brine for subsequent use in Indian style curries etc", *Ind Food Pack.*, **34**(6), 9–16 (1980).

[99] Beerh O.P., Saxena A.K., Manan J.K., "Improvement of the traditional method of manufacture of carrot murrabba", *Ind Food Pack.*, **38**(4), 59–6 (1984).

[100] Jayaraman K.S., Dasgupta D.R., "Development and storage ability of intermediate moisture carrot", *J Food Sci.*, **43**, 1880–1881 (1978).

[101] Sampathu S.R., Chakraberty S., Kamal P., Bisht H.C., Agrawal N.D., Saha N.K., "Standardization and preservation of carrot halwa—an Indian sweet", *Ind Food Pack.*, **35**(6), 60–67 (1981).

22

Okra: Preservation and Processing

DEVINDER KAUR[1]*, MUDITA VERMA[1], URVASHI SHARMA[1] AND NEETU MISHRA[1]

ABSTRACT

Okra (Abelmoschus esculentus (L.) Moench. family of Malvaceae a native of tropical Africa, a plant grown in the tropics, sub–tropics and warmer areas of the temperate zones for its immature seed pods which are consumed as a vegetable. Fresh okra is available in the market for a limited duration and the shelf life does not exceed more than a week. This slimy pod and its seeds are a powerhouse of various nutrients like protein and Vitamin C. Drying and dehydration are most widely used primary methods of food preservation. Drying allows longer periods of storage with minimized packaging requirements and reduced shipping weights. Many studies has been done on the effect of drying okra with or without any pretreatments like blanching, boiling etc followed by vacuum, microwave, solar and hot air oven drying. Various studies on okra packaging by using MAP and CAP is also done on okra varieties. It has various health benefits against diabetes, digestive disorders, impaired cholesterol metabolism etc. Immuno–modulatory action, detoxifying properties, plasma glucose and cholesterol lowering effect have been well established by the researchers especially owing to its nutritive value, antioxidant content, and mucilaginous nature.

Key words: Okra, Protein, Vitamin, Drying, Packaging, Pretreatments, Health benefits, Antioxidants, Chilling, Blanching, Boiling, Millard reaction, Deteriorates, Humidity, Moisture, Modified Atmospheric Packaging, Medicinal properties, Mucilage, Cholesterol, Plasma replacer, Fibre, Hepato protective, Cancer, Diabetes, Constipation, Kidney disease, p 38 MAPK, TGF-1.

[1] Centre of Food Technology University of Allahabad, Allahabad, UP, India
**Corresponding author* E-mail: devi_sonu@yahoo.com

INTRODUCTION

Okra (*Abelmoschus esculentus* (L.) Moench, also known as *Lady's finger, Bhindi* and *Gumbo*; is a flowering plant valued for its edible fruits and belongs to the family of *malvaceae*. This African species was mentioned for the first time in 1216 B.C. in Egypt[1] and its culture has subsequently spread throughout Asia, America and in Europe. Okra is of African origin where it has been cultivated for more than 4000 years but now is also grown in different tropical and warm temperature regions of the world like–Greece, Iran, Egypt, India, Japan, southern United States Turkey and Philippines[2]. Nowadays, the most important okra–producing countries are India, Nigeria, Pakistan, Ghana and Egypt[3]. India ranks top in the Okra production with 6350 Metric tonnes in the year 2013[4]. Okra is grown in the tropics, sub–tropics and warmer areas of the temperate zones for its immature seed pods which are consumed as one of the important vegetables with tremendous nutritional values.

It grows best in hot weather where temperature ranges above 26°C. In the peak season, it is produced in large quantities much more than what the local populace can consume. Commercially, fresh pods are usually marketed in open street markets or supermarkets without any kind of temperature or humidity control. Consumers identify the loss of fruit quality by the external yellowing and by the toughening or darkening on tips or lack of tissue rupture of the apical pod when twisted with the fingers. Proper processing, preservation, marketing and utilization of okra are necessary to arrest the wastage experienced during the peak season. Such efforts should involve the development of appropriate technologies for processing and preserving okra to produce products of high market value. This will enhance the standard of living of the farmers and processors. It is however important to assess the level of okra production, processing and preservation to serve as a guide in selecting appropriate methods and techniques.

Composition

The edible portion (fresh fruits) contains 86.10% moisture, 9.70% carbohydrates, 2.25% protein, 1.00% fibre, 0.20% fat and 9.00% ash in addition to vitamins A, B, C and iodine[5]. The fruits are consumed as vegetables, raw, cooked or fried in stews, gumbos and creole dishes together with other vegetable as salad[6]. Okra fruits contains high amount of calcium, vitamin B_6 and folic acid. The dried and powdered or dehydrated okra is used in thickening soups, as emulsifier for salad dressing and as flavoring in preparing food products[7]. Okra is a source of protein, folic acid, vitamin B6, C and A, iron, calcium[8,9] and dietary fiber[10]. Okra contains large quantities of glycans, which are responsible for the viscosity of aqueous suspension[11] and the stringy, gum–like consistency that is particularly desirable in soups. However, the data is quite limited regarding the polyphenolic profile of Okra, its identification and qualitative analysis. Only four quercetin derivatives have been identified till date[12, 13]. Qualitative and quantitative analysis demonstrated that this vegetable is rich in polyphenolic compounds, especially the seeds which are mainly composed by quercetin derivatives and catechins and skins by quercetin and hydroxycinnamic acid derivatives.

Drying

Drying is one of the most widely used primary methods of food preservation. The purpose of drying food products is to allow longer periods of storage with minimized packaging requirements and reduced shipping weights[14]. Fresh okra is available in the market for a limited duration and the shelf life does not exceed more than a week. Large quantities of okra fruits produced during the main production season are usually left to deteriorate, as they cannot be kept for long. The fresh pods of Okra have an extremely short shelf life mainly due to high respiration and water loss rates[15]. Okra is highly perishable because of its high moisture content and respiratory activities; thus, it is necessary to preserve the commodity. In fresh produce like Okra, water loss is influenced by internal factors (morphological and anatomical characteristics, surface area–to–volume ratios, surface injuries, and maturity stage) and external or environmental factors (temperature, relative humidity, air movement, and atmospheric pressure)[16]. Producers are forced under the circumstances to give their commodities out at very low "take–away" prices.

Moisture content of okra at the time of its harvest ranges from 1300 to 1400% (dry basis) whereas its safe moisture content for storage in dried form is about 10% (dry basis). There is a considerable demand for the processed vegetables, especially during the off–season, and the dried products have tremendous potential to meet this demand. Owing to its high moisture content okra is susceptible to rapid deterioration. Traditional methods for preserving Okra involve slicing and sun drying of the fruits until they become brittle[17]. Traditionally, okra fruits have been processed by drying to extend the shelf life well beyond the few weeks when they are in season[17]. In most cases fruits are either sliced or smaller whole fruits are sun dried, on racks, trays, concrete floors and on roof tops till they become brittle. This method however implies problems associated with lack of pretreatment, non uniformity of slice thickness (resulting in uneven drying), direct exposure to dirt, insects and other pests, thereby affecting the nutritional and sensory attributes of the final product[17, 18]. The final product obtained from drying usually has a brown or dark– brown colour which is quite non–appealing to the consumer. Like various other vegetables, both fresh and dried okra are packed in baskets for retailing purposes which results in an obvious over exposure to weather. However, these practices lead to quick deterioration as well as contamination of produce[19].

Time taken for drying of okra from the initial moisture contents of 88.7% (w.b.) to final moisture content of around 15 ± 0.5% (w.b.) was 100 hours in open sun drying[20]. Although sun drying may be cheap, its drawbacks are well documented in literature[20]. Conventional sun drying often yields poor quality because it fails to protect the produce against dust, rain and wind, or even insects, rodents, birds and domestic animals while drying[21]. Drying of okra with solar and hot–air driers is definitely an effective means of overcoming the above mentioned problems faced in preservation. It has been reported the air–drying characteristics and kinetics of okra which advocated that solar drying provides a better means of drying okra. Solar drying still uses the sun as the heat source, but a specially designed container increases the temperature and air current to speed up drying[20]. Results indicated that drying took place in the falling rate period. Fick's diffusion model was applied to describe 'Moisture transfer' from okra and the 'Effective diffusivity' was calculated which was found to increase with increasing temperature. An Arrhenius relation with an activation energy value of 51.26 kJ/mol expressed the effect of temperature on the diffusivity. The Page model gave the best results for describing single layer drying of Okra[20]. Pretreatment of Okra before solar drying provided no apparent advantage and did not improve the quality of the final product[22].

Chilling Injury

Okra pods are highly perishable due to a high respiration rate, moisture content and chilling sensitivity and therefore okra has a short shelf life. Okra pods have increased membrane permeability with chilling injury, and the degree of chilling injury may differ with cultivar.A fresh pod can be stored for 7–10 days at 7–10°C and 90–95% RH. When stored at temperatures less than 7°C for an extended period, it gets chill damaged very rapidly[23]. The symptoms of chilling injury include surface and internal discoloration (browning), pitting, water soaked areas, off–flavor development, and accelerated incidence of surface moulds and decay especially organisms not usually found growing on healthy tissue[24]. It is well established that surface browning is associated with loss of membrane integrity which occurs during stress and senescence in plant tissues[25]. Storage at low temperatures and delayed fruit senescence extended postharvest life of okra, but the beneficial effects may be limited by the development of chilling injury–associated disorders, including skin browning, translucency, or pitting[26]. Thus, inhibition or reduction of chilling injury development is definitely an effective way to extend storage life and maintain quality of okra at low temperature. Calcium dips and modified atmospheres have been reported to reduce chilling symptoms. Ethylene is likely involved in the development of chilling injury symptoms in some commodities[27, 28]. An inhibitor of ethylene perception, 1–MCP, is thought to interact with ethylene receptors and thereby prevents ethylene–dependent responses[29]. Researchers have shown that 1–MCP inhibited the chilling injury symptoms of some horticultural products, such as apple[30, 31] avocado[28], pineapple[32], persimmon[33], 'Fallglo' tangerine and grapefruit[36], 'Nova' and 'Ortanique' mandarins[35], plum[36] and loquat[37].

Application of 1–MCP effectively alleviated the development of chilling injury disorders (browning and pitting) in okra fruit stored at 7 °C, which was found to be associated with lower membrane permeability and lipid peroxidation. In addition, the light green color of okra fruit was well maintained by 1–MCP. Furthermore, okra fruit treated with 1–MCP showed less intensive decrease in chlorophyll fluorescence and increase in respiration rate during cold storage[38].

Effect of Processing

Literatures have shown that the processing temperature can result in a significant decline in flavor and nutritional value of okra especially bioactive compounds such as ascorbic acid and carotenoids[39, 40]. Blanching and boiling are the major processes involved while cooking of Okra. Process of wet heating at the 80–100°C for several minutes inactivate the enzymes causing color, sensorial and nutritional changes in fruit and vegetable products[41]. They not only cause a drastic depreciation in the vitamin C content but also lower the antioxidant properties in green leafy vegetables[42]. Heating process with hot water allows the removal of minerals, water soluble vitamins and other soluble compounds such as soluble protein and sugar from the Okra[39]. Though, Vitamin A is usually retained during drying but being light sensitive entity, storage in dark places becomes essential. Losses of heat labile Vitamin C may be decreased *via* practicing suitable pretreatment methods and may even also causes an increase in the Vitamin C content[43]. Processing treatments of blanching and sulphiting in combination with frozen storage led to better retention of ascorbic acid content, physicochemical properties and sensory attributes of sliced okra than refrigerated storage and solar drying, followed by ambient temperature storage[44, 45]. Frozen storage has been shown to promote retention of visco–elastic properties of okra[46]. Vacuum frying is used to improve the quality of fried food by allowing minimal exposure to oxygen and by using pressure to evaporate water at lower than 100°C[47] . The main objective of frying is to develop characteristic colors flavor and aromas in the crust of the fried food. During frying process, the Maillard reaction, the outer crust of fried food, the removal moisture, oil absorption and the loss of nutrient occurred. Therefore, after vacuum frying process, organic and conventional okra resulted in a loss of the chemical qualities with the exception of the beta–carotene and fat contents. The significant increase in β (beta) –carotene and fat content owing to oil absorption and entrainment[39,40,41,42,43,44,45,46,47,48].

Packaging

Fresh Okra deteriorates quickly during storage due to tenderness and a high respiration rate[49]. In order to reduce respiration and weight loss, most vegetable commodities are kept at temperatures below 10°C. However, Okra is chilling– sensitive and develops water–soaked lesions, dark ribs, and surface pitting when stored at temperatures below 10°C for extended period[50, 51]. Postharvest dips, packaging, and controlled–atmosphere (CA) storage have been somewhat successful in extending Okra shelf life. Perforated packaging has been found to be more beneficial over non–perforated packaging as the former produces less off odor than those in non perforated packaging. Ventilation is substantially significant in storage as it avoids buildup of carbon dioxide and heat, as compared to non perforation which permits slow gas exchange combined with vapor and heat buildup leading to rapid deterioration. Water loss is influenced by internal factors (morphological and anatomical characteristics, surface area–to–volume ratios, surface injuries, and maturity stage) and external or environmental factors (temperature, relative humidity, air movement, and atmospheric pressure)[52]. Treatments to overcome these factors may include waxes, other surface coatings and wrapping with plastic films, or manipulating the environment such as maintenance of high relative humidity and control of air circulation[53].

Storage of okra at room temperature in 400–gauge polyethylene (PE) bags increased shelf life by 5 to 6 days compared to unpackaged okra[54]. The use of 40–µm PE bags in conjunction with low–temperature storage (4.4°C) reduced weight loss and prevented color changes without causing chilling injury in okra[55]. Okra pods were subjected to different combination of packaging and storage temperature. Three levels of packaging were used consisting of perforated (punched) polyfilm bags, non–perforated polyfilm bags (0.03 mm thickness) and unpackaged to modify the storage atmosphere. Pods weighing 100g were put in the various packages and the unpackaged pods were placed on flat cartons. Four levels of storage temperature, 4°C, 8.5°C, 13°C and room temperature storage were used for storing okra pods. Pods stored at 4°C were left unsuitable for marketing after 21 days of storage as they got frozen and had the highest electrolyte leakage due to high chilling injury. All unpackaged Okra pods stored at all temperatures were also left unmarketable after 21days of storage as they were dry and shriveled. These pods had the greatest weight loss (79%) at 13°C compared to weight loss of less than 30% observed in the pods stored at the same temperature (13°C) but kept in perforated packages. The best marketable pods that were not frozen and had the least off odor were those kept at 13°C storage temperature treatment in both non perforated and perforated packaging. Blackening was inhibited under these storage conditions resulting in the best visual appearance of okra pods. Atmospheric modification was instrumental in controlling pod provided by packaging thus resulting in enhanced quality and marketability of the pods after 21 days[56].

It has been found that okra dipped in a solution of 500 ppm ascorbic acid and stored in cellophane bags at 2°C showed no chilling injury and had an improved shelf life of 28 days[56]. Reduced chilling injury on Okra dipped in solutions of calcium and potassium salts before storage has also been established[58]. Okra held at 5°C in CA (5% to 10% CO_2) had reduced chilling injury[50].

MAP

Rate of weight loss is very high in unripe okra pods hence a relative humidity of about 95–100% is required to reduce dehydration, pod toughening and loss of fresh appearance. Modified atmosphere, a technique of sealing actively respiring produce in polymeric film packages to modify oxygen and carbon dioxide concentrations within the package has been associated with improved moisture retention and enhanced metabolism and activities of decay–causing microorganisms. The technique also minimizes browning of damaged surfaces of produce thereby increasing storability of a wide range of fresh produce. Due to high respiration rate, okra pods bleach and may develop off–flavors when stored in poorly ventilated containers for prolonged periods (over 24 hours). Factors like excessive heating (heat of respiration) and high relative humidity (condensation) promotes decay in stored okra[59]. In addition, temperature inside the package affects storability by influencing the film permeability and hence composition of oxygen and carbon dioxide. PVC film is effective in minimizing the weight loss and maintains the water content of okra irrespective of the storage temperature used. Modified atmosphere induced by the PVC film reduces the chilling injury development in fruits stored at a damaging temperature of 5°C. Storing fruits at 25°C induces quick chlorophyll degradation which is avoided by lowering the temperature to 5 or 10°C and wrapping them with PVC film. The appearance of severe chilling symptoms is associated to lower chlorophyll content in the fruit pericarp[15].

Okra Seed

Nowadays, as there is an exponential increment in the demand for vegetable oils due to the rising human population and the expanding oleo–chemicals industry, the exploration of some nonconventional and newer resources of vegetable oils is gaining much concern. Okra, which is currently grown mainly as a vegetable crop, has a great potential for cultivation as an oilseed crop as its mature pods contain high quantity of seeds containing considerable amount of oil which could be characterized and harnessed for commercial purposes. Some previous studies revealed the potential of Okra seed as a source of oil and protein for both the temperate and tropical regions[60,61,62]. Oil concentration of Okra seeds from Greece was found to be 15.9 to 20.7%, depending on the extraction method. The oil was found to contain high level of linoleic acid (up to 47.4%) and tocopherols isomers[64]. The percentage of respective fatty acids revealed came out to be myristic (0.2%), palmitic acid (30.2%), stearic acid (40%), palmitoleic acid (22.4%), oleic acid (24.2%). Okra seed oil is a rich source of linoleic acid, a polyunsaturated fatty acid which plays an essential role in human nutrition[64]. Protein content of Okra seed is as high as 45% after extraction of oil[65].

Medicinal Properties of Okra

Okra's mucilage has various nutritional, pharmacological and industrial applications. It has been used medically as a blood plasma replacer or blood volume expander which binds cholesterol and bile acid carrying toxins, dumped into the liver[66,67]. The anti–adhesive qualities of Okra were assumed to be due to a combination of glycoproteins and highly acidic sugar compounds making up a complex three–dimensional structure[68,69,70].

The slimy characteristic of the mucilage soothes and facilitates a comfortable passage and elimination of the endogenous toxins and excess cholesterol via anal route. Fiber content of Okra has been found active in stabilizing blood sugar levels as well as in controlling the rate at which sugar is absorbed from the intestinal tract. Nearly half of the fibre is soluble in the form of gums and pectins and this soluble fiber lower serum cholesterol thereby reducing the risk of heart diseases. The other half which is insoluble cleanses the intestinal tract by absorbing water, other soluble secondary metabolites and toxins thereby decreasing the risk of some forms of cancer, especially colorectal cancer. Thus, this action of insoluble fibers plays a key role in preventing and mitigating constipation[67,68,69,70,71]. Okra is used as a mucilaginous food additive against gastric irritative and inflammatory diseases, due to its high carbohydrate content. The plant has been used to treat various disorders, such as cancer, diabetes, liver diseases, constipation, urine retention, inflammation, bacterial and fungal infections[72,73]. The antioxidant activity of okra is predominantly attributed by flavonoids and phenolics[74,75]. The hepatoprotective activity of *Abelmoschus esculentus* (Linn.) *Moench* root through its antioxidant capacity has been studied[76]. The peel and seeds of *A. esculentus* has been studied for their *in vivo* antioxidant and anti–diabetic properties in rat models. Supplementation of okra significantly increased liver, kidney and pancreas SOD, CAT, GSH, GPx levels and decreased TBARS levels in diabetic rats compared to diabetic control rats[77]. *Abelmoschus esculentus* may significantly reduce the risk of blindness or its major causes among type 2 diabetes mellitus in Africa[78].

Abelmoschus manihot (L.) extract has been proved clinically effective in chronic kidney diseases by ameliorating renal inflammation and glomerular injury, by reducing the infiltration and the activation of macrophages in glomeruli and TNF–β protein expression in the kidney, in addition to the inhibition of p38MAPK signaling pathway activity through the down–

regulation of p–p38MAPK and TGF–β1 protein expressions *in vivo*[79]. *A. esculentus* L. polysaccharides has been shown to reveal stimulatory effects on rat dendritic cells and promotes the secretion of T (H) 1 cytokines that supports its immunomodulatory action[80]. *A. esculentus* polysaccharides are recommended to be used in future as a new carrier for nasal drug delivery[81].

CONCLUSION

Okra is one such food crop which in spite of having a massive production scale needs to score a lot in food technological aspects. Whether it be processing, preservation, transport and various types of storage practices, each entity is a major concern of consistent research and development issues because the best breakthrough in these arena are yet to come and of course traditional contemporary methods are quite lacking in maintaining the demand pull of okra by the growing population. Nonetheless, similar conscious efforts are needed in tapping the nutritional, medicinal, commercial and industrial viabilities of okra, so as to both support the previously established health benefits and promulgate significant innovative beneficial strategies in the light of new researches.

REFERENCES

[1] Ames J.M., Macleod G., "Volatile components of okra", *Phytochem.*, **29**(4), 1201–1207 (1990).

[2] Karakoltsidis P.A., Constantinides S.M., "Okra seeds: A new protein source", *J Agric Fd Chem.*, **23**, 1204–1207 (1975).

[3] FAO STAT, (downloaded from http://faostat.fao.org/ on September 2, 2015) (2013).

[4] Hand Book on Horticulture Statistics, Government of India, Ministry of Agriculture Department of Agriculture and Cooperation, New Delhi, 2014. (Accessed from www.agricoop.nic.in on September 6, 2015),

[5] Kochhar S.L., "Tropical Crops", *A text Book of Economy Botany*, pp. 263–264 (1981).

[6] Kashif S.R., Yaseen M., Arshad M., Ayub M., "Response of okra (*Hibiscus esculentus L.*) to soil given encapsulated calcium carbide", *Pak J Bot.,* **40**, 175–181 (2008).

[7] Nonneck I.L., *"Vegetable production pub",* Nan No. 1, Sotram Reinhold Company, N.Y. pp. 608–112 (1989).

[8] Aworh O.C., Olorunda A.O., Akibo O., "Quality attributes of frozen okra as influenced by processing and storage", *J Fd Tech.,* **15**, 429–433 (1980).

[9] Ihekoronye A.I., Ngoddy P.O., "Integrated Food Science and Nutrition for the Tropics", Macmallian, London, England, pp. 144–306 (1985).

[10] Adom K.K., Dzogbefia V.P., Ellis W.O., "Combined effect of drying time and slice thickness on the solar drying of okra", *J Sci Fd Agric.,* **73**, 315–320 (1996).

[11] Owoeye A.L., Laurie M.C., Allaghen N.N., Onyezili F., "Chemical and physical parameters affecting the viscosity of the mixed okra and tomato homogenate", *J Sci Fd Agric.*, 53, 283–286 (1990).

[12] Huang Z., Wang B., Eaves D.H., Shikany J.M., Pace R.D., "Phenolic compound profile of selected vegetables frequently consumed by African American in the Southeast Union States", *Food Chem.,* **103**, 1395–1402 (2007).

[13] Shui G., Peng L.L., "An improved method for the analysis of major antioxidants of Hibiscus esculentus Linn", *J Chromatogr.,* **1048**, 17–24 (2004).

[14] Okos M.R., Narasimhan G., Singh R.K., Witnauer A.C., "Food dehydration" *In*: Heldman D.R. and Lund D.B. (*eds*.), *Handbook of Food Engineering*, Marcel Dekker, New York, (1992).

[15] Finger F.L., Dlla–justina M.E., Casali V.W., Puiatt M., "Temperature and Modified atmosphere affect the quality of Okra", *Sci Agric.,* **65**(1), 312–313 (2008).

[16] Crisosto C.H., "Postharvest factors affecting fruit quality and postharvest deteriorations". *Perishables handling Newsletter* No **75**, 2–5 (1993).

[17] Kordylas J.M., "Processing and Preservation of Tropical and Sub tropical Foods", *ELBS with Macmillan*. pp. 3–4 & 58–73 (1991).

[18] Tindal N.D., "Vegetables in the Tropics", *Macmillian* Press Ltd., London & Basingstoke, pp. 325–328 (1983).

[19] Adegoreye A.S., Eniayeyi D., Ajibola O.O., "Tensile and comprehensive strength of two horticultural packaging materials: Cane and palm frond," *Int J Fd Sci Tech.,* **25**, 226–229 (1990).

[20] Doymaz I., "Drying of green bean and okra under solar energy", *Chem Indus Chem Engin Quarterly.,* **17**(2), 199–205 (2011).

[21] Jain D., Tiwari J.N, "Thermal aspects of open sun drying of various crops", *Energy.,* **28**, 37– 54 (2003).

[22] Adom K.K., Dzogbefia V.P., Ellis W.O., Simpson B.K., "Solar drying of okra–effects of selected package materials on storage stability", *Fd Res Int.,* **29**(7), 589–593 (1996).

[23] Perkins–Veazie P.M., Collins J.K., "Cultivar, packaging and storage temperature differences in storage temperature differences in postharvest shelf life of okra", *Hort Technology.,* **2**, 350–352 (1992).

[24] Mitchel F.G., Kader A.A., "Postharvest Technology of Horticultural Crops", 2nd Edition, *University of California Publication* p. 3311 (1992).

[25] Jiang H., Solyom A.M., Timmermann B.N., Gang D.R., "Characterisation of gingerol –related compounds in ginger rhizome (*Zingiber officinale* Rosc.) by high performance liquid chromatography/electrospray ionization mass spectrometry", *Rapid Comm Mass Spect.,* **19**, 2957–2964 (2005).

[26] Boontongto N., Srilaong V., Uthairatanakij A., Wongs–Aree C., Aryusuk K., "Effect of methyl jasmonate on chilling injury of okra pod.". *Acta Hortic.,* **746**, 323–327 (2007).

[27] Ben–Amor M., Flores B., Latche A., Bouzayen M., Pech J.C., Fomojaro F., "Inhibition of ethylene biosynthesis by antisense ACC oxidase RNA prevents chilling injury in Charentais cantaloupe melons", *Plant Cell Environ.,* **22**, L1579–1586 (1999).

[28] Pesis E., Ackerman M., Ben–Arie R., Feygenberg O., Feng X.Q., Apelbaum A., Goren R., Prusky D., "Ethylene involvement in chilling injury symptomds of avocado during cold storage", *Postharvest Biol Technol.,* **24**, 171–181 (2002).

[29] Watkins C.B., "The use of 1–methylcyclopropene (1–MCP) on fruits and vegetables", *Biotechnol Adv.,* **24**, 389–409 (2006).

[30] Arquiza J., Hay A.G, Nock J.F., Watkins C.B., "l–Methylcyclopropene (l–MCP) interactions with diphenylamine degradation, superficial scald metabolism, and polyphenol oxidase and peroxidase activity in apples", *J Agric Fd Chem.,* **53**, 7565–7570 (2005).

[31] Fan X.T., Mattheis J.P., "Development of apple superficial scald, soft scald, core flush, and greasiness is reduced by MCP", *J Agric Fd Chem.,* **47**, 3063–3068 (1999).

[32] Selvarajah S., Bauchot A.D., John P., "Internal browning in cold–stored pineapples is suppressed by a Internal browning in cold–stored pineapples is suppressed by a post–harvest application of 1–methylcyclopropene", *Postharvest Biol Technol.,* **23**, 167–170 (2001).

[33] Salvador A., Arnal L., Monterde A., Cuquerella J., "Reduction of chilling injury symptoms in persimmon fruit cv. 'Rojo Brillante' by 1–MCP", *J Postharvest Biol Technol.,* **33**, 285–291 (2004).

[34] Dou H., Jones S., Ritenour M., "Influence of 1–MCP application and concentration on post–harvest peel disorders and incidence of decay in citrus fruit", *J Hortic Sci Biotechnol.*, **80**, 786–792 (2005).

[35] Salvador A., Carvalho C.P., Monterde A., Martinez–Javega J.M., "1–MCP effect on chilling injury development in 'Nova' and 'Ortanique' mandarins.", *Food Sci Technol Int.*, **12**, 165–170 (2006).

[36] Candan A.P., Graell J., Larrigaudiere C., "Postharvest quality and chilling injury of plums: Benefits of 1–methylcyclopropene", *J Agric Res.,* **9**, 554–564 (2011).

[37] Cao S.F., Zheng Y.H., Wang K.T., Rui H.J., Shang H.T., Tang S.S., "effect of 1–methcyl cyclopropene on chilling and cell wall metabolism in loquat fruit", *J Hortic Sci Biotechnol.,* **85**, 147–153 (2010).

[38] Huang S., Li T., Jiang G., Xie W., Chang S., Jiang Y., Duan X., "1–Methylcyclopropene reduces chilling injury of harvested okra (*Hibiscus esculentus L.*) pods", *Scientia Horticulturae.*, **141**, 42–46 (2012).

[39] Fellows P.J., Food processing technology–Principles and practice Second edition (ISBN: 1 85573 533 Elsevier Publishing, (2000).

[40] Onayemi O., Badifu G., "Effect of blanching and drying methods on the nutritional and sensory quality of leafy vegetables", *Plant Fds Hum Nutri.,* **37**, 291–298 (1987).

[41] Beckett S.T., "Physico–chemical aspects of food processing", Springer Science & Business Media; First Edn. (1995).

[42] Ganiyu O., "Effect of blanching on the antioxidant properties of some tropical green leafy Vegetables", *LWT Food Sci Tech.*, **38**, 513–517 (2005).

[43] David E., Whitfield V., "Solar Drying, Solar Dryer Systems, important resource to improve food preparation", *International conference on solar, Kimberly South Africa.,* pp. 1– 4 (2000).

[44] Nwanekezi E.C., Okorie S.U., "Effects of Holding Methods and Time on the Vitamin Contents of Five Tropical Leafy Vegetables', *Nigerian Food Journal.*, **31**(1), 53–63 (2013).

[45] Inyang U.E., IIke C., "Effect of blanching, dehydration method temperature on the ascorbic acid, colour, sliminess and other constituent of okra fruit", *Int J Food Sci Nutr.*, **49**, 125–130 (1998).

[46] Aworh O.C., Olorunda A.O., Akibo O., "Quality attributes of frozen okra as influenced by processing and storage", *J Fd Tech.,* **15**, 429–433 (1980).

[47] Olorunda A.O. and Tung M.A., "Rheology of fresh and frozen okra dispersions", *J Fd Tech.,* **12**, 593–598 (1977).

[48] Edem D.O., "Palm oil: Biochemical, physiological, nutritional, hematological, and toxicological aspects: A review", *Plant Fds Hum Nutri.,* **57**, 319–341 (2002).

[49] Hardenburg R.E., Vatada A.L., Wang C., "The Commercial Storage of Fruits, Vegetables, and Florist and Nursery Stocks", *Agriculture Handbook 66*, U.S. Department of Agriculture, Agricultural Research Service, Washington, DC (1986).

[50] Ilker Y., "Physiological manifestations of chilling injury and its alleviation in okra fruits (*Abelmoschus esculentum (L) Moench*)", *PhD Dissertation, Univ. of California* (1976).

[51] Perkins–Veazie P., Collins J.K., "Cultivar, packaging, and storage temperature differences in postharvest shelf life of okra", *Hort Technology*., **2**(3), 350–352 (1992).

[52] Crisosto C.H., "Postharvest factors affecting fruit quality and postharvest deteriorations". *Perishables Handling Newsletter* No **75**, 2–5 (1993).

[53] Sargent S.A., Ritenour M.A., Brecht J.K., "Handling cooling and sanitation techniques for maintaining post–harvest quality", University of Florida Cooperative Extension Service, Institute of Food and Agriculture Sciences, EDIS (2000).

[54] Singh B.P., Dhankhar B.S., Pandita M.L., "Effect of pre–packaging materials on storage life of fresh okra *(Abelomschus [sic] esculentus(* L) Moench) fruits", Haryana, *J Hort Sci.,* **9**, 175–179 (1980).
[55] Fontenot J.F., Wilson P.W., Butts K., Shuh D.M., Brewer H.M., "Extending the shelf life of okra pods", *Louisiana Agr.,* **30**, 16–18 (1987).
[56] Ngure J.W., Aguyoh J.N., Gaoquiong L., "Interactive effects of packaging and storage temperatures on the shelf–life of okra", *J Agrl Biol Sci.,* **4** (3), 44–49 (2009).
[57] Aderiye B.I., "Effects of ascorbic acid and prepackaging on shelf life and quality of raw and cooked okra", *Fd Chem.,* **16**, 69–77(1985).
[58] Ilker Y., Morris L.L., "Alleviation of chilling injury of Okra", *Hort Science.,* **10**, 324 (1975).
[59] Medlicot A., "Post–harvest handling of okra: Product specifications and postharvest handling for fruits, vegetables and root crops exported from the Caribbean", Fintrac (1990).
[60] Crossley A., Hilditch T.P., "The fatty acid and glyceride of okra seed oil", *J Sci Fd Agric.,* **2**, 251–255 (1951).
[61] Rubatzky V.E., Yamaguchi M., "World vegetables", *In: Principles production and nutritive values; 2nd edition,* pp. 681–686 (1997).
[62] Oyelade O.J., Ade–Omowaye B.I.O., Adeomi V.F., "Influence of variety of protein, fat content and some physical characteristics of okra seeds", *J Food Engg.,* **57**, 111–114 (2003).
[63] Andras C.D., Simandi B., Orsi F., Lambrou C., Tatla D.M., Panayiotou C., Domokos J., Doleschall F., "Supercritical carbon dioxide extraction of okra (*Hibiscus esculentus* L) seeds", *J Sci Fd Agric.,* **85**, 1415–1419 (2005).
[64] Savello P.A., Martin F.W., Hill J.M., "Nutritional composition of okra seed meal", *J Agric Food Chem.,* **28**, 1163–1166 (1980).
[65] Oyeuga Y.A., "*Nigeria's Food and Feeding Stuffs, their chemistry and nutritive value", 3rd Edn., Ibadan University Press,* Nigeria (1968).
[66] Siemonsma J.S., Kouame C., "*Abelmoschus esculentus* (L.) Moench", Wageningen Agricultural University, Wageningen", Netherlands (2000).
[67] Zook S.W., "All about Okra", www. physiology.wise. edu/ravi/okra p. 6 (2004).
[68] Lengsfeld C., Faller G., Hensel A., "Okra polysaccharides inhibit adhesion of Campylobacter jejuni to mucosa isolated from poultry *in vitro* but not *in vivo*", *Animal Feed Sci Technol.,* **135**, 113–125 (2007).
[69] Lengsfeld C., Titgemeyer F., Faller G., Hensel A., "Glycosylated compounds from okra inhibit adhesion of *helicobacter pylori* to human gastric mucosa", *J Agric Fd Chem.,* **52**, 1495–1503 (2004).
[70] Wittschier N., Faller G., Hensel A., "An extract of Pelargoniun sidoides (Eps7630) inhibits in situ adhesion of *helicobacter pylori* to human stomach", *Phytomed.,* 14, 285–288 (2007).
[71] Wolford R., "Okra", www.*urbanex.Uiiuc.Edu/veggies/okra1.html.* University of Illinoise Extension
[72] Soares de Sousa Ferreira G., Assreuy A.M., Gadelha C.A., Gomes de Morais V., "Purification and biological activities of *Abelmoschus esculentus* seed lectin", *Protein J.,* **31**, 674–680 (2012).
[73] Klafke G.B., Moreira G.M., Monte L.G, Pereira J.L, Brandolt T.M., Xavier M.O., Santi–Gadelha T., Dellagostin O.A., Pinto L.S., "Assessment of plant lectin antifungal potential against yeasts of major importance in medical mycology", *Micropathologica.,* **175**(1–2), 147–51 (2013).
[74] Liao H., Liu H., Yuan K., "A new flavonol glycoside from the *Abelmoschus esculentus* Linn", *A Pharmocognosy Magazine*, **8**, 12–15 (2012).
[75] Ai G., Liu Q., Hua W., Huang Z., Wang D.J., "Hepatoprotective evaluation of the total flavonoids extracted from flowers of *Abelmoschus manihot* (L.) Medic: *In vitro* and *in vivo* studies", *Ethnopharmacol.,* **146**(3), 794–802 (2013).

[76] Saravanan S., Pandikumar P., Pazhanivel N., Gabriel P.M., Ignacimuthu S., "Hepatoprotective role of *Abelmoschus esculentus* (Linn.) Moench on carbon tetrachloride induced liver injury", *Toxicology Mechanisms Methods.*, **23**(7) 528–536 (2013).

[77] Sabitha V., Ramachandran S., Naveen K.R., Panneerselvam K., "Investigation of *in vivo* antioxidant property of *Abelmoschus esculentus* (L) moench. fruit seed and peel powders in streptozotocin–induced diabetic rats", *J Ayurveda Integr Med.*, **(4)**, 188–93 (2012).

[78] Moïse M.M., Benjamin L.M., Doris T.M., Dalida K.N., Augustine N.O., "Role of Mediterranean diet, tropical vegetables rich in antioxidants, and sunlight exposure in blindness, cataract and glaucoma among African type 2 diabetics", *Int J Ophthalmol.*, **5**(2), 231–237 (2012).

[79] Tu Y., Sun W., Wan Y.G., Che X.Y., Pu H.P., Yin X,J., Chen H.L., Meng X.J., Huang Y.R., Shi X.M., "Huangkui capsule, an extract from *Abelmoschus manihot* (L.) medic, ameliorates adriamycin–induced renal inflammation and glomerular injury via inhibiting p38MAPK signaling pathway activity in rats", *J Ethnopharmacol.*, **20**, 147(2), 311–320 (2013).

[80] Sheu S.C., Lai M.H., "Composition analysis and immuno–modulatory effect of okra (*Abelmoschus esculentus* L.) extract", *Food Chem.*, **134**(4), 1906–1911 (2012).

[81] Sharma N., Kulkarni G.T., Sharma A., Bhatnagar A., Kumar N., "Natural mucoadhesive microspheres of *Abelmoschus esculentus* polysaccharide as a new carrier for nasal drug delivery", *J Microencapsul.*, **30**(6), 589–598 (2013).

Subject Index

D

E

F

G

U

V

W

—